629.892 MacC
McComb, Gordon.
Robot builder's
    sourcebook

$ 24.95

⊲ **W9-ASK-559**

ROBOT BUILDER'S SOURCEBOOK

ROBOT BUILDER'S SOURCEBOOK

Gordon McComb

McGraw-Hill

New York | Chicago | San Francisco | Lisbon | London | Madrid | Mexico City | Milan
New Delhi | San Juan | Seoul | Singapore | Sydney | Toronto

The **McGraw·Hill** Companies

Cataloging-in-Publication Data is on file with the Library of Congress

Copyright © 2003 by Gordon McComb. All rights reserved. Printed in the United States of America. Except as permitted under the United States Copyright Act of 1976, no part of this publication may be reproduced or distributed in any form or by any means, or stored in a data base or retrieval system, without the prior written permission of the publisher.

1 2 3 4 5 6 7 8 9 0 PBT/PBT 0 9 8 7 6 5 4 3 2

ISBN 0-07-140685-9

The sponsoring editor for this book was Scott Grillo and the production supervisor was Pamela A. Pelton. It was set in New Baskerville by TopDesk Publishers' Group.

Printed and bound by Phoenix Book Technology.

Throughout this book, trademarked names are used. Rather than put a trademark symbol after every occurrence of a trademarked name, we use names in an editorial fashion only, and to the benefit of the trademark owner, with no intention of infringement of the trademark. Where such designations appear in this book, they have been printed with initial caps.

Information contained in this work has been obtained by The McGraw-Hill Companies, Inc. ("McGraw-Hill") from sources believed to be reliable. However, neither McGraw-Hill nor its authors guarantee the accuracy or completeness of any information published herein and neither McGraw-Hill nor its authors shall be responsible for any errors, omissions, or damages arising out of use of this information. This work is published with the understanding that McGraw-Hill and its authors are supplying information but are not attempting to render engineering or other professional services. If such services are required, the assistance of an appropriate professional should be sought.

 This book is printed on recycled, acid-free paper containing a minimum of 50% recycled, de-inked fiber.

DEDICATION

For Lane.
Continuing the McComb saga.

CONTENTS

Introduction	*ix*
How to Use the Source Listings	*xiii*
Acknowledgements	*xv*
Actuators	1
Actuators-Motion Products	2
Actuators-Motors	14
Actuators-Other	38
Actuators-Pnuematic	39
Actuators-Shape Memory Alloy	42
Barcoding	46
Batteries and Power	51
Books	62
Books-Electronics	62
Books-Robotics	63
Books-Technical	64
Communications	83
Communications-Infrared	83
Communications-RF	85
Competitions	93
Competitions-Combat	95
Competitions-Entrant	98
Competitions-Maze	100
Competitions-Other	100
Competitions-Soccer & Ball Playing	103
Competitions-Sumo	104
Computers	106
Computers-Data Acquisition	106
Computers-I/O	108
Computers-Single Board Computers	113
Distributor/Wholesaler	120
Distributor/Wholesaler-Industrial Electronics	120
Distributor/Wholesaler-Other Components	127
Electronics	130
Electronics-Circuit Examples	130
Electronics-Connectors	135
Electronics-Display	135
Electronics-Miscellaneous	138
Electronics-Obsolete	140
Electronics-PCB-Design	141
Electronics-PCB-Production	144
Electronics-Soldering	147

VIII CONTENTS

Electronics-Sound & Music	149
Electronics-Specialty	154
Entertainment	156
Entertainment-Art	156
Entertainment-Books & Movies	158
Fasteners	163
Fests and Shows	175
Internet	177
Internet-Bulletin Board/Mailing List	177
Internet-Calculators & Converters	180
Internet-Edu/Government Labs	181
Internet-Informational	190
Internet-Links	198
Internet-Personal Web Page	202
Internet-Plans & Guides	209
Internet-Research	212
Internet-Search	218
Internet-Usenet Newsgroups	227
Internet-Web Ring	231
Journals and Magazines	234
Kits	240
Kits-Electronic	240
Kits-Robotic	247
LEGO	250
LEGO-General	250
LEGO-Mindstorms	252
Machine Framing	259
Manufacturer	266
Manufacturer-Components	266
Manufacturer-Glues & Adhesives	268
Manufacturer-Semiconductors	270
Manufacturer-Tools	278
Materials	283
Materials-Fiberglass & Carbon Composites	288
Materials-Foam	292
Materials-Lighting	297
Materials-Metal	302
Materials-Other	310
Materials-Paper and & Plastic Laminates	317
Materials-Plastics	320
Materials-Store Fixtures	333
Materials-Transfer Film	334
Microcontrollers	339
Microcontrollers-Hardware	339
Microcontrollers-Programming	359
Microcontrollers-Software	362

CONTENTS

Motor Control	366
Outside-of-the-Box	380
Portal	382
Portal-Other	382
Portal-Programming	383
Portal-Robotics	384
Power Transmission	387
Professional Societies	405
Programming	407
Programming-Examples	407
Programming-Languages	408
Programming-Platforms & Software	412
Programming-Robotic Simulations	413
Programming-Telerobotics	415
Programming-Tutorial & How-to	416
Radio Control	418
Radio Control-Accessories	427
Radio Control-Hardware	428
Radio Control-Servo Control	429
Radio Control-Servos	432
Retail	436
Retail-Armatures & Doll Parts	438
Retail-Arts & Crafts	440
Retail-Auctions	447
Retail-Automotive Supplies	450
Retail-Discount & Department	451
Retail-Educational Supply	454
Retail-General Electronics	457
Retail-Hardware & Home Improvement	477
Retail-Office Supplies	481
Retail-Opticals & Lasers	482
Retail-Other	485
Retail-Other Electronics	488
Retail-Other Materials	490
Retail-Robotics Specialty	493
Retail-Science	500
Retail-Surplus Electronics	508
Retail-Surplus Mechanical	515
Retail-Train & Hobby	519
Robots	526
Robots-BEAM	526
Robots-Educational	528
Robots-Experimental	530
Robots-Hobby & Kit	530
Robots-Industrial/Research	539
Robots-Personal	543

✗ CONTENTS

Robots-Walking	544
Sensors	546
Sensors-Encoders	551
Sensors-GPS	557
Sensors-Optical	562
Sensors-Other	567
Sensors-RFID	572
Sensors-Strain & Load Cells	575
Sensors-Tilt & Accelerometer	577
Sensors-Ultrasonic	581
Supplies	587
Supplies-Casting & Mold Making	588
Supplies-Chemicals	600
Supplies-Glues & Adhesives	601
Supplies-Paints	605
Test and Measurement	607
Tools	613
Tools-Accessories	618
Tools-CNC	619
Tools-Hand	625
Tools-Machinery	628
Tools-Power	631
Tools-Precision & Miniature	633
Toys	638
Toys-Construction	640
Toys-Electronics	649
Toys-Robots	649
User Groups	653
Video	660
Video-Cameras	660
Video-Imagers	667
Video-Programming & APIs	668
Video-Transmitters	670
Wheels and Casters	672
Appendix A: Yellow Pages—First Line of Defense	*683*
Company Reference	*689*
Where to Find It	*707*
About the Author	*712*

INTRODUCTION

Imagine...

...an ocean—the Internet—so vast and deep the information contained in its more than 40 million Web sites would fill tens of thousands of libraries. Now imagine trying to find something in any of those thousands of libraries. Enter the word "robotics" at the popular Google.com search engine, and you'll get a listing of some *one million* Web sites and pages. If each "hit" were a piece of paper, the stack would reach some 30 stories—325 feet—into the air!

Imagine...

...you're trying to find some specific information about robotics in this stack, and you're not sure what to look for.

That's precisely what faces a growing legion of amateur and educational robotics enthusiasts. The boom in robotics has brought with it a mountain of suppliers and information. Thousands of companies worldwide provide parts, plans, kits, and other material for building robots; thousands more support the robotic craft with programming languages, operating systems, and computers.

Imagine...

...the most relevant information contained in one centralized clearinghouse. This is what I imagined, and it's why I compiled the book you now hold in your hands. *The Robot Builder's Sourcebook* is designed to serve as a compendium for the amateur robotics enthusiast: *what it is, where to get it,* and *how to get started.*

What You'll Find in this Book

This book goes far beyond any search engine or links page you'll find on the Internet. In *The Robot Builder's Sourcebook* you will find:

- *Over 2,500 robot resources,* including mail order suppliers, online retailers, and informational Web sites. Each listing is placed in a category, such as *Actuators - Motors,* and include detailed information about the resource—including (when available) address and phone number.

- *Sources for unusual parts and supplies* you might not have known existed. How about thin luminous "neon rope," in vibrant colors, to dress up your robot; or maybe adhesive tape that conducts electricity one way, but not another; or perhaps a special casting material that softens in a commonly available solvent, then sets to any shape you can imagine for your robot.

- *Dozens of "sidebars" with additional information* to help you understand critical robotics technologies, such as motor types, sensor designs, and choice of materials.

- *Over 200 articles of relevant advice to both beginner and experienced reader,* on various robot building topics.

The Robot Builder's Sourcebook is the Yellow Pages for amateur and educational robotics. It is designed to be a constant reference for all robotics enthusiasts, as it contains resources for both common and uncommon parts and supplies.

About the Listings

Not every company or individual having to do with robotics is listed in this book. That would be counterproductive—it would mean tens of thousands of listings, and you'd once again drown in a sea of too much information.

Those that are listed in this book are either primary resources that no robot maker should do without, or are representative of a large group of similar sellers. They are included to give you an idea of the product offerings. The same goes for schools with robotics programs, and for people who want to show off their robot creations; it's impossible to list them all.

In most cases, resources listed have Web pages, so you can easily find out more information about them. This is not just a pro-Internet bias, but a reflection of the realities of commerce.

Who's Not Listed

All this said, certain companies who conduct business on the Web have been intentionally left out. For commercial sites (sites that collect your money and sell you things through the mail), no listing is provided for companies:

- That are only on a free server (such as Yahoo, Geocities, or AOL). There are rare exceptions to this, when I either know the company well, or they have been in business for so long the possibility of fraud is virtually non-existent. In such cases I'll tell you why they were excepted.
- That fail to provide at least a phone number *or* a mail address (at the minimum, a post office box). This is to help protect you against fly-by-nighters—shady characters who set up quick storefronts, take your money, then hide in the anonymity of the Internet.
- Whose site is "under construction" with little worthwhile information contained on it.

For all pages, no listing is given if the site:

- Includes content, links, or advertising to "adult" material, or to online casinos and gambling. No, I'm not a prude, but I recognize many amateur robotics enthusiasts are under 18. They don't need me pointing them to porn on the Internet, thank-you-very-much.
- Contains excessive pop-up or pop-under windows. I tolerate one or two pop-up windows, but a gaggle of them, especially one right after the other, is total baloney, and they don't earn the right to be listed here.
- Merely frames other Web sites. No posers and wannabees!
- Is a simple "grab-bag" link lists. Search engines like Google.com do a better job.
- Tries to add itself (or other sites) to your Favorites list.
- Has not been updated within the last 12 months.
- Provides little or nothing of interest to the amateur robot builder. This applies to some of the very high-end industrial robot manufacturers.

These Aren't Paid Ads!

No way! The Internet is full of enough ads as it is.

No one paid me, or the publisher (cash, freebies, beer, dancing girls, whatever), to be listed in this book. And they especially didn't pay for the special "highlights" that focus on specific products or companies. Those companies, groups, and individuals listed in this book are included because I felt they contributed to the art and science of amateur robots.

Be Included in the Next Edition!

We (the publisher and I) welcome your submissions if your company, school, or personal Web site is related to amateur (and not industrial) robotics. If accepted, your submission will appear in the next edition of *Robot Builder's*

Sourcebook. Send a short e-mail, describing your company, school, or Web site to:

listings@robotoid.com

If you're still working on your Web site, please complete it before submitting your request. Submissions from non-US entities are especially encouraged.

Updates and Changes

People on the Internet move, change addresses, or plain go out of business. Some of the listings in *Robot Builder's Sourcebook* are bound to change over time. To help reduce the frustration of dead-end links, we regularly survey the companies and Web sites included in this book, and provide updates at the following:

http://www.robotoid.com/sourcebook/

Here, you'll find:

- New additions to the sourcebook.
- News of major changes, like Amazon.com going bankrupt (nah...).
- Searchable database of all the links provided in *Robot Builder's Sourcebook.*

The searchable database is useful if you try a Web address printed in the book, and it's no longer working. To reduce wear-and-tear on your mouse and keyboard, we use a special coding feature for our searchable database. If you find a Web address is no longer functional, locate the six-digit ID number included with its listing. A typical ID number looks like this:

012345

Enter the ID into the Search box, and click the Go button. The latest Web address we have for that listing is displayed.

Report Changed or Dead Links

Though we make every effort to look for, change, and remove bad links, some fly under our radar. If you find a link that is no longer working, please report it using the Changed Links button on the main *Robot Builder's Sourcebook* page.

What You Need to Use The Internet

Many of the resources listed in this book rely on contact via the Internet. In fact, some sellers, like Amazon, only do business through the Internet. They discourage the "older fashioned" methods of mail order buying, and may not even provide a mailing address.

Odds are you already have what you need to make use of the Internet, but to recap:

- *Web browser.* Any reasonably recent version of your favorite should be fine. For some Web pages, you'll want a browser that can display graphics and run Java and JavaScript code.
- *Shockwave.* Some sites use Shockwave, an add-in program that provides animation and sound effects. Shockwave is a "plug-in" that works with your Web browser. You can download it at http://www.macromedia.com/.
- *E-mail reader.* If you use a Web-based e-mail service (like Hotmail or Yahoo), then your browser is your e-mail reader. Otherwise you will need an e-mail program, such as Microsoft Outlook Express (comes with all new versions of Windows). Additionally, there are several free and nearly free e-mail readers you can try.

- *Newsgroup reader.* A newsgroup reader allows you to read and post Usenet newsgroup messages. Newsgroup readers comes as part of Windows (Outlook Express), and are included with many Web browsers, such as Netscape.

HOW TO USE THE SOURCE LISTINGS

Each listing in *The Robot Builder's Sourcebook* provides basic contact information. In most cases, the address and/or phone number is included for sellers of products and services. Internet-only resources, such as search engines, include the name of the site, its URL address, and a description.

Here's a sample listing for a seller of products or services:

Robots R Us (1) 🏅   909090 *(4)*

123 Main St.
Anytown, XY 99999 *(2)*
USA *(3)*

📞   (123) 555 1212 *(5)*

📠   (123) 555-9876 *(6)*

🚫   (800) 555-1122 *(7)*

✉   info@robotzrus *(8)*

🌐   http://www. robotzrus (9)

Manufacturer and seller of lots and lots of robot goodies. Everything is given away for free! *(10)*

1. *Name* of the resource.
2. *Mailing address* for the resource.
3. *Country* the resource is located in.
4. *Six digit ID number* that can be used at the support site for this book to look up the latest known URL for the resource. See the Introduction for more information.
5. *Voice phone.* Outside North America the country code is included.
6. *Fax phone.* As with the voice number, the country code is included for listings outside North America.
7. *Toll free phone.* In almost all cases, these numbers are good only when dialing within the seller's country. In North America, many toll free numbers do not function if calling locally, or within the same state.
8. Main *e-mail* contact.
9. Web site *URL address.*
10. *Description* of products or services provided.

Note: Not all listings include all of the above information. For example, some companies wish to only publish their toll free phone, and not their local access phone...sounds crazy, but that's their approach to doing business.

Many of the listings also include one or more icons to help you identify special features of the resource. Here's what the icons mean:

🏅   My personal pick, because I know the company well or have ordered from them and had good results.

📄   Resource provides a printed catalog or sales brochure. The catalog may not be free, so be sure to check.

xv

Resource provides at least rudimentary product descriptions on their Web site. Simple listings of product lines or manufacturers do not a Web catalog make.

Listing is a premium resource, known for competently providing products to robot builders and other hobbyists.

$ Seller requires a minimum order of more than $20 (for sellers in the United States only). Note: Some sellers will accept orders of any size, but will tack on a handling charge if it's under a certain amount. This is not the same as a minimum order requirement, where your order is refused if it doesn't measure up.

Listing is for a manufacturer or wholesaler who may, or may not, sell directly to individuals. If the resource does not sell directly, the Web site typically indicates where and how products can be purchased. In some instances, a resource will sell directly only if a regional distributor or representative is not available, so be sure to check with the company for exact policies.

Resource sells online, typically through an e-commerce shopping cart, but may also accept phone, fax, or mail orders from product listings available on the site.

Resource provides useful information only, and no sales.

Resource conducts local business in a bricks-and-mortar retail store. Some listings are for companies that sell both online and locally.

ACKNOWLEDGEMENTS

I'm indebted to the countless robot enthusiasts on the comp.robotics.misc newsgroups, who are always willing to share their ideas, hints, tips, and sources. You were the inspiration for this book, and I hope it'll be useful to you.

Special thanks to Matt Wagner (my agent) at Waterside Productions, and to Scott Grillo, my publisher at McGraw-Hill. Scott actually pays me to write about things I love doing—building robots. Could life be any better?

This book wouldn't have been possible without the hard and compassionate work of Chuck Wahrhaftig and Judy Allan of TopDesk Publishers' Group, tamers of the wild Quark beast. Thanks, guys.

When the call went out for photos, dozens of kind folks responded on short notice, and I'm grateful for the product photos you've provided. A special kudos to Ed Sparks, for his wonderful CAD drawings. Check out FirstCadLibrary.com for some first-rate 3D illustrations of motors, gears, and more.

Finally, a heartfelt thanks to my family—wife Jennifer, daughter Mercedes, son Max, and grandson Lane—for letting me stay up past my bedtime to finish this book.

ACTUATORS

Actuators are mechanisms that produce motion from some energy source, such as electricity or air pressure. Common actuators are electric motors and air cylinders. Actuation mechanisms are used to control how that motion is applied.

The companies in this main section make, distribute, or sell a very broad line of actuators and mechanisms and are listed here as generic resources. Additional companies and resources are listed in the following subcategories:

ACTUATORS-MOTION PRODUCTS: Specializes in mechanisms (though may also sell motors and other components)

ACTUATORS-MOTORS: DC geared and nongeared motors, servomotors, stepper motors (but not motors for radio-controlled models)

ACTUATORS-OTHER: Mechanical actuators (e.g., solenoids) that don't neatly fit elsewhere

ACTUATORS-PNEUMATIC: Air cylinders, control values, pumps, and other fittings

ACTUATORS-SHAPE MEMORY ALLOY: Material that contracts when heat or electricity is applied.

Locomotion Systems

Mobile robots use wheels, tracks, or legs to move around—there are exceptions, of course: some snake-like robots crawl. We'll just concentrate on the usual designs.

Wheels: The Primary Moto-vator

Wheels are by far the most popular method of providing robot mobility. Wheels can be just about any size, from an inch or two in diameter to over 10 or 12 inches. Tabletop robots have the smallest wheels, less than 2 to 3 inches in diameter. Wheel size is critical for smaller 'bots because larger wheels weigh more. Robots can have just about any number of wheels, although two is the most common. In a two-wheel robot, the machine is balanced by one or two casters on either end.

Wheeled robots are the most popular.

Legs: I'm Walkin' Here, I'm Walkin'!

More and more amateur robots have *legs*. Legs are often preferred for robots that must navigate over uneven terrain. Most amateur robots are designed with six legs, which affords *static balance*—the ability of the robot to be balanced at all times because a minimum of three legs (in tripod arrangement) are touching the ground at any one time. Robots with fewer legs must take more careful steps, or use *dynamic balance* (shifting of weight) to keep from falling over.

Despite their looks, six-legged walking robots are not difficult to build.

Tracks: O' My Tears

Tracks (or treads) are similar to what tanks use. The tracks, one on each side of the robot, act as giant wheels. The tracks turn, and the robot lurches forward or backward. Track drive is best for robots used only outdoors and only over soft ground, like dirt.

SEE ALSO:

MOTOR CONTROL: Electronic circuits for controlling motors

POWER TRANSMISSION: Gears, bearings, belt, chain, and other components

RADIO CONTROL-SERVOS: Servo motors used with radio-control models

Danaher Motion MC   203507

45 Hazelwood Dr.
Amherst, NY 14228
USA

- ☏ (716) 691-9100
- 📠 (716) 691-9181
- ⊘ (800) 566-5274
- 🌐 http://www.danahermcg.com/

U.S.-based manufacturer and distributor of several top-quality motion product brands. The company is the corporate parent of:

Ballscrews and Actuators—
http://www.ballscrews.com/

PMI (pancake motors)—
http://www.kollmorgen.com/

Superior Electric (motors)—
http://www.superiorelectric.com/

Portescap (miniature gearmotors)—
http://www.portescap.com/

Warner Linear (linear actuators, ballscrew products)—http://www.globallinear.com/

The Danaher Motion site provides some handy online tools, include a linear actuator selector, a ballscrew selector, and a units converter. Online catalogs and datasheets are in Adobe Acrobat PDF format.

Ballscrew nut, with reciprocating ball bearing channels.

Invensys Plc   204159

Carlisle Place
London
SW1P 1BX
UK

- ☏ +44 (0) 2078 343848
- 📠 +44 (0) 2078 343879
- 🌐 http://www.invensys.com/

Invensys is a large corporate parent of many motion control and automation brands, many of which may be familiar to you. These include:

- Barber-Colman (motors)—http://www.barber—colman.com/
- Clarostat Sensors and Controls (potentiometers)—http://www.speed-position.invensys.com/
- Hansen Transmissions (gearboxes)—http://www.hansentransmissions.com/
- Lamda Electronics (power supplies)—http://www.lambdapower.com/
- Rexnord (power transmission components)—http://www.rexnord.com/
- W M Berg (retail power transmission components)—http://www.wmberg.com/

MSC Industrial Direct Co., Inc.   202826

75 Maxess Rd.
Melville, NY 11747-3151
USA

- ☏ (516) 812-2000
- ⊘ (800) 645-7270
- 🌐 http://www.mscdirect.com/

See listing under **Materials**.

⚙ ACTUATORS-MOTION PRODUCTS

Motion products are mechanisms and components designed primarily for use in motion control equipment, particularly linear motion assemblies (such as ACME screws, ballscrews, X-Y translation tables, and linear actuators). Motion products may be sold individually, or as a complete system, including motor, all mechanical parts, electrical control, and feedback sensors.

Of the companies listed in this section, nearly all are geared toward industrial motion control applications; therefore, their prices for parts tend to be high-a typical 18-inch-long ballscrew mechanism might cost in excess of $300. However, they are good sources for mechanical information, and many do sell low-cost versions suitable for amateur robotics. Their wares can often be found on the surplus market.

Companies involved in so-called motion mechanicals are usually also manufacturers, distributors, or sellers of motors and motor control devices.

SEE ALSO:

DISTRIBUTOR/WHOLESALER-INDUSTRIAL ELECTRONICS: For buying new motors

MOTOR CONTROL: Circuitry for operating motors found in this section

POWER TRANSMISSION: Individual components; gears, belts, etc.

RETAIL-SURPLUS MECHANICAL: Mail order and local retail for buying salvaged motors

Allied Devices   202122

325 Duffy Ave.
Hicksville, NY 11801
USA

((516) 935-1300
📠 (516) 937-2499
✉ info@allieddevices.com
🌐 http://www.allieddevices.com/

Manufacturers and distributors of high-precision motion products and mechanical components. Offerings include:

Delta Computer Systems, Inc.

http://www.deltacompsys.com/
Hydraulic and servo motion control systems

Linear Bearings

http://www.linearbearings.com.au/
What else—linear bearings (and other motion mechanicals)

THK

http://www.thk.com/
Linear motion products

- Rotary motion assemblies (gearheads, speed reducers, differentials, etc.)
- Rotary motion components (shafts, couplings, shaft adapters, etc.
- Gears (including metric)
- Linear motion assemblies (racks, pinions, linear slides, ACME screws and leadnuts, etc.)
- Assembly hardware (screws, nuts, hangars, set screws, springs, etc.)

Anaheim Automation   203517

910 East Orangefair Ln.
Anaheim, CA 92801-1195
USA

((714) 992-6990
📠 (714) 992-0471
✉ webmaster@anaheimautomation.com
🌐 http://www.anaheimautomation.com/

Industrial motor control systems (motors, controllers, and tachs); X-Y tables.

Applied Industrial Technologies   203445

One Applied Plaza
Cleveland, OH 44115-5053
USA

((216) 426-4189
📠 (216) 426-4820
⊘ (877) 279-2799
✉ products@apz-applied.com
🌐 http://www.appliedindustrial.com/

Industrial bearings, linear slides, gears, pulleys, pneumatics, hydraulics, and other mechanical things. Also hosts Maintenance America, online reseller of industrial maintenance supplies and general industrial supplies (wheels, casters, fasteners, and more), tools, paints, and adhesives.

Applied Motion Products   203879

404 Westridge Dr.
Watsonville, CA 95076
USA

((831) 761-6555

(831) 761-6544

(800) 525-1609

info@applied-motion.com

http://www.appliedmotionproducts.com/

Applied Motion is a major manufacturer of industrial motion control products, including stepper and servo motors, gearheads, and motor control electronics. Products are available through distributors and are also quite common on the surplus market. For this reason, visit the Web site for its technical reference materials.

A selection of servo motors. Photo Applied Motion Products.

Servo motor, driver, and support software. Photo Applied Motion Products.

Arrick Robotics   202558

P.O. Box 1574
Hurst, TX 76053
USA

((817) 571-4528

(817) 571-2317

info@robotics.com

http://www.robotics.com/

In the words of the Web site, Arrick Robotics specializes in PC-based automation products including stepper motor control systems, linear and rotary positioning tables, robotic workcells, and automation accessories such as pulley reducers and grippers. These products are used in a variety of settings including laboratories, factories, machine shops, and education.

Products include:

- Gantry CNC components
- Arobot
- Trilobot

For the average amateur builder, the ARobot is the most affordable product sold by Arrick Robotics. The ARobot is a small mobile robot experimenter's platform designed to be programmed using an onboard Basic Stamp II microcontroller.

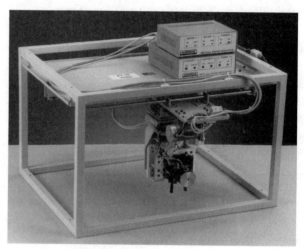

Robotic workcell, available in various sizes. Photo Arrick Robotics.

Automationdirect.com   202829

3505 Hutchinson Rd.
Cumming, GA 30040
USA

((770) 889-2858

(770) 889-7876

(800) 633-0405

sales@automationdirect.com

http://www.automationdirect.com/

Online e-commerce and catalog mail-order source for motion control products, including shaft encoders, servos, and driver electronics. Much of the product offering is for high-end industrial and not for small home-brew robots. Plan your budget accordingly.

The Web site also sports a large selection of user manuals and documentation for products available for download (most are in Adobe Acrobat PDF format).

B&B Motor & Control Corp.   203514

21-21 41st Ave.
Long Island City, NY 11101
USA

((718) 784-1313
    (718) 784-1930
   sales@bbmotor.com
   http://www.bbmotor.com/

Motors and motion mechanicals. Products include AC/DC and stepper motors and motor controllers, encoders, linear slides, linear actuators, precision gearheads, rotary tables, servos, and steppers.

Ball Screws & Actuators Company, Inc.   202524

3616 Snell Ave.
San Jose, CA 95136-1305
USA

((408) 629-1132
    (408) 629-2620
    (800) 882-9957
   sales@ballscrews.com
   http://www.ballscrews.com/

High-end, high-quality ballscrews, threaded rod, linear stages, and other motion mechanicals. Full product lines include:

- Supernuts and leadscrews
- Ballscrews
- Complete screw assemblies
- Stock stages
- Rails and bearings
- Linear actuators
- Accessories (such as couplings and mounting flanges)

The company's ActiveCam line of antibacklash nuts is designed to reduce costs considerably from traditional ballscrews and is designed for use when driving light loads. The BS&A Web site also provides a number of useful technical articles on designing with and using motion mechanicals. An example article is "Straight Talk on Leadscrews." They sell direct if there isn't a distributor near you.

Barrington Automation   202494

780 Tek Dr.
Crystal Lake, IL 60014
USA

((815) 477-1400
    (815) 477-9818
   info@barrington-atn.com
   http://www.barringtonautomation.com/

Barrington Automation manufactures industrial motion products including translation stages, linear slides, elevating tables, and grippers, as well as other high-end products.

Also see their machine-framing subsidiary at:

http://www.frame-world.com/

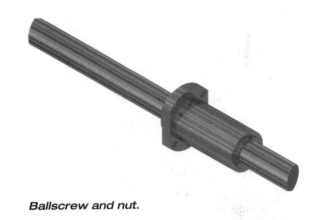

Ballscrew and nut.

Bayside Automation Systems and Components   202478

27 Seaview Blvd.
Port Washington, NY 11050
USA

((516) 484-5353
    (800) 305-4555
   http://www.baysidemotion.com/

Precision motion products: bearings, linear-positioning slides, gear reducers, and servo motor amplifiers.

BEI Technologies   202700

One Post St.
Ste. 2500
San Francisco, CA 94104
USA

((415) 956-4477
    (415) 956-5564

sales@bei-tech.com

 http://www.bei-tech.com/

In the words of the Web site, BEI Technologies is leading the development of intelligent sensors across many markets. The sophisticated sensor technologies available and in development at BEI provide new levels of performance for our customers. Our intelligent sensors are enabling new technologies for automation products and systems around the world.

Offers optical encoders, pressure sensors, motors, motor control circuits, inertial sensors (gyros), and position sensors.

See also BEI Industrial Encoder Division:

http://www.beiied.com/

Bishop Wisecarver   203483

(see addresses below)

 http://www.boschframing.com/

Machine framing, motion mechanics (linear actuators, linear guides, and bushings).

For linear motion products:

14001 South Lakes Dr.
Charlotte, NC 28273
(704) 583-4338
Fax: (704) 583-0523
Toll free: (800) 438-5983

For machine-framing products

816 East Third St.
Buchanan, MI 49107
(616) 695-0151
Fax: (616) 695-5363
Toll free: (800) 32-BOSCH

Bodine Electric Company   203880

2500 West Bradley Pl.
Chicago, IL 60618-4798
USA

((773) 478-3515

℘ (773) 478-3232

⊘ (800) 726-3463

 http://www.bodine-electric.com/

Bodine is a well-known and respected manufacturer of motors of all shapes, sizes, and kinds. Of Bodine's product line, most robotics folk are interested in their DC motors and gearmotors, and brushless DC motors and gearmotors. The Bodine product catalog is on CD-ROM, and the Web site contains numerous technical articles that relate to Bodine's products, as well as general information about motors from many other companies. Their white paper, "Brushless DC Motors Explained," is particularly interesting.

Compumotor / Parker Hannifin Corp. 203928

5500 Business Park Dr.
Rohnert Park, CA 94928 7904
USA

((707) 584-7558

℘ (707) 584-2446

⊘ (800) 358-9068

🌍 http://www.compumotor.com/

Compumotor manufacturers and sells (online or through catalog sales) a wide range of high-end industrial motors and motion control products. Lines include:

- Controllers
- Servo and stepper drives and drive/controllers
- Motors: linear servo, rotary servo, linear stepper, and rotary stepper
- Incremental encoders

The Compumotor.com Web site contains a copious amount of free technical literature, such as white papers on servo design, DC and stepper motor construction basics (written for the engineer), and datasheets. Their product catalogs are available on CD-ROM, online, or as printed books.

Compumotor is a division of Parker Hannifin; visit the main Web site:

http://www.parker.com/

Cross Automation   203930

2020 Remount Rd.
P.O. Box 1079
Gastonia, NC 28053-1079
USA

((704) 867-4401

℘ (704) 866-9525

⊘ (800) 866-4568

info@crossco.com

 http://www.cross-automation.com/

Distributor of motors, motor control, sensors, and mechanical systems (gearheads, actuators, translation tables, etc.) for industrial automation.

Del-Tron Precision 204003

5 Trowbridge Dr.
Bethel, CT 06801
USA

 (203) 778-2721

○ (800) 245-5013

 deltron@deltron.com

🌐 http://www.deltron.com/

Precision linear translation tables and slides, bearings, rollers, and guides.

Directed Perception, Inc. 203460

1485 Rollins Rd.
Burlingame, CA 94010
USA

((650) 342-9399

(650) 342-9199

info@DPerception.com

🌐 http://www.dperception.com/

Camera pan-tilt mechanism.

Emerson Power Transmission Manufacturing 204030

8000 W. Florissant Ave.
St. Louis, MO 63136
USA

((314) 553-2000

(314) 553-3527

🌐 http://www.emerson-ept.com/

Mondo major manufacturer of power transmission and motion products. Brands include:

Browning—World leader in V-belt drives

Morse—Roller chain drives

SealMaster—Bearings, rod ends

US Gearmotors—Fractional horsepower AC and DC gearmotors

Rollway—2,000 types of bearings

Kop-Flex—Industrial shaft couplings

Hiwin Technologies Corporation 204252

520 Business Center Dr.
Mount Prospect, IL 60056
USA

((847) 827-2270

(847) 827-2291

🌐 http://www.hiwin.com/

Providers of ball, ACME, and leadscrews; DC motors, linear actuators, linear bearings, rails and guides; positioning tables, motor control circuits, and stepper motor drives.

igus GMBH 203444

igus GmbH
Spicher Straße 1 a
D-51147 Köln
Germany

(+49 (0) 2203 96490

+49 (0) 2203 96492

🌐 http://www.igus.de/

Makers of polymer (plastic) bearings, chain, linear slides, and other mechanicals. Web site is available in many languages, including English and German.

Rod end.

What's a Leadscrew?

Leadscrew is a generic term for a threaded rod that translates rotary motion to linear motion. The rod is connected to the shaft of a motor. Situated along the length of the rod is a *nut*, which is connected to some platform, or *carriage*. When the rod turns, the nut travels forward or back, which causes the carriage to likewise move.

Leadscrews translate rotational motion to linear motion.

There are many types of leadscrews, and some folks reserve the term for a specific type of rod that has a specific kind of thread. However, the point remains that the rod has threads machined along its length, and that it couples with a nut with compatible machining so the threads will properly engage.

Threaded Rod, Acme Rod, Ballscrew Rod

Rod stock is often used in robotics (and machinery tools such as CNC) to translate rotational motion from a motor to linear motion. The motor shaft is attached to the rod; a nut is threaded onto the rod, and the nut is connected to some carriage that slides back and forth.

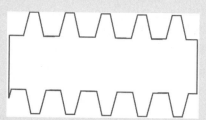

Acme leadscrew rod.

The generic term for these devices is *translation table*, or *translation stage*. When two such translation tables are connected to form a cross or T, then they are called an X-Y table, because they provide linear movement along both the X and Y axes.

There are many types of rods:

Ballscrew rod.

- *Threaded rod* can be found at hardware and home improvement stores. It's like a long version of a bolt, but without the bolt head. The rod is threaded down its entire length. The threads are V shaped, just like a machine bolt. Threaded rod is inexpensive and is suitable for applications where accuracy is not critical.

- *Acme rod* is especially made for building translation tables. The threads are more squared off, and, of course, a special nut is required to couple with the rod. Acme rod is an ideal "middle ground" that offers reasonable price with reasonable accuracy.

- A *ballscrew rod* (or coil rod) is intended to couple with a ballscrew nut. These nuts use recirculating ball bearings that run in tiny channels. The ball bearings ride against shallow, rounded threads in the rod. Ballscrew rods and nuts are the most expensive of them all, because of the mechanicals that go into the recirculating ball bearings. They are seldom needed for amateur robotics applications.

Uses for Linear Motion Products in Amateur Robotics

Translation tables and leadscrew mechanisms are the primary domain of manufacturing and factory automation, where machines do the work of humans—only in less time and often with greater precision. Their application in amateur robotics is less common, though no less compelling.

For starters, you can always build a translation table for use with your wood router in order to build robot parts. Used with a computer, you can cut out and mill parts with accuracies far exceeding that of a skilled craftsman. You might make a set of legs—all exactly alike—for your walking robot, for example. The possibilities are endless.

Linear motion is also used in many forms of robotic arms and grippers. For instance, a roving wheeled robot might be equipped with a kind of forklift arm in the front that can grab objects. With a linear mechanism, the forklift raises up or down to carry objects around the room.

This Rokenbok radio-control "robotic" toy is outfitted with a linear motion forklift gripper.

A walking robot can use linear motion to lift its legs for each step. Photo Jim Frye, Lynxmotion, Inc.

Finally, certain advanced legged robots use linear motion mechanicals to replicate walking gaits. One motor swings the leg back and forth, and another motor lifts the footpad of the leg. To walk forward, the leg is lifted up and positioned forward. The leg is lowered, and then moved to the back. The process repeats itself for each step of each leg.

Build Your Own Linear Translation Table—For Under $20

Industrial linear translation tables cost big bucks because they are engineered to provide high accuracy under fairly heavy loads. But for many home-shop and amateur robotics applications, supreme accuracy under heavy-duty use is not critical. This makes it possible to build a workable linear motion translation table for little spending money. Nearly all of the materials are available at hardware stores. Here's how:

A clamping-style coupler can be used to connect the all-thread to a motor shaft.

Start with a length of 3/8-inch all-threaded rod. The rod is available in 1-, 2-, and 3-foot lengths. Get the length you need for your application—if you're building a translation table that requires an 18-inch travel, you'll need a 2-foot length. The threaded rod will act as the leadscrew.

For the leadscrew nut purchase one 3/8-inch coupler. The coupler looks like a very long hex nut, and in fact, that's exactly what it is. You will need to glue this nut to the moving carriage of your translation table. Use a fast-setting epoxy. Don't use superglue or its ilk, as it won't stand up to the mechanical stress.

You will need at least one guide shaft, and two is preferable. Look for a pair of 1/2- or 3/8-inch-diameter steel cold-rolled shafts at the same place in the hardware store as you found the all-thread.

The carriage transport can be made out of a block of wood or plastic. Plastic is marginally better as it doesn't swell with moisture or heat, but a good hardwood such as birch will also work (don't use oak as it's too heavy, and don't use pine because it absorbs moisture like a sponge). Drill three holes in a triangular configuration, with the center hole near the top. The two holes for the shafts should be drilled carefully and should be only slightly larger than the diameter of the shafts. Drill the larger center hole so that it accommodates the 3/8-inch coupler. Measure the coupler at its widest point, and select a drill bit accordingly.

The front of the translation table requires careful drilling as well. Drill two holes for the guide shafts just large enough so that you can firmly seat the shafts into the wood. You can always apply epoxy if there's some play. You may wish to drill the holes for the car-

What's a Leadscrew (continued)

riage and the front panel by sandwiching them together, as this will help ensure proper alignment.

The two holes for the guide shafts in the rear of the translation table should be larger so the shafts will freely "float" by a few millimeters. This helps prevent the carriage from binding. Do not glue these shafts into position.

The one item you might not be able to find at the hardware store is a suitable coupler to connect your motor to the all-thread leadscrew. A clamping-type coupler can be used to attach shafts of different diameters together. Select a coupler for the diameter of the all-thread (in this case, 3/8-inch) and the diameter of your motor. Suppose your motor has a 1/4-inch shaft; you need a 3/8-inch to 1/4-inch reducing coupler. Sources for couplers are found in the **Power Transmission** section.

Kerk Motion Products, Inc.   203505

1 Kerk Dr.
Hollis, NH 03049
USA

((603) 465-7227
    (603) 465-3598
✉   info@kerkmotion.com
🌐   http://www.kerkmotion.com/

Linear mechanicals; leadscrews, acme nuts, linear rails, and spline rails.

Linear Industries, Ltd.   203511

1850 Enterprise Way
Monrovia, CA 91016
USA

((626) 303-1130
    (626) 303-2035
🚫   (800) 821-2875
🌐   http://www.linearindustries.com

Motion control mechanicals. Serving California and the Northwestern, U.S.

Machine Systems Ltd   202818

7974 Jackson Rd.
Ann Arbor, MI 48103
USA

    (734) 424-0202
🚫   (800) 386-6404
✉   sales@machsysinc.com

🌐   http://www.machsysinc.com/

Sellers of ballscrews, leadscrews, X-Y translation tables, linear guides, and other motion mechanicals.

SEE ALSO:

http://www.ballscrews.ws/

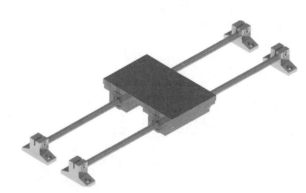

Two-rod linear slide.

Merlin Systems Corp, Ltd.   202086

ITTC Tamar Science Park
1 Davy Road
Derriford, Plymouth
PL6 8BX
UK

(+44 (0) 1752 764205
    +44 (0) 1752 772227
✉   info@merlinsystemscorp.co.uk
🌐   http://www.merlinsystemscorp.co.uk/

Makers of the Humaniform Muscle, a lightweight actuator technology ideal for robotics. Other products include:

- MIABOTS-Intelligent autonomous micro robots
- LEX Sensor-Digital absolute position sensor
- Humaniform robotics and control systems technology
- Stretch sensor

Minarik Corporation 203510

905 East Thompson Ave.
Glendale, CA 91201
USA

- (818) 637-7500
- (818) 637-7509
- (800) 427-2757
- http://www.minarikcorp.com/

Full-line mechanical (bearings, shafts, gears, chain, etc.); electronics (PWM drives and sensors); online ordering plus many local warehouses within the U.S.

Motion Systems Corporation 204136

600 Industrial Way West
Eatontown, NJ 07724
USA

- (732) 222-1800
- (732) 389-9191
- appengineer@motionsystem.com
- http://www.motionsystem.com/

Electromechanical linear actuators. Offers technical data for all models.

Nook Industries 203506

4950 East 49th St.
Cleveland, OH 44125-1016
USA

- (216) 271-7900
- (216) 271-7020
- (800) 321-7800
- nook@nookind.com
- http://www.nookindustries.com/

Nook specializes in linear mechanicals (leadscrews and ACME screws), linear bearings, worm-gear actuators, and ballscrew products.

Linear slide.

Parker Hannifin, Inc. 203929

6035 Parkland Blvd.
Cleveland, OH 44124
USA

- (216) 896-3000
- (440) 266-7400
- (800) 272-7537
- http://www.parker.com/

Parker Hannifin is a multifaceted manufacturer of industrial automation components. Their main lines include linear actuators; pneumatic rotary drives; motors; structural framing; linear tables and slides; gearheads; and gantry robots. Most products are available only through distributors; however, some (like Compumotor division motors) can be purchased directly.

PIC Design 202483

86 Benson Rd.
P.O. Box 1004
Middlebury, CT 06762
USA

- (203) 758-8272
- (203) 758-8271
- (800) 243-6125
- sales@pic-design.com
- http://www.pic-design.com/

Precision mechanical components, motion control mechanicals, X-Y translation tables, leadscrews, belts, pulleys, and gear products.

Rockford Ball Screw 204250

3450 Pyramid Dr.
Rockford, IL 60119
USA

(815) 874-9532

(815) 874-7897

(800) 475-9532

Open-style linear bearing.

http://www.rockfordballscrew.com/

Rockford makes leadscrews, rated by load (such as 0 to 1,000 pounds, 1,001 to 2,000 pounds, etc.), screw diameter, and pitch.

Rockwell Automation   203871

Firstar Building
777 East Wisconsin Ave.
Ste. 1400
Milwaukee, WI 53202
USA

(414) 212-5200

(414) 212-5201

http://www.rockwellautomation.com/

Rockwell Automation manufactures a broad line of automation electronics and components: stepper motors, servo motors, encoders, linear motion products-about 500,000 items in all . . . a little too many to list here. Products are available from distributors and through company online stores.

Covers the following brands:

Allen Bradley (http://www.ab.com/) and Electro-Craft

Reliance Electric (http://www.reliance.com/)

Dodge PT (http://www.dodge-pt.com/)

Rockwell Software
(http://www.software.rockwell.com/)

Ballscrew assembly.

Secs, Inc.   202123

520 Homestead Ave.
Mt. Vernon, NY 10550
USA

(914) 667-5600

(914) 699-0377

http://www.prosecs.com/

Gears and such:

- Gears, gearheads, and gear assemblies
- Bearings
- Belts and pulleys
- Shafts
- Differentials
- Speed reducers
- Sprockets
- Linear ball sides
- Linear actuators
- Clamps
- Couplings
- Clutches

Seitz Corp.   202484

212 Industrial Ln.
Torrington, CT 06790-1398
USA

(860) 496-1949

(800) 261-2011

http://www.seitzcorp.com/

Plastic gears, gears, and motion control mechanicals.

Servo Systems Co.   202599

115 Main Rd.
P.O. Box 97
Montville, NJ 07045-0097
USA

((973) 335-1007
 (973) 335-1661
○ (800) 922-1103
✉ info@servosystems.com
🌐 http://www.servosystems.com/

Servo Systems Co. is a full-service motion control distributor and robotic systems integrator. The Web site contains copious descriptions and technical data on their industrial components. The company also sells motion mechanicals, such as linear stages. Be sure to check out their "surplus bargains" pages for affordably priced servos, as well as other gear.

Specialty Motions, Inc.   204251

22343 La Palma Ave.
Ste. 112
Yorba Linda, CA 92887
USA]

((714) 692-7511
 (714) 692-7510
○ (800) 283-3411
🌐 http://www.smi4motion.com/

Specialty Motions makes and sells (online or through distributors) pro-grade linear motion mechanicals: leadscrews, ACME screws, ballscrews, linear bearings, guides, rails, and more. The Web site contains plenty of technical documents on building and using linear motion gear, and the online e-store provides spec information and (usually) a picture or mechanical drawing.

Ballscrew.

Techmaster Inc.   203481

N93 W14518 Whittaker Way
Menomonee Falls, WI 53051
USA

((262) 255-2022

 (262) 255-4052
🌐 http://www.techmasterinc.com/

Techmaster is a distributor of aluminum structural framing (machine framing) and motion control products.

Thomson Industries, Inc.   202827

2 Channel Dr.
Port Washington, NY 11050
USA

((516) 883-8000
 (516) 883-7109
○ (800) 554-8466
✉ Thomson@thomsonmail.com
🌐 http://www.thomsonindustries.com/

Thompson manufactures a broad line of linear bearings, X-Y translation stages, linear actuators, ballscrews, and other products for linear motion contraptions. The company offers "designer guide" literature, datasheets, and application notes (in Adobe Acrobat PDF format).

Thomson linear slide.

Warner Electric   203437

449 Gardner St.
South Beloit, IL 61080
USA

((815) 389-3771
 (815) 389-6425
○ (800) 234-3369
✉ info@warnerelectric.com
🌐 http://www.warnernet.com/

Warner Electric manufactures industrial power transmission products including clutches and brakes, as well as contact and noncontact sensors.

ACTUATORS-MOTORS

Motors that turn under electric power. Motor types included in this section are permanent magnet (PM), brushed, brushless and coreless continuous DC motors; servo motors (but not R/C servomotors); stepper motors; motors with gearheads attached.

Many of the listings in this section are for manufacturers of motors, and the majority do not sell directly to the public. However, links to distributors and customer representatives are provided on the manufacturers' Web sites. And, let's not forget that manufacturer Web sites are gold mines for datasheets (including for motors you might have gotten surplus), application notes, and technical white papers.

SEE ALSO:

ACTUATORS-PNEUMATICS: Air power

ACTUATORS: SHAPE MEMORY ALLOY: Electrically driven wire "muscles"

DISTRIBUTOR/WHOLESALER-INDUSTRIAL ELECTRONICS: For buying new motors

MOTOR CONTROL: Circuitry for operating motors found in this section

POWER TRANSMISSION: Mechanical drive components like gears and sprockets

RADIO CONTROL-SERVOS: R/C servo motors for model airplanes and cars

RETAIL-GENERAL ELECTRONICS: Mail order and local retail for buying new motors

RETAIL-SURPLUS MECHANICAL: Mail order and local retail for buying salvaged motors

SENSORS-ENCODERS: Feedback mechanisms for motors

A. O. Smith   203948

11270 W. Park Pl.
Milwaukee, WI 53224-9508
USA

✆ (414) 359-4000

📠 (414) 359-4064

🌐 http://www.aosmithmotors.com/

A. O. Smith manufactures small DC, fractional and sub-fractional-horsepower motors. Motor catalogs and spec sheets are available at the Web site.

Anaheim Automation   203517

910. East Orangefair Ln.
Anaheim, CA 92801-1195
USA

✆ (714) 992-6990

📠 (714) 992-0471

✉ webmaster@anaheimautomation.com

🌐 http://www.anaheimautomation.com/

Industrial motor control systems (motors, controllers, and tachs) and X-Y tables.

Animatics Corporation   203133

3050 Tasman Dr.
Santa Clara, CA 95054
USA

✆ (408) 748-8721

📠 (408) 748-8725

✉ response@animatics.com

🌐 http://www.smartmotor.com/

Industrial servomotors and controllers.

ARSAPE   202510

Rue Jardinire 33
2306 La Chaux-de-Fonds
Switzerland

✆ +41 32 910 6050

📠 +41 32 910 6059

✉ info@arsape.ch

🌐 http://www.arsape.ch/

Manufacturer of miniature drive systems, such as stepper motors, servo motors, precision gearheads, and drive electronics. The company is based in Switzerland, but the products are available worldwide through distributors and manufacturer's representatives; see the Web site for a list of distributors.

The Web site is in English.

Astro Flight Inc.   203303

13311 Beach Ave.
Marina Del Rey, CA 90292
USA

((310) 821-6242

(‍ (310) 822-6637

✉ info@astroflight.com

🌐 http://www.astroflight.com/

Astro Flight's high-powered gear motors are meant for model airplanes, cars, and boats, but they are also used in Robot Warriors robots. These top-quality motors are available with planetary gearboxes. The company also sells a line of high-performance gearboxes and electronic speed controls.

Astro Flight cobalt motors. Photo Astro Flight Inc.

Aveox, Inc.   204126

31324 Via Colinas
Ste. 103
West Lake Village, CA 91362
USA

((818) 597-8915

(‍ (818) 597-0617

🌐 http://www.aveox.com/

Aveox provides high-performance brushless DC permanent magnet brushless servo motors for the remote-control hobbyist, as well as for industrial motion control. Sells motors and controllers.

Baldor Electric Company.   203222

5711 R.S. Boreham, Jr. St.
P.O. Box 2400
Fort Smith, AR 72901
USA

((501) 646-4711

(‍ (501) 648-5792

⊘ (800) 828-4920

🌐 http://www.baldor.com/

Baldor is a manufacturer of industrial motors. Product suitable for robotics include DC, DC gearhead, and servomotors. The Web site sports a nice product cross-reference style of product selection. Prices are provided. You won't be using these to build a small tabletop bot, but they're good candidates for your next combat robot. Products are sold through local distributors.

Barber Colman   300005

🌐 http://www.eurotherm.com/

See also the listing for Invensys Plc (in the Actuators section).

Bison Gear & Engineering   300001

3850 Ohio Ave.
St. Charles, IL 60174
USA

((630) 377-4327

(‍ (630) 377-6777

⊘ (800) 282-4766

🌐 ttp://www.bisongear.com/

Bison Gear distributes industrial motors for medium-duty applications. Several of their 12-volt parallel shaft and right-angle gearmotors might be useful in larger robots, especially combat style. They offer online buying, spec sheets, technical information, and a regular newsletter.

Buehler Motor GmbH   300003

P. O. Box 45 01 55
90212 Nuernberg
Germany

(+49 (0) 9114 5040

(‍ +49 (0) 9114 54626

✉ marketing@buehlermotor.de

🌐 http://www.buehlermotor.com/

Buehler manufactures (and sells through distributors and representatives) compact permanent magnet DC motors, with and without gearheads. Their products are occasionally found on the surplus market, and the Web site is handy for its repository of technical specs.

Sales offices are in Germany and the U.S.; Web site is in German, English, and French.

Compumotor / Parker Hannifin Corp. 203928

5500 Business Park Dr.
Rohnert Park, CA 94928 7904
USA

- ((707) 584-7558
- (707) 584-2446
- Ø (800) 358-9068
- 🌐 http://www.compumotor.com/

Compumotor manufactures and sells (online or through catalog sales) a wide range of high-end industrial motors and motion control products. Lines include:

- Controllers
- Servo and stepper drives and drive/controllers
- Motors: linear servo, rotary servo, linear stepper, and rotary stepper
- Incremental encoders

The Compumotor.com Web site contains a copious amount of free technical literature, such as white papers on servo design, DC and stepper motor construction basics (written for the engineer), and datasheets. Their product catalogs are available on CD-ROM, online, or as printed books.

Compumotor is a division of Parker Hannifin; see the main Web site:

http://www.parker.com/

Servo motor. Photo Compumotor-Parker Hannifin Corp.

Densitron Technologies plc 202175

Unit 4, Airport Trading Estate
Biggin Hill
Kent
TN16 3BW
UK

- (+44 (0) 1959 542000
- +44 (0) 1959 542001
- ✉ europe.sales@densitron.net
- 🌐 http://www.densitron.com/

Densitron is a maker of electromechanical components-including rotary solenoids, stepper motors, and electric clutches, as well as computer-related product such as displays.

Donovan Micro-Tek 203980

67 W. Easy St.
Ste. 112
Simi Valley, CA 93065
USA

- ((805) 584-1893
- (805) 584-1892
- Ø
- ✉ info@dmicrotek.com
- 🌐 http://www.dmicrotek.com/

Itty-bitty stepper motors-8, 10, and 15mm. Also carries encoders, drive electronics, and gearboxes. Everything is on the small side.

EA Electronics 204090

8 Maple St.
Ajax, ON
L1S 1V6
Canada

- ((905) 619-1813
- ✉ email@eaelec.com
- 🌐 http://www.eaelec.com/

Purveyors of motors and motor control electronics for remote-control model ships. But what's a robot if not a ship with wheels instead of a rudder? In other words, just about everything for R/C model ships will work in a robot, too. Their speed-control boards are designed for some high-current DC motors and will handle 10 to 20 amps.

Among the products offered:

- Multifrequency speed control
- Optically isolated smart control
- DC PM Motors (from Johnson, Coleman, and Pittman, the big names in small DC motors)

Electronics are sold though distributors; motors sold directly.

Edmond Wheelchair Repair & Supply 203777

1604 Apian Way
Edmond, OK 73003
USA

☏ (405) 359-5006
⊘ (888) 343-2969
✉ Sales@edmond-wheelchair.com
🌐 http://www.edmond-wheelchair.com/

Wheelchair and scooter parts, including motors, wheels, and batteries.

Electric Motor Warehouse   300002

G 1460 E. Hemphill
Burton, MI 48529
USA

☏ (810) 744-1240
☏ (810) 744-1424
⊘ (877) 986-6867
✉ sales@electricmotorwarehouse.com
🌐 http://www.electricmotorwarehouse.com/

Online retailer of motors for home and light industry. They have a selection of Leeson and Dayton brand fractional and subfractional-horsepower DC motors (geared and nongeared) for the combat robot builders out there. They also offer a nice range of 12-volt parallel and right-angle shaft Dayton gearmotors. Prices are reasonable.

Elite Speed Products   203305

3923 East Mound St.
Columbus, OH 43227
USA

☏ (614) 231-4170
✉ sales@elitespeedproducts.com
🌐 http://www.elitespeedproducts.com/

Ni-cad batteries, electric motors, and motor replacement parts (brushes, armatures, etc.) for R/C applications.

Ellis Components   204216

Stonebroom Industrial Estate
Stonebrook
Alfreton
Derbyshire
DE55 6LQ
UK

☏ +44 (0) 1773 873151
☏ +44 (0) 1773 874645
✉ sales@ellis-components.co.uk
🌐 http://www.elliscomponents.co.uk

Resellers of Bosch 12- and 24-volt DC automotive motors.

Enigma Industries   204214

P.O. Box 27522
Anaheim, CA 92809-0117
USA

✉ info@EnigmaIndustries.com
🌐 http://www.enigmaindustries.com/

Enigma sells a unique real-world simulation program for developing drive trains for coaxially driven (wheel on either side) robots and especially larger robots for combat. You enter the technical details of the motor and the details of your robot design, including gear ratios, tire diameters, motor voltage and capacity, and weight. The program then simulates how the robot will perform given the motor you've selected.

The technical specs of common motors are already in the program's database, and you can edit those figures if needed, or create new entries. The program simulates battery voltage, state of charge, internal resistance, as well as over- and under-volting of the motor (e.g., the performance of motor driven at 12 volts, but whose faceplate rating is 19 volts).

Another unique aspect is that the database includes the approximate price and sources for the motors.

Large motor, with keyway shaft.

Estimation of Motor Torque   204058

🌐 http://www.westernelectric.com.au/m_select/select1.htm

From Western Electric in Australia, here is a semitechnical brief on estimating motor torque requirements..

See other technical papers and datasheets at the following site:

http://www.westernelectric.com.au/

EVdeals   204218

9 South St.
Plainville, MA 02762
USA

☎ (508) 695-3717
🖷 (508) 643-0233
✉ scott@EVdeals.com
🌐 http://www.evdeals.com/

Electric motors for bikes and scooters, 12- to 48-volt batteries and battery chargers.

Faulhaber Group   204239

Postfach 1146
D-71094 Schnaich
Germany

☎ +49 (0) 7031 6380
🖷 +49 (0) 7031 6381
🌐 http://www.faulhaber.com

Maker of miniature and microminiature motors. The Faulhaber Group consists of the following motor manufacturers:

Faulhaber GMBH (Germany)—
http://www.faulhaber.de/

MicroMo Electronics, Inc. (U.S. sales)—
http://www.micromo.com/

MiniMotor SA (Europe)—
http://www.minimotor.ch/

GE Industrial Systems   203947

41 Woodford Ave.
Plainville, CT 06062
USA

☎ (860) 747-7111
🖷 (860) 747-7393
🌐 http://www.geindustrial.com/

GE Industrial Systems supplies a wide range of products for residential, commercial, industrial, institutional, and utility applications. Mostly high-end industrial components; their product is also a common find on the surplus market. The GE Industrial Web site provides downloadable catalogs and technical references. Will sell industrial product direct, limited to territory.

Among some of the cogent GE products for larger robots:

- Contactors
- Controllers and I/O
- Drives
- Embedded computers
- Motion control
- Motor control centers
- Motors
- Sensing solutions
- Sensors, solenoids, and limit switches

Globe Motors   203882

2275 Stanley Ave.
Dayton, OH 45404-1249
USA

☎ (937) 228-3171
🖷 (937) 229-8531
🌐 http://www.globe-motors.com/

In the words of the Web site, "Globe Motors designs, manufactures, and distributes precision, subfractional horsepower motors and motorized devices throughout the world."

Main product line is industrial DC and DC gearhead motors, including both brushed and brushless. Most products are described in Web catalogs and consist of individual Adobe Acrobat PDF files. Products are sold by a network of distributors.

The company also supplies a line of specialty products including rotary actuators and linear actuators (screwjacks).

Globe motor.

Motors for Amateur Robots

Electric motors are used to "actuate" something in your robot: its wheels, legs, tracks, arms, fingers, sensor turrets, or weapon systems. There are literally dozens of types of electric motors (and many more if you count gasoline and other fueled engines), but for amateur robotics, the choice comes down to these three:

A variety of different motors types suitable for small robots.

- In a *continuous DC motor*, application of power causes the shaft to rotate continually. The shaft stops only when the power is removed, or if the motor is stalled because it can no longer drive the load attached to it. These motors are listed in the section you are reading now, **Actuators-Motors**.

- In a *stepping motor*, applying power causes the shaft to rotate a few degrees, then stop. Continuous rotation of the shaft requires that the power be pulsed to the motor. As with continuous DC motors, there are subtypes of stepping motors. Permanent magnet steppers are the ones you'll likely encounter, and they are also the easiest to use. Stepper motors are also listed in this section.

- A special subset of continuous motors is the *servo motor*, which in typical cases combines a continuous DC motor with a "feedback loop" to ensure accurate positioning. There are many, many types of servo motors; a common form is the kind used in model and hobby radio-controlled cars and planes. These are called R/C (radio-controlled) servos. R/C servo motors are listed in **Radio Control-Servos**.

Choosing a Motor Type

With three common motor types for amateur robots to pick from—DC, stepper, and servo—it can be hard to know which one is best. The answer is not simple, because each motor type has its own pros and cons.

Motor Type	Pros	Cons
Continuous DC	• Wide selection available, both new and used. • Easy to control via computer with relays or electronic switches. • With gearbox, larger DC motors can power a 200-pound robot.	• Requires gear reduction to provide torques needed for most robotic applications. • Poor standards in sizing and mounting arrangements.
Stepper	• Does not require gear reduction to power at low speeds. • Low cost when purchased on the surplus market. • Dynamic braking effect achieved by leaving coils of stepper motor energized (motor will not turn, but will lock in place).	• Poor performance under varying loads. Not great for robot locomotion over uneven surfaces. • Consumes high current. • Needs special driving circuit to provide stepping rotation.
R/C servo*	• Least expensive nonsurplus source for gear motors. • Can be used for precise angular control or for continuous rotation (the latter requires modification). • Available in several standard sizes, with standard mounting holes.	• Requires modification for continuous rotation. • Requires special driving circuit. • Though more powerful servos are available, practical weight limit for powering a robot is about 10 pounds.

*The discussion is limited here to R/C servo motors; there are other types of servo motors, but most are mondo expensive.

Bear in mind that all motors are available in different sizes.

- Small motors are engineered for applications where compactness is valued over torque. While there are small high-torque motors, these tend to be expensive because they use rare earth magnets, high-efficiency bearings, and other features that add to their cost.

- Large motors may produce more torque, but also require higher currents. High-current motors require larger capacity batteries and bigger control circuits that won't overheat and burn out under the load. Therefore, match the size of the motor with the rest of the robot. Don't overload a small robot with a large motor when big size isn't important.

- When decided on the size of the motor, compare available torque *after* any gear reduction. Gear reduction always increases torque. The increase in torque is proportional to the amount of gear reduction: If the reduction is 3:1, the torque is increased by about three times (but not quite, because of frictional losses).

Motor Specifications

Motors are rated by their voltage, current draw, output speed, and torque.

Operating Voltage

Operating voltage specifies the nominal (normal) voltage the manufacturer recommends for the motor. Most small DC motors are designed for 1.5- to 12-volt operation, with the majority in the 3- to -6-volt range. Larger DC motors designed for heavy-duty applications usually require 12 to 24 volts, with some needing upward of 90 volts. Often, but not always, the higher the voltage, the more powerful the motor (this does not apply to stepper motors, where very low voltages—on the order of just a few volts—are common for heavy-duty motors). Most motors can be run at operating voltages higher or lower than the specified rating.

- Lower voltages reduce torque and speed.
- Higher voltages increase torque and speed.

(Note: For stepper motors, speed is not greatly altered by changing the voltage.)
Avoid applying excessively high voltages to a motor, or else it could overheat and burn out. R/C servo motors contain their own electronics which are not designed for use over about 7.2 volts.

Current Draw

Current draw is the amount of current, specified in milliamps or amps, that the motor requires to produce a certain amount of torque. Motors consume different amounts of current depending on how they are operated:

- *No-load*. A motor that doesn't have anything attached to its shaft isn't doing any work and is said to be free-running. No-load current tends to be very low.

- *Load*. As the motor does work, its load and current draw increases. Manufacturers rate the current draw under load using different standards, making it hard to judge a motor using this specification alone.

- *Stalled*. When the motor shaft stops rotating, it "stalls" and draws as much current as will flow through the windings. This specification is useful for "worst-case scenario" engineering planning.

- *Shorted*. Maximum current flows into the motor when the coils are shorted out. The motor will not run, and like any short circuit, if operated in this way for any length of time, serious damage can result to other systems on the robot.

Torque

Torque is the way the strength of the motor is measured. It is typically calculated by attaching a lever to the end of the motor shaft and a weight or gauge on the end of that lever. The length of the lever usually depends on the unit of measurement given for the weight. Examples:

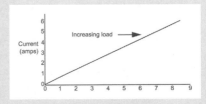

Example motor curve.

- Metric measurements use Newton meters (Nm), kilograms-force meters (kgf-m), or sometimes gram-centimeters (gm-cm).
- Standard measurements use ounce-inches (oz-in), or pound-feet (lb-ft), or pound-inches (lb-in). It's common to reverse the nomenclature and call it foot-pounds and inch-pounds.

Speed

The *speed* of the motor indicates how fast its shaft is turning. DC motors without a gearbox spin at 3,000 to over 12,000 rpm (revolutions per minute). With a gearbox, the speed can vary from under 1 rpm on up.

Stepping motors are not rated in rpm, but pulses (or steps) per second. The speed of a stepper motor is a function of the number of steps required to make one full revolution times the number of steps applied to the motor each second. Typical values are 200 or 300 pps.

Motor Specs: Comparing Apples to Oranges

Motor torque specifications are not standardized, both in the way the torque is measured and the way that torque is specified. This makes it difficult to adequately compare the torque of motors to determine which one(s) is best suited for the job.

Why the disparity? Many motors are designed expressly for a specific application, and the specifications are tailored to be meaningful for that application. Amateur robot builders don't typically buy 10,000+ motors at a time, so we have to buy "off the rack" rather than have something specially made for us. Naturally, we also have to accept whatever

A high-current high-power gear motor commonly used in heavy-duty robot applications.Photo National Power Chair.

type and format of specifications that are provided with the motor. Torque may be provided in ounce-inches for one motor and Newton meters for another.

Basic Torque Conversion Factors

Fortunately, it's possible to compare these disparate measurement schemes with simple math. Here are some basic conversion equivalents:

1 pound-foot = 1.35 Newton meter

1 Newton meter = 8.851 pounds-inches

1 pound-foot = 12 pounds-inches

16 ounce-inches = 1 pound-inch

1 Newton = 1 kilogram-meter per second squared

Cross Comparison Chart

The following table shows the relationships between the most commonly used torque specifications: ounce-inches, pound-inches, pound-feet, Newton meters, and kilograms per force meters.

oz-in	lb-in	lb-ft	Nm	kgf-m
1	0.0625	0.0052	0.0071	0.0007
16	1	0.0833	0.113	0.0115
192	12	1	1.356	0.1383
141.6	8.851	0.7376	1	0.1
1416	86.8	7.231	9.807	1

Specifications in Horsepower?

AC-operated motors are often specified in horsepower (hp) rather than torque. While both horsepower and torque represent the strength of the motor, they are not the same thing. (Specifically, horsepower calculations also include the amount of work performed during a period of time, such as one second, while torque does not.)

Small DC-operated motors for amateur robotics are not rated in horsepower, because the ratings are less than 1/50 or even 1/100 hp. Manufacturers rely on horsepower ratings for larger motors; you may see it for big brute motors intended for electric wheelchairs.

For the curious, 1 hp is equal to 550 lb-ft per second, or 76.04 kgf-m per second.

Converters on the Web

Several motor manufacturers and other Internet resources offer torque converter calculators on their Web sites. Check the following:

Bodine Electric

http://www.bodine-electric.com/

Online Conversion

http://www.onlineconversion.com/

Convert Me

http://www.convert-me.com/

Anatomy of a Permanent Magnet DC Motor

By far, the most common motor used in amateur robotics is the small *permanent magnet* (PM) variety. It runs off direct current (DC) and comes in thousands of shapes and sizes. But despite all the variations of PM motors, they all work about the same.

A Peek Inside

At the center of the motor is the *central shaft*; wrapped around this shaft is a core of windings, called the *armature*. The armature turns within a cavity of the motor that is made of *magnets*. Though multiple sets of magnets may be used, the fundamental design of PM motors uses two *field magnets*, one with its north pole facing the armature, and one with its south pole facing the armature. The two magnets are placed on opposite sides of one another inside the motor.

Electricity is applied to the windings, and this causes the motor to turn within the magnetic field. The direction of the motor is determined by the polarity of the DC voltage applied to the two terminals of the motor.

Brushed and Brushless

Permanent magnet motors can be brushed or brushless. The *brushed* variety uses small contacts, called *brushes*, that make intermittent contact with the windings on the armature. As the armature spins, the brushes touch an insulated metal collar (the *commutator*) that is positioned on the end of the armature. The commutator is typically split in three electrically insulated sections, which each third going to windings on the armature. As the motor spins, the electrical field is then alternated between the windings.

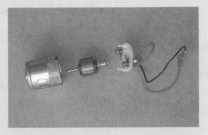

What the inside of a PM motor looks like.

In the least expensive motors, the brush is really just a piece of flexible copper wire, but on fancier motors, real metal brushes or pieces of carbon are used. These brushes can be replaced as part of regular motor maintenance. You won't find replaceable brushes on smaller DC motors, but they are common on larger units, such as wheelchair drives.

Brushes can wear out over time; *brushless* motors are a variation on the theme where the alternating field through the coils comes from electronic control, rather than mechanical contact via brushes. Brushless motors are more complicated to drive, but can last longer and are often more efficient.

Speed Control

The speed of the motor is primarily determined by its input voltage. Put through more voltage, and the armature spins faster. The actual speed depends on several factors, however, including the load on the motor. The speed is the fastest when the motor spins freely (no load). Speed decreases for a given voltage as the load on the motor shaft increases.

A simple method of controlling the speed of a motor is to vary its voltage with a rheostat. Turning the dial increases or decreases the voltage reaching the motor, and so the motor changes speed. This is a rather inefficient, however; most motor speed controllers use a technique known as *pulse width modulation* (PWM) where voltage to the motor is rapidly turned on and off several hundred or even several thousand times per second.

PWM works by varying the ratio of the "on" time to the "off" time. The longer the off time, the slower the motor, because the motor will receive less voltage. Likewise, a PWM period of 100% means the motor is fully on and runs the fastest. A PWM period of 0% means the motor is off and is stationary.

Selecting the Right Motor Supply Voltage

The design of the average permanent magnet motor does not require exact operating voltages. A motor "designed" for 12-volt operation will usually work just fine at 8 or 10 volts, or even at 14 to 16 volts. Operating the motor at a voltage higher or lower than its rating proportionally affects the speed of the motor, its current draw, and its torque.

- With a lower voltage, the motor will run slower, draw less current, and provide less torque.
- With a higher voltage, the motor will run faster, draw more current, and provide more torque.

There is some danger in operating a motor at too high a voltage. You'll want to carefully experiment with the motor to determine if it can withstand a higher voltage. Things to consider when "overvolting" a motor:

- Because the motor is turning faster, it can develop more internal friction, wearing out the bearings or bushings sooner than normal.
- With more current flowing through the windings, the motor can get hot. If it gets very hot, the windings may burn out.
- Excessive heat may partially demagnetize the magnets in the motor, resulting in diminished performance.
- With more current flowing to the motor, the control circuit or relay operating the motor may overheat and cook to extra crispy.

Typical Problems with Permanent Magnet Motors

Permanent magnet motors of all types, styles, designs, makes, and models are prone to failure, simply because they are fast-moving mechanical devices. Over time, motors can develop a variety of problems that can cause them to fail.

For small DC motors (those under about 1.5 inches in diameter and without replaceable brushes), repair is not practical; replace them with identical or similar units. Larger DC motors may be repairable, given the proper tools. However, bear in mind that disassembling a PM motor may result in diminished performance.

- *Shorted windings.* Applying too much current to the motor can overheat it, and the windings can short. When this happens the motor ceases to turn. A short can be determined with a volt-ohm meter. A reading of 0 ohms is a clear indication of a short. If the reading is nonzero, but low (1-2 ohms), spin the shaft of the motor and watch for a variation in the resistance. Also check for a short between the terminals and the metal case of the motor. It should read very high ohms—virtually an open circuit.
- *Dirty brushes or commutator.* These can be cleaned with alcohol, but if they are worn, replacement is necessary.
- *Dry or worn bearings.* The cheapest DC motors don't use internal bearings, but they may use washers that can crack or disintegrate. Repair or throw away. If the motor uses bearings, they must be replaced with the exact same kind.
- *Gummed lubrication, loss of lubrication.* Whether the motor uses brushes, washers, or nothing at all, the lubricant added at the factory may dry up, thin out over time, or gel up into a gooey mess. Applying a *very small* dab of synthetic grease fixes this situation, but be sure to keep the brushes and commutator clean! The motor will not work if these are mucked up with grease.

Be a Motor Mouth

Knowing the lingo of motors helps you understand their specifications, which can in turn help you select the right one for the job. Here's a short recap of the most common terms you'll encounter in motor specifications.

- *Breakdown torque.* The maximum torque of the motor that doesn't cause an abrupt change in either speed or output power.
- *Braking.* Any means of slowing down the motor. Brakes can be completely electronic, produced by grounding or shorting out the terminals of the motor. They can also be mechanical, similar in function to the brakes in a car. Brakes are more common in larger AC-operated motors.

- *Duty Cycle.* The amount of time the motor can be operated between off periods. Continuous duty cycle motors can be run 24/7; intermittent duty cycle motors are intended to be operated for periods of a few minutes at a time.

- *Full-load Amps* or *Full-load Current.* The amount of current, in amps or milliamps, the motor draws when it is operating at its rated output torque and voltage.

- *Full-load Torque.* See Torque Load.

- *Mounting.* The means by which the motor is mounted. Some motors are intended to be mounted by the face (the end where the motor shaft is). Machine screws hold it in place. Other motors are intended to be mounted by the gearbox. Others have no mounting holes at all and are intended to be clamped into place.

- *No-load Speed.* The speed of the motor, in revolutions per minute (rpm) when there is nothing attached to its shaft. The no-load speed is always faster than the load or output speed

- *Output Speed.* The speed, in revolutions per minute (rpm) of the motor in full-load condition, or free running (not turning any load), and at a given voltage. Output speed is affected by voltage to the motor and the load on the motor shaft.

- *Overhung load.* Also called radial force or radial load, the force applied at right angles to the motor shaft. This force may be the weight of the robot (if the wheels are attached directly to the motor shaft) or the force caused by a pulley, sprocket, or gear. Exceeding the overhung load can cause premature death of the motor.

- *Reversible.* Specifies if the motor can run in either clockwise or counterclockwise direction. Most DC motors are reversible, but not all.

- *RPM.* See Output Speed.

- *Sleeve bearings.* Specifies the type of bearings used for the output shaft of the motor. Cheap motors have no bearings at all, but instead use fiber, plastic, or metal shims. Well-made mechanical bearings make the motor quieter and last longer under heavier loads.

- *Stall Torque.* The amount of torque exhibited at the output of the motor when the shaft is prevented from moving. This torque is the result of the maximum amount of current possible flowing through the motor windings. Stall torque is specified without assumptions made regarding physical damage that can be caused by operating the motor in this condition (locking a shaft in a gear motor may actually tear the gearbox apart).

- *Torque.* The twisting force of the shaft of the motor (or of the output shaft of the gearbox, if the motor is attached to one). Torque can be specified in a number of different units, with inch-pounds being the most common in North America.

- *Torque Load.* The maximum torque produced by the motor without an abrupt change in speed or output power.

- *Voltage.* The specified operating voltage of the motor. Most motors can be operated at higher or lower voltages, though extremes should be avoided.

Of Fish, Weight Scales, and Torque

First a story. Thomas Edison was not only an inventor, he was a businessman. When he learned he was losing money because his phonograph players were being broken in rail shipment, he asked his engineers to come up with a stronger box, one that could withstand being dropped off a loading dock at a train station. Edison was also quite frugal, and given the number of phonographs he sold, he didn't want to make the box *too* strong, or else he'd spend money unnecessarily.

The engineers set off to work. A week later, Edison returned to find them still grappling with the problem. They had built a box, but were arguing over their calculations on whether it had the strength to withstand the drop. Disgusted, Edison put a phonograph into the box and shoved it off the loading dock. The box splintered into pieces. The box wasn't strong enough for the job—back to the drawing board!

The moral: Empirical discovery—testing variables under real-life conditions—is sometimes the fastest and most economical approach to solving a problem. You can use the same techniques with robots, like when determining the torque needs of the motors. Rather than complex calculations and math, a perfectly suitable approach is to directly measure the force needed to pull a stationary robot along the ground.

Measuring Torque with a Fish Scale

Here's how it's done: Attach a 6- to 8-foot length of string to a hanging spring scale, like the kind used for fishing. A scale that reads up to a couple pounds is sufficient. Attach temporary wheels, of the same diameter you plan to use, to a wooden box. The box should be the approximate weight and dimensions as your robot. Secure the end of the string to the tread of the wheel, along the top.

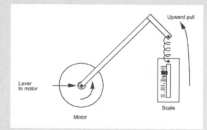

Use a fishing scale to measure the torque requirements of your wheeled robot.

Holding the scale in your hand, and with the string nearly parallel to the floor, pull the robot until it just starts to move. Note the reading on the scale. Then, do a simple bit of math:

Torque = Reading on the scale times the radius of the wheel

The result, *torque*, is in inch-pounds (or pound-inches, same thing).

Example: If the scale reads 1 pound, and the wheel is 10 inches in diameter (radius is half the diameter, or 5 inches), then the minimum torque needed to get the robot moving with one motor is 5 inch-pounds.

Note: You will need to convert ounces to decimal in order do the calculations. Each ounce is approximately 0.065 of a pound. So, if the scale reads 1 pound, 7 ounces, multiply 7 by 0.065 (result: 0.45), then add in the number of pounds (1.45). The scale, in decimal pounds, reads 1.45. It does *not* read 1.7!

Yet another method is to use a torque wrench, a common tool in the automotive garage. Securely attach the wrench to the hub of the wheel, and turn the wrench until the wheel begins moving. Don't turn the wrench more than you need to take a quick reading; you may ruin the wrench otherwise.

Interpreting the Results

Consider that most robots use at least two motors for locomotion, one on each side. For forward momentum, the motors will share the load; a reading of 5 inch-pounds represents the total torque to move the robot forward, but two motors need only develop (at minimum) half that amount.

In reality, however, it is not uncommon to steer a robot with just one motor, and therefore, you should rely on the single-motor torque test as the minimum torque needed to propel the robot. You will also want to take additional readings over various surfaces that the robot will travel. This includes carpet as well as carpet/tile thresholds (the "bump" that separates carpet from a tile, wood, or linoleum floor). If the robot is expected to run over uneven ground, take incline measurements as well. You can set up incline boards using lumber and stacks of books.

When It's Time for Math

Though empirical testing is a quick and fairly reliable method of determining the torque requirements for your robot, you may still want to learn how to use the various inertial formulas intended for these calculations. Most any engineering book has them. The work goes faster if you have a scientific calculator that can perform trig functions. The Microsoft Windows Calculator, when switched to Scientific mode, is suitable.

Selecting the Right Stepper Motor

There are many types of stepper motors, but the two encountered most often are the unipolar and bipolar. On the outside, both motors look the same; inside, the motors differ by their electrical connections and windings. Both unipolar and bipolar motors are like two motors sandwiched together and have two sets of windings. In a bipolar motor, the windings attach to the external power supply or driver circuit at their end points only; in a unipolar motor there is an extra set of taps at the center of the two windings.

Though variations exist, the typical bipolar stepping motor has four leads; unipolar steppers have 5, 6, or 8 leads, depending on how the windings are connected internally.

Unipolar and bipolar motors require different actuation techniques. Most self-contained stepper motor control ICs are designed for unipolar motors, and therefore tend to be less expensive. Bipolar motors must always be powered in bipolar mode. Unipolar motors can be powered in unipolar or bipolar mode. When used in bipolar mode, the center taps of the unipolar motor are left unconnected.

Some additional considerations:

- Bipolar motors provide slightly higher torque, but can't always reach the same top speeds of unipolar motors.
- Unipolar motors are more common on the surplus and used markets, making them less expensive.

Motor Sizes

Stepper motors come in lots of sizes, but many are a standardized *frame size*, with a standardized mounting plate. The following NEMA (National Electrical Manufacturers Association) frame sizes are for the height and width of the motor; the length is undetermined and can vary. The following sizes are for the mounting flange, which provides holes at each corner for attaching the motor.

Motor Frame Size	Mounting Flange Dimension	Flange Hole Centers
NEMA 17	40mm, 1.57 inches	1.22 inches
NEMA 23	56mm, 2.22 inches	1.856 inches
NEMA 34	3.25 inches	2.739 inches

Bigger NEMA frame sizes exist, of course, but are used infrequently in small mobile robots.

The length of the motor varies depending on its internal construction. Stepper motors are often said to be single-, double-, or triple-stacked, which denotes the number of "submotors" inside. The more stacks, the longer the motor. Additional stacks are used to increase the torque of the motor.

Single or Double Shaft

It is not uncommon for stepper motors to have a shaft on both ends of its frame. The shafts are usually the same diameter and length. One shaft is typically used to drive whatever load is required by the motor, and the other shaft is used for connection to some kind of feedback mechanism, such as a visible indicator or an optical encoder. In the case of the latter, the optical encoder provides information to a control circuit on the position of the motor shaft.

Stepper Phasing

A stepper motor requires a sequence of pulses applied to its various windings for proper rotation. This is called *phasing*. By their nature, all stepper motors are at least two-phase. Unipolar motors can be two- or four-phase, and some are six-phase.

Step Angle

Step angle is the amount of rotation of the motor shaft each time the motor is pulsed. Step angle can vary from as small as 0.9 degrees (1.8 degrees is more common) to 90 degrees. The step angle determines the number of steps per revolution.

Pulse Rate

Stepper motors have an upper limit to the number of pulses they can accept per second. Heavy-duty steppers usually have a maximum *pulse rate* (or step rate) of 200 or 300 steps per second, which equates to 60 to 180 rpm. Some smaller steppers can accept a thousand or more pulses per second.

It is important to note that stepper motors can't be motivated to run at their top speeds immediately from a dead stop. To achieve top speeds, the motor must be gradually accelerated (also called *ramped*).

Running Torque

Steppers motors provide more torque at slow speeds. This is opposite of DC motors, which develop increased torque the faster they turn. The *running torque* of a stepper motor determines the amount of work it can perform. The higher the running torque, the larger the mass the motor can move.

Note that stepping motors are also rated by their *holding torque*, which the amount of force the motor exerts when its windings are energized, but not pulsed. This imparts a kind of braking effect where the motor will resist turning when you don't want it to.

Voltage and Current Ratings

Steppers for 5-, 6-, and 12-volt operation are not uncommon. Unlike DC motors, however, using a higher voltage than specified doesn't result in faster operation, but more running and holding torque. "Overvolting" stepper motors is a common technique for increasing its torque. It's not uncommon to apply voltages of 100 to 500% over the rating on the faceplate of the motor. However, in doing this the motor can get quite hot, and care must be exercised to prevent overheating and damage.

The current rating of a stepper is expressed in amps or milliamps per *energized phase*. The power supply must be able to deliver at least as much current as the specified rating for the motor. For unipolar motors, two windings are powered at a time, requiring the supply to deliver at least twice as much current as that specified. If, for example, the current per phase is 1 amp, the minimum current requirement is 2 amps.

Golf Car Catalog, The   202317

Mountaintop Golf Cars, Inc.
9647 Hwy 105 South
Banner Elk, NC 28604
USA

☎ (828) 963-6775

📠 (828) 963-8312

🚫 (800) 328-1953

✉ FindIt@GolfCarCatalog.com

🌐 http://www.golfcarcatalog.com/

All replacement parts for golf cars, including motors and batteries.

Grainger (W.W. Grainger)   202928

100 Grainger Pkwy.
Lake Forest, IL 60045-5201
USA

☎ (847) 535-1000

📠 (847) 535-0878

🌐 http://www.grainger.com/

Can you imagine rummaging through shelves holding 5 million different products? If you're doubtful, take a look at Grainger, one of the world's leading retail industrial supply companies. Their printed catalog is thicker than a phone book and offers everything for your robot building from plastic rods to fractional-horsepower gearmotors. Obviously, that leaves 4,999,998 other products, which I won't describe here.

In addition to Grainger's online presence, they have some 600 local outlets that stock core merchandise. Catalogs are available in printed form or on CD-ROM.

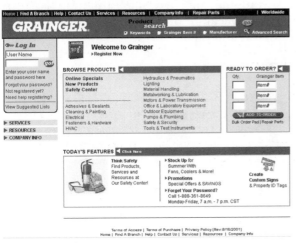

Grainger Web site.

Hansen Corporation   204004

901 South First St.
Princeton, IN 47670-2369
USA

☎ (812) 385-3415

📠 (812) 385-3013

✉ sales@hansen-motor.com

🌐 http://www.hansen-motor.com/

Manufacturers of miniature precision DC motors (with/without encoders and gearheads) and stepper motors.

Haydon Switch and Instrument, Inc   204005

1500 Meriden Rd.
Waterbury, CT 06705
USA

☎ (203) 756-7441

📠 (203) 756-8724

🚫 (800) 243-2715

✉ info@hsi-inc.com

🌐 http://www.hsi-inc.com/

Makers of rotary and linear stepper motors. Available through distributors or online.

Hiwin Technologies Corporation   204252

520 Business Center Dr.
Mount Prospect, IL 60056
USA

☎ (847) 827-2270

📠 (847) 827-2291

🌐 http://www.hiwin.com/

Providers of ball, ACME, leadscrews, DC motors, linear actuators, linear bearings, rails and guides, positioning tables, motor control circuits, and stepper motor drives.

Hurst Manufacturing   300004

1551 East Broadway
Princeton, IN 47670
USA

☎ (812) 385-2564

📠 (812) 386-7504

🚫 (888) 225-8629 Ext. 244

http://www.myhurst.com/

Hurst manufactures brushless DC, brushed DC, synchronous, and stepping motors, as well as linear actuators and gearboxes. Their motor/gearbox combos are seen frequently on the surplus market; the Web site provides helpful technical data and spec sheets if you should find you're the owner of one of these motors. You can also buy some products online.

Servo motor.

Johnson Electric Holdings Limited 203946

6-22 Dai Shun St.
Tai Po Industrial Estate
Tai Po, New Territories
Hong Kong

(+85 2 2663 6688
(+85 2 2663 6110
 http://www.johnsonmotor.com

Johnson Electric makes motors. You probably have a half-dozen of them in your house, car, and computer. They are commonly used in automotive, power tools, home appliances, compact disc players, VCRs, printers, faxes, and other business equipment, and toys.

You likely won't buy directly from Johnson Electric, but odds are you'll run across their motors when buying surplus. The Johnson Web site provides technical details on its motors (requires Adobe Acrobat Reader).

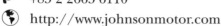

KidsWheels 203771

13266 Pond Springs Rd.
Austin, TX 78729
USA

((512) 257-2399
((512) 257-0093
 Alf@KidsWheels.com
 http://www.kidswheels.com/

Monster trucks for kids: Perego, Fisher-Price Power Wheels, and other makes. Can be ordered from the

Web site. Also offers parts, such as gearboxes, wheels, and batteries.

LEESON Electric Corporation 203975

2100 Washington St.
P.O. Box 241
Grafton, WI 53024-0241
USA

((262) 377-8810
((262) 377-9025
 leeson@leeson.com
 http://www.leeson.com/

Leeson sells big motors and small, with 4,000 off-the-shelf models to choose from. Most are for industrial applications and are available with or without a gearhead. Check out their extensive technical reference guide.

Offices in the U.S., Canada, Italy, and China.

Mabuchi Motor America Corp. 202321

430 Matsuhidai
Matsudo
Chiba 270-2280
Japan

(+81 47 384 1111
(+81 47 389 5299
 slsinq@mabuchi-motor.co.jp
 http://www.mabuchi-motor.co.jp/

Mabuchi makes motors, primarily for the small toy and appliance market. Their Web site is useful for the spec sheets available on current motor models. You can search by application or part designation. The Web site is in English and Japanese.

Magmotor Corporation 202849

7 Coppage Dr.
Worcester, MA 01603
USA

((508) 929-1400
((508) 929-1401
⊘ (866) 246-6867
 http://www.magmotor.com/

High-performance DC servomotors (brush and brushless); the products are also sold through RobotBooks.com. These are high-quality motors for performance robotics, such as combat bots.

Magmotor. Photo Robotbooks.com

Maxon Motor AG   202920

Brnigstrasse 220
P.O. Box 263
CH-6072 Sachseln
Switzerland

☎   +41 41 666 1500
📠   +41 41 666 1650
✉   info@maxonmotor.com
🌐   http://www.maxonmotor.com/

Nice precision gearmotors. You probably can't afford them new, but they're neat to look at! The company also sells DC servos, encoders, motor control units, and gearheads (spur and planetary). Spec sheets and technical white papers are available at the Web site.

Maxon gearmotors occasionally become available on the surplus market, and some are current product. You can use the datasheets on the Maxon Web site to obtain technical information on the motor.

McMaster-Carr Supply Company ⚇   202121

P.O. Box 740100
Atlanta, GA 30374-0100
USA

☎   (404) 346-7000
📠   (404) 349-9091

✉   atl.sales@mcmaster.com
🌐   http://www.mcmaster.com/

Literally everything you need under one roof: materials (plastic and metal), fasteners, hardware, wheels, motors, pipe and tubing, tools, bits and other tool accessories, power transmission (gears, belts, and chain), ball transfers, ball casters, spherical casters, regular casters, and many more. Walk-in stores are located in Chicago, Ill.; Cleveland, Ohio; Los Angeles, Calif., and New Jersey.

McMaster-Carr Web site.

Merkle-Korff Industries ⚇   202821

1776 Winthrop Dr.
Des Plaines, IL 60018
USA

☎   (847) 296-8800
📠   (847) 699-0832
✉   sales@merkle-korff.com
🌐   http://www.merkle-korff.com/

Manufacturer of motors, including DC and DC gearhead. You can choose from among stock motors that are ready to ship and available online. Or, if you need something specific, you can mix and match motors, gearboxes, shafts, encoders, and other accessories from among standardized parts.

See also MK Koford for motor drives, encoders, and brush/brushless motors:

http://www.koford.com/

Micromech

202189

5-8 Chilford Court
Braintree, Essex
CM7 2QS
UK

☏ +44 (0) 1376 333333
📠 +44 (0) 1376 551849
🌐 http://www.micromech.co.uk/

Purveyors of stepper and servo motors and their associated control circuitry. Also: DC gearhead motors, gearheads, and X-Y translation tables. Online sales, including clearance items of old, discontinued, and demo products.

MicroMo Electronics, Inc.

202921

14881 Evergreen Ave.
Clearwater, FL 33762-3008
USA

☏ (727) 572-0131
📠 (727) 573-5918
🚫 (800) 819-9516
✉ mmeweb@micromo.com
🌐 http://www.micromo.com/

MicroMo Electronics manufactures high-quality micro and miniature motors. This stuff ain't cheap, but if you need precision and power, this is the way to go. The company's product line includes:

- Coreless DC motors
- Permanent magnet and brushless DC motors
- Gearheads
- Encoders and tachometers
- Brakes
- Controllers and amplifiers
- Stepping gearmotors and drivers

SEE ALSO:

Micro-Drives for miniature gearhead motors—
http://www.micro-drives.com/

Parent company Faulhaber (based in Germany)—
http://www.faulhaber.com/

The Web site provides a noncommerce product catalog, plus short technical articles on such topics as "What Is a Permanent Magnet DC Motor?" Many articles include links to datasheets, application notes, and even exploded views of motor innards. Sells direct in North America; through distributors elsewhere.

Small motor.

Minarik Corporation

203510

905 East Thompson Ave.
Glendale, CA 91201
USA

☏ (818) 637-7500
📠 (818) 637-7509
🚫 (800) 427-2757
🌐 http://www.minarikcorp.com/

Full-line mechanical (bearings, shafts, gears, chain, etc.); electronics (PWM drives and sensors); online ordering plus many local warehouses throughout the U.S.

Motion Group, Inc., The

202993

P.O. Box 669
Clovis, CA 93613-0669
USA

📠 (559) 325-7117
🚫 (800) 424-7837
✉ sales@motiongroup.com
🌐 http://www.motiongroup.com/

Motion Group supplies industrial-strength stepper motors and motion control products, particularly the components need for CNC retrofits of metalworking lathes and mills.

Motor Selection Primer

204125

http://www.aveox.com/primer.html

Which motor to choose? This short primer answers the basic questions.

National Power Chair

202169

4851 Shoreline Dr.
P.O. Box 118
Mound, MN 55364
USA

((952) 472-1511

℘ (952) 472-1512

⃠ (800) 444-3528

✉ info@npcinc.com

🌐 http://www.npcinc.com/

What do wheel chairs and battling robots have in common? They both use high-power gearmotors from National Power Chair. This company began rebuilding motors and gearboxes for the wheelchair industry in 1981. Their product line includes over 100 different models manufactured since the year The Beatles released the song "Ticket To Ride."

See also NPC Robotics—
    http://www.npcinc.com/robots/

See also battery rechargers—
    http://www.npcinc.com/rchargers.html

NPC's robotics page also includes wheels and hubs, as well as specially selected motors for large robots.

Model 41250. Photo National Power Chair Inc.

NPC Robotics

See listing for National Power Chair (this section).

Example power window motor.

Oriental Motors   202923

2570 W. 237th St.
Torrance, CA 90505
USA

((310) 784-8200

℘ (310) 325-1076

⃠ (800) 816-6867

🌐 http://www.orientalmotor.com/

Oriental makes stepper and DC motors for industrial process control. They are common on the surplus market. Spec sheets are available at the Web site.

Pacific Scientific   203436

4301 Kishwaukee St.
P.O. Box 106
Rockford, IL 61105-0106
USA

((815) 226-3100

℘ (815) 226-3148

✉ customer_service@atg.pacsci.com

🌐 http://www.pacsci.com/

High-performance motors. Stepper, servo, and DC.

Picard Indistries   202360

4960 Quaker Hill Rd.
Albion, NY 14411
USA

((716) 589-0358

℘ (716) 589-0358

✉ jcamdep4@iinc.com

🌐 http://www.picard-industries.com/

Picard specializes in miniature smart motors and sensors. Their product line includes programmable solenoids, motor control, and sensors.

Pittman   202511

343 Godshall Dr.
Harleysville, PA 19438
USA

((215) 256-6601

℘ (215) 256-1338

✉ info@pittmannet.com

🌐 http://www.pittmannet.com/

Pittman is a major U.S. manufacturer of brush and brushless DC motors (standard and servo), many of them with built-in gearboxes. A number of the motors you may buy surplus were manufactured by Pittman, and you can find many datasheets on Pittman product on the Web site. Pittman motors are available through distributors, but most regular folks won't be able to afford them new. Keep an eye out for them in surplus stores.

Flatted shaft vs...

...a keyway shaft.

Pontech 202837

9978 Langston St.
Rancho Cucamonga, CA 91730
USA

((413) 235-1651
⊘ (877) 985-9286
✉ info@pontech.com
🌎 http://www.pontech.com/

Pontech produces low-cost servo and stepper motor controllers. Their SV203 product can control up to eight R/C servos and has the ability to be commanded by computer, joystick, or infrared remote control (different versions of the SV203 are available with unique features).

The company's Windows-based animation software is used to prepare a series of actions for the servos and is useful for "show bots" that repeat a series of moves. Pontech offers for sale standard R/C servos and stepper motors that have been tested with their other products.

For an example walking robot that uses the SV203 controller, see Six-Legged Walking Robot at:

http://www.pontech.com/files/hex01.htm

Portescap 202924

157, rue Jardiniere
La Chaux-de-Fonds, CH-2301
Switzerland

(+41 32 925 6111
↻ +41 32 925 6596
🌎 http://www.portescap.com

Precision Swiss-made ironless DC and brushless DC motors, gearmotors. Very nice. Very compact. Very expensive. Distributed in the U.S. by Danaher Motion.

Portescap also manufactures disc stepper motors, planetary gearheads, and spur gearheads.

RAE Corp. 203972

4615 W. Prime Parkway
McHenry, IL 60050-7037
USA

((815) 385-3500
↻ (815) 363-1641
⊘ (800) 323-7049
✉ raesales@raemotors.com
🌎 http://www.raemotors.com/

RAE offers fractional-horsepower, permanent magnet, DC brush-type motors; parallel shaft and right-angle gearmotors; as well as DC motor drives. RAE motors are often found on large combat robots; models range from "small" 1/70 horsepower to 1/3 horsepower. Gearboxes are heavy-duty, with many right-angle worm drives, capable of propelling a robot weighing several hundred pounds. Motors can be purchased through distributors, and several are available from Grainger under the Dayton brand-name.

Automotive window motor, with gearbox.

RMB Roulements Miniatures SA 204011

Eckweg 8
Box 6121
CH-2500 Biel-Bienne 6
Switzerland

✆ +41 32 344 4300
📠 +41 32 344 4301
✉ info@rmb-group.com
🌐 http://www.rmb-ch.com/

Makers of incredible small miniature bearings and Smoovy motors. For the latter, see:

http://www.smoovy.com/

Robotic Power Solutions 204215

305 9th St.
Carrollton, KY 41008
USA

✆ (502) 639-0319
✉ sjslhill@bellsouth.net
🌐 http://www.battlepack.com/

Specializing in combat robot parts, the company sells ni-cad and NiMH battery packs, chargers, and cobalt AstroFlight gearmotors. Their "Battlepack Kits" include batteries, wires, support bars, heat-shrink tubing, and padding foam.

Rockwell Automation 203871

Firstar Building
777 East Wisconsin Ave.
Ste. 1400
Milwaukee, WI 53202
USA

✆ (414) 212-5200
📠 (414) 212-5201
🌐 http://www.rockwellautomation.com/

Rockwell Automation manufactures a broad line of automation electronics and components: stepper motors, servo motors, encoders, linear motion products—about 500,000 items in all . . . a little too many to list here. Products are available from distributors and through company online stores.

Covers the following brands:

Allen Bradley (http://www.ab.com/) and Electro-Craft

Reliance Electric (http://www.reliance.com/)

Dodge PT (http://www.dodge-pt.com/)

Rockwell Software
(http://www.software.rockwell.com/)

Smoovy/RMB Group 202508

Eckweg 8, case postale 6121
CH 2500 Biel-Bienne 6
Switzerland

✆ +41 32 344 4300
📠 +41 32 344 4301
✉ info@rmb-group.com
🌐 http://www.smoovy.com

Manufacturer of Swiss-made precision motors, gearmotors, linear actuators, and controllers. The Web site also provides several technical white papers and technical briefs on motor selection and control. Smoovy motors are tiny. Did I say tiny? Make that tiny! Their model SPE39004 gearmotor is only 3.4mm wide at its thickest and is but 15mm long, including the shaft. You could swallow the thing and not even know it.

Sullivan Products 204217

1 North Haven St.
P.O. Box 5166
Baltimore, MD 21224
USA

✆ (410) 732-3500
📠 (410) 327-7443
✉ sales@sullivanproducts.com
🌐 http://www.sullivanproducts.com/

Sullivan manufactures accessories for performance R/C aircraft. They offer a number of hardware-related items like pushrods and landing gear, but also of interest is their electric starter for gasoline-powered airplane engines. Starter motors exhibit high torque and are ideal for use in robot drives, and they tend to be affordably priced compared to many other motors in the same class

Motors from cordless drills make for good robot power plants.

Superior Electric 202628

383 Middle St.
Bristol, CT 06010
USA

☎ (860) 585-4500

📠 (860) 589-2136

🌐 http://www.superiorelectric.com/

Makers of industrial DC motors, brushed and brushless.

SEE ALSO:

http://www.slosyn.com/

Tamiya America, Inc.-Educational 203143

Attn: Customer Service
2 Orion
Aliso Viejo, CA 92656-4200
USA

☎ (949) 362-6852

🚫 (800) 826-4922

🌐 http://www.tamiyausa.com/
product/educational/

Tamiya is one of the largest plastic model manufacturers in the world. Just about every model maker has built at least one Tamiya tank, car, or other vehicle. Lesser known is that Tamiya sells a broad line of educational mechanical components, including robot kits, motor kits, pulleys, wheels, and more. Tamiya is based in Japan, and they have offices all over the world. I picked Tamiya's U.S. distribution office, but their products are available everywhere. See the company's Web site for locations of other distributors and dealers:

http://www.tamiya.com/

Tamiya's Educational Kits includes these useful projects. Almost all are constructing with a precut plywood base.

- Light-sensing electronic robot-item 75013
- Dung beetle electronic robot-item 75014
- Powered forklift (with 2-channel wired remote)-item 70070TA
- Earth-moving dump truck (with 3-channel wired remote)-item 70067TA
- Wall-hugging mouse-item 70068
- Remote-controlled bulldozer-item 70104
- Power shovel/dozer-item 70107

- Tracked vehicle chassis-item 70108
- 4WD chassis-item 70113

Tamiya's Robots & Construction line includes a number of motors. These are made with molded plastic, and many can be built using a number of reduction ratios.

Tamiya Inc.

Tamiya is one of the world's largest makers of plastic models. They also made a complete line of educational mechanical components, including motorized gearboxes, wheels, plastic chain, rubber tracks, and more.

A Tamiya motor, ready for assembly

One of the most useful is a dual-gear motor, which consists of two small motors and independent drive trains. You can connect the long output shafts to wheels, legs, or tracks.

One of their most popular kits is a tracked tractor/shovel that is controlled via a switch panel. You can substitute the switch panel with computerized control circuitry (relays make it easy, but you could also use power transistors or an H-bridge), providing for full forward and backward movement of the tank.

Tamiya can be found at the following Web sites:

http://www.tamiya.com/
http://www.tamiyausa.com/

- 3-speed crank axle gearbox-item 70093
- Planetary gearbox-item 72001
- Twin motor gearbox-item 70097
- High-speed gearbox-item 72002
- High-power gearbox-item 72003
- 4-speed crank axle gearbox-item 70110
- Worm gearbox-item 72004 (this is one of my personal favorites)
- 6-speed gearbox-item 72005

And tires/wheels (always sold in pairs). These tires are made to work with the above motors.

- Off-road tires-item 70096
- Truck tire set (two pair)-70101
- Sports tire set-item 70111
- Narrow tire set-item 70145

Additional mechanical items include:

- Ladder-chain and sprocket set-item 70142
- Ball caster (set of two)-item 70144
- Pulley unit set-item 70121
- Pulley set/small-item 70140
- Pulley set/large-item 70141
- Track and wheel set-70100

Finally, they sell a line of battery boxes, switches, 2- and 4-channel wired remote-control boxes, mechanical switches, and sound-activated switches.

Fine products, indeed, but they can sometimes be hard to find. Few retail stores, even well-stocked hobby shops, carry the Educational product line, let alone everything. Your best bet is online sales such as Towerhobbies.com. Try this Google.com search phrase for starters:

tamiya educational motor

Tamiya mouse kit.

Team Delta Engineering   202174

1035 North Armando St., Unit D
Anaheim, CA 92806
USA

 (208) 692-4502

 dan@teamdelta.com

http://www.teamdelta.com/

Home of Team Delta robot combat group. Also sells components and other goods in support of combat robotics.

- Switches, wire, and passives
- R/C electronic switch interfaces
- R/C electronic specialty interfaces
- IFI Victor 883 power controller
- Antennas, power converters, and power supplies
- Stock books, T-shirts, stickers, and videos
- Mechanics
- The Team Delta bookshelf
- Kits, learning, and services

Gear motor. Photo Team Whyachi.

Team Whyachi LLC   204219

814 E. 1st Ave.
P.O. Box 109
Dorchester, WI 54425
USA

 info@teamwhyachi.com

http://www.teamwhyachi.com/

Team Whyachi is both an entrant in combat-style robot contests and an online retailer of high-performance parts for combat bots. Their small but still quite useful product line includes high-power gearmotors, wheels, and motor speed controls.

High powered Etak motor motor.
Photo Team Whyachi.

Wilde EVolutions, Inc.   203576

18908 Hwy 99, Ste. A
Lynnwood, WA 98036-5218
USA

- ((425) 672-7977
- ((425) 672-7907
- ⊘ (800) 327-8387
- ✉ info@wilde-evolutions.com
- ⊕ http://www.wilde-evolutions.com/

Electric vehicles and parts, such as high-capacity batteries, motors, high-current relays, and speed controllers, for golf carts and other small electric vehicles and scooters.

ZapWorld.com   204220

117 Morris St.
Sebastopol, CA 95472
USA

- ((707) 824-4150
- ((707) 824-4159
- ⊘ (800) 251-4555
- ✉ zap@zapworld.com
- ⊕ http://www.zapworld.com/

ZapWorld sells motors, batteries, and parts for small electric vehicles, namely bikes, ground scooters, and aquatic scooters (as well as some odd stuff, such as personal hovercrafts). Use the stuff for your own performance robots.

ACTUATORS-OTHER

Electrical motors and pressure systems (pneumatic and hydraulic) make up the bulk of actuators. But other types—primarily electrically driven solenoids—exist and are included here.

SEE ALSO:

ACTUATORS-MOTORS: DC motors

ACTUATORS-SHAPE MEMORY ALLOY: Wire that contracts when heat/electricity is applied

ACTUATORS-PNEUMATIC: Air-driven cylinders and motors

RETAIL-GENERAL ELECTRONIC: Where to buy solenoids

BICRON Electronics Company   203671

50 Barlow St.
Canaan, CT 06018
USA

- ((860) 824-5125
- ((860) 824-1137
- ✉ info@solenoid.com
- ⊕ http://www.solenoid.com

Distributor of solenoids. Tubular and frame, pull and push. Downloadable technical guide in Adobe Acrobat PDF format.

Open frame solenoid.
Photo Bircon
Electronics Co.

Densitron Technologies plc   202175

Unit 4, Airport Trading Estate
Biggin Hill
Kent
TN16 3BW
UK

- (+44 (0) 1959 542000
- (+44 (0) 1959 542001
- ✉ europe.sales@densitron.net

http://www.densitron.com/

Densitron is a maker of electromechanical components-including rotary solenoids, stepper motors, and electric clutches, as well as computer-related product such as displays.

Google Search: Relays & Solenoids 202692

http://directory.google.com/Top/Business/
Industrie/Electronics_and_Electrical/
Relays_and_Solenoids/

Google.com's category list for relays, solenoids, and related electromechanical devices.

Cylindrical solenoid.

Use the clevis end of a solenoid shaft to attach to linkages.

Ledex 204187

801 Scholz Ave.
Vandalia, OH 35377-0427
USA

((937) 454-2345
(937) 898-8624
http://www.solenoids.com/

Ledex makes solenoids-rotary and linear. They're sold through distributors or online; you can obtain spec sheets and technical data about their products, and solenoids in general, at the Web site.

ACTUATORS-PNEUMATIC

Pressure systems perform work from the pressure of gas (pneumatic) or liquids (hydraulic). For the most part, hydraulic systems are too large, heavy, and expensive for use in amateur robotics, and they can be messy if there is a leak. Pneumatic systems, using compressed air, are common and affordable. So, we'll cover just pneumatic systems.

The resources in this section specialize in pneumatic systems, including pumps, air reservoirs, air cylinders, air motors, and pressure-control valves. Not listed here are pneumatic tools, which are air powered, and include the components that can be cannibalized for use in a robot, such as the air motor from a drill or retrofitting a 12-volt solenoid to operate the trigger valve from an old ratchet wrench.

SEE ALSO:

ACTUATORS-MOTION PRODUCTS: Mechanical motion devices, such as linear stages

ACTUATORS-MOTORS: DC motors

ACTUATORS-SHAPE MEMORY ALLOY: Wire that contracts when heat/electricity is applied

TOOLS AND TOOLS-POWER: Pneumatic power tools

Air America, Inc. 202173

2240 Elmhurst Road
Elk Grove Village, IL 60007
USA

((847) 545-9999
(847) 545-0275
Sales@airamerica.com
http://www.airamerica.com/

High-pressure air systems and regulators for paintball. Some of the components can be used for air-powered robots.

In Europe the address is:

Air America UK
Sprawls Farm, Bures Road
West Bergholt, Colchester
Essex CO6 3DN. UK

Tel1: +44 (0) 1206 240831

Tel2: +44 (0) 1206 243450

Fax: +44 (0) 1206 240821

sales@airamerica-uk.com

Airpot Corporation   204249

35 Lois St.
Norwalk, CT 06851
USA

℡   (203) 846-2021

℡   (203) 849-0539

   (800) 848-7681

✉   service@airpot.com

🌐   http://www.airpot.com/

Airpot makes what's known as air-damping dashpots-basically, miniature shock absorbers. Dashpots provide resistance to linear motion and are often used to reduce the sudden jarring of a lever or a bar moving in one direction or the other. The company also sells a line of small air cylinders using the trade name Airpel.

Applied Industrial Technologies

203445
One Applied Plaza
Cleveland, OH 44115-5053
USA

℡   (216) 426-4189

Pressure Actuation Systems

Motors are amazingly powerful for their size. Imagine: Something the size of a walnut can propel a 1-pound robot across the room. But if that seems incredible to you, consider the amount of work that a pressure actuation system can produce. Pressure is provided in the form of air or other gas, or some kind of liquid. Air-pressure systems are called pneumatic; liquid-pressure systems are called hydraulic. Both are more bulky than motor actuators, but they can also provide a lot more power.

Pneumatic cylinders use air to provide powerful linear motion.

You've seen *hydraulic* power at work if you've ever watched a bulldozer go about moving dirt from pile to pile. And, while you drive, you use it every day when you press down on the brake pedal. Similarly, *pneumatic* power uses air pressure to move linkages. Pneumatic systems are cleaner than hydraulic systems, but all things considered, aren't as powerful.

Hydraulic and pneumatic systems are pressurized using a pump. The pump is driven by an electric motor, so in a way, robots that use hydraulics or pneumatics are fundamentally electrical. The exception to this is using a pressurized tank, like a SCUBA tank, to provide air pressure in a pneumatic robot system. Eventually, the tank becomes depleted and must either be recharged, using some pump on the robot, or removed and filled back up, using a compressor.

Hydraulic and pneumatic systems are rather difficult to effectively implement, but they provide an extra measure of power over DC and AC motors. With a few hundred dollars in surplus pneumatic cylinders, hoses, fittings, solenoid valves, and a pressure supply (battery-powered pump, air tank, regulator), you could conceivably build a hobby robot that picks up chairs, bicycles, even people.

If you wish to experiment with pressure systems, I recommend first starting with pneumatics. The components are cheaper, and leaking air requires no cleanup, whereas leaking hydraulic fluid (the same basic stuff as power steering fluid) is very messy.

(216) 426-4820

(877) 279-2799

products@apz-applied.com

http://www.appliedindustrial.com/

Industrial bearings, linear slides, gears, pulleys, pneumatics, hydraulics, and other mechanical things. Also hosts Maintenance America, online reseller of industrial maintenance supplies and general industrial supplies (wheels, casters, fasteners, and more), tools, paints, and adhesives.

Tie-rod air cylinder.

Big Boys Toys   203865

1554 West Branch St.
Arroyo Grande, CA 93420
USA

(805) 481-0873

(888) 520-7458

info@bbtpaintball.com

http://www.bbtpaintball.com/

Online sellers of paintball air pumps, air regulators, and other components. With some ingenuity, these can be adapted for use in pneumatically powered robots.

Bimba Manufacturing Company   204113

P.O. Box 68
Monee, IL 60449-0068
USA

(708) 534-8544

(708) 235-2014

(800) 442-4622

support@bimba.com

http://www.bimba.com/

Bimba is one of the most recognized brands of pneumatic cylinders and associated components. Basically, if it exists in Pneumatic Land, Bimba makes and sells it. Their product is sold through distributors, but because of its popularity among designers, you'll regularly see

Bimba cylinders- new and used-at surplus stores. Check surplus stores first, and if it's not available, consider ordering it from a Bimba representative.

Double-acting cylinder.

Clippard Minimatic, Inc.   203037

7390 Colerain Ave.
Cincinnati, OH 45239
USA

(513) 521-4261

(513) 521-4464

(877) 245-6247

http://www.clippard.com/

Manufacturer of pneumatic and electronic control devices-cylinders, solenoid valves, and control electronics. Available online or through distributors.

Fiero Fluid Power, Inc.   202172

5280 Ward Rd.
Arvada, CO 80002
USA

(303) 431-3600

(800) 638-0920

fiero@fierofp.com

http://www.fierofp.com/

Distributor of automation, motion control, and machine-framing products. Their product line includes:

- Bimba-pneumatics
- 80/20-machine framing
- Tol-O-Matic -pneumatics
- Watts-air valves
- Foster-air fittings
- Robohand-translation stages
- Skinner Valve

- Mosier-pneumatics
- Turn Act-translation stages

Offices in Denver, Colo., and Salt Lake City, Utah.

Ball joint on end of air cylinder for attaching to linkages.

Close-up of an air cylinder shaft.

McKibben Artificial Muscles 202018

http://rcs.ee.washington.edu/BRL/devices/mckibben/

The McKibben Artificial Muscle is a pneumatic actuator that its developers say exhibits properties found in real muscle. It's basically a rubber balloon in a plastic mesh, but while it looks simple enough, to make it work right requires some planning and engineering.

Paintball-Online.com 203866

7340 SW Hazelfern Rd.
Portland, OR 97224
USA

- ((503) 620-4847
- ✆ (503) 620-5737
- ⊘ (800) 875-4547
- ✉ sales@paintball-online.com
- 🌎 http://www.paintball-online.com/

Air pumps, CO2 and N2 air systems, fill stations, expansion chambers and regulators, and other components for paintball. Use as well for making pneumatically powered robots.

Pass thru air manifold.

Tee air splitter.

 ## ACTUATORS-SHAPE MEMORY ALLOYS

A shape memory alloy (SMA) is a metal material that undergoes molecular change when heated. Most SMAs are manufactured so that they contract mainly in one direction when heated. The material is electrically conductive, so passing a current through it causes it to heat up. The most common application for SMAs in robotics is as "artificial muscles." The SMA wire acts as a kind of small solenoid when electricity is applied.

SMA material is sold by a variety of companies worldwide. This section lists both the manufacturers of shape memory alloys, as well as mail-order retailers who will sell to consumers. Most retailers sell sample packs so you can experiment with SMA. The material is available in bulk, if you happen to need a lot of it.

SEE ALSO:

BOOKS-ROBOTICS: Some books have been written about using SMA

RETAIL-ROBOTICS SPECIALTY: Sellers of consumer robot kits and parts often sell SMA kits

ROBOTS-HOBBY & KIT: Robotic kits that use SMA

Dynalloy, Inc. 203094

3194-A Airport Loop Dr.
Costa Mesa, CA 92626-3405
USA

- ((714) 436-1206
- ✆ (714) 436-0511

sales@dynalloy.com

 http://www.dynalloy.com/

Dynalloy is a manufacturer of shape memory alloys specially made to be used as actuators. Offers wire by the meter, sample kits, and precrimped Flexinol (for ease in attaching it to things).

Memory-Metalle GmbH   204047

Am Kesselhaus 5
D-79576 Weil am Rhein
Germany

(+49 (0) 7621 799121

📞 +49 (0) 7621 799244

 info@memory-metalle.de

🌐 http://www.memory-metalle.de/

Shape memory alloy (SMA) products, services, and technical details. Web site is in English.

Memry Corp.   204045

3 Berkshire Blvd.
Bethel, CT 06801
USA

((203) 739-1100

📞 (203) 798-6606

🚫 (866) 466-3669

✉ productinfo@memry.com

🌐 http://www.memry.com/

From the Web site, "Memry Corporation is a recognized leader in the development, manufacturing and marketing of semi-finished materials (wire, strip and tubing), components and assemblies utilizing the properties exhibited by shape memory alloys, in particular nickel titanium (Nitinol or NiTi)."

Mondo-tronics Inc.   203735

PMB-N, 4286 Redwood Hwy.
San Rafael, CA 94903
USA

((415) 491-4600

📞 (415) 491-4696

✉ info@musclewires.com

🌐 http://www.musclewires.com/

Reseller of shape memory alloy; books, and kits. Sold through distributors or the company's RobotStore.com Web site.

Nanomuscle, Inc.   202825

2545 West Tenth St.
Ste. A
Antioch, CA 94509
USA

((925) 776-4726

📞 (925) 755-9572

 sales@nanomuscle.com

 http://www.nanomuscle.com/

Nanomuscle is a specially manufactured shape memory alloy that does the job of a miniature solenoid. Apply voltage, and the Nanomuscle actuator contracts several millimeters; remove voltage, and the device relaxes. A developer's kit is available, and the company provides onsite purchasing in quantities of 25 or more units, only.

My personal note: When I checked their prices I felt they wanted too much money for these things. Wait until the price becomes more reasonable- under $15.

Nitinol Devices & Components   202975

47533 Westinghouse Dr.
Fremont, CA 94539
USA

((510) 623-6996

📞 (510) 623-6995

 sales@nitinol.com

 http://www.nitinol.com/

In the words of the Web site, "NDC is a leading supplier of Nitinol materials (wire, tube, sheet, strip, and bar) and components to the medical and commercial industries worldwide." Datasheets of products are provided for download.

Shape Memory and Superelastic Technologies   204046

http://www.smst.org/

Volunteer organization of industry professionals dedicated to disseminating technical education of shape memory and superelastic properties, especially Nitinol

Building Robots with Shape Memory Alloy

As early as 1938, scientists observed that certain metal alloys, once bent into odd shapes, returned to the original form when heated. This property was considered little more than a laboratory curiosity because the metal alloys were weak, difficult and expensive to manufacture, and broke apart after just a couple heating/cooling cycles.

The Stiquito robot uses tiny SMA wires for walking "muscles." Photo James Conrad, Stiquito.com.

Research into metals with memory took off in 1961, when William Beuhler and his team of researchers at the U.S. Naval Ordnance Laboratory developed a titanium-nickel alloy that repeatedly displayed the memory effect. Beuhler and his cohorts developed the first commercially viable shape memory alloy, or SMA. They called the stuff Nitinol, a fancy-sounding name derived from *Ni*ckel *Ti*tanium *Na*val *O*rdnance *L*aboratory.

Since its introduction, Nitinol has been used in a number of commercial products—but not many. For example, several Nitinol engines have been developed that operate with only hot and cold water. In operation, the metal contracts when exposed to hot water, and relaxes when exposed to cold water. Combined with various assemblies of springs and cams, the contraction and relaxation (similar to a human muscle) causes the engine to move.

In 1985, a Japanese company, Toki Corp., unveiled a new type of shape memory alloy specially designed to be activated by electrical current. Toki's unique SMA material, trade named BioMetal, offers all of the versatility of the original Nitinol, with the added benefit of near-instant electrical actuation. BioMetal and materials like it—Muscle Wire from Mondo-Tronics or Flexinol from Dynalloy—has many uses in robotics, including as novel locomotive actuation. From here on out we'll refer to this family of materials generically as shape memory alloy, or simply SMA.

At its most basic level, SMA is a strand of nickel titanium alloy wire. Though the material may be very thin (a typical thickness is 0.15mm—slightly wider than a strand of human hair), it is exceptionally strong. In fact, the tensile strength of SMA rivals that of stainless steel: The breaking point of the slender wire is a whopping six pounds. Even under this much weight, SMA stretches little. In addition to its strength, SMA also shares the corrosion-resistance of stainless steel.

Shape memory alloys need little support paraphernalia. Besides the wire itself, you need some type of terminating system, a bias force, and an actuating circuit. Companies that sell SMA materials provide these components, or information on them.

Not all sources of shape memory alloy sell to individuals or in small quantities. The premier resellers of SMA to the public are:

DYNALLOY, Inc.
http://www.dynalloy.com/

Mondo-tronics Inc.
http://www.musclewires.com/

Stiquito
http://www.stiquito.com/

alloys. Conference procedures and links to companies and other organizations involved in shape memory alloys.

Shape Memory Applications, Inc.   202985

1070 Commercial St.
Ste. 110
San Jose, CA 95112
USA

📞 (408) 727-2221
📠 (408) 727-2778
 http://www.sma-inc.com/

Manufacturer of shape memory alloy, include tube, sheet, and foil.

🏭

Special Metals Corporation   204048

3200 Riverside Dr.
Huntington, WV 25705-1771
USA

📞 (304) 526-5100
🚫 (800) 334-4626
✉ info@smcwv.com
🌐 http://www.specialmetals.com/

Makers of shape memory alloy materials. Technical documents available for download in Adobe Acrobat PDF format.

🌐   🏭

Stiquito   202248

Micro Robotics Supply, Inc.
101 Pendren Pl.
Cary, NC 27513-2225
USA

✉ stiquito@stiquito.com
🌐 http://www.stiquito.com/

Stiquito is a small and simple robot that uses shape memory alloy (SMA) wire for movement. This is official Stiquito page, maintained by author Jim Conrad, and supports the product and several books written about it.

 💻

TiNi Alloy Company   204044

1619 Neptune Dr.
San Leandro, CA 94577
USA

📞 (510) 483-9676
📠 (510) 483-1309
✉ info@tinialloy.com
🌐 http://www.sma-mems.com/

TiNi is involved with R&D to produce a thin-film shape memory alloy to activate microminiature (MEMS) actuators, including "millivalves" and "microvalves."

Toki Corp.   204043

✉ biometal@toki.co.jp
http://www.toki.co.jp/BioMetal/_index.html

From the Web site, "BioMetal is one of Ti-Ni based Shape Memory Alloys; however, its properties are specially arranged for use in our own-manufactured actuators. The material being metal, it provides smooth and living creature-like (biological) movements, thus it has been named 'BioMetal.' BioMetal is offered in the form of a thin wire (BioMetal Fiber) which, facilitating electrical current passage, performs best in tensile-directional usage."

BAR CODING

Bar codes are used to uniquely identify objects and are so commonplace in our society that we take them for granted. For robotics, bar codes can be used as a way to program them, as well as for navigation and object detection.

A typical bar code system includes a reader (usually a hand-held wand, but also a hand-held reader that uses a CCD or laser element), and a bar code interface to translate the black-and-white hashes into usable data. Specialty printers are available for preparing bar codes, though they are not strictly needed for most robotics applications. Bar code printing software allows you to print bar codes on standard laser or ink-jet printers.

Most manufacturers of bar code equipment listed in this section will not sell directly to the general public; however, they provide useful technical details not only about their products, but also about the science of bar coding. There are numerous online retailers specializing in bar code hardware and software; the ones listed in this section are representative of the hundreds that are out there.

Bar Code Discount Warehouse, Inc. 202727

2950 Westway Dr.
Ste. 110
Brunswick, OH 44212
USA

✆ (330) 220-3699
📠 (330) 220-3099
⊘ (800) 888-2239
✉ sales@bcdw.com
🌐 http://www.bcdw.com

Online reseller of bar-coding products and supplies. Products listed by category or manufacturer.

Barcode Direct 202725

Ste. 2, 40 Radnor Rd.
Galston, NSW 2159
Australia

✆ +61 2 9653 3030
📠 +61 2 9653 3130
✉ info@barcodedirect.com.au
🌐 http://www.barcodedirect.com.au/

Bar code wands and systems; CCD and laser readers; and bar code label printers.

Of special interest are the articles on bar code technologies and applications. Among the gems:

- "Sizing Applications for 2D Symbols"—a detailed overview of 2D codes, like Maxicode, used by UPS.
- "Introduction to Barcoding Technology"—fairly in-depth basics on bar codes.
- "An Introduction to Radio Frequency"—short discourse on RFID identification.

BARCODE Island 203727

http://www.barcodeisland.com/

Barcode Island is an informational Web site for developers interested in using bar codes. Technical information includes symbology sets; a user-to-user forum allows visitors to ask and answer questions.

Barcode Mall 202718

503 Marten Rd.
Princeton, NJ 08540
USA

✆ (908) 359-7023
📠 (908) 359-1203
✉ a1barcode@barcodemall.com
🌐 http://www.barcodemall.com/

Bar code wands, CCD and laser scanners, printers, and label supplies.

Barcode Store 202720

c/o HALLoGRAM Publishing
14221 E 4th Ave.
Ste. 220
Aurora, CO 80011
USA

✆ (303) 340-3404
📠 (303) 340-4404
✉ sales@barcodestore.com
🌐 http://www.barcodestore.com/

Online and mail-order retailer of bar code wands, CCD scanners, laser scanners, printers, labels, software, and supplies.

Practical Robotics Applications for Bar Coding

Bar codes are used to identify packages in shipment, mail in the post office, food at the grocery store, even the book you're reading now. The same technology behind bar coding for product identification can be used by robots. The components needed for bar coding include a printed bar code (which you can make with any printer), a reader or wand, and a code translator.

There are many types of bar codes, based on different coding sequences. A common bar code is Code 39 (also known as Code 3-of-9), which can encode both text and numbers. Most commercial bar code printing and scanning hardware/software can work with Code 39.

For a robotic application, you can interface a bar code wand to the computer or microcontroller of a robot. Scan the code, and the information is fed into the robot. Here are some ideas for using bar codes with robotics:

- Prepare coded programming strips, such as "wander" or "seek can of soda." The robot reads the code and is programmed for action.

- Place bar codes along the wall baseboard, for a wall-following robot. Attach a bar code wand (or better yet a CCD scanner) to the robot so the wand faces the baseboard and sees the codes. The bar codes can be used to identify where in the room the robot is. A similar technique can be used for a maze-solving robot.

- A similar technique can be used in a line-following robot, only in addition to a line, the robot reads bar code information as it travels. It could use this information for additional navigation instruction.

- Given a wide-field bar code scanner (the kind used in warehouses), your robot can see bar codes from several feet away. Such a robot could identify not only where in the room it is, but in which room.

Bar coding is now a mainstream product, with numerous online and local retailers selling all the components you need to get started. And don't forget surplus, as new bar-coding technologies come out all the time and companies exchange old equipment for new.

Bar Code Translation Hardware

The latest bar code wands and CCD scanners incorporate software to read and interpret common bar codes. These connect to a PC via the keyboard connector or through a USB or serial port. The products are ideal if your robot is based on a PC-compatible computer. Zebra Technologies, at **http://www.zebra.com/,** is a premier manufacturer of all-in-one bar code wands and CCD scanners.

Other wands are simply optical readers and must be used with an interface. The interface connects to the wand and outputs a signal (usually via RS-232) for reading by a computer. Older model interfaces tended to be quite bulky, so if your robot is small, the interface will dwarf the rest of the machine.

Plain optical reader wands (no interface electronics) can be used with a bar code translation chip. One such product, available in single or multiple quantities is:

BarcodeChip.com
http://www.barcodechip.com/

The bar code chip provides data via a simple serial connection and can be readily interfaced to microcontrollers.

Using a Brother P-Touch PT-2300 to Print Bar Codes

The Brother P-Touch PT-2300.

I admit I'm a label freak. I use labels to identify every drawer and bin in my robotics lab. That way, I don't have to guess what's in each one. In the old days, the labels were made by the Dymo Labeler, a mechanical device that embossed text on plastic strips. I still use my old Dymo occasionally, but these days, most of my labels are created on a Brother P-Touch electronic labeling system. This self-contained machine uses special label cartridges and sports a miniature typewriter keyboard for entering the text.

There are numerous P-Touch labelers, and they vary mostly in the width of the labels they can print, the size of the type, and number of lines that can be printed. Obviously, wider labels can print more lines.

The P-Touch PT-2300 prints in eight type sizes, in four fonts, with up to six lines of text, and takes different size tapes (maximum 1-inch wide). But the feature I like the most is that it connects to a computer via a standard USB cable; you have to supply the cable, but it's available anywhere. With included software, you can print using any available system font, as well as bar codes.

Several bar code formats are available, including Code 39, a good generic format that most every bar code wand or reader will recognize. The PT-2300 is one of several higher-end labelers from Brother than will print bar codes.

One use of the bar-coding print feature: create small strips of instructions to give the robot. This is similar to the popular Rumble Robots, which are "programmed" using cards with bar codes printed on them. You can make your own bar code cards for your robot, with each card a separate behavior. For instance, one card might tell the robot to shy away from light; another might tell it to go toward the light.

Bear in mind that the bar codes themselves do not include the programming code to instruct the robot. The core code is contained in the robot, and the bar code is used to activate a desired state.

Barcode Warehouse   202726

Telford Drive
Newmark Industrial Estate
Newmark Notts
NG24 2DX
UK
☎   +44 (0) 1636 602000
📠   +44 (0) 1636 602001
✉   ross@thebarcodewarehouse.co.uk
🌍   http://www.thebarcodewarehouse.co.uk/

Retailer of bar code wands, CCD/laser scanners, printers, labels, and bar-coding supplies.

📄   🌍   🖥

---

barcode-barcode / pmi   202722

11150 Woodward Ln.
Cincinnati, OH 45241

USA
☎   (513) 782.5050
📠   (513) 782.5051
⊘   (800) 325.7636
✉   info@barcode-barcode.com
🌍   http://www.barcode-barcode.com/

Discount bar code hardware and supplies.

🌍   🖥

BarcodeChip.com   202523

☎   (765) 287-1987
📠   (765) 287-1989
✉   dvanhorn@cedar.net
🌍   http://www.barcodechip.com/

Self-contained bar code decoding chips with serial interfaces. Connect a bar code wand or amplified infrared detector to the input of the chip, scan any of several types of commonly used bar codes, and the bar code data is available for serial downloading from the chip. Easy connection to microcontrollers such as the Basic Stamp, PICmicro, and Atmel AVR via three-wire serial.

Barcode chip.

BarcodeHQ/Data Worth 202723

623 Swift St.
Santa Cruz, CA 95060
USA

✆ (831) 458-9938
🖶 (831) 458-9964
🚫 (800) 345-4220
✉ wds@barcodehq.com
🌍 http://www.barcodehq.com/

Full line of bar code hardware (wands, CCD scanners, laser scanners, printers, etc.), labels, and software.

SEE ALSO:

http://www.pcbarcode.com/—in the U.K.
http://www.codesbarres.com/—in France
http://www.strichcode.com/—in Germany

Custom Sensors Inc. 202721

30 York St.
Auburn, NY 13021
USA

✆ (315) 252-3741
🖶 (315) 253-6910

✉ info@csensors.com
🌍 http://www.csensors.com/

Bar code hardware and software. Decoded and undecoded wands. Web site includes help/informational pages.

Dissecting the CueCat 203004

http://cipherwar.com/info/tools/cuecat/

How to take apart a (free) CueCat bar code wand and use it for general bar-coding applications

IDAutomation.com, Inc. 203884

550 N. Reo St.
Ste. 300
Tampa, FL 33609
USA

✆ (813) 261-5064
🖶 (813) 354-3583
✉ admin@idautomation.com
🌍 http://www.idautomation.com/

Supplier of bar code products, software, and supplies.

RACO Industries / ID Warehouse 203843

5480 Creek Rd.
Cincinnati, OH 45242
USA

✆ (513) 984-2101
🖶 (513) 792-4272
🚫 (800) 446-1991
✉ info@racoindustries.com
🌍 http://www.idwarehouse.com/

Resellers of various bar code and RFID tagging systems.

Symbol Technologies, Inc. 203728

One Symbol Plaza
Holtsville, NY 11742-1300
USA

✆ (631) 738-5200
🖶 (631) 738-5990
🚫 (800) 722-6234

info@symbol.com

http://www.symbol.com/

Symbol is a major manufacturer of bar code readers and printers. Products are available through distributors.

Wasp Bar Code Technologies 202724

1400 10th St.
Plano, TX 75074
USA

((214) 547-4100

✆ (214) 547-4101

✉ sales@waspbarcode.com

🌐 http://www.waspbarcode.com/

Manufacturer and online retailer of bar-coding hardware, labels, and software.

ZEBEX America, Inc. 202719

Ilene Court, Building 11
Unit #2
Hillsborough, NJ 08844
USA

((908) 359-2070

✆ (908) 359-1272

✉ sales@zebex.com

🌐 http://www.zebex.com/

Manufacturer of hand-held and stationary bar code readers, bar code printers, and bar code scanners. Products are available through distributors.

Zebra Technologies Corporation 203842

333 Corporate Woods Pkwy.
Vernon Hills, IL 60061-3109
USA

((847) 634-6700

✆ (847) 913-8766

⊘ (800) 423-0442

🌐 http://www.zebra.com/

Zebra manufactures bar code readers (wands, CCD and laser scanners), RFID readers and tags, and bar code label printers. Products are available through distributors and retailers.

One interesting Zebra technology is the printing of radio frequency ID tags as peel-off labels. An integrated circuit is embedded in the label, which facilitates the RF identification. Such "smart labels," as Zebra calls them, do not need to be in the line of sight of the reader (RFID will work behind obstructions and often at a greater distance than standard optical bar coding). Versions of the labels also contain a digital memory that can be programmed and reprogrammed many times.

The Web site contains several technical white papers on bar coding and RFID technologies.

---

eScan Technologies
http://www.e-scan.com/
Barcode readers, printers

Genesis POS
http://www.genesispos.com/
Barcode hardware, software, and labels

National Barcode
http://www.nationalbarcode.com/
Barcode hardware and software

PCBarcode
http://www.pcbarcode.com/
Barcode wands, readers, printers, and software

Progressive Microtechnology, Inc.
http://www.scanpmi.com/
Barcode wands and other hardware, software

PSC Inc.
http://www.pscnet.com/
Manufacturer of barcode readers

Scan Technology, Inc.
http://www.scantec.com/
Barcode hardware

ScanSmart
http://www.scansmart.com/
Barcode hardware, printers, PDA barcode add-ons

BATTERIES AND POWER

Mobile robots require a self-contained power source, and most robots use a battery or battery pack for this purpose. As common as batteries are, alkaline batteries available at the local supermarket are not ideal for use in robotics; your robot will drain its batteries in short order, and you'll have to replace them frequently-a costly proposition.

For this reason, the ideal robot power source is the rechargeable battery. Two rechargeable battery technologies stand out as being easy to get and relatively inexpensive. The older nickel-cadmium (ni-cad) batteries, as well as the newer nickel metal hydride (NiMH) batteries, are available at hobby stores, home improvement outlets, even some discount stores like Target or Wal-Mart.

Unusual power sources, such as experimental fuel cells, are also included in this section, as well as battery holders and battery chargers.

SEE ALSO:

MANUFACTURER-TRAIN & HOBBY: Makers of radio control components, including batteries

RADIO CONTROL-GENERAL R/C COMPONENTS: batteries, battery packs, and holders

RETAIL-HARDWARE & HOME IMPROVEMENT: Good retail battery source

RETAIL-GENERAL ELECTRONICS: Batteries and battery packs

RETAIL-SURPLUS ELECTRONICS: Used (but still good) batteries

Advanced Battery Systems
204116

300 Centre St.
Holbrook, MA 02343
USA

☎ (781) 767-5516
📠 (781) 767-4599
🚫 (800) 634-8132
✉ periphex@aol.com
🌐 http://www.advanced-battery.com/

Batteries: sealed lead acid, ni-cad, NiMH, lithium, lithium-ion, lithium polymer, alkaline; in all traditional sizes and capacities. Also rechargers and battery packs (custom and stock) for cell phones, laptops, cordless phones, and other electronic devices.

All Effects Company, Inc.
203443

17614 Lahey St.
Granada Hills, CA 91344
USA

☎ (818) 366-7658
📠 (818) 366-3768
✉ eric@allfx.com
🌐 http://www.allfx.com/

All Effects makes an ingenious Makita battery power adapter. Makita batteries are available at most any hardware or home improvement store and come in 7.2 volts, 9.6 volts, and 12 volts. The adapter allows for convenient use of these batteries for other applications.

All Effects is perhaps best known as a mechanical special effects company for motion pictures and television. The company is headed by Eric Allard, one of the most creative mechanical effects engineers working in the movie business today. His company's credits include the Johnny 5 robot from the Short Circuit films (the second of which he coproduced), Ghostbusters, Alien Resurrection, and Stuart Little, and the Energizer Bunny TV commercials.

Batteries America
204112

2211-D Parview Rd.
Middleton, WI 53562
USA

☎ (608) 831-3443
📠 (608) 831-1082
🚫 (800) 308-4805
✉ ehyost@chorus.net
🌐 http://www.mrnicd-ehyostco.com/

Batteries (including sealed lead acid, ni-cad, NiMH) and rechargers; single-cell and packs.

🌐 💻

Batteries Plus
202176

925 Walnut Ridge Dr.
53029, WI Hartland
USA

🚫 (800) 274.9155
✉ sales@batteriesplus.com
🌐 http://www.batteriesplus.com/

U.S. nationwide battery retail chain, sells product for retail and commercial applications. Check the Web site for a store location.

Battery Mart   202850

1 Battery Dr.
Winchester, VA 22601
USA

- (540) 665-0065
- (540) 665-9623
- (800) 405-2121
- ✉ info@batterymart.com
- 🌐 http://batterymart.com/

Mail order batteries, large and small, all types (including sealed lead acid, motorcycle, ni-cad, NiMH); chargers. Online sales; local stores in Winchester, Va. and Martinsburg, W. Va. See their section of Robot Batteries-a selection of sealed lead acid batteries in sizes from single cells to large 12-volt packs.

Battery Mart Web site.

Battery Specialties   203149

3530 Cadillac Ave.
Costa Mesa, CA 92626
USA

- ☏ (714) 755-0888
- ☏ (714) 755-0889
- ⊘ (800) 854-5759
- ✉ sales@batteryspecialties.com
- 🌐 http://www.batteryspecialties.com/

Resellers of alkaline, ni-cad, sealed lead acid, and lithium batteries, and battery holders.

DC-DC Power Supply   204053

http://www.jarcom.com/inmotion/power.htm

How to build a DC-DC power supply for powering such devices as a PC motherboard from a single battery. Uses a slew of Maxim power control chips for the regulation and conversion.

eBatts.com   202579

703 Rancho Conejo Blvd.
Newbury Park, CA 91320
USA

- ☏ (805) 499-4332
- ⊘ (800) 300-1540
- ✉ info@eBatts.com
- 🌐 http://www.ebatts.com/

Rechargeable batteries. In single cells or packs for camcorders, laptops, cordless phones, and cellular phones.

Edmond Wheelchair Repair & Supply 203777

1604 Apian Way
Edmond, OK 73003
USA

- ☏ (405) 359-5006
- ⊘ (888) 343-2969
- ✉ Sales@edmond-wheelchair.com
- 🌐 http://www.edmond-wheelchair.com/

Wheelchair and scooter parts, including motors, wheels, and batteries.

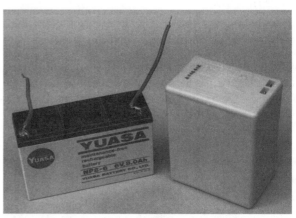

Sealed lead acid batteries.

Elite Speed Products   203305

3923 East Mound St.
Columbus, OH 43227
USA

((614) 231-4170

 sales@elitespeedproducts.com

http://www.elitespeedproducts.com/

Ni-cad batteries, electric motors, and motor replacement parts (brushes, armatures, etc.) for R/C applications.

Energizer Holdings, Inc. 203228

533 Maryville University
St. Louis, MO 63141
USA

⊘ (800) 383-7323

http://www.eveready.com/

Energizer either makes bunny toys or batteries; it's hard to tell from the commercials. In any case, the Web site has some interesting technical information about batteries, including an overview of battery chemistry, charge and discharge curves, and other engineering data. Haven't found any technical info on rabbits yet....

Energy Sales 204063

355 E. Middlefield Rd.
Mountain View, CA 94043
USA

((650) 969-0800

(650) 961-2000

⊘ (800) 963-6374

 info.ca@energy-sales.com

 http://www.energy-sales.com/

Distributors of name brand primary and rechargeable batteries. Locations also in Oregon and Washington.

Acme Model Engineering Co.

http://www.acmemodel.com/

Battery holders for AAA, N, AA, C, D and 9-volts

Battery City

http://www.battery-city.com/

Consumer-packaged batteries: ni-cads, NiMH, specialty; local to Southern California

---

EVdeals 204218

9 South St.
Plainville, MA 02762
USA

((508) 695-3717

(508) 643-0233

 scott@EVdeals.com

 http://www.evdeals.com/

Electric motors for bikes and scooters; 12- to 48-volt batteries; and battery chargers.

Fuel Cells 2000 204211

http://www.fuelcells.org/

Online resource for fuel cells-how they work, who makes them, where you can find them (including small demonstrator kits, possibly useful in experimental robots).

FuelCellStore.com 204210

P.O. Box 4038
Boulder, CO 80306-4038
USA

((303) 881-8343

 info@fuelcellstore.com

 http://www.fuelcellstore.com/

Online retailer of exotic fuel cells and fuel cell demonstrator kits. Be the first on your block to power your robot with one. Also offers DC-DC converters (most small fuel cells don't develop much voltage), fuel, and how-to books.

Gillette Company, The 203226

Berkshire Corporate Park
Bethel, CT 06801
USA

⊘ (800) 551-2355

http://www.duracell.com/

Duracell makes batteries. Billions of dollars worth every year. You'll buy them at your local store or through the mail, but the Web site provides some interesting semi-technical articles in the Technical/OEM section.

Chemical Makeup of Batteries

While there are hundreds of battery compositions, only a small handful of them are regularly used in amateur robots.

- *Carbon-zinc* batteries are also known as garden-variety "flashlight" cells, because that's the best application for them—operating a flashlight. They're a simple battery with relatively low current capacities. While they can be "rejuvenated" to bring back some power, they are not rechargeable, and they end up being expensive for any high-current application (like running a robot).

- *Alkaline* batteries offer several times the current capacity of carbon-zinc and are the most popular non-rechargeable battery used today. They cost several times more than carbon-zinc. Robotics applications tend to discharge even alkaline batteries rather quickly, so a bot that gets played with a lot will run through it's fair share of cells. Good performance, but at a price.

- *Rechargeable alkaline* batteries are the mass-merchandizing answer to the high cost of regular alkaline batteries used in high-demand applications—robotics is certainly one such application, though battery makers had things like portable CD players in mind when they designed rechargeable alkalines. These cells require a recharger designed for them and can be recharged dozens or hundreds of times before discarding. All things considered, rechargeable alkalines are probably the best choice as direct replacements for regular alkaline cells. More about why this is a factor in the article titled "Mixing and Matching Battery Voltage."

- *Nickel-cadmium* (or *ni-cad*) rechargeable batteries are an old technology and unfortunately, one that has caused considerable poisoning of the environment—cadmium is extremely toxic. So, battery makers have been weaning consumers off ni-cads, favoring instead the battery formulation that follows. While you can still get ni-cads, there's little reason to, so we'll ignore them as a choice.

- *Nickel-metal hydride* (*NiMH*) rechargeable batteries not only offer better performance than ni-cads, they don't make fish, animals, and people (as) sick when they are discarded in landfills. They are the premier choice in rechargeable batteries today, but they're not cheap. Like rechargeable alkalines, they require a recharger made for them. (Many of the latest rechargers will work with rechargeable alkalines, ni-cads, and NiMH; just don't use a ni-cad recharger with NiMH.)

- *Lithium-ion* (*Li-ion*) cells are frequently used in the rechargeable battery packs for laptop computers and high-end camcorders. They are the Mercedes-Benz of batteries, and are surprisingly lightweight for the current output they provide. However, Li-Ion cells require specialized rechargers and are frightfully expensive.

- *Lithium* (without the *-ion*) batteries are also available. These are non-rechargeable and are used for long-life applications, such as for smoke detectors. They are also used as memory backups and are commonly available in 3-, 6-, and 9-volt cells.

- *Sealed lead-acid* (*SLA*) batteries are similar in makeup to the battery in your car, except that the electrolyte is in gel form, rather than a sloshy liquid. SLAs are "sealed" to prevent most leaks, but in reality, the battery contains pores to allow oxygen into the cells. SLA batteries, which are rechargeable using simple circuits, are the ideal choice for very high current demands, such as battle bots or very large robots.

- *Polymer* batteries are among the latest in rechargeable technology. They are used for medium- to high-current electronics applications like cellular phones. These batteries use lithium as a component, but they are not quite the same as the lithium and Li-ion cells mentioned earlier. Polymer batteries can be manufactured with thicknesses as small as 1mm wafers.

Batteries at a Glance

Battery	Volts/cell*	Application	Recharge**	Notes
Carbon-zinc	1.5	Low demand, flashlights	No	Cheap, but not suitable for robotics or other high-current applications.
Alkaline	1.5	Small appliance motors and electric circuits	No	Available everywhere; can get expensive when used in a high-current application like robotics.
Rechargeable alkaline	1.5	Substitute for non-rechargeable variety	Yes	Good alternative to non-rechargeable alkalines.
Ni-cad	1.2	Medium- and high-current demand, including motors	Yes	Being phased out because of their toxicity.
NiMH	1.2	High-current demand, including motors	Yes	High capacity; still a bit pricey.
Li--ion		High-current demand, including motors	Yes	Expensive, but lightweight for their current capacity.
Lithium	3	Long life, very low current demand	No	Best used as battery backup for memory circuits.
SLA	2.0	Very high current demand	Yes	Heavy for their size, but very high capacities available.
Polymer	3.8	Long life, medium-current demand for electronics	Yes	Cells can be made to most any size and shape; very high price; voltage varies widely over discharge.

*Nominal volts per cell for typical batteries of that group. Higher voltages can be obtained by combining cells.

**Many non-rechargeable batteries can be "revitalized" by zapping them with volts for a few hours. However, such batteries are not fully recharged with this method and are re-discharged very quickly.

Battery Voltage and Current

The two most critical aspects of batteries are their voltage and their current. The importance of *voltage* is obvious: The battery must deliver enough electrical juice to operate whatever circuit it's connected to. A 12-volt system is best powered by a 12-volt battery. Lower voltages won't adequately power the circuit, and higher voltages may require voltage reduction or regulation, either of which entails some loss of efficiency.

If voltage is akin to the amount of water going through a pipe, then *current* is the pressure of that water. The higher the pressure, the more forceful the water is when it comes out. Similarly, current in a battery determines the ability of the circuit it's connected with to do heavy work. Higher currents can light bigger lamps or move bigger motors.

Because batteries cannot hold an infinite amount of energy, the current capacity of a battery is often referred to as an *energy store* and is referred to simply as *C*, for capacity.

Battery current is rated in amp-hours, or roughly the amount of amperage (a measure of current) that can be delivered by the battery in a one-hour period. In actuality, the amp-hour rating is an idealized specification: It's really determined by discharging the battery over a 5- to 20-hour period. Few batteries can actually deliver their rated amp-hour currents throughout that stated hour.

Smaller batteries are not capable of producing high currents, and their specifications are listed in milliamp-hours. There are 1,000 milliamps in an amp. Therefore, a battery that delivers half an amp is listed with a capacity of 500 milliamp-hours (abbreviated *mAh*).

Common Battery Sizes

If you've ever changed the batteries in an electronic device, you already know there are different sizes available. There's no point in telling you batteries are available in N, AAA, AA, C, D, and 9-volt ("transistor") size.

But with the notion of different sizes fresh in our minds, its worth noting that the size of the battery directly affects its capacity. For comparison purposes, following is the typical capacity ratings for rechargeable ni-cad or NiMH batteries:

Some common battery sizes.

Cell Size	Diameter	Height	Weight	Capacity in mAh
N	12.0	30.0	5	150
AAA	10.5	44.5	12	650
1/3 AA*	14.0	14.0	7	50
1/2 AA*	14.0	17.0	14	110
2/3 AA*	14.0	28.3	14	600
4/5 AA*	14.0	42.2	23	1200
AA	14.0	50.0	25	1500
A*	17.0	50.0	35	2200
1/2 C	23.0	26.0	23	2100
C	25.2	49.2	80	3500
D	32.2	60.0	150	7000
9-volt	25.7 _ 17.4	48.2	45	160–200

Diameter and weight are in millimeters; weight is in grams.

*Typically used in specialty battery packs, and not available in traditional consumer packaging. They are available from battery specialty retailers as replacement cells, however.

Mixing and Matching Battery Voltage

Elsewhere we've noted that not all battery cells provide the same voltage. Alkaline cells provide 1.5 volts nominal (which means normal or average) per cell. Ni-cads and NiMH batteries provide 1.2 volts. Most cells are used in battery packs, where they are connected in series. This acts to add the volts from each cell to provide a higher voltage of the whole pack. Therefore, four 1.5-volt batteries will yield 6 volts; but four 1.2-volt batteries will yield only 4.8 volts.

Because of this variance in cell voltage, it is not always possible to simply substitute one battery type for another in a battery pack. Circuits designed to work with a 6-volt pack may not function, or may function erratically, if used with a 4.8-volt pack. If you've been using your robot with alkalines, it may not work if you substitute lower voltage NiMH batteries.

When it doubt, check the manufacturer's datasheet (if one is available) or test the device with the lower voltage pack. Be sure to verify operation throughout the whole discharge period of the batteries. The robot may work fine when the batteries are fresh; but the voltage at the terminal of the battery is reduced as it is discharged. This means your robot could stop working after a short period of time, before the batteries are fully discharged.

The Right Voltages for Your Robot

Standards are a wonderful thing. You can buy a TV at your local electronics boutique and know that it'll work when you plug it into the socket at home. Don't count on electronics for robots to be as accommodating. There is no standard for operating electronic equipment: Some require 5 volts, others need 3.3 volts, and yet still others need 12, 15, 24, or 48 volts, and everything in-between.

Providing the proper voltages to the various subsystems in your robot requires careful planning. Obviously, the easiest way to manage the power requirements of your robot is to chose components that operate at a single voltage—say, 5 volts. That's not always possible, especially for a mechanical device like a robot, which uses a wide variety of systems.

There are three basic approaches to powering the various components in your robot. Each one is discussed below.

Single Battery; Multiple Voltages

Most of the electrical equipment in your home or office is operated from a single power supply (such as wall current). Each piece of equipment, in turn, uses this voltage as is (as in the case of an electrically powered fan), or it converts the incoming current via a transformer or other device to the voltage required. This is the natural approach because each piece of equipment is a stand-alone unit and doesn't depend on any other to operate.

This same approach can be used in your robot. A single battery—delivering, say, 12 volts—powers different subsystems. A voltage regulator or DC-DC converter is used to provide each subsystem with the precise voltage it requires.

While this approach sounds good in theory, in practice it can be expensive and/or difficult to implement properly.

- *Linear voltage regulators*, the most common variety, are cheap but relatively inefficient. In effect, they "step down" voltage from one level to another; the difference in voltage is dissipated as heat. The heat can be dealt with; the real problem is the unnecessary drain on the battery. It's better to conserve battery power for productive tasks, like running the robot's motors.

- *Switching voltage regulators* are more efficient—some offer efficiencies of up to 80%—but they are more expensive to implement, and many require additional components and design consideration. Like linear voltage regulators, switching regulators "step down" one voltage to provide another.

- *DC-DC converters* are self-contained voltage changers. They are the most expensive of the lot, but they require no additional components. DC-DC converters can step down or step up voltages and can provide negative voltages. The disadvantage of many DC-DC converters (besides cost) is that they require high input voltages in order to supply adequate current at the output. For instance, the input voltage may be on the order of 24 to 48 volts, in order to provide reasonable current at 5 or 12 volts.

Multiple Batteries; Multiple Voltages

A potential alternative to voltage regulation or conversion is to add separate battery packs to your robot. One battery pack may power the main electronics of the robot; another may power the motors. This often works out well because the electronics probably need regulation, and the motors do not. The pack for the electronics can be 6 or 7.2 volts (regulated to 5 volts), and the pack for the motors can deliver 12 volts.

The trick to making this work is to tie all the ground (negative terminal) wires of the battery packs together. Each subsystem receives the proper voltage from its battery pack, but the shared grounds ensure that the various parts of your robot work together.

The exception to tied grounds is if you use optoisolators. A typical application of optoisolators in robotics is to control the drive motors. The electronics and the motors are on completely separate circuits, and their grounds are not tied. Rather than connect wires directly from the electronics to the motor control circuitry, the electronics instead power optoisolators, which contain a light emitting diode (LED) and a phototransistor. The link between electronics and motor control is therefore made of light, not wire.

Single Battery; Single Voltage

Depending on the subsystems of your robot, you may be able to use a single battery and single voltage for everything. Example: If your electronics do not contain any 5-volt TTL parts, you might be able to run all circuitry at 12 volts, along with the motors of your robot. Since applying excessive voltage to electronics can damage them, always check the specifications first.

The disadvantage of using a single battery for both electronics and motor is that DC motors—especially large ones—produce a lot of electrical noise that can disrupt the operation of microcontrollers and computers. If you plan on operating your robot from a single, nonregulated battery pack, be sure to add noise suppression to the motors. One effective noise suppression technique is to solder 0.1-uF nonpolarized disc capacitors across the terminals of the motor and/or from each terminal to the ground case of the motor.

Using Ready-made Rechargeable Battery Packs

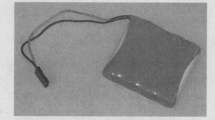

Many of today's consumer products use rechargeable batteries. It's a fair bet that the majority of these use specially made battery packs, rather than individual cells. Manufacturers must often specify so-called sub sizes for the cells in the pack because of size issues. It's hard to get a couple of AAs, or even AAAs, in the handset of a cordless phone, for instance.

A typical battery pack for a consumer electronics device (in this case, a cordless phone).

Replacement battery packs are available for most popular brands and makes of cordless phones, cell phones, personal CD players, and other consumer electronics products. While these replacement packs carry a premium price, there are some advantages:

- Most packs are smaller than ones you can make yourself and are handy if space is a problem.
- Purchasing the replacement pack for a discarded device allows you to hack the recharger electronics from it. This saves you the cost and trouble of buying or building a recharger.
- If the battery pack of a discarded product is still good, you can recharge and use it and save yourself the cost of buying new batteries.

Hawker Energy Products Inc. 203439

617 N. Ridgeview Dr.
Warrensburg, MO 64093-9301
USA

- (660) 429-6437
- (660) 429-6397
- (800) 964-2837
- info.US@hawker.invensys.com
- http://www.hepi.com/

Hawker claims to be the world's largest battery manufacturer. I can believe it, considering how many products and stores carry their wares. Their main product line is sealed lead acid (SLA), which are available in single cells (usually 2 volts per) and in packs up to 24 and 48 volts.

The Web site offers no direct purchasing (Hawker representatives and distributors handle that), but there are plenty of battery spec sheets and technical papers to print out and take to bed with you.

Interesting tidbit: Hawker is the battery of choice among many battling robot builders.

House of Batteries 203440

16512 Burke Ln.
Huntington Beach, CA 92647
USA

- (714) 375-0222
- (714) 375-0235
- (800) 432-3385
- sales@houseofbatteries.com
- http://www.houseofbatteries.com/

House of Batteries is a distributor and online retailer of batteries, battery packs, and chargers. Most name brands are carried, in common consumer and industrial sizes. Not all products are available for online ordering.

Jbro Batteries / Lexstar Technologies 204087

1938 University Ln.
Lisle, IL 60532-2150
USA

- (630) 964-9081
- (800) 323-3779

 http://www.jbro.com/

Manufacturer and distributor of rechargeable batteries and battery chargers. Brands include Panasonic and Hawker.

KidsWheels 203771

13266 Pond Springs Rd.
Austin, TX 78729
USA

- (512) 257-2399
- (512) 257-0093
- Alf@KidsWheels.com
- http://www.kidswheels.com/

Monster trucks for kids: Perego, Fisher-Price Power Wheels, and other makes. Can be ordered from the Web site. Also offers parts, such as gearboxes, wheels, and batteries.

Nexergy 204062

1909 Arlingate Ln.
Columbus, OH 43228-9331
USA

- (614) 351-2191
- sales.columbus@nexergy.com
- http://www.nexergy.com/

Alkaline, sealed lead acid, and other battery packs; online ordering in bulk for a discount.

NiCad Lady Company, The 204064

20585 Camino del Sol
Unit B
Riverside, CA 92508
USA

- (909) 653-8868
- (909) 653-5189
- nicdlady@nicdlady.com
- http://nicdlady.com/

Ni-Cad Lady sells ni-cad, NiMH, and sealed lead acid batteries in all sizes. Also carries rechargers and battery packs for cell phones and laptop computers.

Pico Electronics, Inc. 204008

143 Sparks Ave.
Pelham, NY 10803-1837
USA

✆ (914) 738-1400
📠 (914) 738-8225
⊘ (800) 431-1064
✉ info@picoelectronics.com
🌐 http://www.picoelectronics.com/

Manufacturers and sellers of miniature and micro-miniature transformers, inductors, DC-DC converters, and DC power supplies.

Planet Battery 203977

46 Baker St.
Providence, RI 02905
USA

📠 (401) 781-1340
⊘ (877) 528-1117
✉ pb@planetbattery.com
🌐 http://www.planetbattery.com/

Batteries and battery packs. Sells sealed lead acid battery packs in different voltages and capacities.

Power Sonic Corp. 203976

9163 Siempre Viva Rd.
Stes. A-F
San Diego, CA 92154
USA

✆ (619) 661-2030
📠 (619) 661-3648
✉ national-sales@power-sonic.com
🌐 http://www.power-sonic.com/

Rechargeable batteries and rechargers.

- Sealed lead-acid batteries, ranging from 0.5 to 100 Ah
- Nickel-cadmium and nickel-metal hydride batteries
- Automatic battery chargers

The Web site hosts numerous spec sheets, datasheets, and materials safety sheets on battery technologies and compositions.

Robotic Power Solutions 204215

305 9th St.
Carrollton, KY 41008
USA

✆ (502) 639-0319
✉ sjslhill@bellsouth.net
🌐 http://www.battlepack.com/

Specializing in combat robot parts, the company sells ni-cad and NiMH battery packs, chargers, and cobalt AstroFlight gearmotors. Their "Battlepack Kits" include batteries, wires, support bars, heat-shrink tubing, and padding foam.

Sanyo Energy 204060

🌐 http://www.sanyobatteries.net/

Industrial and consumer batteries: ni-cad, NiMH, lithium, and lithium-ion. Product specifications and technical white papers are provided at the Web site.

Super Battery Packs 203297

B. Furgang
3125 Rockgate
Simi Valley, CA 93063
USA

✉ bfurgang@ez2.net
🌐 http://www.superbatterypacks.com/

High-performance R/C battery packs, made into packs. Uses NiMH batteries.

Tadiran U.S. Battery Division 204059

2 Seaview Blvd.
Port Washington, NY 11050
USA

✆ (516) 621-4980
📠 (516) 621-4517
⊘ (800) 537-1368
✉ sales@tadiranbat.com
🌐 http://www.tadiranbat.com/

Providers of lithium batteries. Product descriptions, spec sheets, technical briefs.

Thomas Distributing   203148

128 East Wood
Paris, IL 61944
USA

- ((217) 466-4210
- ␂ (217) 466-4212
- ⊘ (800) 821-2769
- 🌐 http://www.nimhbattery.com/

Batteries (specializing in NiMH) and battery holders. Very large selection. Also provides semitechnical backgrounders on battery technologies.

TNR Technical   204061

3400 West Warner Ave., #K
Santa Ana, CA 92704
USA

- ((714) 427-5175
- ␂ (714) 427-5187
- ⊘ (800) 490-8418
- ✉ pat@tnrtechnical.com
- 🌐 http://www.batterystore.com/

Batteries of all types and sizes; major manufacturers. Alkaline, ni-cad, NiMH, sealed lead acid, lithium, and lithium-ion. In separate cells or packs (e.g., for camcorders, cordless phones, and laptops). Many cells available in round and rectangular/prismatic packages, with size and weight specifications provided in handy cross-reference tables.

Offices also in Florida and North Carolina.

Wilde EVolutions, Inc.   203576

18908 Hwy. 99, Ste. A
Lynnwood, WA 98036-5218
USA

- ((425) 672-7977
- ␂ (425) 672-7907
- ⊘ (800) 327-8387
- ✉ info@wilde-evolutions.com
- 🌐 http://www.wilde-evolutions.com/

Electric vehicles and parts, such as high-capacity batteries, motors, high-current relays, and speed controllers, for golf carts and other small electric vehicles and scooters.

ZapWorld.com   204220

117 Morris St.
Sebastopol, CA 95472
USA

- ((707) 824-4150
- ␂ (707) 824-4159
- ⊘ (800) 251-4555
- ✉ zap@zapworld.com
- 🌐 http://www.zapworld.com/

ZapWorld sells motors, batteries, and parts for small electric vehicles, namely, bikes, ground scooters, and aquatic scooters (as well as some odd stuff like personal hovercrafts). Use the stuff for your own performance robots.

BOOKS

Books are still the premier source of specialty information. The following sections list online bookstores, especially those that carry technical books. Some of these bookstores are also open for walk-in business through one or more retail stores. Also listed are publishers and support Web sites for books of interest to the robot builder.

The Books section is divided into several subcategories:

BOOKS: Online "heavyweights" that carry most every book in print

BOOKS-ELECTRONICS: Books on electronics, such as electronics theory and construction

BOOKS-ROBOTICS: Special-interest books on robots, robotics, and artificial intelligence

BOOKS-TECHNICAL: General-purpose books on technical topics

Amazon.com 202586

http://www.amazon.com/

The ubiquitous online retailer of books and more. Separate stores sell books, DVD and VHS movies, kitchen and household appliances, toys (online Toys'R'Us), computers, software, and magazine subscriptions. Amazon.com is strictly Internet-based and is not a catalog mail-order company; phoning or writing in your order is discouraged.

Amazon Books Web page.

Barnes & Noble.com 204024

http://www.bn.com/

Barnes & Noble is the largest bookseller in the world, with retail stores and an online e-commerce Web site.

They have hundreds of books on robotics and other technical subjects.

Fatbrain.com recently teamed up with B&N.com, and consequently, a larger selection of engineering, computer and technical books can be found at its Web site:

http://btob.barnesandnoble.com/

Barnes & Noble Web page.

BOOKS-ELECTRONICS

Art of Electronics, The 203285

http://www.artofelectronics.com/

The Art of Electronics and its companion student workbook are the standard texts for electronics in many schools. Fairly laden with theory and math, the books are nevertheless aimed at a general academic and hobbyist audience. This is the support Web site for the book.

Lakeview Research 202154

5310 Chinook Ln.
Madison, WI 53704
USA

((608) 241-5824

((608) 241-5848

✉ jan@lvr.com

🌎 http://www.lvr.com/

Web site of author and technical guru Jan Axelson, who provides information and tools relating to USB, parallel ports, RS-232 and RS-485 serial communications, 8052-Basic microcontrollers, and making printed circuit boards.

USB Complete, Second Edition.
Photo Jan Axelson.

McGraw-Hill Professional 300010

http://www.books.mcgraw-hill.com/

Publishers of this book and many others on various technical topics. The Web site is divided into subjects, and you can search for relevent books by keyword, title, or author. Books can be purchased online or through most any book retailer.

Square One Electronics 202491

P.O. Box 501
Kelseyville, CA 95451-0501
USA

((707) 279-8881
((707) 279-8883
✉ sqone@pacific.net
🌍 http://www.sq-1.com/

Publishers of how-to books on PIC microcontroller programming and stepper motors. Available from the company or from many online booksellers, such as Amazon.com. Titles include:

- Easy PIC'n
- PIC'n Up the Pace
- PIC'n Techniques
- Serial PIC'n

BOOKS-ROBOTICS

Personal Robot Technolgies, Inc. 203713

P.O. Box 612
Pittsfield, MA 01202
USA

((413) 684-5220
🚫 (800) 769-0418
🌍 http://www.smartrobots.com/

Support Web site for the book, Personal Robot Navigator, published by AK Peters.

Robotbooks.com 203537

✉ webmaster@robotbooks.com
🌍 http://www.robotbooks.com/

Robotbooks.com is a "partnered" Web reseller of books, toys, movies, and other robotic artifacts. (By partnering, the Web site doesn't do any fulfillment of their own, but rather provides affiliate links to various e-commerce Web sites on the Internet.) The Web site is run by Carlo Bertocchini, a well-known BattleBots contestant-his robots include Biohazard.

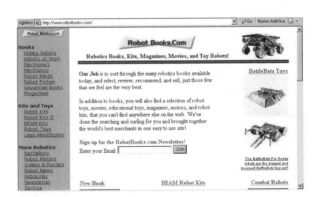

Robotbooks.com Web site.

Robotics Universe 202332

✉ gort@robotoid.com
🌍 http://www.robotoid.com/

Online support Web site for my robotics books, including the one you're reading now.

Robotics Universe Web site.

Stiquito 202248

Micro Robotics Supply, Inc.
101 Pendren Pl.
Cary, NC 27513-2225
USA

 stiquito@stiquito.com

 http://www.stiquito.com/

Stiquito is a small and simple robot that uses shape memory alloy (SMA) wire for movement. This is official Stiquito Web site, maintained by author Jim Conrad, and supports the product and several books written about it.

BOOKS-TECHNICAL

A K Peters, Ltd. 203374

63 South Ave.
Natick, MA 01760-4626
USA

📞 (508) 655-9933
📠 (508) 655-5847
🌐 http://www.akpeters.com/

AK Peters publishes intermediate- and advanced-level books on engineering, math, physics, robotics, and other disciplines. Many of their books are college textbooks, so don't expect easy bedtime reading. Their Mobile Robots: Inspiration to Implementation (Jones, Flynn, and Seiger) has been a standard text for many years. AK Peters also publishes the popular (though somewhat expensive) RugWarrior kit. The books are available through booksellers; the Web site includes a browsable catalog of current and upcoming books.

A. G. Tannenbaum 202904

P.O. Box 386
Ambler, PA 19002
USA

📞 (215) 540-8055
📠 (215) 540-8327
✉ k2bn@agtannenbaum.com
🌐 http://www.agtannenbaum.com/

A unique source for service manuals of old but not forgotten electronic components: TVs, VCRs, amateur radio rigs, fax machines, microwaves, audio gear, and more. The company also buys and sells old vintage radios, TVs, tube hi-fi sets, test equipment, hard-to-find tubes and high-voltage capacitors. Manuals are available as copies, and, depending on availability, some are sold as originals.

What do service manuals for old stuff have to do with robots? First and foremost, yesterday's electronic components often contain useful mechanical and electrical parts, such as gears, pulleys, and motors. The service manuals help identify odd-ball parts you may find and can assist you in adapting them to your robot. This is especially true of motors in VCRs. One brand of VCR will use the same parts for several years, so one manual covers a lot of VCR models.

Second, it's fun. Maybe you could build a robot from a vintage Philco radio.

Lindsay's Technical Books 🎖 203271

P.O. Box 538
Bradley, IL 60915-0538
USA

📞 (815) 935-5353
📠 (815) 935-5477
✉ lindsay@lindsaybks.com
🌐 http://www.lindsaybks.com/

Old (reprints) and new books on electronics, how-to (woodworking, metalworking, etc.), high voltage, and other unusual topics. Lindsay's sells a large assortment of metalworking and foundry books, some old and some new. Also books on plastic injection molding and plastic vacuum forming.

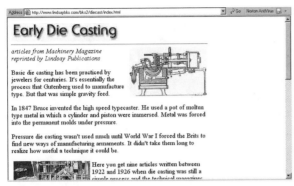

Lindsay's Technical Books Web site.

Nerd Books 203441

911 Washington Blvd.
Ste. 201
Roseville, CA 95678
USA

((916) 677-1400

 info@nerdbooks.com

 http://www.nerdbooks.com/

Online technical bookstore. Also has local store in Sacramento, Calif., area.

OPAMP Technical Books 203744

1033 N. Sycamore Ave.
Los Angeles, CA 90038
USA

((323) 464-0977

 (800) 468-4322

 http://opamp.com/

Mail-order technical books. Retail bookstore in Los Angeles, Calif. (actually, Hollywood). OPAMP calls itself "The Technical Person's Discount Technical Bookstore."

PID Without the Math 203081

http://members.aol.com/pidcontrol/booklet.html

Informational Web site about PID Without the Math, a semitechnical overview of controlling servo motors.

Powell's Books 203812

((503) 228-0540 Ext. 482

 (800) 291-9676

 help@powells.com

 http://www.powells.com/

Online and local store in Oregon.

ProfBooks Robotics 202408

http://www.profbooks.com/Robotics/

U.K.-based bookseller of high-level books. Most of the robotics they carry are textbooks for university study, though there are hobbyist-level books thrown in for good measure.

SAMS Technical Publishing 204248

5436 W. 78th St.
Indianapolis, IN 46268-3910
USA

((317) 334-1256

 (800) 428-7267

customercare1@samswebsite.com

http://www.samswebsite.com/

SAMS is perhaps best known for their Photofact service guides. In the days before they plastered "no user service-able parts inside" over every consumer electronics device, people actually fixed things when they broke. SAMS was there with a Photofact, which included full schematics, troubleshooting, and alignment information.

SAMS still sells Photofacts for many current and older model products, including TVs, printers, and VCRs, the latter of which is useful for robot builders because video recorders are a gold mine of parts. You can use the information in the Photofact as a way to hack parts for your robot. Because most parts are reused for many models of VCRs, a small handful of Photofacts will serve you for a number of VCRs you may find on the garage sale and thrift store circuit.

For the Photofact products, which can be purchased online, see:

http://www.samswebsite.com/photofacts.html

U.S. Government Printing Office 203573

http://www.access.gpo.gov/

One of the world's biggest (if not the biggest) publishers is the U. S. government. All sorts of pamphlets and books are for sale here. Online bookstore; GPO bookstores are situated in major cities across the country. Among the riveting titles:

- Dielectric and Magnetic Properties of Printed Wiring Boards and Other Substrate Materials

- Inventory of Electric Utility Power Plants in the United States (in case your robot needs to plug in)

- Questions and Answers About EMF

- Stages to Saturn: A Technological History of the Apollo/Saturn Launch Vehicles

At one time, GPO books were very inexpensive compared to other books. That trend seems to have changed. The typical "meaty" GPO book is well over $35, and many are over $50. Also includes some CD-ROM titles.

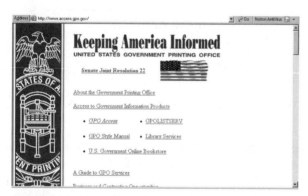

Web site for the United States Government Printing Office.

Workshop Publishing 204001

272 Morgan Hill Dr.
Lake Orion, MI 48360
USA

((248) 391-2974

(248) 391-8290

WorkshopPublishing@Hotmail.com

http://www.build-stuff.com/

A "How to Build Stuff" emporium.

Plans include:

- Table Top Vacuum Forming Machine -For molding plastic parts

- Secrets of Building a Vacuum Forming Machine

- Secrets of Building an Injection Molding Machine

Books include:

- Vacuum Forming for the Hobbyist -Low-budget plastic molding for the hobbyist

- Understanding Thermoforming-An overview of commercial vacuum forming

- Mold Making and Casting Guides-four-book set, with sources for materials

- Prop Builders Molding and Casting Book-450 photos, covers many materials.

- How to Cast Your Own Plastic Parts-Cast urethane resin into silicone molds

- Cutting Costs in Short Run Injection Molding-Make your own short-run tooling

A completed plastic vacuum former. Photo Doug Walsh, Workshop Publishing.

Further Reading

Nothing beats books. They're relatively cheap, and the better ones pack more information than you'll find on a galaxy of Web pages. For the robot-minded, here are some books you'll want to consider adding to your reading list. These are available for purchase, new or used, but don't forget the local library. If a book you want isn't carried by your library, ask the librarian if it is available from another branch. Or, ask that the book be considered for purchase.

I don't list every single robotics book, just the ones I think are most relevant to amateur robotics. Same goes for electronics and engineering texts.

Online or Local

Only the more general-interest robotics books are carried in stock by local bookstores. You'll need to special order the books on higher-level robotics or else find them online. The premier sources for online books are:

Store Name	URL	Notes
Amazon	http://www.amazon.com/	Best-known of the online bookstores; books can be ranked by publication date (great for seeing what's new), bestsellers, author, title, and customer reviews.
Barnes and Noble	http://www.bn.com/	The online subsidiary of the Barnes & Noble bookstore chain. Not as many robotics titles as Amazon, but the ones they have are likely to be the ones you want. See also http://btob.barnesandnoble.com/ for their technical titles.
Bookpool	http://www.bookpool.com/	Not many books on robotics (or electronics or engineering), but nice discounts.
Brown Technical Bookshop	http://www.brownbookshop.com/	Specializes in highly technical books. Few robotics titles, but several engineering, math, electrical, and associated books.
OpAmp Technical	http://opampbooks.com/	Online and local Books (Los Angeles, Calif.) retailer of technical books. Limited selection of robotics books, but offers to beat Amazon's price. Only lists books alphabetically.
Powells	http://www.powells.com/	Offers a fair number of robotics titles; displays the list in alphabetical order, by author, or by price. Occasionally offers discounts. Local stores in Oregon.
San Diego Technical Books	http://www.booksmatter.com/	Online and local (San Diego, Calif.) book retailer. Good selection, and you have the ability to list books by publication date, author, title, or selling rank. Discounts are rarely given, however.
Stacey's	Booksellers http://www.staceys.com/	Online and local (San Francisco, Calif., Bay Area) book retailer. Good selection, but not discounted and browsing options are limited.

Google.com search:
technical bookstore
   Additionally, several roboticscentric Web sites cater to the robot reader.
Robotbooks.com—http://www.robotbooks.com/
GoRobotics.net—http://www.gorobotics.net/

   In the descriptions that follow, you'll see this icon, which means I personally recommend this book for any well-stocked robotics library. This *does not* mean the other books are not recommended, useful, or good. But I realize not everyone has the cash to purchase, or the time to read, *every single book* on robotics. I've limited myself to voting for just two "recommended" books in each category (exception: children's nonfiction robotics books, of which there are really no exceptional ones). If nothing else, consider these books as starting points for building your own robotics reading list.

Hobby Robotics

 Applied Robotics

Edwin Wise
Delmar Learning, 1999
ISBN: 0790611848
Grab bag of amateur robotics, with a strong emphasis on hardware design (using an Atmel AVR microcontroller). Keynote: Included on the CD-ROM with book is "Fuzbol," developed by the author's company, a *fuzzy-logic* robotics operating system.

Build A Remote-Controlled Robot

David R. Shircliff
Tab Books, 2002
ISBN: 0071385436
Construction plans for a remote-controlled robot vehicle.

Build Your Own Combat Robot

Pete Miles, Tom Carroll
Osborne McGraw-Hill, 2002
The first in what will likely be a long line of books on building battling robots, like the ones on *BattleBots* and *Robotica*. Provides handy construction tips for heavy-duty robots, even if you don't want to attach lawnmower blades to the thing.

Build Your Own Robot!

Karl Lunt
A K Peters Ltd., 2000
ISBN: 1568811020
Reprint of the Karl's columns in *Nuts & Volts* magazine; the columns appeared in the early to mid-1990s, so the text is a bit on the old side. But much of it is still useful. Portions of the book get *very* technical, so don't try this one if you're a beginner.

Introduction to Robotics, An

Harprit S. Sandhu
Nexus Special Interest Ltd., 1997
ISBN: 1854861530
Nice entry-level guide to robotics, including the construction of a servo-operated robot.

Muscle Wires Project Book

Roger G. Gilbertson
Mondo-Tronics, 2000

ISBN: 1879896141

Book and kit about "Muscle Wires," a brand of shape memory alloy available from the author's company (Mondo-tronics). Projects include Boris, a walking robot.

Personal Robot Navigator, The

Miller, Winkless, Phelps, Bosworth
A K Peters Ltd., 1999
ISBN: 188819300X

Mostly theory, this book also contains a software CD-ROM of robot navigation simulator (Robonav).

PIC Robotics: A Beginner's Guide to Robotics Projects Using the PICMicro

John Iovine
McGraw-Hill, 2002
ISBN: 0071373241

Robot project book using the PIC microcontroller as the central brain.

Practical Robotics: Principles and Applications

Bill Davies
CPIC Technical Books, 1997
ISBN: 096818300X

General book on different mechanical construction ideas, with practical details. Not much actual robot building in this one, but the pieces can readily be put together to make one.

Robot Builder's Bonanza, Second Edition

Gordon McComb
McGraw-Hill, 2000
ISBN: 0071362967

My book on beginning and intermediate robotics. I'm proud to say it's the most popular and longest-selling book on amateur robotics ever published. (Yes, I get to put a "must have" symbol next to my own book. So sue me!)

Robot Building for Beginners

David Cook
APress, 2002

For raw beginners with little or no experience. You need some shop tools like a drill press and tapping set to build some of the designs, though.

Robot DNA (series)

McGraw-Hill; 2002

A unique series of books on robotics construction. Each book, aimed at the intermediate to advanced roboteer, details a specific aspect of construction. I acted as series editor (with Myke Predko) and contributed to several of the books.

Robots, Androids and Animatrons, Second Edition

John Iovine
McGraw-Hill, 2001
ISBN: 0071376836

Several entry-level projects, including a small six-legged walking robot.

Stiquito for Beginners: An Introduction to Robotics

James M. Conrad, Jonathan W. Mills
IEEE Computer Society Press, 2000
ISBN: 0818675144

Stiquito: Advanced Experiments with a Simple and Inexpensive Robot

James M. Conrad, Jonathan W. Mills
Institute of Electrical and Electronic Engineers, 1997
ISBN: 0818674083
Beginner and advanced books on constructing small, legged robots using shape memory alloy wire (e.g., BioMetal, Nitinol, Dynalloy, or Muscle Wires). Both books contain the basic parts you'll need to build the sample robot.

LEGO Robotics and LEGO Building

Building Robots with Lego Mindstorms

Mario Ferrari, Giulio Ferrari, Ralph Hempel
Syngress Media Inc, 2001
ISBN: 1928994679
Highly recommended guide to intermediate- and advanced-level LEGO Mindstorms robotics. Lots of mechanics; many programming examples are in Not Quite C.

Creative Projects with LEGO Mindstorms

Benjamin Erwin
Addison-Wesley Pub Co, 2001
ISBN: 0201708957
Using color illustrations, this book demonstrates over a dozen fun and unusual LEGO Mindstorms creations. A great book for the classroom.

Dave Baum's Definitive Guide to LEGO Mindstorms

Dave Baum
APress, 2000
ISBN: 1893115097
Part on LEGO Mindstorms mechanics, and part on using the author's Not Quite C (NQC) programming language, which he wrote especially for the Mindstorms platform.

Extreme Mindstorms: An Advanced Guide to Lego Mindstorms

Dave Baum et al.
APress, 2000
1893115844
Collection of intermediate-level topics on LEGO Mindstorms. The chapters are written by a variety of well-known Mindstorms experts.

Jin Sato's Lego Mindstorms: The Master's Technique

Jin Sato
No Starch Press, 2002
ISBN: 1886411565
Advanced-level LEGO Mindstorms robots, such as a robotic dog.

Joe Nagata's Lego Mindstorms Idea Book

Joe Nagata
No Starch Press, 2001
ISBN: 1886411409
Intended to both inspire and inform, this book is aimed at the intermediate- and advanced-LEGO roboteers. Few of the designs in Joe's book could be considered "the usual stuff."

Unofficial Guide to LEGO Mindstorms Robots

Jonathan B. Knudsen
O'Reilly & Associates, 1999
ISBN: 1565926927
Author Jonathan Knudsen captures the early enthusiasm and awe of the LEGO Mindstorms and provides a potpourri of ideas and concepts. Some fun designs.

Technical Robotics, Theory and Design

Computational Principles of Mobile Robotics

Gregory Dudek, Michael Jenkin
Cambridge University Press, 2000
ISBN: 0521568765
Math-heavy with an emphasis on computation and algorithms, intended for advanced undergraduate and graduate students studying mobile robotics.

Mobile Robots: Inspiration to Implementation

Joseph L. Jones, Anita M. Flynn, Bruce A. Seiger
A K Peters Ltd, 1999
ISBN: 1568810970
Hands-on guidebook to constructing mobile robots. Offers two main projects: Tutebot, which introduces basic mobile robot concepts, and Rug Warrior, based on the Motorola MC68HC11 microcontroller. As presented, Tutebot is too expensive to build for what it does (the authors recommend motors that retail for about $30 each), but you can build it with other components for a lot less.

Robot Evolution: The Development of Anthrobotics

Mark Rosheim
John Wiley & Sons, 1994
ISBN: 0471026220
Technical overview on robots and robot systems. For the hard core.

Robotic Explorations: An Introduction to Engineering Through Design

Fred Martin
Prentice Hall
ISBN: 0130895687
This is not an inexpensive book, but it's one of the best for teaching the fundamentals of robotics in the classroom. Written by a respected robotics professor, this is a textbook suitable for first- or second-year robotics courses in high school and college. The author provides explicit examples and exercises using LEGO Technic pieces and the MIT Handy Board controller, but you can apply what you learn to most any robotics platform.

Sensors for Mobile Robots: Theory and Application

H. R. Everett
A K Peters Ltd, 1995
ISBN: 1568810482
Robots are nothing without sensors to allow them to see their way. This is an overview book designed to help familiarize you in the role of sensors in robotics and the available technologies. Written by one of the experts in the field.

Robotics Essays and Overviews

Age of Spiritual Machines, The

Ray Kurzweil
Penguin USA, 2000
ISBN: 0140282025
Inventor Ray Kurzweil sees a mixed future for robots. Though his time frames may not hold up, he offers a scintillating look at the thinking (and perhaps feeling) machine of tomorrow.

Flesh and Machines: How Robots Will Change Us

Rodney Allen Brooks
Pantheon Books, 2002
ISBN: 0375420797
Penned by a respected robotics pioneer from MIT, *Flesh and Machines* explores the coming of sentient beings and how human beings will be forever changed by it. Parts of Dr. Brooks's vision are chilling, others parts are inspiring.

Managing Martians

D. Shirley, D. Morton
Bantam Doubleday Dell, 1998
ISBN: 0767902408
The chronicle of the birth and life of the Mars Pathfinder Sojourner robot. Written by the team leader of the Sojourner project.

Personal Robotics: Real Robots to Construct, Program, and Explore the World

Richard Raucci
A K Peters Ltd, 2000
ISBN: 156881089X
The subtitle of this book may be a tad misleading to some; there aren't any true construction plans in this book. Instead it shows you several ready-made robots and robot kits, as well as finished robots built by others. This is an inspirational overview and buyer's guide-book.

Robo Sapiens

Peter Menzel, Faith D'Aluisio
MIT Press, 2000
ISBN: 0262133822
Beautiful full-color encyclopedia on robotics. Available as a hardcover "coffee table" book or softbound. You'll want keep this one for a long time, so get the hardcover.

Robot in the Garden, The

Ken Goldberg
MIT Press, 2000
ISBN: 0262072033
A collection of essays on remotely controlled robotics: telerobotics and telepistemology, controlled via communications links like the Internet. See also the author's *Beyond Webcams* (MIT Press, 2001; ISBN: 0262072254).

Robot Riots: The Good Guide to Bad Bots

Alison Bing, Erin Conley
Dorset Press, 2001
ISBN: 0760730008

Quick review of combat robots, of the BattleBots ilk.

Artificial Intelligence and Behavior-Based Robotics

Artificial Intelligence and Mobile Robots

D. Kortenkamp, R. P. Bonasso, R. Murphy
MIT Press
ISBN: 0262611376
A review of functional systems and techniques in artificial robots used by researchers and companies.

Behavior-Based Robotics

Ronald C. Arkin
MIT Press, 1998
ISBN: 0262011654
Overview text of various behavior-based robotics techniques. The book contains references to robots that have used the various AI techniques described, with moderate detail of their implementation.

Cambrian Intelligence: The Early History of the New AI

Rodney Allen Brooks
MIT Press, 1999
ISBN: 0262522632
A collection of Rodney Brooks's (he's a professor at MIT) earlier works on robotics intelligence.

Introduction to AI Robotics, An

Robin R. Murphy
MIT Press, 2000
ISBN: 0262133830
A roundup of several popular approaches to artificial intelligence in robotics. Fairly technical, but still readable.

Mobile Robotics: A Practical Introduction

Ulrich Nehmzow
Springer Verlag, 2000
ISBN: 1852331739
An introductory textbook aimed at the college student studying mechatronics or robotics. Emphasis is on mobile robots and is heavy on concepts and theory, but not actual construction.

Robot: Mere Machine to Transcendent Mind

Hans Moravec
Oxford University Press, 2000
ISBN: 0195116305
A speculative but informed look at one possible future with intelligent robots. Written by a professor at Carnegie Mellon University. See also the author's *Mind Children: The Future of Robot and Human Intelligence* (Harvard Univ Press, 2000; ISBN: 0674576187).

Vehicles: Experiments in Synthetic Psychology

Valentino Braitenberg
MIT Press, 1986
ISBN: 0262521121

A small but highly influential book on how simple machines can mimic living organisms.

Robot Books for Children

Artificial Intelligence: Robotics and Machine Evolution

David Jefferis
Crabtree Pub, 1999
ISBN: 0778700569
A look at robotics through the lens of artificial intelligence. Ages: 9–12

Inventor's Handbook: Robots

Bobbi Searle
Silver Dolphin, 2000
ISBN: 1571454187
Thirty-two-page book and kit for building a cardboard and plastic robot. Ages: 9–12

Robots Among Us: The Challenges and Promises of Robots

Christopher W. Baker
Millbrook Press, 2002
ISBN: 0761319697
Picture book about robots in everyday life. Ages: 9–12.

Mechanical Design

Five Hundred and Seven Mechanical Movements

Henry T. Brown
Astragal Press, 1995
ISBN: 1879335638
Short but sweet, this book covers the foundations of basic mechanics, such as the gear, pulley, and windlass. The mechanisms are shown in self-explanatory illustrations. The book is a reprint from the late nineteenth century; I have a version of the book published in the 1950s.

Home Machinist's Handbook

Doug Briney
McGraw-Hill, 1984
ISBN: 0830615733
Covering basic machine shop practice for the home, this perennial bestseller is the ideal companion for the intermediate or advanced robot builder. The emphasis is on machining in the home "lab," and the example tools are common Sherline desktop mills and lathes.

Illustrated Sourcebook of Mechanical Components

Robert O. Parmley (editor)
McGraw-Hill, 2000
ISBN: 0070486174
Gargantuan book on mechanical design and components. The typical and atypical are covered.

Mechanical Devices for the Electronics Experimenter

Britt Rorabaugh
Tab Books, 1995
ISBN: 0070535477

Overview of mechanical construction of common components, such as motors and linkages.

Mechanisms and Mechanical Devices Sourcebook

Neil Sclater, Nicholas P. Chironis
McGraw-Hill, 2001
ISBN: 0071361693
Over 2,500 mechanical devices elements and 1,200 illustrations. Never wonder how something works; look it up in this terrific resource. Special chapters on robotic systems, couplings and linkages, gears, and pneumatics.

The Prop Builder's Molding & Casting Handbook

Thurston James
Betterway Pub; 1990
ISBN: 1558701281
You don't have to be a Hollywood movie prop builder to enjoy this book (though it's really written for any kind of prop builder, from blockbuster motion pictures to the community theater production). Various materials and finishing techniques are covered, including the use of papier-mâché, plaster, and resins.

Tabletop Machining

Joe Martin
Sherline Products Inc., 1998
ISBN: 0966543300
Written by the head honcho of desktop machine manufacturer Sherline, this handy tome teaches you the basics of using small lathes and mills. The book is most suitable for owners of Sherline products, as you would imagine, but is applicable to most any other brand.

Microcontroller/Microcontroller Programming

AVR Risc Microcontroller Handbook

Claus Kuhnel
Newnes, 1998
ISBN: 0750699639
Using the Atmel AVR 8-bit microcontroller.

Basic Stamp

Claus Kuhnel, Klaus Zahnert
Newnes, 2000
ISBN: 0750672455
Covers the Parallax Basic Stamp I, II, and IIsx. Gets pretty technical and is a good read for intermediate and advanced users.

Basic Stamp 2:Tutorial and Applications, The

Peter H. Anderson
ISBN: 0965335763
Example-driven tutorial on programming the Basic Stamp.

Design with PIC Microcontrollers

John B. Peatman
Prentice Hall, 1997
ISBN: 0137592590

Entry-level text for hardware designers on how to use the Microchip PICmicro microcontrollers.

Easy Pic'N: A Beginner's Guide to Using PIC16/17 Microcontrollers

David Benson
Square One Electronics, 1999
ISBN: 0965416208

Entry-level book on using the Microchip PIC microcontroller. Programming examples are in assembly language. See also *PIC'n Up the Pace: An Intermediate Guide to Using PIC Microcontrollers* (1999, ISBN: 0965416216) and *PIC'n Techniques, PIC Microcontroller Applications Guide* (1999, ISBN: 0965416232), both by the same author, and *Serial PIC'n: PIC Microcontroller Serial Communications* (Roger L. Stevens, 1999, ISBN: 0965416224). All are published by Square One Electronics.

Microcontroller Application Cookbook, The

Matt Gilliland, Ken Gracey
Woodglen Press, 2000
ISBN: 0615115527

Real-world applications for the Basic Stamp II. Most of the application projects are simple and down-to-earth, and all include sample code.

Microcontroller Idea Book, The

Jan Axelson
Lakeview Research, 1994
ISBN: 0965081907

Though older than most books on microcontrollers, this is one of the better ones. The projects center around the 8052 chip with built-in Basic. Schematics and code fill the book. Jan is an engineer, yet the book is readable; useful to both beginner and midlevel microcontroller programmers.

Microcontroller Projects Using the Basic Stamp, Second Edition

Al Williams
CMP Books, 2002
ISBN: 1578201012

Using the Basic Stamp microcontroller.

PIC Microcontroller Project Book

John Iovine
Tab Books, 2000
ISBN: 0071354794

Project-oriented book on using the Microchip PIC microcontroller. Projects include stepper motor control, counters, and robotic sensing.

Programming & Customizing PICmicro Microcontrollers

Myke Predko
McGraw-Hill, 2000
ISBN: 0071361723

How to program the PIC microcontroller. Includes CD-ROM. See also *PICmicro Microcontroller Pocket Reference* (2000; ISBN: 0071361758) and *Programming and Customizing the 8051 Microcontroller* (1999, ISBN: 0071341927) by the same author and publisher.

Programming and Customizing the AVR Microcontroller

Dhananjay V. Gadre

Tab Books, 2000
ISBN: 007134666X
Entry-level volume on using the Atmel AVR 8-bit microcontrollers. Includes a CD-ROM with program examples, with special sections on interfacing the AVR via serial, USB, and IrDA.

Programming and Customizing the Basic Stamp

Scott Edwards
Tab Books, 2001
ISBN: 0071371923
The definitive all-purpose guide on using the Parallax Basic Stamp microcontroller (covers the Basic Stamp I, II, and IIsx).

Programming and Customizing the HC11 Microcontroller

Thomas Fox
McGraw-Hill Professional Publishing, 1999
ISBN: 0071344063
All about using the venerable Motorola MC68HC11 microcontroller.

Electronics How-to and Theory

Art of Electronics, The

Paul Horowitz, Winfield Hill
Cambridge Univ Press, 1989
ISBN: 0521370957

Student Manual for the Art of Electronics

Paul Horowitz, T. Hayes
Cambridge Univ Press, 1989
ISBN: 0521377099
The Art of Electronics, and its companion student workbook, are the standard texts for electronics in many schools. Fairly laden with theory and math, the books are neverthe-less aimed at a general academic and hobbyist audience.

Bebop to the Boolean Boogie

Clive Max Maxfield
LLH Technology Pub, 1995
ISBN: 1878707221
Unconventional approach (including humorous illustrations) to teaching logic circuits. Readable and quite engrossing. An extensive index makes it easy to locate important topics.

Beginner's Guide to Reading Schematics

R. J. Traister, A. L. Lisk
Tab Books. 1991
ISBN: 0830676325
Like the title says: how to read schematics. For beginners.

Build Your Own Low-Cost Data Acquisition and Display Devices

Jeffrey Hirst Johnson
Tab Books, 1993
ISBN: 0830643486
Specializes in data acquisition on PC-compatible computers.

CMOS Cookbook

Don Lancaster
Newnes, 1997
ISBN: 0750699434
One of two desk references no electronics hobbyist or engineer should be without. The other is *TTL Cookbook* (Sams; 1980 ISBN: 0672210355), also by Don Lancaster. You can read much more from Don at his Web site: ***http://www.tinaja.com/.***

Electronic Circuit Guidebook (various volumes)

Joseph J. Carr
Delmar Learning
  Volume 1: Sensors; 1997, ISBN: 0790610981
  Volume 2: IC Timers; 1997, ISBN: 0790611066
  Volume 3: Op Amps; 1997, ISBN: 0790611317
  Volume 4: Electro Optics; 1997, ISBN: 0790611325
  Volume 5: Digital Electronics; 1998; ISBN: 0790611295
Reference books on various electronics topics of interest to all engineers.

Forrest Mims Engineer's Notebook, The

Forrest M. Mims, Harry L. Helms
LLH Technology Pub, 1993
ISBN: 1878707035
Most anyone involved in electronics knows Forrest Mims, a prolific writer who has authored numerous magazine articles, columns, and books since the 1960s. *The Forrest Mims Engineer's Notebook* is a compilation of how-to and circuit design guidance and is suitable for all students of electronics. See also from the same author *The Forrest Mims Circuit Scrapbook* (2000; ISBN: 1878707493).

IC Op-Amp Cookbook

Walter G. Jung
Prentice Hall PTR, 1998
ISBN: 0138896011
Various ways op-amps can be used in circuits. More of a reference.

Logicworks 4: Interactive Circuit Design Software for Windows and Macintosh

Addison-Wesley Pub Co, 1999
ISBN: 0201326825
Book and CD-ROM with a student version of the LogicWorks circuit design and simulation software.

Making Printed Circuit Boards

Jan Axelson
McGraw-Hill, 1993
ISBN: 0070027994
Covers various techniques for constructing prototype circuit boards.

Practical Electronics for Inventors

Paul Scherz
Tab Books, 2000
ISBN: 0070580782

A unique book with a unique angle, this text demonstrates the hows and whys of electronics circuits for people who may not want to learn all the nitty-gritty details. The book still has some technical formulas, but the emphasis (like the title says) is on practical electronics.

Printed Circuit Board Materials Handbook

Martin W. Jawitz (editor)
McGraw-Hill, 1997
ISBN: 0070324883
Intended for the electronics professional, this book covers various materials and techniques used in printed circuit board manufacture.

Teach Yourself Electricity and Electronics

Stan Gibilisco
Tab Books, 2001
ISBN: 0071377301
Introduction to electronics, with self-test exercises.

Interfacing to IBM PC (and compatibles)

Parallel Port Complete

Jan Axelson
Lakeview Research, 1997
ISBN: 0965081915
Everything you ever wanted to know about interfacing with the parallel port on your PC compatible. Jan is arguably the leading author on port interfacing. See also from the same author and publisher:
>   *Serial Port Complete*; 1998, ISBN: 0965081923
>   *USB Complete*; 2001, ISBN: 0965081958

PC PhD: Inside PC Interfacing

Myke Predko
Tab Books, 1999
ISBN: 0071341862
Hardware and software for interfacing with a PC, including serial, parallel, and bus-level boards.

Programming the Parallel Port

Dhananjay V. Gadre
CMP Books, 1998
ISBN: 0879305134
Using the parallel port of the PC-compatible computer to control relays and other devices and to accept input from sensors. Program examples in the C language.

Real-World Interfacing with Your PC

James Barbarello
Delmar Learning, 1997
ISBN: 0790611457
Entry-level guide on using the serial and parallel ports of the PC compatible to interface with various things.

Use of a PC Printer Port for Control & Data Acquisition

Peter H. Anderson
ISBN: 0965335704
and

Parallel Port Manual Vol. 2, The

Peter H. Anderson
ISBN: 0965335755
Fairly in-depth coverage of parallel port interfacing on the PC, with a number of unique topics not found in similar books, such as using an ultrasonic sensor and interfacing to a compass module. Most of the projects are useful in robotics.

Robotics and Electronics Magazines

 ### Circuit Cellar

http://www.circuitcellar.com/
Monthly magazine on professional-level electronics projects. Heavy emphasis on embedded systems.

Elektor

http://www.elektor-electronics.co.uk
General-interest electronics magazine.

Everyday Practical Electronics

Wimborne Publishing Ltd.
Allen House
East Borough, Wimborne
Dorset BH21 1PF
UK
General-interest electronics project magazine.

 ### Nuts & Volts Magazine

430 Princeland Court
Corona, CA 91719
http://www.nutsvolts.com/
Combining great articles and columns with classified ads, this grassroots magazine is a gold mine every month. Be sure to check out the regular robotics and Basic Stamp columns.

Poptronics

Gernsback Publications
500 Bi-County Blvd.
Farmingdale, NY 11735
http://www.gernsback.com/
General-interest monthly magazine on electronics and electronic projects, with regular features and columns on robotics. I wrote the Robotics Workshop column for a while; check back issues in the 2000–2001 time frame.

Robot Science and Technology

3875 Taylor Road, Ste. 200
Loomis, CA 95650
http://www.robotmag.com/

An excellent but irregularly published magazine on amateur robotics. Subscription is for six issues, but there's no guarantee when the issues will come out. Perhaps it's best to buy back issues, and the current issue at the regular cover price, as it comes available.

Robot Fiction

Many robot builders are also interested in fictional tales about robots. As judging fiction is far more subjective than nonfiction, and because some of the best works of robot fiction are old and/or out of print, I won't presume to list specific books here for adult fiction. Instead, I'll just summarize some of the more relevant works and authors.

- *Isaac Asimov.* The stories in *I, Robot* ushered in a new era of robotics storytelling. Asimov's famous "Three Laws of Robotics" were first introduced in these stories (the book is a compilation of shorts published in magazines such as *Astounding*). The bulk of robot fiction before this time painted robots as evil, a manifestation of the out-of-control mechanized world people were living in during the first half of the twentieth century. Asimov wrote extensively about robots; see *The Rest of the Robots*, *The Bicentennial Man*, *The Robots of Dawn*, *The Complete Robot*, and many others.

- *Brian Aldiss.* One of my personal favorites, Aldiss forces a darker look at robotics and our relationship with them. A seminal work is *But Who Can Replace a Man*, a collection of shorts published in 1958.

- *Philip K. Dick.* Where Asimov was optimistic about robotics, Philip K. Dick was pessimistic (he arguably might have more realistic, as well). The movie *Bladerunner* was inspired by a Philip K. Dick story, *Do Androids Dream of Electric Sheep*, a tale about "replicants"—engineered biological creatures who look just like humans, but aren't allowed rights. While replicants aren't robots per se, the story deals with many of the same ethical questions being raised today about sentient machines: Do they have the same rights as humans, even though they aren't "real"? Dick also wrote about robots in *Second Variety, Imposter, Martian Time Slip, Autofac*, and several other novels and short stories.

- *Jack Williamson.* In a career than spanned many years, Williamson wrote two of the most influential fictional works on robots: *The Humanoids* (1949) and *With Folded Hands* (1947). Both were about a benign race of androids that take over the work, and spirit, of their human subjects.

- *Stanislaw Lem.* Lem's 1974 collection of robot short stories, *The Cyberiad: Fables for the Cybernetic Age*, is part fiction, part satire. Nothing is sacred, not even Asimov's Three Laws of Robotics.

- *Clifford D. Simak.* Of particular interest, look for *City* (1952), where the Earth is inhabited by old robotic cats, dogs, and other machines, and *Time and Again* (1951), an allegorical tale of robot slaves seeking to free themselves.

- *Ron Goulart.* Robots that don't work right. Goulart's books on faltering contraptions are hilarious, but I also understand they are an acquired taste. Try one out to see if you like it.

- *Ray Bradbury.* One of the best-known science fiction authors of all time, Bradbury has written a number of short stories and novels about robots, often taking a look at their dark side. A number of thought-provoking robot stories can be found in *I Sing the Body Electric!* (1969); more can be found in *The Illustrated Man* and *Long After Midnight* story collections.

- *David Gerrold.* A gifted novelist and television writer (he penned the *Trouble with Tribbles* episode for the original *Star Trek* TV series), David's work has spanned a wide gamut of topics. His novel *When Harlie Was One* (also noted with variations, such as *When H.A.R.L.I.E. Was One*) is about a self-learning machine, growing and maturing with age just as humans do. A unique aspect of this book is that like computer software, it's gone through several releases since its original 1972 version.

These books are suitable for younger readers interested in robots and their interaction with humans. Many were written decades ago, so they have no sex, graphic violence, or drug themes. The local library is the best source for these.

- *Robots of Saturn*, by Joseph Greene. This young adult novel (now out of print) was published in 1962 and is remarkable in that it involves telerobotic machines that mind-meld with humans. An unusual premise over 40 years ago, the idea is just now becoming practical.

- *The Runaway Robot*, by Paul W. Fairman (writing as Lester Del Ray). Published in 1965, this book concerns a family robot that runs away from its young master while on the Mars colony. Also out of print.

- *Ricky Ricotta's Mighty Robot vs. the Mecha-Monkeys from Mars* is part of the popular Ricky Ricotta Giant Robot series for first readers. These books are in print and available in bookstores and in libraries.

COMMUNICATIONS

The following sections list resources involved with compact wireless communications using either infrared or radio frequency (RF) signals. The communications products selected for these sections are adaptable to robotics, are relatively low powered, and can be operated at 5 volts.

Infrared communications links are used primarily for hand-held remote controls. RF links have a greater distance and (depending on the power output of the transmitter) can be operated around corners or through walls. Note that some of the RF transmitters listed in the section Communications-RF may require special licensing, depending on local laws.

SEE ALSO:

ELECTRONICS-CIRCUIT EXAMPLES: Plans for homemade communications links

KITS-ELECTRONIC: Infrared and RF construction kits

MICROCONTROLLERS-HARDWARE: Kits and products for connecting controllers to communications devices

RADIO CONTROL: Receivers and transmitters for model R/C

RETAIL-GENERAL ELECTRONICS: Additional sources for infrared and RF products

VIDEO-TRANSMITTERS: RF transmitters for short-range video

COMMUNICATIONS-INFRARED

Innotech Systems Inc. 203785

320 Main St.
Port Jefferson, NY 11777
USA
✆ (631) 262-1260
📠 (631) 262-0294
✉ sales@innotechsystems.com
🌐 http://www.innotechsystems.com/

Innotech Systems provides infrared and RF remote controls and remote-control systems. They also offer voice-operated remote controls and "Parallel Port Universal Remote Control Integrated Circuit," a microcontroller

MOVERS AND SHAKERS

Cynthia Breazeal

http://www.ai.mit.edu/people/cynthia/cynthia.html

"Dr. Cynthia" is currently a postdoctoral fellow at MIT and is best known for her pioneering work with Kismet, a self-learning robot and one of many in her studies of socially intelligent humanoid machines. She has also written a book on this subject.

IC for use with computer or microcontroller-based system.

Some parts have a minimum order of 100 or 1,000 units. Datasheets in Adobe Acrobat PDF format.

PC Remote Control 204148

http://www.pcremotecontrol.com/

Products to control your PC via a hand-held remote control. Interface circuit examples and downloadable communications software.

Xilor Inc. 202454

1400 Liberty St.
Knoxville, TN 37909
USA
✆ (865) 546-9863
📠 (865) 546-8324
🚫 (800) 417-6689
✉ info@rfmicrolink.com
🌐 http://www.rfmicrolink.com/

Wireless remote controls (both RF and infrared), pressure switches, and conductive rubber. See also listing under **Sensors-Other**.

Evation

http://www.evation.com/

Manufacturer of Irmand IR remote control device and software: control your PC (or robot) from a universal remote control

Help—My Remote Control Won't Work!

A number of robot toys and kits come with an infrared hand-held remote control, the same kind of thing you use to channel surf on your TV. Their operation is simple: Press a button and infrared light flashes from an emitter on the front. A sensor on the robot receives the flashes, decodes them, and responds to the command you have given.

Alas, for something so simple, remote controls don't always work as expected. And it's not always because the batteries are dead—though that's a leading cause. So, if the remote control for your robot has stopped functioning, review these points to get it back on track.

- *Dead or corroded batteries.* This has to be mentioned, simply because it's so obvious. Open the battery compartment and look for corroded battery contacts. If a battery has leaked, clean the battery contacts with a pencil eraser or battery contact cleaner. Wash your hands thoroughly to remove residue battery electrolyte. Replace with known, fresh batteries.
- *Batteries inserted incorrectly.* Be sure the batteries are not reversed when you insert them in the remote. Otherwise the remote will not work, or it will work erratically.
- *Bent battery contacts.* Replacing batteries all the time in the remote can bend the battery contacts inside. Be sure they make a good physical connection to the terminals on the battery.
- *Dirty or gummed up contacts.* Spilling a drink into the remote control is a sure-fire way of ruining it. Water and diet sodas will usually dry, and the remote will come back to life. Sugared sodas, coffee, and milk will leave a residue, requiring the remote be disassembled and cleaned or thrown away.
- *Obstruction or dirty LED.** Same for the infrared receiver. Look for tape, a broken piece of plastic, dirt, or crud covering the LED on the remote control. Do the same for the infrared receiving module on the robot.
- *Bad infrared LED.* This is less common, but it can happen if the remote is used a lot. You can try replacing the LED. Most any high-intensity LED will work, but be sure to observe correct polarity.
- *Bad IR sensor.* This can happen too, though it is unlikely. The sensor can be replaced with a new one. This is much easier if the robot was built from a kit, where you can purchase a direct replacement from the kit maker.

(*Some of you may object to using the term *LED*—light emitting diode—for an infrared emitter. While they technically do not emit light visible to the human eye, it is indeed visible light to other animals, such as cats, as it is in the near-infrared region of the electromagnetic spectrum. Therefore, LED is the correct term.)

Testing the Remote Control

There are several simple ways to test if a remote control is working. One easy approach is to place the remote near an AM radio. Dial to a position where there is no station (a distant station is okay). Press buttons on the remote. You'll be able to hear a high-pitched "trill" sound as you press the buttons. If you do not hear the sound, the remote is not working at all.

In instances where the basic electronics of the remote are working, but its infrared LED is not, you can test this with an LED sensor, available at RadioShack and many other electronics stores. The sensor is a passive device: It glows after exposure to a bright infrared light source.

COMMUNICATIONS-RF

A3J Engineering, Inc. 203197

15344 E. Valley Blvd.
Ste. C
City of Industry, CA 91746
USA

- (626) 934-7600
- (626) 934-7609
- info@3jtech.com
- http://www.3jtech.com/

Wireless and PC/104 modems:

- Pegasus II-56K
- Universal "pocket" modem
- Pegasus III-Infrared 56K V.90 data/fax modem
- PC/104 56K and 33.3K data modem

Web cams:

- uCAMit USB camera
- CAMit remote-controllable security camera with built-in modem

ABACOM Technologies 202035

32 Blair Athol Crescent
Etobicoke, ON
M9A 1X5
Canada

- (416) 236-3858
- (416) 236-8866
- abacom@abacom-tech.com
- http://www.abacom-tech.com/

ABACOM Technologies manufactures and distributes low-power miniature (and I do mean miniature) radio transmitters and receivers, for both audio and data transmission. Modules are available for AM and FM; the AM versions are quite affordable and ideally suited for short-range, low-data-rate robotics communication. Available in 418, 433.92, 868.35 and 916.5 MHz versions; data rates are up to 2,400 bits per second. Depending on the model, maximum range is about 300 feet.

The FM modules are more expensive, but they offer faster data rates (up to 9,600 bps on most units, and 19,200 bps on specialty modules) and a range of up to 500 feet. Available frequencies for the various FM modules are 418 MHz, 433.92 MHz, and 403 MHz.

Compatible AM and FM receivers are offered in small form factors for inclusion in your own projects.

Also available:

- Keyfob enclosures
- Audio/video transmitter receiver modules
- Data encoder and decoder modules
- Power amplifiers
- Antennas

Downloadable catalogs in Adobe Acrobat PDF format.

Model TX-EVAL. Photo ABACOM Technologies.

AeroComm 204247

10981 Eicher Dr.
Lenexa, KS 66219
USA

- 913-492-2320
- 913-492-1243
- 800-492-2320
- sales@aerocomm.com
- http://www.aerocomm.com/

AeroComm makes and sells a broad line of high-frequency and microwave data communications modules, intended primarily for OEMs, but also sells smaller quantities and consumer products. Products include:

- ConnexRF instant wireless modules—2.4 GHz radios in a credit card-sized form factor
- Wireless RS232 links-self-contained, with whip antennas
- Wireless printer-sharing module

Bluetooth   204107

http://www.bluetooth.com/

Industrywide standard for computer-to-computer wireless communications.

Communications Specialists, Inc.   202876

426 West Taft Ave.
Orange, CA 92865

 (800) 854-0547

 http://www.com-spec.com/

Tone-signaling products for two-way radios. Check out their 64-tone encoders and decoders.

Computronics Corporation Ltd.   203662

Locked Bag 20
Bentley
Western Australia 6983
Australia

(+61 8 9470 1177

📠 +61 8 9470 2844

✉ kdare@computronics.com.au

🌏 http://www.computronics.com.au

Industrial electronics: electronic displays, tools, and components. Includes soldering stations, high-brightness LEDs, chemicals, RF transmitter and receiver modules.

Data Hunter   203025

5132 Bolsa Ave.
Ste. 102
Huntington Beach, CA 92649
USA

((714) 892-5461

📠 (714) 892-9768

✉ info@datahunter.com

🌏 http://www.datahunter.com/

Data Hunter sells a number of RF communications systems:

- "The Finger" includes all RF circuitry (baseband processor, MAC, RF power amplifier) inside of a tuned antenna.

- "Tiny Radios" support the most popular industry data communications standards and proprietary standards.

- "The Tick" is an RS-232 parasitically powered data radio.

DC Electronics   202⬛

P.O. Box 3203
Scottsdale, AZ 85271-3203
USA

((480) 945-7736

📠 (480) 994-1707

🚫 (800) 467-7736

✉ clifton@dckits.com

🌏 http://www.dckits.com/

DC sells a variety of middle- to high-end kits, many of them useful in robotics. Their main product line is transmitters and receivers.

Diverse Electronics Services   202282

Carl A. Kollar
1202 Gemini St.
Nanticoke, PA 18634-3306
USA

((570) 735-5053

✉ carl@diverseelectronicservices.com

🌏 http://www.diverseelectronicservices.com/

PIC-based motor controllers, radio-controlled device controllers, transmitter/receiver sets.

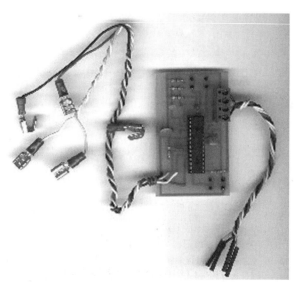

Model RCIC-2; a dual channel radio control interface with mixing. Photo Diverse Electronics Services.

Elsema Pty Ltd 204106

Unit 3, 10 Hume Rd.
Smithfield, NSW 2164
Australia

 +61 2 9609 4668
 +61 2 9725 2663
 support@elsema.com
 http://www.elsema.com/

RF transmitters and receivers (1 to 16 channel), keychain FOBs, relay output receivers.

Ewave, Inc./Electrowave 203390

7419 Gracefield Ln.
Dallas, TX 75248
USA

(972) 248-2931
(972) 931-6996
sales@electrowave.com
http://www.electrowave.com/

Ewave specializes in "RF data modems with extremely low latency and high throughput for harsh industrial environments and demanding applications." The company makes the Ewave Stamper, used in the FIRST competitions, and resold by Parallax Inc.

Glolab Corp. 203040

307 Pine Ridge Dr.
Wappingers Falls, NY 12590
USA

(845) 297-9772

Data-Linc

http://www.data-linc.com/

License-free spread spectrum data modems

MicroDAQ.com, Ltd.

http://www.spread-spectrum-radio.com/

Sellers of spread spectrum data modems; specializes in small data logging devices

Omnispread Communications Inc.

http://www.omnispread.com/

Low-cost spread spectrum data modems

---

 kits@glolab.com
 http://www.glolab.com/

Glolab manufactures and sells multichannel wireless transmitters and receivers, encoder and decoder modules (to permit controlling more than one device through a wireless link). They also provide pyroelectric infrared sensors and suitable Fresnel lenses. An amplifier and hookup diagram from the PIR sensor is available on the Web site.

Model KR4A four-channel receiver. Photo Glolab Corp.

Model KR4B four-channel receiver, with relays. Photo Glolab Corp.

Hamtronics, Inc. 202897

65 Moul Rd.
Hilton, NY 14468-9535
USA

(716) 392-9430
http://www.hamtronics.com/

Modules to make amateur radio exciters and receivers. If you have the proper operating license (depending on unit), these are useful as communications links for your robot. Products are available assembled and/or as a kit.

Robot Radio Links

Radio communications can be used in robotics for two primary purposes:

Wireless modules can be used to communicate with your robot.

- To command the robot, either completely for all its discrete functions, or to provide general commands for basic operations, such as Run or Stop. General commands may also be used to select and activate programs already resident in the robot's computer.

- To receive data from the robot, usually either a video signal or some form of telemetry.

Radio links are common for discrete function control in combat robotics. The operator of the robot uses an R/C transmitter (outfitted with a frequency crystal for land use, rather than airplane use) to operate the motors and weapons of the robot. Typical transmitters have four or five channels, with each channel operated by the twin joysticks and other knobs on the transmitter. At a minimum, three channels are used: one each for the right and left motors, and one for the weapon.

Video is a typical application for receiving a radio signal from a robot. Video transmitters and receivers that operate in the 2.4 GHz microwave range are common and fairly inexpensive—under $150 or $200 for the pair. Range is limited to under 200 feet outdoors, or from 20 to 50 feet when used indoors.

Wireless data modems are used for sending digital data to a robot or for receiving data. High-speed wireless data modems are expensive, so most applications call for relatively low speeds from 2,400 bits per second (bps) to a ceiling of 19,200 bps. This is considerably slower than even today's slowest computer modems, but the low speed is necessary to preserve data integrity.

When selecting a receiver and transmitter for wireless data between you and your robot, consider the following:

- *Power output determines range.* Depending on your country's laws, higher power outputs may require certification of the device or even licensing. In the U.S., most wireless data modems operate at a power output that does not require licensing.

- *Range contributes to maximum data rate.* Data rates can be fastest over shorter distances, because the received signal is clearer. Over longer distances, the data rate must be reduced in order to reduce or eliminate errors.

- *The right antenna can greatly increase range.* Radio-frequency signals radiating from a properly designed and mounted antenna will travel further than signals from a transmitter without an antenna. Be sure to use an antenna properly matched for the transmitter you are using—sometimes, it's just a simple wire, but consult the documentation on how to position or wrap the wire.

- *Use a compatible antenna on the receiver.* The same rules apply to the receiver as to the transmitter. Be sure to consider the orientation of the antennas on the receiver and the transmitter—if the units have stick antennas, avoid having one point up, while the other points sideways.

Alternatives to RF Modules

Purchasing an RF transmitter and receiver module is one way to provide a communications link between you and your robot. In additional, several ready-made products can be hacked for their RF systems and pressed into use as radio links between you and your bot.

- *Walkie-talkie.* Many toy walkie-talkies include a "code sender" button for transmitting Morse code. By connecting the receiver to an AC-coupled interface and 567-tone decoder, you can add simple on/off control of your robot.

- *Garage door opener.* Try to find a used one that's being discarded; the electronics—the part you want—last longer than the mechanics. Hack the receiver to work as an on/off control for your robot.

- *"Key-ring" appliance control.* You can purchase a radio-controlled powered outlet at many department and home improvement stores. Hack the module to work with your robot. The transmitter is a key ring, with one or two buttons (some control several modules).

A key-ring transmitter from a store-bought wireless appliance control module.

- *Wireless car alarm kit.* Two- and three-function wireless car alarm kits can be retrofitted for controlling a robot. You can find them at auto parts stores—see **Retail - Automotive Supplies** for leads—and are fairly popular at weekend swap meets. Price is lower at the swap meets than at car parts stores.

Lemos International

202183

48 Sword St.
Auburn, MA 01501
USA

☎ (508) 798-5004

📠 (508) 798-4782

✉ sales@lemosint.com

🌐 http://www.lemosint.com/

Lemos sells the Radiometrix line of RF modules and data modems, miniature video transmitters, and various electronic components. Will sell in low quantities, but most sales are for 200+ units.

Linx Technologies, Inc.

202008

575 S.E. Ashley Place
Grants Pass, OR 97526
USA

☎ (541) 471-6256

📠 (541) 471-6251

🚫 (800) 736-6677

✉ info@linxtechnologies.com

🌐 http://www.linxtechnologies.com/

Linx sells RF modules, antennas, and connectors. Boasts easy-to-interface implementations because of

their modular designs; products are available from Digi-Key, among other distributors. Evaluation kits are also offered.

MaxStream, Inc.

204246

P.O. Box 1508
Orem, UT 84059-1508
USA

☎ (801) 765-9885

📠 (801) 765-9895

🚫 (866) 765-9885

✉ info@maxstream.net

🌐 http://www.maxstream.net/

Manufactures and sells small RF data modules and developer kits. Their 24XStream 2.4 GHz wireless OEM development kit includes two XStream wireless modules (these are compact, about the size of a book of matches) with wire antennas, a pair of PCB serial interface boards (RS-232/422/485 interface board schematic), cables, power supplies, development diskette with software examples, and datasheets. 900 MHz developer kits and modules are also available. Products support data rates of 9,600 or 19,200 bps.

North Country Radio   202461

P.O. Box 53 Wykagyl Station
New Rochelle, NY 10804-0053
USA

  (914) 235-6611

  (914) 576-6051

✉ Rgraf30832@aol.com

🌐 http://www.northcountryradio.com/

RF transmitter/receiver kits. The company has been selling RF, video, and specialty electronics kits for amateur and experimental use since 1986.

Radiometrix Ltd   204177

Hartcran House
Gibbs Couch
Carpenders Park
Hertfordshire
WD19 5EZ
UK

📞 +44 (0) 2084 281220

📠 +44 (0) 2084 281221

✉ info@radiometrix.co.uk

🌐 http://www.radiometrix.co.uk/

Radiometrix specializes in the design and manufacture of low-power radio modules for cable-free data links. Choice of UHF and VHF frequencies. Their TXx/RXx (such as the TX1 or RX3) modules are easily interfaced on homemade circuits and need only power and a small antenna to operate. Depending on the antenna, terrain, and data rate, range is from a few hundred meters to well over 10 kilometers. Lower data rates can be used over longer distances, and, of course, obstructions or use indoors reduces the range.

Model TX2/RX2 RF receiver and transmitter modules. Photo Radiometrix Ltd.

Model BiM3, 869/914MHz high speed FM radio transceiver module. Photo Radiometrix Ltd.

Radiotronix   204166

207 Industrial Blvd.
Moore, OK 73160
USA

📞 (405) 794-7730

📠 (405) 794-7477

✉ sales@radiotronix.com

🌐 http://www.radiotronix.com/

Radiotronix designs, manufactures, and markets RF transmitter, RF receiver, and RF transceiver modules. Units operate in the 433.92 MHz and 902-928 MHz range. Evaluation kits available. The company sells products in single quantity for individual experimenters and in much higher quantities for OEMs. When I checked their Web site, prices were posted, and they seemed quite reasonable. Datasheets and application notes (in Adobe Acrobat PDF format) are available for download from the site.

Model EWM-900-FDTC, 56-channel, full-duplex, 902-928 MHz data and audio transceiver module. Photo Radiotronix Inc.

Ramsey Electronics, Inc.   202353

793 Canning Pkwy.
Victor, NY 14564
USA

-   (716) 924-4560
-   (716) 924-4886
-   (800) 446-2295
- OrderDesk@ramseyelectronics.com
- http://www.ramseyelectronics.com/

Ramsey makes and sells a wide variety of electronics kits, including RF transmitters and receivers, miniature video cameras, digital sound recorders, DC motor speed controllers, tone encoder and decoder, and plenty more. Also sells test equipment and tools. The company's downloadable catalog is in Adobe Acrobat format and is available in one chunk or divvied up into sections.

Reynolds Electronics   202009

3101 Eastridge Ln.
Canon City, CO 81212
USA

-   (719) 269-3469
-   (719) 276-2853
- support@rentron.com
- http://www.rentron.com/

Reynolds offers kits and ready-made products for the electronics enthusiast and robotmeister, including PicBasic and PicBasic Pro compilers, Basic Stamp, Microchip PICmicro, Intel 8051 microcontrollers, remote controls, tutorials, projects, RF components, RF remote-control kits, and infrared kits and components.

Reynolds Electronics Web site.

Reynold Electronics

Though I've never met the proprietors of Reynold Electronics, I've always imagined them with thick horn-rimmed glasses, pocket protectors, and slide rules hanging off their belts—you know, the classic IQ-of-200-nerd-always-wins-the-science-fair stereotype. They're in the business of selling components for amateur robotics, but they're also very free with the application notes, how-tos, and tutorials; I tried counting them, then gave up.

Reynolds specializes in certain areas of the intermediate and high-end robotics hobby, particularly for remote control, wireless communications, and servo control. Example products and Web tutorials (complete with schematics, code, color photos, and descriptions) are as follows:

- 900 MHz wireless embedded systems programming
- Ruf-Bot—Build an RF remote-control system for robotics control
- Build a "10-key serial keypad"
- Quick and simple infrared schematics
- Build a custom RS-232 analog voltage meter

There are, in addition, numerous articles furnished by author and robot experimenter Myke Predko on using the PIC.

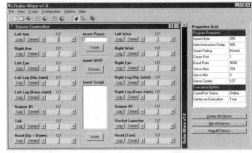

Robo-Ware, one of Rentron's many products.

Another unique product is Robo-Ware, software that generates code for the Basic Stamp and Pic Basic for running multiple servos, using the Scott Edwards Mini SSC II serial servo controller. The idea is that you create animations or moves, then export the resulting code for use with a Basic Stamp I/II or PIC microcontroller (when using a PIC, you need to export to Pic Basic or Pic Basic Pro, which is a Basic-language compiler for the PIC microcontrollers).

Reynold Electronics is at:

http://www.rentron.com/

RF Digital Corporation 202162

2029 Verdugo Blvd.
Ste. 750
Montrose, CA 91020
USA

((818) 541-7622
(℘ (818) 541-7644
(🕸) http://www.rfdigital.com/

RF Digital sells radio frequency transmitters and receivers (including those from Linx, such as the RFD24002 transmitter module), antennas, encoders, decoders, and Basic Stamp microcontrollers. They sell in single quantities and high quantities for OEMs (in case you're building a thousand robots-for taking over the world maybe?).

RF Monolithics, Inc. 203018

4347 Sigma Rd.
Dallas, TX 75244-4589
USA

((972) 233-2903
(℘ (972) 387-8148
⊘ (800) 704-6079
(🕸) http://www.rfm.com/

Maker of RF transmitter and receiver modules. Available through distributors.

Smarthome, Inc. 202330

17171 Daimler St.
Irvine, CA 92614-5508
USA

((949) 221-9200
(℘ (949) 221-9240
⊘ (800) 762-7846
(🕸) http://www.smarthome.com/

Smarthome specializes in home automation: transmitters and receivers, X10 devices, video cameras. An example product is the 7426AX single-channel long-range transmitter, able to send 27.255 MHz data signals up to 10 miles. A four-channel version is also available, as are companion receivers.

Telelink Communications 203663

P.O. Box 5457
North Rockhampton
Queensland 4702
Australia

(+61 7 4934 0413
(℘ +61 7 4934 0311
(🕸) http://www.telelink.com.au/

Telelink distributes low-power transmitters and receivers for embedded, microcontroller, and similar applications. The company offers a Universal Evaluation Kit, miniature UHF FM data transmitter and receiver modules, among others.

Xilor Inc. 202454

1400 Liberty St.
Knoxville, TN 37909
USA

((865) 546-9863
(℘ (865) 546-8324
⊘ (800) 417-6689
✉ info@rfmicrolink.com
(🕸) http://www.rfmicrolink.com/

Wireless remote controls (both RF and infrared), pressure switches, and conductive rubber. See also listing under Sensors-Other.

 **COMPETITIONS**

Robotic competitions are all the rage these days. The televised combat matches are but one kind of robotic competition; there's a large number of robots that compete in a variety of engineering challenges, including robot sumo, where two robots try to push each other outside of a playing mat; soccer and ball playing; maze solving; and even simulated firefighting. The common attribute of these competitions is that, in almost all cases, the robot is acting completely on its own. In the TV combat games, the robot is commanded by a human, using a remote-control link.

The following sections list many of the most popular robot competitions worldwide. The first section is filled with general competitions and varied challenges. Subsequent sections highlight specific competition events. Most competitions are held annually, and some are not open to the general public.

SEE ALSO:

INTERNET-BULLETIN BOARD/MAILING LIST: Hangouts for competition entrants

INTERNET-USENET NEWSGROUPS: Post and read messages about competitions

INTERNET-WEB RING: Listings of sites dedicated to various robotics and electronics interests

6.270 Autonomous Robot Design Competition   203329

http://web.mit.edu/6.270/

This is the home page for MIT's famous robot competition. In the words of the site: "6.270 is a hands-on, learn-by-doing class open only to MIT students, in which participants design and build a robot that will play in a competition at the end of January. The goal for the students is to design a machine that will be able to navigate its way around the playing surface, recog-

So You Want to Enter a Competition

Competitions help hone important robot-building skills. They can also be fun; and if you lose, disappointing. Keep the following in mind if you're planning on entering a robot competition.

- *Don't wait until the last minute to start!* Beginners especially grossly underestimate the time it takes to build a working robot. Start *months* before the competition event. Ideally, don't register for the event unless your robot is done or nearly done.
- *Match your robot with the competition.* Just because you built a vacuum-cleaning buddy doesn't mean it'll kick robutt in a sumo competition. Know the limitations of your robot and what it's best designed to do.
- *Start small, and work up.* Don't try to build C3P-0 for your first robot. Start with a simple project, and enter a less-challenging competition. Go up from there.
- *Read the rules before building your robot.* If you're targeting a specific competition for a new robot, download and read the rules for the event before you start construction. It will save you from costly mistakes.
- *Read the rules before entering the competition.* If your robot is already built, you still need to read the rules to make sure it qualifies. Measure and weigh your robot to be sure it's not over- or undersized for the competition.
- *Enter the proper event class.* Many competitions have several events, differentiated by class: The class may be by weight, size, age (yours, not the robot's), or school grade.
- *Don't try to bend the rules.* You and your robot will probably be disqualified and all your work will be for nothing.
- *Above all, have fun.* Competitions are meant to be a challenge. If you don't like to be challenged, competitions aren't for you. Even if you don't win, you'll still learn a lot in the process, and you'll meet others who share your passion for robotics.

nize other opponents, and manipulate game objects. Unlike the machines in Introduction to Design (2.70), 6.270 robots are totally autonomous, so once a round begins, there is no human intervention (in 2.70 the machines are controlled with joysticks)."

Web site for 6.270 competition.

AAAI Mobile Robot Competition   203330

http://www.cc.gatech.edu/~tucker/aaairobot/

AAAI stands for American Association for Artificial Intelligence. They sponsor an annual robot contest with such as these: Robot Rescue, service robot for serving food to humans, and BotBall, a program for high schoolers to build robots that play tabletop sports. The thrust of all of the AAAI competitions is to demonstrate autonomy in robotics. Human intervention or remote control is not part of the concept. These are challenging competitions.

SEE ALSO:

http://www.botball.org/

All Japan MicroMouse Contest   203331

http://www.bekknet.ad.jp/~ntf/mouse/mouse-e.html

Home of the MicroMouse competition, sponsored by New Technology Foundation.

BEST   203971

http://www.bestinc.org/

From the Web site: "BEST Robotics Inc. was established in 1997 as a national non-profit organization whose purpose is boosting engineering, science, and technology among pre-college students. It is managed by a hub council and steering team composed of volunteer leaders representing the BEST hubs and Texas BEST regional play-off championship."

The events change each year, but most are designed for a playing field that is 12 by 48 feet. See also:

http://www.brazosbest.org/

Carnegie Mellon Mobot Races   203335

http://www-2.cs.cmu.edu/~mobot/

MObile roBOTs at Carnegie Mellon (Pittsburgh, Pa.). Competitions include a robot slalom and joust.

DPRG RoboRama   203348

http://www.dprg.org/dprg_contests.html

Robot competition sponsored by the Dallas Personal Robotics Group. Contests include line following, sumo, and firefighting.

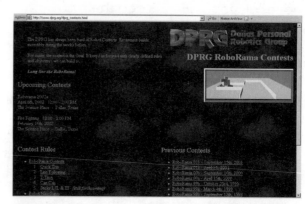

RoboRama from the Dallas Personal Robotics Group.

FIRST   202096

http://www.usfirst.org/

According to the Web site: "The First Robotics Competition is an annual design competition that brings professionals and young people together in teams to solve an engineering design problem in an intense and competitive way. The Competition is a program of FIRST (For Inspiration and Recognition of Science and Technology), a nonprofit organization founded in 1989 by inventor and entrepreneur Dean Kamen.… The Competition aim is to show students not only that the technological fields hold many varied opportunities for success and are accessible and rewarding, but also that the basic concepts of science, math, engineering, and invention are exciting and interesting."

Intelligent Ground Vehicle Competition   203341

http://www.secs.oakland.edu/SECS_prof_orgs/PROF_AUVSI/

At Oakland University in Rochester, Mich.

International Festival of Sciences and Technologies   203343

http://www.robotik.org/defaultuk.htm

France's big robot competition includes walking machines and robo-soccer. Web site is in English and French.

Manitoba Robot Games   203072

http://www.scmb.mb.ca/mrg.html

Variety of robotic competitions held in Manitoba, Canada.

Robot Competition FAQ   202021

http://www.robots.net/rcfaq.html

Frequently asked questions (FAQ) on robot contests and competitions.

Robothon   202015

http://www.seattlerobotics.org/robothon/

In the words of the Web site: "The Robothon is a national event that showcases the capabilities and technological developments in robotics from the amateur robotics community. The Robothon is an event where people from around the world can come together to present new robotic technologies, share ideas, meet fellow robotic enthusiasts, show off their robotic creations, and compete in several robotic competitions. The

Robothon is a public event to help promote and educate the general public that science and technology is fun and exciting for all age groups."

The contest is held in sunny Seattle, Wash. Sponsored by the Seattle Robotics Society. See:

http://www.seattlerobotics.org/

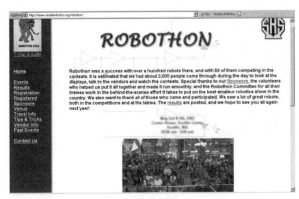

Robothon Web site.

Singapore Robotic Games   203354

http://guppy.mpe.nus.edu.sg/srg/

Eleven different competitions including legged robot race, wall climbing, and "robot battlefield."

Techno Games   203956

http://www.technogames.net/

Sponsors of robotic and mechanical challenges, such as robot soccer, shot put, and MicroMouse. Held in the U.K.

COMPETITIONS-COMBAT

Typical robot combat involves human operators controlling a remotely operable "robot" (I put "robot" in quotes because there are some who feel these aren't real robots). The most famous combat robots have so far appeared on television, but regional—and untelevised—combat events are starting to appear as well. Many of the combat robots are outfitted with various weapons, such as saw blades, pickaxes, and hammers.

Also included in this section are some support sites for builders and fans of combat robots, including how-tos for better robot battle.

SEE ALSO:

COMMUNICATIONS-RF: Radio control

COMPETITIONS-ENTRANT: Robots who fight

INTERNET-PERSONAL WEB PAGE: Owners show off their robots, some of which do battle

RADIO CONTROL: Transmitters and receivers for remotely operated robots

BattleBots   202836

701 De Long Ave.
Unit K
Novato, CA 94945
USA

☎ (415) 898-7522

📠 (415) 898-7525

✉ info@battlebots.com

🌐 http://www.battlebots.com/

BattleBots is an American television bot-bashing event with a decided comic flavor. The robots are anything but humorous, with such denizens as BioHazard that rip, shred, and punch, metallic-gladiator style. BattleBot events are open to all contestants, though your robot must qualify in prematches before you're allowed on TV.

Battlebricks   202240

http://www.battlebricks.com/

Battling LEGO: "When Good Plastic Goes Bad." You can imagine the rest. The Battlebricks team is located in Albany, N.Y.

Bot Bash   203334

http://www.botbash.com/

People who like to build robots and then go out and smash them into each other.

Central Jersey Robo Conflict   203336

http://users.rcn.com/ljstier/rules.html

Remote-controlled devices in competitive and combat-oriented games.

DragonCon Robot Battles   203339

http://www.scenic-city.com/robot/

Immobilize your opponent. Kill, kill, kill. The competition is staged at Atlanta's Dragon*Con science fiction convention.

*Web site for annual robotics competition staged at Atlanta's Dragon*Con.*

MechWars   203709

http://www.tcmechwars.com/

Big brute robots that like to bash one another.

Robot Conflict   203349

http://robotconflict.com/

Home page of the Robot Conflict series of events and also the home of the Northeast Robotics Club (NERC).

Robot Dojo   202167

http://www.robotdojo.com/

Help and history about, and for, combat robotics.

Robot Wars   203351

http://www.robotwars.co.uk/

A TV show produced in the U.K. about robots that want to tear each other apart. And they seem like such nice robots, too.

RobotCombat.com

202978

http://www.robotcombat.com/

RobotCombat.com (operated by ro-battler Jim Smentowski) is a portal for robotics, where the specialty of the house is machines that bash up each other.

A very useful feature is their continually updated links pages at:

http://www.robotcombat.com/links.html

Robotcombat.com Web site.

Care and Feeding of the "Combot"

Combat robots (combots) are designed to inflict damage on an opponent. That makes them potentially dangerous to humans, too. So, if you're planning on building a combat robot—whether or not you enter competitions—remember, safety is Rule One. Ignoring safety procedures, or lapsing into carelessness, may cost you fingers, hands, arms, eyes, even your life.

  Here are some simple pointers to keep in mind when building and experimenting with a combat robot:

1. Don't even think about building a combat robot unless you're experienced. They are *not* for first-timers.

2. Don't try to skimp on the proper materials, especially for the weapons systems of the robot. A fast spinning lawnmower blade needs a proper bearing and housing. Trying to make do with thin sheet metal, plastic, or other lightweight material may cause damage to your robot, and you!

3. Always work with a friend. Have your friend standing by while you try out the remote controls of the robot.

4. Because of their weight, and the need for good speeds, most combat robots use motors requiring 30 to 100 amps of battery current. Batteries that deliver this much current are dangers in themselves if they are not treated with respect. *Never* short out the terminals of the battery just to "see what would happen." I'll tell you what will happen: You'll melt the metal used to short the terminals, you'll get severe—and possibly life-threatening—shock and burns.

5. Work in an adequately sized workshop. *Never, ever* operate the weapon system of the robot while in an enclosed room (do it outside, in the open) or while you or others are nearby. There's a reason they have thick, unbreakable plastic surrounding the arenas in televised robot bouts.

6. The machinery and tools used to build a heavy-duty robot can be just as dangerous as the robot. Wear protective gear and always use safe materials-handling procedures. If you're welding the robot frame, be sure you know all about welding safety. Yes, people have burned down their houses because of careless welding practices. Don't be the next!

7. Finally, if you doubt your construction abilities, pass on the combat robot and build something else.

MOVERS AND SHAKERS

Mark Thorpe

http://www.marcthorpe.com/

Mark Thorpe, an industrial designer and artist, is generally attributed to be the "father" of remotely controlled robotic combat.

Robotica 203015

http://tlc.discovery.com/fansites/robotica/robotica.html

The Learning Channel's answer to BattleBots. Not as funny, but the competitions are more varied, with added obstacles like the Gauntlet and the Labyrinth. The Web site also includes Q&A, short articles, and a user-to-user forum.

Robots@War 203716

http://www.robotsatwar.com/

Open-air live robot battle from the U.K.

Robotwars 203014

http://www.robotwars.co.uk/

Main site of the venerable Robot Wars television combat show, seen in over two dozen countries. The show may be from the U.K, but these robots are no English gentlemen!

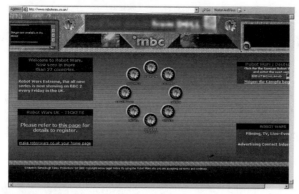

RobotWars Web site.

RobotWars MAD 203717

http://www.robotwarsmad.co.uk/

Says the Web site: "Getting sick and tired in being behind in the world of Robot Wars? What happens when Robot Wars is not on TV? Come to Robot Wars MAD.co.uk for all the latest news, gossip on all the Robots, teams and also, lots of Quiz's, Polls and more things to keep you active."

Society of Robotic Combat (SORC) 202412

http://www.sorc.ws/

The WWE of battling bots. Not as much sweat, but the same amount of swearing.

Technical Guide to Building Fighting Robots 202299

http://homepages.which.net/~paul.hills/

Several useful and insightful articles on intermediate and advanced robot-building techniques by Paul Hills. Examples include:

- Making a high-power servo-theory and circuits
- "The physics of axe weapons"
- Using DC motors in fighting robots

Fairly technical and in-depth.

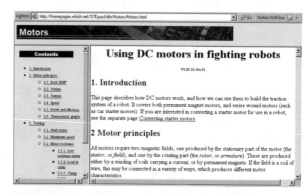

Web site for serious combat robot builders.

COMPETITIONS-ENTRANT

Some robot builders like to show off their creations. This section highlights robots made to fight other robots. Most are metallic gladiators and are something of "TV stars," having appeared on televised robot bouts.

SEE ALSO:

COMPETITIONS-COMBAT: Where to find fighting robots

INTERNET-EDU/GOVERNMENT LAB: Some schools sponsor robot challenges

INTERNET-PERSONAL WEB PAGE: Owners show off their robots, some of which do battle

Anvil   203384

http://www.focalpoint.freeserve.co.uk/

Anvil combat robot. With building diary and pictures. Also how-to pages using the OOPic microcontroller.

Automatum   203395

http://www.automatum.com/

Several BattleBots entrants, such as Pressure Drop and Complete Control, with lots of construction details and full-color photographs.

Iceman   203382

http://www.myth.demon.co.uk/IceMan/

Combat robot for U.K. Robot Wars.

Jim Struts/Miss Struts   203383

http://www.eyeeye.demon.co.uk/

"Arnold Terminegger" and other robots for combat.

Jon's Robot Wars pages   203380

http://www.use-the.net/robots/

Jon's battle robots, advice, datasheets, and more.

KillerBotZ   203377

http://www.killerbotz.org

Builder of death match combat robots.

KillerHurtz and TerrorHurtz   203381

http://www.johnreid.demon.co.uk/

KillerHurtz and TerrorHurtz are combat robot entrants. The site also provides some excellent how-to articles, calculators, and datasheets. Be sure to check out the following subdirectories and pages at the site:

- Motor characteristics: /howto/motorgraph.htm
- Power train calculator: howto/calculator.htm
- Pneumatics page: /howto/pneumatics.htm

Web site for KillerHurtz and TerrorHurtz

Liverdyne Robotics   202817

http://www.usswarrior.co.uk/

Combat robot design team. These guys make a mean cup of English tea.

M5 Industries   203397

http://www.m5industries.com/

A visual effects company in San Francisco that also builds combat robots.

Puppetmaster Combat Robotics   203388

http://www.puppetmaster-robotics.com/

Home page of combat robot entrant Scarab. Also useful FAQ on entering (and maybe winning) a robot combat competition.

RabidLabs   202843

http://www.rabidlabs.com/

"Home of the Rabid Lab Rat." Show-off site for the robot combat team RabidLabs.

Rage 203396

http://www.clineworks.com/rage.html

Construction photos of this BattleBots contender.

Team Boltz 203435

http://www.teamboltz.com/

Combat competition entrant: Bad Cow, Psycho Chicken.

Team Minus Zero 203398

http://www.tmz.com/

Combat robot entrant. Interesting and useful technical info pages. Also plenty of color photos of the 'bots.

Team Saber 204066

http://www.teamsaber.com/

Home page of Team Saber, BattleBots contestant. Nice tech articles on topics such as using servos.

Team Whyachi LLC 204219

814 E. 1st Ave.
P.O. Box 109
Dorchester, WI 54425
USA

Street Wheeler. Photo Team Whyachi LLC.

 info@teamwhyachi.com

 http://www.teamwhyachi.com/

Team Whyachi is both an entrant in combat-style robot contests and an online retailer of high-performance parts for combat bots. Their small but still quite useful product line includes high-power gear motors, wheels, and motor speed controls.

COMPETITIONS-MAZE

Maze—following robots navigate a maze without human intervention. There are literally dozens of such competitions, but nearly all are variations on the same theme. They are adequately represented by the following listings.

Micromouse Bilby 202413

http://www.usq.edu.au/users/billings/bilby/

Maze contest, but with robotic mice instead of real ones. Held in Australia.

Micromouse Competition 203345

http://www.ece.ucdavis.edu/umouse/

The MicroMouse maze contest page from the University of California at Davis.

COMPETITIONS-OTHER

Here you'll find additional competitions that don't fit neatly into the other categories, and because of their uniqueness, they deserve special attention.

Association for Unmanned Vehicle Systems International (AUVSI) 203876

http://www.auvsi.org/

In the words of the Web site: "The Association for Unmanned Vehicle Systems International (AUVSI) is the world's largest non-profit organization devoted exclusively to advancing the unmanned systems community. AUVSI, with members from government organ-

MOVERS AND SHAKERS

Dean Kamen

http://www.usfirst.org/about/bio_dean.htm

Dean Kamen makes news even when he keeps his mouth shut. One of his latest inventions is the once-top secret Segway Human Transporter, a type of scooter that looks a lot like a push-type rotary lawn-mower, but is significantly more advanced. Mr. Kamen is also well known for his establishment of FIRST (For Inspiration and Recognition of Science and Technology), a nonprofit group that promotes the pursuit of technological studies for young people.

izations, industry and academia, is committed to fostering, developing, and promoting unmanned systems and related technologies."

AUVSI (Autonomous Underwater Vehicle Competition)   202019

http://www.auvsi.org/competitions/water.cfm

Home page of the Autonomous Underwater Vehicle Competition. The competition is held annually.

Co-Evolutionary Robot Soccer Show   202258

http://www.legolab.daimi.au.dk/cerss/

Here's what the Web site says: "The Co-Evolutionary Robot Soccer Show is a game that allows you to develop robot soccer players by using the concept of co-evolution. You can develop the robot soccer players in the software provided for free at this Web site. You evolve different robot soccer players by changing the parameters for the evolution (the population parameters and the fitness formula parameters). When you have evolved a good robot soccer player, you can send the player to our server. Each night (European time), the server will play 50,000 matches between the uploaded players and generate a new Highscore List every morning. So you can keep track on how your player(s) is/are doing by going to this site every morning. At the end of the competition, the first players on the Highscore List will win the sponsored prizes. These prizes include a LEGO Mindstorms Robotic Invention System."

First Internet Robot Contest (FIRC)   203782

http://www.roboticspage.com

Online robotics contents-send in your robot to be judged. Look through the entrants to see who won.

Knex K-Bot World Championships   203344

http://www.livingjungle.com/

Competition robots, K'NEX style.

LEGO League International   204228

http://www.firstlegoleague.org/

The FIRST LEGO League is an international group of LEGO enthusiasts interested in fostering the fields of robotics, science, and technology to school-age children. See also FIRST:

http://www.usfirst.org/

Web site for LEGO League International.

Microrobot NA Inc.   203939

P.O. Box 310
451 Main Street
Middleton, NS B0S 1P0
Canada

☎   (902) 825-1726

📠   (902) 825-4906

🚫   (866) 209-5327

✉   info@microrobotna.com

http://www.microrobotna.com/

Robot kits, microcontroller boards, and parts for sumo, soccer, and line-following competition 'bots. For example, the company's Robo-Lefter is a maze solving MicroMouse. (Its name is derived from the left-turn maze-solving algorithm it uses.) The products are available through distributors or directly from Microbot NA.

RoboFesta-Europe 203951

http://www.robofesta-europe.org/

RoboFesta-Europe is part of an international movement to promote interest throughout Europe in science and technology, including robotics. Sponsors Olympics-style competition events.

See also the international RoboFesta page:

http://www.robofesta.net/

RoboFesta-International 203952

http://www.robofesta.net/

RoboFesta is an international movement to promote interest throughout the world in science and technology, including robotics. Sponsors Olympics-style competition events. Web page in Japanese, English, and French.

SEE ALSO:

http://www.robofesta-europe.org/

RoboFlag 202215

http://roboflag.carleton.ca/

Autonomous mobile robots compete against one another to capture a flag. See also:

http://roboflag.carleton.ca/gallery/

http://roboflag.carleton.ca/competition/

http://robotag.carleton.ca/

Hosted by Carleton University (Ottawa, Ontario, Canada).

Robot Vacuum Cleaner Contest 203352

http://www.botlanta.org/Rally/index.html

Fastest robot to vacuum up a half pound of rice wins.

Robotag.com 203995

http://www.robotag.com/

Pictures, movies, and descriptions of Team SCUD's Robotag entrants.

Robotag.com Web site.

RSSC Robot Talent Show 203353

http://www.dreamdroid.com/talentshow.htm

A regular talent show put on by the Robotics Society of Southern California; judging is by software, hardware, and how good your 'bot looks in a swimsuit.

Team SCUD 203778

http://www.robotag.com

Tag-playing entrant. Plenty of design photos.

Trinity College Fire-Fighting Home Robot Contest 202095

http://www.trincoll.edu/events/robot/

World-famous firefighting contest. Robots roll or walk through a scaled-down model of a house looking for a candle to extinguish. Harder than it sounds.

Walking Machine Challenge 202561

http://www.sae.org/students/walking.htm

The Society of Automotive Engineers sponsors a challenge in college-level engineering to design, build, and test a walking machine with a self-contained power source. Many of the resulting designs are quite sophisticated.

Western Canadian Robot Games 202093

http://www.robotgames.com/

The Western Canadian Robot Games is one of the oldest robot competitions, with events that include sumo wresting, something called atomic hockey, a hallway navigation game for walking robots, and a series of challenges specially designed for BEAM robots. The competitions are held annually in Alberta, Canada.

Web site for the Western Canadian Robot Games.

COMPETITIONS-SOCCER & BALL PLAYING

Robotic soccer is played either by two robots against one another, or one team of robots against another. This form of robot competition is among the most challenging, especially for multirobot teams. In the multiple-robot version of the game, each robot plays a certain position on the team and is programmed accordingly. The robots must communicate with one another in order to play successfully. In some soccer tournaments, an overhead machine vision system provides a view of the game field, and this view is electronically interpolated to remotely command the players.

The RoboCup soccer competition is the most famous, and there are local versions of it for those who cannot travel to Japan, where the main event is staged.

Soccer is one form of robot game that involves balls. Other forms include volleyball and ball collecting.

BotBall 203969

http://www.botball.org/

Competition sponsored by the KISS Institute for Practical Robotics (KIPR).

Canada First Robotic Games 202094

http://www.canadafirst.org/

Canada First is a team-based robotics competition sponsored by various Canadian companies to motivate students in the fields of math, science, and technology. The competitions generally involve soccer and ball handling of some type.

Web site is in English and French.

FIRA Robot World Cup 203340

http://www.fira.net/

Robot Soccer in China.

Intelligent Robot Contest Festival 203342

http://www.robotics.is.tohoku.ac.jp/inrof.html

The rules and regulations of the Intelligent Robot Contest Festival in Sendai, Japan.

Web site is in Japanese and English.

Jerry Sanders Creative Design Competition 203333

http://dc.cen.uiuc.edu/

A competition of autonomous or radio control ball collectors. Or maybe it's "bill collectors," in which case, I don't like this competition at all!

KISS Institute for Practical Robotics (KIPR)

202540

http://www.kipr.org/

In the words of the Web site: "KISS Institute for Practical Robotics (KIPR) is a private non-profit community-based organization that works with all ages to provide improved learning and skills development through the application of technology, particularly robotics. We do this primarily by providing supplementary, extra-curricular and professional development classes and activities. KISS Institute's activities began in 1993."

KIPR also sponsors the annual Bot Ball tournament for middle and high school students.

RoboCup

202560

http://www.robocup.org/

RoboCup is an international project to promote the fields of artificial intelligence and robotics. This is accomplished with soccer competitions-two teams of robots square off against one another and play a game of soccer without human intervention. RoboCup is held annually in Japan.

Web site for RoboCup, held yearly in Japan.

RoboCup Junior

203935

http://www.robocupjunior.org.au/

RoboCup Junior, based in Melbourne, Australia, is for the design, construction, and competition of autonomous soccer-playing robots. The competitions are intended for school-age students.

Sony Robotbox

202057

http://www.sony.co.jp/en/SonyInfo/dream/robotbox/

Online magazine for RoboCup competitions.

Trinity LEGO Cybernetics Challenge

203355

http://www.cs.tcd.ie/research_groups/cvrg/lego/index.html

A game of robot volleyball between teams of two robots, which were built using LEGO Mindstorms.

 # COMPETITIONS-SUMO

Robot sumo involves two contestants on a round playing mat. The object is for one contestant to push its opponent off the mat before it gets pushed out. In all but a few cases, the robot fighters are autonomous and are not controlled remotely by a human operator.

The playing mat is typically 6 feet in diameter and is painted black. A white stripe is painted near the outside rim of the mat and is used by the robot to help it determine the boundary of the playing field.

Sumo competitions are separated into weight classes. All robots within a class must weigh less than the stipulated amount, such as 2, 10, or 20 pounds. No weaponry is typically allowed in a sumo competition—it's all about pushing.

CIRC Autonomous Sumo Robot Competition

203337

http://www.circ.mtco.com/

By the Central Illinois Robotics Club.

Critter Crunch

203338

http://www.milehicon.org/critrule.htm

A robotic combat in which the object is to immobilize your opponent or to push it out of the arena. Two weight classes: 2 pounds and 20 pounds.

How to Build a Sumo Wrestling Robot 202848

http://www.cercot.demon.co.uk/sumo/build.htm

Discussion and pictures for building a sumo-style competition robot.

International Robot Sumo Wrestling Competition 203356

http://www.chibashoten.com/robot/

Japanese champions come to the U.S.! First prize is $2,000. Much grunting.

Northwest Robot Sumo 203346

http://www.sinerobotics.com/sumo/

One of the biggest American sumo competitions.

Robot sumo, that is. The sponsor of the competition is Sine Robotics:

http://www.sinerobotics.com/

OCAD Sumo Robot Challenge 203347

http://www.student.ocad.on.ca/info/sumo/

Bashing/crashing/smashing robots sponsored by the Ontario College of Art & Design.

Robot Sumo 203350

http://www.robots.org/events.htm

An annual sumo competition held at the Exploratorium in San Francisco. Sponsored by the San Francisco Robotics Society of America.

Tips for Effective Sumo Robot Wrestling

If you've ever watched a real sumo-wrestling match, then you know that at least part of the sport relies on weight and girth. Same with sumo robot wrestling, where the aim is to push your opponent outside the game field—usually a round, black circle. All things considered, a heavier robot will push a lighter robot aside, but there is more science involved than you might think.

- In most competitions, the advantage of weight is minimized by putting contestants into various classes. For the most part, you'll want your robot to be on the heavier side of its class, for any advantage that might give you.
- Still, bear in mind that the heavier the robot, the harder it might be for its motors to provide adequate speed and torque. Don't blindly sacrifice agility for weight. A fast-moving robot can more readily get out of trouble or position itself for a sideways push.
- Wheel traction is critical. Be sure to read the rules for the competition, as you may be restricted on the size, material, and thickness of the wheels you use. Soft rubber provides better traction and helps prevent your opponent from pushing you off the play field. Wider wheels offer increased surface area and therefore better traction.
- A low center of gravity ensures your robot won't be easily turned over. A low, squat robot is usually the best design.
- Many sumo robots use a scoop-like shovel at the front to partially lift their competitors off the play field. Once lifted, robots are easy prey. See if the competition rules allow for a shovel (most do not permit a weapon).

COMPUTERS

Listings in this section deal with computers for robot control, as well as computer interfacing and data acquisition. The emphasis is on small computers, particularly those that are self-contained and can be powered by a single 5- or 12-volt DC source.

COMPUTERS-DATA ACQUISITION

Data acquisition involves circuitry that converts an analog signal, such as temperature or vibration, to digital form so that it can be processed by a computer. This section lists data acquisition boards for PC-compatible computers as well as general-purpose data acquisition modules. Data acquisition modules typically provide their data through a standard interface, such as PC parallel port, RS-232, or USB.

DATAQ Instruments, Inc.   202667

241 Springside Dr.
Ste. 200
Akron, OH 44333
USA

((330) 668-1444
⅋ (330) 666-5434
∅ (800) 553-9006
✉ info@dataq.com
🌎 http://www.dataq.com/

Dataq makes and sells data acquisition hardware and accessories, data acquisition software, signal-conditioning components, and data acquisition starter kits (some low-cost bargains here).

DATEL, Inc.   204205

11 Cabot Blvd.
Mansfield, MA 02048-1151
USA

((508) 339-3000
⅋ (508) 339-6356
∅ (800) 233-2765
✉ websales@datel.com
🌎 http://www.datel.com/

Manufacturer and distributors of:

- DC-DC converters
- Sampling A/D converters
- Data acquisition boards

Datasheets available for the products in Adobe Acrobat PDF. Web site is in English and Japanese.

Embedded Acquisition Systems   203059

c/o Kin Fong
2517 Cobden St.
Sterling Heights, MI 48310
USA

⅋ (240) 266-4252
✉ sales@embeddedtronics.com
🌎 http://embeddedtronics.com/

Makers of MiniDaq, a small data acquisition module for the PC. Also offers the EAS Finger Board II; scaled-down Handy Board. The Web site includes pics of prototype robots the company has made using their products.

LabJack Corporation   203845

3112 S. Independence Ct.
Lakewood, CO 80227-4445
USA

((303) 942-0228
⅋ (720) 294-0550
🌎 http://www.labjack.com/

Products include an affordable USB-based data acquisition module.

LabJack U12. Photo LabJack Corp.

Giving Your 'Bot a Brain

Even the Scarecrow from *The Wizard of Oz* wanted a brain. You want your robot to be at least as smart as the Scarecrow, right? You can be your own wizard of Emerald City by giving out brains to your robotic creations. Instead of some phoney-baloney diploma, the brains you give your robots will be electronic in nature. The only trick is—which electronics?

Noncomputer Electronic Components

Noncomputer discrete components—transistors, resistors, capacitors, and integrated circuits—can be used to control a robot. Such robots typically are "hard-wired" to perform some task, such as follow a line or seek out light. BEAM robots are a good example of machines that have noncomputer brains: No number crunching is going on in their electronics. Rather, simple electronic circuits compel the robot to move.

A popular design uses a 555 timer IC to control a motor. Separate 555 chips control each motor of a two-wheeled robot. The 555 timer outputs a series of short pulses, and the duration of these pulses determines the speed of the motor. If one motor goes a little slower than the other, the robot will move in wide, sweeping arcs. Add some bumper switches to the robot to back it up and another 555 to time how long the robot reverses direction, and the machine will do a remarkable job of navigating around a room.

Watching the robot, you'd think the thing were controlled by a computer, but its "brain" is a simple collection of parts you can buy at Radio Shack. (If you're interested, the design for such a robot can be found in the book, *Mobile Robots: From Inspiration to Implementation*, among other sources.)

Computers and Microcontrollers

As attractive as noncomputer brains are, they are basically "hard-wired" as circuitry, and making changes to them requires lots of work. A robot controlled by a computer can be "rewired" simply by changing the *software* running on the computer.

There is an almost endless variety of computers that be used as robot brains. The three most common are:

- *Microcontroller*, programmed either in assembly language or a high-level language such as Basic or C. The LEGO Mindstorms RCX is a good example of a robot run from a microcontroller. Microcontrollers are available in 4-, 8-, 16-, and 32-bit versions (plus a few others, for special purposes). The Basic Stamp, PICmicro, AVR, OOPic, and BasicX are good examples of microcontrollers commonly used in amateur robotics.

- *Single board computer*, also programmed either in assembly language or a high-level language, but generally with more processing power than a microcontroller. Single board computers (SBCs) are a lot like "junior PCs," but on a single circuit board. In fact, many SBCs are IBM PC-compatible and use Intel microprocessors capable of running any Intel-based program, including the MS-DOS operating system. A common SBC form factor is PC/104, which gets its name from "Personal Computer" (originally of IBM fame) and the number of pins (104) used to connect two or more PC/104-compatible boards together.

- *Personal computer*, such as a PC compatible or an Apple Macintosh, or even an older model such as the venerable Commodore 64. A good option for robotics is a PC-compatible laptop, particularly the older monochrome models that required less battery power. They're also fairly inexpensive on the used market.

Prairie Digital, Inc.   202208

920 Seventheenth St.
Industrial Park
Du Sac, WI 53578
USA

☎   (608) 643-8599

📠   (608) 643-6754

✉   sales@prairiedigital.com

🌐   http://www.prairiedigital.com/

Prairie Digital sells I/O and interface boards:

- General-purpose data acquisition system for PCs
- 8-bit analog-to-digital converter
- 12-bit analog-to-digital converter
- Serial port data acquisition and control module
- Low-cost 5 1/2-digit A/D with RS-232 serial control
- Relay board

COMPUTERS-I/O

I/O stands for input/output, circuitry designed to interface a computer to external circuitry. On a PC, I/O is used to connect the computer to a printer, for example. The I/O performs important functions of conditioning the data and acting as a barrier to problems caused by the external device. The idea here is that it's cheaper to replace an I/O board or module than the complete computer.

The I/O listed in this section is particularly suited for robotics and other embedded applications. Depending on the configuration, I/O can be used with single board computers (SBCs) or with microcontrollers. Examples of I/O modules include relay bays (the computer activates the relays from program control), self-

contained Web servers, and generic interfaces to stepper motor or H-bridge drivers.

SEE ALSO:

COMPUTERS-SINGLE BOARD COMPUTERS: Something to connect the I/O to

INTERNET-CIRCUIT EXAMPLES: Plans for home-made I/O

LEGO-MINDSTORMS: Ideas for I/O expansion for the Mindstorms robot

MICROCONTROLLERS-HARDWARE: Using a microcontroller instead of an SBC

ActiveWire, Inc.   202346

895 Commercial St.
Ste.700
Palo Alto, CA 94303
USA

☎   (650) 493-8700

📠   (650) 493-2200

✉   salesinfo@activewireinc.com

🌐   http://www.activewireinc.com/Manufacturers and sellers of ActiveWire-USB, a low-cost module designed to provide programmable I/O pins that can interface to anything, via a USB serial port on a PC or Macintosh. Programming examples are provided for Macintosh, Win95/98/2000/ME, Linux, FreeBSD, and LabView.

The company offers add-on interfacing modules, including:

- Motor control board, for controlling up to six small DC motors (maximum 600mA to 1A)

All USB

http://www.allusb.com/
USB converters - USB to serial, USB to parallel, etc.

B&B Electronics

http://www.bb-elec.com/
Products includes USB converters

MOVERS AND SHAKERS

Steve Mann

http://www.eecg.toronto.edu/~mann/

Professor Steve Mann, of the University of Toronto, literally wears his work. He's widely known as the first person to actively and continually wear implanted electronic gizmos—the "Six-Million-Dollar Teacher." His research involves the "wearable computer," a silicon brain that assists people in their everyday lives, possibly (at some point in the future) becoming a part of them. Of course, shades of *The Borg* and a lot of other sci-fi themes, but Professor Mann doesn't seem to have domination of the galaxy on his mind.

- LCD character display module
- Opto-isolator module, for connecting to relays, H-bridges, SCRs, and other high-current devices

ActiveWire-USB

ActiveWire-USB is a module for connecting any PC, Macintosh, or Linux box that has a USB port. It sports 16 input/output (I/O) lines for connecting to real-world devices, such as motors, relays, and other digital devices. Size of the module is small: 3.3 _1.9 inches, and no separate battery is needed (unless you need to drive heavier loads), as it derives its power from the USB line.

You need a USB A-to-B cable, sold separately (and every one) to connect the module to your computer. Because the ActiveWire-USB module is plugged into your PC, it means your robot will be tethered to it while your PC controls it.

With the proper driver loaded on your computer, you can use your favorite programming language (JavaScript, Visual Basic, C, etc.) to command the ActiveWire-USBis 16 port pins, configuring them individually as either inputs or outputs.

Robotic applications for the ActiveWire-USB module include motor control. You need to attach a motor driver circuit, such as an H-bridge, to the pins of the module, in order to drive the motor. (If you donit wish to create an H-bridge circuit yourself, the ActiveWire company also sells a ready-made motor bridge add-in board that can power up to six small DC motors.)

Because the I/O pins on the ActiveWire-USB module can be configured as an input or as an output, the device can be used to interface switches to the computer. The company provides "tech notes" for these typical applications, as well as downloadable copies of the hardware and software manuals, which you can browse before purchasing the product.

For more information on ActiveWire-USB, visit:

http://www.activewireinc.com/

Athena Microsystem Solutions 202522

10624 Rockley Rd.
Houston, TX 77099
USA

✆ (281) 418-5631

📠 (281) 256-3851

✉ info@athenamicrosystems.com

🌐 http://www.athenamicro.com/

Athena sells microcontrollers and single board computer peripherals. Product highlights include:

- AMS-HE/DE-Precision Hall-effect DC current sensing module
- AMS876-SIMMStick plug-in module based on Microchip's PIC16F876 Flash memory microcontroller
- SLI-OEM-Serial LCD controller
- AMS-900PA/232-Spread spectrum wireless transceiver

JKmicrosystems, Inc. 202333

1403 Fifth St.
Ste. D
Davis, CA 95616
USA

✆ (530) 297-6073

📠 (530) 297-6074

✉ jkmicro@jkmicro.com

🌐 http://www.jkmicro.com/

JKmicrosystems makes single board computers, peripheral boards (A/D converter, I/O, relay), keypads, and LCD displays.

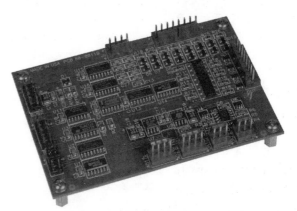

Model uIO Input output expansion board. Photo JKmicrosystems, Inc.

Of Inputs and Outputs

Number Five (from the movie Short Circuit) was always shouting about "more input." Robots need both input and output—ways to get data from the outside world (input), and ways to control external devices, like motors (output)

There are several specialized forms of I/O found on computers and microcontrollers that robots can use. Many are proprietary to a given circuit architecture: Microcontrollers from National may favor one approach; chips from Philips may favor another. In the end, many do similar jobs.

Serial Communications

I2C—Inter-Integrated Circuit, a two-wire serial network protocol used by Philips to allow integrated circuits to communicate with one another. With I2C you can install two or more microcontrollers in a robot and have them communicate with one another. One I2C-equipped microcontroller may be the "master," while the others are used for special tasks, such as interrogating sensors or operating the motors.

Microwire—A serial synchronous serial communications protocol used in National Semiconductor products, and popular for use with the PICMicro line of microcontrollers from Microchip Technologies. Most Microwire-compatible components are used for interfacing with microcontroller/microprocessor support electronics, such as memory and analog-to-digital converters.

SCI—Serial communications interface, an enhanced version of the UART, detailed later.

SPI—Serial peripheral interface standard used by Motorola and others to communicate between devices. Like Microwire, SPI is most often used for interfacing with microcontroller/microprocessor support electronics, especially outboard EEPROM memory.

Synchronous serial port—Data is transmitted one bit at a time, using two wires. One wire contains the transmitted data, and the other wire contains a clock signal. The clock serves as a timing reference for the transmitted data. Note that this is different from asynchronous serial communication (see the following), which does not use a separate clock signal.

UART—Universal asynchronous receiver transmitter, used for serial communications between devices, such as your PC and the robot's computer or microcontroller. *Asynchronous* means that there is no separate synchronizing system for the data. Instead, the data itself is embedded with special bits (called *start* and *stop* bits) to ensure proper flow. The USART (Universal Synchronous/Asynchronous Receiver Transmitter) can be used in either asynchronous or synchronous mode, providing for faster throughput of data.

Data Conversion

ADC—Analog-to-digital conversion transforms analog (linear) voltage changes to binary (digital). ADCs can be outboard, contained in a single integrated circuit, or included as part of a microcontroller. Multiple inputs on an ADC chip allow a single IC to be used with several inputs (4, 8, and 16 input ADCs are common).

DAC—Digital-to-analog conversion transforms binary (digital) signals to analog (linear) voltage levels. DACs are not as commonly employed in robots; rather they are commonly found on such devices as compact disc players.

Pulse and Frequency Management

Input capture—An input to a timer that determines the frequency of an incoming digital signal. With this information, for example, a robot could differentiate between inputs, such as two different locator beacons in a room. Input capture is similar in concept to a tunable radio.

PWM—Pulse width modulator, a digital output that has a square wave of varying duty cycle (e.g., the "on" time for the waveform is longer or shorter than the "off" time). Often used with a simple resistor and capacitor to approximate digital-to-analog conversion, to create sound output, and to control the speed of a DC motor.

Pulse accumulator—An automatic counter that counts the number of pulses received on an input over *x* period of time. The pulse accumulator is part of the architecture of the microprocessor or microcontroller and can be programmed autonomously; that is, the accumulator can be collecting data even when the rest of the microprocessor/micro-controller is busy running some other program.

Special Functions

Hardware interrupts—Interrupts are special input that provides a means to get the attention of a microprocessor or microcontroller. When the interrupt is triggered, the micro-processor can temporarily suspend normal program execution and run a special sub-program.

Comparator—An input that can compare a voltage level against a reference; the value of the input is then lower (0) or higher (1) than the reference. Comparators are most often used as simple analog-to-digital converters where HIGH and LOW are represented by something other than the normal voltage levels (which can vary, depending on the kind of logic circuit used). For example, a comparator may trigger HIGH at 2.7 volts. Normally, a digital circuit will treat any voltage over about 0.5 or 1 volt as HIGH; anything else is considered LOW.

Analog/mixed-signal (A/MS)—Inputs (and often outputs) that can handle analog or digital signals, under software guidance. Many microcontrollers are designed to handle both analog and digital signals on the same chip, and to even mix -and match analog/digital on the same pins of the device.

External reset—An input that resets the computer or microcontroller so that it clears any data in RAM and restarts its program (the program stored in EEPROM or elsewhere is not erased).

Switch debouncer—Cleans up the signal transition when a mechanical switch (push button, mercury, magnetic reed, etc.) opens or closes. Without a debouncer, the control electronics may see numerous signal transitions and could interpret each one as a separate switch state. With the debouncer, the control electronics sees just a single transition.

Input pullup—Pullup resistors (5 to 10K) are required for many kinds of inputs to control electronics. If the source of the input is not actively generating a signal, the input could "float" and therefore confuse the robot's brain. The pullup resistors, which can be built into a microcontroller and activated via software, prevent this floating from occurring.

J-Works, Inc.

12328 Gladstone St.
Unit 1
Sylmar, CA 91342
USA

☏ (818) 361-0787

202196

☏ (818) 270-2413

✉ sales@j-works.com

🌐 http://www.j-works.com/

PC and USB I/O products, including:

- A/D converters

- Opto I/O
- Relay
- Digital I/O
- Temperature
- Counter

NetMedia Inc. / Siteplayer   202151

10940 N. Stallard Pl.
Tucson, AZ 85737
USA

((520) 544-4567
((520) 544-0800
✉ info@siteplayer.com
🌐 http://www.siteplayer.com/

Siteplayer is a Web server the size of a postage stamp. Really, a 1-inch-square postage stamp! The Siteplayer module has a built-in Web server and Ethernet adapter, allowing it to be used on any TCP/IP connection. SitePlayer sports eight I/O pins, which can be individually programmed via Web commands.

The use in robotics is obvious, including remote control of a robot connected via the Internet. Add a video camera, and you-or anyone else-can control a robot from around the globe.

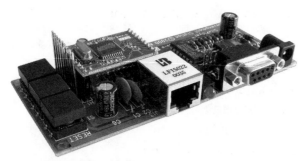

SitePlayer, on the SitePlayer development kit board. Photo NetMedia, Inc.

Sealevel Systems   204069

155 Technology Pl.
Liberty, SC 29657
USA

((864) 843-4343
((864) 843-3067
✉ support@sealevel.com
🌐 http://www.sealevel.com/

Makes and sells I/O converters and interfaces for USB, RS-232 and RS-422 serial, and PCI slot.

Weeder Technologies   202949

1710-B Brighton Cove
Ft. Walton Beach, FL 32547
USA

((850) 863-5723
((850) 863-5723
🌐 http://www.weedtech.com/

Weeder manufactures and sells a line of "stackable modules" for a variety of RS-232 applications. Each module has its own address to facilitate communications. The modules are individually addressable; the address is set simply by using a DIP switch. You can stack up to 32 modules (hence, peripherals) on the same RS-232 cable connected to a host PC. Reasonable prices.

- Solid state relay module
- Analog input module
- Analog output module
- Stepper motor driver
- Pulse counter/timer
- Multidrop peripheral interface

Analog to digital interface. Photo Weeder Technologies.

Winford Engineering   203048

4169 Four Mile Rd.
Bay City, MI 48706
USA

((989) 671-2941
((989) 671-2941
⊘ (877) 634-2673
✉ sales@winfordeng.com
🌐 http://www.winfordeng.com/

Winford produces computer I/O cards and accessories for the PC. Their CRD155B is an 8-bit ISA card that provides 24 digital I/O lines. The CRD155B card can be programmed and controlled by using Winford Engineering's Portal API.

COMPUTERS-SINGLE BOARD COMPUTERS

The term single board computer is a holdover from days when most computers consisted of a main board (called the motherboard) and additional peripheral boards (called daughter cards) for extra features such as I/O of disk drive interface. Today, even desktop PCs use a single printed circuit board, all basic functions are built in-including I/O, disk interface, sound, and display adapter.

Today the term single board computer has evolved to mean a compact computer board, especially one that can be powered by a single 5- or 12-volt voltage source and that is intended primarily for use as a controller in a piece of hardware-a so-called embedded application. Most of the single board computers (SBCs) in this section are smaller than 4 inches square.

SEE ALSO:

MICROCONTROLLERS-HARDWARE: Computers on a single chip

PROGRAMMING-LANGUAGES: Programming languages, such as C and Basic, for computers

PROGRAMMING PLATFORMS & SOFTWARE: Operating systems

Aaeon Electronics, Inc.   202888

3 Crown Plaza
Hazlet, NJ 07730
USA

((732) 203-9300

⊠ (732) 203-9311

✉ sales@aaeon.com

🌐 http://www.aaeon.com/

Manufacturer and distributor of PC/104 modules, LPX-size SBCs, and media SBCs (compact computers designed for media applications). Offices in the U.S., Korea, and Germany.

🌐 🏭

Advanced Digital Logic   203198

4411 Morena Blvd.
Ste. 230
San Diego, CA 92117
USA

((858) 490-0597

⊠ (858) 490-0599

✉ general@adlogic-pc104.com

🌐 http://www.adlogic-pc104.com/

Embedded and stand-alone PC/104 single board computer (SBC) modules. Also sells Flash memory, audio network cards, and other peripheral add-in cards for PC/104.

All Industrial Systems, Inc.   202889

672 Still Meadows Circle East
Palm Harbor, FL 34683
USA

((727) 786-1009

✉ sales@usattro.com

🌐 http://www.usattro.com/

Makers of a broad line of single board computers, embedded 386/486/Pentium systems, and SBC peripherals.

Arcom Control Systems, Inc.   202890

7500 West 161st St.
Stilwell, KS 66085
USA

((913) 549-1000

⊠ (913) 549-1001

⊘ (888) 941-2224

✉ sales@arcomcontrols.com

🌐 http://www.arcomcontrols.com/

PC/104 single board computers and peripherals, developer's kits, and embedded boards. Operating systems include Linux and Windows. Sales locations worldwide.

Axiom Manufacturing, Inc.   202891

2813 Industrial Ln.
Garland, TX 75041
USA

(972) 926-9303
 (972) 926-6063
 sales@axman.com
 http://www.axman.com/

In the words of the Web site: "Axiom Manufacturing is a diverse microcontroller company specializing in single board computers, embedded controllers, custom design, and manufacturing solutions."

Products include single board computers based on the Motorola 68HC11 and 68HC12 microcontrollers, 80CXX microprocessor, MPC555 PowerPC, and MMC2001 Mcore microcontroller.

Diamond Systems Corporation   203199

8430-D Central Ave.
Newark, CA 94560
USA

 (510) 456-7800
 (510) 456-7878
🚫 (800) 367-2104
✉ techinfo@diamondsystems.com
🌐 http://www.diamondsystems.com/

Makers of PC/104 form-factor single board computers with analog I/O and serial ports.

EMAC, Inc.   202198

2390 EMAC Way
Carbondale, IL 62901
USA

📞 (618) 529-4525
📠 (618) 457-0110
✉ info@emacinc.com
🌐 http://www.emacinc.com

Manufacturer of PC-compatible SBCs,; SBC microcontrollers, embedded servers, microprocessor trainers, remote-access devices.

EMJ Embedded Systems   203065

220 Chatham Business Dr.
Pittsboro, NC 27312
USA

📞 (919) 545-2500

 (919) 545-2559
🚫 (800) 548-2319
✉ emjembedded@emj.com
🌐 http://www.emjembedded.com/

Products include serial-interface LCD displays, PC/104 single board computers, embedded modems for PC/104, miniature hard drives, and Flash memory.

Industrologic, Inc.   203849

3201 Highgate Ln.
St. Charles, MO 63301
USA

📞 (636) 723-4000
📠 (636) 723-6000
🚫 (800) 435-1975
✉ info@industrologic.com
🌐 http://www.industrologic.com/

Industrologic sells single board computers and I/O modules (including relay and LED readout), Atmel-based AT89 MCU board-level systems, RS-232 data acquisition.

JKmicrosystems, Inc.   202333

1403 Fifth St.
Ste. D
Davis, CA 95616
USA

📞 (530) 297-6073
📠 (530) 297-6074
✉ jkmicro@jkmicro.com

Model 386Ex single board computer. Photo JKmicrosystems, Inc.

http://www.jkmicro.com/

JKmicrosystems makes single board computers, peripheral boards (A/D converter, I/O, relay), keypads, and LCD displays.

Micro Computer Specialists, Inc.   202892

1070 Joshua Way
Vista, CA 92083
USA

📞 (760) 598-2450

🚫 (800) 542-9662

🌐 http://www.mcsi1.com/

Embedded computers, including 386 XT Size 386SX-40 single board computers.

Micromint, Inc.   202197

902 Waterway Pl.
Longwood, FL 32750
USA

📞 (407) 262-0066

📠 (407) 262-0069

🚫 (800) 635-3355

✉️ sales@micromint.com

🌐 http://www.micromint.com/

Micromint is a leading supplier of single board computers and industrial embedded controllers. Their products include:

- Domino (80C52 with BASIC)
- PicStic micro modules
- Answer MAN (network-based data acquisition)

PicStick. Photo Micromint, Inc.

- TI01 (based on Polaroid ranger)
- Embedded modem module
- 80C52-BASIC chip
- Power line interface for X-10

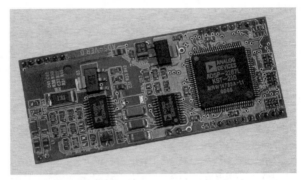

Model 2400EMM data modem. Photo Micromint, Inc.

PC/104 Consortium   203051

http://www.pc104.org/

According to the site, "We are a Consortium of over 100 members worldwide who have joined together to disseminate information about PC/104 and to provide a liaison function between PC/104 and standard organizations." Fair enough.

PC104.COM   203050

http://www.pc104.com/

Information and links to providers of PC/104-related products and services. Also includes sections on what PC/104 is, a PC/104 FAQ, and PC/104-controlled systems.

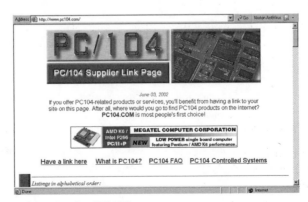

Web site for PC104.com

Single Board Computers

Not long ago, the notion of a computer on a single, small circuit board was science fiction. Now, they're so plentiful you'll find them at hundreds of sources. In this book we provide a handful of sources to get you started; there are many others, particularly those that specialize in computers for particular applications, such as controlling an elevator, operating a construction crane, or adjusting the fuel-air mixture in your car's engine.

These applications are often referred to as *embedded*, because a computer is "embedded" as part of the overall functionality of the device. For this reason, *single board computers* are almost universally used for some embedded application, though they are also of premier interest to robot builders, because of their power, small size, and low energy requirements.

With the proliferation of single board computers (or *SBCs*) has also come a dizzying array of types, sizes, and styles. Fortunately, a form -factor known as *PC/104* provides a handy standard that makes it easier to select and use SBCs and peripherals, even from among different manufacturers.

PC/104 boards measure about 3.5 inches square (specifically 90 by 96mm). The "104" comes from the number of interconnection pins used to stack the boards together. This stacking allows you to add to a PC/104 SBC without using back planes (like in a PC-compatible computer) or connectors and cables. There are both 8- and 16-bit versions for the PC/104, and there are different options within the specification.

More information on the PC/104 standard can be found at the PC/104 Consortium at:

http://www.pc104.org

While PC/104 is perhaps the most common standard form factor for SBCs, it's not the only one. A fairly popular SBC is the PCI or ISA "daughter card," designed to fit into an expansion slot of a PC-compatible (PCI and ISA are connector standards; PCI is the current type used in PCs, and ISA is the older 8- and 16-bit style). Though intended to be used inside another computer, daughter card SBCs are in fact independent computers. They are for applications that require extra processing power, when you don't want to tax the main processor of the computer.

Should You Use a PC as a Robot Brain?

The average PC may be desk-bound, but that doesn't mean you can't mount it on your robot and use it in a portable environment. That said, you may not want to use the computer for your robot's brain. Some PCs are more suited for conversion to mobile robot use than others. Consider the following qualities of a computer pressed into service as the brain of an untethered mobile robot:

- *Small size*. In this case, *small* means that the computer can fit in or on your robot. A computer small enough for one robot may be a King Kong to another. Generally speaking, however, a computer larger than about 12 inches by 12 inches is too big for any reasonably sized 'bot.

- *Standard power supply requirements*. Some computers need only a few power supply voltages, most often +5, and sometimes +12. A few, like the IBM PC-compatible, requires* negative reference voltages of -12 and -5. (*Required* is a nebulous thing: Some PC-compatible motherboards will still function if the -12 and -5 voltages are absent, though functions such as RS-232 serial may not operate correctly.)

- *Accessibility to the microprocessor system bus or an input/output port*. The computer won't do you much good if you can't access the data, address, and control lines. The IBM PC architecture provides for ready expansion using "daughter" cards that connect to the motherboard. It also supports a variety of standard I/O ports, including parallel and serial.

- *Uni- or bidirectional parallel port.* If the computer lacks access to the system bus, or if you elect not to use that bus, you should have a built-in parallel port. This allows you to use 8-bit data to control functionality of your robot. The Commodore 64, no longer made but still available in the used market, supports a fully bidirectional parallel port.

- *Programmability.* You must be able to program the computer using either assembly language or a higher-level language such as Basic, C, Logo, or Pascal.

- *Mass storage capability.* You need a way to store the programs you write for your robot, or every time the power is removed from the computer, you'll have to rekey the program back in. (Recall that microcontrollers and SBCs equipped with Flash or EEPROM memory retain their programs even when power is removed.) Floppy disks or small, low-power hard disk drives are possible contenders here.

An old 486-class motherboard. One possible use for this relic is as a robot brain.

- *Availability of technical details.* You can't tinker with a computer unless you have a full technical reference manual. The reference manual should include full schematics, or at the very least, a pinout of all the ports and expansion slots. Some manufacturers do not publish technical details on their computers, but the information is usually available from independent book publishers. Visit the library or a bookstore to find a reference manual for your computer.

Using an AC Inverter for a PC Power Source

You've read elsewhere in this section about using a PC motherboard as the brains for your 'bot. They're cheap, reliable, and very competent computers, but a disadvantage is their power requirements. Many require different voltages, namely +/-5 and +/-12 volts.

Another option is to power your robot from a 12-volt battery connected to an AC inverter. These are available at auto supply stores and many department stores that carry automotive supplies. They are designed to work with the 12-volt system of a car and can provide enough operating juice to power the typical later-model PC (older models may use very power-hungry components). The PC plugs into the inverter as if it were a wall outlet.

AC inverters are available in different wattages. As you might imagine, the higher the wattage, the more expensive the inverter. You do not need a 3,000-watt inverter; try a 400- to 800-watt version first.

Protean Logic   204040

11170 Flatiron Dr.
Lafayette, CO 80026
USA

☎ (303) 828-9156

📠 (303) 828-9316

🌐 http://www.protean-logic.com/

Makers of single board computers and microcontroller boards, many using the TICKit interpreter engine, said to offer faster processing than the Basic Stamp. Products include:

- TICKit 63 processor IC
- TICKit 63 computer module
- RSB509b serial data buffer IC
- TICKit 63 single board computer

R.L.C. Enterprises, Inc.   202893

2985 Theatre Dr.
Paso Robles, CA 93446
USA

(805) 239-9737

(805) 239-9736

 http://www.rlc.com/

Makers of embedded single board PCs with the Windows CE operating system, touch screens, and I/O interfaces built in. Intended mainly for original equipment manufacturers (OEMs) purchases, but sells in smaller quantities.

Tern, Inc.   202005

1724 Picasso Ave.
Ste. A
Davis, CA 95616
USA

(530) 758-0180

(530) 758-0181

salester@tern.com

http://www.tern.com/

Tern provides 16- and 32-bit microcontrollers (Intel-based or NEC V25) and software development packages for C, C++, and x86 assembly language.

Vesta Technology   202199

11465 West I-70 Frontage Rd. North
Wheat Ridge, CO 80033
USA

(303) 422-8088

(303) 422-9800

Micro/Sys

http://www.embeddedsys.com/
PC/104 SBCs and embedded systems

Midwest Micro-Tek

http://www.midwestmicro-tek.com/
Embedded and single board computers

VersaLogic Corp.

http://www.versalogic.com/
PC/104 single board computers

---

sales@vestatech.com

http://www.vestatech.com/

In the words of the Web site: "Vesta Technology Inc. develops programmable controllers, for use in embedded systems, machine control, and OEM industrial applications. These tiny, powerful computers are based on the PIC16C62, PIC16C74, 80C188EB, and 68332G processors, and come in a variety of models depending on the system level resources (such as memory and ports) that your project requires."

Product line highlights include:

- Single-tasking and multitasking BASIC
- A/D interface
- TTL digital I/O
- Optically isolated inputs, relay outputs
- D/A output amplifiers

Vikon Technologies   202338

6 Way Rd.
Middlefield, CT 06455
USA

(860) 349-7055

(860) 349-7088

info@vikon.com

http://www.vikon.com/

Vicon makes and sells embedded systems, including single board computers, Atmel AVR programming development boards, PIC development kits and boot loaders, PIC prototyping boards, and SimmStick bus-compatible products.

Wilke Technology GmbH   204041

Krefelder Str. 147
52070 Aachen
Germany

+49 (0) 2419 18900

+49 (0) 2419 189044

info@wilke-technology.com

http://www.wilke-technology.com/

Single-tasking and multitasking single board computers. Products include BASIC-Tiger professional software.

Win Systems

202894

P.O. Box 121361
Arlington, TX 76012
USA

 (817) 274-7553

 (817) 548-1358

info@winsystems.com

http://www.winsystems.com/

Win Systems is a manufacturer of single board computers, PC/104, PC/104-Plus, and STD bus products.

Z-World, Inc.

202895

2900 Spafford St.
Davis, CA 95616
USA

 (530) 757-3737

 (530) 757-3792

zworld@zworld.com

http://www.zworld.com/

Single board computers (with or without built-in Ethernet connectivity), I/O boards, embedded control systems, C compiler engineered for Z-World embedded products.

RabbitCore. Photo Z-World, Inc.

Zykronix, Inc.

202896

357 Inverness Dr. South
Ste. C
Englewood, CO 80112
USA

 (303) 799-4944

(303) 799-4978

sales@zykronix.com

http://www.zykronix.com/

Little Monsters single board computers, stackable PCMCIA/PCI, USB, and sound cards.

∿ DISTRIBUTOR/WHOLESALER

In the retail chain, a distributor is a company or other organization that purchases specialized product for resale, either to other distributors, retailers, or end users. Manufacturers often prefer working strictly through a distributor for cost reasons: it's cheaper to sell their product to a small handful of "go-betweeners" (other distributors or retailers) than thousands-if not tens of thousands-of end users.

Wholesalers serve a similar purpose as distributors, but the term is meant to be restricted to those who buy and sell in quantities. When buying through a distributor, you can often purchase in single quantities (though minimum order amounts and extra handling charges may apply). Wholesalers want you to buy in bulk, and they offer discounts to make that a more attractive option.

The companies listed in the Distributor/Wholesaler sections run the gamut from high-quantity wholesalers to what are in effect retail sales of individual components. Most of the distributors listed are for industrial electronics and may serve the needs of garage shop tinkers, schools, and companies.

It's important to note that the Internet is changing the way manufacturers view their distributor/customer relationships. Thanks to e-commerce and online buying, the costs of servicing customers are reduced greatly, allowing more and more manufacturers to deal directly with the buying public. In some cases, such direct sales are cheaper than going through a distributor; in most instances, buying online direct from the manufacturer is for convenience only.

∿ DISTRIBUTOR/WHOLESALER- INDUSTRIAL ELECTRONICS

Distributors and wholesalers of industrial electronics supply a much broader range of product than the average electronics retailer. The product may also be available in larger quantities (possibly representing a savings, if you need that many), with greater numbers of options and choices. Most of the companies listed in this section sell directly, either online or in retail establishments. Minimum order amounts may apply, so you'll want to save up your orders to avoid any additional fees.

Industrial electronics distributors and wholesalers are used to dealing with purchasing departments of companies, where X part is specified by the customer, and X part is shipped. For this reason, most distributors and wholesalers publish a "line card" showing either the products they carry or the brands they represent (or both). If online ordering is not offered, you will need to locate copies of catalogs from various manufacturers and order by part number.

SEE ALSO:

RETAIL-GENERAL ELECTRONICS: Resellers of electronic components

RETAIL-SURPLUS ELECTRONICS: Used or overstock electronics at considerable savings

---

Active Electronics 202870

One of the names for Future-Active, mail-order, and retail stores selling general electronics merchandise.

Active Electronics Components Depot— http://www.activestores.com/

Future Electronics— http://www.futureelectronics.com/

Future Active—http://www.future-active.com/

Future Global

Web site for Future Active, a business unit of Active Electronics.

All American Semiconductor, Inc. 203907

16115 NW 52nd Ave.
Miami, FL 33014
USA

(305) 621-8282

(305) 620-7831

 http://www.allamerican.com

60,000 electronic components from more than 75 suppliers.

Allied Electronics 202571

7410 Pebble Dr.
Fort Worth, TX 76118
USA

(817) 595-3500

(817) 595-6444

(800) 433-5700

http://www.alliedelec.com/

Allied Electronics is a prime source for all electronics (components, chemicals, tools, you name it), in single or multiple quantities. Their prices are often lower than the competition's. Do note the minimum order ($50 at the time of this writing). The online catalog is database driven and is searchable.

Datasheets are provided for many components, even fairly obscure semiconductors. You need Adobe Acrobat Reader to access the datasheets or catalog-page details.

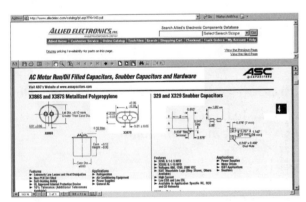

Allied Electronics Web site.

America II Corp. 203909

2600 118th Ave. North
St. Petersburg, FL 33716
USA

(727) 573-0900

(727) 572-9696

(800) 767-2637

sales@aiie.americaii.com

http://www.americaii.com/

Specializes in both prime (new) semiconductors and other electronics components, as well as excess inventory (surplus).

Appleton Electronic Distributors, Inc. 203603

205 W. Wisconsin Ave.
Appleton, WI 54911
USA

(920) 734-5767

(920) 734-5172

(800) 877-8919

sales@aedwis.com

http://www.aedwis.com/

Distributor of industrial and general electronics. Local store in Appleton, Wisc.

SEE ALSO:

http://www.marshelectronics.com/

AG Electrónica
http://www.agelectronica.com/
Electronics distributor in Mexico

Dalbani Corporation
http://www.dalbani.com/
General electronics and tools

Philmore & Datak
http://www.philmore-datak.com/
Wide range of active and passive components; connectors, cables, switches, tools, accessories; kits LED dice, RF transmitter, PIR movement detector

Polykom
http://www.polykom.com/
Industrial electronics supplier in Australia

Reptron Electronics, Inc.
http://www.reptron.com/
45,000 electronic component products from over 60 vendors

Arrow Electronics, Inc.   202872

25 Hub Dr.
Melville, NY 11747
USA

☎   (516) 391-1300
⊘   (877) 237-8621
✉   onlineservice@arrow.com
🌐   http://www.arrow.com/

Arrow distributes a full line of electronic components to industry. Products can be located by manufacturer, part number, or category. Sales offices are located worldwide. Before ordering, note the handling fee added to each shipment.

Avnet Inc. (Avnet Electronics)   203457

2211 South 47th St.
Phoenix, AZ 85034
USA

☎   (480) 643-2000
📠   (480) 643-7240
🌐   http://www.avnet.com/

Full-line distributor with local offices worldwide. Refer to the Web site for locations.

SEE ALSO:

Avnet Kent, interconnect, passive and electromechanical products:

http://www.avnetkent.com/

Avnet Cilicon, handling of semiconductors:

http://www.cilicon.com/

Avnet Tools and Test, test and measurement gear:

http://www.etoolsandtest.com/

B.T.W. Electronic Parts   202572

560 Denison Street, #2
Markham, ON
L3R 2M8
Canada

☎   (905) 479-0797
📠   (905) 479-3601
⊘   (800) 719-8948
✉   info@btw-electronics.com
🌐   http://www.btw-electronics.com/

General electronics; searchable product list by part number, description, and manufacturer.

📄   🌐 $ 💻

Digi-Key   202358

701 Brooks Ave. South
Thief River Falls, MN 56701
USA

☎   (218) 681-6674
📠   (218) 681-3380
⊘   (800) 344-4539
🌐   http://www.digikey.com/

Digi-Key is one of the largest mail-order retailers/distributors of electronic components in North America. See the listing in Retail-General Electronics.

📄   🌐   📦   💻

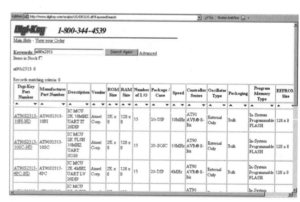

Digi-Key Web site, showing found components.

Electrocomponents plc   203910

International Management Centre
5000 Oxford Business Park South
Oxford
OX4 2BH
UK

☎   +44 (0) 1865 204000
📠   +44 (0) 1865 207400
✉   queries@electrocomponents.com
🌐   http://www.electrocomponents.com/

Electrocomponents has 27 operating companies-two in the U.K., 11 in the rest of Europe, and 14 in the rest of the world, including Japan and the U.S. In all, the company distributes some 300,000 products to over 1,500,000 technical and industrial professionals throughout the world.

Allied Electronics serves Canada and the U.S. The RS International Export Service can supply RS products anywhere in the world.

Electronic Depot Inc. 204236

1301 Buttercup Ct.
Lawrenceville, GA 30044-2113
USA

((770) 237-3088
📠 (770) 962-8812
⊘ (888) 453-2707
🌐 http://www.electronicdepotinc.com/

General electronics distributor: passive and active components, wire and cable, tools, soldering stations, fasteners, connectors; much of their line is geared toward electronics production.

Electronic Supply Center 203611

Skagit Whatcom Electronics
620 West Division
Mount Vernon, WA 98273
USA

((360) 336-3073
📠 (360) 336-5214
✉ sales@electronic-supply.com
🌐 http://www.electronic-supply.com/

General and specialty electronics. Product line includes miniature cameras, electronic kits, and electronic learning labs.

Future Electronics 202567

237 Hymus Blvd.
Pointe-Claire, Quebec
H9R 5C7
Canada

((514) 694-7710
📠 (514) 695-3707
🌐 http://www.futureelectronics.com/

Mondo catalog and retail sales of all kinds of electronic parts. Web site is in English and French.

Other Future Electronics Web sites:
http://www.future-active.com/

http://www.future-active.com/
http://www.activestores.com/

Hdb Electronics 202908

2860 Spring St.
Redwood City, CA 94063
USA

((650) 368-1388
📠 (650) 368-1347
⊘ (800) 287-9432
✉ info@hdbelectronics.com
🌐 http://www.hdbelectronics.com/

Full-line electronics: active and passive components, chemicals, transformers, switches, relays, solenoids, and dozens of other product groups. The Web site provides links to spec sheets on manufacturers' sites.

Inland Empire Components 203878

601-C Crane St.
Lake Elsinore, CA 92530-2722
USA

((909) 245-6555
📠 (909) 245-6556
⊘ (800) 566-5427
🌐 http://www.lookic.com/

Inland Empire is a general stocking distributor, principally for OEMs, and provides passive and active components, tools, optical sensors, specialty semiconductors (just from the As: Analog, Agilent, Allegro, Advanced Linear Devices, and many others). They provide a continually updated stock list in Excel, comma-delimited, and other formats.

Marsh Electronics Inc. 203874

1563 S. 101st
Milwaukee, WI 53214
USA

((414) 475-6000
📠 (414) 771-2847
🌐 http://www.marshelectronics.com/

Industrial electronics distributor. Local stores in Appleton, Wisc; Cleveland and Columbus, Ohio; and Indianapolis, Ind.

Marshall Electronics, Inc.   202474

1910 E. Maple Ave.
El Segundo, CA 90245
USA

  (310) 333-0606

  (310) 333-0688

  (800) 800-6608

✉ sales@mars-cam.com

🌐 http://www.mars-cam.com/

Specialty electronics with two divisions of possible interest to robot builders:

- Cable & Connectors Division
- Optical Systems Division

Marshall makes itty-bitty CMOS black-and-white and color imagers that are used extensively in other companies' products.

Newark Electronics   202500

4801 N. Ravenswood
Chicago, IL 60640
USA

  (773) 784-5100

  (800) 463-9275

🌐 http://www.newark.com/

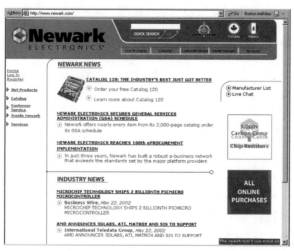

Newark Web site.

Prime-component distributor to business and industry; also caters to hobbyist market, but minimum orders may apply. Sells large quantities, when available. U.S. subsidiary of Premier Farnell, based in the U.K. Printed catalog also available on CD-ROM.

📄 🌐 $ 💻

NTE Electronics, Inc.   202627

44 Farrand St.
Bloomfield, NJ 07003
USA

  (973) 748-5089

  (973) 748-6224

✉ general@nteinc.com

🌐 http://www.nteinc.com/

NTE is a "master distributor" of semiconductors and other electronics components. They provide a cross-reference catalog to transistors and other semiconductors, available in print or searchable CD-ROM or on the Web site.

📄 🌐 🏭

Nu Horizons Electronics Corp.   203911

70 Maxess Rd.
Melville, NY 11747
USA

  (631) 396-5000

(631) 396-5050

(888) 747-6846

✉ sales@nuhorizons.com

🌐 http://www.nuhorizons.com/

Nu-Horizons is a full-line electronics distributor, offering the major brands-and some not-so-major. Source for Allegro, Exar, Winbond, and Oki Semiconductor. All of these have products worth the robot builder's attention.

🌐 💻

PartMiner Inc.   204050

80 Ruland Rd.
Melville, NY 11747
USA

(631) 501-2800

(800) 969-2000

✉ support@freetradezone.com

🌐 http://www.freetradezone.com/

PartMiner is a components distributor that also operates as a kind of gateway to researching components and obtaining technical data, best pricing, and purchasing information.

Pioneer-Standard Electronics, Inc. 203886

6065 Parkland Blvd.
Cleveland, OH 44124
USA

✆ (440) 720-8500
📠 (440) 720-8501
🌐 http://www.pioneerstandard.com/

Pioneer-Standard's Industrial Electronics Division is a full-line electronics distributor, specializing in electronic components in all quantities, including 1,000+ quantities for OEMs. Sales offices are located throughout North America, and you can buy by mail order.

Premier Farnell plc 203908

25/28 Old Burlington St.
London
W1S 3AN
UK

✆ +44 (0) 2078 514100
📠 +44 (0) 2078 514110
✉ information@premierfarnell.com
🌐 http://www.premierfarnell.co.uk

Parent company to Farnell, CPC, and Buck & Hickman in the U.K.; Newark Electronics and MCM in the U.S. See:

http://www.farnell.com/
http://www.cpc.co.uk/

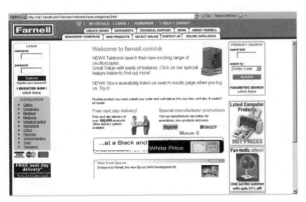

Web site of UK-based Farnell

http://www.buckhickman.co.uk/
http://www.newark.com/
http://www.mcmelectronics.com/

R & D Electronic Parts, Inc. 202909

370 Montague Expressway
Milpitas, CA 95035-6832
USA

✆ (408) 262-7144
🚫 (800) 675-1177
✉ info@randdelectronicparts.com
🌐 http://www.randdelectronicparts.com/

General electronics line, aimed primarily at electronics production in California's Silicon Valley, includes passive and active components, solder tools and solder rework stations, metal prototyping (K&S Engineering), chemicals, connectors, construction tools, and static control.

Rapid Electronics 203664

Severalls Lane
Colchester
Essex
CO4 5JS
UK

✆ +44 (0) 1206 751166
📠 +44 (0) 1206 751188
✉ sales@rapidelec.co.uk
🌐 http://www.rapidelec.co.uk/

Industrial electronics supplier. Products include DC-to-DC converters, crimp connectors, and some 26,000 other product lines from over 320 suppliers.

R.P. Electronics 203061

2060 Rosser Ave.
Burnaby, BC
V5C 5Y1
Canada

✆ (604) 738-6722
📠 (604) 738-3002
🚫 (888) 921-7770
✉ info@rpelectronics.com
🌐 http://www.rpelectronics.com/

R.P. Electronics is an electronics parts distributor stocking a wide selection of electronics components and test instruments. The company distributes locally and around the world (based in Burnaby, B.C., Canada).

Sager Electronics 203912

97 Libbey Industrial Pkwy.
Weymouth, MA 02189
USA

((781) 682-4844
🖰 (781) 682-4819
🚫 (800) 724-3780
🌐 http://www.sager.com/

Full-line electronics distributor: active and passive components, electromechanical (solenoids, relays), connectors, wire and cable, hardware for electronics, etc. Databooks available on the site (in Adobe Acrobat PDF format).

SAYAL Electronics 203620

1-3791 Victoria Park Ave.
Toronto, ON
M1W 3K6
Canada

((416) 494-8999
🖰 (416) 494-9721
✉ sales@sayal.com
🌐 http://www.sayal.com/

General electronics distributor. Locations in Canada and will ship worldwide.

Standard Supply Electronics 202574

3424 South Main St.
Salt Lake City, UT 84115
USA

((801) 486-3371
🖰 (801) 466-2362
🚫 (800) 453-7036
🌐 http://www.standardsupply.com/

Full-line electronics distributor.

Tri-State Electronics 204237

200 West Northwest Hwy.
Mount Prospect, IL 60056
USA

((847) 255-0600

Where to Get Stuff: Wholesalers and Manufacturers

Some wholesalers and manufacturers will sell direct to individual users. If the company has a Web site, their sales terms are usually indicated, and you can find out if they will sell to you.

- Wholesalers provide parts in quantity to industry, typically product made by other firms. They offer attractive discounts because they make up for the low prices with higher volume. Wholesalers seldom deal with individuals or in low quantities.

- Distributors sell smaller quantities to industry, schools, and sometimes individuals. Check with the companies near you and ask for their "terms of service."

- Some electronics manufacturers are willing to send samples of their products —some free, some at a small surcharge.

If you belong to a local robotics club or user's group, you may find it advantageous to go through the club to establish a relationship with a local electronics parts distributor. Buy in bulk to save.

📠 (847) 445-0896

✉ sales@tselectronic.com

🌐 http://www.tselectronic.com/

Distributor/online e-tailer of passive and active components (resistors, capacitors, transistors, ICs, etc.), batteries, relays, switches, solder and soldering stations, tools, and wire and cable.

TTI Inc. 203914

2441 Northeast Pkwy.
Fort Worth, TX 76106
USA

📞 (817) 740-9000

📠 (817) 740-9898

⊘ (800) 225-5884

✉ information@ttiinc.com

🌐 http://www.ttiinc.com/

Industrial products including Murata rotary position sensor, connectors, passives.

Vitel Electronics 203851

969 Derry Road East
Suite 110
Mississauga, ON
5T 2J7
Canada

📞 (905) 564-9720

📠 (905) 564-5719

🌐 http://vitelelectronics.com/

Canadian-based full-line electronics online/mail-order retailer. Retail stores across Canada.

∿ DISTRIBUTOR/WHOLESALER-OTHER COMPONENTS

Organizations in this section cater mainly to nonelectronic components (these are often referred to as passive components). This may include hardware used to build electronic devices, battery holders, wire, cabling, soldering equipment and supplies, mechanical repair parts (belts, gears, etc. for VCRs). I have also included in this section some specialty electronic components, such as high-end linear and rotary potentiometers, microphones, speakers, and electron tubes.

Note that some of the companies listed in this section are wholesalers and deal only with distributors or retailers. However, they make available their catalog, either online or in printed form, which you can study to determine the scope of what's available.

SEE ALSO:

RETAIL-GENERAL ELECTRONICS: Resellers of electronic components

RETAIL-SURPLUS ELECTRONICS: Save money with used or overstock components

---

Caltronics 202575

335 Maple Ave.
Torrance, CA 90503
USA

📞 (310) 787-0782

📠 (310) 787-1096

⊘ (800) 346-4936

✉ sales@caltronix.com

🌐 http://www.caltronix.com/

Caltronics is a master distributor/wholesaler of electronics accessories. Their line includes electronic com-

---

Ace Hobby Distributors, Inc.
http://www.acehobby.com/
R/C distributor/manufacturer

Air Electro, Inc.
http://www.airelectro.com/
Industrial connectors

CID Inc.
http://cidonline.com/
Industrial components, relays, connectors, enclosures, etc.

IQC International
http://www.iqc.co.uk/
Industrial connectors

Peters-de Laet
http://www.pdel.com/
Industrial components: connectors, relays, chemicals

ponents, AC adapters, connectors and sockets, cables, wire, chemicals, batteries, cable clamps and fasteners, tools, solder and soldering equipment, relays, and switches. The company sells wholesale to retailers and direct to consumers (at regular price).

Contact East, Inc. 202951

335 Willow St.
North Andover, MA 01845-5995
USA

  (978) 682-9844

  (800) 225-5370

  http://www.contacteast.com/

Contact East specializes in production-floor electronics supplies, including adhesives, tools, soldering stations and solder, tools, and test gear.

Electronics Parts Center 202902

1019 S. San Gabriel Blvd.
San Gabriel, CA 91776
USA

  (626) -286-3571

  (775) 257-1375

  (800) 501-9888

  http://www.electronicsic.com/

Specializes in replacement/service parts for electronics products (power supplies, monitors, TVs, you name it). Look up parts by part number or function. Includes mechanical VCR parts, such as rollers, gears, and belts. This is one way to get mechanical components for cheap, though the engineering selection is somewhat limited.

ElextronixOnline 203259

1 Herald Sq.
Fairborn, OH 45324
USA

  (937) 878-1828

  (937) 878-1972

  (800) 223-3205

  sales@electronix.com

  http://www.electronix.com/

Electronics parts, specializing in repair and replacement parts.

International Components Corporation 202880

175 Marcus Blvd.
Hauppauge, NY 11788
USA

  (631) 952-9595

  (631) 952-9597

  (800) 645-9154

  http://www.icc107.com/

Manufacturer of electronics parts, mainly microphones, speakers, and piezo elements. But also capacitors, electron tubes, cables, connectors, and accessories. Available through distributors.

International Hobbycraft Company 202135

2890 Dundee Rd.
Northbrook, IL 60062-2502
USA

  (847) 564-9945

  (847) 564-9951

  ihc@hobby-exporter.com

  http://www.hobby-exporter.com/

International Hobbycraft is an export trading company that represents hobby-related manufacturers for international export. The company represents many of the recognized brand names in the hobby business.

Keystone Electronics Corp. 203312

31-07 20th Rd.
Astoria, NY 11105
USA

  (718) 956-8900

  (718) 956-9040

  (800) 221-5510

  kec@keyelco.com

  http://www.keyelco.com/

Keystone sells electronics hardware-such items as assembly hardware, knobs, panel covers, standoffs, battery holders, clamps. Their products can be acquired through local retailers and distributors.

Mode Electronics Ltd. 203313

6830 Burlington Ave.
Burnaby, BC
V5J 4H1
Canada

((604) 435-6633

℘ (604) 435-8890

✉ info@mode-elec.com

🌐 http://www.mode-elec.com/

AC adapters, sockets, zillions of things. Serves all of North America. Catalog pages are in Adobe Acrobat PDF format.

RAF Electronic Hardware 203822

95 Silvermine Rd.
Seymore, CT 06483-3995
USA

((203) 888-2133

℘ (203) 888-9860

🌐 http://www.rafhdwe.com/

RAF sells hardware for use in electronics equipment: handles, spacers, and fasteners in both English and metric dimensions. Product listings are in English, French, and Spanish. Same stuff-just different words.

As an example of how these products might be used, metal or plastic standoffs can be used to separate the "decks" of a small robot that makes use of individual circuit boards or mounting plates, stacked one on top of the other. Standoffs are different from spacers; the latter tend to be quite short (under a quarter of an inch); standoffs are available in lengths up to several inches and come with or without threaded ends. Depending on your application, you can choose male or female threads. A standoff with one male and one female threaded end can be joined end to end to make a longer standoff.

The company offers a free sampler kit; their products are available from local distributors, which are listed on the Web site.

State Electronics, Inc. 203325

36 Rte. 10
East Hanover, NJ 07936
USA

((973) 887-2550

℘ (973) 887-1940

⊘ (800) 631-8083

🌐 http://www.potentiometers.com/

State Electronics makes and sells stock (off-the-shelf) and custom potentiometers, including precision linear pots. They carry rotary encoders and Hall-effect potentiometers. The product line includes Ohmite, Potter & Brumfield, and Clarostat. Minimum quantities usually apply when ordering online.

ELECTRONICS

This section encompasses sources and resources for electronic circuits, design, and production. Specialty components that enhance a robot with unique user interface features are located in this section as well. This includes LCD displays, which a robot can use to communicate with you, as well as sound input and output.

ELECTRONICS-CIRCUIT EXAMPLES

The Internet is a grand repository of circuits ideas and plans that you can use to construct your robot. This section details a number of useful circuits (typically in schematic form) that are well suited to robotics, including using comparator ICs, constructing interfaces to PC-compatible computers, and building your own microcontrollers.

SEE ALSO:

INTERNET-INFORMATIONAL: More sharing of electronics ideas

INTERNET-PLANS & GUIDES: Step-by-step tutorials for electronics/robotics construction

KITS-ELECTRONIC: Ready-made electronic kits you put together

ROBOTS-KIT & HOBBY: Robots in kit form; assembly required

Al's Robotics 203718

http://alsrobotics.botic.com/

Gallery of Al's 'bots, including EVO (BEAM robot), Psycho Mantis, and KTX1, a small PIC-based mobile robot. The site also includes a number of useful beginner tutorials on robotics and electronics:

- Tutorial for hacking/modifying servos for continuous motion
- General information on sensors
- LM339 Comparator IC

AVR-based Robot Hardware 204115

http://homepage.ntlworld.com/seanellis/avrrobot_hw.htm

Circuit and construction details on:

- Processor module
- Eyes
- Motor driver
- RS232 buffer
- Programming
- Laying out strip board with Eagle

BasicElectronics.com 203787

http://www.basicelectronics.com/

Information, tutorials, circuit examples, FAQ, and online calculators on various electronics topics. Calculators include:

- 555 timer calculator (astable or monostable)
- Resistor color calculator (written by my friend and fellow author Danny Goodman)
- Ohm calculator

Most of the calculators require a browser capable of running JavaScript.

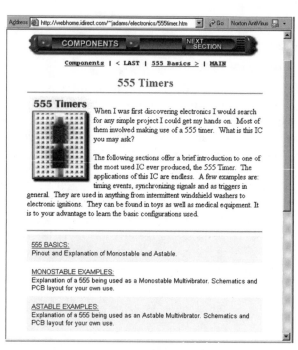

One of many tutorials at BasicElectronics.com

Bill's Homemade Electronics Emporium 202047

http://www.mnsi.net/~boucher/emporium.htm

"Information for making your own electronic projects at home." The basics, building your lab, using tools, creating printed circuit boards, project photos.

Bob Blick Technical Reference   202077

http://www.bobblick.com/

Home page of hardware designer Bob Blick, including a free "tech reference" section containing numerous circuits of interest to the robot builder. For example, the H-bridge page describes a circuit (with schematic) for a DC motor H-bridge of about 100 watts. Also:

- Servo pulse to PWM
- Servo pulse to dual H-bridge
- PIC programmer
- DTMF tone decoder
- LCD serial terminal
- 555 timer as an A/D converter

Boondog Automation   202212

c/o Paul Oh
3500 Powelton Ave.
Ste. B-402
Philadelphia, PA 19104
USA

paul@cs.columbia.edu
http://www.boondog.com

Products and tutorials centering on using the PC as a host for various automation tasks. Be sure to check out the following tutorials, written by Professor Paul Y. Oh:

- 8255 PC Interface Card (PCB artwork available)
- TRIPOD: Template for Real-Time Image Processing Development
- Controlling Devices over the Internet
- Wireless FM Transmitter
- Parallel Port Interface Box
- ADC/DAC PC Interface Card
- Quadrature Encoder Card
- Using IRQs: Hardware Interrupt Interfacing
- PC-based DC motor speed control
- 8254 Timer/Counter Card
- Infrared Emitter and Detector
- Long-Range Infrared Emitter and Detector
- Visual Basic (DLLs) and PC Interfacing
- DTMF Touch-Tone Generator and Decoder
- Hacking a Mouse for Encoders

Google Searches for Circuit Examples

You can use Google.com to find thousands of schematics that have been posted online. Some circuit examples even come with printed circuit board layouts. Of course, you can never be sure if the examples you find on some Web page have been tested. If in doubt, contact the author to determine if there are any known issues with the schematic

The basic search phrase is *schematic*; by placing it first in the search string, Google will base its returned hits with the highest relevancy to schematic diagrams. Here are some examples:

schematic motor h-bridge

schematic mosfet transistor h-bridge

schematic "infrared sensor"

schematic "dc-dc converter"

schematic adxl202

Also, don't forget to check the Web sites of semiconductor manufacturers. Many provide application notes with working schematics. See **Manufacturer-Semiconductors** for a list of several makers of semiconductors popular in robotics.

Bowden's Hobby Circuits 202932

http://ourworld.compuserve.com/homepages/
Bill_Bowden/

Site includes over 100 circuit diagrams, as well as links to related sites, commercial kits and projects, newsgroups, and educational areas.

Connecting a PC Keyboard to the BS2
203069

http://www.emf-design.com/bs2/reader.htm

Instructional article on how to connect a PS/2 keyboard to a Basic Stamp. Programming code included.

DesignNotes.com 202193

82 Walker Ln.
Newtown, PA 18940
USA

((215) 860-6867
🖶 (215) 860-8085
✉ info@designnotes.com
🌐 http://www.designnotes.com

Free interactive site aimed at electronics design engineers, programmers, and anyone interested in electronics. Features include:

- Designing for Dollars contest
- Design Notes Archives
- Design Forum
- Peer Review
- Designer's School
- Designer's Store (Design, Development, and Test Equipment-software, hardware)

DesignNotes.com Web site.

Droid Maker's Workshop 202130

http://www.geocities.com/droidmakr/

From robot enthusiast Clifford Boerema, robot-building help, especially for first timers (but some electronics skill or knowledge is handy). Some very nice semi-technical articles, including schematics.

FC's Solar Circuits 202043

http://www.solorb.com/elect/solarcirc/

Plans, circuit diagram, and description for a variety of solar energy projects, including an AA-size battery solar charger. Additional nonsolar projects include:

- Battery low-voltage beeper
- PWM DC motor speed control
- Seven-component-regulated LED lamp

Hans Wedemeyer's Projects, Code, and More 203705

http://hans-w.com/

Hans shows us various electronics projects, many of them with schematics and building details. Several are ideally suited to robotics:

- Seven-port RS-232 multiplexer
- Eight-channel 12-bit serial ADC
- Dial-up remote monitoring using serial ADC
- DC motor speed control
- Single-channel 24-bit USB ADC
- Bar code reader connection
- Humidity-measuring project
- CNC stepper motor driver

HMBOTS 203433

http://www.hmbots.homestead.com/

BEAM 'bots spoken here. The site provides a number of BEAM-ish circuits and BEAM robot construction ideas. There's also a very well illustrated guide-with closeup color pictures-to modifying an R/C servo for continuous rotation. One of the better guides I've seen.

How to Control a HD44780-based Character LCD
202156

http://home.iae.nl/users/pouweha/lcd/lcd.shtml

How to control an industry-standard character LCD panel. General info and code examples (for 8051 and PIC controllers).

Laurier's Handy Dandy Little Circuits
203049

http://members.shaw.ca/roma/

Includes projects like DC power supply and electronic train whistle, and also schematics for a countdown timer, signal function generator, motor controls, battery charger, and more.

Lud's Open Source Corner
204056

http://drolez.com/hardware/

How-to articles with code and programming examples on:

- Atmel AVR-Open Source Software: PWM/servo controllers with GPIO and serial interface
- Palm Cybot-Control a cybot from your PalmOS-based device and SmallBASIC

Mobot Building Info Pages
202214

http://www.mobots.com/makingMobots/

From Mobots.com; a small handful of useful tutorials on motor drive and motion control topics:

- PWM (Pulse-Width Modulation) DC Motor Speed Control with 555 Timer
- Stepper Motor Tester Using the AVR AT90S1200
- MCUs or Controller Boards?
- Quadrature Decoding Demo

Products and Information for the Student and Hobbyist in Electronics and Robotics
202930

http://www.hawkeselectronics.com/page_4.htm

Among the hands-on projects (descriptions, circuit examples, code examples), you'll find:

- 8051 microcontroller-based projects and software featuring The Zip 51 project board
- Robotics (including DTMF and remote control)
- Wireless
- PIC projects

RC Electronics Projects of Ken Hewitt
203392

http://www.welwyn.demon.co.uk/

Reprints from Ken's articles in the U.K. magazine Radio Control Models and Electronics, such as a motor speed controller V-tail mixer. Parts for some projects available for sale.

Robobix
203706

http://www.geocities.com/robobix/

Circuits include ultrasonic distance measurement, light reflection distance measurement, and simple infrared object detection.

Robot Projects
202066

http://www.robotprojects.com/

This site provides hands-on examples of a variety of interesting robotics projects, most of which revolve around using the OOPic microcontroller (the site is maintained by Scott Savage, the developer of the OOPic). Projects include:

- Racing Rover-Collision avoidance sensors on a high-speed robot
- Big-O-Trak-Retrofitting a Milton Bradley Big Trak with an OOPic

Retrofitting a Big Trak toy, one of many ideas at RobotProjects.com.

- WilbyWalker-CADD drawings and source code for a six-legged walker
- Contactless Angular Measurement-Measure the angle your robot is to a wall
- Recycling the sonar unit from a Polaroid camera
- Experiments with the SP0256 speech synthesizer
- Controlling 21 servos from your PC

Robotics Information and Articles 204120

http://www.leang.com/robotics/

Example 'bots and online articles (good ones) on such subjects as:

- Controlling servo motors with various microcontrollers
- RF serial communication for the MIT Handy Board
- H-bridge motor driver circuit
- Infrared proximity sensor

For Kam Leang's past and current robot projects, see also:

http://www.leang.com/robotics/

Serial LCD Interface Using AVR 90s2313 202155

http://members.tripod.com/Stelios_Cellar/AVR/SerialLCD/serial_lcd_interface_using_avr.htm

Tutorial and code examples (for the Atmel AVR controller) on interfacing the chip to an LCD panel using an Hitachi HD44780A LCD controller.

Small PC Board for the 68HC812A4 202253

http://www.rdrop.com/users/marvin/other/otherprj.htm

Circuit schematic and PCB layout for a 68HC812A4-based microcontroller board.

Steve Curtis: Robotics Experiments 202254

http://www.freenetpages.co.uk/hp/SteveGC/index.htm

Steve Curtis shares his robot designs, with schematics and programming examples.

Tomi Engdahl: Main Page 202838

http://www.hut.fi/~then/

Circuit examples, how-tos, tutorials, and more on many different aspects of electronics, with an emphasis on PC interfacing and architecture.

University of Alberta 202025

http://nyquist.ee.ualberta.ca/html/cookbook.html

Gateway page to the University of Alberta Circuit Cookbook archive.

Wayne Gramlich's Personal Projects 203792

http://gramlich.net/projects/index.html

Here, we learn that Wayne has far too many interests for even one lifetime, but somehow he finds time to play with all sorts of interesting things, including electronics and robotics. What I like about Wayne's home page is that it's very much a work-in-progress, regularly updated, a mix of finished with half-finished ideas, and presented for no other reason than to share with others.

This is not the usual hardware hacker's page: Wayne doesn't just show a picture of his projects with a short description. Instead, he treats us (at least for the more completed projects) with circuit diagrams, board layouts, even Gerber and Excellon PCB manufacturing files, should we wish to produce our own prototypes.

Be sure to check out Wayne's "RoboBricks"—a concept like LEGO Mindstorms, but consisting of more sophisticated interconnecting modules.

Wenzel Associates, Inc.: Technical Library 203268

http://www.wenzel.com/documents/library.html

Technical articles and tutorials. Including:

- Time and Frequency Circuits and Articles
- Crystal Oscillator Tutorial Articles

- Handy Spreadsheets-Calculate PLL response, phase noise under vibration

Wiresncode 202218

http://www.wiresncode.com/

Modular Ethernet interfaces for 8051 and HC11 controllers, LCD display module controller, Simmstick-compatible CS8900-based Ethernet interface.

 ## ELECTRONICS-CONNECTORS

This section details specialty connectors ideally suited for high-current applications in robotics.

SEE ALSO:

COMPETITIONS-ENTRANT: Photos and discussion of how combat robots are built

DISTRIBUTOR/WHOLESALER-OTHER COMPONENTS: Additional source for connectors

MOTOR CONTROL: Subsystems for motors

Anderson Power Products 203438

13 Pratt's Junction Rd.
P.O. Box 579
Sterling, MA 01564-0579
USA

((978) 422-3600
ℰ (978) 422-3700
🌐 http://www.andersonpower.com/

Power connectors of all sizes and colors. Connectors are available in single-pole or multipole versions (basically, one wire or many). Technical datasheets are provided in Adobe Acrobat PDF format.

Methode Electronics, Inc. 204007

7401 W. Wilson Ave.
Chicago, IL 60706
USA

((708) 867-6777
ℰ (708) 867-6999
🌐 http://www.methode.com/

Manufacturer of connectors (signal, power, RF) and heat-shrinkable tubing.

 ## ELECTRONICS-DISPLAY

Compact display units, typically liquid crystal display (LCD) panels, are used in robotics as a way for the machine to talk to its master. Many display modules for robotics use a serial communications scheme that requires only one or two I/O lines from the robot's microcontroller or computer. This conserves the robot's available I/O lines for other tasks.

SEE ALSO:

ELECTRONICS-CIRCUIT EXAMPLE: How to build your own serial LCD interface

KITS-ELECTRONIC: Additional LCD circuits

RETAIL-GENERAL ELECTRONICS: Source for small LCD panels

RETAIL-SURPLUS ELECTRONICS: Source for used LCD panels

Bitworks Inc. 202159

#1 Bitworks Way
Prairie Grove, AR 72753
USA

((501) 846-5777
ℰ (501) 846-5016
✉ lenny@bitworks.com
🌐 http://www.bitworks.com/

Bitworks produces display products, including:

- 320_240 LCD touch screen
- Scalable flat panel monitor card for DOS/VGA/SVGA/XGA
- Compact Flash to IDE adapter

Control a Serial LCD from a PIC 202147

http://www.mastincrosbie.com/mark/
electronics/pic/lcd.html

BiPOM Electronics,Inc.
http://www.bipom.com/
Products include serial LCD displays

Matrix Orbital
http://www.matrixorbital.com/
Serial interface displays

Sample code for interfacing a PIC to an LCD via a serial line. Requires a serial LCD, such as those sold by Scott Edwards Electronics.

CrystalFontz America Inc. 203459

15611 East Washington Rd.
Valleyford, WA 99036
USA

- (509) 291-3514
- (509) 291-3345
- (888) 206-9720
- sales@crystalfontz.com
- http://www.crystalfontz.com/

CrystalFontz makes LCD displays and drivers. Their products include the usual 1x16 and 2x16 line LCD, as well as models with graphical display, backlighting, and serial connections (most LCDs are for parallel connections, which takes up more I/O lines on your microcontroller or computer).

2x16 LCD display, from CrystalFontz. Photo Crystalfontz America, Inc.

Graphics LCD display, from CrystalFontz. . Photo Crystalfontz America, Inc.

Decade Engineering 203064

5504 Val View Drive, SE
Turner, OR 97392
USA

- (503) 743-3194
- (503) 743-2095
- info@decadenet.com
- http://www.decadenet.com/

Manufacture of low-cost OSD (on-screen display) modules and other video function blocks for use by system designers. Easily connected to microcontrollers or computers. The products serve as character overlay generators for video overlay and can be used on robots with analog video transmitters.

Designtech Engineering Co. 202363

2001 S. Blue Island Ave.
Chicago, IL 60608
USA

- (312) 243-4700
- (312) 243-4776
- http://www.designtechengineering.com/

Designtech makes graphical LCD panels with touch control.

Earth Computer Technologies, Inc. 202192

32701 Calle Perfecto
San Juan Capistrano, CA 92675
USA

- (949) 248-2333
- (949) 248-2392
- lcdking@earthlcd.com
- http://www.earthlcd.com

Makers of display electronics:

- LCD panels: text, graphics, industrial LCD monitors
- Low-cost LCD kits for popular single board computers
- LCD controllers
- Touch screen controllers

E-Lab Digital Engineering, Inc. 202490

Carefree Industrial Park
1600 N. 291 Hwy. Ste. 330
P.O. Box 520436
Independence, MO 64052-0436
USA

- (816) 257-9954
- (816) 257-9945
- support@elabinc.com
- http://www.elabinc.com/

E-Lab makes a series of building block ICs and modules that support various microcontrollers, including the Microchip PIC or Atmel AVR. Among their products are:

- Serial text LCD controller IC
- Octal seven-segment LED decoder
- Unipolar stepper motor controller
- Bipolar stepper motor controller
- Serial to parallel-printer IC

EMJ Embedded Systems — 203065

220 Chatham Business Dr.
Pittsboro, NC 27312
USA

((919) 545-2500

🖰 (919) 545-2559

🚫 (800) 548-2319

✉ emjembedded@emj.com

🌐 http://www.emjembedded.com/

Products include serial-interface LCD displays, PC/104 single board computers, embedded modems for PC/104, miniature hard drives, and Flash memory.

HDS Systems, Inc. — 203538

P.O. Box 42767
Tucson, AZ 85733
USA

((520) 325-3004

🚫 (877) 437-7978

✉ Sales@hdsSystems.com

🌐 http://www.hdssystems.com/

Makers of super-duper bright LEDs in LED arrays and housings.

Interfacing an LCD Display with the 8051 — 202931

http://www.hawkeselectronics.com/lcd.htm

How to marry a parallel LCD to an 8051 microcontroller.

LedVision Holding, Inc. — 202348

303 Sherman Ave.
Ackley, IA 50601
USA

((641) 847-3888

🖰 (641) 847-3889

✉ sales@ledvision.com

🌐 http://www.ledvision.com/

LedVision makes large LED displays, the kind used by retail shops and movie theaters.

Lumitex, Inc. — 203043

8443 Dow Cir.
Strongsvill, OH 44136
USA

((440) 243-8401

🖰 (440) 243-8402

🚫 (800) 969-5483

✉ info@lumitex.com

🌐 http://www.lumitex.com/

Lumitex designs, develops, and manufactures custom backlighting for LCDs, machine vision systems, and other applications. Some good technical notes and fact sheets.

RS232-interfaced LCD Display — 202161

http://www.geocities.com/mdurller/lcd.html

In the words of the Web site: "This is the schematic for the serial interface board. It consists of a MAX232, ATMEL AT90S2313 microcontroller and some support circuitry for the IR receiver module and LCD panel. The design goal of this project is to make a small, embedded RS232 TTY that has an alphanumeric LCD as monitor output and an IR remote control as the 'keyboard' input."

Scott Edwards Electronics, Inc. — 202179

1939 S. Frontage Rd. #F
Sierra Vista, AZ 85635
USA

((520) 459-4802

🖰 (520) 459-0623

info@seetron.com

 http://www.seetron.com/

Scott Edwards Electronics (otherwise known as Seetron) manufactures and sells serial LCD and VFD displays that easily interface to a computer or micro-controller. The company also offers the Mini SSC II interface to control up to eight R/C servos from a single serial connection.

Serial-interface LCD from Scott Edwards Electronics. Courtesy of Seetron.com.

⋀ ELECTRONICS-MISCELLANEOUS

Here, you'll find electronics odds and ends, including online calculators for electronics formulas, sources for electronics-related swap meets, and electronic pet containment systems that can be retrofitted to work with robotics.

SEE ALSO:

FESTS AND SHOWS: More electronics swap meets

INTERNET-REFERENCE AND INTERNET-SEARCH: Where to find additional electronics resources

ARRL: Database of Hamfests   203549

http://www.arrl.org/hamfests.html

Database of upcoming hamfests, swap meets set up and patronized by ham radio enthusiasts. But you'll find more than radio gear at most hamfests; many are good sources for general surplus electronics and surplus mechanicals, including motors, solenoids, relays, soldering stations, tools, and more. Hamfests are usually held on Saturdays, and many start early in the morning. To get the best deals, get there as early as possible, or else the good stuff will already be gone.

Electronics2000.com   204172

http://www.electronics2000.com/

Resource page for electronics information and calculators. The Web site also provides an online store for selling inexpensive CD-ROM compilations of small electronics-related utility programs.

Innotek, Inc.   203870

One Innoway
Garrett, IN 46738
USA

((219) 357-3148

📠 (219) 467-5102

⊘ (800) 826-5527

 support@innotek.net

🌐 http://pet.innotek.net/

Innotek manufactures a line of pet containment receivers and transmitters, suitable for dogs and wandering robots. On a robot, the transmitter could be used as a homing beacon or as an invisible fence to keep the robot from leaving an area.

PetSafe   203869

10427 Electric Ave.
Knoxville, TN 37932
USA

⊘ (800) 732-2677

📧 freedog@pet-containment.com

🌐 http://www.petsafe.net/

Makers of indoor pet containment systems. The "fence" is an electronic transmitter that's located at the exit point of the house. A lightweight receiver goes around the neck of your Golden Retriever; if your dog tries to escape, a warning signal is sounded on the receiver (the receiver worn by the retriever). Consider the uses in robotics.

Radio Fence Distributors, Inc.   203868

1133 Bal Harbor Blvd.
Ste. 1151
Punta Gorda, FL 33950
USA

Makers of electronic pet containment and training products. These devices are used to keep Bowser from leaving the yard. One application in robotics is to provide a radio "fence" for a security or lawn-mowing 'bot. Put the perimeter wire around an outside yard or inside room, and hack the collar receiver to interface with your robot's microcontroller. If the robot wanders near the perimeter wire, the receiver will signal its proximity.

Electronic Pet Fences—For Robots

You may have seen them advertised on TV or in one of those "everything for your home" catalogs. They're electronic pet containment fences, and they're made to keep a wandering dog inside your house or yard. The idea behind them is simple. When you know how they work, you'll instantly see their application for robotics.

A wire is placed around some perimeter—say, your back yard. Attached to the wire is a transmitter that produces a low-level radio-frequency signal. The receiver is attached to your dog's collar. As the dog approaches the perimeter wire, the signal radiating from it gets stronger. At a certain point, when your dog is right beside the perimeter wire, the signal is at its strongest. The collar receives this strong signal and produces either a tone and/or a mild shock as a means of controlling the dog's behavior. The tone isn't loud, and the shock isn't painful; rather, they are meant to help condition the dog that going beyond the perimeter is not allowed.

You can use a pet containment system for robotic control, as well. One potential use is for a robotic lawnmower. Place the perimeter wire at the edges of your lawn. Then retrofit the collar receiver so that instead of emitting a tone, or producing a shock, it sends a signal to the robot's microcontroller.

Another method leaves the collar receiver intact, but involves more experimentation on your part. A telephone pickup coil—the kind designed to attach to the handset of a phone to record a conversation—makes for a good induction pickup for most pet containment systems.

Consider that while electronic pet containment systems may be designed to prevent an animal from going beyond a perimeter, for robotics, they can be used to keep a robot within a specified zone. Merely reverse the logic condition of the robot's programming: Keep the robot near the perimeter, instead of avoiding it. In this way, the robot can follow an invisible electronic "guide fence," a kind of track that keeps the machine on course.

Pet containment fences can be purchased online (several are listed in the **Electronics-Miscellaneous** category), as well as from most pet supply stores and from a number of discount department stores such as Target or Wal-Mart.

IC Cabinet Kits

A number of electronics specialty retailers, such as Jameco and Digi-key, offer preselected kits of common TTL and CMOS integrated circuits. The kits include a labeled cabinet. Before buying any IC kits, be sure to do the math: TTL and CMOS chips are among the least expensive electronic components, so there's no need to pay more than you have to.

Figure the cabinet costs no more than $10. If the kit retails for $140 and contains 350 pieces, then you are paying an average of 37 cents per chip. That's not bad. But if the kit contains 200 parts, then the price per chip skyrockets to 65 cents each. You might as well make up your own IC selection. It'll be a lot cheaper.

Of course, you're not limited to using perimeter fences to keep a robot from crossing over. Just as easily, the fence can be used as a guiderail, keeping a robot on course as it rolls around the house or side yard. A side benefit of adapting an electronic pet containment fence to your robot is that, unlike your dog, your 'bot won't leave behind unpleasant little surprises for you to step in.

⌁ ELECTRONICS-OBSOLETE

Here, you'll find sources for obsolete, old, and outdated electronics components. "Old" and "outdated" are in the eyes of the beholder, and many of the sources listed here are dealers in surplus components that are still in production and use today. The listings that follow are good sources for hard-to-find parts that no one else seems to carry anymore. Most offer an online search database so you can quickly find parts you need.

SEE ALSO:

DISTRIBUTOR/WHOLESALER-INDUSTRIAL ELECTRONICS: Not everything they carry is new

RETAIL-SURPLUS ELECTRONICS: Additional sources for older components

America II Corp. 203909

2600 118th Ave. North
St. Petersburg, FL 33716
USA

☎ (727) 573-0900
📠 (727) 572-9696
🚫 (800) 767-2637
✉ sales@aiie.americaii.com
🌍 http://www.americaii.com/

Specializes in both prime (new) semiconductors and other electronics components, as well as excess inventory (surplus).

American Microsemiconductor 202905

P.O. Box 104
133 Kings Rd.
Madison, NJ 07940
USA

☎ (973) 377-9566
📠 (973) 377-3078
✉ info@americanmicrosemi.com
🌍 http://www.americanmicrosemi.com/

Scavenging Parts

Among the best sources for obsolete parts are garage sale discards. Here is just a short list of the electronic and mechanical items you'll want to be on the lookout for and the primary robot-building components they have inside:

- *VCRs* are perhaps the best single source for parts, and they are in plentiful supply. You'll find motors, switches, LEDs, cable harnesses, and IR receiver modules on many models.

Old VCRs are among the best sources for parts.

- *CD players* have optical systems you can gut out if your robot uses a specialty vision system. Inside you'll find a laser diode, focusing lenses, miniature multicell photodiode arrays, diffraction gratings, and beam splitters, plus microminiature motors and a precision leadscrew-positioning device.

- *Fax machines* contain numerous motors, gears, miniature leaf switches, and other mechanical parts.

- *Mice, printers, old scanners, disk drives, and other discarded computer peripherals* contain valuable optical and mechanical parts.

- *Mechanical toys*, especially those that are motorized, can be used either for parts or a robot base.

American Microsemiconductor specializes in obsolete and discontinued parts-diodes, transistors, triacs, SCRs, and integrated circuits. Many of their products are intended for quantity purchases by manufacturers who need a replacement source of old or discontinued parts, but smaller quantities are available as well.

Not all products carried by American Semiconductor are old or outdated or no longer needed. A good portion is actually still sold by other retailers and is still made by one or more manufacturers.

Also found on the Web site are some sweet and simple tutorials on common semiconductors: transistors, diodes, zeners, MOSFETs, and others.

Web site of American Microsemiconductor.

Lansdale Semiconductor

http://www.lansdale.com/

Obsolete ICs - "today's technology, tomorrow"

Metco Electronics

http://www.metcoelectronics.com/

Specializes in "vintage" components, for repair of antique electronics

Online Technology Exchange

http://www.onlinetechx.com/

Clearing house for obsolete, discontinued, and hard-to-find ICs

Midcom (UK) Ltd. 203428

Unit B
Rigby Close
Warwick
CV346TH
UK

☎ +44 (0) 1926 420772
📠 +44 (0) 1926 420095
🌐 http://www.midcom-uk.com/

Specializes in obsolete and military components: capacitors, connectors, diodes, resistors, semiconductors, etc.

Zygo Systems Limited/Chips2Ship 203429

The Old Forge
Nuneham Courtenay
Oxford
OX44 9NX
UK

☎ +44 (0) 1865 343951
✉ salesi@chips2ship.com
🌐 http://www.chips2ship.com/

Supplies obsolete ICs, transistors, diodes, and other components.

SEE ALSO:

http://www.factorydirect.co.uk/

Ⅶ ELECTRONICS-PCB-DESIGN

This section contains software for designing printed circuit boards (PCBs). The design process may include creating the schematic and layout out the PCB. The more advanced PCB layout tools generate industry standard files for automated manufacture of PCBs. Most of the products that follow are for professionals, or at least very serious amateurs, and their cost reflects this.

See also the following section, Electronics-PCB-Production.

Altium Limited/CircuitMaker 203722

12A Rodborough Rd.
Frenchs Forest NSW
2086
Australia

+61 2 9975 7710

+61 2 9975 7720

http://www.microcode.com/

Publishers of the CircuitMaker 2000 Virtual Electronics Lab software, allowing you to design, simulate, and out-put printed circuit board designs. Student versions and demos are available for download.

SEE ALSO:

http://www.protel.com/

Breadboard Your Circuits!

A typical electronics breadboard. They come in various sizes and styles.

Prior to soldering together any electronic circuit, you should test it with a *solderless breadboard*. This allows you to con-firm that circuit works, and it lets you experiment before you commit it to a permanent circuit board. Solderless bread-boards consist of a series of holes with internal contacts spaced one-tenth of an inch apart, just the right spacing for ICs. You plug in ICs, resistors, capacitors, transistors, and 20- or 22-gauge wire in the proper contact holes to create your circuit.

Solderless breadboards come in many sizes. For the most flexibility, get a double-width board, one that can accommodate at least 10 ICs. A typical double-width model is shown in the figure. Smaller boards can be used for simple projects; circuits with a high number of components require bigger boards. While you're buying a breadboard, purchase a set of prestripped wires. The wires come in a variety of lengths and are already stripped and bent for use in breadboards. The set costs $5 to $7, but you can bet the price is well worth it.

Tips to Reduce Static

Static shock—like the kind you get after scuffling your feet on the carpet and touching a doorknob—can destroy sensitive electronic circuits. Poof! There goes weeks of work on your latest robot. Here are some simple steps to minimize static.

- *Wear low-static clothing and shoes.* Wear natural fabrics, such as cotton or wool. Avoid wearing polyester and acetate clothing.

- *Use an antistatic wrist strap.* The wrist strap grounds you at all times and prevents static buildup.

- *Before touching the robot, discharge the static.* Go ahead: Touch the doorknob (or other metal) before picking up the robot.

- *Avoid playing with your robots on nylon carpets.* Nylon builds up a static charge in most everything. Wood, title, or linoleum floors are better. So are polypropylene (Olefin) car-pets, which don't build up as much static.

- *If possible, avoid the use of acrylic plastic for the body of the robot.* Of all plastics, acrylics tend to build up the most static charge. If you want to use plastic, opt for ABS, PVC, or styrene.

- *During construction of the robot, ground your soldering iron.* A grounded iron not only helps prevent damage from electrostatic discharge, but it lessens the chance of a bad shock, should you accidentally touch a live wire.

- *Use component sockets.* When building your projects that use ICs, install sockets first. This reduces the chance of damaging static when soldering, but it also makes it easier to replace zapped ICs!

Altium Limited / Protel 202839

12A Rodborough Rd
Frenchs Forest NSW
2086
Australia

☎ +61 2 9975 7710
🖷 +61 2 9975 7720
🌐 http://www.protel.com/

Protel is high-end printed circuit board design software. Intended for the electronics professional, the software supports numerous advanced features: multi-sheet, hierarchical schematic entry; mixed-mode simulation; PLD design; rules-driven board layout; shape-based autorouting; signal integrity simulation; integrated document management; and design team collaboration.

SEE ALSO:

http://www.altium.com/

CadSoft Computer GmbH 202527

Hofmark 2
D-84568 Pleiskirchen
Germany

☎ +49 (0) 8635 6989-10
🖷 +49 (0) 8635 6989-40
✉ sales@cadsoft.de
🌐 http://www.cadsoft.de/

Home page for EAGLE Layout Editor. Freeware "lite" version available. Worldwide distributors.

Douglas Electronics Inc. 203655

2777 Alvarado St.
San Leandro, CA 94577
USA

☎ (510) 483-8770
🖷 (510) 483-6453
✉ info@douglas.com
🌐 http://www.douglas.com/

Cadence Design Systems, Inc.
http://www.orcad.com/
Makers of Orcad PCB layout software

Pro-level PCB design software for Macintosh and Windows.

The company also sells stock breadboards in various shapes and sizes (such as PC/104 and PC PCI), as well as Miniboard and Handy Board PCBs.

ECD, Inc./PCBexpress 202205

13626 S. Freeman Rd.
Mulino, OR 97042
USA

☎ (503) 829-9108
🖷 (503) 829-5482
🌐 http://www.pcbexpress.com/

ECD (Electronic Controls Design) provides PCB production in small quantities or large.

PCBexpress is designed to give you the lowest prices for your small-quantity circuit board needs. Options are limited to facilitate the short production runs.

SEE ALSO PCBPRO:

http://www.pcbpro.com/

ECD, Inc./PCBpro 202203

13626 S. Freeman Rd.
Mulino, OR 97042
USA

☎ (503) 829-9108
🖷 (503) 829-5482
🌐 http://www.pcbpro.com

ECD (Electronic Controls Design) provides PCB production in small quantities or large.

PCBpro offers large quantities of boards and a variety of lead times suited for production runs. Options are available for board thickness, copper weight, number of layers, gold fingers, and testing.

SEE ALSO PCBEXPRESS:

http://www.pcbexress.com/

ExpressPCB/Engineering Express 202356

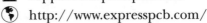
✉ support@expresspcb.com
🌐 http://www.expresspcb.com/

Custom (and proprietary) PCB layout software for creating short-run PCB boards through ExpressPCB service. Free downloadable PCB layout program for Windows PC.

Note two different entities: ExpressPCB and PCB Express.

Ivex Design International, Inc.   202303

P.O. Box 7156
Beaverton, OR 97007
USA

☏ (503) 848-6520
📠 (503) 848-7552
✉ info@ivex.com
🌍 http://www.ivex.com/

Makers of the WinDraft, WinBoard, Ivex Spice circuit board software. Together these programs are used to design the schematic, test it, and produce a circuit board layout. The board layout uses industry standard output files so that you can have it machine produced.

ELECTRONICS-PCB-PRODUCTION

This section contains sources for companies that produce printed circuit boards (PCBs) using standard automation files you provide to them. These automation files are prepared using compatible PCB design software (see the previous section, Electronics-PCB-Design). The majority of the companies listed here will produce short runs of PCBs-two or three boards-for a minimum setup fee. The more boards you order, the less each board costs, because it is the setup fee that represents the single largest cost in short-run PCB production.

Santa Barbara PCB

http://www.916design.com/
Custom CAD library and layout services for the electronics community

Techniks, Inc.

http://www.techniks.com/
Press n Peel PCB transfer film

Also listed here are materials for producing your own PCBs in your workshop, typically using a laser or ink-jet printer to prepare PCB artwork on transfer film.

AP Circuits   204182

Unit 3, 1112- 40th Ave. NE
Calgary, AB
T2E 5T8
Canada

☏ (403) 250-3406
📠 (403) 250-3465
✉ staff@apcircuits.com
🌍 http://www.apcircuits.com/

Printed circuit boards by mail. Small quantities or large.

Basic Soldering Guide, The   203184

http://www.epemag.wimborne.co.uk/solderfaq.htm

Guide to soldering: how to solder, types of iron, desoldering, troubleshooting.

Cimarron Technology, Inc.   204200

1611 S. Utica
PMB 276
Tulsa, OK 74104
USA

☏ (918) 747-3874
📠 (918) 747-4849
✉ browndog@cimarrontechnology.com
🌍 http://www.cimarrontechnology.com/

Cimarron sells a line of surface mount adapters for prototyping and breadboarding. You solder the surface mount IC or other component on the adapter board, then plug the board into a standard breadboard. Products include:

- SO8 to 8-pin DIP adapter
- SOT-23 to 6-pin DIP adapter
- Narrow/wide SO to 28-pin DIP adapter

The company also sells some unique "ProtBlock" premade circuits mounted on miniature boards. The (surface-mounted) components are not included, so that you can design your own circuit using the parts and values that best fit your needs. One interesting product is the amplifier module using a common 8-pin op-amp IC.

Dealing with Failures

Failure design is a type of engineering that assumes nondestructive errors in a system that self-corrects for problems. In many cases, the specific nature of the problems is not known ahead of time. Software programs that "catch" unknown errors before they cause early termination of a program are a good example of simple failure design.

Knowing that things will always go wrong can help in reviewing the problems inherent in robotics. Failure is common in all forms of engineering, and building robots is no exception. As these failures occur, you'll learn from them, and you'll build better robots.

Robots can fail in any number of ways: mechanical, electrical, or programming.

Mechanical Failure

You already know that mechanical problems are when something breaks or falls off the robot. But they're also design faults, where a robot is engineered or constructed in such a way that it causes problems. A robot that's top-heavy is one example: If the center of gravity is too high, the robot may tip over easily.

The quality of construction and the sturdiness of the materials you use determine how readily a robot will "come apart at the seams." Obviously, a robot made with cardboard and electrical tape will not last as long as a robot made with wood, plastic, or metal, and that has proper fasteners to hold everything together.

You can use the "pull test" to determine if your robot construction methods are sound. Once you have attached something to your robot—using glue, nuts and bolts, or whatever system—give it a good tug. If it comes off, the construction isn't good enough. Look for a better way.

Electrical Failure

Electronic failures are not as obvious as mechanical ones. Unless your circuit has gone up in smoke, it may not be readily apparent what's wrong with it. Too, circuits that functioned properly in a solderless breadboard during the early design phase may no longer work once you've soldered the components in a permanent circuit, and vice versa. There are many reasons this can happen, including mistakes in wiring, odd capacitive effects, even variations in tolerances due to heat transfer.

- *Circuit never worked.* Review your wiring and make necessary repairs. Double-check the accuracy of the schematic or drawing you're working from. Is there a typo?
- *Circuit worked before, and now doesn't.* Look for a short circuit or a broken wire. Try working backward to find the problem. For example, apply power directly to the motors. If the motors work, you know they're okay, so try next to manually control (using jumper wires) the H-bridge motor driver circuit you're using. If the motors turn, then the H-bridge is fine, and it's time to look for problems in the microcontroller or computer section.

Other electrical problems may be caused by errors in programming (see the following section); sensors that are stuck, dirty, or simply unreliable (this is the case with many proximity sensors using infrared); or weak batteries. Always try a new or freshly charged set of batteries before you tear your robot apart. Never assume batteries are okay. Test them under load (that is, connect them up, then take a meter reading).

Programming Failure

More and more robots use computers or microcontrollers as central processing units, and these are operated by software you give them. If the software contains programming bugs, the robot may not work at all or it may work only erratically. There are three basic kinds of programming "bugs"; in all cases, the fix is to review the program, find the error, and revise the code.

- *Compile bugs* are caused by bad syntax. Look for misspelled commands, missing characters (end-of-line semicolons in a C program are a good example), or bad formatting. Programming languages that compile your code will usually flag compile time bugs, but may still let you execute the program on your robot. Be sure to first fix the problem before you transfer the program to the robot's microcontroller or computer.

- *Run-time bugs* are caused by a disallowed condition. A run-time bug isn't caught by the compiler. Rather, it occurs when the microcontroller/computer attempts to run the program. A typical run-time bug is using a data type that is too small to hold the value expected of it. Example: A byte data type holds values up to 255; a run-time bug may occur if you attempt to store a value of 256 (in which case you'd chose a word data type, which can hold numbers to 65,535).

- *Logic bugs* are when a program doesn't work the way you expected it to. These are caused by improper math, incorrect assumptions of how sensors interact with the environment, or any other conditions based on false information.

Dyna Art 203789

1947 Sandalwood Pl.
Clearwater, FL 33760-1713
USA

- (727) 524-1500
- (727) 524-1225
- mail@dynaart.com
- http://www.dynaart.com/

Makes and sells a PCB transfer system; uses laminator-like machine to fuse artwork (prepared by laser printer) onto copper clad.

Easy Printed Circuit Board Fabrication 203963

http://www.fullnet.com/u/tomg/gooteepc.htm

Well-researched and detailed article on using a laser printer and standard white paper to create transfer sheets for making your own printed circuit boards.

ECD, Inc./PCBexpress 202205

13626 S. Freeman Rd.
Mulino, OR 97042
USA

- (503) 829-9108
- (503) 829-5482
- http://www.pcbexpress.com/

ECDElectronic Controls Design) provides PCB production in small quantities or large.

PCBexpress is designed to give you the lowest prices for your small-quantity circuit board needs. Options are limited to facilitate the short production runs.

SEE ALSO PCBPRO:

http://www.pcbpro.com/

ECD, Inc./PCBPro 202203

13626 S. Freeman Rd.
Mulino, OR 97042
USA

- (503) 829-9108
- (503) 829-5482
- http://www.pcbpro.com

ECD (Electronic Controls Design) provides PCB production in small quantities or large.

PCBpro offers large quantities of boards and a variety of lead times suited for production runs. Options are available for board thickness, copper weight, number of layers, gold fingers, and testing.

SEE ALSO PCBEXPRESS:

http://www.pcbexress.com/

ExpressPCB / Engineering Express 202356

✉ support@expresspcb.com

🌍 http://www.expresspcb.com/

Custom (and proprietary) PCB layout software for creating short-run PCB boards through ExpressPCB service. Free downloadable PCB layout program for Windows PC.

Note two different entities: ExpressPCB and PCB express.

PCBexpress 202349

🌍 http://www.pcbexpress.com/

See ECD (this section).

PCB-Pool/Beta LAYOUT GmbH 203661

Feldstraße 2
D-65326 Aarbergen
Germany

☎ +49 (0) 6120 907010

📠 +49 (0) 6120 907014

✉ verkauf@pcb-pool.com

🌍 http://www.pcb-pool.com

PCB layout service, offering low-cost prototypes. The service combines jobs, based on size and type, to save money. A number of popular PCB layout software formats are supported, including Eagle, Protel, and Orcad.

Web site is in German and English. Offices also in Ireland.

Printed Circuit Board Fabricators 203989

http://www.pcbfab.com/

A handy reference with advice and insights on the process for producing a printed circuit board. Sponsored by RD Chemical.

RS-274X Gerber Format 202204

http://www.barco.com/ets/data/rs274xc.pdf

Adobe Acrobat PDF file that explains the RS-274X Gerber Format. Fairly technical, with over 50 pages of info.

Think & Tinker, Ltd. 202992

P.O. Box 1606
Palmer Lake, CO 80133
USA

☎ (719) 488-9640

📠 (719) 481-0464

🚫 (800) 392-4941

🌍 http://www.thinktink.com/

Products for printed circuit board production, including dry film lamination, film imaging, PCB drills, bubble tanks, and tinning.

Vector Electronics and Technology, Inc. 202625

11115 Vanowen St.
N. Hollywood, CA 91605
USA

☎ (818) 985-8208

📠 (800) 423-5659

🚫 (800) 423-5659

✉ inquire@vectorelect.com

🌍 http://www.vectorelect.com/

Maker of printed circuit boards and wire wrapping and point-to-point prototyping products, including Slit-n-Wrap wire-wrapping tool. Available through distributors.

⋀ ELECTRONICS-SOLDERING

Sources for soldering stations and supplies, particularly those used for production (such as surface mount and rework and repair), are listed in this section. Some companies listed here do not sell directly to consumers, but may provide technical information about their products on their Web sites and may offer a replacement parts service.

SEE ALSO:

DISTRIBUTOR/WHOLESALER SECTIONS:
Additional sources for solder and tools

TEST AND MEASUREMENT: Additional sources for soldering stations

TOOLS: Soldering stations

Computronics Corporation Ltd.   203662

Locked Bag 20
Bentley
Western Australia 6983
Australia

(+61 8 9470 1177

✆ +61 8 9470 2844

✉ kdare@computronics.com.au

🌎 http://www.computronics.com.au

Industrial electronics: electronic displays, tools, and components. Includes soldering stations, high-brightness LEDs, chemicals, RF transmitter and receiver modules.

Cooper Industries, Inc.   202618

3535 Glenwood Ave.
Raleigh, NC 27612
USA

((919) 781-7200

✆ (919) 783-2116

🌎 http://www.coopertools.com/

Weller solder tools, wire-wrapping prototyping products, Xcelite hand tools.

Weller WES50, from Cooper Tools.
Photo Courtesy of Cooper Tools.

Elenco Electronics   202139

150 W. Carpenter Ave.
Wheeling, IL 60090
USA

((847) 541-3800

✆ (847) 520-0085

✉ elenco@elenco.com

🌎 http://www.elenco.com/

From the Web site: "Elenco is a major supplier of electronic test equipment and educational material to many of the nation's schools and hobbyists. We also have a network of distributors selling our products from coast to coast and abroad."

The company sells electronic kits under the Amerikit brand.

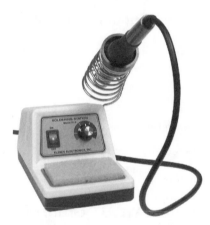

Elenco's SL-5 soldering station.
Photo Elenco Electronics Inc.

Techni-Tool, Inc.   202605

1547 N. Trooper Rd.
P.O. Box 1117
Worcester, PA 19490-1117
USA

🚫 (800) 832-4866

✉ sales@techni-tool.com

🌎 http://www.techni-tool.com/

Techni-Tool sells lubricators, testers, hand tools, and soldering stations, among other products.

Wahl Clipper Corp.   202619

2900 Locust St.
Sterling, IL 61081
USA

((815) 625-6525

🌎 http://www.iso-tip.com/

In addition to Trim N Vac hair clippers, personal massagers, and other household appliances, Wahl makes

---

Kester Solder

http://www.kester.com/

Manufacturer of solder; see the Web site for spec sheets, materials safety documents, and information about lead-free solder

Midwest Tech Services, Inc.

http://www.midwesttech.com/

Tools and supplies for electronics production

---

soldering irons and soldering tools for industry. Sold through distributors.

WASSCO   202568

12778 Brookprinter Pl.
Poway, CA 92064
USA

((858) 679-8787
♥ (858) 679-8909
⊘ (800) 492-7726
✉ sales@wassco.com
🌐 http://www.wassco.com/

WASSCO distributes production soldering materials and supplies. Offered are solder, cleaning chemicals, abrasives, soldering tools, static control products, test and measurement gear, and hand tools.

∿ ELECTRONICS-SOUND & MUSIC

This section is an eclectic mix of products and services for adding sound to your robot. The sound may be music or musical effects, recorded voice, speech synthesis, or even speech recognition.

American Musical Supply Inc.   202616

P.O. Box 152
Spicer, MN 56288
USA

((320) 796-2088
♥ (320) 796-2080
⊘ (800) 458-4076

🌐 http://www.americanmusical.com/

Musical instruments and amplifiers; sound effects.

Angela Instruments   203286

10830 Guilford Rd.
Ste. 309
Annapolis Junction, MD 20701
USA

((301) 725-0451
♥ (301) 725-8823
✉ steve@angela.com
🌐 http://www.angela.com/

Electronics (new, surplus), with emphasis on musical instruments. Also sells microphones, tubes, capacitors, and variacs. Consider some of the guitar and musical effects products for use in adding unusual prerecorded sounds and voices to your robot.

AT&T Labs Natural Voices
Text-to-Speech Engine   203079

http://www.naturalvoices.att.com/

AT&T's Natural Voices software is among the best-sounding text-to-speech synthesis ever developed. You may not even be aware that it's a computer speaking to you! Variations include male and female voices, and vocabularies (with the proper accents) are available for English, neutral Spanish, and German. These vocabularies are implemented as "fonts," and at the time of this writing, additional voice fonts for British English, Parisian French, Castilian Spanish, and others were on the way.

The software, which runs under Windows, Linux, and various flavors of Unix, is intended for interactive applications such as speech-enabled Web pages. However, the company (smartly) offers the software at very low cost for desktop use, and that would include using it to provide your robot with a voice. If your robot operates from a microcontroller, you can always record little snips of voices for playback with electronic voice chips.

Atmel: Dream Sound Synthesis
Datasheet   203327

http://www.atmel.com/atmel/products/prod209.htm

Datasheet for Atmel's Dream sound chips; check out SAM9773, a single-chip music synthesizer with effects. It

Making Your Robot Sound Off

Robots shouldn't be mute. Even basic sounds and sound effects can be accomplished using relatively inexpensive sound-recording chips. One of the best known, and most common, is the ChipCorder line of digital sound recorders, now manufactured by Winbond. (You may know them from their previous owner, Information Storage Devices, or ISD). The ChipCorder ICs are readily available to the electronics hobbyist and amateur robot builder via a number of mail-order retailers, such as Digi-Key and Jameco.

ChipCorders vary in duration of sound they can record. Versions are available that record from 30 seconds to 2 minutes of sound. The more sophisticated chips can be controlled via a microprocessor or microcontroller so that individual segments of sound can be individually played back. In this way, you can have (say) a minute of recorded sound, but can play back discrete portions to produce various sound effects or vocal phrases.

Prices for the chips vary depending on feature and recording time, but most cost under $15. While there are certainly other makers of sound storage/playback integrated circuits, the ChipCorder chips are by far the most widely used and among the most affordable.

A rich assortment of datasheets and application notes is available for the ChipCorder products on the Winbond Web page at ***http://www.winbond-usa.com/*** (Winbond is based in Taiwan, and this is their U.S. Web site; however, this site contains most of the ChipCorder downloadable information).

Quadravox (***http://www.quadravox.com/***) provides value-added circuits based on the ChipCorder line. One product is a ChipCorder prototyping board, the QV400D, that lets you experiment with different ChipCorder variations. It comes with everything you need, including power supply and support software, to create, edit, and implement recorded sounds on the ChipCorder IC.

Voice Synthesis Text-to-Speech Internet Sites

Not long ago, integrated circuits for the reproduction of human-sounding speech were fairly common. With the proliferation of digitized recorded speech, however, unlimited speech synthesizers have become an exception instead of the rule. The companies that made stand-alone speech synthesizer chips either stopped their manufacture or were themselves sold to other firms that no longer carry the old speech parts.

If you plan on using recorded (mechanical or electronic) sound with your robot, you may want to consider any of the several text-to-speech Internet sites, such as the Bell Labs TTS (Text-to-Speech) project. Software running on the Internet server lets you type in the text you want to synthesize. A sound file (WAV, AU, or AIFF format) is returned to you. Save the file and use it for producing a sound sample with a cassette tape or recording chip.

There are several interactive online text-to-speech processing sites, including the following:

AT&T Labs Natural Voices Text-to-Speech Engine

http://www.naturalvoices.att.com/demos/

http://www.research.att.com/projects/tts/

Bell Laboratories (Lucent Technologies) TTS system

http://www.bell-labs.com/project/tts/

Speech Technology Group: Text-to-Speech Demo for Spanish

http://www.gth.die.upm.es/research/synthesis/synth-form-concat.html

is programmed serially, meaning that it's ideally suited for use with microcontrollers and robotics, where input/output pins are always in short supply. The chip is surface mount, however; so you'll need to make a circuit board if you want to breadboard.

See also the datasheets for the SAM9707 "Integrated Sound Studio" and the SAM9743 "Single-Chip Music System."

Bell Laboratories (Lucent Technologies) TTS System   203757

http://www.bell-labs.com/project/tts/

Interactive test site for experimenting with text-to-speech. Type in text, specify how you want it to sound (man, woman, big man, child, even a gnat!), and the spoken speech is returned as an audio clip. You can save the audio clips to your computer. If your robo contains a digital sound recording chip, like those from ISD, you can transfer the recorded voice segments to the chip, and your creation will now have a voice.

Can be slow at times, but worth it.

Big Briar, Inc.   202202

554-C Riverside Dr.
Asheville, NC 28801
USA

((828) 251-0090
✆ (828) 254-6233
⊘ (800) 948-1990
✉ info@bigbriar.com
🌐 http://www.bigbriar.com/

Home of music synthesizer guru Bob Moog. According to the Web site, "We are a community of musicians, business professionals, and technicians who work together to bring you some of the finest electronic music gear on today's market." The synthesizer products are useful in creating recorded sound effects for your robot.

Carvin Guitars   202615

12340 World Trade Dr.
San Diego, CA 92128
USA

((858) 487-1600
✆ (858) 487-8160

⊘ (800) 854-2235
🌐 http://www.carvin.com/

Build a guitar bot, or . . . I was thinking of the musical products (amplifiers, recorders, etc.) for producing sound effects or background music for your robotic creations.

Fonix Corporation   203760

180 W Election Rd.
Draper, UT 84020
USA

((801) 553-6600
✆ (801) 553-6707
🌐 http://www.acuvoice.com/

Speech synthesis for Web pages. They provide a service called i-Speak, which is voice synthesis for e-mail, Web pages.

SEE ALSO:

http://www.speakthis.com/

Gyrofrog: The Theremin   202713

http://www.gyrofrog.com/theremin.html

Information about the Theremin electronic music device.

PAiA Electronics, Inc.   202201

3200 Teakwood Ln.
Edmond, OK 73013
USA

((405) 340-6378
✆ (405) 340-6300
✉ sales@paia.com
🌐 http://www.paia.com/

PAiA Electronics is a well-respected leader in hobby electronics kits for the musician with big dreams, but little bread. They provide low-cost, high-quality, user-assembled kits of innovative electronic products. You can adapt many of their products to robotics, especially if you're into sound and music effects. Some of the kits that you might find useful include:

- Modular Synthesizer
- Theremax Theremin/Gestural Controller

In Focus: PAiA Electronics

PAiA makes low-cost kits for those into making electronic music. Several of their kits can be used for creative robotic projects. Let's look more closely at two of the products, the theremin and the vocoder.

A theremin (named after Leon Theremin, its inventor) is an electronic instrument that uses capacitance to change the pitch and volume of sound. You've heard it if you've ever watched an old 1950s science fiction movie; theremins were used constantly for sound effects for such things as flying saucers and robots, and for the eerie background music. Consider how a theremin works: by body capacitance. As you move your hand (or other body part) closer to or further away from a metal plate on the theremin, the pitch changes accordingly.

Modern theremins, like PAiA's Theremax, produce complex waveforms as their output, so they sound like music. But you can use the same principles of the theremin, simplified, to create a robot sensor that detects proximity to people, animals, and many other objects. The theremin circuits would detect relative closeness, but also the rate of change in proximity.

The vocoder is another electronic device from another era that is still in popular use (several techno and Euro pop groups actively use the vocoder in song after song). The device was originally created to compress human speech into a smaller frequency range, without sacrificing intelligibility. (In fact, the concept behind the vocoder, which was developed in the late 1930s, is still in use today for telephone communications.) In a vocoder for music and sound effects, a person's voice is modulated by another tone—a buzz or a rasp is common—to produce something very robotic or mechanical.

Your robot may be equipped with a solid state recorded sound chip, where you can record snippets of your voice for automated playback. But because it's your voice, it doesn't much sound like a robot—it sounds like you. By processing your voice through a decoder (PAiA's is an inexpensive analog unit and is quite capable), you can transform your voice to that of an automaton.

- FatMan Analog MIDI Synth
- Vocoder
- TubeHead Vacuum Tube Preamp
- Headphone Distribution Amplifier
- Stereo Compressor
- MIDI Drum Brain

Phonemes and Allophones   202835

http://www.umanitoba.ca/faculties/arts/linguistics/russell/138/feb3/allophon.htm

Informational page about the linguistic parts of speech-what makes up the sounds we say. Phonemes and allophones are used in computerized speech generation.

Quadravox, Inc.   202182

1701 N.Greenville Ave.
Ste. 608
Richardson, TX 75081
USA

((972) 669-4002

(972) 437-6382

(800) 779-1909

info@quadravox.com

http://www.quadravox.com/

Add sound to your robots. Quadravox offers a number of application modules, including:

- QV531-MP3 player module
- QV301-Parallel control sound module
- QV306-RS232-controlled sound module

Check out their sound module prototyping boards, designed to work with the Winbond ISD series of digital

Model QV306m4, from Quadravox. Photo Quadravox, Inc.

sound recorders. In fact, much of the company's product line is based on the ISD series of digital recording chips.

SEE ALSO QUADRAVOX TEXT-TO-SPEECH SYNTHESIS WHITE PAPER:

http://www.quadravox.com/qvdiph.pdf

Model QV306m1, from Quadravox. Photo Quadravox, Inc.

RC Systems, Inc.   203883

1609 England Ave.
Everett, WA 98203
USA

✆ (425) 355-3800

📠 (425) 355-1098

✉ info@rcsys.com

🌎 http://www.rcsys.com/

RC Systems manufactures and sells text-to-speech modules, called DoubleTalk, that can be attached to computers or microcontrollers. Their V8600A voice synthesizer is a modular text-to-speech processor that automatically converts plain English text into a high-quality male voice.

RC Syste,s V8600 speech synthesizer module. Photo RC Systems, Inc.

Sensory Inc.   204039

1991 Russell Ave.
Santa Clara, CA 95054-2035
USA

✆ (408) 327-9000

📠 (408) 727-4748

🌎 http://www.sensoryinc.com

Sensory makes and sells inexpensive voice recognition modules and software. Popular units include the Voice Direct 364 module (also available as a developer's kit), as well as the SC-6x product line of speech synthesis ICs. The company also offers the Voice Extreme developer's kit (under $150), a programmable voice recognition module that supports (as an optional mode) speaker independence, meaning that no speaker training is required.

Note that the company's Voice Direct Speech Recognition Kit, resold by a number of online retailers, is discontinued and has been replaced by the Voice Direct 364.

Developer's kits are available directly from the company or through distributors.

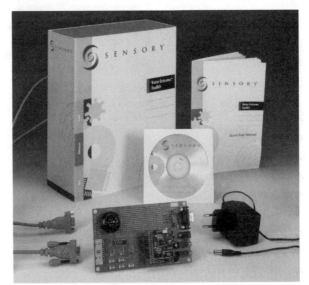

Voice Extreme Developer's kit, from Sensory. Photo Mario Galizia, Grafiche Galizia; Sensory, Inc.

SoftTalk Text-to-Speech System   203754

http://members.aol.com/softtalk/

Informational page on SoftTalk, a software text-to-speech system for embedded systems, including the Motorola MC68HC11 family of microprocessors.

Voice Direct 364 from Sensory, Inc.

Robots that listen to your voice commands and obey? The Voice Direct, from Sensory, Inc. can do just this. It's relatively easy to set up and use. The unit consists of a small double-sided circuit board, which is ready to be connected to a microphone, speaker (for auditory confirmation), battery, and either relays or a microcontroller.

The Voice Direct board recognizes up to 15 words or phrases (in so-called external host mode, the device can recognize up to 60 words) and is said to have a 99%+ recognition accuracy. Phrases of up to 3.2 seconds can be stored, so you can tell your robot to "come here," or "stop, don't do that!" The product comes with full circuit and connection diagrams.

Keep the following in mind when using a voice recognition system:

• You must be reasonably close to the microphone in order for the system to accurately understand your commands. The better the quality of the microphone, the better the accuracy of the recognition.

• If using a voice recognition system on a mobile robot, you may wish to extend the microphone away from the robot so that motor noise is reduced. For best results, you'll need to be fairly close to the robot and speak directly and clearly into the microphone.

• Consider using a good-quality RF or infrared wireless microphone for your voice recognition system. The receiver of the wireless microphone is attached to your robot; you hold the microphone itself in your hand.

The Web site for Sensory, Inc. is:

http://www.sensoryinc.com

SoftVoice Text-to-Speech System   203758

✉ info@text2speech.com

🌐 http://www.text2speech.com/

SoftVoice is speech synthesis software for Windows. The product is available for commercial license. Says the Web site: "Implemented as Windows DLL's, SoftVoice TTS is a state-of-the-art expert system for the conversion of unrestricted English text to high quality speech in real time. The SoftVoice system is the most versatile in the industry, with 20 different preset voices that the user can quickly select according to his or her preference."

Ʌ ELECTRONICS-SPECIALTY

In this section you'll find specialty electronics, components, and modules of practical use in robotics.

SEE ALSO:

ELECTRONICS-CIRCUIT EXAMPLES: Others share their favorite electronic circuits

KITS-ELECTRONICS: Circuits you put together

RETAIL-GENERAL ELECTRONICS: Additional sources for unique electronics components

RETAIL-SCIENCE: Interesting and unusual science-related kits

High Tech Chips, Inc.   203047

631E Windsor Rd. #8
Glendale, CA 91205
USA

🌐 http://www.hightechips.com/

Source code (released under GPL) of single-purpose specialty ICs based on microcontrollers. These include:

• LED flashers
• Monostable multivibrator
• Digital version of classic 555
• Musical sound generator
• Programmable musical sound generator
• PWM controller

Logiblocs Ltd.   203936

P.O. Box 375, St.
Albans
AL1 3GA
UK

📞 +44 (0) 1727 763700

📠 +44 (0) 1727 763700

✉️ feedback@logiblocs.com

🌐 http://www.logiblocs.com/

Logiblocs are electronic building blocks for elementary and junior high school students. They plug together to make complete and working circuits.

📄 💻

Photon Micro-Light 202716

PhotonLight.Com
200 W. 38th Ave.
Eugene, OR 97405
USA

📞 (541) 927-3552

📠 (541) 484-6898

🚫 (877) 584-6898

✉️ bryan@photonlight.com

🌐 http://www.photonlight.com/

High-brightness LED. Blindingly bright.

 🏭

Pico Electronics, Inc. 204008

143 Sparks Ave.
Pelham, NY 10803-1837
USA

📞 (914) 738-1400

📠 (914) 738-8225

🚫 (800) 431-1064

✉️ info@picoelectronics.com

🌐 http://www.picoelectronics.com/

Manufacturers and sellers of miniature and micro-miniature transformers, inductors, DC-DC converters, DC power supplies.

Piezo Systems Inc. 204009

186 Massachusetts Ave.
Cambridge, MA 02139
USA

📞 (617) 547-1777

📠 (617) 354-2200

✉️ sales@piezo.com

🌐 http://www.piezo.com/

Piezoelectric transducers, actuators, and drivers:

- Transducer elements (benders and extenders)
- Motion instruments (one- and two-axis mirror tilter, ultrasonic rotary motor)
- Electronics
- Fans and resonators
- Design and prototyping
- Acoustic sources

Plus tech info on piezo transducers.

 💻

SimmStick 202337

✉️ webmaster@simmstick.com

🌐 http://www.simmstick.com/

Informational page about SimmStick (available through dealers; see Dealers link). The SimmStick is a 30-pin simm board that is used as a standard form factor for a variety of devices, including PIC and Atmel AVR microcontrollers.

Valiant Technology Ltd. 203813

Valiant House
3 Grange Mills, Weir Road
London
SW12 0NE
UK

📞 +44 (0) 2086 732233

📠 +44 (0) 2086 736333

✉️ info@valiant-technology.com

🌐 http://www.valiant-technology.com/

Makers and sellers of unique electronic building block components. See listing under Toys-Construction.

ENTERTAINMENT

Work and no play make Tobor a bored robot. That goes for robot builders,
too. Don't forget the entertainment side-books, movies, and art-on robotics themes. Many
of today's best robot builders were influenced by fictional robots in movies and
books-who hasn't wanted to build their own R2-D2 after seeing *Star Wars?*

The two categories that follow take a look at the entertainment value of robots:

ART-Robots intended as artistic forms of expression, rather than functional machines. Includes posters, framable art, cinema special effects, specialty magazines, "standees" (life-size cardboard models), museums, and more.

BOOKS & MOVIES-The fictional side of robots, as presented in storytelling.

ENTERTAINMENT-ART

ABoyd Company, LLC, The   202772

P.O. Box 4568
Jackson, MS 39296
USA

 (601) 948-3479
Ø (888) 458-2693
✉ info@aboyd.com
🌐 http://www.aboyd.com/

Science fiction and monster movie memorabilia. See listing under Toys-Robots.

Acme Vintage Toys & Animation Gallery
202773

9976 Westwanda Dr.
Beverly Hills, CA 90210
USA

☎ (310) 276-5509
✆ (310) 276-1183
✉ acmetoys@yahoo.com
🌐 http://www.acmetoys.com/

Toys and animation art. Tin toy robots, Robby the Robot, Star Wars and Star Trek character sets, Mr.

Web shopping cart for Acme Vintage Toys.

Atomic, Gigantor, and other robot figurines. Most are collectables, and most are very expensive.

All Effects Company, Inc.   203443 🎖

17614 Lahey St.
Granada Hills, CA 91344
USA

☎ (818) 366-7658
✆ (818) 366-3768
✉ eric@allfx.com
🌐 http://www.allfx.com/

All Effects makes an ingenious Makita battery-powered adapter. Makita batteries are available at most hardware or home improvement stores and come in 7.2 volts, 9.6 volts, and 12 volts. The adapter allows for convenient use of these batteries for other applications.

All Effects is perhaps best known as a mechanical special effects company for motion pictures and television. The company is headed by Eric Allard, one of the most creative mechanical effects engineers working in the movie business today. His company's credits include the Johnny 5 robot from the Short Circuit films (the second of which he coproduced), Ghostbusters, Alien Resurrection, Stuart Little, and the Energizer Bunny TV commercials.

🏭

Celebrity Standups   202765

436 Highway 31 South
Alabaster, AL 35007
USA

☎ (205) 663-2755
✉ celebritystandup@aol.com

http://celebritystandups.com/

Standees: Star Wars (including C3PO, R2-D2, Battle Droid), Star Trek (original and newer shows, including Captain Kirk, Data, and Seven of Nine), Marvin the Martian, T-2 Endoskeleton, George Jetson.

Cinefex
300011

P.O. Box 20027
Riverside, California 92516
USA

☎ (909) 781-1917
📠 (909) 788-1793
✉ website@cinefex.com
🌐 http://www.cinefex.com/

Cinefex is the premier magazine for special effects in movies and television. The magazine, which comes out four times a year, is beautifully produced in full color. Each issue concentrates on the digital, optical, and mechanical special effects of at least one, and sometimes two or three, motion pictures or television productions. The articles are semitechnical, and most film jargon is explained.

Movies are gravitating to doing more and more special effects using only computers, but productions with mechanical effects (other than explosions and car crash ups) are still fairly common. The articles on mechanical effects work are especially notable for robot builders, and you can learn quite a few secrets and techniques from them.

Back issues are available; I recommend you thumb through the index at the site to search for topics and films that interest you. The soldout issues are provided as xerographic copies. For these, the cover is in color, but the articles are black and white.

Cinefex is the leading publication for and about special effects in movies and TV.

Dimensional Designs
202771

Dept. GITG
1845 Stockton St.
San Francisco, CA 94133-2908
USA

☎ (415) 788-0138
📠 (415) 956-9262
✉ DimDesigns@aol.com
🌐 http://www.dimensionaldesigns.com/

Reproductions of popular science fiction figures, including tin Robby the Robot, and latex masks of Outer Limits TV series aliens. Don't expect cheap stuff here. Monsters can be expensive.

H. I. Gosses: Robots and Special Effects
202233

http://go.to/robots-sfx

A truly radical personal Web page of robotic and special effects creations. Now, if they only made real robots that looked like these. . . .

Jeff's Robots-Toy Robots
202074

http://www.jeffbots.com/

Jeff collects robot toys and models and displays them here. Plus plenty of links.

Movie Goods
202974

6601 Center Dr. West
Ste. 500
Los Angeles, CA 90045
USA

☎ (310) 342-8295
📠 (310) 342-8296
🚫 (866) 279-2403
🌐 http://www.moviegoods.com/

Originals and reproductions of famous movie posters and lobby cards. Yep, Day the Earth Stood Still, Logan's Run, Silent Running, Forbidden Planet, Lost in Space, Star Wars . . . they're all there.

Movie Poster Shop, The 202766

#9, 3600 21 St. NE
Calgary, AB
T2E 6V6
Canada

 (403) 250-7588
(888) 905-7588
(403) 250-7589
mail@moviepostershop.com
http://www.moviepostershop.com/

Movie posters (original and reproduction), theater lobby cards, standees.

Museum of Unnatural Mystery 202775

http://www.unmuseum.org/

The Web site says it best: "Welcome to the Museum of Unnatural Mystery, a slightly bizarre, cyberspace, science museum for all ages. Are there really flying saucers? What killed the dinosaurs? Is there something ancient and alive in Loch Ness? The Museum takes a scientific look at these, and other, questions."

Some fun stuff here.

NEET-O-RAMA 202768

93 West Main St.
Somerville, NJ 08876
USA

(908) 722-4600
neetstuff@mindspring.com
http://www.neetstuff.com/

Movie posters, standees, and video. Check out the Twonky Video category for thousands of unusual cult movies, including hard-to-find 1940s through 1960s science fiction titles.

Retrofire-Robot & Space Toy Collectibles 202067

http://www.retrofire.com/

Interesting "high-tech" toys, including tin robots, space ray guns, and more.

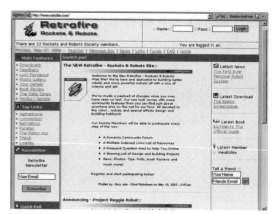

Retrofire's Web site.

SCI FI 202770

http://www.scifi.com/

Online site for SciFi Channel; also reviews and discusses robot toys.

Standees.com 202767

303 Lippincott Dr. Ste. 220
Marlton, NJ 08053
USA

Info@standees.com
http://www.standees.com/

"Standees" are cardboard cutouts of famous people, like movie actors. That includes movie robots, like Robot B-9 from Lost in Space (talking and nontalking versions) and, of course, Star Trek and Star Wars characters. Most standees are full size, but fold at strategic places for storage.

ENTERTAINMENT-BOOKS & MOVIES

ABoyd Company, LLC, The 202772

P.O. Box 4568
Jackson, MS 39296
USA

(601) 948-3479
(888) 458-2693
info@aboyd.com
http://www.aboyd.com/

Science fiction and monster movie memorabilia. See listing under Toys-Robots.

Amazon.com   202586

http://www.amazon.com/

The ubiquitous online retailer of books and more. Separate stores sell books, DVD and VHS movies, kitchen and household appliances, toys (online Toys'R'Us), computers, software, and magazine subscriptions.

Barnes & Noble.com   204024

http://www.bn.com/

Barnes & Noble is the largest bookseller in the world, with retail stores and an online e-commerce site. They have hundreds of books on robotics and other technical subjects.

Jeff's Robots-Toy Robots   202074

http://www.jeffbots.com/

Jeff collects robot toys and models and displays them here. Plus plenty of links.

Retrofire-Robot & Space Toy Collectibles   202067

http://www.retrofire.com/

Interesting "high-tech" toys, including tin robots, space ray guns, and more.

Rad Robot Movies

These movies are out on VHS or DVD. Find them at your local video store. Some are considered special interest, so they may be available only for sale and not rent.

The Day the Earth Stood Still (1951)

The quintessential robot flick, and yet the robot (Gort) is only a minor character. But what a character! Tick him off and he'll open his visor and disintegrate you with a piercing laser beam. Cool stuff.

Gort was played by Lock Martin, one of the tallest actors (7'7") ever to work in Hollywood—first as the doorman of the Graumans Chinese Theater, then as host of a Los Angeles children's TV show named *The Gentle Giant*. The Gort costume (there were two, with laced "zippers" on the front or back, depending where the camera was placed) was latex rubber and measured 8'2".

Despite its age, *TDTESS* ranks as one of the best science fiction movies ever made. The special effects are low-tech by today's standard—the sound effect of Gort's destructive ray was created by playing a tape backward—but the story is top-notch, and the action thrilling. *From out of space . . . a warning and an ultimatum. Strange power from another planet menaces the earth! What is this invader from another planet . . . Can it destroy the earth?*

Robot Monster (1953)

Directed by Phil Tucker, *Robot Monster* is one of those "so bad it's good" movies. Tucker directed a number of sex and teen exploitation films in the 1950s (including *Dance Hall Racket*, written by legendary comic Lenny Bruce). This flick stars George Nader, a popular B-picture actor of the 1950s and 1960s.

For whatever reason, the robot in this picture looks like a gorilla in a diving helmet, probably because the actor playing the robot wore a gorilla suit wearing a deep-sea diving helmet. The robot manages to kill off all but a half-dozen people on earth, then falls in love. Jeepers! *Moon monsters launch attack against earth! How can science meet the menace of astral assassins?*

Forbidden Planet (1956)

Before Leslie Nielson got silly and starred in all those *Naked Gun* movies, he was a dashing leading man in many 1950s and 1960s movies. Of his most famous is *Forbidden Planet*, where he and his crew of space voyagers come to fetch a scientist and his pretty daughter and return them to Earth.

Only the scientist doesn't want to go, and besides he doesn't need to, because he's built his very own mechanical butler, a robot named Robby. Robby is polite, couldn't hurt a fly (his programming follows Isaac Asimov's Three Laws of Robotics, and therefore forbids it), and can make smooth, aged whiskey by the gallons. Needless to say, he becomes a favorite of many, including the audience.

An interesting story and some fine acting, but parts are a tad scary for the little kids. Robby the Robot would later make a number of guest appearances in other films, on the *Twilight Zone* show, and even on the *Superman* television series.

Kronos (1957)

What do Kronos and mothers-in-law have in common? They both zap the energy out of everything they come into contact with. Low-budget but still engaging, *Kronos* the movie is the story of an alien race that has used up all the electricity of their home planet (sounds like they're from California). So, they send a giant robot to Earth to collect all of *our* energy.

Kronos is one of many science fiction flicks that takes place predominantly in California's high desert—because it was so cheap for Hollywood producers to film there. To this day, whenever I drive through the Anza-Borrego or Mohave deserts of the U.S. Southwest, I get an eerie feeling Kronos is right around the next bend.

Parts of *Kronos* are really cheesy. Don't expect multi-million-dollar special effects. This is 1957, after all, the same year Russia launched Sputnik. *World-destroying monster!*

The Outer Limits, Vol. 41: I, Robot (1963)

An episode of the original *Outer Limits* television series, this is an adaptation from the Isaac Asimov short story of the same name. Stars Leonard Nimoy, before he donned his pointed ears in *Star Trek*. The robot stands accused of murder, but is it a machine or a man?

This prototypical episode of the *Outer Limits* was directed by Leslie Stevens, who would later turn out the movie *Incubus*, starring William Shatner, another *Trek* alumnus. The movie is memorable on many counts, the least of which it is entirely in the Esperanto language! No robots, though—probably because robots don't speak Esperanto.

Westworld (1973)

Written and directed by Michael "Jurrasic Park" Crichton, *Westworld* explores what happens when humans put too much trust into machines. It's a story line that Crichton would explore many times in his films, using various technologies that "go worng."

Westworld is one of three "worlds" in an interactive resort for rich folks. The guests are real, but just about everything else is a machine, including the gunslingers and the hookers. What goes *worng* in *Westworld* is a gunbot, played deftly by Yul Brenner. For some unknown reason, the programming for Yul and the rest of the robots get scrambled (they must be running Microsoft Windows!), and they end up killing all but one guest.

There are few special effects with robot-like mechanisms, except for a fairly unrealistic shot of the inside of Yul Brenner's head. Enjoy this movie more for its "be careful what you ask for" mid-1970s message about high-tech. *Westworld . . . where robot men and women are programmed to serve you for . . . romance . . . volence . . . anything.*

The Terminator (1984)

Forever frightening all of the women in the world unlucky enough to be named Sarah Conner, *The Terminator* put actor Arnold Schwarzenegger on the Map of the Stars. Great effects for such a low-budget film.

The Terminator robot is a cyborg: human-looking on the outside, but high-tech machine on the inside. Except for some brief shots at the beginning of the film, it's not until the end of the movie do we see the innards of the Terminator: chrome-plated "bones" operated by powerful hydraulic systems. In reality, "it's only a model" (in the words of Monty Python). No working robot was ever created for the film; what you see is a cleverly designed puppet operated by cables. It looks real enough!

The Terminator skeleton was designed by veteran Hollywood mechanical and makeup effects artist Stan Winston. Winston was also responsible for the mechanical and creature effects for such films as *Aliens, Inspector Gadget, Small Soldiers, Galaxy Quest*, and *Terminator III*. He's one of the best.

Short Circuit (1986)

Let's forget the human characters in this film. Please. Instead, just concentrate on the robots, designed by conceptual artist Syd Mead, and expertly crafted by effects supervisor Eric Allard.

Briefly, *Short Circuit* is about a group of five battlefield robots. One robot is accidentally electrocuted, and this somehow fuses his computer, causing him to become alive. Now a sentient creature, this robot—referred to as Number Five—fears for his life and escapes. The rest of the movie is about Number Five being chased by the security department of the Big Bad Defense Company that created him.

Actual working (remotely controlled) robots were created for the movie, which makes *Short Circuit* one of the few films that employed real robotic actors. (For some shots, bits and pieces of the Number Five robot were controlled by cable using off-screen puppeteers.)

If you can find it, get a copy of issue 28 of the magazine *Cinefex* (see **http://www.cinefex.com/** for starters). This issue contains an in-depth article covering the construction of the *Short Circuit* robots by Allard and his All Effects Company. Enlightening reading. *Something wonderful has happened. . . . Number Five is alive!*

Robot Jox (1990)

Little story but cool special effects. The robots are huge, legged vehicles used in combat death matches. An early incarnation of the MechWarriors concept. *The ultimate killing machine, part man, part metal.*

The Iron Giant (1999)

Agree or disagree, this movie is not only one of the best robot films ever made, but also one of the best animated pictures, *ever!*

The Iron Giant takes place amid the fear and suspicion of the late 1950s (1957, to be precise). The Soviet Union had just launched *Sputnik* (see *Kronos*, above), and Americans were sure the next "man-made moon" would contain a nuclear bomb, dropped into the middle of their sleepy little town.

Something *does* fall from the sky, but it's not what folks are expecting; it's a giant robot sent to earth for some unknown reason. The robot feeds itself by eating metal. Before you can say "pass the ketchup, please," the hulking 'bot becomes entangled in some high-voltage power lines and is on the brink of destruction. Enter our hero, nine-year-old Hogarth Hughes, who saves the robot and ends up becoming best pals with it.

What might have been just another "kid and his big robot buddy" film, *The Iron Giant* is unique in how it deals with humanity, even when that humanity comes from a metal leviathan. Vin Diesel plays the voice of the robot.

Two end effectors up! *It came from outer space!*

Honorable Mentions

***Doctor Who* (1975)**—British television series co-starring the Daleks, a race of really, really bad robots. There are some older *Doctor Who* movies that you might enjoy, as well. Daleks are in many of them.

***Gigantor* (1965)**—An early Japan animation series where a little boy controlled his very own giant robot. They fought crime together.

***Lost in Space* (1965-1968)**—Sold to the CBS network as "the Swiss Family Robinson in outer space," this television series started out as serious, but is best viewed today as the ultimate in camp. When the show was first on, I had a major crush on Angela Cartwright (she played Penny Robinson), so I didn't pay much attention to The Robot (sometimes referred to by fans as Robot B9 or Robot YM-3).

Others noticed the machine over the cute brunette, and some of them have even created their own scratch-built LIS robots, faithful down to the red grippers.

Fun fact: The original pilot for *Lost in Space* lacked both The Robot and the miserable Dr. Smith. Good thing producer Irwin Allen reshot the pilot with these two great characters added.

***Lost in Space* (1998)**—Not everyone's cup of tea, but a personal favorite for me, this full-feature version of the 1960s television show is packed with modern mechanical and digital effects. The Robot is a radically updated version of the TV model. After the movie came out, there were truckloads of models and toys based on this robot.

***Metropolis* (1927)**—Considered the first science fiction epic film, *Metropolis* is perhaps best known for the shapely metallic robotrix created by the mad scientist, Dr. Rotwang. Look carefully at the background walls of his laboratory: You'll see an upside down pentagram; this symbol is often considered the sign of witchcraft and the occult. The symbology is not accidental. *Metropolis* came out during a period when the world was reeling from the effects of the industrial revolution. Suspicion of mechanics outstripping humanity helped set the notion that people who created artificial life were evil. Isaac Asimov's robot stories, and his Three Laws of Robotics, were aimed at reversing this long-held stereotype.

***Silent Running* (1972)**—An allegory of the many ills besetting the United States during the Vietnam War era, *Silent Running* stars Bruce Dern as a sort of interstellar gardener. With the aid of three small robots, he tends to Earth's last remaining agriculture, which is now limited to several "pods" on a space ship. The anthropomorphic robots are a highlight of the film, but they are not mechanical; each one has a person inside it.

***Star Wars* (1978 to ??)**—Honorably mentioned here because the films, and the robots in them, get mentioned so much everywhere else. Who doesn't know about C-3PO and R2-D2? 'Nuff said.

How to Search for More Movie Robots

Use the Internet Movie Database at:

http://www.imdb.com/

to search for movies—past, present, and near future, on any topic. For example, to find movies about robots, enter *robot* in the search box, and choose *Plot* for the search type.

FASTENERS

Fasteners are the often forgotten element of robot construction. The smallest and simplest robots don't need fasteners, of course-you can just tape or glue the parts together. But larger 'bots need more substantial hardware to keep everything in its place, and that means an assortment of screws, nuts, washers, bolts, and other fasteners.

The local hardware store is a great source for fasteners. It'll have most everything you need, in a variety of sizes. But while hardware stores are prime sources for fasteners, they are neither the only source, nor the best. Not all sizes are available, and retail hardware stores typically sell small packages of fasteners at a huge markup. You'll go broke buying packages of four or five machine screws; bulk packages will save lots of money. Alas, though many hardware stores sell some bulk fasteners, they only offer the more popular types and sizes in such packaging.

The listings that follow are fastener specialty outlets. With few exceptions, they conduct business by mail order or online, and a few have local stores. Most fastener retailers sell in quantity, and pass on the savings to you. They also provide more variety, including specialty nylon and plastic fasteners (which weigh less than steel), unique drive styles, unusual sizes or lengths, and metric.

Before purchasing fasteners, determine which sizes you use the most. If possible, settle on a few standard sizes; you'll save money. Consider using smaller hardware when practical. It's cheaper and lighter. Case in point: The smallest fasteners of my first robots were 6/32, because that's all the local hardware stores sold in bulk. Yet 6/32 is oversized for many applications for small robots. These days, I gravitate toward 4/40 fasteners, purchased from a specialty fastener retailer in bulk. Not only does 4/40 hardware weigh about 50 percent less than 6/32 hardware, it's cheaper to boot.

SEE ALSO:

DISTRIBUTOR/WHOLESALER-OTHER COMPONENTS: More specialty fasteners and hardware

MACHINE FRAMING: Build robots with little or no fastener hardware

RETAIL-AUTOMOTIVE SUPPLIES: Fasteners for the car

RETAIL-HARDWARE & HOME IMPROVEMENT: Retail (online, local) outlets for fasteners

12 Volt Fasteners 202630

USA

⊘ (800) 310-9152

✉ sales@12voltfasteners.com

🌐 http://www.12voltfasteners.com/

Fasteners: Nuts, bolts, screws, and washers in metric, black oxide, stainless steel, and zinc plated. Cable ties. Sold in bulk (1,000 qty.). Specializing in self-drill and self-tapping screws, such as drywall and truss screws.

Aaron's General Store 202661

http://www.aaronsgeneralstore.com/

This is a portal to a number of online specialty fastener stores. Among the offerings are:

http://www.AaronsPushNuts.com/

http://www.AaronsMachineScrews.com/

http://www.AaronsCapScrews.com/

http://www.AaronsSelf-TappingScrews.com/

http://www.AaronsMetricScrews.com/

http://www.AaronsWoodScrews.com/

http://www.AaronsMilitaryScrews.com/

http://www.AaronsScrewdrivers.com/

http://www.AaronsSecurityScrews.com/

http://www.AaronsTorxScrews.com/

For complete company information, see:
http://www.AaronsMachineScrews.com/

Aaron's Machine Screws 202634

111 Pacifica
Ste. 130
Irvine, CA 92618-7421
USA

☎ 714) 838-3575

📠 (714) 838-3165

⊘ (877) 838-3575

✉ sales@AaronsMachineScrews.com

🌐 http://www.aaronsmachinescrews.com/

Machine screws with Phillips, slotted, hex, Torx, and Pozidriv; SEMS screws, bolts, SAE and USS washers and nuts. Steel and stainless steel industrial fasteners. Hard-to-find items are a specialty. The site offers a very nice search feature with which you can systematically select the physical traits of the fasteners you want, such as head type, drive type, size, finish, thread type, and more. You can buy in low quantity, but the savings really kick in when you buy in bulk.

One important note: The site often lists a "carton quantity" that may be lower than the same quantity sold as bulk. This happens when the carton quantity is the same as an offered bulk quantity (for example, bulk quantity of 1,000 pieces, and a carton with a 1,000 pieces). It's easier to sell a precounted box of fasteners than it is to count the number of pieces (usually by weight) sold in bulk; you pay for the extra handling then.

SEE ALSO:

http://www.aaronswoodscrews.com/ for wood screws

http://www.aaronsgeneralstore.com/ for general portal to hardware

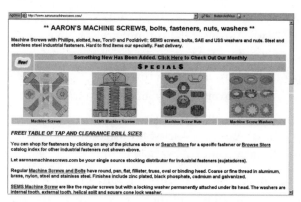

One of Aaron's online fastener depots.

Airparts, Inc. 203153

2400 Merriam Ln.
Kansas City, KS 66106
USA

  (913) 831-1780

  (913) 831-6797

  (800) 800-3229

  airparts@airpartsinc.com

  http://www.airpartsinc.com/

Metal supplies: aluminum and steel sheets, rods, tubes, etc.; fasteners and hardware. An excellent source for aluminum, chromolly, and other homebuilt-aircraft construction supplies. You don't need to build an airplane; a robot will do.

Large selection of aluminum extruded pieces, including rods, angles, bars, and tubing. For many of the products, the aluminum alloy (such as 6063/T52 or 6061/T6) is specified.

All Metals Supply, Inc. 202823

600 Ophir Rd.
Oroville, CA 95966
USA

  (530) 533-3445

  (530) 533-3453

  (888) 668-2220

  sales@allmetalssupply.com

  http://www.allmetalssupply.com/

All Metals is a distributor of ferrous and nonferrous metals, fasteners, and industrial hardware. Metal products include aluminum, copper, zinc, and others in bar, tube, extruded shapes, sheets, and plate. Fastener products include machine screws and bolts of all sizes, in stainless, zinc, and nylon. The Web site is for reference only; order directly from company.

Allmetric Fasteners, Inc. 204082

3790 Yale St.
Houston, TX 77018
USA

  (713) 695-2220

  (713) 695-3311

  (800) 424-7705

  sales@allmetric.com

  http://www.allmetric.com/

Fasteners, metric style.

Manufacturers Fastener World
http://www.m-f-w.com/
Surplus fasteners

Promptus Electronic Hardware
http://www.promptusinc.com/
Fasteners, handles, spacers, and other hardware for electronics gear

Fascinated by Fasteners

One way to put a robot together is with chewing gum and bailing wire. It works until the gum dries out and the bailing wire breaks off. For a more permanent construction, use fasteners—screws, nuts, bolts, and assorted hardware. Most fasteners are made so they can be applied and later removed. Removing fasteners is handy if you

Nuts, screws, and washers make up the most common fasteners used in robotics.

need to rebuild your robot or want to disassemble parts so you can reuse them. Construction techniques using permanent glues and adhesives don't allow for easy disassembly.

There are many kinds of fasteners. The most common for robotics construction is the machine screw. It's like a regular garden-variety household screw, except it does not have a tapered end. The end is flat, because the screw is meant to be secured with a nut or other retainer. Machine screws are also sometimes called bolts, but the current trend is to reserve "bolt" for large fasteners. Cars are put together with bolts; amateur robots are put together with machine screws.

Machine screws vary by size and number of threads per inch (or per millimeter, for metric fasteners) and also by the head type and the driver type. Pan and flat-head types are common finds at any hardware store, though also available are round, oval, truss, and Fillister. Common drive types are slotted and Phillips; Torx, hex, and Pozidrive can be purchased from fastener specialty outlets. The alternative drive types are recommended for high-torque applications or if a fastener needs to be removed often.

Nuts are used to hold machine screws in place. There are hex (six-sided) nuts, blind nuts that dig into soft material, self-locking nuts that have a nylon insert, Tinnerman nuts (these are flat pieces of metal that act as retainers), wing nuts, and many others. For obvious reasons, the nut must match—in both size and threads per inch—with the machine screw it's intended for.

Flat washers are used to spread the tension of the screw and nut across a larger area. Lock washers (teeth on the inside, outside, or both; as well as split) help prevent the nut from working itself loose from the screw.

While most of us are familiar with standard zinc-plated steel fasteners, there are in fact many other materials used to make screws, nuts, and washers.

- Stainless steel offers added strength and resistance against rusting or corrosion.
- Brass is a softer metal that's most often used for looks.
- Nylon is considerably lighter than steel or brass and is advantageous when weight is a concern.
- Aluminum is used when a metal fastener is desired and weight is a consideration. It's also the preferred fastener for aluminum structure, as it will not cause corrosion (which can happen if you use steel fasteners with aluminum framing).

Depending on the source, steel fasteners are available in plain zinc finish and also black oxide, hot-dipped galvanized (useful for applications where water or the elements might cause rust or other corrosion), and colored zinc—green and yellow are common choices.

When shopping for fasteners, note that you can save considerably by purchasing in quantity. At the hardware store, a package of ten #8 machine screws may cost 99 cents, yet a package of 100 may be priced at $3.99. The price difference is buying in bulk. If you think you'll make heavy use of a certain size fastener in your robots, invest in the bigger box, and pocket the savings.

American Bolt and Screw Manufacturing Corporation
202633

601 Kettering Dr.
P.O. Box 51300
Ontario, CA 91761
USA

✆ (909) 390-0522
📠 (909) 390-0545
🚫 (800)-325-0844
🌐 http://www.absfasteners.com/

Suppliers of fasteners (bolts, screws, nuts, washers, you name it) to industry. Caters to high-quantity purchases.

Additional locations in Arizona, Oregon, Texas, Georgia, and Indiana.

Applied Industrial Technologies
203445

One Applied Plaza
Cleveland, OH 44115-5053
USA

✆ (216) 426-4189
📠 (216) 426-4820
🚫 (877) 279-2799
✉ products@apz-applied.com
🌐 http://www.appliedindustrial.com/

Industrial bearings, linear slides, gears, pulleys, pneumatics, hydraulics, and other mechanical things. Also hosts Maintenance America, online reseller of industrial maintenance supplies and general industrial supplies (wheels, casters, fasteners, and a lot more), tools, paints, and adhesives.

Atlantic Fasteners
203825

49 Heywood Ave.
P.O. Box 1168
West Springfield, MA 01090-1168
USA

✆ (413) 785-1687
📠 (413) 785-5770
🚫 (800) 800-2658
✉ info@atlanticfasteners.com
🌐 http://www.atlanticfasteners.com/

Fasteners: 67,847 varieties, with pictures. A downloadable catalog in Adobe Acrobat PDF format is available.

Barnhill Bolt Co., Inc.
203830

2500 Princeton NE
Albuquerque, NM 87107
USA

✆ (505) 884-1808
📠 (505) 888-1559
🚫 (800) 472-3900
🌐 http://www.barnhillbolt.com/

Fasteners for all occasions, including all-thread, threaded couplers, thumbscrews, roll pins, rings, retailer, and the usual nuts, bolts, and washers. In zinc, steel, stainless, brass, nylon. Metric and standard.

Bolt Depot
204110

286 Bridge St.
North Weymouth, MA 02191
USA

✉ customer-service@boltdepot.com.
🌐 http://www.boltdepot.com/

Wood screws, sheet metal screws, machine screws, hex bolts, carriage bolts, lag bolts, socket head cap screws, nuts, washers-standard and metric sizes. Sales by individual pieces or small-quantity boxes.

BoltsMART
202636

Kinetic Information Technologies
1502 109th St.
Grand Prairie, TX 75050
USA

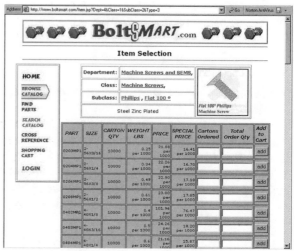

BoltsMART's Web site.

(972) 606-1544
(972) 606-1545
info@boltsmart.com
http://www.boltsmart.com/

Nuts, bolts. You can browse their online catalog or search by keyword. Standard, metric, zinc, steel, stainless, hot-dipped galvanized, nylon, and brass. Sold in bulk high quantities (e.g., 1,000 for the smaller items) and cartons.

Brikksen Company 204109

11470 Hillguard Rd.
Dallas, TX 75243
USA

(214) 343-5703
(214) 348-6990
(800) 962-1614
sales@brikksen.com
http://www.brikksen.com/

Specializes in metric stainless steel fasteners.

Du-Mor Service & Supply Co. 202635

10693 Civic Center Dr.
Rancho Cucamonga, CA 91730
USA

(909) 483-3330
(909) 483-3123
info@du-mor.com
http://www.du-mor.com/

Fasteners (including stainless steel), steel stock, shop tools, ferrous and nonferrous pipe and fittings, cutting

tools and abrasives, hardware (swivels, brackets, etc.), and equipment.

EL-COM 203827

12691 Monarch St.
Garden Grove, CA 92841
USA

(714) 230-6200
(714) 230-6222
(800) 228-9122
http://www.elcomhardware.com/

Fasteners and hardware, mainly for cabinetry. Also casters, aluminum extrusions (squares, channels, bars), plastic laminates, and foam products.

Fastenal Company 202640

2001 Theurer Blvd.
Winona, MN 55987
USA

(507) 454-5374
(507) 453-8049
sales@fastenal.com
http://www.fastenal.com/

Industrial components and parts (casters, etc.). Local outlets in many U.S. states.

Fastener Barn, LLC 202631

436 South 100th Ave.
Zeeland, MI 49464
USA

Understanding Thread Sizes

What do the values 4/40, 6/32, 8/32, and others mean? These are thread sizes. For fasteners under 1/4 inch, thread sizes are denoted using this fractional nomenclature.

The first number represents a standard diameter of the screw using a standard sizing metric (akin to shoe sizes). The second is the number of threads per inch (also called *pitch*). Given a 6/32 machine screw, for example, the screw is a size #6 (9/64-inch, or 0.140625-inch), and there are 32 threads to the inch. Pitch is coarse or fine, so not all #6 screws are 32 pitch.

((616) 748-1246

♨ (616) 748-1960

⊘ (800) 509-2276

 surplus@fastenerbarn.com

🌐 http://www.fastenerbarn.com/

Surplus nuts and bolts.

Fastener-Express   203164

25581 Paseo De La Paz
San Juan Capistrano, CA 92675
USA

((949) 661-9630

♨ (949) 661-9445

⊘ (877) 546-1148

✉ fastenerexpress@yahoo.com

A Cornucopia of Machine Screw Drive and Head Styles

Machine screw fasteners come in a variety of drive and head styles. Here is an overview of the several commonly available drive and head types, and what makes them unique.

 Slotted — Made for general fastening and low torque drive; screwdriver may slip from the slot.

 Phillips — Cross-point drive resists drive slippage, but lots of slots become easily stripped out when using an improperly sized driver.

 Hex, Torx, Pozidrive — Specific size and type of driver required, which minimizes stripping. Disadvantage: You must have the proper tool to fasten and unfasten.

 Pan — Good general-purpose fastener; shallow head provides less grip for the driver.

 Round — Taller head protrudes more than pan head, but provides greater depth for driver. Good for higher-torque applications.

 Countersunk — Used when head must be flush with the materials surface. Requires countersunk hole. Also called flat head.

 Fillister — Extra deep head for very high torque.

Hex bolt — Uses no slot, and requires wrench to tighten. For highest-torque applications.

http://www.fastener-express.com/

Fastener assortments, socket screws, metric fasteners, aluminum fasteners, servo and flange screws, machine screws, sheet metal screws, nuts, washers, nylon fasteners.

Fastenerkit.com   202641

60 Fairview Ave.
Poughkeepsie, NY 12601
USA

((845) 454-7010

((845) 454-0070

 http://www.fastenerkit.com/

Fastener kits. Bolts, nuts, washers, clips.

Fuller Metric Parts   202642

9652-188th Street
Surrey, BC
V4N 3M2
Canada

((604) 882-9202

((604) 882-9240

∅ (800) 665-4825

✉ surreysales@fullermetric.com

 http://www.fullermetric.com/

Metric fasteners; all sizes and styles, including pins, threaded spacers, and socket head screws. Check out their tech info pages (most in Adobe Acrobat PDF format) for such documents as "Thread Identification Charts," "Stainless Steel Material Data," and "Torque Figures." Offices in Vancouver and Toronto, Canada.

K-Surplus Sales Inc.   202644

1403 Cleveland Ave.
National City, CA 91950
USA

((619) 474-6177

((619) 474-3521

✉ kplus@pacbell.net

 http://www.ksurplus.com/

Surplus fasteners and hardware.

Maryland Metrics   204105

P.O. Box 26
Owings Mills, MD 21117-0261
USA

((410) 358-3130

((410) 358-3142

∅ (800) 638-1830

✉ sales@mdmetric.com

 http://www.mdmetric.com/

Something of a one-stop shop, Maryland Metrics carries bearings, linear bearings, fasteners, rods, gears, pneumatic and hydraulic fittings, and a variety of power transmission items. Good assortment of technical info.

Standard American Threads at a Glance

UNC (coarse)	UNF (fine)
1/4-20	1/4/-28
5/16-18	5/16-24
3/8-16	3/8-24
7/16-14	7/16-20
1/2-13	1/2-20
5/8-11	5/8-18
3/4-10	3/4-16
7/8-9	7/8-14
1-8	1-14

McFeely's Square Drive Screws 203826

1620 Wythe Rd.
P.O. Box 11169
Lynchburg, VA 24506-1169
USA

⊘ (800) 443-7937

✉ tech@mcfeelys.com

🌐 http://www.mcfeelys.com/

Fasteners, tools, adhesives. Check out the technical information about screws.

Metric Specialties, Inc. 204084

622 S. Flower St.
Burbank, CA 91502
USA

☎ (818) 848-6696

📠 (818) 848-3951

⊘ (800) 800-6696

🌐 http://www.metricspecialties.com/

Specializes in metric fasteners: screws, washers, nuts, inserts, pins, rings, and other hardware. Also metric tools.

Micro Fasteners 204080

110 Hillcrest Rd.
Flemington, NJ 08822
USA

☎ (908) 806-4050

📠 (908) 788-2607

⊘ (800) 892-6917

Micro Fasteners specializes in small fastener hardware.

Screw Sizes at a Glance

Screw Size	Inch/Fraction	Inch/Decimal	Millimeter
#1	1/16	0.0625	1.58750
#2	5/64	0.078125	1.98437
#3	3/32	0.09375	2.38125
#4	7/64	0.109375	2.77812
#5	1/8	0.125	3.17500
#6	9/64	0.140625	3.57187
#8	5/32	0.15625	3.96875
#9	11/64	0.171875	4.36562
#10	3/16	0.1875	4.76250
#11	13/64	0.203125	5.15937
#12	7/32	0.21875	5.55625
#13	15/64	0.234375	5.95312
#14	1/4	0.250	6.35000
#16	17/64	0.265625	6.74687
#18	19/64	0.296875	7.54062
#20	5/16	0.3125	7.93750
#24	3/8"	0.375	9.52500

info@microfasteners.com

http://www.microfasteners.com/

Fasteners (machine screws, nuts, lock washers, rivets, etc.) predominately in petite sizes. U.S. and metric threads. Volume discounts. Most fasteners are sold in packs of 100 pieces or less. Materials include zinc, stainless, nylon, and brass.

Micro Plastics, Inc.   204081

Hwy. 178 N.
Flippin, AR 72634
USA
((870) 453-2261

 (870) 453-8676

 mpsales@microplastics.com

http://www.microplastics.com/

Major manufacturer and seller of plastic fasteners, including clips, cable ties, hose clamps, plastic stand-offs, panel fasteners, whole plugs, threaded rod, and the usual screws, nuts, and washers. Products are available in standard or metric sizes.

MSC Fasteners   204077

104 Oakdale Dr.
Zelienople, PA 16063
USA

Metric Conversion Table

Metric fasteners don't use the same sizing nomenclature as their SAE cousins. Screw sizes and pitches are defined by a standardized diameter notation, the thread pitch (number of threads per millimeter) followed by length—all in millimeters. For example:

M2-0.40-5mm
Means the screw is 2mm in diameter, has a pitch of 0.40 threads per millimeter, and has a length of 5mm. Note that most metric screws use standard threads, so the pitch may be omitted:

M2-5mm
Use the following table to compare metric screw sizes.

Diameter	mm	Inch	Diameter	mm	Inch
M1	1	0.0393	M20	20	0.7874
M1.1	1.1	0.0433	M22	22	0.8661
M1.2	1.2	0.0472	M24	24	0.9448
M1.4	1.4	0.0551	M27	27	1.0629
M.17	1.7	0.0669	M30	30	1.181
M1.8	1.8	0.0708	M33	33	1.299
M2	2	0.0787	M36	36	1.417
M2.2	2.2	0.0866	M39	39	1.535
M2.3	2.3	0.0905	M42	42	1.654
M2.5	2.5	0.0984	M45	45	1.772
M3	3	0.1181	M48	48	1.890
M3.5	3.5	0.1378	M52	52	2.047
M4	4	0.1574	M56	56	2.205
M4.5	4.5	0.1771	M60	60	2.362
M5	5	0.1968	M64	64	2.520
M6	6	0.2362	M68	68	2.677
M7	7	0.2755	M72	72	2.835
M8	8	0.3149	M76	76	2.992
M10	10	0.3937	M80	80	3.150
M12	12	0.4724	M85	85	3.346
M14	14	0.5511	M90	90	3.543
M16	16	0.6299	M95	95	3.740
M18	18	0.7086	M100	100	3.937

☏ (724) 452-8003

📠 (724) 452-1145

🚫 (800) 359-7166

✉ mscfasteners@mscfasteners.com

🌐 http://mscfasteners.com/

Fasteners: body washers, button head, socket cap screws, lag screws, carriage bolts, levis pins, cotter pins, drive screws, flat-head socket cap screws, hex-head cap bolt, unslotted machine screws, wood screws, Neoprene-backed flat washers, acorn nuts. Carries stainless steel, brass, and metric fasteners.

Pacific Fasteners   202482

3934 East 1st Ave.
Burnaby, BC
V5C 5S3
Canada

☏ (604) 294-9411

📠 (604) 294-4730

✉ PacFast@pacificfasteners.com

🌐 http://www.pacificfasteners.com/

Fastener hardware: screws, bolts, nuts, washers, pins, rivets, socket caps, U bolts.

Phoenix Fastener Company, Inc.   202637

2501 West Homer St.
Chicago, IL 60647-4309
USA

☏ (773) 276-9661

📠 (773) 276-9680

🚫 (800) 621-1905

✉ sales@phoenixfastener.com

🌐 http://www.phoenixfastener.com/

Fasteners (steel, zinc, brass), all-thread, anchors, staples, screws, tape, and more.

Reid Tool Supply Co. 🏆   203820

2265 Black Creek Rd.
Muskegon, MI 49444
USA

🚫 (800) 253.0421

✉ mail@reidtool.com

🌐 http://www.reidtool.com/

Reid is an all-purpose industrial supply resource. See listing under Power Transmission.

Screwfix Direct Ltd.   203857

FREEPOST
Yeovil
Somerset
BA22 8BF
UK

🚫 0500 41 41 41

✉ online@screwfix.com

🌐 http://www.screwfix.com/

E-tailer of fasteners, tools, hardware, and other home improvement items.

Small Parts Inc. 🏆   202120

@listing-address:13980 N.W. 58th Court
P.O. Box 4650
Miami Lakes, FL 33014-0650
USA

☏ (305) 557-7955

📠 (305) 558-0509

🚫 (800) 220-4242

✉ parts@smallparts.com

🌐 http://www.smallparts.com/

Small Parts is a premier source for-get this!-small parts. All jocularity aside, Small Parts is a robot builder's dream, selling most every conceivable power transmission part, from gears to sprockets, chains to belts, and bearings to bushings. Product is available in a variety of materials, including brass, steel, and aluminum, as well as nylon and Delrin. Rounding out the mix is a full selection of raw materials: metal rod, sheets, tubes, and assorted pieces, as well as a huge assortment of fasteners.

Now about prices. Small Parts is for the serious builder, both amateur and pro. A little brass gear might cost $6, but what you pay for (apart from the precision, of course) is the selection of being able to find just about everything you need.

You can browse through their online catalog at

http://www.engineeringfindings.com/ or get their printed catalog.

Smith Fastener Company · 202987

3613 East Florence Ave.
Bell, CA 90201
USA

((323) 587-0382

✆ (323) 587-8712

⊘ (800) 764-8488

✉ sales@smithfast.com

🌐 http://www.smithfast.com/

Fasteners of all types: nuts, bolts, screws, and washers; threaded inserts; blind rivets; automotive fasteners and electrical; roll pins, spring pins, split pins, and slotted pins.

Specialty Tool & Bolt · 204065

108A Aero Camino
Santa Barbara, CA 93117-3198
USA

((805) 968-3581

✆ (805) 968-3385

⊘ (800) 722-6587

✉ info@SBBOLTS.com

🌐 http://specialtytoolandbolt.com/

Fasteners: stainless steel, miniature screws, machine screws, bolts, nuts, washers.

Tower Fasteners Co. Inc. · 202639

1690 North Ocean Ave.
Holtsville, NY 11742
USA

((631) 289-8800

✆ (631) 289-8810

⊘ (800) 688-6937

🌐 http://www.towerfast.com/

Master distributor, with online ordering, for several fastener brands, 3M adhesives, hardware for electronics, clamps and couplers, and power transmission. Distribution centers located along the East Coast.

Common Metric Coarse Thread Pitches

Size	Pitch
M2	0.40
M3	0.50
M4	0.70
M5	0.80
M6	1.00
M8	1.25
M10	1.50
M12	1.75
M14	2.00
M16	2.00
M18	2.50
M20	2.50
M22	2.50
M24	3.00
M27	3.00
M30	3.50
M33	3.50
M36	4.00
M39	4.00
M42	4.50
M45	4.50
M48	5.00
M52	5.00
M56	5.50
M64	6.00

Wicks Aircraft Supply   203160

410 Pine St.
Highland, IL 62249
USA

((618) 654-7447

↻ (618) 654-6253

⊘ (800) 221-9425

✉ info@wicksaircaft.com

🌐 http://www.wicksaircraft.com/

Small aircraft parts; specialty fasteners. Note: This is not little model aircraft stuff, but stuff for small aircraft- ones people can climb into. Products of particular interest to robobuilders are:

- Composite materials (epoxy, foam, cloth)
- Steel, aluminum, plastic
- Hardware (bolts, nuts, washers, etc.)
- Control system accessories
- Wheels, brakes, tires

You can order most products online, either by browsing or by index search.

Nearest Equivalent SAE Screw Size

This table shows the nearest equivalent SAE screw size for the purpose of substituting metric screws when plans or instructions call for SAE.

Metric	SAE
M.17	#1
M1.8	
M2	#2
M2.2	
M2.3	#3
M2.5	
M3	#4
M3.5	$6
M4	#8
M4.5	#9
M5	#11
M6	#13
M7	#16
M8	#20
M10	#22

FESTS AND SHOWS

In this section you'll find listings for amateur radio fests (called hamfests) and general electronics swap meets. These events, held on a regular basis-some monthly, some yearly-are good sources for new and used components for robot building.

ARRL: Database of Hamfests 203549

http://www.arrl.org/hamfests.html

Database of upcoming hamfests, swap meets set up and patronized by ham radio enthusiasts. But you'll find more than radio gear at most hamfests; many are good sources for general surplus electronics and surplus mechanicals, including motors, solenoids, relays, soldering stations, tools, and more. Hamfests are usually held on Saturdays, and many start early in the morning. To get the best deals, get there as early as possible, or else the good stuff will already be gone.

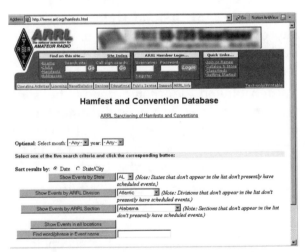

Web page of the ARRL, where you can locate Ham fests near you.

Bara Hamfest 202366

http://www.bara.org/

Annual hamfest, held in Washington Township, N.J.

Crown Amateur Radio Convention 202365

http://www.nofars.org/hamfest.htm

Annual hamfest in Jacksonville, Fla.

Dayton Hamvention 204201

http://www.hamvention.org/

This is the Big Kahuna of ham conventions. Held every year, people come from all over the world-literally!-to attend this thing. Held in Dayton, Ohio.

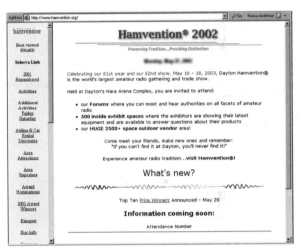

The Dayton Hamvention.

Denton Hamfest 202369

http://dentonhamfest.org/

From the Web site: "The Denton Hamfest is a one day event promoting amateur radio, cool technologies, good Tamales (we promise) and all the coffee you can drink."

Not to be confused with the Dayton Hamvention.

Hamfest Minnesota 202370

http://www.hamfestmn.org/

Information on the annual hamfest in St. Paul, Minn.

Hoosier Hills Hamfest 202367

http://www.hoosierhillshamfest.org/

Annual ham fest near Bedford, Ind.

Pacificon 202368

http://www.pacificon.org/

Ham fest held annually in the fall. Refer to the Web site for place and time.

Robot Show, The 203953

http://www.robotshow.co.uk/

The Robot Show is a U.K.-based symposium on personal, educational, industrial, and experimental robotics. Check the Web site for a current calendar of events.

Going as a Seller

You don't have to be a full-time retailer to sell at a swap meet or ham fest. All you need is something to sell, and a willingness to stand in one place for a few hours while potential buyers rummage through your stuff. Policies vary, but most swap meets and ham fests require only an up-front booking fee for a space. The more regular the show, the more likely you will be required to have a state resale license, for sales tax purposes. Call the show organizer for specifics.

The space rented to you may simply be a marked-off area of a room or parking lot; you bring your own table, chairs, and display cases. If it's an outdoor event, be sure to bring sun block and maybe an umbrella chair or table -- you need the shade. It can get hot standing out in the sun all day.

Tip: Go with a helper. That way you can sneak away when nature calls, and not worry that someone will steal all your merchandise.

Getting the Best Deals

Veteran swap meet buyers know that the best stuff goes fast. So they're the first in line when the show opens. Sure, that may mean getting a little less sleep, but you might be rewarded with a really good deal.

Here are some more tips for getting the very best at swap meets and ham fests:

• Do a quick run-through before buying anything. That way you won't overpay for something that another exhibitor is selling for less.

• Exhibitors like to "pre-shop" to pick up the best one-of-a-kind items, which they can resell at the show, or on eBay. You have to beat these folks to the punch.

• When buying electronic goods, be sure you can return it for a replacement or refund if it doesn't work. Exhibitors who have Monday through Friday store hours someplace in town are the best.

• When possible, pay cash. While credit card fraud is not common at swap meets and ham fests, the nature of the event does attract the occasional unscrupulous dealer.

• Don't leave the seller's booth without a complete receipt (emphasis on complete). You may need it to return goods, or even to get out of the exhibit hall.

• It pays to know the going rates before attending any show. Use the Internet to find general prices of the things you're interested in.

INTERNET

Like the opening remarks introducing a famous speaker at a banquet, "The Internet needs no introduction." This global resource brings together buyer and seller, student and teacher, database and researcher, and much more. Since the early 1990s, when the Internet was opened for commercial exploitation, it's become the number one resource for finding robotics information and products.

The sections that follow break up the Internet into cohesive units that are of primary importance and relevance to robot builders. These units include bulletin boards and mailing lists, educational and government labs where high-end robotics research takes place, informational sites, search engines, Usenet newsgroups, and plenty more.

INTERNET-BULLETIN BOARD/MAILING LIST

Before the Internet came along, robot builders had to settle for "talking shop" while sitting around a big pickle barrel every Thursday night at Sam's Feed and Hay store. Still, it was a great way to share construction ideas, trade horror stories, and ask for advice. These days, we have the Internet as the virtual equivalent of Sam's hospitality, with thousands upon thousands of bulletin boards and mailing lists that cater to a variety of special interests.

Though bulletin boards and mailing lists serve about the same purpose—they let individuals read and post messages—they work differently from each other, and each has its pros and cons.

- Bulletin boards (also called forums) are Web sites where you can read messages left by others and, if the mood strikes you, post replies of your own. In most cases, the bulletin board is software than runs on the Web server, and you need only a Web browser to use it. A few bulletin boards require add-in browser software, namely, Java and JavaScript.

- Mailing lists are conducted via e-mail. You sign up to receive messages, and every message posted to the list is sent to you. If you want to reply to a message, you just dash off an e-mail. The e-mail is sent to the list, not directly to the person who posted the original message.

Mailing lists are great for low-volume messaging; when the messages get to 20 or 30 a day, you may wish to get the digest version of the mailing list (if one is made available). Bulletin boards require a special trip to the appropriate Web site, but you can read only those messages with topics that interest you.

Among the most popular bulletin boards/mailing lists are the Yahoo eGroups. Each group specializes in a certain topic, such as programming with the Atmel AVR microcontroller or general robotics. You have a choice of reading all messages on the Yahoo site (this would be the "bulletin board setting)" or receiving messages via e-mail. You can opt to receive all messages or just daily digest versions.

SEE ALSO:

INTERNET-USENET NEWSGROUPS: Additional user-to-user help

PORTAL-ROBOTICS: All-encompassing Web sites that usually include message forums

USER GROUPS: Local and virtual groups meet; some variations on the pickle barrel thingie

---

Art & Robotics Group (ARG)   202535

http://www.interaccess.org/arg/

From Toronto, Canada. User group, discussion board, and latest news on the artistic side of robotics.

AVR Forum   203310

http://www.avr-forum.com/

Stomping grounds for geeks who are into the Atmel AVR line of 8-bit microcontrollers. Includes links, sample code, user-to-user forums, and an AVR FAQ.

AVRFreaks   203022

http://www.avrfreaks.org/

User-to-user forums, code examples, resources, application notes, articles, and links for those involved with programming the Atmel AVR line of 8-bit microcontrollers. Be sure to check out the free code library (requires free registration).

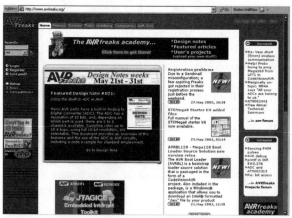

Web page for AVRFreaks, for users of the Atmel AVR microcontroller.

Genesis Robotics   204108

http://www.genesis-robotics.vze.com/

General-interest robotics forum.

Industry Community   204192

http://www.industrycommunity.com/ee/

Bulletin boards for those into electronics.

PICList   203292

http://www.piclist.com/

The PICList is a bulletin board/mailing list for folks interested in the Microchip PICmicro and similar processors.

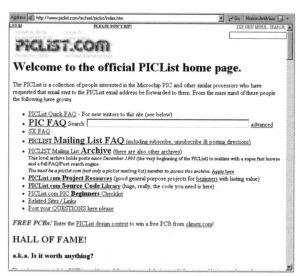

Main PICList homepage.

Slashdot.org   204020

http://slashdot.org/

Billing itself as "News for Nerds," Slashdot is a premier hangout spot for news and views regarding the high-tech world. Its discussions are always lively (and sometimes a little rough). Though most of the discussions are on computing topics, and there's a strong anti-Microsoft sentiment, there are plenty of articles and discussion threads on robotics, LEGO, and artificial intelligence.

Most discussion threads begin with a posting about some article, news story, or Web page. Thousands of readers flock to the link and often overwhelm it-this is called "being slashdotted." If you try to visit a link listed in a current discussion thread and it's unavailable, try later; the Web server has exhausted its resources by the intense rush of interest.

If you participate in the discussions, be sure to sign in, or your posting will list you as an "Anonymous Coward." Yes, /. can be a tough place sometimes!

Yahoo Groups: AVR-Chat   203926

http://groups.yahoo.com/group/AVR-Chat/

For anyone into programming the Atmel AVR microcontroller.

Yahoo Groups: BasicX   203710

http://groups.yahoo.com/group/basicx/

BasicX microcontroller from NetMedia.

Yahoo Groups: BEAM Robotics   202997

http://groups.yahoo.com/group/beam/

BEAM robotics builders. See also a similar group at:

http://groups.yahoo.com/group/beamrobotics/

Yahoo Groups: CAD_CAM_EDM_DRO   203695

http://groups.yahoo.com/group/
CAD_CAM_EDM_DRO/

Discussion list for CAD, CAM (CNC), EDM, and DRO. For those who are acronym impaired, this means:

    CAD—Computer-aided design

CAM—Computer-aided manufacturing
CNC—Computerized numeric control
EDM—Electrical discharge machining
DRO—Digital readout

Yahoo Groups: CerfCube 203026

http://groups.yahoo.com/group/CerfCube/

Hardware, software, and hacking of the Intrinsyc CerfCube.

Yahoo Groups: Directory List-Artificial Intelligence 203747

http://dir.groups.yahoo.com/dir/Science/
Computer_Science/Artificial_Intelligence

Top-level list of groups about artificial intelligence.

Yahoo Groups: Directory List-Hobbies and Crafts 203749

http://dir.groups.yahoo.com/dir/
Hobbies___Crafts/Models/Radio-Controlled

Top-level list of groups covering radio-controlled models. Because of the weirdo triple underscores in the URL name, you may want to start at

http://groups.yahoo.com/ then drill down to the Hobbies & Crafts, then Models, then Radio-Controlled subgroups.

Yahoo Groups: Directory List-Robotics 203748

http://dir.groups.yahoo.com/dir/Science/
Engineering/Mechanical/Robotics

Top-level list of groups involved in robotics.

The FIRST group on Yahoo Groups.

Yahoo Groups: FIRST 203922

http://groups.yahoo.com/group/FRCtech2002/

FIRST Robotics Competition message board.

Yahoo Groups: Legged Robots 203763

http://groups.yahoo.com/group/legged-robots/

Robots that walk on one or more legs.

Yahoo Groups: OOPic 203711

http://groups.yahoo.com/group/oopic/

OOPic microcontroller from Savage Innovations.

Yahoo Groups: Open Source Motor Controllers 204227

http://groups.yahoo.com/group/osmc/

Discussion of design issues affecting the Open Source Motor Controller Project. See also:

http://www.dmillard.com/osmc/

Yahoo Groups: PARTS 203921

http://groups.yahoo.com/group/PARTS/

Online meeting place for members of the Portland Area Robotics Society (PARTS).

SEE ALSO:

http://www.portlandrobotics.org/

Yahoo Groups: Piclist 203924

http://groups.yahoo.com/group/piclist/

Catering to the PICMicro microcontroller user.

Yahoo Groups: Rabbit-Semi 203923

http://groups.yahoo.com/group/rabbit-semi/

For users of the Rabbit Semiconductor C-programmable microcontroller.

Yahoo Groups: Robotics 203648

http://groups.yahoo.com/group/Robotics/

General robotics.

Yahoo Groups: Rug Warrior 203746

http://groups.yahoo.com/group/rugwarrior/

For fans of the Rug Warrior robotics kit.

Yahoo Groups: San Diego Robotics Society 202055

http://groups.yahoo.com/group/sdrs-list

Online discussion group for the San Diego Robotics Society.

Yahoo Groups: Seattle Robotics 203920

http://groups.yahoo.com/group/SeattleRobotics/

Message area for the Seattle Robotics Society.

SEE ALSO:

http://www.seattlerobotics.org/

Yahoo Groups: SXtech 203925

http://groups.yahoo.com/group/sxtech/

For those involved with the Scenix SX microcontroller and Parallax SX-Key.

Yahoo Groups: Wires and Circuits 202345

http://groups.yahoo.com/group/wiresandcircuts/j

For the electronics hobbyist, a place to share circuits and project ideas.

INTERNET-CALCULATORS & CONVERTERS

How many drams are in a scruple? Find out with these free online converters and calculators. The specialty Web sites that follow do away with lengthy formulas and allow you to simply enter your data into a text box, and out comes the answer. Converters and calculators are available for electronics design formulas (like calculating the components for use in a 555 timer IC), conver-

sion of weights and measures to other weights and measures, torque calculators, and weight calculators for plastic and metal pieces.

BasicElectronics.com 203787

http://www.basicelectronics.com/

Information, tutorials, circuit examples, FAQ, and online calculators on various electronics topics. Calculators include:

- 555 timer calculator (astable or monostable)
- Resistor color calculator (written by my friend and fellow author Danny Goodman)
- Ohms calculator

Most of the calculators require a browser capable of running JavaScript.

Online Conversion 203385

http://www.onlineconversion.com/

Online converters for 5,000 units and 30,000 conversions.

Includes the following converters: Length/Distance; Temperature; Speed; Volume Liquid and Dry; Weight; Metric Weight; Computer; Date/Time; Cooking; Angles; Area; Power; Energy; Density; Force; Pressure; Astronomical; Numbers; Finance; Miscellaneous; Fun Stuff; Clothing; Light; Torque; Viscosity; Frequency; Flow Rate; Acceleration.

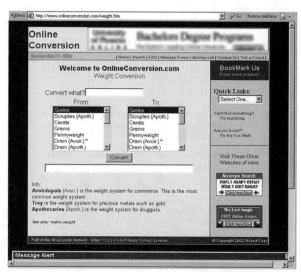

Access some 30,000 conversions at Online Conversion.

Robot Powertrain Calculator 203399

http://www.killerhurtz.co.uk/howto/calculator.htm

Calculate torque, efficiency, and other important motor data using this handy JavaScript calculator. See also the Java applets and JavaScript programs at:

http://www.johnreid.demon.co.uk/howto/

Torque Conversion Calculator 204111

http://www.bodine-electric.com/MotorCalculators/
    orqueConversionCalculator.htm

A simple calculator for converting between units of torque measurement. Thanks to the folks at Bodine.

Torque Speed Applet 204075

http://www.pmdi.com/calculator/
    tsp/tspApplet.html

Java applet for calculating torque/speed and power/speed of a DC motor. You need to input several values (which you obtain from the specification sheet for the motor). Requires Java.

From Precision MicroDynamics; see:

http://www.pmdi.com/

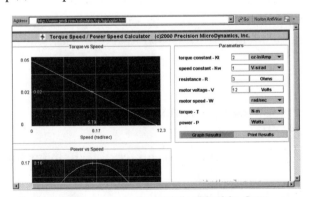

Calculate torque and speed with this Java applet.

Weight Calculator for Metal and Plastic Shapes 203386

http://www.matweb.com/weight-calculator.htm

Online automatic weight calculator for a variety of metals and plastics. Input the material (brass, aluminum, ABS plastic, etc.) and/or the density of the material—which you can get from manufacturer spec sheets—the profile shape, and the dimensions.

Provided by MatWeb, a materials information database with data on 26,000+ materials, including metals, plastics, ceramics, and composites.

INTERNET-EDU/ GOVERNMENT LABS

Some of the best robotics research is taking place in educational institutions and government labs. Much of it is open to public inspection (since taxpayer money funds it), so educational and government lab sites are a treasure trove of ideas for robotics.

There is a lot to see at many of the educational and government labs listed in this section, and you'll want to spend time poking around to see what you can find. For example, the Massachusetts Institute of Technology (MIT) site contains dozens of subsites, with perhaps thousands of pages and images directly related to robotics.

Aerial Robotics Club 202549

http://www.rose-hulman.edu/Users/groups/
RC/Public/HTML/index.html

According to the Web site: The task of the Aerial Robotics Club is to design, build, and maintain an aerial vehicle which is capable of interacting with its environment. Annual competition. Sponsored by Rose-Hulman Institute of Technology, in Terre Haute, Ind. Be sure to check out the photo and video galleries, and the (usually) live lab cam.

MOVERS AND SHAKERS

Marvin Minsky

http://web.media.mit.edu/~minsky/

Marvin Minsky is a true pioneer of modern robotics, going back to 1951 with his invention of the first neural network simulator. His work includes endowing machines with a human capacity for commonsense reasoning. His book *The Society of Mind* presents a novel engineering of the intelligent thought that is derived from thousands, even millions, of simple processes.

Australia Telerobot
203418

http://telerobot.mech.uwa.edu.au/

Control a robot in Western Australia from your Web browser! Just remember robots spin backward Down Under.

BARt-UH
202024

http://www.irt.uni-hannover.de/~biped/

BARt-UH is a bipedal autonomous walking robot designed at the University of Hannover in Germany. Web site is in English.

Biorobotics
202279

http://biorobots.cwru.edu/

Bots at the Biologically Inspired Robotics Lab at Case Western Reserve University. According to the site, the lab "is dedicated to the advancement of the field of robotics using insights gained through the study of biological mechanisms."

BIP2000 Anthropomorphic Biped Robot
202973

http://www.inrialpes.fr/bip/Bip-2000/

Fancy two-legged walking robot Laboratoire de Mecanique des Solides and INRIA Rhne-Alpes. Web site is in English and French.

MOVERS AND SHAKERS

Seymour Papert

http://www.papert.com/

Professor Papert envisioned kids learning by computer when computers were the size of Buicks. He spearheaded the development of the Logo programming language, which became a de facto standard in schools across the country in the 1970s and early 1980s. The LEGO Mindstorms line of robotic construction kits is named after his book on using technology to teach children: *Mindstorms: Children, Computers and Powerful Ideas.*

Bruno Jau Robotic Hand
203091

http://uirvli.ai.uiuc.edu/tlewis/pics/hand.html

Finally-a robot that will give you a hand.

Carnegie Mellon, Robotics Institute
203012

http://www.ri.cmu.edu/home.html

All about the Robotics Institute at Carnegie Mellon University in Pittsburgh. Pa.

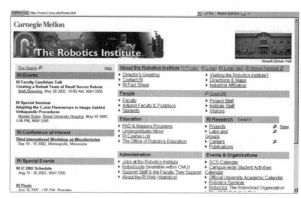

Robotics at Carnegie Mellon University

Carnegie Mellon, Robotics Institute
202016

http://www.frc.ri.cmu.edu/

Information about the Field Robotics Center at Carnegie Mellon University (Pittsburgh, Pa.).

Carnegie Mellon University: Minerva
203090

http://www-2.cs.cmu.edu/~minerva/tech/index.html

Minerva is an autonomous tour guide—"We're walking, we're walking, we're stopping. . . ."

Case Western Reserve University-IGERT
203092

http://neuromechanics.cwru.edu/

About the Neuro-Mechanical Systems program at Case Western Reserve University (Cleveland, Ohio).

Cognitive Architectures   204122

http://ai.eecs.umich.edu/cogarch2/

Online articles compare a variety of current proposed cognitive architectures and a workable structure for classifying and comparing future proposed cognitive architectures.

SEE ALSO:

http://ai.eecs.umich.edu/cogarch0/

http://ai.eecs.umich.edu/cogarch0/subsump/

So, What Does Your Robot Do?"

If robot makers had a dollar for every time someone asked them that, they'd be neck and neck with Bill Gates as the world's richest people. But to be fair, the uses of amateur robots aren't always clear.

Here's a short list of common robotic functions, should you be wondering what your robot should do—whether you've built it already or not.

- *Wanderer.* This robot does nothing special, except explore. Entire counties have been built and populated thanks to the explorers, so this is a great job for any robot!
- *Line tracer.* A simple robot that traces a line is a precursor to many types of worker machines that dutifully follow a track. Includes delivery robots, sentry robots, and many others.
- *Wall follower.* Like a line tracer, the wall follower seeks to navigate its realm guided by the perimeters of a room. All rooms have walls or other boundaries of some type, and they can be used by simple machines for simple navigation.
- *Maze solver.* The typical maze-solving robot is an enhanced version of the wall follower. Mazes can be solved using various algorithms that are based on following the walls of a maze in a certain sequence.
- *Vacuum cleaner.* This group encompasses any class of janitorial robot, whether it cleans the carpets, waxes the floor, or mows the lawn. The issues are the same: Do some work within a confined space, without knocking over people, pets, and things.
- *Burglar alarm/sentry.* If a robot is going to wander or trace a line or follow a wall, it might as well do something constructive at the same time. Robots make for perfect "mobile burglar alarms" or sentries. Equip yours with a video camera and it can record crooks in the act. Given the proper sensors, the robot might also sniff out noxious fumes, like carbon monoxide.
- *Drink server.* Mobile or stationary robots can serve drinks, appetizers, and other food. The robot must have a strong arm that can lift the glass or food, which makes them among the harder machines to build.
- *Robot gladiator.* Whether the robot fights sumo style or head to head in a death match, robot gladiators are either autonomous or remote-controlled virtual combatants.
- *Firefighter (simulation).* There are real firefighting robots, and most are teleoperated by a human. For a self-governing firefighting robot, simulation is almost as good. Most firefighting contests are designed to prove a design and are staged in miniature "houses" with candles as the fire.
- *Personal assistant.* We're talking simple stuff here, like alarm clocks for waking up in the morning.
- *Educational experience.* Let's not forget that building any robot teaches important mechanical and scientific principles to the maker. That ought to be worth something today.
- *Conversation piece.* A robot of your very own, especially one you built, is a great ice-breaker at parties.

Cool Robot of the Week   202012

http://ranier.hq.nasa.gov/telerobotics_page/coolro-bots.html

According to the Web site, "The honor of being listed as 'Cool Robot of the Week' is bestowed upon those robotics-related web sites which portray highly innovative solutions to robotics problems, describe unique approaches to implementing robotics systems, or present exciting interfaces for the dissemination of robotics-related information or promoting robotics technology."

1998 archive:
http://ranier.hq.nasa.gov/telerobotics_page/coolrobots98.html

1999 archive:
http://ranier.hq.nasa.gov/telerobotics_page/coolrobots99.html

2000 archive:
http://ranier.hq.nasa.gov/telerobotics_page/coolrobots00.html

2001 archive:
http://ranier.hq.nasa.gov/telerobotics_page/coolrobots01.html

Cornell Robotics and Vision Laboratory   203420

http://www.cs.cornell.edu/Info/Projects/csrvl/csrvl.html

Past and current projects at Cornell Robotics and Vision Laboratory (Ithaca, N.Y.).

DEMO-Dynamical & Evolutionary Machine Organization   204124

http://www.demo.cs.brandeis.edu/

From the Web site: "DEMO attacks problems in agent cognition using complex machine organizations that are created from simple components with minimal human design effort." From Brandeis University (Waltham, Mass.).

See also The Golem Project:

http://www.demo.cs.brandeis.edu/golem/

Field Robotics Center   202546

http://www.frc.ri.cmu.edu/

At Carnegie Mellon University.

In the words of the Web site: "Research into sun-synchronous navigation will discover, express, and exhibit the importance of reasoning about sunlight as it pertains to robotic exploration."

Franklin Institute's Robotics   203190

http://www.fi.edu/qa99/spotlight2/

Robotics at the Franklin Institute (Philadelphia, Pa.).

Georgia Tech Intelligent Systems & Robotics   203191

http://www.cc.gatech.edu/isr/

From the site: "The goal of the Intelligent Systems and Robotics group in the College of Computing at Georgia Tech is to understand and design systems which use intelligence to interact with the world, making computer controlled systems more autonomous and ubiquitous."

Georgia Tech Mobile Robot Lab   202547

http://www.cc.gatech.edu/ai/robot-lab/

From Georgia Tech: "The Mobile Robot Laboratory's charter is to discover and develop fundamental scientific principles and practices that are applicable to intelligent mobile robot systems."

Hexplorer 2000   203761

http://real.uwaterloo.ca/~robot/

The Hexplorer is a six-legged walking robot at the University of Waterloo, located in Ontario, Canada. Construction details and programming overview are provided.

See also the main page for the Motion Research Group:

http://real.uwaterloo.ca/

Image Science and Machine Vision Group
203192

http://www-ismv.ic.ornl.gov/

Says the Web site: "The Image Science and Machine Vision Group is currently involved in three programmatic areas: measurement and controls for industry, biological sciences, and surveillance and security."

Intelligent Systems and Robotics Center (ISRC)
203135

http://www.sandia.gov/isrc/home.html

Among other projects, ISRC contemplates robots for warfare and national security. Research includes Surveillance and Reconnaissance Ground Equipment (SARGE), Miniature Autonomous Robotic Vehicle (MARV), Accident Response Mobile Manipulator System (ARMMS), and a Robot that Makes Up Acronyms (RTMUA).

Iowa State University Robotics Club
202548

http://www.ee.iastate.edu/~cybot/

At Iowa State University.

From the Web site: "Project Cybot is a unique combination of a continuous senior design project and a club open to all students at ISU."

IRIDIA Projects and Activities
204121

http://iridia.ulb.ac.be/Projects/

Artificial intelligence white papers and project summaries. From IRIDIA, the artificial intelligence research laboratory of the Université Libre de Bruxelles.

JPL Rover and Telerobotics
203194

http://robotics.jpl.nasa.gov/

If it walks on another celestial body—like Mars—and was launched by NASA, JPL built it. Here, you can read about JPL's past, present, and future projects. Be sure to check out the Robotic Vehicles Group page.

Laboratory for Perceptual Robotics (LPR)
203877

http://www-robotics.cs.umass.edu/lpr.html

University of Massachusetts Perceptual Robotics laboratory.

LEGO: Distributive Intelligence with Robots
202246

http://www.ceeo.tufts.edu/me94/

How they got LEGO Mindstorms robots to work together. Demonstrations include Whistling Brothers, Travel by Beacon, and Wandering Cyclops.

Lenox High School Bot Club
202059

http://www.loganbot.com

Here's what the Web site has to say: "This site details the efforts of the Lenox High School Bot Club towards the construction of a super-heavyweight Battlebot named Logan. We feel that real life engineering projects that incorporate many scientific applications are a great way to bring technology into the classroom. Helpful information, links and 12 pages of combat robot building tips are also available."

Machine Intelligence Laboratory
203073

http://www.mil.ufl.edu/

The goings-on at the Machine Intelligence Laboratory at the University of Florida (Gainesville).

Massachusetts Institute of Technology
202083

77 Massachusetts Ave.
Cambridge, MA 02139-4307
USA

📞 (617) 253-1000

🌐 http://web.mit.edu/

This is the main Web site for the Massachusetts Institute of Technology (MIT) in Cambridge, Mass. Links on the main page take you to various labs and research centers at the campus. Spend some time on this one.

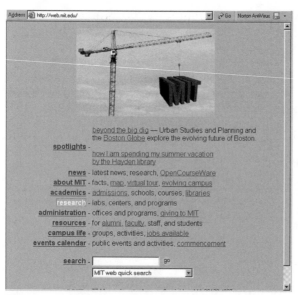

Robotics at Massachusetts Institute of Technology.

MIT: Artificial Muscle Project 202446

http://www.ai.mit.edu/projects/muscle/muscle.html

The Artificial Muscle Project at the MIT (Cambridge, Mass.) Artificial Intelligence Laboratory plays around with linear actuators using a substance known as polymer hydrogel. This material is said to have characteristics similar to human muscle.

MIT: FTP site 203294

ftp://cherupakha.media.mit.edu/pub/

Downloadable files of various projects, research papers, and doctoral theses from MIT.

Note that this is an FTP (File Transfer Protocol) site, and can be used with a specialized FTP program or with most browsers.

MIT: Leg Laboratory 203256

http://www.ai.mit.edu/projects/leglab/

The Leg Lab is world-renowned for its designs of various single- and multipedal robots. Movies are available for many of the designs.

MIT: Logo Foundation 203119

http://el.www.media.mit.edu/logo-foundation/

Informational page about the Logo programming language, originally developed by professors at MIT (Cambridge, Mass.).

MIT: MindFest 203101

http://www.media.mit.edu/mindfest/

MindFest is a yearly gathering of LEGO-heads. You can see pictures of past events and read up on upcoming ones.

Mobile Robots at Loughborough 202266

http://www.lboro.ac.uk/departments/el/robotics/

There be robots at the Department of Electronic & Electrical Engineering at Loughborough University (Loughborough, England).

NASA 202325

http://www.nasa.gov/

The main home page of the National Aeronautics and Space Administration, the outfit in the U.S. that launches the space shuttle and the occasional robot.

Daily update of things at the National Aeronautics and Space Administration.

NASA JPL: Mars Pathfinder 202275

http://mars.jpl.nasa.gov/MPF/index1.html

Informational site about the Mars Pathfinder mission. The mission may be over, but the interest in it is not.

Navy Center for Applied Research in Artificial Intelligence 202559

http://www.aic.nrl.navy.mil/

From the Web site: "The Navy Center for Applied Research in Artificial Intelligence (NCARAI) has been involved in both basic and applied research in artificial intelligence since its inception in 1982. NCARAI, part of the Information Technology Division within the Naval Research Laboratory, is engaged in research and development efforts designed to address the application of artificial intelligence technology and techniques to critical Navy and national problems."

Poly-PEDAL Lab 204158

http://polypedal.berkeley.edu/

The Poly-PEDAL Lab studies motion in animals and insects. The walk (gait) and balance studies often help in designing legged robots.

Polypod 203401

http://robotics.stanford.edu/users/mark/polypod.html

From the Web site: "Polypod is a bi-unit modular robot. . . . This page presents work done in 1993 and 1994. Work on the next generation, called 'PolyBot' started mid 1998 at Xerox PARC as part of the modular robotics project under the smart matter theme."

Robotics and Computer Vision Laboratory 203419

http://www-cvr.ai.uiuc.edu/

The way they see things as the Robotics and Computer Vision Laboratory at the University of Illinois (Urbana).

Robotics and Intelligent Machines Laboratory 203011

http://robotics.eecs.berkeley.edu/

Research and activities at the Robotics and Intelligent Machines Laboratory at the University of California at Berkeley.

Robotics Group at Columbia University 202263

http://www.cs.columbia.edu/robotics/

Research in robotics—both mobile and stationary—at Columbia University (New York, N.Y.).

Robots at Space and Naval Warfare Systems

See SPAWAR (this section).

Sandia Intelligent Systems & Robotics Center 202556

http://www.sandia.gov/isrc/

In the words of the Web site, "The Intelligent Systems and Robotics Center (ISRC) is a world leader in creating miniature to macro-sized, teleoperated to autonomous, vehicles for military and industrial applications. From environmental clean-up to the battlefield, the ISRC is expert in developing unique intelligent mobile systems."

Side Collision Warning System for Transit Buses 202257

http://www.ri.cmu.edu/projects/project_324.html

Now just imagine the collision system on a big robot instead of a bus. From Carnegie Mellon University.

SPAWAR 202038

http://www.nosc.mil/robots/

A look at robotics at SPAWAR (Space and Naval Warfare Systems Center) in San Diego, Calif. Most of the robots are for military, urban defense, or other applications in which weapons systems—both lethal and nonlethal—are involved.

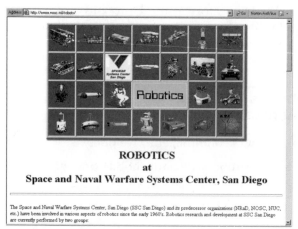

Robotics at Space and Naval Warfare Systems Center.

Stanford Robotics Laboratory 203013

http://www.robotics.stanford.edu/

How robots live at the Stanford Robotics Laboratory in Stanford, Calif. See also Stanford's autonomous helicopter work at:

http://sun-valley.stanford.edu/projects/helicopters/helicopters.html/

Talking Heads 202223

http://www.haskins.yale.edu/haskins/heads.html

From the Web site: "This website provides an overview of the rapidly growing international effort to create talking heads (physiological/computational/cognitive models of audio-visual speech), the historical antecedents of this effort, and related work. Links are provided (where possible) to the sites of many researchers and commercial entities working in this diverse and exciting area."

Tarry Walking Machines 202022

http://www.tarry.de/index_us.html

According to the Web site: "This is the homepage of the Tarry walking machines, which were developed and built by the Department of Engineering Mechanics at the University of Duisburg."

Good detailed look at hexapod designs. Recommended reading.

Toy Robot Initiative 204118

http://www-2.cs.cmu.edu/~illah/EDUTOY/

The Toy Robot Initiative aims to commercialize robotics technologies in education, toys, entertainment, and art. Operated from the Mobile Robot Programming Laboratory in Carnegie Mellon University's Robotics Institute (Pittsburgh, Pa.).

Union College Robotics Club 203366

http://www.vu.union.edu/~robot/
Schenectady, N.Y.

University of Edinburgh AI Machine Vision Unit 202551

http://www.ipab.informatics.ed.ac.uk/mvu/

Overview of the Machine Vision Unit at the University of Edinburgh (that would be in Scotland).

University of Michigan Artificial Intelligence Laboratory 204253

http://ai.eecs.umich.edu/

People and projects at the AI lab at the University of Michigan (Ann Arbor).

University of Michigan Mobile Robotics Lab 203468

http://www.engin.umich.edu/research/mrl/

The Mobile Robotics Lab at the University of Michigan (Ann Arbor). Ho hum? Not quite.

Some special research goes on here in the fields of robot navigation. Be sure to read the details of the mobile robot positioning and obstacle avoidance research. The book Where Am I? (in print, on CD-ROM, and for electronic download), published by the university's Johann Borenstein, is a classic and is required reading in many mechatronics courses.

See also Dr. Borenstein's home page, where he provides links to many more online robotics resources:

http://www-personal.engin.umich.edu/~johannb/

Robotics at the University of Michigan.

University of New Hampshire Robotics Lab

202552

http://www.ece.unh.edu/robots/rbt_home.htm

In the words of the Web site: "The research emphasis of the Robotics Laboratory in the Department of Electrical and Computer Engineering is the application of fast associative memories and other neural network learning techniques (such as CMAC neural networks) to problems in control, pattern recognition, and signal processing."

University of Reading Department of Cybernetics

202082

http://www.cyber.rdg.ac.uk/CIRG/home.htm

The Cybernetic Intelligence Research Group studies intelligence and its real-life applications.

University of Southern California Robotics Research Laboratory

203469

http://www-robotics.usc.edu/

Robotics research at USC spans a large number of labs and projects. These include:

- USC Robotics Research Lab

- Interaction Lab (control and learning in multirobot and humanoid systems)
- Computational Learning and Motor Control Lab
- Laboratory for Molecular Robotics
- Robotic Embedded Systems Lab
- Polymorphic Robotics Laboratory (reconfigurable robotics)

University of Toronto Robotics & Automation

202553

http://www.mie.utoronto.ca/labs/ral/

University of Toronto Robotics and Automation Laboratory and Mechatronics Laboratory.

URBIE Urban Reconnaissance Robot

203193

http://telerobotics.jpl.nasa.gov/tasks/tmr/

According to the Web site: "Urbie's initial purpose is mobile military reconnaissance in city terrain but many of its features will also make it useful to police, emergency, and rescue personnel. The robot is rugged and well-suited for hostile environments and its autonomy lends Urbie to many different applications. Such robots could investigate urban environments contaminated with radiation, biological warfare, or chemical spills. They could also be used for search and rescue in earthquake-struck buildings and other disaster zones."

USC-Robota Dolls

202013

http://www-clmc.usc.edu/~billard/robota.html

Playing with dolls at the University of Southern California.

According to the Web site: "The ROBOTA dolls are a family of mini humanoid robots. They are educational toys. They can engage in complex interaction with humans, involving speech, vision, and body imitation."

USU ECE Center for Self-Organizing and Intelligent Systems

202262

http://www.engineering.usu.edu/ece/projects/csois/

Robot work at Utah State University. Check out past and present projects, including the stair-climbing robot and the omnidirectional wheel designs.

Waterloo Aerial Robotics Group 202554

http://ece.uwaterloo.ca/~warg/

In the words of the Web site: "The Waterloo Aerial Robotics Group [at the University of Waterloo in Canada] is a team of engineers who are developing a series of fully autonomous vehicles (both air and ground). The goal is to have a fleet of robots that can work cooperatively toward some predefined goal without the slightest bit of help from any human crew."

 INTERNET-INFORMATIONAL

What could be better than information about robotics? Try free information about robotics! The resources that follow provide, at no cost, a bevy of useful information about robotics, electronics, mechanics, programming, computers, or some other subject directly related to robot building.

In many cases, the sites listed provide additional informational resources, so be sure to check each Web site's table of contents or home page. Note that a few of the resources listed here are provided in PDF format and therefore require the Adobe Acrobat program, which

you can download for free from Adobe. See the introduction for more details.

SEE ALSO:

INTERNET-EDU/GOVERNMENT LABS: Research robots

INTERNET-PERSONAL WEB PAGE: Others share their robot creations with you

INTERNET-PLANS & GUIDES: Full plans and guides for electronics and robotics

MANUFACTURER (VARIOUS): Freebie datasheets, application notes, and product designs for electronic components and products

---

Ackerman Steering and Racing Oval Tracks 204127

http://www.auto-ware.com/setup/ack_rac.htm

What Ackerman Steering is all about. The article talks about steering for full-size racing cars, but the concepts are the same for any size vehicles, even robots.

All About MEMS 204142

http://www.analog.com/imems/

Gateway to MEMS (microelectromechanical systems) technology, published by Analog Devices, a leader in the field.

Alphadrome Robots and Space Toys 202072

http://www.alphadrome.com/

For collectors of tin robots and space toys. Includes a discussion board for collectors.

Antique Radio-Phil's Old Radios 202662

http://antiqueradio.org/

Information, want ads, and beginner's info on working with old-time radio, antique circuits, and electron tubes.

Art of Motion Control, The 204184

http://www.taomc.com/

Motion control techniques in art. Entertaining reading.

Animal Makers
http://www.animalmakers.com/
Animatronics company specializing in life-like animated animals for motion pictures and television

Jim Fuller's Resource Site
http://www.southwest.com.au/~jfuller/
Robot and programming resources

MacRobotics
http://www.macrobotics.com/
Robotics and the Macintosh

Motion Control, Inc.
http://www.utaharm.com/
About the Utah arm, "robotic" prosthetic arm

Automation Sensors   202112   (🌐) http://www.automationsensors.com/

6550 Dumbarton Cir.
Fremont, CA 94555
USA

(📞) (435) 753-7300
(📠) (435) 753-7490
(🚫) (888) 525-7300

Makers of self-contained ultrasonic sensors and pressure products. Check out the technical reference section for a number of application notes, as well as handy white papers (in Adobe Acrobat PDF format) on such things as dielectric constants, bulk densities, engineering unit abbreviations, and thread specifications.

Self-Contained versus Tethered Robots

What's a real robot? One commonly accepted definition says that it's a *self-contained, autonomous* (self-governed) machine that needs only occasional instructions from its master to set it about its various tasks. A self-contained robot includes its own power system, brain, wheels (or legs or tracks), and manipulating devices such as claws or hands. The robot does not depend on any other mechanism or system to perform its tasks. It's complete, in and of itself.

However, this definition ignores the legions of factory robots that are, in effect, mechanical arms connected to a computer someplace else in the room. They are considered "robots" partly because of convention: The term was used by their early creators, and it's stuck ever since.

Self-contained robots are those that incorporate all of the necessary ingredients for a self-governed machine—sensors, processing, and mechanical action—all in one box. Such a robot may be mobile or stationary. A mobile robot has wheels, legs, or some other form of locomotion. The typical mobile robot is designed either for exploring or delivering; the typical stationary robot is designed for manipulating objects, such as for construction or handling dangerous materials.

Tethered robots, on the other hand, are robots in form, but not necessarily in intelligence. The mechanism that does the actual task is the robot itself; the support electronics or components may be separate. The link between robot and control components might be a wire, a beam of infrared light, or a radio signal. Though only marginally considered robots, the tethered variety comprise the bulk of all robots in existence today.

Reality versus Fantasy

Where does reality end and fantasy begin? When building robots, the line isn't always clear. Separating reality from fantasy helps avoid overreaching designs and lost effort. Fantasy is a *Star Wars* R2-D2 robot projecting a hologram of a beautiful princess. Reality is a home-brew robot that rambles down the hallway, maybe even hitting the walls as it goes. Fantasy is a giant killer robot what walks on two legs and shoots a death ray from a visor in its head. Reality is a foot-tall "trashcan" robot that offers houseguests a diet soda.

Sure, everyone wants to build a robot that fully replicates human intelligence and abilities, but the reality is that such a robot is far away, even for engineering teams spending millions of dollars. It's important to be wary of impossible plans. Don't attempt to give your robot features and capabilities that are beyond your technical expertise or budget.

Here's an idea to help keep you on the reality track: When designing your robot, write some notes about what you want it to do, then put the notes away. Let them gel in your brain for a week or two. Quite often, when you review your original design, you will realize that some of the features and capabilities are mere wishful thinking and beyond the scope of your time, finances, or skills. Make it a point to refine, alter, and adjust the design of the robot before, and even during, construction.

Autonomous Robot Controller 202468

http://indai.com/robot/

In the words of the Web site: "ARC is a simulator of multiple mobile robots. It is used to test controllers devised by the user in order to streamline the design phase previous to their installation in real robots. This simulation is, of course, only an approximation of a real situation and in some cases a valid controller in simulation will not be so in the real world."

AutoPilot UAV project 202036

http://autopilot.sourceforge.net/

The AutoPilot UAV project is intended to develop a nonpiloted "drone" helicopter. According to the Web page, "The goal is to produce an autonomous aerial vehicle that can stay aloft for over two hours, carrying over 50 kg of payload and cost less than $10,000."

All is done with free open source software.

Bill Ruehl 202061

http://www.robotdude.com/

Microcontroller info and projects:

- Hardware hacks
- Robot-building info
- Links

Chuck Rosenberg: Robot Pages 203475

http://www-2.cs.cmu.edu/~chuck/infopg/

Articles, photos, and resources. Check out "Practical Robot Building Lessons"—some excellent advice on robotics, including connectors and repairs.

Chuck's Robotics Notebook 202654

http://www.mcmanis.com/chuck/robotics/

View of robotics from robomeister Chuck McManis. Semitechnical.

Droid Maker's Workshop 202130

http://www.geocities.com/droidmakr/

From robot enthusiast Clifford Boerema, robot-building help, especially for first timers (but some electronics skill or knowledge is handy). Some very nice semi-technical articles, including schematics.

Dustbots 202845

http://www.dustbots.com/

Vacuum cleaner and cleaning robots.

Fred Barton Productions, Inc. 202683

P.O. Box 1701
Beverly Hills, CA 90213-1701
USA

📞 (310) 234-2956
📠 (310) 234-0956
✉ tobor1701@earthlink.net
🌐 http://www.the-robotman.com/

Information on robotic props by Fred Barton; check out the Robot Museum.

Fred also sells a limited number of full-sized, licensed Robby the Robot (from Forbidden Planet fame) robots.

According to the Web site: "Robby, the Robot, manufactured by Fred Barton Productions, Inc. of Hollywood, California, is an exact 1:1 scale replica of the famous movie robot as seen in MGM's classic sci-fi thriller Forbidden Planet. The Collector's Edition DX of Robby is computerized, remote controlled, and incorporates a digital audio sound-track from the movie that lights the nine mercury-vapor neon tubes in his mouth synchronously with the robot's original voice as heard in the film."

Fred Barton and his robot re-creations. Photo Fred Bardon Productions, Inc.

Fred's Robby is over 7 feet tall, weighs just over 100 pounds, and is made of fiberglass. The robot is officially licensed through the current owners of the original film.

Furby Autopsy   204156

http://www.phobe.com/furby/

See the insides of a Furby. Clear photographs and close-ups. Not for the squeamish.

Amazingly, there are also Furby Autopsy T-shirts and coffee mugs. Buy them here.

G. W. Lucas: Differential Steering   203745

http://rossum.sourceforge.net/papers/DiffSteer/

A detailed and technical paper on differential steering, especially as it relates to robots.

Generating Sony Remote Control Signals with a BASIC Stamp II   203068

http://www.whimsy.demon.co.uk/sircs/index.html

As the title says.

Generic Algorithms for Gait Synthesis in a Hexapod Robot   202969

http://www.iguana-robotics.com/people/tlewis/publications/rodney2.PDF

Just like the title says. The document requires Adobe Acrobat Reader.

Handhelds.org   204146

http://www.handhelds.org/

As published on the Web site: "Our goal is to encourage and facilitate the creation of open source software for use on handheld and wearable computers."

Hero-1 Robot   202219

http://irobot.org/hero/default.htm

Projects, FAQ, and information about the ever popular HERO robot, once sold by Heathkit.

Hints and Tips for Prototyping with SMD   203044

http://www.geocities.com/vk3em/smtguide/smtguide.htm

Informational guide on how to work with surface mount components.

History of Robotics   204123

http://cache.ucr.edu/~currie/roboadam.htm

A history of robotics, as told by Adam Currie. Not a bad overview.

HwB: Connector Menu   202088

http://www.cc86.org/~pjf/hwb/menu_Connector.html

Pinout descriptions of dozens of connector types: MIDI, PCMCIA, parallel, serial, etc.

ICybie Hacking   203029

http://www.aibohack.com/icybie/hacking.htm

Steps to hack a Tiger Electronics ICybie pet dog robot.

Imaginerobots.com   202657

http://www.imaginerobots.com/

Danh Trinh's ever-most-excellent repository of robotic creations. Danh builds some of the best robots I've seen on the Web; check out his Robot Bug hexapod, which he built from scratch using plastic parts created with his desktop CNC machine (he uses a retrofitted Sherline desktop mill). A particular interest of Danh's is aerial robotics.

The RoboBug, from Imaginerobotics.com. Photo Mike Potter

Robo forklift, from Imaginerobotics.com. Photo Mike Potter.

Inexpensive Homebrew Inertial Guidance System   202334

http://www.precision3d.org/IGS/

Plans, schematic, and description of building a homebrew inertial guidance system based on a Tokin CG-16D gyro sensor, designed to function for aerial navigation.

Infra Red Remote Controls-How They Work   203046

http://www.cim.mcgill.ca/~arlweb/mechatronics/p5/homeworks/hw1p5.htm

Good semitechnical details on how infrared remote controls work. From the homework notebook of Pedro Serrano, an undergraduate student at McGill University. (Gee, my homework never looked this good. . . .)

SEE ALSO:

http://www.cim.mcgill.ca/~arlweb/

Introduction to Capacitors   203005

http://www.execpc.com/~endlr/index.html

Learn all you wanted to know about capacitors here.

Joe Mehaffey and Jack Yeazel's GPS Information   203430

http://www.gpsinformation.net/

Joe and Jack yack about global positioning satellite (GPS) receivers and how to interface with them using computers.

KISS Institute for Practical Robotics (KIPR)   202540

http://www.kipr.org/

In the words of the Web site: "KISS Institute for Practical Robotics (KIPR) is a private non-profit community-based organization that works with all ages to provide improved learning and skills development through the application of technology, particularly robotics. We do this primarily by providing supplementary, extra-curricular and professional development classes and activities. KISS Institute's activities began in 1993."

KIPR also sponsors the annual Bot Ball tournament for middle and high school students.

Klatt's History of Speech Synthesis   204222

http://www.cs.indiana.edu/rhythmsp/ASA/Contents.html

Speech synthesis through the ages, with downloadable sample clips, including the infamous "Bicycle Built for Two" used in 2001: A Space Odyssey.

LED FAQ Pages, The 203783

http://www.pioneernet.net/optoeng/LED_FAQ.html

Informational Web page on light-emitting diodes.

Line Follower 202470

http://filebox.vt.edu/users/afalck/www/research/Controls-2.html

Lots of technical information and math about creating a line-following robot. Many design formulas.

Max's Little Robot Shop 202302

http://www.users.qwest.net/~kmaxon/page/

Robo guru Kenneth Maxon shares some information on a variety of robot-related technologies, including robot construction, controllers, and injection molding. Some nice illustrations of projects.

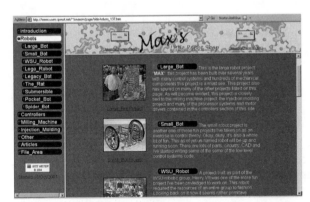

Kenneth Maxon Web page.

Mechatronics Tutorial Information 202939

http://www.engr.sjsu.edu/bjfurman/courses/ME106/mechatronicstutorials.htm

Some useful lecture notes on things like basic electronics, op amps, and electronics symbols. From San Jose State University (San Jose, Calif.).

Modular Reconfigurable Robotics 203402

http://www.parc.xerox.com/spl/projects/modrobots/

Xerox has developed some ideas for robots that remake themselves into various shapes.

Murray McKay 202336

http://home.midsouth.rr.com/mmckay/

Murray is a robot builder, and he shares with us his designs and how-to tutorials. Included for such robots as Weevil and Aardvark are schematics, example programs, and descriptions. See also his page on constructing an infrared proximity detector.

NASA: Robotics Education Project 203707

http://robotics.nasa.gov/

From the Web site: "The NASA Robotics Education Project (REP) is dedicated to encouraging people to become involved in science and engineering, particularly robotics. REP works to capture the educational potential of NASA's robotics missions by supporting educational robotics competitions and events, facilitating robotics curriculum enhancements at all educational levels, and maintaining a web site clearinghouse of robotics education information."

Newsgroups: Comp.robotics FAQ 202098

http://www.truegift.com/robots/

Revised (and for now somewhat abridged) version of the comp.robotics.misc FAQ.

Norwich Robotics Project 203791

http://www.norwichroboticsproject.org/

From the site director, Nick Sheldon: "In December 1998 an application was made . . . to design and build eight mobile robots, each to be controlled by a Pentium PC, and use these to teach free courses on robotics to the general public. The proposed courses would consist of morning workshops, to be held at weekends, and cover all levels from basic skills to A level equivalent."

This Web site reviews the courses, worksheets, and other materials that comprise the project. Some interesting robotics tidbits to be found here, including a self-assessment quiz for those interested in pursuing higher education in robotics and mechatronics.

Omniscience Futureneering 203770

http://www.webcom.com/sknkwrks/

Various remote-control application notes, including radio-controled mowers and hacks to give servos more power. Winner of the Best Tagline on a Web Page Award: "Everything is dangerous if you're stupid."

ParalleMIC 203136

http://www.parallemic.org/

Online resource for "parallel mechanisms," used extensively in robotic arms and hands and also in the legs of some walking robots.

Paul's Cheap Sonar Range Finder Design 203466

http://www.hamjudo.com/sonar/

How Paul built an inexpensive ultrasonic sonar system using a PIC16F84 microcontroller.

Project 64 202509

http://project64.c64.org/index.htm

Collection of software and documentation for the Commodore 64, including interface plans and schematics.

Puppet-Building Information 204134

http://www.puppetbuilder.com/info/

Brief details on the puppet-making art, from puppeteer Nick Barone. The armatures, control linkages, and even foam bodies of puppets can be used to create robots.

Puppetry.info 204135

http://www.puppetry.info/

Technical aspects of puppetry, including construction techniques.

RepairFAQ: Basic Testing of Semiconductor Devices 202899

http://www.repairfaq.org/REPAIR/F_semitest.html

From Sam Goldwasser's RepairFAQ: how to test semiconductors; test most with a multimeter, but other test tools are discussed.

Main page for Repair FAQ.org.

RepairFAQ: Capacitor Testing 202900

http://www.repairfaq.org/REPAIR/F_captest.html

From Sam Goldwasser's RepairFAQ: how to safely test and discharge capacitors. Hint: It's not done by licking the terminals.

RepairFAQ: Salvaging Interesting Gadgets, Components, and Subsystems 202868

http://www.repairfaq.org/REPAIR/F_gadget.html

From Sam Goldwasser's RepairFAQ: where to find useful parts by grubbing only junk.

Robot Maxamilian 203088

http://www.howtoandroid.com/

Billed as "A website dedicated to showing others how to build their own android robots." Very good information on mechanics and animatronics, and includes a parts list with estimated costs.

Robot Room 202076

http://www.robotroom.com/

By the author of *Robot Building for Beginners*; provides extra projects, tips, techniques, resources and links, and an errata for the book.

Robotics for Sculptors 203144

http://www.sculptor.org/3D/Cutting/robotics.htm

Links for sculptors interesting in integrating their art with machine control (such as 3D routers).

Robotics Frequently Asked Questions List 202097

http://www.frc.ri.cmu.edu/robotics-faq/

From the Web site: "This is the Frequently Asked Questions (FAQ) list for the internet robotics newsgroups comp.robotics.misc and comp.robotics.research. This list provides a resource of answers to commonly (and some uncommonly) asked questions regarding robotic systems, organizations, periodicals, and pointers to numerous other resources on the net."

This FAQ is rather old (1996 is the latest revision at this time of this writing). Most of the information contained within it is out-of-date.

Robotics Mini-FAQ for Beginners 202029

http://www.acroname.com/robotics/info/mini_faq.html

Robotics and engineering instructor John Piccirillo (jpicciri@eb.uah.edu) offers us the Mini-FAQ, which he keeps updated on a fairly regular basis. A great source of information for getting started.

Robotics Universe 202332

✉ gort@robotoid.com

🌐 http://www.robotoid.com/

Online support site for my robotics books, including the one you're reading now.

RobotsLife.com 203471

http://www.robotslife.com/

According to the Web site, RobotsLife.com has a unique mission: to "help people understand what's happening with these electronic creatures." Includes news and links.

Sharp Sensor Hack for Analog Distance Measurement 202475

http://www.cs.uwa.edu.au/~mafm/robot/sharp-hack.html

Reengineering a Sharp GPIU5 infrared detector module to determine distance.

Tech Toys Today 203138

http://www.techtoystoday.com/

Tech Toys Today is a free informational site hosted by amateur robotics enthusiast and book author Dennis Clark. There are a number of very handy advanced-beginner and intermediate-level robotics projects, including:

- Serial PWM chip
- Serial compass
- Serial servo controller
- PIC-based IRPD (infrared proximity detector)

Technical Guide to Building Fighting Robots 202299

http://homepages.which.net/~paul.hills/

Several useful and insightful articles on intermediate and advanced robot-building techniques by Paul Hills. Examples include:

- Making a high-power servo-theory and circuits
- "The physics of axe weapons"
- Using DC motors in fighting robots

Fairly technical and in-depth.

Trains.com 202395

http://www.trains.com/

All aboard trains. Trains.com is a one-stop shop for model trains of all gauges. Included are links, articles, how-tos, product reviews, and more. The site is the centerpiece of specialty magazines Model Railroader, Trains, Classic Toy Trains, and Garden Railways and Classic Trains, and there are links to these publications.

Triboelectric Charging of Common Objects 202668

http://www.ece.rochester.edu:8080/~jones/demos/charging.html

All about triboelectric charging (causing static electricity) of various common materials.

Twysted Pair 203272

http://www.twysted-pair.com/

Electronics how-to and theory. Says the site: "Tools to help you work with resistors, capacitors, inductors, FETs, transformers, diodes, transistors, digital logic, TTL, and CMOS devices."

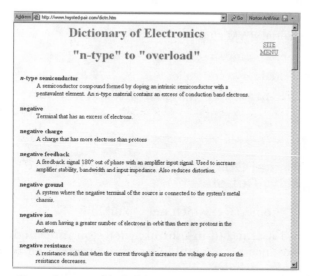

One of several helpful pages at Twysted Pair.

United States Library of Congress 203281

http://lcweb.loc.gov/

The U.S. Library of Congress is the world's largest library. Use it to research just about anything, including the history of robotics. While much of the material is viewable (or at least indexed) on the Web site, serious research requires that you visit Washington D.C. Of course, books cataloged by the Library of Congress are indexed on the Web page, but are for viewing at the indicated reading room.

Walt Noon's Show Design 204157

http://www.pe.net/~magical/robots/

Overview of mechanical animations for special events and promotions.

INTERNET-LINKS

Link pages lead you to other sites on the Web. Most such pages have links that are categorized by subject, and others are a kind of specialized search engine for technical subjects. The listings that follow are but a small selection of link pages, and they were selected for their relevancy to robotics.

SEE ALSO: INTERNET-SEARCH.

ArtsAndCrafts.co.uk 203189

http://www.artsandcrafts.co.uk/

Here, you'll find a links directory for arts and crafts in the U.K.

Bowden's Hobby Circuits 202932

http://ourworld.compuserve.com/
    homepages/Bill_Bowden/

Site includes over 100 circuit diagrams, as well as links to related sites, commercial kits and projects, newsgroups, and educational areas.

Chris Hillman's Robotics/Animatronics/SPFX links 202053

http://members.aol.com/c40179/

Very extensive set of links to Web pages about robotics, animatronics, and mechanical special effects (materials and providers).

Links from Chris Hillman

Craft Site Directory 203531

http://www.craftsitedirectory.com/

Categorized links to various craft-oriented Web sites, large and small. Categories include woodworking, general crafts, metal crafts, and supplies.

DMOZ Open Directory Project-AI 202439

http://dmoz.org/Computers/Artificial_Intelligence/

Open Directory Project for artificial intelligence links.

DMOZ Open Directory Project-AI Robotics 202433

http://dmoz.org/Computers/
Artificial_Intelligence/Robotics/

Open Directory Project for artificial intelligence and robotics.

DMOZ Open Directory Project-Artificial Life 202438

http://dmoz.org/Computers/Artificial_Life/

Open Directory Project for artificial life.

DMOZ Open Directory Project-Electronics 202442

http://dmoz.org/Science/Technology/Electronics/

Open Directory Project for electronics.

DMOZ Open Directory Project-Embedded Hardware 202440

http://dmoz.org/Computers/Hardware/Embedded/

Open Directory Project for embedded hardware (e.g., microcontrollers or single board computers).

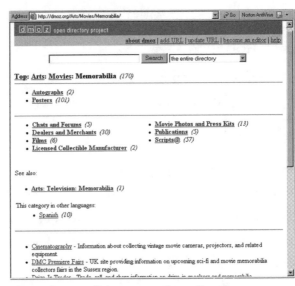

One of many links pages from the Open Directory Project pages.

DMOZ Open Directory Project-Home Automation 202436

http://dmoz.org/Computers/Home_Automation/

Open Directory Project for home automation topics.

DMOZ Open Directory Project-Instruments and Supplies 202443

http://dmoz.org/Science/
Instruments_and_Supplies/

Open Directory Project for lab and scientific instruments and supplies.

DMOZ Open Directory Project-Materials 202441

http://dmoz.org/Science/Technology/Materials/

Open Directory Project for various types of materials, such as polymers and metals.

DMOZ Open Directory Project-Programming 202437

http://dmoz.org/Computers/Programming/

Open Directory Project for programming topics.

DMOZ Open Directory Project-Robotics 202434

http://dmoz.org/Computers/Robotics/

Open Directory Project for general robotics.

DMOZ Open Directory Project-Speech Technology 202435

http://dmoz.org/Computers/Speech_Technology/

Open Directory Project for speech technology, both recognition and synthesis.

Dontronics: PIC List 204117

http://www.dontronics.com/piclinks.html

Don McKenzie's listing of useful PIC sites.

ePanorama.net 202647

http://www.epanorama.net/

Basically, a big huge links page to a variety of electronics topics, including this laundry list of subjects near and dear to any robot builder:

- Bar code technology
- Basics
- Books
- Cabling
- PCB design and making
- Component dealers
- Data communications
- GPS
- IC pinouts

- IR remote control
- Laser
- Motor control
- Optics
- Optoelectronics
- Power supplies
- Prototyping
- Remote control
- Repair information
- Robotics
- Soldering
- Wiring information

Global Sources 204037

http://www.globalsources.com/

Intended for importers and exporters, Global Sources provides thousands of links to manufacturers, distributors, and wholesalers for toys/games/hobbies, electronic components (active, passive, and electromechanical), industrial machines and supplies, and many other topics.

Hero MegaLinks 202217

http://www.doorbell.com/yatu/heros.html

Once upon a time, there was a robot named HERO. Links to friends of HERO are listed on this fan page.

IndustryLink.com 204183

http://www.industrylink.com/

Over 1,000 companies, by categories. Includes categories for Electronics, Machine Tools, Metals, Plastics & Polymers, and Wireless.

Internet FAQ Archives 202431

http://www.faqs.org/

FAQs (frequently asked questions) of all colors and descriptions.

Macintosh Robotics Resource List 203080

http://www.concentric.net/~Jjlee/robotics/
maclist.shtml

Links . . . for using the Macintosh to program robots. Somewhat unique, as (for whatever reasons) the PC dominates the field of amateur robotics.

Robot Shapes, Styles, and Sizes

Amateur robots come in all sizes, from no larger than a deck of cards, to well over the size of a refrigerator. Most are on the smaller end of this scale, with robots weighing in at 2 to 10 pounds.

Robots come in a variety of shapes. The shape is dictated mainly by the internal components that make up the machine and also by the intended application. Most 'bot designs fall into one of the following categories:

A "turtle-size" desktop robot. Photo Jim Frye, Lynxmotion.

- *Turtle.* Turtle robots are simple and compact, designed primarily for "tabletop robotics." Turtlebots get their name because their body somewhat resembles the shell of a turtle and also from early programming, with Logo turtle graphics, which was adapted for robotics use in the 1970s.

- *Miniature vehicle.* These are small automatons with wheels or tracks. In hobby robotics, they are often built using odds and ends like used compact discs, extra LEGO parts, or the chassis of a radio-controlled car. LEGO robots fit this category.

A walking robot. Photo Jim Frye, Lynxmotion.

- *Rover.* Any of a larger group of rolling or tracked robots designed for applications that require some horsepower, such as vacuuming the floor.

- *Walker.* A walking robot uses legs, not wheels or tracks, to move about. Most walker 'bots have six legs, like an insect, as the six legs provide excellent support and balance.

- *Arms.* Arm designs are used by themselves in stationary robots or can be attached to a mobile robot.

- *Android.* Android robots are specifically modeled after the human form: a head, torso, two legs, and possibly one or two arms.

What Happened to About.com Listings?

Knew you'd ask. You may know of About.com, a gigantic links site. About.com is divided into many different categories, including hobbies, art, robots, artificial intelligence, and other relevant sections, and is maintained by compensated "guides" who troll the Internet looking for links and other resources.

About.com is not listed in this book, despite the number and usefulness of its resource listings, because—quite simply—they are pop-up fiends. Visiting an About.com listing can be sheer hell, with three and sometimes four pop-up windows appearing. After trying to include some of their resources for this book, I decided I had had enough.

Feel free to investigate About.com on your own. But be prepared to close a lot of pop-up windows as you go.

Meccano Sources 202309

http://www.freenet.edmonton.ab.ca/meccano/mecsou.html

Where to get Meccano (Erector Set) kits and parts.

R/C Web Directory Index 202165

http://www.towerhobbies.com/rcweb.html

Where to find R/C information, parts, and manufacturers on the Web. The site is maintained by Tower Hobbies, a major online R/C retailer.

RepairFAQ: Sam's Neat, Nifty, and Handy Bookmarks 202869

http://www.repairfaq.org/REPAIR/F_sambook.html

From Sam Goldwasser's RepairFAQ: awholelottalinks.

RoboMenu 202407

http://www.robotics.com/robomenu/

Gallery of robots from all over the world.

SEE ALSO:

http://www.robots.net/

Robot Directory, The 203461

http://www.robotdirectory.org/

Robot showcase. Robots are listed by category, such as flying, walking, or wheeled.

Robotica.pagina.nl 202328

http://robotica.pagina.nl

Robotics links. Some are to pages in Dutch.

Roger's Embedded Microcontrollers Home Page 202938

http://www.ezl.com/~rsch/

Land o' links for microcontrollers.

Slashdot.org 204020

http://slashdot.org/

Billing itself as "News for Nerds," Slashdot is a premier hangout spot for news and views regarding the high-tech world. See the listing under Internet-Bulletin Board/Mailing List.

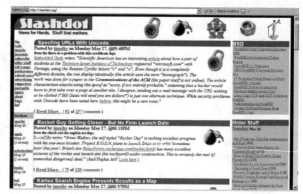

Slashdot: "News for Nerds."

Woodworker Online 203220

http://www.woodworker-online.com/

Woodworker Online provides links and resources for the home woodworking enthusiast. Links are organized by type.

INTERNET-PERSONAL WEB PAGE

The Internet lets you show off what you've done and share with others your discoveries, projects, plans, and objectives. This section provides numerous personal Web pages of robotics and electronics experimenters. Most contain photos of the owners' projects, as well as construction notes or diaries. And many provide schematics, programming code, and other practical examples.

SEE ALSO:

INTERNET-INFORMATIONAL: More sharing of ideas

INTERNET-PLANS & GUIDES: More sharing of schematics and how-tos

LEGO (VARIOUS): Sharing of LEGO creations

PROGRAMMING-EXAMPLES: Programming code you can try

Alan E. Kilian Home Page   202044

http://bobodyne.com/web-docs/

Alan let's us examine a large assortment of his robotics and electronics projects. Highlights include:

- Trippy the robot
- Paper about range finding
- PIC-SERVO software
- PIC 12C509 experiments
- Tryclops and Trippy ranger experiments

See also Twin Cities Robotics Group:

http://www.tcrobots.org/

Alex Brown's Robotics Page   202003

http://abrobotics.tripod.com/

Web page of robot builder Alex Brown, who provides details of several of his projects and writings, including:

- Snuffy firefighting robot-winner of Trinity 2001/senior division
- Ebo-plans and explanation of building small turtle robot
- PID-informational page on PID-based control equations

Al's Robotics   203718

http://alsrobotics.botic.com/

Gallery of Al's 'bots, including EVO (BEAM robot), Psycho Mantis, and KTX1, a small PIC-based mobile robot. The site also includes a number of useful beginner tutorials on robotics and electronics:

- Tutorial for hacking/modifying servos for continuous motion
- General information on sensors
- LM339 Comparator IC

Aquabots   202224

http://www.geocities.com/SoHo/Exhibit/
8281/aquabots.html

Descriptions and pictures of a school of aquatic robots. Also some land-based BEAM robots.

Arvid Animatronics   203056

http://www.backinsf.com/j1432/animatronics/arvid.html

All about Arvid, a homebuilt animatronic creature. The projected was created for a special effects class for the Academy of Art in San Francisco, Calif.

Atomic Zombie   203291

http://www.atomiczombie.com/

The Web page of BioHazard, RoadKill, and D. J. Dogster at Atomic Zombie, all of whom are into cutting-

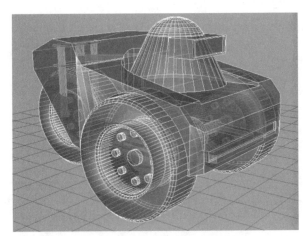

A CAD drawing of AtomicZombie's The Sentinel. Courtesy Brad and Devon Graham.

Beginnings of The Sentinel, with hacked-up General Motors differential. Courtesy Brad and Devon Graham.

edge robotics. Nicely done personal Web page of a person who knows how to wield a mean blowtorch.

BEAM Hexapod 203009

http://members.tripod.com/sparkybots/hexapod.htm

A relatively heavy-duty BEAM walking robot. Circuit example and construction pictures.

BEAM Robotics 202376

http://www.nis.lanl.gov/projects/robot/

This is the main site to Mark Tilden's LANL (Nonproliferation and International Security) page. Mark is responsible for the BEAM concept, and here we see his philosophy of BEAM—though in fairly typical Tilden—speak: "The idea is to improve robo-genetic stock through stratified competition and have an interesting time in the process." I think he means scientists just want to have fun.

You'll also find links to other Web sites, where to go for more information, and news of upcoming BEAM competitions.

Ben's RCX Robots 203110

http://www.ben.com/LEGO/rcx/

About Ben's LEGO Mindstorms robots, including Micro Rover, a tiny robot with differential steering and a single front bumper, and Miniature R2-D2.

BiPed Robot ⅋ 203035

http://members.chello.at/alex.v/

Alex has gone and built himself a bipedal (two-legged) walking robot that exhibits dynamic balance. Operated by aircraft servos. Watch the MPEG movies to see the machine in action.

The Web site provides hardware and software design overview, including a 3D exploded view of the robot's parts. Notice the two servos in the ankles of both legs; they are critical in allowing the robot to balance.

Bob Greiner's Lost in Space B9 Robot Project 203404

http://members.tripod.com/bobgreiner/

Bob's Lost in Space Robot B9. Very detailed photographs and step-by-step building diary makes this one a great read.

BoneyNet 204052

http://www.boney.clara.net/

Home of MABLE, a robot built for a final-year university project, PIC info, and more interesting stuff.

Chip Shults Home Page ⅋ 203039

http://home.cfl.rr.com/aichip/

Home page of robotics and automation expert Charles (Chip) W. Shults III. Examples, how-tos, history of robots and artificial intelligence.

Check out a custom-made "brain chip" that implements a simple Braitenberg vehicle at:

http://home.cfl.rr.com/aichip/CIR02.htm

Chris Renton: Fred ⅋ 202058

http://members.optusnet.com.au/~renton/
       fred/index.htm

Fred means "Free Roaming Electronic Dream." It's Chris Renton's version of a personal robot he made from parts collected from the local Dick Smith's and

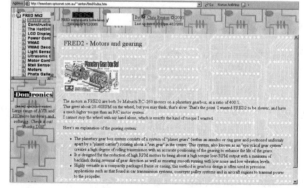

Information about Fred at Chris Renton's Web page.

other sources. Plenty of circuit designs, plans, theory of operation, and even printed circuit board layouts. This one's A+.

Creaturoides 202323

http://www.creaturoides.com/

Gallery and descriptions of several robotics designs. Web pages are in Spanish.

Dale's Homemade Robots 202864

http://www.wa4dsy.net/robot/

Some of Dale's robots include the Suckmaster II vacuum 'bot. Some well-made stuff here.

Daniel Livingston-B9 Today 203405

http://www.b9robotresource.com/

News and diary of Daniel's Robot B9 (from Lost in Space). Done in a USA Today newspaper motif. A fun read.

Robot B-9 Journal.

Dave Novick's Robots 203756

http://www.me.ufl.edu/~dkn/robots/

Check out Dave's RoboBug walking robots.

Dizzy-An Aware Kind of Robot 202464

http://www.xs4all.nl/~sbolt/edz.htm

Parts, plans, and kit (you can order it from The Netherlands from the site). Some interesting ideas.

DreamDroid Robotics 202329

http://www.dreamdroid.com/

As said by the Web site: "DreamDroid Robotics is dedicated to research and implementation of various robotic applications. In here you will find information on all the projects, competitions, and more. We believe that robots have a future in areas such as autonomous artificial intelligence driven robots, combat robotics, personal and entertainment robots."

E-Bot: The Educational Robot 202276

http://home.earthlink.net/~apendragn/ebot/

Says the Web site: "The E-Bot is a bare-bones educational robot with lots of room for expandability. It is a variation of the B-Bot idea developed by Marvin Green and Cricket by Henry Arnold, modified to use a PIC 16C66 microcontroller and with a built-in software package. It is part of an introductory robotics course being developed for the Robotics Society of Southern California."

Extensive documentation (in Adobe Acrobat PDF format), development details, circuit examples, and software examples are provided by the author, Arthur Ed LeBouthillier.

Forbidden-planet.org 202769

http://www.forbidden-planet.org/Robby/

Fan site for the movie Forbidden Planet.

Frank Scott's Hexapod Robots 203755

http://www.frasco.demon.co.uk

Hexapod robots: pictures, descriptions, parts lists, and background information.

G-Bot 203406

http://www.g-botproject.com/

Another homemade Robot B9 (from Lost in Space fame). This one is quite good.

Greg's Robots 203087

http://www.elnet.com/~gad/Robots.htm

Greg's projects include a Nerf Missile Launcher.

Havinga Software Robot Pages 203720

http://www.havingasoftware.nl/robots/robots.htm

Projects include autonomous robot Snuf, Bi-Ped Robot, and video capture and pattern recognition.

Hioxz Robotics 203000

http://members.home.nl/gamesz/index.html

Web page of "Hioxz from the Netherlands, a student very busy with robotics."

Goodly number of robot project descriptions, some with schematics.

Imaginerobots.com 202657

http://www.imaginerobots.com/

Danh Trinh's ever-most-excellent repository of robotic creations. Danh builds some of the best robots I've seen on the Web; check out his Robot Bug hexapod, which he built from scratch using plastic parts created with his desktop CNC machine (he uses a retrofitted Sherline desktop mill). A particular interest of Danh's is aerial robotics.

Jeff's Robots-Robot Menagerie 202655

http://home.pacbell.net/jkkroll/bots.html

Robots by Jeff Kroll. Nice construction pictures.

Karl Lunt, Author 203465

http://www.seanet.com/~karllunt/

Karl is the author of *Build Your Own Robot!* (AK Peters), and he wrote a long-running column in Nuts & Volts magazine on robotics. Be sure to check out Karl's sbasic and tiny6th compilers for the Motorola HC11/12 microcontrollers.

Ken Boone's Robotic Home Page 203416

http://users.aol.com/kensrobots/kensrobots.html

Ken's been building robots since the mid-1980s, and he discusses his creations on this Web page.

Mark's B9 Robot Resources 203408

http://homepage.mac.com/markthompson1/B9/

Where to find parts, plans, and help for building your own Robot B9 of Lost in Space fame.

MHEX-My Six-Legged Walking Robot 202971

http://www.geocities.com/viasc/mhex/mhex.htm

Very nicely done 12-servo hexapod, created out of machined aluminum.

Micro Robots 204079

http://www.geocities.com/acicuecalo/

Examples of walking and rolling robots, circuits, and programming code. Mostly in Spanish, with some in English.

Murray McKay 202336

http://home.midsouth.rr.com/mmckay/

Murray is a robot builder, and he shares with us his designs and how-to tutorials. Included for such robots as Weevil and Aardvark are schematics, example programs, and descriptions. See also his page on constructing an infrared proximity detector.

Otis's Basic Stamp Robot Page   204181

http://home.epix.net/~iracerc/stamp.html

Description, pictures, and circuit layout of an R2-D2 robot, running on Topo-style extreme camber wheels, is controlled from a Basic Stamp. Talk about an identity crisis! See other links to additional project pages.

Oualid Burström's Robots   204153

http://www.fatalunity.com/~oualid/

Pictures, blueprint pictures, software examples, and descriptions of several interesting robots, including walkers and robotic arms.

Patrick Innes Robotics Page   202968

http://home.earthlink.net/~pkinnes/bots.html

Gallery of interesting robots, including a rover made with discarded CDs.

Phil Pemberton's Website   202976

http://www.philpem.f9.co.uk/

SPO256 replacement project, 6502 Web ring, the 6502 Appreciation Group.

Richfiles: The Robotics Page   204140

http://richfiles.calc.org/RobotTopics.html

Pictures and short descriptions of robots, mostly BEAM style.

Ringo's Robotic Page   202515

http://www.margaritasrus.com/users/ringo/

Ringo's robots include a triangular-shaped tracked vehicle based on Tamiya Educational kit parts.

Robot Hut, The   202068

http://www.geocities.com/robothut/index.html

John Rigg's Toy robot collections and museum, including life-size reproduction versions of Robby the Robot, Robot B9 (from Lost in Space), Gort (from Day the Earth Stood Still), Maria (from Metropolis), and the time machine apparatus from the George Pal version of the movie Time Machine. The museum, located near Spokane, Wash., is open to the public.

The Robot Hut main page

Robot Nut   202070

http://www.robotnut.com/

Gallery of tin and plastic toy robots, most from the 1950s and 1960s.

Robots by Gerhard Schwanz   203024

http://www.gs-roboter.de/

Robots, circuits, and clips. Most is in German. Be sure to check out the interesting MPEG movie files.

Robots I've Built   202982

http://www.chaparraltree.com/robots/

Well, not me, but Raphael Carter.

Roganti's Robotics Zone 202466

http://www.euronet.nl/users/ragman/robotics.html

Contents include:

- Homebuilt Inclinometer
- Robot Turtle-fast-moving, light-seeking, independent turret, mobile base
- Robot Walker-Hexaped-paper tubes make great legs
- Robot Walker-Quadraped-paper tube power
- Robot Walker-Biped-a legged walker using static equilibrium to emulate bipedal locomotion
- Schematic for eight-port serial servo controller for your microcontroller using one output port
- Schematic for 68HC11, 64KB, RS232 port microcontroller board

Seeker Mini Sumo 204176

http://huv.com/miniSumo/seeker/

Construction diary for Seeker, a robot built to compete in the mini sumo class.

Snake Robots 204188

http://www.snakerobots.com/

Eeek! Snake robots. Only these are really cool and don't bite—much. Read about the snakes and their inventor.

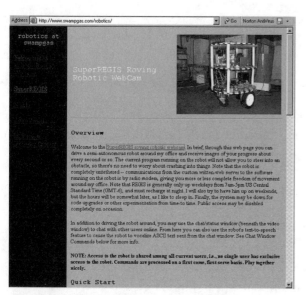

The SuperRegis teleoperated robot at Swampgas.com.

Swampgas Robotics Page 202463

http://www.swampgas.com/robotics/

Be sure to check out SuperRegis, a teleoperated robot. There's also NuBlu and plans for an infrared remote-control decoder chip using PICmicro 16508. The Web page is by Michael Owings, software programmer, book author, and robotics enthusiast.

Tiggerbot 204070

http://jormungand.net/projects/tiggerbot/

Descriptions, programming, and photos of Tiggerbot, a track-based robot that uses sonar for navigation.

Tiny Mobile Robot ♀ 203027

http://home.megapass.co.kr/~cch8960/

Tiny robots. Like the size of a 9-volt battery. Like a chariot for a cockroach.

You'll find pictures, descriptions, circuit schematics, PCB layouts, and parts layouts.

Walking Robots 203470

http://www.walkingrobots.com/

Presented is a collection of walking robots, most of which were made using an abrasive water jet (apparently, this is a kind of machining tool, not a description of a mean boss who spits when yelling at you). Close-up photos but no construction details.

For more information on the manufacturer of specialty parts using the abrasive water jet, see:

http://www.ormondllc.com/

Wayne Gramlich's Personal Projects ♀ 203792

http://gramlich.net/projects/index.html

Here, we learn that Wayne has far too many interests for even one lifetime, but somehow he finds time to play with all sorts of interesting things, including electronics and robotics. What I like about Wayne's home page is that it's very much a work in progress, regularly updated, a mix of finished with half-finished ideas, and presented for no other reason than to share with others.

This is not the usual hardware hackers page: Wayne doesn't just show a picture of his projects with a short description. Instead, he treats us (at least for the more-completed projects) with circuit diagrams, board layouts, even Gerber and Excellon PCB manufacturing files, should we wish to produce our own prototypes.

Be sure to check out Wayne's "RoboBricks"—a concept like LEGO Mindstorms, but consisting of more sophisticated interconnecting modules.

Wilby Walker   203781

http://members.aol.com/wilbywalker/

Description and construction details of the Wilby Walker, a six-legged hexapod. DXF CAM/CAD files are provided.

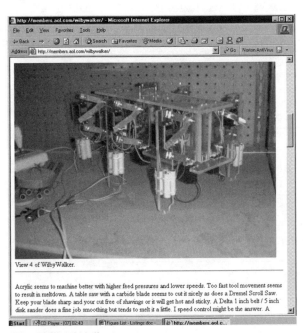

The Wilby Walker

Willy the Robot   204071

http://home.earthlink.net/~apendragn/robot.htm

Robot page of Arthur LeBouthillier, an experienced robot builder who describes the construction of a robot using a PC-compatible computer and 12-volt DC inverter, as well as his unique laser "structured light" vision system.

Wizzes Workshop   203409

http://www.wizzesworkshop.com/

Construction details of a homemade Lost in Space B9 robot.

Wobbly Wheel   202230

http://home1.gte.net/res07vpe/

Modest construction details of a rover-style robot.

 INTERNET-PLANS & GUIDES

This section lists step-by-step how-to guides on robotics and electronics. The information is free, though some sites may also sell supporting kits or products. Subjects include constructing computer interfaces, working with clay and other sculpting materials, and retrofitting a toy into a full animatronic robot.

SEE ALSO:

> **ELECTRONICS-CIRCUIT EXAMPLES:** Circuit schematics, many on a robotics theme
>
> **INTERNET-INFORMATIONAL:** Free information
>
> **INTERNET-PERSONAL WEB PAGES:** Peeks at what other people are doing
>
> **KITS-ROBOTICS:** Robotics kits (usually on the toy end)
>
> **PROGRAMMING-EXAMPLES:** Programming code you can try
>
> **ROBOTICS-HOBBY & KIT:** More robotics kits (typically higher end)

Boondog Automation   202212

c/o Paul Oh
3500 Powelton Ave.
Ste. B-402
Philadelphia, PA 19104
USA

✉ paul@cs.columbia.edu

🌎 http://www.boondog.com

Products and tutorials centering on using the PC as a host for various automation tasks.

Cricket the Robot   203458

http://home.earthlink.net/~henryarnold/

Plans and information on building a nice Basic Stamp-based six-legged walking robot.

Complete plans to build a robot from a 3 1/2-inch floppy drive without taking the drive apart. The floppy drive has all of the motors and electronics you need to get started and compete in a robot contest. Popular site.

Floppy the Robot   202026

http://www.ohmslaw.com/robot.htm

Furby Upgrade   202300

http://www.appspec.net/

What to Name the New Robot

Can't think of what to call your next robot? Here are some common names other proud parents have given their mechanical offspring.

Adam	Your first robot. Be coy and create a female companion robot named Eve. Don't forget to eventually add a SnakeBot.
BugBot or *RoBug*	Robot bug.
Cybot	Apparently a combination of *cybernetic* and *robot*. Even some companies have named their products this.
Fred or *Sam*	No clue why these are so popular.
Junior or *Jr.*	A microchip off the old bloke.
Kelad	*Dalek* (from Dr. Who) spelled backward.
Kirby or *Hoover*	Vacuum cleaner robot.
Max	For *maximum*; or for *Maximillian*, the antagonist robot in the movie *Black Hole*.
Mobot	Mobile robot.
Mouse	And its variations: *MicroMouse, MiniMouse, MickeyMouse, MannyMouse*.
Mowbot	Lawn-mowing robot.
MyBot	Not to be confused with YourBot.
Otto	Word play on *automaton*.
Robot	It worked for Will Robinson
Rover	It does everything but fetch.
Sparky	Hopefully, this is just a love name and isn't what happens when you turn the power switch on.
Spike	Robotic dog; Tom's (from *Tom & Jerry* cartoons) nemesis.
Tobor	Robot spelled backward.

"Upgrade" a Tiger Electronics Furby. Definitely not connected with Tiger in any way, but that shouldn't diminish the interest quotient.

Glass Attic   204133

http://www.glassattic.com/

An "encyclopedia" of polymer clay information. According to the Web site, "Here you'll find over 90 categories (1,300+ pages) of information relating to polymer techniques, lessons, supply sources, tools, and problem solving (as well as to polymer photography, business, teaching, finding inspiration, etc.). Each category contains many links to examples for illustration, and often to lessons as well."

How to Build a Dalek   203415

http://www.steve-p.org/dw/dalek/

JPEG image files of reprints from a Radio Times magazine article originally published in 1973. The construction ideas are suitable for most any kind of larger-scale robot, so even if you don't want to build a Dalek, the plans are still useful.

HowToAndroid.com   203463

http://www.howtoandroid.com/

Plans and ideas for constructing human-form robots. Includes pictures, descriptions, parts lists, simple control schematics.

I.R. BOT   203085

http://www.fictoor.nl/irbot/

Informational on building a robot. Construction details, hardware, software (including a real-time robot OS for the Atmel AVR), and plenty of links.

Jeff Frohwein's Software/Hardware Dev'rs   203045

http://www.devrs.com

Among other topics, GameBoy hacks.

OWI Arm Trainer Interface   202228

http://www.xantz.com/xantz_files/owi_interface.htm

Informational site on interfacing an OWI Arm Trainer kit to a Basic Stamp.

Palm Pilot Robot Kit (PPRK)   202103

http://www-2.cs.cmu.edu/~pprk/

So says the Web site: "The Palm Pilot Robot Kit (PPRK) is a design for an easy-to-build, fully autonomous robot controlled by a Palm handheld computer. This design was created by two Carnegie Mellon Robotics Institute research groups, the Toy Robots Initiative, and the Manipulation Lab, with the intent of enabling just about anyone to start building and programming mobile robots at a modest cost."

Robotics Information and Articles   204120

http://www.leang.com/robotics/

Example 'bots and online articles (good ones) on such subjects as:

- Controlling servo motors with various microcontrollers
- RF serial communication for the MIT Handy Board
- H-bridge motor driver circuit
- Infrared proximity sensor

For Kam Leang's past and current robot projects, see also:

http://www.leang.com/robotics/

WGM Consulting   203019

http://www.wgmarshall.freeserve.co.uk/

See the Robotics link. Several construction plans and programming code for MicroMouse-class robots.

Yoda Project, The   204141

http://www.geocities.com/SouthBeach/8877/yoda_project.html

One of the better instructional pages on the Internet, this one describes refitting a latex Yoda with mechanical innards so that his face moves while he talks. The project, designed by a student at Virginia Tech and his friends, uses R/C servos and a Winbond ISD voice recorder chip. Schematics, parts lists, and programming code are provided. Fun stuff!

 INTERNET-RESEARCH

Look here for Web sites that are particularly well suited for researching technical information, particularly related to robotics and electronics. With few exceptions (which are noted), these resources are free, though some may require registration (your name and e-mail address are typically sufficient).

SEE ALSO:

INTERNET-LINKS: Lists of links related to robotics, electronics, mechanics, and related disciplines

INTERNET-SEARCH: General- and special-interest search engines

Catalogs.com 204094

http://www.catalogs.com/

Catalogs.com is a free search and links site that caters to the inveterate mail-order shopper. It provides short summaries and links to Web sites that provide online catalogs and buying. A few of the subjects are relevant to robobuilders:

- Art-Hobbies-Crafts
- Computer & Office
- Electronics
- Fun Gadgets

ChipCenter-QuestLink 202925

http://www.questlink.com/

Bills itself as "a comprehensive technical resource for the electronic design engineering community." Included is an index of semiconductors, pathways to locate datasheets on ICs and other components, and application notes for many types of circuits.

The http://www.chipcenter.com/ URL takes you to QuestLink, as well.

Locating Chips and Other Electronic Parts

Though it's fun to surf the Internet into the wee hours of the morning, sometimes you have other things to do and want to find something quickly. If you're looking to purchase integrated circuits and other electronic parts, a great time-saver is FindChips.com, a free online resource that is connected into the search engines of more than a dozen electronics wholesalers, distributors, and retailers.

FindChips.com allows you to look for electronic parts and compare availability and pricing.

Just enter the part number of the product you're interested in, and FindChips.com will indicate which of the resources carry the part. Then, descriptions, prices (if available), and other pertinent data are displayed for your review. You can go directly to any of the retailers to make your purchase if you wish.

FindChips.com is located at:

http://www.findchips.com/

While the FindChips.com search engine is primarily intended to find products based on part number, it will also locate many products by general description. For example, enter "4.0 MHz resonator," and odds are it will find several matches.

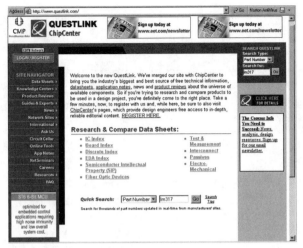

QuestLink ChipCenter.

ChipDocs   204234

http://www.chipdocs.com/

Put this one in the "why didn't they think of this earlier?" file. ChipDocs is a repository of datasheets and application notes for electronic components and semiconductors. Included are datasheets for many obsolete parts, handy because manufacturers often remove these from their sites. There is a monthly charge for the service.

Delphion Inc.   203276

http://www.delphion.com/

Patent search. Nonsubscribers can search through granted U.S. patents; subscribers can search through granted and pending patents in the U.S. and Europe and through other resources.

Don Lancaster/Synergetics   203269

http://www.tinaja.com/

Don Lancaster is the reigning king of technical electronics info on the Web. He is the author of the popular *TTL Cookbook* and *CMOS Cookbook*, and for years he penned various columns in U.S. electronics magazines. Back issues of those columns, plus new material, are available at his site (most are in Adobe Acrobat PDF format). A simple search engine helps you locate subjects of interest.

For the curious, *tinaja* is Spanish for "earthen jar" and is also a village in New Mexico. In many of Don's articles, he refers to "tinaja quests" (giving them away "all expenses paid FOB New Mexico" in return for reader feedback). These are essentially hikes in the Southwestern deserts to find these vessels or jars.

Home page of electronics maven Don Lancaster.

ECG Electronics/NTE Electronics, Inc.
203267

44 Farrand St.
Bloomfield, NJ 07003
USA

☎ (973) 748-5089
📠 (973) 748-6224
✉ general@nteinc.com
🌐 http://www.ecgproducts.com/

Replacement semiconductor and electronic parts.

SEE ALSO:

http://www.nteinc.com/

🏭

EEM.com   202926

http://eem.com/

Electronic Engineer's Master (EEM) offers links to thousands of manufacturers of electronic components, equipment, and services. Search by keyword or browse through categories.

eFunda   202481

http://www.efunda.com/

According to the site: "eFunda is a premier online publisher for original engineering content and software. We offer over 30,000 pages of engineering fundamentals and calculators, authored exclusively by our highly

Searching for Patents

The United States Patent and Trademark Office sponsors a free Web site where you can search for, and browse, any patent going back to 1790. Older patents are shown as graphical TIFF files (for which you need a special add-in for your browser; you can download it from a link available at the USPTO site). Newer patents are in searchable text form.
The USPTO Web search page is at:

http://www.uspto.gov/patft/

Here, you can search for existing patents using the patent number, date of issuance, inventor name, assignee name, and keyword. (You can also search through select patent applications.)
The Quick Search feature is the easiest to use. Here, you can specify up to two search terms and indicate which field of the patent the term(s) relate to. For instance, if you're looking for patents with "robot" in the title, specify *robot* as the search term, and *Title* as the search field. Narrow the found patents by selecting a date range; the system defaults to the most recent years. Patents that match are shown as a page of links.
Note: So-called *design patents* are prefixed with a D. These are fun to look at (you need that TIFF graphics add-in as mentioned previously), as they are patents for the physical appearance of a mechanism. Design patents are often taken out for toys and movie robots and are done to prevent others from making copycat products. Regular (or *utility*) patents are shown as a number, such as 1,234,567 (in case you're curious, this is a patent from 1917 for a shirt collar).
Web searches for patents granted in other countries are also available, but not always for free. Try the following—note that some require registration, and others are by subscription only. While many are government sponsored, others are commercial enterprises, designed for patent attorneys, and subscription fees can be quite high. Most pages are in the native language of the country, though some offer the choice of viewing the site in English and other tongues.

Australia	*http://www.ipaustralia.gov.au/*
Canada	*http://cipo.gc.ca/*
China	*http://jiansuo.com/*
Denmark	*http://www.dkpto.dk/*
Europe	*http://ep.espacenet.com/*
Europe	*http://www.european-patent-office.org/*
Germany	*http://www.patentblatt.de/*
Germany	*http://www.dpma.de/*
Japan	*http://www.jpo.go.jp/*
Korea	*http://www.kipo.go.kr/*
U.K.	*http://www.patent.gov.uk/*
U.S., Europe, Japan, Canada	*http://www.ipsearchengine.com/*
U.S., Europe, Japan	*http://www.delphion.com/*
Worldwide	*http://ipdl.wipo.int/*

educated and experienced staff to give you the most credible engineering information."

Some free; some not.

Electronic Construction from A to Z   203185

http://www.mtechnologies.com/building/atoz.htm

Like the name says.

Electronics Directory, The   203006

http://www.electronicsdir.com/

Searches for datasheets of components.

Electronics Manufacturing Guide, The   202528

http://www.circuitworld.com/

Says the Web page: "Serving over 220,000 requests to more than 10,000 visitors each month, The Electronics Manufacturing Guide offers you the best in online resources." Search for manufacturers, PCB production, distributors, wholesalers.

Electronics2000.com   204172

http://www.electronics2000.com/

Resource page for electronics information and calculators. The site also provides an online store for selling inexpensive CD-ROM compilations of small electronics-related utility programs.

Evolution of the Robot Throughout History   204149

http://cosmics.freeyellow.com/sffile/robothis.html

A history of robots, from 3000 B.C. to present day. Includes a lengthy timeline of robotics history, including real and fictional automatons.

FindChips.com   203328

http://www.findchips.com/

Look up an integrated circuit or other part, and see who sells it. Search queries are performed against the

Look for chips (and other parts) on FindChips.com.

Thomas Register: Business-to-Business

The Thomas Register, once relegated to the reference section of the local library, is now available on the Web (as well as print and electronic versions). With the Thomas Register you can search through some 175,000 listings for businesses that are related to industrial and commercial parts, services, and equipment. You can search by keyword—*shaft encoder*, for example—to look for parts for your next robot.

Most of the firms listed in the Thomas Register are businesses that cater to other businesses. Therefore, minimum orders or quantity purchases may apply.

Visit Thomas Register Online at:

http://www.thomasregister.com/

stock of over a dozen online electronics component resellers.

Fischertechnik Price Watch by Gordy 202426

http://www.primenet.com/~gkeene/fischerprices.html

A listing of Fischertechnik distributors as seen on the Web.

GlobalSpec 204089

http://www.globalspec.com/

An Internet product list with a twist: Everything is for industrial products, like air compressors, motors, power transmissions, and bar code scanners. You can browse categories such as these to find what you need for your advanced robot:

- Sensors, Transducers and Detectors
- Electrical and Electronic Components
- Mechanical Components
- Optics and Optical Components
- Motion and Controls
- Data Acquisition and Signal Conditioning
- Video and Imaging Equipment
- Manufacturing/Fabrication Services

Most categories have one to three dozen listings each and provide company name, product(s), addresses, e-mail, and Web address. Free registration required.

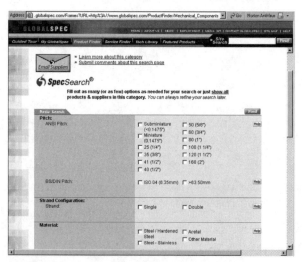

Look for mechanical and electronic components on GlobalSpec.

The site also provides a technical library of downloadable white papers, briefs, and tutorials.

Hardware Book 204179

http://hwb.acc.umu.se/

The Hardware Book, the Internet's largest free collection of connector pinouts and cable descriptions. Handy information to have around the robot lab.

How Stuff Works 203273

http://www.howstuffworks.com/

Free online research guide to how things work. The articles are semitechnical in nature. You can search by keyword or browse interesting topics.

IC Chip Directory/ChipDir.org 202927

http://www.chipdir.org/

The Chip Directory (ChipDir) is an information repository of various integrated circuits, their definitions, functions, and (usually) pinouts. The site is mirrored globally, so you'll find it popping up all over the place.

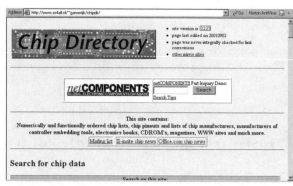

IC Chip Directory Web site.

IC Master 204206

645 Stewart Ave.
Garden City, NY 11530
USA

((516) 227-1300

✆ (516) 227-1901

✉ techsupport@hearstelectroweb.com

http://www.icmaster.com/

Search online (or by book or CD-ROM) by manufacturer or part. Includes 135,000 base part numbers from more than 345 manufacturers. Registration is free for online use; the CD-ROM and book are about $200 and worth the cost if you need to do a lot of component sourcing.

One useful feature is "logo search," where you can match an unknown logo with its corporate owner. Many of the component references also include minimal technical reference data.

PartMiner Inc. 204050

80 Ruland Rd.
Melville, NY 11747
USA

☎ (631) 501-2800
🚫 (800) 969-2000
✉ support@freetradezone.com
 http://www.freetradezone.com/

PartMiner is a components distributor, and they also operate a kind of gateway to researching components, obtaining technical data, best pricing, and purchasing information.

💻

Science Hobbyist 203262

http://www.amasci.com/

Amateur science, cool science, weird science. Links to a large number of off-site articles on science projects, demos, tutorials, and how-tos.

Semiconductor Electronics Resource Center 204190

http://www.dir-electronics.com/

Keep track of semiconductor makers: which ones are in the running, and which aren't. The site also lets you search for distributors, wholesalers, and retailers of semiconductor product.

Silicon Valley Surplus Sources 202929

http://www.kce.com/junk.htm

Listing of Silicon Valley (San Francisco, Calif.) surplus dealers, swap meets, and related retailers.

Thomas Register Online 202111

http://www.thomasregister.com/

Thomas Register is the world's leading resource for information on industrial products and services in North America. Available for free on the Web (registration required for full access), in book form, or on CD-ROM. Most public libraries have the Thomas Register books in their reference sections.

How Stuff Works: An Inside Look at Stuff

Ever wonder how stuff works? You can keep up-to-date with the latest technologies at Michael Brain's HowStuffWorks, the Internet version of the popular series of books.

http://www.howstuffworks.com/

From the home page, you can look up how all kinds of mechanical devices work. For example, to learn more about how optical mice work, you might enter *optical mouse* or *How do optical mice work?* at the main search box. (You don't have to pose searches as questions, but the HowStuffWorks search engine will parse out and understand such queries if you're more comfortable phrasing them that way.)

Matches are shown with short summaries; clicking on a link displays a short article that explains the technology you're interested in—in this case, optical mice. Most of the articles on HowStuffWorks are semitechnical, but lacking are links to more complete information at other sites. You'll need to rely on search engines like Google.com for that.

ToyDirectory.com Inc. 202858

http://www.toydirectory.com/

ToyDirectory.com is a wholesale buyer's guide for the toy and hobby industry. Useful to find out who makes what.

United States Patent & Trademark Office 203275

http://www.uspto.gov/

Search through U.S. patents.

WhatIs.com/TechTarget 204143

http://whatis.techtarget.com/

Look up most any technical term for electronics, mechanics, or computers.

INTERNET-SEARCH

With the vastness of the Internet, search engines are needed to help you find what you need. Otherwise, you'd be slogging through millions of Web sites in order to find the subjects that interest you.

Search engines work by cataloging Web pages and storing critical information about them in databases. With many search engines, you find what you're looking for by entering a search phrase using keywords. A typical keyword search might be as follows:

robotic vision systems

which (depending on the search engine you're using) will show you all the relevant pages having to do with robot vision systems.

While there are many search engines available—and most are free—a small handful are truly useful. Of all of them, Google.com is my favorite, and it gets special attention throughout this book. It's fast, efficient, accurate, and it doesn't bombard you with ads. A number of listings in this section pertain to subject searches you can conduct using Google.com. The searches are geared toward finding online and local suppliers of goods and services.

As good as Google.com is, there are many other special-purpose search engines, including electronic Yellow

Pages that allow you to look for retailers, both local and long distance, based on subject of inquiry.

See also Internet-Links: Pages of Web links on robotics and electronics

AOL Yellow Pages 202520

http://yp.aol.com/

Yellow Pages for the Internet. Search for subject and state; for example, "Electronics, CA" to find electronics specialty stores in California. Important robotics/electronics-related categories:

- Electronic Equipment & Supplies-Manufacturers
- Electronic Equipment & Supplies-Retail
- Electronic Equipment & Supplies-Wholesale
- Robots [wholesale]
- Computer & Equipment Dealers
- Computer Parts & Supplies
- Computer Supplies & Parts-Manufacturers
- Computers-Electronic-Manufacturers

ARRL: Database of Hamfests 203549

http://www.arrl.org/hamfests.html

Database of upcoming hamfests, swap meets set up and patronized by ham radio enthusiasts. But you'll find more than radio gear at most hamfests; many are good sources for general surplus electronics and surplus mechanicals, including motors, solenoids, relays, soldering stations, tools, and more. Hamfests are usually held on Saturdays, and many start early in the morning. To get the best deals, get there as early as possible, or else the good stuff will already be gone.

Buyer's Index 204093

http://www.buyersindex.com/

Buyer's Index bills itself as "The Search Engine for Savvy Shoppers." It's all that. Buyer's Index provides dozens of main categories, such as Tools & Instruments and Toys & Games.

Within these main categories are one or more subcategories, which in turn list companies that sell associated product. Listings include a short description and sometimes a quick recap of such things as the number of items in the catalog, shipping policies, and important restrictions. Not all listings have Web sites, but most do.

The search engine is particularly handy. You can indicate features you want in the returned list: whether they sell retail, wholesale, or both; whether they have a conditional or unconditional return policy; and whether they have a Web catalog or print catalog. Of course, because online retailing and distribution changes rapidly, this information is not always up-to-date, but it's a good starting point.

Useful (to robot building, that is) categories include:

- Art & Craft Supplies & Equipment
- Arts & Entertainment
- Automotive
- Bicycling
- Books
- Collectibles
- Computer Components & Peripherals
- Computer Equipment, Software & Accessories
- Computer Software
- Computer Software for Specific Industries & Professions
- Consumer Electronics
- Electronics & Electrical-Components & Supplies
- Facilities Maint, Repair & Operations
- Hardware & Raw Materials
- Hobbies
- Home Building & Repair
- Household Needs
- Industrial Supplies & Equipment
- Office Supplies & Equipment
- Science, Engineering, & Laboratory
- Security & Safety
- Sewing Fabrics & Textiles
- Tools & Instruments
- Toys & Games

CodeHound 203636

http://www.codehound.com/

CodeHound is an online search engine for examples for programmers and software developers. The site includes coverage of:

- Java
- VB/VB.NET
- Delphi
- C/C++/C#
- SQL
- XML
- Perl
- PHP

The site also sponsors a regular newsletter of interest to the developer community and lists useful third-party resources, including books and upcoming conferences.

Digital City 203289

http://www.digitalcity.com/

Digital City is a localized Yellow Pages, but with classifieds, city guides, and more. Enter your zip code or city and state, and browse through hundreds of shopping categories. To locate, say, electronic surplus stores, enter *electronic surplus* in the search bar after specifying the town.

One or more links may appear; you can continue your search by choosing the *Search for "electronic surplus" in Yellow Pages* link. This starts the AOL Yellow Pages, where you can continue your search.

Use Digital City to find retailers near you.

Google Catalog Search 202578

http://catalogs.google.com/

Search (actual scans of) hundreds of catalogs. Includes sections for Electronics, Computers, and Toys & Games.

Google, the king of all search engines.

Google Search: Adhesives 202685

http://directory.google.com/Top/Business/
Industries/Manufacturing/Chemicals/Adhesives/

Google category list for adhesives: glues, cements, adhesive putties, etc.

Google Search: Art Supplies 202747

http://directory.google.com/Top/Shopping/
Visual_Arts/Supplies/

Google category list for art supplies, including paints, foams and other substrates, and sculpting material.

Google Search: Batteries 202686

http://directory.google.com/Top/Business/
Industries/Industrial_Supply/Batteries/

Google category list for batteries of all descriptions.

Google Search: Chains & Belts 202687

http://directory.google.com/Top/Business/
Industries/Industrial_Supply/Chains_and_Belts/

Google category list for chains and belts, including round, flat, timing, and V belts.

Google Search: Craft Supplies 202746

http://directory.google.com/Top/Shopping/
Crafts/Supplies/

Google category list for craft supplies: ceramics, doll making (including eyes and armatures), metal craft, polymer clay (for mold making), and woodcraft.

Google Search: Cult Movies 202788

http://directory.google.com/Top/Arts/Movies/
Genres/Cult_Movies/

Google category list for cult movies, including sci-fi flicks with robots.

Google Search: DC Motors 202695

http://directory.google.com/Top/Business/
Industries/Electronics_and_Electrical/
Electric_Motors/DC_Motors/

Google category list for DC motors: linear, stepper, permanent magnet, and brushless, as well as exotic types (ultrasonic, piezo, etc.).

Google Search: Displays & Readouts 202705

http://directory.google.com/Top/
Business/Industries/Electronics_and_Electrical/
Displays_and_Readout/

Google category list for LCD, LED, and other forms of displays and readouts.

Google Search: Doll Making 202748

http://directory.google.com/Top/Shopping/Crafts/
Supplies/Doll_Making/

Google category list for doll making; relevant interests are eyes (for "realistic" robots), armatures, and foam.

Searching on the Internet

Both of these statements are true:

- The Internet is a huge sponge that soaks up time, money, and energy.
- The Internet will save you scores of time, money, and energy.

Which statement applies to you depends entirely on how you use the Internet. Careful exploitation of its resources can be rewarding on many levels. With the Internet, you can find most any piece of information, any part, any supplier, but knowing *how* to look is as important as *where* to look. You can waste a lot of time "surfing" the Internet; there are better ways!

Google.com: Window into the Internet

The basic tool for finding things on the Internet is the search engine, a Web-based program that collects data about other pages. Search engines catalog Web sites, assigning "keywords" to them so you can locate what you want among the millions of pages available. Trouble is, most search engines on the Internet are dog-pooh. These days, they've become nothing more than cheap advertising vehicles for companies with the largest budget to spend, and the search results can be both misleading and unproductive.

Google.com is different—or at least it is as of this writing (things on the Internet have a way of changing). In the hopes that Google.com keeps its act clean, throughout this book we use it as the de facto standard search engine. Many of the information articles and sidebars in this book contain references to Google.com "search phrases"—keywords that are used to locate a myriad of associated Web pages.

To use these search phrases, simply type them, exactly as they are printed, in the Google.com search bar. Press the Enter key, and you're off.

Though Google.com is advertising supported, the ads are seldom annoying or overwhelming, nor are they inappropriate for those under 18 years of age. In nearly all cases, the Google.com search engine has an uncanny ability to place the most relevant pages near the top of the list. This alone saves you tremendous time.

Reaching Google.com

The U.S. North American Google.com site is *http://www.google.com/*. However, specialized versions of Google.com are provided for a number of languages and locations. Some of the country-specific Google.com sites are as follows:

Belgium	*http://www.google.be*
Brazil	*http://www.google.com.br*
Canada	*http://www.google.ca*
Deutschland	*http://www.google.de*
France	*http://www.google.fr*
Israel	*http://www.google.co.il*
Italy	*http://www.google.it*
Japan	*http://www.google.co.jp*
Korea	*http://www.google.co.kr*
Nederland	*http://www.google.nl*
New Zealand	*http://www.google.co.nz*
Russian Federation	*http://www.google.com.ru*
Switzerland	*http://www.google.ch*
U.K.	*http://www.google.co.uk*

Searching Wisely

Enter *robotics* into the Google.com search bar, and you'll get over a million Web pages! As you can imagine, robotics is a popular field of endeavor, but using a common search phrase such as this will get you nowhere, fast. Fortunately, Google.com provides a number of ways to narrow your search, so that you have more chance of finding what you want. Here are some tips.

Like most search engines, Google.com uses keywords to match your search phrase with the content of a Web page. Keywords can be contained in the title, a special hidden keywords section of the underlying HTML code that forms the Web page, or the text of the page. How Google.com ranks the relevancy of the keywords it finds on pages is a closely guarded secret, but through observation, it's apparent the search engine favors keywords that appear in the title of the page. Therefore, if a page says " Robotics Stink," it will likely be ranked highly among pages on robots, simply because the phrase "robotics" is in the title.

Knowing this, you can more readily search for companies, products, or information, by using search terms that are most likely to appear in the title of Web pages. Those Web page creators that use non-descriptive titles, or leave them blank, will be ranked much lower in the search results. That's too bad for them.

You can tell Google.com to ignore keywords if they don't appear in the title, thereby limiting the search to *only* words that appear in the titles of pages. You do this with the *intitle* modifier (or its close cousin, *allintitle*). Use

intitle:robotics

to find Web pages on which *robotics* appears only in the title. Note the colon after *intitle*. Also note there is no space between the colon and the keyword.

Ignoring Nonrelevant Pages

Try the *intitle:robotics* search. You'll find that at least some of the so-called hits on the first couple of pages are not about mechanical robots, but about the U.S. Robotics brand of modems, fax modems, and early PalmPilots. Unless you tell it otherwise, Google.com doesn't know the difference between robots that roll or walk on the floor and products made by the old U.S. Robotics company.

There is a simple way to remove these kinds of nonrelevant pages: Use the minus (-) character to tell Google.com to skip any Web page that contains a given keyword. Here's one way:

robotics –us

This fetches all Web pages with *robotics*, then omits all those pages with the keyword *us*. This works, but unfortunately, *us* is a common word. The better way is to use quotes to form a keyword phrase, like this:

robotics –"us robotics"

This time, only those pages that contain *us robotics*, as a complete phrase, are omitted.

Combining Keywords

It's often handy to search for multiple keywords, such as *robotics* and *vision*. With few exceptions, Google.com does not use Boolean searches—*robotics AND vision*, for example—like some other search engines. Google's syntax is a lot easier, yet isn't limited.

+robotics +vision	Finds pages with both of the word forms
robotics vision	Same as above
"robotics vision"	Finds pages with the specific phrase robotics vision

Except for the whole phrase "robotics vision" the keywords in a multiple-keyword search do not have to appear next to each other on the Web pages. That is, *robotics* can appear at the beginning of a paragraph, and *vision* later in the paragraph. That said, Google.com appears to give precedence to those pages where the keywords are found closer together.

Note the + (plus) character in one of the examples. Google.com automatically ignores very common words and does not include them in the results. Adding a + character to the word ensures that Google.com considers it in your search. While both *robotics* and *vision* are not common words and will not be ignored, it's a good practice to include the + character whenever you want to ensure words are considered in the search.

When using the + character, be sure to add a space before it and to type the keyword immediately following:

robot +about	Good
robot+about	Not good
robot + about	Not good

No Wildcards, But . . .

Google.com doesn't support wildcards, special characters—like ? and *—that denote any character or characters. However, the search engine allows you to construct phrases to look for multiple forms of words by using the OR search modifier.

+robotics eye OR eyes

Note *OR* in capitals. Google.com searches are not case sensitive, but you must capitalize OR in order to use it as a search modifier. This phrase looks for all pages with *robotics*, then matches those pages with either *eye* or *eyes*.

Try Multiple Searches

Some searches will require a couple of different alternatives. One simple way of finding pages when you want to look for multiple word forms—*robotics* or *robot*, for instance—is to perform separate searches. For example:

robotics eye OR eyes

robot eye OR eyes

The first search looks for *robotics* with *eye* or *eyes*. The second, for *robot* with *eye* or *eyes*.

Additional Advanced Searches

Google.com supports an Advanced Search page where you can specify a number of special qualifiers, including limiting the found pages to a given language or to those that have been updated within a certain period of time. Similarly, you can also look for just HTML Web pages, Adobe Acrobat PDF files, or many other common file types.

There are several handy Google.com search tricks you'll want to know about. Be sure to read the help pages on the Google.com site for more information.

Reversing the Order of Keywords

Google.com ranks the relevance of pages from the order of the keywords in your search phrase. You get the same pages no matter what order you use, but the pages that appear at the beginning of the list are altered by changing the keyword order. For example:

zebra barcode	Returns pages that favor the Zebra brand.
barcode zebra	Returns pages that favor bar coding in general, followed by those that also contain the keyword *Zebra*.

More than Just Web Pages

The Google.com search engine is not limited to just Web pages. You can also find images (not as useful to robot builders), Usenet newsgroups, and a subject directory.

- The *Groups* search let you sift through years of archives of newsgroup messages.
- The *Directory* search provides Web links to submitted and approved sites. Some of these may be paid advertisements; nevertheless, it's a good way of locating major suppliers.

At the time of this writing, Google.com was experimenting with a new feature called Google Catalogs. This service provides scanned and indexed pages of hundreds of consumer and business mail-order catalogs. If you don't see a Catalogs search option on the main Google.com home page, try the following URL:

http://catalogs.google.com/

En Français Anyone? Page Translations

Robotics isn't limited to just those who speak English. You'll find plenty of Web pages in a variety of languages. Several free services offer text translation from and into the world's most common languages. The translations are not perfect, but they are often close enough to allow you get the gist of what the Web page is all about.

Among the better Web page language translators are Google.com and Altavista.com; my favorite is Google's, so we'll talk about that first. To translate a Web page, go here:

http://www.google.com/language_tools

and in the appropriate text box enter the full URL, including the *http://* portion, of the Web page you want to view. Select the "from" and "to" language choices, such as *Spanish to English*, or *French to German*. The translation takes anywhere from a few seconds to over a minute, depending on the length of the page.

Note that only normal text is translated. Text in a graphic image is not translated and is shown in the original language.

AltaVista's similar offering is available at:

http://babelfish.altavista.com/

It works in a similar fashion to Google.com's translation page and offers additional languages such as Korean, Russian, and Chinese.

Other Search Engines

Of course, there are other search engines on the Web, some good, and some terrible. Here is a listing of some additional general-purpose search engines.

About—*http://www.about.com/* (careful, pop-up window city!)

Alta Vista—*http://www.altavista.com/*

AOL—*http://search.aol.com/*

Buyer's Index—*http://www.buyersindex.com/* (mail-order shopping search engine)

Direct Hit—*http://www.directhit.com/*

Dog Pile—*http://search.dogpile.com/*

Euroseek—*http://www.euroseek.com/*

Excite—*http://www.excite.com/*

Find What—*http://www.looksmart.com/*

Hotbot—*http://www.hotbot.com/*

Infospace—*http://www.infospace.com/*

LookSmart—*http://www.looksmart.com/*

Lycos—*http://www.lycos.com/*

MSN—*http://www.msn.com/*

Overture—*http://www.overture.com/*

Public Internet Library—*http://www.ipl.org/*

Search Hippo—*http://www.searchhippo.com/*

Worldpages—*http://www.worldpages.com/*

Yahoo!—*http://www.yahoo.com/*

Google Search: Dome Squirrel Baffle 203658

This search finds domes for bird feeders. Metal and plastic; many are clear polycarbonate.

Google Search: Electrical/Electronics Hardware 202696

http://directory.google.com/Top/Business/
Industries/Electronics_and_Electrical/Hardware/

Google category list for electronic hardware: connectors, standoffs, fasteners, knobs, etc.

Google Search: Electronics 202791

http://directory.google.com/Top/Science/
Technology/Electronics/

Google category list for general electronics. Check the subsections for narrow fields of interest, such as test and measurement, semiconductors, and design.

Google Search: Fasteners 202688

http://directory.google.com/Top/Business/
Industries/Construction_and_Maintenance/
Materials_and_Supplies/Nails,_Screws_and_Fasteners/

Google category list for fasteners, including nuts, bolts, screws, rivets, clips, and specialty fasteners.

Google Search: Horror Movies 202789

http://directory.google.com/Top/Arts/Movies/
Genres/Horror/

Google category list for horror films, which sometimes also include movies about crazed robots.

Google Search: Industrial Controls 202711

http://directory.google.com/Top/Business/
Industries/Electronics_and_Electrical/
Control_Systems/Industrial/

Google category list for industrial control components, such as single board computers, sensors, actuators, and power transmission products.

Google Search: Industrial Metals 202691

http://directory.google.com/Top/Business/
Industries/Manufacturing/Materials/Metals/

Google category list for industrial metals.

Google Search: Industrial Robotics 202707

http://directory.google.com/Top/Computers/
Robotics/Industrial/

Google category list for factory automation and industrial robotics.

Google Search: Materials Handling 202684

http://directory.google.com/Top/Business/
Industries/Industrial_Supply/Materials_Handling/

Google category list for materials handing, which includes casters and wheels and conveyors.

Google Search: Metal for Crafts 202781

http://directory.google.com/Top/Shopping/
Crafts/Metal/

Google category list for metals and working with it.

Google Search: Metalworking 202749

http://directory.google.com/Top/Shopping/
Crafts/Supplies/Metal_Craft/

Google category list for working with metal, as well as materials and supplies.

Google Search: Motion Control   202708

http://directory.google.com/Top/Business/
Industries/Manufacturing/Factory_Automation/
Motion_Control/

Google category list for factory automation and motion control.

Google Search: Optics   202706

http://directory.google.com/Top/Business/
Industries/Manufacturing/Optics/

Google category list for optics and optical components.

Google Search: Optics Technology   202793

http://directory.google.com/Top/Science/Physics/
Optics/

Google category list for optical design.

Google Search: Optoelectronics   202694

http://directory.google.com/Top/Business/
Industries/Manufacturing/Optoelectronics/

Google category list for industrial optoelectronics (such as LEDs, photosensors).

Google Search: Plastics   202689

http://directory.google.com/Top/Business/
Industries/Manufacturing/Polymers/Materials/
Plastics/

Google category list for plastics: materials and working with it.

Google Search: Plywood   202690

http://directory.google.com/Top/Business/
Industries/Construction_and_Maintenance/
Materials_and_Supplies/Wood_and_Plastics/
Wood_Products/Dimension_Lumber_and_Plywood/

Google category list for plywoods, including those using hardwoods or exotic woods.

Google Search: Relays & Solenoids   202692

http://directory.google.com/Top/Business/
Industries/Electronics_and_Electrical/
Relays_and_Solenoids/

Google category list for relays, solenoids, and related electromechanical devices.

Google Search: Robot Building   203835

http://directory.google.com/Top/Computers/
Robotics/Building/

Google category list for building robots.

Google Search: Robotics   202792

http://directory.google.com/Top/Computers/
Robotics/

Google category list for general robotics topics.

Google Search: Science Fiction Movies   202787

http://directory.google.com/Top/Arts/Movies/
Genres/Science_Fiction_and_Fantasy/

Google category list for science fiction movies, including those in which the robot is the good guy.

Google Search: Science Museums   202790

http://directory.google.com/Top/Reference/
Museums/Science/

Google category list for science museums. Look for ones that specialize in science, physics, and mechanical exhibits.

Google Search: Semiconductors   202794

http://directory.google.com/Top/Science/
Technology/Electronics/Semiconductors/

Google category list for semiconductors.

Google Search: Switches & Encoders 202697

http://directory.google.com/Top/Business/
Industries/Electronics_and_Electrical/
Switches_and_Encoders/

Google category list for switches and encoders (including absolute and incremental, mechanical and optical).

Google Search: Test & Measurement 202693

http://directory.google.com/Top/Business/
Industries/Electronics_and_Electrical/
Relays_and_Solenoids/

Google category list for test and measurement tools.

Google Search: Video 202795

http://directory.google.com/Top/Shopping/
Consumer_Electronics/Video/

Google category list for video topics, including cameras, imagers, recording, transmitting, and programming APIs.

Google Search: Woodcraft 202750

http://directory.google.com/Top/Shopping/
Crafts/Supplies/Woodcraft/

Google category list for woodworking materials.

Google Search: Woodworking 202751

http://directory.google.com/Top/Shopping/
Crafts/Supplies/Woodcraft/Woodworking/

Another Google category list for woodworking materials.

Google Search: Woodworking Tools 202752

http://directory.google.com/Top/Shopping/
Tools/Woodworking/

Google category list for the tools used in woodworking.

Smart Pages 203290

http://www.smartpages.com/

Yellow Pages on the Internet. Look up North American businesses by category or keyword.

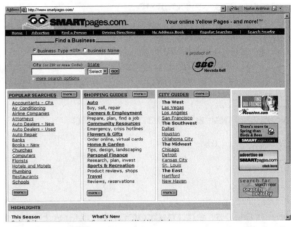

The Smart Pags are the Internet equivalent of printed local Yellow Pages.

 INTERNET-USENET NEWSGROUPS

Usenet is the term used for the portion of the Internet that provides for two-way non-real-time messaging. The more common term is newsgroups, a bit of a misnomer because most of the Usenet groups do not deal with "news," per se. The function of most groups is to share ideas and to post and answer questions.

See also Internet-Bulletin Board/Mailing List

FAQ: rec.toys.lego (LEGO) 202410

http://www.multicon.de/fun/legofaq.html

Support FAQ for the rec.toys.lego newsgroup.

Google Groups: comp.robotics .misc News Archive 202227

http://groups.google.com/groups?
hl=en&group=comp.robotics.misc

Google maintains an archive of most public newsgroups; this URL is for the ever popular comp.robotics.misc (or c.r.m.). Current messages are no more than a day "old," and the archive goes back many years.

Of course, you can view the archive for other newsgroups. Just click the "Group" link near the top of the window, and locate the group of your choice.

Newsgroups: alt.comp. lego-mindstorms

202294

alt.comp.lego-mindstorms

Newsgroups: alt.machines.cnc

203002

alt.machines.cnc

Newsgroups: alt.microcontrollers.8bit

202808

alt.microcontrollers.8bit

Newsgroups: alt.music.makers .theremin

202812

alt.music.makers.theremin

Newsgroups: alt.robotwars

202804

alt.robotwars

Newsgroups: alt.toys.lego

202293

alt.toys.lego

Newsgroups: comp.ai

204173

comp.ai

Newsgroups: comp.ai.life

204174

comp.ai.life

Newsgroups: comp.ai.vision

202291

comp.ai.vision

Newsgroups: comp.arch .embedded

204175

comp.arch.embedded

Newsgroups: comp.arch .embedded.picbasic

202798

comp.arch.embedded.picbasic

Newsgroups: comp.arch .embedded.piclist

202799

comp.arch.embedded.piclist

Newsgroups: comp.arch .fpga

202800

comp.arch.fpga

Newsgroups: comp.home .automation

202344

comp.home.automation

Newsgroups: comp.os.linux .embedded

202805

comp.os.linux.embedded

Newsgroups: comp.robotics .misc

202797

comp.robotics.misc

Newsgroups: Information Beyond the Web

Newsgroups were born in the days long before the Internet became a global marketplace. A newsgroup is a discussion group, a bulletin board, for posting and reading messages. Unlike chat, newsgroups are not real time; you don't directly communicate with others while everyone is online. Newsgroups, which are part of the Internet sometimes referred to as Usenet, can be an excellent source of information and feedback.

As newsgroups are the result of the early Internet, their structure follows something of a gearhead's design of the world, and some aspects of it may be confusing and cryptic. However, it's not complicated once you learn your way around.

What's a Newsgroup?

First and foremost is that newsgroups are divided into two main forms, public and private. Public newsgroups are open to anyone, and most likely, your Internet service provider (ISP) maintains computers just for the purpose of storing newsgroup messages. Your ISP's computers are connected to all the other public newsgroup computers around the globe, and they constantly trade messages back and forth. The end result is that even though you may "connect" to a newsgroup via your local ISP, you are reading the messages of others worldwide; if you post a message of your own, within a few hours it will circulate around the globe.

Private newsgroups are set up by companies or organizations to support their products or agenda. They may or may not be open to the public. In most cases, you must use the newsreader portion of your Web browser to separately log into these private newsgroup servers; they are not part of the public newsgroups provided by your ISP.

Newsgroup Hierarchy

Newsgroups follow a hierarchy, with some one dozen "top-level" categories to choose from. The categories of primary interest to robot builders are highlighted in boldface type:

alt	**Anything and everything**
biz	Business products, services
comp	**Computer and technical hardware, software**
humanities	Fine art and literature, Philosophy 101
misc	What won't fit anywhere else
news	Info about Usenet newsgroups
rec	**Games, hobbies, personal interests, sports**
sci	**Applied science, social science**
soc	Social issues, culture, politics, religion
talk	Current issues and debates

Newsgroups are further divided into one or more additional sublevels, with each sublevel separated from the top-level category name by a period. For example, the main robotics discussion group for amateur and professional robotics is comp.robotics.misc; the group is under the *comp* top-level and is further under the *robotics* sublevel. By the way, there are few sublevels under robotics, so don't expects lots of resources here. This is just how it is for newsgroup subtopics.

Newsgroup Reading and Writing

To read (and optionally post) messages on a newsgroup you need to crank up a newsgroup reader program, which is most often part of the browser you use for surfing the Web, or is part of the operating system. I won't get into exact steps here, as the procedures are documented in your software. Newsgroup readers are supported as part of Outlook Express if you're a Microsoft Internet Explorer user; Netscape has a newsgroup reader built-in.

Newsgroup Shorthand

The categories and subcategories in newsgroups are in a kind of shorthand. Here are what several of the more common ones mean:

alt	Alternative	misc	Miscellaneous
comp	Computer	arch	Architecture
rec	Recreation	ai	Artificial Intelligence
sci	Science		

Newsgroups: rec.crafts.metalworking
202809

rec.crafts.metalworking

Newsgroups: rec.crafts.misc
202810

rec.crafts.misc

Newsgroups: rec.crafts.polymer-clay
202295

rec.crafts.polymer-clay

Newsgroups: rec.models.rc.air
202296

rec.models.rc.air

Newsgroups: rec.models.rc.land
202297

rec.models.rc.land

Newsgroups: rec.models.rc.misc
202298

rec.models.rc.misc

Newsgroups: rec.music.makers.synth
202813

rec.music.makers.synth

Newsgroups: rec.toys.lego
202292

rec.toys.lego

Newsgroups: rec.woodworking
202811

rec.woodworking

Newsgroups: sci.electronics.basics
202801

sci.electronics.basics

Newsgroups: sci.electronics.components
202806

sci.electronics.components

Newsgroups: sci.electronics.design
202803

sci.electronics.design

Newsgroups: sci.electronics.equipment
202807

sci.electronics.equipment

Newsgroups: sci.electronics.misc
202802

sci.electronics.misc

Newsgroups: sci.electronics.repair 203548

sci.electronics.repair

Newsgroups: sci.geo.satellite-nav 203547

sci.geo.satellite-nav

Newsgroups: uk.media.tv .robot-wars 202796

uk.media.tv.robot-wars

INTERNET-WEB RING

Web rings are collections of Web sites—some commercial and some not—that share a common theme or interest. These days, most Web rings are hosted by Yahoo! and similar outfits. Joining and participating are free, but expect advertisements. The following listings are for the hubs of the Web rings. The hub lists the member sites belonging to the ring.

Web Ring: 8051 Microcontroller 202042

http://o.webring.com/hub?ring=80x51

Resources for 80x51 families of microcontroller and related embedded technologies.

Web Ring: Australian Electronics 202084

http://g.webring.com/webring?ring=aering

Electronics retailers and sources in Australia.

Web Ring: Australian Electronics Manufacturers 203795

http://l.webring.com/hub?ring=australianelectr

Australian companies or individuals that manufacture or assemble electronics products in Australia.

Web Ring: AVR Microcontroller 202060

http://r.webring.com/hub?ring=avr

Atmel AVR microcontroller-related sites.

Manners and Etiquette in Newsgroups

Newsgroups can be a horrendous sink of time and energy. Take my advice: Avoid using them as a soapbox, and avoid getting involved in heated discussions. People say things in newsgroups that they'd never say in person, simply because folks are protected sitting behind a computer screen.
Here are some other tips:

- Unless you are an adult, don't reveal your age. If you're old enough to read this book, you're old enough to understand why.

- Sexual harassment is still a problem, particularly in public newsgroups where "joy riders" and "trolls" only stick around to cause trouble. So, if you're female, I suggest you use a male name for your login. Obviously, this isn't a huge problem in groups frequented mostly by women, but harassment is known to be a bigger issue in male-dominated interest groups.

- If you include your e-mail address, use a "throw-away" e-mail account that you can close should the spam (unsolicited commercial mail) get too abundant.

- If you read a message that angers you, don't reply immediately. Flag it, and return to it a day later. If you still want to reply, don't add to any possible "flame war." They consume way too much time.

Web Ring: Basic Stamp SX 202105

http://v.webring.com/hub?ring=stamp

Basic Stamp computers, PIC microcontrollers, and related embedded technologies.

Web Ring: BEAM Robotics 202385

http://v.webring.com/hub?ring=beamring

Dedicated to BEAM robotics.

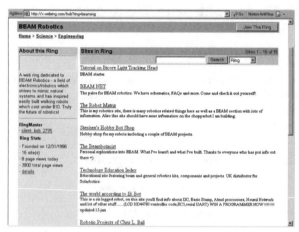

Web Ring: BEAM Robotics.

Web Ring: Circuits and Electronics 202391

http://g.webring.com/hub?ring=circuits

Exploring electronics as a hobby.

Web Ring: Electronic Music 202392

http://w.webring.com/hub?ring=emusic

The Electronic Music Ring (eMusic) features Web sites dealing with the creation and distribution of electronic music.

Web Ring: Electronics Engineering 202402

http://q.webring.com/hub?ring=eewebring

Dedicated to all Web sites and users that participate in or around the world of electronics engineering.

Web Ring: Embedded 202347

http://s.webring.com/hub?ring=embedded

Microcontrollers and embedded RTOS (real-time operating systems)—related sites.

Web Ring: FIRST Robotics 202386

http://f.webring.com/hub?ring=firstrobotics

A collection of FIRST Robotics team sites.

Web Ring: Forth Programming 203793

http://t.webring.com/webring?ring=forth

For people who are interested in the Forth Programming language.

Web Ring: HC11 202104

http://n.webring.com/hub?ring=hc11

Motorola 68HCxx microcontrollers family and related embedded technologies.

Web Ring: Hobby Machinists 203694

http://x.webring.com/hub?ring=hobbymachinists

A collection of personal metalworking Web sites.

Web Ring: LEGO Maniac 203931

http://p.webring.com/hub?ring=lego

For LEGO enthusiasts.

Web Ring: LEGO Mindstorms 202236

http://x.webring.com/hub?ring=legoms

The LEGO Mindstorms Web ring links together sites that have content and/or information about LEGO Mindstorms.

Web Ring: Linux Robotics 202388

http://n.webring.com/hub?ring=linuxrobotics

Noncommercial Linux-based robot projects. Includes sites with mechanical design, electronic circuits, and program code associated with designing and building a Linux-based robot.

Web Ring: Meccano 202310

http://www.meccanoweb.com/meccring/

Caters to the fans of the Meccano (or Erector Set) construction toys.

Web Ring: Model Railroad 202400

http://f.webring.com/hub?ring=modelrailroadele

Circuits and tutorials of interest to do-it-yourselfers of all skill levels.

Web Ring: PalmPilot 204171

http://p.webring.com/webring?ring=geoff

Where Palm users can show off their favorite applications, devices, and Palm accessories.

Web Ring: PICMicro 202107

http://o.webring.com/hub?ring=picmicro

Sites and pages dedicated to the Microchip PIC microcontroller and related projects.

Web Ring: Robot Pets 202389

http://t.webring.com/hub?ring=therobotpetsring

A Web ring for all sites involving Robotics robot pets or animatronics.

Web Ring: Robotics 202384

http://f.webring.com/hub?ring=robotics

For robots, robotics, and robot enthusiasts.

Web Ring: Robotics and EE 202387

http://p.webring.com/hub?ring=roboticsee

The Robotics and Electrical Engineering Web ring is designed to be an easy utility for finding quality hobby electrical and robotics content.

Web Ring: Robotics2 202383

http://h.webring.com/hub?ring=robotics2

Robotics Web pages.

Web Ring: Robots 202071

http://robotwebring.homestead.com/

"Dedicated to all things ROBOT."

Web Ring: Semiconductors 202401

http://p.webring.com/hub?ring=semiconductor

All about semiconductors.

Web Ring: Woodworkers 203230

http://j.webring.com/hub?ring=wood

Collection of woodworking sites.

Web Ring: Zilog 202106

http://v.webring.com/hub?ring=zilog

Resources for Zilog microprocessors and compatibles (e.g., Hitachi and Rabbit), and microcontrollers, and related embedded technologies.

JOURNALS AND MAGAZINES

Whether they are available for free or by subscriptions, journals and magazines offer a tremendous information base on special-interest subjects. This section lists both online and printed journals and magazines (some are available as both). In most cases, the online versions are free, though there may be a limit in the number of current articles posted each issue. In other cases, the online version is just a teaser to the printed magazine.

Several of the listings that follow are for so-called controlled circulation magazines. These are free to qualified readers. The qualification requirements vary depending on the scope of the publication, but most look for readers in a purchasing or recommendation/consulting role. You receive the magazine at no cost if you qualify. Even if you don't, however, the online versions of these magazines are often complete, with all columns and articles intact, or the publication is available for free reading at the local library.

SEE ALSO:

BOOKS-TECHNICAL: More printed information

INTERNET-INFORMATIONAL: Free online sources

MANUFACTURERS (VARIOUS): Look for downloadable datasheets, application notes, and case histories

Advanced Robotics · 203456

http://www.advanced-robotics.org/

Advanced Robotics is the official international journal of The Robotics Society of Japan and is edited by Dr. Hisato Kobayashi. The publication is in English.

See also the home page for The Robotics Society of Japan:

http://www.rsj.or.jp/

Circuit Cellar · 202100

http://www.circuitcellar.com/

Circuit Cellar offers practical, hands-on applications and solutions for embedded-control designers. Recommended reading for the intermediate and advanced robot constructor.

Dr. Dobb's · 202934

http://www.ddj.com/

Dr. Dobb's is an old and trusted journal for software tools for the professional programmer. By paid subscription; content is also available on the Web site. For experienced programmers.

ECN Magazine · 202901

http://www.ecnmag.com/

ECN Magazine contains product news and information for the electronic designer and engineer. Subscriptions to the printed magazine are free to qualified readers.

EDN Access · 202940

http://www.e-insite.net/ednmag/

EDN is a free (for qualified readers) magazine published twice monthly for electronic designers and systems managers. Selected articles are available on the Web site.

Electronic Design · 202941

http://www.elecdesign.com/

Electronic Design magazine is for electronics engineers and students. Each issue contains new product announcements, design articles, tutorials, and how-tos. ED is the home of columnist Bob Pease, who always has something interesting to say. Free to qualified readers; the Web site reproduces content from the magazine in Adobe Acrobat PDF format.

Electronics Products · 202942

http://www.electronicproducts.com/

Electronics Products keeps readers abreast of new product announcements and technologies. The magazine is free to qualified readers.

Electronique Pratique · 202418

Publications Georges Ventillard
2 12, rue Bellevue
Paris 75019
France

 http://www.eprat.com/

French-language magazines for robot and electronics hobbyists:

- Micros & Robots
- Electronique Pratique
- Montages FLASH
- INTERFACES PC

Elektor Electronics 203284

http://www.elektor-electronics.co.uk/

Electronics magazine, published in the U.K. Available by subscription or at newsstands in many countries.

Embedded Systems Programming 202943

http://www.embedded.com/

Embedded Systems is for the microcontroller and the, hmmm, embedded systems developer. Subscriptions are free to qualified readers, and many of the articles are reproduced on the Web site.

Everyday Practical Electronics 202375

http://www.epemag.wimborne.co.uk/

A U.K.-based electronics magazine, catering to hobby construction articles. Subscriptions available internationally.

GPS World 203543

http://www.gpsworld.com/

GPS World specializes in global positioning satellites.

Web site for GPS World magazine. Subscriptions are free for qualified readers.

Home Automator Magazine 203077

http://www.homeautomator.com/

Monthly printed magazine on automating your home. Many of the same principles apply to robots. The magazine is available by subscription only.

IEEE Spectrum 203283

http://www.spectrum.ieee.org/

Spectrum is the official monthly magazine of the IEEE. Selected articles are available online.

Journal of Forth, The 203638

http://www.jfar.org/

Online technical reference for Forth programmers and programming. Serious stuff, and it includes several articles directly relating to robots. An example is "Finite State Machines in Forth."

LinuxDevices.com 202816

http://www.linuxdevices.com/

Online publication specializing on the use of the Linux operating system for embedded applications.

Micro Control Journal 203134

http://www.mcjournal.com/

Micro Control Journal is an online publication with articles, news, and other information on the design and application of industrial automation and control. Many of the articles are in Adobe Acrobat PDF format.

Micro Magazine.com 204010

http://www.micromagazine.com/

From the Web site: "MICRO is the only magazine dedicated to advanced process and equipment control, defect reduction, and yield enhancement strategies for the semiconductor and related advanced microelectronics manufacturing industries. Now in its 20th year of publication, MICRO is recognized by readers as the leader in its field for its technical content as well as its industry news and product coverage."

All that, and it's free to qualified readers.

MMS Online   202109

http://www.mmsonline.com

MMS Online is the online component of the magazine Modern Machine Shop. Subscriptions are free to qualified readers.

Model Airplane News   203299

http://www.modelairplanenews.com/

Model Airplane News covers the R/C flying hobby. Some articles available on the site; the majority is about constructing the air frame and flying the plane, but there are others that deal with electronics and are therefore useful in robotics.

Model Aviation   203300

http://www.modelaircraft.org/mag/index.htm

Official publication of the Academy of Model Aeronautics; available by paid subscription. The magazine includes monthly columns, features, technical and how-to articles.

Modern Materials Handling Online   203257

http://www.manufacturing.net/mmh/

Magazine about materials handling. Before you yawn, this subject includes things like omnidirectional ball casters and conveyor belt parts. In other words, good stuff for a robot.

MRO esource.com   203519

http://www.mro-esource.com/

Online publication specializing in motion control technologies and mechanicals.

For the curious, MRO stands for "maintenance, repair, and operating," and represents the parts and supplies involved in this wide-ranging endeavor. Some of it, like wheels, casters, metal framing, bearings, and roller chain, is useful for robot building.

NASA Tech Briefs   203279

http://www.nasatech.com/

NASA Tech Briefs aren't underwear, but highly readable and informative summaries of the latest in technology from the National Aeronautical and Space Administration. Plenty of tidbits about robotic planetary explorers, and the reading is free.

NASA Tech Briefs provides free articles and reviews of the latest in technology.

New Scientist   203003

http://www.newscientist.com/

Online version of venerable print magazine. New discoveries and breakthroughs are commonly printed here first. Some online content of the print magazine is available; the printed version is available through paid subscription.

Nuts & Volts Magazine   202099

http://www.nutsvolts.com/

The monthly Nuts & Volts magazine is a premier source for robotics and electronics hobbyist information. Articles and columns show how to make printed circuit boards, troubleshoot and repair a circuit, and design a project using integrated circuits or microcontrollers.

The magazine has regular columns on robotics and the Basic Stamp.

Check out their bookstore, including the Nuts & Volts Basic Stamps (reprint of columns).

PC/104 Embedded Solutions Magazine
203200

http://www.pc104-embedded-solns.com/

Monthly online and printed magazine on PC/104 single board computers. Subscriptions are free to qualified readers.

Plastics Technology Online
202110

http://www.plasticstechnology.com/

Plastics Technology Online is a free (to qualified readers) magazine about plastics and plastics manufacturing.

Popular Mechanics
203278

http://popularmechanics.com/

Popular Mechanics is a monthly magazine about understanding and working with mechanical and modern devices. Occasional article or short about robots, but regular science features on such things as new sensor technologies, video systems, and other robot-related topics.

Check out the Technology Watch page:

http://popularmechanics.com/popmech/sci/tech/CURRENT.html

Popular Science
203277

http://www.popsci.com/

Popular Science is the "what's new" magazine. It explains, in lay terms, breakthroughs in science and technology. I wrote lots of articles for PopSci in the early 1980s (though not on robots). Publishes the occasional article on robots and robotics-related topics.

Publications of the American Association for Artificial Intelligence
203455

http://www.aaai.org/Publications/publications.html

AAAI publishes nearly 100 proceedings, technical reports, edited collections, and magazine issues each year, in hard copy, CD-ROM, and electronic form. The AAAI divisions responsible for the bulk of this activity are AI Magazine and AAAI Press.

R/C Car Action
203301

http://www.rccaraction.com/

Monthly magazine, available at newsstands or by subscription, for the R/C enthusiast.

R/C Modeler Magazine
203302

http://www.rcmmagazine.com/

Monthly magazine (available at newsstands or by paid subscription) all about R/C model airplanes. Includes product reviews, how-to articles, and lots of ads. You can use the ads to find R/C parts like servos and hardware.

Real Robots (Cybot)
202529

http://www.realrobots.co.uk/

Real Robots is a twice-monthly ("every fortnight") magazine that combines full-color how-to construction, along with the parts to built a 'bot.

For an independent Web page supporting the Real Robots product, see also:

http://www.cybotbuilder.com/

Robot Science & Technology
202846

3875 Taylor Rd.
Ste. B
Loomis, CA 95650
USA

(916) 660-0730
(888) 510-7728
service@robotmag.com
http://www.robotmag.com/

Quality robotics magazine with an uneven delivery schedule. Subscribe to the online newsletter, and purchase back issues as desired.

Science Daily   203280

http://www.sciencedaily.com/

Science Daily is a digest of the latest happenings in the science world, including new advances in robots, robot sensors, artificial intelligence, and related topics.

Science Magazine   203270

http://www.sciencemag.org/

The "granddaddy" of scientific journals, many break-throughs are published here first. Subscriptions are pricey, but you can read digests and summaries of the main articles online. And this magazine is available at many public libraries. As an alternative, you can purchase a 24-hour pass if you'd like access to a specific article or to do research.

Science News Online   204145

http://www.sciencenews.org/

Science News is a weekly digest-style magazine that covers the latest happenings in all science disciplines. The Science News Online site provides limited access to articles (unless you're a subscriber). Check the weekly table of contents to see if there are any news bits or articles about robots, sensors, or related subjects.

Seattle Robotics Encoder   202988

http://www.seattlerobotics.org/encoder/

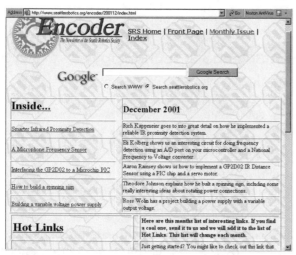

The Encoder is the newsletter of the Seattle Robotics Society.

Semiregular online newsletter with articles, tutorials, and news about amateur robotics. Some of the articles get fairly technical; all of them are good.

Sensors Online   202456

http://www.sensorsmag.com

Sensors Online is dedicated to sensor technology. This free magazine covers sensors for automotive, industrial, commercial, and consumer products. It's a great way to stay abreast of new potential eyes and ears for your robots.

Silicon Chip Magazine   204167

P.O. Box 139
Collaroy, NSW 2097
Australia

☏   +61 2 9979 6503

✉   silchip@siliconchip.com.au

🌐   http://www.siliconchip.com.au/

Silicon Chip is an Australian magazine aimed primarily at professionals and hobbyists interested in electronics and electronics projects. Available by subscription in Australia and New Zealand and by air mail delivery to other parts of the world.

Supply Chain Systems Magazine   203844

http://www.idsystems.com/

Supply Chain is a print and online magazine for "users and implementers of supply chain automation and collaboration." Topics include RFID and bar code. The print version of the magazine is free to qualified readers.

Transactor Online Archive   202991

http://www.csbruce.com/~csbruce/cbm/transactor/

Says the site: "This is an online archive of The Transactor magazine, provided with permission. The Transactor was a popular magazine for the old 8-bit Commodore computers, which was renowned for its overall high quality and the unusually deep technical level of its articles. As a bonus, it was also produced in Canada."

WebElectric Magazine 202998

http://www.webelectricmagazine.com/

Calling itself "the Web magazine for do-ers," the free WebElectric Magazine publishes articles on modern electronics, with a decided slant toward microcontrollers. Read the current issue, or browse back issues. Edited and published by Lawrence Mazza. Some of the projects presented in the magazine are for sale.

WildHobbies 202996

http://www.wildhobbies.com/

WildHobbies bills itself as "the #1 online hobby magazine." Includes a section on robot combat. Free, with user-to-user forums, resource links, news, and classified ads.

Woodworking Pro 202995

http://www.woodworkingpro.com/

A portal for the woodworking professional. News, links, buyer's guide, and a woodworkers magazine (free to qualified readers).

KITS

A great way to learn about something new is to build a ready-made kit. You get all the parts you need and a how-to construction plan to follow. The next sections list resources for both electronics and robotics kits; the two are separated for convenience, though in some cases, a single resource may carry both.

Nearly all of the kits listed here are for hobbyist use and are priced accordingly. You'll want to select a kit based on your skills. Most electronics kits require soldering, so if you need to brush up on your soldering skills, start with a simple low-cost kit. You can learn from it and graduate to bigger and better things.

KITS-ELECTRONIC

These resources manufacture, distribute, or sell electronics kits. For manufacturers and distributors, you can consult the company's Web site for technical documentation, schematics (when available), construction notes, and other pertinent information.

SEE ALSO:

ELECTRONICS-CIRCUIT EXAMPLES: Schematics and plans for electronic circuits you can build

INTERNET-PLANS & GUIDES: Free details on robotics and electronics projects

KITS-ROBOTICS: Robot construction kits, including those for beginners

ROBOTS-HOBBY & KIT: Additional kits, mostly higher end

A-1 Electronics   203062

718 Kipling Ave.
Toronto, ON
M8Z 5G5
Canada

- (416) 255-0343
- (416) 255-4617
- email@a1parts.com
- http://www.a1parts.com/

Kits: electronic, radio, educational lab. Also soldering equipment, surplus, pinhole cameras, tubes, and technical books.

Local store in Toronto, Canada.

Amazon Electronics/Elecronics123   202506

14172 Eureka Rd.
P.O. Box 21
Columbiana, OH 44408-0021
USA

- (330) 549 3726
- (603) 994 4964
- (888) 549-3749
- amazon@electronics123.com
- http://www.electronics123.com/

Product offering includes electronics kits, CMOS (Omnivision) camera modules, video transmitters and receivers, tools, components, microcontroller programmers (PIC and Atmel AVR), books, passive and active components, switches and relays, circuit prototyping, enclosures, and hardware.

Two locations:

USA: Amazon Electronics.

South Africa: Archimedes Products.

C & S Sales   202350

150 W. Carpenter Ave.
Wheeling, IL 60090
USA

- (847) 541-0710
- (847) 541-9904
- (800) 292-7711
- info@cs-sales.com
- http://www.cs-sales.com/

C & S Sales deals with test equipment, soldering irons, breadboards, kits (including OWI robot and Elenco electronic), hand tools, and other products. The Web site regularly lists new, sale, and closeout items.

Carl's Electronics Inc.   202283

P.O. Box 182
Sterling, MA 01564
USA

- (978) 422-5142
- (978) 422-8574 fax
- (800) 439-1417
- sales@electronickits.com
- http://www.electronickits.com/

Carl sells electronics and robotics (mostly OWI) kits.

Component Kits LLC

202206

266 Middle Island Rd.
Unit 15
Medford, NY 11763
USA

((631) 696-1006
∪ (631) 451-9331
 http://www.componentkits.com/

Kits and component assortments. Products include:

- Microcontroller kit
- Stepper motor kit
- Kits (assortments) of resistors, capacitors, crystals

Covington Innovations

203052

285 Saint George Dr.
Athens, GA 30606
USA

((706) 549-4633
✉ Michael@CovingtonInnovations.com
 http://www.covingtoninnovations.com/

Noted book and magazine author; projects, kits, and plans.

See NOPPP, the "No-Parts" PIC Programmer.

Debco Electronics, Inc.

202054

4025 Edwards Rd.
Cincinnati, OH 45209
USA

((513) 531-4499
∪ (513) 531-4455
⊘ 800) 423-4499
✉ debc@debco.com
 http://www.debco.com/

Components (active, passive, cables, connectors, fasteners, etc.), kits, and hand tools.

DLR Kits

202507

2500 E. Imperial Hwy. 201
PMB243
Brea, CA 92821
USA

✉ info@dlrkits.com

 http://www.dlrkits.com/

Electronics kits: infrared controllers, PIC microcontroller experimenter board. The company also offers "component kits" with an assortment of things like capacitors, resistors, crystals, or ICs.

Electronic Goldmine

202652

P.O. Box 5408
Scottsdale, AZ 85261
USA

((480) 451-7454
∪ (480) 661-8259
⊘ (800) 445-0697
✉ goldmine-elec@goldmine-elec.com
 http://www.goldmine-elec.com/

Electronic Goldmine sells new and used electronic components, robot items, electronics project kits, and more.

- General electronic components: capacitors, crystals, displays, fuses, heat sinks, ICs, infrared items, LEDs, potentiometers, resistors, semiconductors (misc.), thermal devices/thermistors, transistors, transformers, voltage regulators
- Passive and electromechanical components: batteries, cables and wire, circuit boards, fans, hardware, keychain lights, knobs, sockets, power supplies, relays, solenoids, switches, test equipment, tools, voltage converters
- Sensor components: bar code scanners, microphones, piezos, sensors, video cameras
- Kits and specialty: alarms and sirens, electronics project kits, motors, optics, robot items, strobe, ultrasonic items

Catalog in PDF format available; printed catalog sent to U.S. addresses only. Be sure to check out the interesting and unusual (and low-cost) robotics kits.

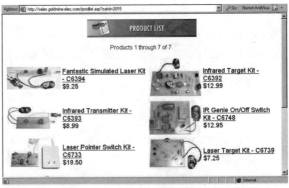

One of many kit product pages at Electonic Goldmine.

Electronic Rainbow, Inc.   202946

6227 Coffman Rd.
Indianapolis, IN 46268
USA

☎ (317) 291-7262
📠 (317) 291-7269
✉ info@rainbowkits.com
🌐 http://www.rainbowkits.com/

Electronic Rainbow makes electronics kits for hobby and education. Kits come with construction details, operation, theory of operation, and testing procedures. PC boards are premarked and have silk-screen layouts to show where the parts go. Some kits available assembled.

Product catalogs are available in Adobe Acrobat PDF format.

Electronic School Supply, Inc. (ESS)   203142

3070 Skyway Dr.
Ste. 303
93455, CA Santa Maria
USA

☎ (805) 922-6383
📠 (805) 928-0253
✉ essinc@esssales.com
🌐 http://www.esssales.com/

ESS manufactures and sells educational kits and lab instructors for electronics and high-technology subjects (such as fiber optics).

Elenco Electronics   202139

150 W. Carpenter Ave.
Wheeling, IL 60090
USA

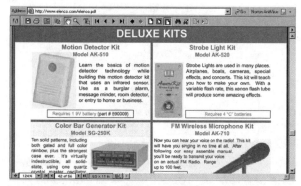

Web catalog page of some of Elenco's many kits.

☎ (847) 541-3800
📠 (847) 520-0085
✉ elenco@elenco.com
🌐 http://www.elenco.com/

From the Web site: "Elenco is a major supplier of electronic test equipment and educational material to many of the nation's schools and hobbyists. We also have a network of distributors selling our products from coast to coast and abroad."

The company sells electronics kits under the Amerikit brand.

Gibson Tech Ed   202404

1216 South 1580 West
Bldg. C
Orem, UT 84058
USA

☎ (801) 434-7664
🚫 (800) 422-1100
🌐 http://www.gibsonteched.com/main.html

Gibson Tech Ed provides electronics and robotics education products to schools, summer camps, home schools, educators, corporations, and individuals. In addition to being a reseller of kits, the company makes their own line of electronics kits; the kits include schematics and theory of operation.

See also http://www.hobbytron.com/

Graymark International, Inc.   202947

Box 2015
Tustin, CA 92681
USA

🚫 (800) 854-7393
🌐 http://www.graymarkint.com/

Electronics project kits, electronics trainers (such as digital/analog), OWI robot kits, some robot books, test instruments, and more.

Henrys Electronics Ltd.   203146

404 Edgware Road
Paddington
W2 1ED
UK

Useful General Electronics Kits for Robots

Ready-made kits are not only an excellent way to hone your electronics construction skills, they also serve as handy modules for enhancing your robotic creations.

While there are many kit makers, two of the most common lines are Kits R Us and Velleman. Their wares are available through a worldwide network of distributors and resellers. (A third line, made by Datak, offers a number of fine kits, but they aren't as universally available; you can locate distributors of this line at *http:// www.philmore-datak.com/*).

Both offer dozens of kits, many of which are directly adaptable to robotics. Here are some of them. (Check the company's Web sites for a list of distributors and resellers.)

Kits R Us—http://www.kitsrus.com/

- 2 x 16 LCD display module
- Sugar cube camera
- Temperature sensor chip
- Ultrasonic movement detector
- Serial temperature sensor
- Walky-talky Kit (27 MHz)
- DC motor speed control
- Voice Direct 364 speech recognition
- UHF transmitter, receiver module, in case
- 20-second record module
- 30-second voice recorder
- Servomotor driver
- 1 W stereo amplifier
- 18 W audio power amplifier
- Unipolar stepper motor driver
- Dual unipolar stepper motor driver
- Atmel AVR programmer
- 4-digit up/down counter
- Electronic heart
- Xenon flasher

Velleman—http://www.velleman.be/

- Electronic strobe light
- FM transmitter
- Analog-to-digital converter card
- Optocoupler input card
- Intelligent motherboard
- Relay card
- 8-to-1 analog multiplexer card
- Speed controller
- 7 W mono amplifier
- Record/playback module
- Sound generator
- 2-channel code lock transmitter/ 1-channel code lock receiver
- Universal relay card
- Stepper motor regulator

Velleman Mini Kits—http://www.velleman.be/

- LED Heart Kit—Beating heart using LEDs; use to dress up a drab 'bot.
- Blinking LEDs Kit
- LED Running Light—8 LED, with different sequence effects (use for show)
- Siren sound generator
- MicroBug—Brightly colored bug-shaped miniature robot

📞 +44 (0) 2072 581831

📠 +44 (0) 2077 240322

🚫 0800 731 6979

🌐 http://www.electronic-gadgets.co.uk/

Reseller of OWI robot kits and electronics kits.

Hobbylinc.com 🎖️ 202715

76 Bay Creek Rd.
Ste. P
Loganville, GA 30052
USA

📞 (770) 466-2667

📠 (770) 466-0650

🚫 (888) 327-9673

✉️ hobbylinc@hobbylinc.com

🌐 http://www.hobbylinc.com/

Hobbylinc sells a wide range of hobby products, including tools, glues and adhesives, and paints. However, of particular merit are the Tamiya Educational kits (gear motors, wheels, tracks, and more) and Elenco electronics kits. Many products lack a description, but have pictures. You should know what you're looking for before shopping. Excellent pricing.

Hobbytron 202403

1216 S. 1580 W.
Ste. C
Orem, UT 84058
USA

📠 (801) 434-9777

🚫 (877) 606-8766

✉️ support@hobbytron.net

🌐 http://www.hobbytron.net/

Online reseller of kits, including electronics and robotics (mostly OWI), toys, and LEGO Mindstorms.

Information Unlimited 202948

P.O. Box 716
Amherst, NH 03031-0716
USA

📞 (603) 673-4730

📠 (603) 672-5406

🚫 (800) 221-1705

✉️ wako2@wavewizard.com

🌐 http://www.amazing1.com/

Information Unlimited is a long-standing developer and seller of unusual science kits. A typical product is their Antigravity Generator, available as plans only, as a kit or ready-made. Some of their plans, kits, and products have uses in the field of robotics. For example, their highly amplified ultrasonic generator might be used for long-distance mapping. Their plasma globes and other light displays are useful to "dress up" an otherwise dull robot.

Jaycar Electronics 203668

P.O. Box 6424
Silverwater. NSW 1811
Australia

📞 +61 2 9741 8555

📠 +61 2 9741 8500

✉️ techstore@jaycar.com.au

🌐 http://www.jaycar.com.au

Jaycar sells a bunch of stuff, including personal electronics, video cameras, test gear, electronics kits, passive and active components, hardware and fasteners, batteries and chargers, and more.

J-Tron Inc. 202505

P.O. Box 378
324 Gilbert Ave.
Elmwood Park, NJ 07407
USA

📞 (201) 398-0500

📠 (201) 398-1010

🚫 (888) 595-8766

✉️ J-Tron@erols.com

🌐 http://www.j-tron.com/

DIY Electronics

http://www.diyelectronics.com/

Electronics and robotics kits and modules; based in Malaysia

SAMPLE Electronics

http://www.sample.co.kr/

Importer and distributor of electronic kits, parts

Datak kits, active and passive electronics components, switches, elecromechanical, wire and cable, fasteners, etc.

Kelvin 202877

280 Adams Blvd.
Farmingdale, NY 11735
USA

- (631) 756-1750
- (631) 756-1763
- (800) 535-8469
- kelvin@kelvin.com
- http://www.kelvin.com/

Kelvin sells educational kits and materials for a high-tech teaching world. Their Technology series includes a number of very useful products for robot building:

- K'NEX
- LEGO Dacta
- Robotics kits (BOE-Bot, MAZER, OWI robots, many others)
- PIC programmers
- Fischertechnik
- Tamiya Educational kits
- Plastics vacuum former
- Plastics injection molder
- Science and chemistry lab components

They also offer project materials in metal, plastic, and wood; magnets; various sizes and types of gearboxes (and motors with and without gearboxes); motor holders; linear actuator motors; wheels; gears; sprockets and sprocket chain; and hundreds of additional products.

Electronics include trainers, board-level solder kits, electronics construction tools, test gear, components (active and passive), and others.

Kelvin's sales are intended for educational institutions. While they will sell to individuals, they say some products may cost more and that some products are only available to schools and teachers. If you're ordering for a school, they accept school POs. Printed catalogs are available to teachers and schools only.

Kit Guy 203141

http://www.kitguy.com/

The Kit Guy loves kits, with information and links to kits makers, including robotics (Lynxmotion, RobotStore).

Kits R Us 202577

Peter Crowcroft
DIY Electronics (HK) Ltd.
P.O. Box 88458
Sham Shui Po
Hong Kong

- +85 2 2304 2250
- +85 2 2729 1400
- http://kitsrus.com/

Electronics kits. Importers and distributors of kits worldwide.

Magenta Electronics Ltd. 203852

135 Hunter Street
Burton-on-Trent
DE14 2ST
UK

- +44 (0) 1283 565435
- +44 (0) 1283 546932
- sales@magenta2000.co.uk
- http://www.magenta2000.co.uk/

Electronics components, kits, PIC programmers, PIC project boards. Also sells parts (e.g., stepper and DC motors) and electronics teaching labs.

Oatley Electronics 202255

P.O. Box 89
Oatley
NSW 2223
Australia

- +61 2 9584 3563
- +61 2 9584 3561
- sales@oatleyelectronics.com
- http://www.oatleyelectronics.com/

Oatley sells test equipment, electronics kits, mechanical components (switches, motors, etc.), active and passive components, lasers and LEDs, RF remote control, FM transmitter kits, and video cameras.

Ozitronics 202680

24 Ballandry Crescent
Greensborough, 3088
Victoria
Australia

- +61 3 9434 3806
- +61 3 9434 3847
- sales@ozitronics.com
- http://www.ozitronics.com/

Electronics kits of all flavors, including servo and stepper motor control, amplifiers, microcontroller programmers, Southern Cross Z80 single board computers, transmitters and receivers, and more. Ships internationally.

Quality Kits/QKits 202200

49 McMichael St.
Kingston, ON
ON K7M 1M8
Canada

- (613) 544 6333
- (613) 544 4944
- (888) 464 5487
- tech@qkits.com
- http://www.qkits.com/

Reseller of Velleman, Minikits, and Kits R Us electronics kits.

Quasar Electronics Ltd. 202681

Unit 14, Sunningdale
Bishop's Stortford
Hertfordshire
CM23 2PA
UK

- +44 (0) 1279 467799
- +44 (0) 7092 203496
- sales@quasarelectronics.com
- http://www.quasarelectronics.com/

Kits (Kits R Us electronics, OWI robot), servo motors, Atmel AVR programmers and products.

According to the site: "The UK's No. 1 Electronic Kit Supplier."

Ramsey Electronics, Inc. 202353

793 Canning Pkwy.
Victor, NY 14564
USA

- (716) 924-4560
- (716) 924-4886
- (800) 446-2295
- OrderDesk@ramseyelectronics.com
- http://www.ramseyelectronics.com/

Ramsey makes and sells a wide variety of electronics kits, including RF transmitters and receivers, miniature video cameras, digital sound recorders, DC motor speed controllers, tone encoder and decoder, and plenty more. Also sells test equipment and tools. The company's downloadable catalog is in Adobe Acrobat format and is available in one chunk or divvied up into sections.

One interesting kit in their current catalog is their "Tickle Stick," a small module that generates high voltage but at low current. For some reason I'm reminded of when R2-D2 gave Salacious Crumb, Jabba the Hutt's annoying couchside pet, the shock of its life on the sail barge in the first sequence of Return of the Jedi. I assume you're reminded of the same thing.

Transtronics, Inc. 203790

3209 W. 9th St.
Lawrence, KS 66049
USA

- (785) 841-3089
- (785) 841-0434
- index@xtronics.com
- http://www.xtronics.com/

Electronics kits, electronic programmers, surplus, and other goodies. Plus lots of useful information and tidbits. Read the explanation of the Jack and Jill rhyme.

Velleman Components NV 203765

Legen Heirweg 33
B-9890 Gavere
Belgium

- +32 (0) 9 384 36 11
- +32 (0) 9 384 67 02
- sales@velleman.be
- http://www.velleman.be

Velleman makes of a wide variety of high-quality electronics kits and components (heat sinks, LCDs, many others). Available through resellers such as Jameco. If from the U.S. use

http://www.vellemanusa.com/

as most products listed on the Belgium Web site are shown as not available in the U.S.

Web site is in English, Dutch, French, German, and Spanish.

Model PK105 signal generator kit, from Velleman.

Wiltronics   202857

P.O. Box 43
Alfredton, 3350
Australia

(+61 3 5334 2513
[] +61 3 5334 1845
✉ sales@wiltronics.com.au
🌐 http://www.wiltronics.com.au/

Wiltronics sells electronics kits, active and passive components, test equipment, tools, and the regular lineup of general electronics. Plus motors, gears, plastic wheels, pulleys, and other mechanicals for robot building and the RoboBall, a functional robot enclosed in a clear plastic ball.

Web site is in English and Japanese.

 # KITS-ROBOTICS

These resources manufacture, distribute, or sell robotics kits. Most of the kits are either for beginners or are aimed at classroom education. Sellers of the OWIKIT and MOVITS robots are listed here.

SEE ALSO:

INTERNET-PLANS & GUIDES: Free details on robotics and electronics projects

KITS-ELECTRONICS: Kits for electronics projects

ROBOTS-HOBBY & KIT: Includes some additional robotics kits, but mostly higher-end

Analytical Scientific, Ltd.   202863

11049 Bandera Rd.
San Antonio, TX 78250
USA

((210) 684-7373
[] (210) 520-3344
🚫 (800) 364-4848
✉ asltd@intersatx.net
🌐 http://www.analyticalsci.com/

Scientific goodies that would make even Mr. Wizard flush with joy. Analytical Scientific carries chemistry, astronomy, anatomy, and biology kits and supplies. They offer the full line of OWI robot kits and other nick-nacks of interest to automaton builders.

C & S Sales   202350

150 W. Carpenter Ave.
Wheeling, IL 60090
USA

((847) 541-0710
[] (847) 541-9904
🚫 (800) 292-7711
✉ info@cs-sales.com
🌐 http://www.cs-sales.com/

C & S Sales deals with test equipment, soldering irons, breadboards, kits (including OWI robot and Elenco electronic), hand tools, and other products. The Web site regularly lists new, sale, and closeout items.

EK Japan Co., Ltd.   203779

http://www.elekit.co.jp/

Manufacturers of the OWIKIT/MOVITS robots, as well as numerous electronics kits. Available through resellers (an online ordering button was nonfunctional at the time of this writing). The site includes interesting information and advance product notice.

You'll find that the prices (in Japanese yen) are often 25 to 50% lower than what the products cost in the U.S. and other countries. The prices shown are suggested retail in Japan; resellers often jack up the price to pay for shipping and duty charges.

Web site is in Japanese and English.

Home page of the exporter of OWIKIT/MOVITS kits.

Graymark International, Inc.   202947

Box 2015
Tustin, CA 92681
USA

⊘ (800) 854-7393

🌐 http://www.graymarkint.com/

Electronics project kits, electronics trainers (such as digital/analog), OWI robot kits, some robot books, test instruments, and more.

Henrys Electronics Ltd.   203146

404 Edgware Road
Paddington
W2 1ED
UK

✆ +44 (0) 2072 581831

📠 +44 (0) 2077 240322

⊘ 0800 731 6979

🌐 http://www.electronic-gadgets.co.uk/

Reseller of OWI robot kits and electronics kits.

Hobbylinc.com ⅄   202715

76 Bay Creek Rd.
Ste. P
Loganville, GA 30052
USA

✆ (770) 466-2667

📠 (770) 466-0650

⊘ (888) 327-9673

✉ hobbylinc@hobbylinc.com

🌐 http://www.hobbylinc.com/

Hobbylinc sells a wide range of hobby products, including tools, glues and adhesives, and paints. However, of particular merit are the Tamiya Educational kits (gear motors, wheels, tracks, and more) and Elenco electronics kits. Many products lack a description, but have pictures. You should know what you're looking for before shopping. Excellent pricing.

Hobbytron   202403

1216 S. 1580 W.
Ste. C
Orem, UT 84058
USA

✆ (801) 434-9777

⊘ (877) 606-8766

✉ support@hobbytron.net

🌐 http://www.hobbytron.net/

Online reseller of kits, including electronics and robotics (mostly OWI), toys, and LEGO Mindstorms.

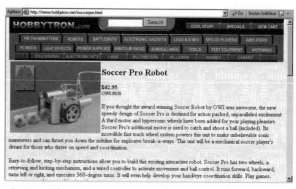

A page from Hobbytron's large selection of OWI/MOVIT line.

Quasar Electronics Ltd.   202681

Unit 14, Sunningdale
Bishop's Stortford
Hertfordshire
CM23 2PA
UK

✆ +44 (0) 1279 467799

📠 +44 (0) 7092 203496

sales@quasarelectronics.com

 http://www.quasarelectronics.com/

Kits (Kits R Us electronics, OWI robot), servo motors, Atmel AVR programmers and products.

According to the site: "The UK's No. 1 Electronic Kit Supplier."

Ready-Made, Kits, or Do It Yourself?

Choice has never been greater for the robot builder. Not only can you construct robots "from scratch," you can buy any of several dozen robot kits. Many can be assembled using a screwdriver and other common tools. Kits are ideal if you don't like the construction aspects of robotics, but instead want to concentrate on the electronics or the programming. You can also purchase ready-made robots.

Whether you choose to buy a robot in ready-made or kit form or to build your own from the ground up, it's important that you match your skills to the project. This is especially true if you are just starting out. While you may want the challenge of a complex project, if it's beyond your present skills and knowledge level, you'll likely become frustrated and abandon robotics before you've given it a fair chance. If you want to build your own robot, start with a simple design, and work up from there.

OWIKITS and MOVITS

The OWIKITS and MOVITS robots are precision-made miniature robots in kit form. A variety of models are available, including manual (switch) and microprocessor-based. The robots can be used as-is, or they can be modified for use with your own electronic systems. For example, the OWIKIT Robot Arm Trainer (model OWI007) is normally operated by pressing switches on a wired control pad. With just a bit of work, you can connect the wires from the arm to a computer interface (via relays, for example) and operate the arm via software control.

Most of the OWIKITS and MOVITS robots come with preassembled circuit boards; you are expected to assemble the mechanical parts. Some of the robots use extremely small parts and require a keen eye and steady hand. The kits are well made and will last for years if they are built properly. I have some OWIKITS robots I purchased almost 20 years ago, and they are still in perfect working order.

OWIKITS ands MOVITS are made by EK JAPAN Co. (Elekit) in Japan; you can see the company's Web site (English and Japanese pages are available) at:

http://www.elekit.co.jp/

The products are available through dozens of distributors and resellers. A major U.S. distributor is RobotKits Direct at:

http://robotikitsdirect.com/

Their Web site separates the robots by skill level and whether soldering is required. Examples of interesting OWIKITS robots include:

- *Navius*—Controlled by a small paper disk, like a player piano. You can make your own disk by filling in the black-and-white patterns.
- *Hyper Line Tracker*—Follows a black line and even remembers the last line it follows.
- *WAO G*—Programmable robot with a microcontroller and two feeler whiskers for detecting objects and table drop-offs.
- *Soccer Jr.*—A basic beginner's robot with wired control.

LEGO

LEGO has become such an important part of the robotics landscape that it deserves its own sections. The listings that follow divide LEGO into two parts:

- General. Basic LEGO constructions, parts identification, online LEGO builder's groups, LEGO Technic parts and creations, and interesting mechanical assemblies made entirely with LEGO.
- Mindstorms. Specifically relating to the LEGO Mindstorms sets, including Robotics Invention System, Robotics Discovery Set, and Droid Developer's Kit.

SEE ALSO:

MICROCONTROLLERS-HARDWARE: Construct your own programmable robotics controller

TOYS-CONSTRUCTION: Other construction toys available

LEGO-GENERAL

Baseplate
203111

http://baseplate.com/

Author and LEGO expert Suzanne D. Rich's Web site on LEGO. Offers some products for sale.

BrickFest
202265

http://brickfest.com/

Annual get-together of LEGO enthusiasts. Check the Web site for time and location.

BrickLink
203129

http://www.bricklink.com/

LEGO marketplace portal.

Brickshelf
202052

http://www.brickshelf.com/

LEGO bricks gallery and pictures. See what other people are making with their LEGO sets. It ain't all scenes of farmers and cows.

Brickshelf offers a gallery of LEGO creations.

Cool LEGO Site of the Week
204160

http://www.lugnet.com/cool/

Just what the name says.

Doug's LEGO Technic Tri-Star Wheel ATV and Robotics page
203103

http://www.visi.com/~dc/

Interesting informational page about unusual locomotion ideas using LEGO Technic.

LEGO Dacta
203127

http://www.lego.com/dacta/

Educational division of LEGO. You can buy online; check the Where to Buy link for a distributor in your country.

LEGO Element Registry
203130

http://w3.one.net/~hughesj/technica/index/index.php

A registry of elements found in various LEGO Technic sets. Covers Gear Wheel, Technical, Expert Builder, Technic, and Mindstorms sets and elements.

LEGO Lexicon 203126

http://members.brabant.chello.nl/~f.buiting/lego/

Glossary of LEGO-related words and phrases.

LEGO Set Inventories 203114

http://www.peeron.com/inv

What things are and which LEGO sets use them. Use it to find out just what that "weird white piece with the spoke-thingies coming out of it" is called.

LEGO Shop-at-Home ⚕ 202583

🚫 (800) 453-4652

🌐 http://shop.lego.com/

LEGO Shop-at-Home is dedicated to all things LEGO. What makes them especially vital is that they sell replacement parts for those that have become damaged or lost or consumed by robotics projects. You can search the online catalog (including by part number), but also be sure to get on the mailing list for the regular catalog.

Availability of replacement parts comes and goes and not all are available in every country. So, if/when you see what you want, be sure to get it. Wait too long, and it may not be available. Among the replacement parts you can get through Shop-at-Home are:

- Pneumatic pump
- Pneumatic cylinder
- Balloon tires
- Technic wheel pack
- Shock absorbers
- Pulleys and rubber belts
- Power pack (battery holder, 9v gear motor)

LEGO Sojourner Mars Lander-Clifford Heath 203105

http://homes.managesoft.com.au/~cjh//lego/

Personal Web page of LEGOmaestro Clifford Heath. Here he makes a LEGO version of the Sojourner Mars Lander.

Lugnet ⚕ 203115

http://www.lugnet.com

Lugnet is the global community of LEGO enthusiasts. They offer news and links about LEGO, but perhaps the most important part is their bulletin boards, which are also available as Usenet newsgroups. Topics include Building, CAD, Technic, Robotics, and Znap.

Expanding the robotics category, we find:

- Education
- Handy Board
- Palm Computing
- Events
- Logo
- Telerobotics
- Code Pilot
- RCX
- Vision Command
- CyberMaster
- Scout
- VLL
- Micro Scout

SEE ALSO:

http://www.lugnet.com/links/

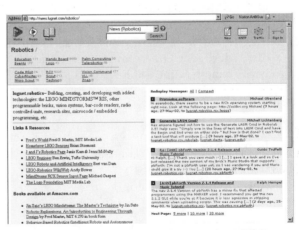

Lugnet provides links and forums for LEGO enthusiasts.

Micro Rover: A Tiny LEGO Rover 202278

http://www.ben.com/LEGO/rcx/micro-rover.html

One tiny LEGO robot. It's powered by a LEGO Micro-Motor. Uses simple actuation control with switches.

LEGO can be used as a basis for making custom robots.

MIT: Kego Catalog
203109

http://el.www.media.mit.edu/groups/el/
Projects/constructopedia/kego.html

Catalog of LEGO and LEGO Technic parts. From our friends at MIT.

Unofficial International Lego Service Catalog
203128

http://members.rogers.com/magundy/lego/
Service.html

Text and graphics lookup for LEGO Technic Service parts. Some of the parts cannot be purchased from LEGO Service, but may be available from other sources such as LEGO Dacto, Pitsco, eBay, or as extra parts from a discarded set. The reference is handy if you need to verify exactly what you have or need.

Important LEGO Web pages:

http://www.lego.com/service/

http://shop.lego.com/

http://www.lego.com/dacta/

http://www.pitsco-legodacta.com/

MOVERS AND SHAKERS

Fred's World—Fred G. Martin, MIT Media Lab

http://web.media.mit.edu/~fredm/

Archival personal Web page of Fred G. Martin, robotics professor, book author, and co-creator of a lot of today's best ideas in robotics and robot education.

LEGO-MINDSTORMS

(MS)2 Expander
202247

http://www.akasa.bc.ca/tfm/lego_ms2.html

Plans and description for the (MS)2 Expander, a homemade ports extender for the LEGO Mindstorms RCX.

- 6 full-featured output ports
- 3 + 1 full-featured output ports and 3 + 2 full-featured input ports
- 6 + 1 full-featured input ports

Web site includes schematics, theory of operation, and construction details.

Android Hand and Arm Prototype
203106

http://www.geocities.com/Colosseum/Dome/
5088/Hand.htm

Description, plans, and pictures for building a robot arm and hand (of a fashion) using LEGO Mindstorms parts.

Bot-Kit
203120

http://www.object-arts.com/Bower/Bot-Kit/

The Bot-Kit is a robot control system for use with the LEGO Mindstorms or CyberMaster. The product uses the Dolphin Smalltalk programming environment.

BrickCommand
203121

http://www.geocities.com/Area51/Nebula/
8488/lego.html

Free Windows-based programming interface for LEGO Mindstorms and LEGO CyberMaster robot controllers.

DosVLL Page
202712

http://www.users.bigpond.com/grant.wells/home-page.htm

Information on how to use a low-level PC to control the LEGO MicroScout brick. The process involves using a simple LED driver circuit which must be constructed and attached to the PC parallel port.

EVAL implementation of DIDABOTS 202244

http://www.idi.ntnu.no/grupper/ai/eval/fdagene/

RCX code and construction examples for LEGO Mindstorms robots made to interact with one another. Here's what the Web site has to say:

"Complex patterns of behavior (such as heap building) can result from a limited set of simple rules that steer the interactions between robots and their environment. We implemented an example of such a behavior originally developed by Maris and te Boekhorst, called DIDABOTS. All robots in the group are the same and perform a very simple behavior: when they run into an obstacle, they back up a bit, turn, and go forward again."

Gordon's Brick Programmer 203122

http://www.umbra.demon.co.uk/gbp.html

Says the Web site, "Gordon's Brick Programmer has been designed as an educational bridge between the very simple visual programming environment provided by LEGO with Mindstorms and the kind of programming environment used by IT professionals. It provides more general access to variables stored inside the RCX than the simple counters in the Mindstorms system."

Hempel Design Group 203100

http://www.hempeldesigngroup.com/lego/

Ralph developed pbForth, a version of Forth for the LEGO Mindstorms. The software is described on this page, with download links. Ralph also provides tips and tricks, such as getting more range from a Mindstorms light sensor and how to build a pressure limit switch.

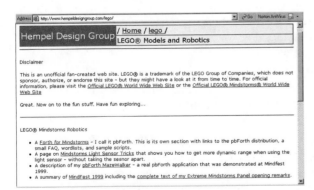

pbForth, and other tidbits, for LEGO Mindstorms robots.

Homebrew LEGO Sensors 203117

http://ex.stormyprods.com/lego/

Descriptions of several homemade sensors for the LEGO Mindstorms RCX, including:

- Temperature sensor
- Light sensor
- Bend sensor
- Rokenbok radio control module

Inchlab 202243

http://www.inchlab.com/

Plenty of projects for those into creating unusual robots with their LEGO Mindstorms. Show-and-tell pages include 2servo Bot, distance tracker, bar code scanner, Cargo Bot, Sharp GP2D02 interface for the RCX, and the LaserBrick.

J and J's Robotics in the Classroom 203118

http://www.occdsb.on.ca/~proj4632/

How robotics, particularly LEGO-based, can be used in education. Includes reviews of products and links.

LEGO League International 204228

http://www.firstlegoleague.org/

The FIRST LEGO League is an international group of LEGO enthusiasts interested in fostering interest in the fields of robotics, science, and technology in school-age children.

SEE ALSO FIRST:

http://www.usfirst.org/

LEGO Mindstorm with Linux Mini-HOWTO 203125

http://www.linuxdoc.org/HOWTO/mini/Lego/

How to get Linux to work with the LEGO Mindstorms, including its infrared programming tower.

Under the Hood of the Mindstorms RCX

The LEGO Mindstorms RCX is the central component of the Mindstorms Robotics Invention System (RIS). The RCX is termed a *programmable brick*; *brick* because it is modeled after the brick-building system employed with the LEGO system, and *programmable* because it contains its own computer.

The RCX is more than a toy; it's a plastic box containing a small microcontroller, as well as interface electronics to connect motors and sensors. At the heart of the RCX is an H8/3292 microcontroller, running at 16 MHz. The H8, and controllers like it, is like a miniature computer, but is designed for "embedded applications" for hardware control.

The LEGO Mindstorms RCX.

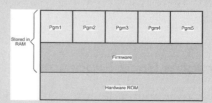

Two kinds of memory are used in the RCX: 16 KROM (read-only memory), which stores the hardware BIOS (basic input/output software) for the device, and 32 K of RAM memory is used to store the operating system (called, erroneously, *firmware*) for the RCX, as well as up to five user programs. The following figure shows the memory layout of the RCX. Note that while the firmware and user programs can be changed since they are stored in RAM, the hardware BIOS cannot be altered.

The memory layout of the LEGO Mindstorms RCX.

A unique aspect of the RCX is that the firmware can be periodically updated. This allows the LEGO Mindstorms programming system to be enhanced, without replacing the RCX unit. Whenever LEGO releases updates for the RCX, you need merely to return to the setup portion of the Mindstorms program and download the new version of the firmware.

Note that the data in RAM is retained by applying a small amount of battery power to the RCX, even when it is turned off. This is why programs and firmware can be lost if the RCX is stored for a length of time and the batteries run dead.

Mindstorms Variations

There are variations to LEGO's Mindstorms and similar product:

- Included in the LEGO Mindstorms Robot Discovery Set (RDS) is the *Scout*, a programmable brick that supports two motors and two sensors. The Scout can also be programmed via a computer with the addition of an infrared programming tower (like the one that comes with the RIS).

- Several specialty construction sets, like the LEGO Mindstorms Droid Developer's Kit, come with the *Micro Scout*. The Micro Scout contains its own motor, and its basic programming is built-in. However, it can also be programmed through a system called VLL—for Visible Light Link. The latest versions of the RCX can send signals to the Micro Scout; one robot programs another robot!

The LEGO Mindstorms Scout is a simpler programmer brick than the RCX.

- The *RoboLab* construction set comes with the similar programmable brick as the RCX in the Mindstorms Robotics Invention System, but the package is sold with different programming software. The RoboLab is intended for educational applications, and its software is more suitable for classroom environments and science experiments.

- The *CyberMaster* is an earlier incarnation of the RCX, and while it shares many of the same features, it is not as flexible. It was designed to be used in the classroom.

LEGO Mindstorms-Home Page   203017

http://www.legomindstorms.com/

Home page for LEGO Mindstorms. You'll find construction ideas, example robots others have built, advanced software for Mindstorms development, and links to various places to spend more money on LEGO.

LEGO Mindstorms home page.

LEGO Mindstorms Compatible Resistor Switch Pad 203099

http://www.akasa.bc.ca/tfm/lego_resistor.html

How to make a multiplexing switch pad that will work on one sensor input of a LEGO Mindstorms RCX robot.

LEGO Mindstorms Compatible Temperature Sensor 203097

http://www.akasa.bc.ca/tfm/lego_temp.html

How to build a LEGO Mindstorms compatible temperature sensor.

HiTechnic Products

http://www.hitechnicstuff.com/

Standard and custom-made LEGO Mindstorms add-on sensors; products include ultrasonic sensor in "brick" shape

Techno-Stuff Robotics

http://www.techno-stuff.com/

Custom-made sensors and other add-on products for LEGO Mindstorms for sale

---

LEGO Mindstorms Internals 202414

http://www.crynwr.com/lego-robotics/

A technical overview of the internal hardware of the LEGO Mindstorms RCX.

The LEGO Mindstorms RCX robotics controller.

LEGO Mindstorms SDK 203123

http://mindstorms.lego.com/sdk/

The LEGO Mindstorms (and its older brother, the LEGO Technic CyberMaster) are basically programmable microcontrollers. The programming environments that come with the LEGO products are but one way of commanding your Mindstorms robot. LEGO offers a free download of a software developer's kit (SDK) that describes a more robust programming environment, one that permits far greater flexibility than the standard environments.

Two SDKs are available:

http://mindstorms.lego.com/sdk/ Covers RCX 1.x
http://mindstorms.lego.com/sdk2/ Covers RCX 2.x

The versions relate to the firmware that has been downloaded to the Mindstorms RCX (this firmware can be replaced). There's little reason for you to still be using the first version of the firmware, so use the /sdk2 download.

The technical documentation is in Adobe Acrobat PDF format.

LEGO Mindstorms: GP2D12 Distance Sensor 202241

http://philohome.free.fr/sensors/gp2d12.htm

How to retrofit a Sharp GP2D12 infrared distance sensor to work with the LEGO Mindstorms RCX.

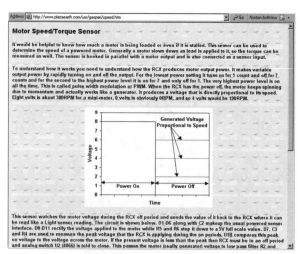

A page from Michael Gasperi's site on alternative LEGO Mindstorms sensors.

LEGO Proximity Detector
203113

http://www.mop.no/~simen/legoproxi.htm

Informational page and example code on how to best use the infrared proximity detector that comes with the LEGO Mindstorms set.

LEGO: Distributive Intelligence with Robots
202246

http://www.ceeo.tufts.edu/me94/

How they got LEGO Mindstorms robots to work together. Demonstrations include Whistling Brothers, Travel by Beacon, and Wandering Cyclops.

LEGO: Zero Force Limit Switch, Rotation and Linear Motion Sensor
202245

http://www.umbra.demon.co.uk/sensor1.html

Instructions and pictures for building several types of homebrew sensors, including a rotation sensor and an infrared optical slot sensor.

LEGOBots at Indiana University
204163

http://www.indiana.edu/~legobots/

How college kids spend their time these days.

LEGO-Robotics WikiWeb
203102

http://www.object-arts.com/wiki/html/
    Lego-Robotics/FrontPage.htm

A Wiki is a place for visitors to contribute their own articles, comments, and reviews. As per the Web site, "The aim of this site is to act as a central repository for all things to do with robotics experimentation using the popular LEGO construction sets. In general this will revolve around the LEGO Mindstorms and CyberMaster sets but information about other LEGO based robotics activities is also welcome here."

legOS
203674

http://www.noga.de/legOS/

legOS is an open source embedded operating system for the LEGO Mindstorms. Getting full use from it requires some experience with the C programming language. Versions are available for Unix and Windows operating systems.

Machina Speculatrix
203417

http://www.plazaearth.com/usr/gasperi/walter.htm

An overview of W. Grey Walter's Machina Speculatrix robotic machine, as well as how similar machines might be constructed with LEGO.

Mario Ferrari's Personal Web Pages
203098

http://www.geocities.com/CapeCanaveral/
    Galaxy/9449/

Home page of Mario Ferrari, LEGO Mindstorms tinkerer and book author.

Mindstorms RCX Sensor Input Page
202237

http://www.plazaearth.com/usr/gasperi/lego.htm

Michael Gasperi's well-received Web page on various sensors and sensor improvements for the LEGO Mindstorms RCX.

NQC (Not Quite C)   203078

http://www.enteract.com/~dbaum/lego/nqc/

Home page for the NQC (Not Quite C) programming language for the LEGO Mindstorms. Versions are available for the PC and Macintosh. You can download it here, along with documentation, a handy FAQ, and links to support sites and add-in software.

pbForth   202267

http://www.hempeldesigngroup.com/lego/pbFORTH /pbForthGUI.html

Information about pbFORTH, a Forth language alternative for the LEGO Mindstorms. Free download.

RCX and Palm   203108

http://members.rotfl.com/vadim/rcx/

Informational page about using Palm Pilot with the LEGO Mindstorms Robotic Invention System.

RCX Internals   203107

http://graphics.stanford.edu/~kekoa/rcx/

Want to know what's under the hood of the LEGO Mindstorms RCX "brick"? This page tells you the details.

Mindstorms Alternative Programming Languages

Though the LEGO Mindstorms Robotics Invention System comes with a means to program the RCX, you are not limited to it to take command of your LEGO robots. Among the more common alternative programming languages and environments for the RCX are the following:

- *Not Quite C (NQC)*. This is perhaps the most popular replacement language for the RCX. It uses C-like programming statements and is fairly robust. Versions are available for the PC, Linux, and Macintosh. Some NQC fans prefer to add a graphical user interface, and many of these are available.

  http://www.enteract.com/~dbaum/nqc/
- *pbForth*. Using a variant of the Forth language, pbForth is replacement firmware for the RCX, rather than just a replacement programming language (NQC, above, uses the standard RCX firmware). The system can be used with the PC, Macintosh, and Linux.

  http://www.hempeldesigngroup.com/lego/pbForth/
- *legOS*. Perhaps the most difficult to use, but one of the most robust replacement programming packages is legOS. Like pbForth, legOS is a new firmware, and therefore works around some of the limitations (real and perceived) in the stock RCX firmware. legOS is supported under Linux and Windows, and to use it effectively, you need C programming experience.

  http://www.noga.de/legOS/
- *Java*. Never to be left out, there are several Java implementations of RCX programming alternatives. Among them is leJOS (pronounced Ley-J-oss) and TinyVM. Most Java implementations work under Linux and Windows.

  http://lejos.sourceforge.net/
  http://tinyvm.sourceforge.net/

RCX IO Extender 204164

http://www.pressroom.com/~wesm/mStorms/
IOExtender/IOExtender.html

Construction details on expanding the interface ports on a LEGO Mindstorms RCX.

Robert Munafo's LEGO Creations 204165

http://home.earthlink.net/~mrob/pub/lego/

Examples of exotic LEGO Mindstorms robots.

The Shop 203112

http://www.ozbricks.com/bobfay/

Bob Fay's everything LEGO Web site, including project construction plans, sensor tweaks, service and repair procedures.

TinyVM 204221

http://tinyvm.sourceforge.net/

Open source Java for LEGO Mindstorms.

Trinity LEGO Cybernetics Challenge 203355

http://www.cs.tcd.ie/research_groups/cvrg/
lego/index.html

A game of robot volleyball between teams of two robots built using LEGO Mindstorms.

Tufts University: Robolab 203124

http://www.ceeo.tufts.edu/graphics/robolab.html

Robolab LEGO Mindstorms/CyberMaster programming, from Tufts University (Boston, Mass.).

VLL Library for NQC 203104

http://www.mi-ra-i.com/JinSato/
MindStorms/DDK/vll-e.html

Information on using VLL with the Not Quite C programming language.

VLL Transmit for legOS 204161

http://www.ben.com/LEGO/rcx/vll.html

Code (for legOS) for transmitting a VLL code sequence from the LEGO Mindstorms RCX to the LEGO MicroScout.

VLL: Programming the LEGO MicroScout 204162

http://eaton.dhs.org/lego/

Programming the LEGO MicroScout in VLL using bar codes on transparencies.

Web Ring: LEGO Mindstorms 202236

http://x.webring.com/hub?ring=legoms

The LEGO Mindstorms Web ring links together sites that have content and/or information about LEGO Mindstorms.

More LEGO Resources

See also these excellent online resources for LEGO Mindstorms:
Lugnet—*http://www.lugnet.com/*

RCX Internals—*http://graphics.stanford.edu/~kekoa/rcx/*

LEGO Mindstorms Internals—*http://www.crynwr.com/lego-robotics/*

MACHINE FRAMING

The person who thought up the concept for machine framing must have played with Play-Doh as a kid: Put some Play-Doh clay into a little machine, push down on the lever, and out comes various shapes. Change the orifice on the machine and out comes different shapes. Now imagine, instead of clay, you do the same thing with liquid aluminum. When the aluminum cools, it creates what's known as an extrusion; the profile of the extrusion matches the orifice that the hot, molten metal was pushed through.

Now assume the orifice is shaped so that the aluminum extrusion has four or more sides, and each side has a "T-slot" in it. Within this T-slot can be placed a variety of attachments, and they tighten against the groove of the T-slot to make a secure connection.

Such is machine framing, where aluminum rod is extruded into various shapes for the express purpose of creating "construction set frames" for building machinery. With the appropriate connectors and accessories, it's possible to assemble a frame with aluminum profile extrusions that has few, if any, nuts and bolts. Everything is held together with connectors that cinch against the T-slots.

Whether or not the inventor of machine framing began his or her career with Play-Doh, you can use these pieces of aluminum to create strong and lightweight frames for larger robots. With the proper connectors, your robot frames can be reconfigurable. There are swivel joints, for example, that allow you to change the shape of the frame by loosening a set screw, moving the joint to a new position, and retightening the screw.

A few sources will sell a few pieces at a time, but most expect a big order and prefer working with industrial clients and factories. Check minimum-order requirements, and note special shipping charges. Unless the seller will cut the pieces for you, there may be extra freight charges because the extrusions normally come in 8-foot (or longer) lengths.

SEE ALSO:

MATERIALS-METAL: More metal to choose from

MATERIALS-STORE FIXTURES: PVC and aluminum tubing that you can make roboframes with

RETAIL-HARDWARE & HOME IMPROVEMENT: Aluminum Ls and Us

80/20 Inc. 202447

1701 South 400 East
Columbia City, IN 46725
USA

- (219) 248-8030
- (219) 248-8029
- info@8020.net
- http://www.8020.net/

80/20 bills itself as the "manufacturer of The Industrial Erector Set." They make a well-engineered modular T-slotted aluminum framing system. With this versatile product, you can construct robot frames and mechanical systems without welding, drilling, tapping, or other labor-intensive fastening. Framing components are available in fractional (inch) or metric sizes. A full catalog is provided in Adobe Acrobat PDF format on the Web site. Separate PDF files list prices.

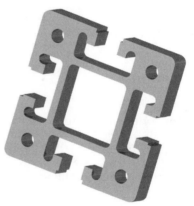

Aluminum profile extrusion from 80/20.

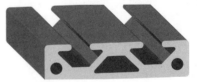

Another aluminum profile extrusion from 80/20.

Alloy Frame Systems 203485

P.O. Box 66118
Seattle, WA 98166
USA

- (206) 431-9168
- (800) 301-5561
- http://www.alloyframesystems.com/

Makes aluminum extrusions and assembly hardware for machine framing. Nice online descriptive catalog (no e-commerce), with downloaded DFX AutoCAD files

Futura Industries

http://www.futuraind.com/

Aluminum extrusions for frame construction (T-Slots brand name)

Textube Corporation/Creform

http://www.textube.com/creform/

Material handling system of pipes, joints, and accessories; normally used to build industrial frames, but can also be used to build sturdy robot bodies

(should you be interested in that sort of thing). Order from the company or one of their local distributors.

Automat   203255

Talstraße 64
D-69198 Schriesheim
Germany

- +49 (0) 6203 61954
- +49 (0) 6203 68561
- http://www.compact-technik.de/

The "construction set for R&D," Automat manufactures and sells a high-tech prototype system with construction beams, gears, and other mechanicals. Similar to Fischertechnik.

Web site is in German and English.

Bishop Wisecarver   203483

(see addresses below)

- http://www.boschframing.com/

Machine framing, motion mechanics (linear actuators, linear guides, bushings).

For linear motion products:
14001 South Lakes Dr.
Charlotte, NC 28273
(704) 583-4338
Fax: (704) 583-0523
Toll free: (800) 438-5983

For machine-framing products:
816 East Third St.
Buchanan, MI 49107

(616) 695-0151
Fax: (616) 695-5363
Toll free: (800) 32-BOSCH

Bosch Automation Products   202496

816 E. Third St.
Buchanan, MI 49107
USA

- (616) 695 0151
- (616) 695-5363
- (877) 282-6724
- marketing-bap@us.bosch.com
- http://www.boschautomation.com/

Manufacturer of aluminum structural framing.

SEE ALSO:

http://www.boschframing.com/

Aluminum profile extrusion from Bosch.

Corner gussets are used to add strength.

compact technik gmbh   202493

D-69198 Schriesheim/Heidelberg
Talstraße 64
Germany

+49 (0) 6203 61954

+49 (0) 6203 68561

http://www.compact-technik.de/

Automat prototype construction kits. Web site is in English and Deutsch.

Fiero Fluid Power Inc. 202172

5280 Ward Rd.
Arvada, CO 80002
USA

(303) 431-3600

(800) 638-0920

fiero@fierofp.com

http://www.fierofp.com/

Distributor of automation, motion control, and machine-framing products. Their product line includes:

- Bimba-pneumatics
- 80/20-machine framing
- Tol-O-Matic-pneumatics
- Watts-air valves
- Foster-air fittings
- Robohand-translation stages
- Skinner Valve
- Mosier-pneumatics
- Turn Act-translation stages

Local offices in Denver and Salt Lake City.

Flexible Industrial Systems, Inc. 203480

2284 Paragon Dr.
San Jose, CA 95131
USA

(408) 437-1600

contact@goflexible.com

http://www.goflexible.com/

Flexible Industrial Systems manufactures and distributes industrial machine framing, assembly tools, conveyors, workstations, enclosures, and material-handling devices. They are resellers of Bosch aluminum framing.

Frame World 202495

780 Tek Dr.
Crystal Lake, IL 60014
USA

(815) 477-1400

(815) 477-9818

info@frame-world.com

http://www.frame-world.com/

Maker and sellers of aluminum structural framing components. They have an extensive catalog online. Technical details include both a DXF and a DWG file, should you wish to design your robot frames on AutoCad or other program that accepts these drawing file formats. Technical details about each part are provided in Adobe Acrobat PDF files.

Division of Barrington Automation.

Selected Sellers and Manufacturers of Machine Framing

80/20 Inc.
http://www.8020.net/

Bosch Automation Products
http://www.boschautomation.com/

Flexible Industrial Systems, Inc.
http://www.goflexible.com/

Frame World/Barrington Automation
http://www.frame-world.com/

Kee Industrial Products, Inc.
http://www.keeklamp.com/

Techmaster Inc.
http://www.techmasterinc.com/

Textube Corporation/Creform
http://www.textube.com/creform/

Aluminum extrusions are commonly used with T-nuts to build versatile machine frames.

Industrial Profile Systems   202497

Corporate Headquarters
6703 Theall Rd.
Houston, TX 77066-1215
USA

 (281) 893-0100
 (281) 893-4836
🌐 http://www.industrialprofile.com/

Extruded aluminum profiles, connectors, fastener hardware.

Kanya AG/SA/Ltd.   203484

Neuhofstrasse 9
CH-8630 Rti
Switzerland

📞 +41 55 251 5858
📠 +41 55 251 5868
✉ info@kanya.com
🌐 http://www.kanya.com/

Machine-Framing Alternatives

If machine framing is too rich for your blood, but you like the idea of assembling framing using tubing and connectors, consider retail display fixture hardware designed for retail stores. Available in plastic, aluminum, or steel, this round or square material can be cut to length, then joined using various connectors.

One common style is 1 1/2-inch PVC tubing, available in white, gray, black, and other colors. The tubing connects with various shapes of corner connectors. For example:

- A two-way L joins two pieces of pipe at a right angle.
- A three-way T joins three pieces of pipe: two pipes connected end to end and a third as a riser.
- A four-way T joins four pieces of pipe in an intersection arrangement.

. . . and so on. The pipe is made to fit snugly into the connectors. PVC is easy to cut and drill, so you can readily attach other components to the frame. One popular retail store fixture system is Kee Klamp, made by Kee Industrial Products.

Yet another idea is to using so-called furniture-grade PVC pipe for small robot frames. This pipe, available in sizes ranging from 3/4 to 2 inches, is intended to make outdoor furniture using PVC pipe and fittings. It is furniture grade because the pipe has been treated with an ultraviolet-resistant chemical (ordinary PVC becomes brittle in time from exposure to ultraviolet).

As with retail store fixture materials, you can cut the pipe to length and assemble it using various connectors. Use PVC solvent cement to prevent the pipe from working loose (or you can use self-tapping screws if you want to be able to disassemble the frame in the future). Cut holes in the pipe to mount other components.

Use PVC pipe and fittings to make lightweight robot frames.

Manufacturer of machine framing. Web site is in multiple languages, including English, Deutsch, French, and Spanish. Available through agents worldwide.

Kee Industrial Products, Inc.   202498

100 Stradtman St.
Buffalo, NY 14206
USA

- (716) 896-4949
- (716) 896-5696
- (800) 851-5181
- info@keeklamp.com
- http://www.keeklamp.com/

Kee makes round tube structures (play structures, scaffolds, etc.) and fittings. Their Kee Klamp system can be used to produce exceptionally strong heavy-duty robots without welding, drilling, or bolting (you need to cut the pieces to length). Offices worldwide. Their online catalog includes clear color pictures of all the arts so you can more readily visualize how you might construct something with their system.

Reid Tool Supply Co.   203820

2265 Black Creek Rd.
Muskegon, MI 49444
USA

- (800) 253.0421
- mail@reidtool.com
- http://www.reidtool.com/

Reid is an all-purpose industrial supply resource. See listing under Power Transmission.

Techmaster Inc.   203481

N93 W14518 Whittaker Way
Menomonee Falls, WI 53051
USA

- (262) 255-2022
- (262) 255-4052
- http://www.techmasterinc.com/

Techmaster is a distributor of aluminum structural framing (machine-framing) and motion control products.

Techno Profi-Team   203482

2101 Jericho Turnpike
Box 5416
New Hyde Park, NY 11042-5416
USA

- (516) 328-3970
- support@techno-profi.com
- http://www.techno-profi.com/

Seller of aluminum extruded profiles and connectors for machine framing. You can view their catalogs online (they're in both Adobe Acrobat PDF and HTML formats).

Selected Sellers of Retail Store Fixtures

Alpha Store Fixtures, Inc.
http://www.storefixtures2000.com/

Display Warehouse
http://www.displaywarehouse.com/

Kee Industrial Products, Inc.
http://www.keeklamp.com/

Outwater Plastics Industries Inc.
http://www.outwater.com/

PVC Store, The
http://www.thepvcstore.com/

Tebo Store Fixtures
http://tebostorefixtures.com/

Machine Framing: Erector Sets for Big Boys and Girls

Ever watch a scaffold being erected at a building site? With nothing more than some pipe and connectors, the scaffold can be built to most any shape and any size—within the limits of physics, of course. The scaffold is a kind of large-scale construction toy, not unlike the basic concept of the Erector set, LEGO, Construx, K'NEX, and a long line of similar toys. The difference is simply one of scale.

Imagine something in-between toy construction sets and scaffolding. That's the idea behind machine framing. The system employs lengths of plastic or aluminum rod, connected together with various elbows, tees, and hinges. Rather than a simple rounded rod, the material is extruded into unique profiles: hexagonal or square shapes, with flanges and ribs that make it easy to secure other components.

With machine framing you can more quickly design the body and other structures of your robots. The framing pieces are cut to length with a hacksaw, then assembled with connectors and other hardware. The typical machine-framing construction does not require drilling, taping, or welding, and this is the main attraction of the system. Everything can be fastened tightly together using the supplied connectors and hardware.

Aluminum Leads the Parade

The most common form of machine framing is made with aluminum and goes by the more mundane descriptor of aluminum profile extrusions. Several manufacturers offer aluminum extrusions, each one with slightly different shapes and connector types (so they aren't always compatible with one another).

An aluminum profile extrusion.

Framing profiles comes in a variety of sizes and usually in two "classes": fractional and metric. The choice depends on what measurement system you're most comfortable working in. We'll assume fractional for our discussion, but note that similar products are available in equivalent metric sizes as well.

Machine-framing components are first measured by their cross-section dimensions, such as 1 by 1 inch, or 1 1/2 by 2 inch. For most robots, the smaller material is sufficient, and it's less expensive all around. Sizes of up to 3 by 6 inches are available, when maximum structural strength is required.

Literally hundreds of profiles (extrusion shapes) are available, with the "T-slot" the most common. In this profile, a T-shaped slot is extruded on all four sides of a square tube. With the proper hardware, other components can be fastened anywhere along the length of the slot. Components can be moved to another position along the length simply by loosening their fastener, sliding it up or down, and retightening. No drilling! No welding!

Frames are constructed using connectors, fasteners, or hinges. Depending on the system, a square frame can be made using four lengths of profile extrusion and four L-shaped connectors. Hinges offer added flexibility by allowing you to set the angle of the joined pieces. The hinges can be tightened to stay in place.

Now, How About Cost?

While machine framing offers a fast and efficient way to build the frames and other mechanical systems of medium and large robots, the cost is not cheap. Most extrusions are sold by the foot or the inch. A 6-foot length of 1 by 1 inch T-slot costs on average about $15 to $20. A similar length of extruded aluminum channel stock, available at the home improvement store, is barely half that amount; of course, to use it you must drill mounting holes as needed.

Costs can be controlled by carefully selecting the pieces you need, in the exact lengths you need, so there is no waste. A foot of wasted extrusion can cost several dollars, so it's wise to make your investment stretch as far as it can.

Somewhat less expensive is plastic extrusion. This material is not as strong as aluminum extrusion, and supplementary hardware is not as extensive. However, it costs considerably less. Like aluminum extrusions, the common style in plastic is the T-slot, either with a square shape or an octagon (eight-sided) profile.

Plastic extrusions are cheaper, but not quite as strong.

XBeams, Inc.   204026

130 W. Union St.
Pasadena, CA 91103
USA

🚫   (800) 765-2535

✉   Info@XBeams.com

🌏   http://www.xbeams.com/

Kits, projects, and spare parts of aluminum profile extrusions, connectors, hinges, and other components. Nice prices, and the extra lengths of aluminum framing are precut to standard sizes.

See also Evolution Robotics:

http://www.evolution.com/

🌏

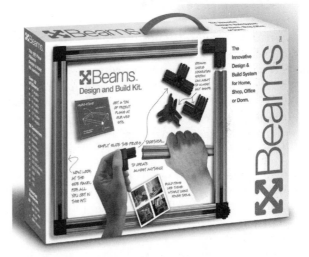

X-Beams packages pre-cut extrusions and connectors. Photo XBeams, Inc.

MANUFACTURER

The following sections list manufacturers of products of keen interest to the robot constructionist. These include components, semiconductors, tools, and hobby products.

Most manufacturers won't sell directly to the public. But they are nevertheless good resources for product information, datasheets, and application notes.

Plus, a growing number of manufacturers are selling refurbished gear and replacement parts directly from their Web sites. Example replacement parts include motors, gearboxes, and batteries for cordless drills and screwdrivers, or wheels and drive shafts for power garden equipment. Scan the company's parts catalog and you're sure to find items that can be used on a robot.

MANUFACTURER-COMPONENTS

Manufacturers listed here produce electronic parts that aren't classified as semiconductors. Most of what you'll find here are used in robotics for sensors, such as linear and rotary potentiometers for position sensors and LED and infrared components for proximity sensors. Also included here are manufacturers of connectors, wiring harnesses, relays, switches, and other electronics hardware.

Most of the manufacturers in this section do not sell directly to individuals, but a few will sell in small quantities via phone orders or the Internet.

SEE ALSO:

DISTRIBUTOR/WHOLESALER-OTHER COMPONENTS: Source for buying these components

MANUFACTURER-SEMICONDUCTOR: Makers of active components, such as transistors

RETAIL-GENERAL ELECTRONICS: Another source for buying this stuff

SENSORS-DISTANCE & PROXIMITY: Self-contained sensors using technologies found in this section

---

Bourns, Inc. 202830

1200 Columbia Ave.
Riverside, CA 92507-2114
USA

((909) 781-5690

((909) 781-5273

✉ trimcus@bourns.com

🌐 http://www.bourns.com/

Switches, encoders, potentiometers, linear motion potentiometers. Some of product line is carried by Allied.

Products of merit include:

- Chip resistors and chip arrays
- Pointing devices (strain gauge analog joysticks, now handled by Synaptics)
- Encoders
- Precision rotary potentiometers
- Resistor and R/C networks
- Linear motion potentiometers
- Switches
- Trimming potentiometers

Bourns Web site

CTS Corporation 202832

905 West Blvd. North
Elkhart, IN 46514
USA

((574) 293-7511

((574) 293-6146

🌐 http://www.ctscorp.com/

CTS makes industrial encoders, potentiometers, piezo transducers, and connectors. Some of product line is carried by Digi-Key and other mail-order distributors.

CUI Inc. 202831

9615 SW Allen Blvd.
Ste. 103
Beaverton, OR 97005
USA

(503) 643-6129

 (800) 275-4899

 http://www.cui.com/

Encoders, switches, sensors (pressure, current), speakers, connectors, piezo transducers. Some product line carried by Digi-Key.

Danaher Controls 202709

1675 Delany Rd.
Gurnee, IL 60031-1282
USA

(847) 662-2666

(847) 662-6633

(800) 873-8731

http://www.dancon.com/

High-end encoders, switches, counters, timers, and relays.

Gardner Bender 203233

6101 N. Baker Rd.
Milwaukee, WI 53209
USA

(414) 352-4160

http://www.gardnerbender.com/

Tool and electrical parts (e.g., wire connectors and switches) manufacturer.

Heyco Products, Inc. 202620

1800 Industrial Way North
P.O. Box 517
Toms River, NJ 08754
USA

(732) 286-1800

(732) 244-8843

http://www.heyco.com/

Maker of connectors and terminals. Some cool products include:

- Heyco Flex II-liquid tight tubing and fittings
- Heyco Flex V-slit tubing
- Shorty, Universal, and Armbor bushings
- Wire clips-J, C, and U style
- Aluminum solderless connectors

Sold through distributors or direct from the manufacturer.

ICO RALLY 202621

2575 East Bayshore Rd.
Palo Alto, CA 94303-3210
USA

(650) 856-9900

(650) 856-8378

http://www.icorally.com/

Distributor of connectors, harnessing, solderless terminals.

Leviton Mfg. Company Inc. 203249

59-25 Little Neck Pkwy.
Little Neck, NY 11362-2591
USA

(718) 229-4040

(800) 824-3005

http://www.leviton.com/

Manufacturer of switches, electrical outlets, and other electrical products. The company provides a number of spec sheets on their Web site. Product is available in retail stores and regional distributors.

Web site for Leviton.

Lite-On, Inc. 203769

720 South Hillview Dr.
Milpitas, CA 95035
USA

(408) 946.4873

(408) 941.4597

 http://www.liteon.com/

Infrared receiver modules, LEDs, phototransistors, photointerrupters. Useful site for app notes and spec sheets. Datasheets available on the Web site; products are available from distributors such as Digi-Key.

Midori America Corporation 203442

2555 E. Chapman Ave.
Ste. 400
Fullerton, CA 92831
USA

((714) 449-0997

 (714) 449-0139

✉ midoriusa@aol.com

Lumex, Inc.

http://www.lumex.com/

Manufacturer and distributor of Opto- and Photo-Electronic Components and devices

Mallory

http://www.nacc-mallory.com/

Makers of capacitors, "Sonalert" piezo annunciator, and other electronics components

Molex

http://www.molex.com/

Maker of connectors, including for power

Nemco

http://www.nemcocaps.com/

Maker of capacitors

NKK Switches

http://www.nkkswitches.com/

Manufacturer of switches

Sharp Electronic Components Group

http://www.sharp.co.jp/ecg/index-e.html

Optoelectronics, laser diodes, infrared devices

Switchcraft

http://www.switchcraft.com/

Maker of switches; all types

---

 http://www.midoriamerica.com/

Manufacturer of linear potentiometers and inclinometers.

Ohmite Mfg. Co. 202626

3601 Howard St.
Skokie, IL 60076
USA

((847) 675-2600

 (847) 675-1505

✉ ohmite@wwa.com

🌐 http://www.ohmite.com/

Major manufacturer of resistors, inductors, and potentiometers.

 MANUFACTURER-GLUES & ADHESIVES

Look here for companies that specialize in glues and adhesives with particular merit for robot building. There are actually hundreds of makers of various glues, adhesives, and cements, and I've listed only a few of them here as representative of what's available.

The obvious use for glues and adhesives (and their close cousin, cements) is to hold robot parts together. But depending on the material, you may also use them to create such things as sensors.

SEE ALSO:

RETAIL-ARTS & CRAFTS: Sells a variety of sticky materials

RETAIL-HARDWARE & HOME IMPROVEMENT: Glues and adhesives for the home

SUPPLIES-GLUES & ADHESIVES: Retail sources for these products

---

Ambroid/Graphic Vision 202128

61 Katie Ln.
Swanzey, NH 03446
USA

((603) 352-2794

 (603) 352-8356

🚫 (800) 242-2794

info@graphicvisioninc.com

 http://www.ambroid.com/

IPS Plastic Solvent Cements

IPS Corp. is a major manufacturer of solvent cements for joining plastic pipe. Their consumer and industrial products are available at hardware, home improvement, and plastics specialty retailers everywhere. The company's solvent cements are available for rigid and flexible PVC, ABS, styrene, acrylic, and other plastics, and they come in a number of consistencies and formulations.

The products are called solvent cements because they actually dissolve a little bit of the plastic during the cementing process. As the cement cures, the plastic pieces are then mutually joined, and the bond is very strong—almost as strong as the rest of the plastic.

IPS sells a number of WELD-ON brand solvent cements that have practical use in robotics using PVC, acrylic, or other plastic parts.

- **WELD-ON 2007**. Water-thin and clear, this cement is intended for joining PVC pieces (standard and foamed). It cures very quickly and can be applied through a needle applicator and a squeeze bottle.

- **WELD-ON 4052**. Medium-bodied, this cement can join PVC (standard and foamed), ABS, acrylic, styrene, polycarbonate, and many other plastics. This cement is termed "low VOC"—VOC stands for "volatile organic compounds." Such cements may take a little longer to cure, but they release fewer bad things into the air.

- **WELD-ON 1707**. Medium-bodied, this cement is intended solely for joining ABS plastics. ABS is a little heavier and stronger than PVC and is also available in sheet and extruded forms.

- **WELD-ON 3**. Designed for bonding acrylic sheets, leaving a water-clear and barely visible seam.

IPS Corp. is located at:
http://www.ipscorp.com/

Ambroid makes glues: Ambroid liquid cement (general adhesive), styrene plastic cement, Proweld (for ABS, styrene, butyrate and acrylic plastics), EZ Mask peel-away liquid, and Tac-N-Place. The products are available in many hobby stores and in arts and crafts stores.

3M Worldwide 203526

3M Product Information
3M Center, 515-3N-06
St. Paul, MN 55144-1000
USA

(651) 737-6501

(651) 737-7117

(800) 713-6329

http://www.3m.com/

Manufacturer of a variety of industrial and consumer products. Many application notes are available on the Web site. Way too many to list here, so see http://search.3m.com/ to search the product database.

Among the most useful products:

- Engineered adhesives (spray, tape, putty, reclosable fasteners-like Velcro, only more aggressive stickiness)

- Paints and protective coatings

- Abrasives (sandpaper, Scotch-Bright)

- Foils and tapes (metal, plastic, paper, duct, foam)

Oe of the many products pages at the 3M site.

tesa AG   203867

Quickbornstraße 24
20253 Hamburg
Germany

 http://www.tesa.com/

Makers of unusual adhesive tapes, including transfer tape, which is just the gooey part, without the backing. It's used heavily in manufacturing, because the lightweight adhesive keeps parts in place when ordinary tapes will not.

Additional offices around the world. Check the Web site for locations, addresses, and phone numbers. Main Web site is in English and German.

Tesa products are available from retailers and distributors. Try the following Google.com search phrase:

tesa adhesive transfer tape

⋀ MANUFACTURER-SEMICONDUCTORS

Semiconductors include integrated circuits, transistors (signal and power), diodes, and related circuitry. The semiconductor manufacturers listed here run the gamut from large multi-billion-dollar companies with fabrication factories around the world to small specialty designers that "send out" their ICs, transistors, and other parts to be built by someone else.

As it turns out, the smaller outfits often have the most interesting specialty products. Look especially for motor controllers, power drivers, switching regulators, DC-DC converters, analog-to-digital converters, solid-state accelerometers and tilt sensors, optical sensors, LCD drivers, and microcontrollers.

Most companies listed in this section do not sell directly to individuals. The main benefit of the manufacturers listed here is the product datasheets and application notes you can obtain from them. These are routinely published as PDFs, which require the free Adobe Acrobat Reader program to view (see http://www.adobe.com/ for downloading details).

Note that a number of semiconductor makers are only too happy to send out free samples. When samples are available, they are typically for current product only (no legacy products kept in stock for perpetuity), and package styles are limited—for instance, ICs only in surface mount and not in DIP packages. Check the companys' Web sites for their samples policies.

SEE ALSO:

DISTRIBUTOR/WHOLESALER-INDUSTRIAL ELECTRONICS: Purchase semiconductors

INTERNET-RESEARCH: Look for semiconductors by part number (e.g., Findchips.com and IC Chip Directory)

ELECTRONICS-DISPLAY: LCD panels to go along with LCD driver ICs

MICROCONTROLLERS-HARDWARE: Includes makers of microcontroller ICs and subsystems

RETAIL-GENERAL ELECTRONICS: Purchase semiconductors

Advanced Micro Devices (AMD)   203071

One AMD Pl.
P.O. Box 3453
Sunnyvale, CA 94088
USA

☏ (408) 732-2400

⊘ (800) 538-8450

🌐 http://www.amd.com/

Makers of the AMD Athlon and XP processor, Flash memory, and embedded processors. Technical datasheets available on the Web site-look for the TechDocs logo. Additional technical documentation may be ordered through AMD Literature Support.

Agilent Technologies, Inc.   202010

SPG Technical Response Center
3175 Bowers Ave.
Santa Clara, CA 95054
USA

☏ (408) 654-8675

📠 (408) 654-8575

⊘ (800) 235-0312

✉ SemiconductorSupport@agilent.com

🌐 www.semiconductor.agilent.com

Agilent manufactures a wide variety of semiconductor and electronics products. While some are intended strictly for volume OEM purchasing, others can be obtained through industrial electronics distributors, such as Arrow and Newark (and yet others can't be found anywhere—they're listed on the Agilent site, but no purchasing information is available). Copious amounts of datasheets and application notes are available for download from the site (most are in Adobe Acrobat PDF format).

Some of Agilent's product lines include:

- Fiber optics
- Infrared and bar code
- LED
- Motion control
- Optical navigation
- Optocoupler

Among key products for robotics include optical mouse ICs and sensors for optical navigation. One particularly useful chip is the ADNS-2051 optical mouse sensor. This chip works by taking thousands of digital pictures each second. It has a resolution of up to 800 counts per inch (CPI) and can track motion of up to 14 inches per second. One possible application is robot odometry over nonfibrous (i.e., not carpeted) surfaces. Specifics regarding this chip can be found at:

http://www.agilent.com/view/opticalnavigation/

Note the staggered DIP package design; this is not your standard IC, and it requires special prototyping boards or custom circuit boards.

In addition, Agilent offers a number of low- and medium-cost optical linear and rotary encoders for motion control. To round out the line, the company provides various versions of quadrature encoder interface and counter ICs, including several that are in a DIP (dual-inline pin) IC package, making them easy to breadboard.

Finally, check out Agilent's line of CMOS vision sensors at:

http://www.agilent.com/view/Imaging/

Optical mouse sensor. Photo Agilent Technologies.

Allegro Micro Systems, Inc. 202911

115 Northeast Cutoff
Worcester, MA 01606
USA

☎ (508) 853-5000

🖷 (508) 853-7861

✉ sales@allegromicro.com

Advanced Linear Devices, Inc.
http://www.aldinc.com/

Specializes in linear ICs, including power MOSFETs, analog switches, voltage comparators, analog-to-digital converters, op-amps

Linear Technologies
http://www.linear.com/

Makes a variety of analog and digital ICs

Mitsubishi Semiconductors
http://www.mitsubishichips.com/

Semiconductor manufacturer

Pericom Semiconductor Corp.
http://www.pericom.com/

Interface electronics

Sharp Microelectronics
http://www.sharpsma.com/

Semiconductors, CCD imagers, and sensors

Smart Modular Technologies, Inc.
http://www.smartm.com/

Flash memory

🌐 http://www.allegromicro.com/

Manufacturers of Allegro and Sanken semiconductors. Among Allegro's product offerings useful in robotics are the following:

- Motor control ICs for servo, DC, brushless, and stepper motors

Web page for Allegro Micro Systems.

- H-bridge ICs
- Hall-effect sensors

Allegro provides a stunning amount of technical information; look for their datasheets and application notes (in Adobe Acrobat PDF format). A line of reference design demo boards allows you to easily test and experiment with various products.

Altera Corp.   203070

101 Innovation Dr.
San Jose, CA 95134
USA

  (408) 544-7000

🚫 (800) 767-3753

🌐 http://www.altera.com/

Makers of programmable logic devices (PLDs).

Analog Devices, Inc. ⚲   202912

One Technology Way
P.O. Box 9106
Norwood, MA 02062-9106
USA

📞 (781) 461-3333

📠 (781) 461-4482

🚫 (800) 262-5643

🌐 http://www.analog.com/

Analog Devices is a key manufacturer of precision linear semiconductors. Among their product line is the low-cost ADXL series of accelerometer/tilt sensors, which can be readily used in amateur robotics. Their Web site contains copious amounts of datasheets, white papers, application notes, and other documentation. Samples can be ordered directly from the site; a number of industrial electronics distributors, such as Allied Electronics and Newark, carry the Analog line for resale.

The company's main product lines include the following:

- Single- and dual-axis accelerometers
- Analog-to-digital and digital-to-analog converters
- Op-amps
- LCD drivers
- Temperature sensors
- Digital signal processing

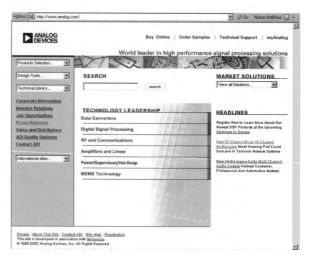

Analog Web site.

Analog Devices ADXL202 accelerometer module.

Applied Micro Circuits Corp.   202958

6290 Sequence Dr.
San Diego, CA 92121-4358
USA

📞 (800) 935-2622

📠 (858) 450-9333

✉ product support@amcc.com

🌐 http://www.amcc.com/

Semiconductor maker; analog and digital.

ARM Ltd.   204152

110 Fulbourn Road
Cambridge
CB1 9NJ
UK

📞 +44 (0) 1223 400400

📠 +44 (0) 1223 400410

🌐 http://www.arm.com/

ARM is an industry leader providing 16/32-bit embedded RISC microprocessors. The company licenses its

RISC processors, peripherals, and system-chip designs to leading international electronics companies.

Atmel 202374

2325 Orchard Pkwy.
San Jose, CA 95131
USA

((408) 441-0311

℘ (408) 436-4200

🕲 http://www.atmel.com/

Atmel manufactures a number of semiconductor lines of interest to the robot experimenter. First and foremost is their AVR 8-bit microcontrollers. The AVRs compete with Microchip's PICmicro controllers. The AVR line was designed to be programmed using a high-level language compiler, such as C or Basic (several such compilers—some free, some not—are available for the AVR). Much of the available AVR line is available from online retailers such as Digi-Key and Inland-Empire.

AVR also manufactures 8051-architecture microcontrollers, 32-bit ARM microcontrollers, Flash memory, CCD imagers, RF identification chips, and field programmable gate arrays (FPGA).

Datasheets and application notes are provided for most product in Adobe Acrobat PDF format.

Web site for Atmel microcontrollers.

Dallas Semiconductor 202913

http://www.dalsemi.com/

A branch of Maxim Semiconductor. Among other products, Dallas produces the 1-Wire and iButton special-function microcontrollers. Web site is in English, Chinese, and Japanese. See the entry for Maxim (in this section) for corporate address and phone number.

SEE ALSO:

http://www.ibutton.com/

Where to Get Stuff: Freebies

Shhhh! I'm not supposed to tell you this, but a number of electronics component makers will send you free samples, just for the asking. While it helps if you're an instructor at a school, or an engineer working for a company building millions of units of some gizmo, the truth is, most semiconductor and electronics manufacturers will send free samples to most anyone. Just don't be greedy, and observe limits to their kindness (most companies will send a "handful" of parts—usually under two each, or a total of 10 per request).

Availability of freebies is especially useful, as the newest semiconductors may not be carried by distributors and wholesalers, making it hard to purchase these products. You can get a jump ahead by asking for free samples and incorporating them in your work.

Sample kits are also sometimes available from wholesalers, distributors, and resellers. Look for kits of parts, fasteners, or various types of hardware. The kits are limited and are obviously meant to entice you into buying more, but they're free, and you never know when some piece from a kit will come in handy.

Elantec Semiconductor, Inc. 202959

675 Trade Zone Blvd.
Milpitas, CA 95035
USA

☎ (408) 945-1323
📠 (408) 945-9305
🚫 (888) 352-6832
✉ tech@elantec.com
🌐 http://www.elantec.com/

Products include video amplifiers, laser diode drivers, DC-DC converters, and MOSFET drivers. Datasheets available in Adobe Acrobat PDF format.

Exar Corp. 202960

48720 Kato Rd.
Fremont, CA 94538
USA

☎ (510) 668-7000
📠 (510) 668-7001
🌐 http://www.exar.com/

Maker of specialty ICs, including video and imaging, interface and timing (their XR2206 function/waveform generator is an all-time classic), and serial UARTs. The usual datasheets, of course, are provided in Adobe Acrobat PDF format.

Fairchild Semiconductor 202903

82 Running Hill Rd.
South Portland, ME 04106
USA

☎ (207) 775-8100
📠 (207) 761-6020
🚫 (800) 341-0392
🌐 http://www.fairchildsemi.com/

Fairchild manufactures just about every kind of semiconductor there is, including linear, logic, microcontroller, and interface chips. Copious amounts of datasheets are provided on the site.

Infineon Technologies AG 203784

Infineon Technologies AG
St.-Martin-Str. 53
81669 Mnchen
Germany

☎ +49 (0) 8923 40
📠 +49 (0) 8923 424694
🌐 http://www.infineon.com/

Manufacturer of microcontrollers, memory, Hall-effect and optical (infrared reflective and slotted) sensors, magnetorestrictive sensors, and other electronics components. Very good assortment of datasheets and application notes.

Intel Corporation 203913

2200 Mission College Blvd.
Santa Clara, CA 95052-8119
USA

☎ (408) 765-8080
📠 (408) 765-6284
🚫 (800) 628-8686
🌐 http://www.intel.com/

Intel makes the microprocessor used in over 300 million computers worldwide—but who's counting, right? They provide copious amounts of datasheets on their products on their Web site, as well as research papers and white papers on the latest technologies, including wireless.

International Rectifier 203743

233 Kansas St.
El Segundo, CA 90245
USA

☎ (310) 322-3331
📠 (310) 322-3332
🌐 http://www.irf.com/

Semiconductor manufacturer; most notably power transistors. Provides a good amount of technical information and application notes. Product is available through resellers. Their power MOSFETs are labeled IRFxxx (the xxx is a three- or four-digit number), and as such are easy to identify.

At the IR site you'll find numerous application guides, fact sheets, and datasheets (in Adobe Acrobat PDF format). The company sells some product directly, mainly its reference design developer kits. Individual components are available from distributors, and the online catalog pages for the components include direct links to distributor stock.

Web site for International Rectifier.

Intersil   202957

7585 Irvine Center Dr.
Ste. 100
Irvine, CA 92618
USA

(949) 341-7000

(949) 341-7123

(888) 468-3774

 http://www.intersil.com/

Intersil specializes in power-management ICs (including power MOSFET components), analog/mixed signals ICs, and interface ICs. Direct ordering is available through the Web site, as are datasheets and application notes (in Adobe Acrobat PDF format).

Linear Integrated Systems   202181

4042 Clipper Ct.
Fremont, CA 94538-6540
USA

(510) 490-9160

(510) 353-0261

sales@linearsystems.com

http://www.linearsystems.com/

Makers of specialty linear semiconductors: bipolar transistors, DMOS switches, MOSFETs, A/D and D/A converters, and voltage-controlled resistors.

Maxim Integrated Products   202914

120 San Gabriel Dr.
Sunnyvale, CA 94086
USA

(408) 737-7600

(408) 737-7194

(800) 998-8000

http://www.maxim-ic.com/

Maxim is a semiconductor maker, perhaps best known to hobbyists for their power-management and interfacing ICs. The company makes several compact switching regulator chips that offer both ease of use and relatively high currents. The company also sells sensors, microcontrollers, and linear ICs. Datasheets (in Adobe Acrobat PDF format) are available on the Web site. Online ordering is available.

Microchip Technology   202371

2355 W. Chandler Blvd.
Chandler, AZ 85224
USA

(480) 792-7200

(480) 899-9210

http://www.microchip.com/

Microchip makes a broad line of semiconductors, including the venerable PICmicro microcontrollers. Their Web site contains many datasheets and application notes on using these controllers, and you should be sure to download and save them for study.

The company is also involved with radio-frequency identification (RFID), selling readers and tags, as well as developer's kits.

 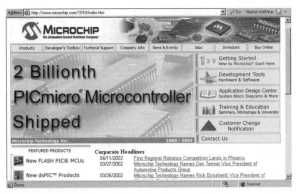

Home of Microchip.

Motorola Semiconductor Products Sector   204054

http://mot-sps.com/

Motorola makes thousands of products, and navigating through their offerings is a full-time job in itself. The online Products Selector lets you narrow your search to fields of interest within the company's semiconductor businesses, which include sensors, microcontrollers, analog ICs, and digital ICs. You'll find plenty of datasheets and application notes along the way.

National Semiconductor Corp.   202961

2900 Semiconductor Dr.
Santa Clara, CA 95052-8090
USA

  (408) 721-5000

  (408) 739-9803

  http://www.national.com/

Maker of all kinds of semiconductors, including linear and digital ICs, specialty ICs (e.g., motor drive), and COP8 microcontrollers. Datasheets available at the Web site.

ON Semiconductor Corporation   203916

5005 E. McDowell Rd.
Phoenix, AZ 85008
USA

  (602) 244-6600

  (800) 282-9855

  ONlit@hibbertco.com

  http://www.onsemi.com/

Lots of stuff available, including:

Analog chips: amplifiers and comparators; analog switches; mux/demux; battery management; interface ICs; motor control; power management; thermal management

Logic chips: arithmetic; buffers/drivers; flip-flops; gates; inverters; latches; mux/demux/encoders/decoders; octals and bus interface; 16-bit/18-bit functions; receivers/transceivers; translators; specialty functions

Technical documents online (in Adobe Acrobat PDF format).

Philips Semiconductor   202915

Hurksestraat 19, Postbus 218
5600 MD Eindhoven
The Netherlands

  +31 31-40-279-1111

  +31 40-27-248-25

  http://www.semiconductors.philips.com/

Philips makes linear, digital, and power semiconductors, plus sensors such as the KMZ10 magnetorestrictive device used in compasses. Plenty of datasheets (in Adobe Acrobat PDF format) for download. Some items can be purchased online through distributors, but the minimum quantity may be in the hundreds or even thousands!

Siliconix Incorporated   203915

2201 Laurelwood Rd.
Santa Clara, CA 95054
USA

  (408) 988-8000

  (408) 567-8950

  http://www.siliconix.com/

Silconix (now a brand of Vishay) makes power MOSFETs and power ICs. Their site contains a number of useful datasheets and application notes regarding the selection and use of power MOSFET devices. The company also sells low-cost demonstrator and developer boards based on their products.

ST Microelectronics   202916

20 Route de Pre-Bois
ICC Bldg.
CH-1215 Geneva 15
Switzerland

  +41 22 929 2929

  +41 22 929 2900

  http://www.st.com/

Manufacturer of a broad line of semiconductors, including the venerable L293D 4-channel power driver (typically used in controlling the motors in small robots), CMOS imager, and control chipsets. An almost endless amount of datasheets and technical white papers is provided on the site.

Texas Instruments   202917

12500 TI Blvd.
Dallas, TX 75266-0199
USA

  (972) 995-3773

(972) 995-4360

 (800) 336-5236

 http://www.ti.com/

TI is a multifaceted electronics company with many products useful to the robot builder. These include standard TTL and CMOS logic, microcontrollers, op-amps, linear devices such as the venerable LM555, motor H-bridge circuits, so-called power management (DC-DC converters, voltage regulators, MOSFET and power drivers), ADC and DAC chips, and audio amplifiers.

Among TI's most used nongeneric products useful in robotics is the L293D quad half-H driver (functionally equivalent to the ST Micro L293D) and the SN754410 quad half-H driver (basically the same as an L293D, but with greater current capacity). These two chips are used extensively in small robot designs to control motors via a microcontroller or computer.

TI provides extensive datasheets and application notes (in Adobe Acrobat PDF format) and even supplies the approximate U.S. price (per 1000 quantity) for many of their products. Products are available through distributors.

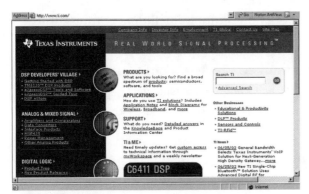

Main Web page for Texas Instruments.

Ubicom / Scenix   204178

635 Clyde Ave.
Mountain View, CA 94043
USA

 (650) 210-1500

 (650) 210-8715

 http://www.ubicom.com/

Makers of the Scenix microcontroller.

Unitrode Corp.

See Texas Instruments (this section).

Vishay Intertechnology, Inc.   203906

63 Lincoln Hwy.
Malvern, PA 19355-2120
USA

 (610) 644-1300

 (610) 889-9429

 http://www.vishay.com/

Vishay is a manufacturer of discrete semiconductors, diodes, and rectifiers. The company is also a maker of infrared data communication devices (IRDCs), strain gauges, and load cells.

SEE ALSO:

http://www.siliconix.com/ -Power ICs and MOSFETs

http://www.vishay.com/brands/
measurements_group/ -Strain gauges

Winbond Electronics Corporation   202316

2727 North First St.
San Jose, CA 95134
USA

 (408) 943-6666

 (408) 544-1789

 (800) 677-0769

 info@winbond.com

 http://www.winbond.com.tw/

Winbond is a semiconductor manufacturer. Their line includes the ISD series of digital-recording ICs (they bought out the ISD company that introduced these), the PowerSpeech text-to-speech synthesizer chip, an unlimited vocabulary text-to-speech IC, and microcontrollers. Datasheets available on the site.

For the company's U.S. subsidiary's Web site, see:

http://www.winbond-usa.com/

Winbond parent company Web site.

Zarlink Semiconductor Inc. 202962

400 March Rd.
Ottawa, ON
K2K 3H4
Canada

☏ (613) 592-0200

🖷 (613) 592-1010

🌐 http://www.zarlink.com/

Zarlink's old name was Mitel Corp. Their product line is only indirectly related to amateur robotics and is listed here more for their technical documentation and white papers than for stuff you can buy. Zarlink makes semiconductor chips for such applications as global positioning satellite, Bluetooth computing controllers, embedded pacemakers, hearing aids, and wireless networks. Check the Product & Design Data link for some interesting reading on their products.

Zilog 203021

532 Race St.
San Jose, CA 95126
USA

☏ (408) 558-8500

🖷 (408) 558-8300

✉ info@zilog.com

🌐 http://www.zilog.com/

Zilog makes microprocessors and support components. They helped usher in the personal computer with the now—famous Z80 microprocessor—which is still available in updated packages and is often used in embedded systems and single board computers. Zilog also offers general and special-use microcontrollers, wireless infrared controllers, and other components.

An extensive array of datasheets, application notes, and developer tools is provided on the site. Most are in Adobe Acrobat PDF format. Products are available for sale from distributors, including Digi-Key and Pioneer-Standard.

MANUFACTURER-TOOLS

The single greatest thing that separates us humans from animals is not that we enjoy seeing robots tear each other's guts out, but that we use tools to help us do our work. Tools make other tools, so the only limits on the tools we use are their size and practicality. In this section are listed manufacturers of a broad array of tools used in robotics construction. This category spans from soldering irons to woodworking saws to computer-controlled mills.

Very few of the companies listed here will sell directly to individuals; they prefer instead that you buy your tools from retailers. That's okay, because many manufacturer

---

Ingersoll-Rand Company/Tool and Hoist

http://www.irtools.com/

Automotive and pneumatic tools

Jet/WMH Tool Group

http://www.jettools.com/

Industrial power tools and machinery

Johnson Level & Tool

http://www.johnsonlevel.com/

Leveling tools

Makita U.S.A., Inc.

http://www.makita.com/

Maker of power tools, including cordless drills (provides replacement batteries and other parts)

Milwaukee Electric Tool Corporation

http://www.mil-electric-tool.com/

Makers of power tools, such as keyhole saws and drills

Olympia Group, Inc.

http://www.olympiaweb.com/

Tool manufacturer: brands include Olympia-Tools, Village Blacksmith, Disston, Roughneck, and Euro-Cut

Ryobi Power Tool

http://www.ryobitools.com/

Manufacturer of power tools (saws, drills, more)

Skil Power Tools

http://www.skil.com/

Power saws, drills and other tools

Snap-on Inc.

http://www.snap-on.com/

Maker of hand and power tools

Web sites contain datasheets and specifications of their products, safety sheets, even repair guides. And a few sell replacement parts online. You don't have to own the tool to make use of the replacement parts, which include motors, batteries, gears and gearboxes, sprockets, and other mechanical components.

SEE ALSO:

DISTRIBUTOR/WHOLESALER-INDUSTRIAL ELECTRONICS: Sellers of electronic tools

RETAIL-GENERAL ELECTRONICS: Hand tools and soldering equipment for electronics construction

RETAIL-HARDWARE & HOME IMPROVEMENT: Places to buy tools

TOOLS (VARIOUS): SOURCES AND RESOURCES ABOUT TOOLS OF ALL TYPES

Arrow Fastener Company, Inc. 203221

271 Mayhill St.
Saddle Brook, NJ 07663
USA

((201) 843-6900

📠 (201) 843-3911

🌐 http://www.arrowfastener.com/

Makers of staplers, industrial stapler guns, nail guns, and glue guns.

Black & Decker

203223
626 Hanover Pike
Hampstead, MD 21074
USA

🚫 (800) 544-6986

🌐 http://www.blackanddecker.com/

Black & Decker makes power tools for home and small shop use. Some basic specification information on products available at the site.

Channellock Inc. 203225

1306 South Main St.
Meadville, PA 16335
USA

🚫 (800) 724-3018

✉ pliers@channellock.com

🌐 http://www.channellock.com/

Channellock makes pliers and other hand tools. Some technical details about each tool is provided in the online store, where you can also buy products. Most of the time, though, you'll just pick these up at the corner hardware store.

Cooper Industries, Inc. 202618

3535 Glenwood Ave.
Raleigh, NC 27612
USA

((919) 781-7200

📠 (919) 783-2116

🌐 http://www.coopertools.com/

Weller solder tools; wire-wrapping prototyping products, Xcelite hand tools.

DEWALT Industrial Tool Co. 203227

701 E. Joppa Rd.
TW425
Baltimore, MD 21286
USA

🚫 (800) 433-9258

🌐 http://www.dewalt.com/

Maker of power tools, including a line of cordless drills and screwdrivers. Replacement batteries and chargers are available.

Web page for DEWALT Industrial Tool Co.

Dremel 202136

4915 21st Street
Racine, WI 53406
USA

 (262) 554-1390

(800) 437-3635

 dremelcs@execpc.com

http://www.dremel.com/

Makers of the Dremel multipurpose motorized rotary tool.

The do-just-about-anything Dremel tool.

FEIN Power Tools Inc. 203229

1030 Alcon St.
Pittsburgh, PA 15220
USA

 (412) 922-8886

(412) 922-8767

(800) 441-9878

 cmorriso@feinus.com

http://www.fein.com/

Drills, cordless screwdrivers, power sanders, vacuum cleaners. Offices worldwide.

Great Neck Saw Manufacturers, Inc. 203232

165 East 2nd St.
Mineola, NY 11501
USA

Tools of the Trade: Soldering Iron

A *soldering iron* is a requirement for any robot building that requires you to construct or repair electronic circuits or wiring.

- For maximum flexibility, invest in a *modular soldering pencil*. This style of soldering iron lets you change the heating element. A 25- or 30-watt heating element is preferred.
- *Stay away from "instant-on" soldering irons.* They put out far to much heat for most any application other than soldering large-gauge wires.
- Always use a *soldering stand* to keep the soldering pencil in a safe, upright position.
- Get one or two *small soldering tips* for intricate printed circuit board work, but invest in a single medium- or large-size tip for routine soldering chores.
- You need *solder* to feed the soldering iron. And not just any kind of solder, but the rosin or flux core type. Acid core and silver solder should never be used on electronic components.
- A s*ponge* is used to clean the soldering tip during use. Keep the sponge damp and wipe the tip clean every few joints. An ordinary kitchen sponge is fine, but don't use the kind that contains soap.

Extra soldering tools you may want to add to your shop:

- *Heat sink*, for attaching to sensitive electronic components during soldering. The heat sink draws the excess heat away from the component and helps prevent damage to it.
- *Desoldering vacuum tool*, to soak up molten solder. Used to get rid of excess solder, to remove components, or redo a wiring job.
- *Dental picks*, for scraping, cutting, forming, and gouging into the work.
- *Rosin cleaner* (spray or brush-on). Apply the cleaner after soldering is complete to remove excess resin.
- *Solder vise* or "third hand." The vise holds together pieces to be soldered, leaving you free to work the iron and feed the solder.

(516) 746-5352

sales@greatnecksaw.com

http://www.greatnecksaw.com/

Great Neck manufactures hand tools, including hammers, saws, wrenches, cutting tools, threading tap and die, battery rechargers, and more.

Jepson Tool Power   203234

20333 S. Western Ave.
Torrance, CA 90501
USA

 (310) 320-3890

(310) 320-1318

(800) 456-8665

info@jepsonpowertools.com

http://www.jepsonpowertools.com/

Maker of power tools. Moderate technical details provided about the tools.

Laser Products   203235

1335 Lakeside Dr.
Romeoville, IL 60446
USA

(630) 679-1300

(877) 679-1300

info@lasersquare.com

http://www.lasersquare.com/

Tool manufacturer (laser lines and levels). Interesting possibilities for robots.

Porter Cable Corporation   203251

4825 Hwy. 45 North
P.O. Box 2468
Jackson, TN 38302-2468
USA

(731) 668-8600

(731) 660-9525

(800) 487-8665

http://www.porter-cable.com/

Manufacturer of power tools.

Porter Cable is well known for their routers, and other tools as well.

Roto Zip Tool Corporation   203252

1861 Ludden Dr.
Cross Plains, WI 53528
USA

(608) 798-3737

(608) 798-3739

(877) 768-6947

customer_service@rotozip.com

http://www.rotozip.com/

Power tool manufacturer; makers of the RotoZip spiral saw. The RotoZip is both a drill and a saw and can be used to cut out shapes in wood or plastic (or very lightweight metal). Think of it as a larger version of the Dremel, but with one bit that is used for the drilling and cutting. Check out the User Tips & Techniques at the company's Web site.

The Roto Zip rotary saw, ready for action. Photo Roto Zip Tool Corporation.

Stanley Works, The   203224

200 Moody St.
Waltham, MA 02453
USA

(781) 899-7900

(781) 899-6485

 (866) 899-7900

 foreman@constructiontoys.com

🌐 http://www.stanleyworks.com/

Maker of hand tools of all types. Web site includes product brochures and some technical information on the company's tools.

📄 🌐 🏭

Toro Company, The   202305

Consumer Division
8111 Lyndale Ave. South
Bloomington, MN 55420
USA

🚫 (800) 348-2424

🌐 http://www.toro.com/

Toro makes gas- and electric-powered garden tools. The Web site lacks meaningful technical details (in case you want to hack a gas or electric motor for your robot). But the site does provide links to nearby Toro parts outlets, where you can pick up parts either for unusual uses in your robot or to put your lawn mower back together after trying to hack it for your robot.

🏭

Zircon Corporation   202308

1580 Dell Ave.
Campbell, CA 95008
USA

📞 (408) 866-8600

 (408) 866-9230

🚫 (800) 245-8265

📧 info@zircon.com

🌐 http://www.zircon.com/

Manufacturer of electronic measuring tools, including an ultrasonic tape measurer, laser levels, electronic metal detector, stud finder (metal proximity or density change), electronic water-level sensor. Some hacking potential.

🏭

⬛ MATERIALS

Listings in this section are general sources for a wide range of mechanical materials useful in robot construction. Many of the companies could best be described as "materials superstores," general stores that offer everything from pieces of metal and plastic to power transmission items (gears, sprockets, roller chain, bearings, etc.), construction tools, motors, glues and adhesives, specialty tapes, fasteners, and much more.

The catalogs of many general materials outlets are huge, with some spanning several thousand pages. Personally, I prefer the printed catalogs for reference, but tend to do my actual buying online; those that sell online provide searchable databases of components, making it easier to find things when you're in a hurry.

SEE ALSO:

FASTENERS: Specialty sources for fasteners of all types

POWER TRANSMISSION: Gears, sprockets, you name it

TOOLS (VARIOUS): Hand, power, and machinery tools

Enco Manufacturing Co. ⚙   203672

400 Nevada Pacific Hwy.
Fernley, NV 89498
USA

📞 (770) 732-9099
🚫 (800) 873-3626
✉ info@use-enco.com
🌐 http://www.use-enco.com/

Enco is a premier mail-order source for shop tools, power tools, hand tools, production tools (lathes, mills, hydraulic presses, metal brakes, you name it), bits, saws, casting materials, plastics, hardware (door and cabinet), fasteners, tooling components, ACME rods and nuts, welding equipment and supplies, and lots more. They print a master catalog and send out sales catalogs on a regular basis. The sales catalogs contain some real bargains.

You can also use their shopping cart system. Their online catalog is basically scans of their master catalog in Adobe Acrobat PDF format. The printed catalog is definitely easier to use, and you can "quick order" any part by entering its catalog number in the shopping cart.

Enco provides printed and on-line catalogs.

Grainger (W. W. Grainger) ⚙   202928

100 Grainger Pkwy.
Lake Forest, IL 60045-5201
USA

📞 (847) 535-1000
📠 (847) 535-0878
🌐 http://www.grainger.com/

Can you imagine rummaging through shelves holding 5 million different products? If you're doubtful, take a look at Grainger, one of the world's leading retail industrial supply companies. Their printed catalog is thicker than a phone book, and they offer everything for your robot building from plastic rods to fractional horsepower gearmotors. Obviously, that leaves 4,999,998 other products, which I won't describe here.

In addition to Grainger's online presence, they have some 600 local outlets that stock core merchandise. Catalogs are available in printed form or on CD-ROM.

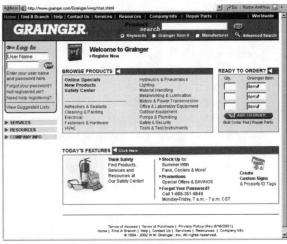

Search for thousands of products on the Grainger site.

Graybar Electric Company, Inc. 203973

34 N. Meramec Ave.
Clayton, MO 63105
USA

☎ (314) 512-9200

☏ (314) 512-9453
⊘ (800) 472-9227
🌐 http://www.graybar.com/

Graybar Electric sells tens of thousands of electronic, electromechanical, and mechanical products. While

Material for Robot Bodies

Robots can be easily constructed from aluminum, steel, tin, wood, plastic, paper, foam, or a combination of them all. Here are some common materials used in constructing robot bases and other components.

- *Plastic.* Pound for pound, plastic has more strength than many metals, yet is easier to work with. You can cut it, shape it, drill it, even glue it. Effective use of plastic requires some special tools, and availability of extruded pieces might be somewhat scarce unless you live near a well-stocked plastic specialty store. Use mail order instead.

- *Foam board.* Art supply stores stock what's know as foam board, also called "Foam Core." Foam board is a sandwich of paper or plastic glued to both sides of a layer of densely compressed foam. The material comes in sizes from 1/8 inch to over 1/2 inch, with 1/4 to 1/3 inch being fairly common. The board can be readily cut with a small hobby saw (paper-laminated foam board can be cut with a sharp knife; plastic-laminated foam board should be cut with a saw). Foam board is especially well suited for small robots for which lightweightedness is of extreme importance.

- *Rigid expanded plastic sheet.* Expanded sheet plastics are often constructed like a sandwich, with thin outer sheets on the top and bottom, and a thicker expanded (air-filled) center section. When cut, the expanded center section often has a kind of foam-like appearance, but the plastic itself is stiff. Rigid expanded plastic sheets are remarkably lightweight for their thickness, making them ideally suited for small robots.

- *Aluminum.* Aluminum is a great robot-building material for medium and large machines because it is exceptionally strong for its weight. Aluminum is easy to cut and bend using ordinary shop tools. It is commonly available in long lengths of various shapes, but it is somewhat expensive.

- *Steel.* Although sometimes used in the structural frame of a robot because of its strength, steel is difficult to cut and shape without special tools. Stainless steel is sometimes used for precision components, like arms and hands, and also for parts that require more strength than a lightweight metal (such as aluminum) can provide. Expensive.

- *Tin, iron,* and *brass.* Tin and iron are common hardware metals, often used to make angle brackets, sheet metal (various thickness from 1/32 inch on up), and (when galvanized) nail plates for house framing. Brass is often found in decorative trim for home construction projects and as raw construction material for hobby models. All three metals are stronger and heavier than aluminum. Cost is fairly cheap.

- *Wood.* Wood is easy to work with, can be sanded and sawed to any shape, doesn't conduct electricity (avoids short circuits), and is available everywhere. Disadvantage: Ordinary construction plywood is not recommended; use the more dense (and expensive) multi-ply hardwoods for model airplane and sailboat construction. Common thicknesses are 1/4 to 1/2 inch—perfect for most robot projects.

their emphasis is on industrial controls and systems, they offer plenty of tools, supplies, PVC, plastics, couplers, hardware, fasteners, and more.

In addition to online and catalog sales, Graybar operates nearly 300 offices and distribution facilities in Canada, Mexico, Puerto Rico, Singapore, and the U.S.

HST Materials, Inc.   202145

777 Dillon Dr.
Wood Dale, IL 60191
USA

   (630) 766-3333
   (630) 766-6335
   info@hstmaterials.com
   http://www.hstmaterials.com/

HST Materials is a stocking distributor, fabricator, and die cutter of foam tapes, sponge and dense rubber gasketing, heat-shrink tubing, fiberglass and expandable sleeving, cable ties, tapes, adhesives, and specialty products. Local retailer in Wood Dale, Ill.

Notable products include:

- Foam tape and sheet
- Extruded seals and gaskets
- Tubing/sleeving
- Cable ties
- Adhesives

Maintenance America   203875

http://www.maintenanceamerica.com/

Online ordering facility for Applied Industrial Technologies.

SEE:

http://www.appliedindustrial.com/

MatWeb   203704

http://www.matweb.com/

MatWeb is billed as "Your Source for Materials Information." That it is. The site provides information on more than 26,000 materials, including metals, plastics, ceramics, and composites.

In the words of the Web site: "MatWeb's database includes thermoplastic and thermoset polymers such as ABS, nylon, polycarbonate, polyester, and polyolefins; metals such as aluminum, cobalt, copper, lead, magnesium, nickel, steel, superalloys, titanium and zinc alloys; ceramics; plus a growing list of semiconductors, fibers, and other engineering materials."

Materials property data is taken directly from manufacturers, suppliers, and distributors.

McMaster-Carr Supply Company   202121

P.O. Box 740100
Atlanta, GA 30374-0100
USA

   (404) 346-7000
   (404) 349-9091
   atl.sales@mcmaster.com
   http://www.mcmaster.com/

Literally everything you need under one roof: materials (plastic, metal), fasteners, hardware, wheels, motors, pipe and tubing, tools, bits and other tool accessories, power transmission (gears, belts, chain), ball transfers, ball casters, spherical casters, regular casters, and many more. Walk-in stores in Chicago, Cleveland, Los Angeles, New Jersey.

MSC Industrial Direct Co., Inc.   202826

75 Maxess Rd.
Melville, NY 11747-3151
USA

   (516) 812-2000
   (800) 645-7270
   http://www.mscdirect.com/

Let's put it this way: MSC doesn't sell hamburgers or french fries. However, they sell just about everything else, and they have distribution points and offices all across North America. If you're looking for gears, tools, tool supplies (saws, bits, abrasives), glues and adhesives, couplers, hydraulics, pneumatics, wheels, casters, fasteners, paints, bearings, belts, linear slides, leadscrews, gearmotors, machinable wax, casting urethanes . . . well, you get the picture. Their catalog is 5,000 pages, available in print, CD-ROM, or online form, and contains over a half million items.

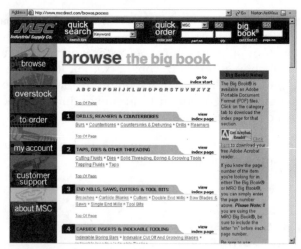

The MSC Web page.

Outwater Plastics Industries Inc.   202532

4 Passaic St.
Wood-Ridge, NJ 07075
USA

☎ (973) 340-1040
📠 (973) 916-1640
🚫 (800) 688-9283
✉ outwater@outwater.com
🌐 http://www.outwater.com/

Outwater Plastics offers a wide variety of materials, including casters; metal and plastic extrusions; machine framing; retail store fixtures (PVC and metal pipe, clamps, and connectors); and kitchen, bath, and furniture hardware. Online and printed catalog; get the printed catalog, as it's more complete.

Small Parts Inc.   202120

13980 N.W. 58th Ct.
P.O. Box 4650
Miami Lakes, FL 33014-0650
USA

☎ (305) 557-7955
📠 (305) 558-0509
🚫 (800) 220-4242
✉ parts@smallparts.com
🌐 http://www.smallparts.com/

Small Parts is a premier source for-get this!-small parts. All jocularity aside, Small Parts is a robot builder's dream, selling most every conceivable power transmission part, from gears to sprockets, chains to belts, and bearings to bushings. Product is available in a variety of materials, including brass, steel, and aluminum, as well

as nylon and Delrin. Rounding out the mix is a full selection of raw materials: metal rod, sheets, tubes, and assorted pieces, as well as a huge assortment of fasteners.

Outwater Plastics

With a catalog the size of a phone book, Outwater Plastics offers a dizzying array of products, from tiny fasteners to fiber optics to huge sheets of metal-laminated decorative sheets. A main thrust of the company is parts and supplies for furniture-manufacturing businesses; another is architectural flourishes for builders and interior designers. These products are of only minor interest to robot builders, unless you're building a 'bot that looks just like an 1850s two-story Colonial.

Other product lines in their catalog, however, are notable for constructing automatons. I'll summarize some of the more useful stuff here:

- Plastic trim, moulding, rods, and bar stock—available in acrylic, PVC, and several other common plastic types, in a variety of colors.
- Engineering plastic—bars, blocks, and slabs for milling and turning. Materials include polyethylene, Hydex, Zytel nylon, Kynar, Hytrel, ABS, PEEK, cast nylon, Delrin, and Teflon.
- Spherical casters and ball casters—for use as omnidirectional wheel casters on your robots.
- Display framing—in metal and PVC; tubes, squares, connectors, and fastening hardware. Use it to make robot frames.
- Lazy Susan turntables, from 3 to 12 inches—for making rotating joints for stationary robots or for robot heads.
- Retail store fixtures, including slatwall—think "out of the box" for making robot bases.
- Kerfkore flexible foam laminate.
- 3M industrial adhesives, in aerosol spray cans or pump sprays.
- Medium- to large-size fasteners (most start at 1/4-inch 20), in stainless, brass, zinc, and nylon.
- Aluminum and plastic channel—for making robot fames, bodies, and other parts.

The company's Web site lists some of the products for sale, but you really do need to get a copy of the catalog. You can also order free sample kits of several products.

The Outwater Plastics catalog is as big as a phone book and infinitely more fun to read.

Now about prices. Small Parts is for the serious builder, both amateur and pro. A little brass gear might cost $6, but what you pay for (apart from the precision, of course) is the selection, being able to find just about everything you need.

You can get their printed catalog, or browse through their online catalog; see:

http://www.engineeringfindings.com/

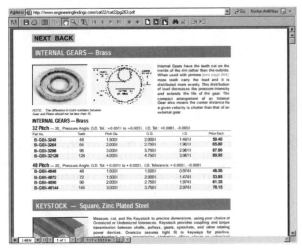

One of many parts pages at the Small Parts online ordering site.

Special Effect Supply Co. 203414

164 East Center St. #A
North Salt Lake, UT 84054
USA

((801) 936-9762

✉ spl_efx@xmission.com

🌐 http://www.fxsupply.com/

See listing under Supplies - Casting & Mold Making.

Supply Depot Inc. 202613

2830 Via Orange Way
Spring Valley, CA 91978
USA

((619) 660-2813

✆ (619) 660-2829

⊘ (800) 286-1699

✉ info@supplydepotinc.com

🌐 http://www.supplydepotinc.com/

Supply Depot sells some 190,000 industrial products, including machine tools, abrasives, carbide, cutting tools, clamping and fixturing, fluids, hand tools, machinery, material handling, precision tools, power tools, workholding, and assorted sundries.

Travers Tool Co., Inc. 202956

128-15 26th Ave.
Flushing, NY 11354
USA

⊘ (800) 221-0270

🌐 http://www.travers.com/

Source for metalworking tools and supplies, including fasteners, fluid power, hand tools, machine tools and accessories, metal, power and hand tools, power transmission equipment, materials handling, and machinable wax.

Robotic Skeletal Structures

In nature, and in robotics, there are two general types of support frames: endoskeletons and exoskeletons. Neither is inherently better; it all depends on the function of the robot and the ways in which you want to use it.

- *Endoskeleton* support frames are the kind found in many critters—including humans, mammals, reptiles, and most fish. The skeletal structure is on the inside; the organs, muscles, body tissues, and skin are on the outside of the bones. The endoskeleton is a characteristic of vertebrates.

- *Exoskeleton* support frames are the "bones" on the outside of the organs and muscles. Common exoskeletal creatures are spiders, all shellfish such as lobsters and crabs, and an endless variety of insects.

MRO Stands for . . . ?

You may see the acronym *MRO* used by online retailers and Web sites. MRO stands for "maintenance, repair, and operating," and it represents the parts and supplies involved in both factory upkeep as well as its day-to-day operation.

Some MRO materials are useful for robot building. For example, MRO suppliers often carry heavy-duty wheels and casters, replacement DC motors, metal stock and other raw materials, bearings, drive chain, and similar products. Much of it is aimed at industrial applications, so the materials will tend to be for heavy-duty jobs.

WESCO International, Inc.　　　　203974

4 Station Sq.
Pittsburgh, PA 15219
USA

📞 (412) 454-2200
📠 (412) 454-2505
🌐 http://www.wescodist.com

Choose from over 60,000 industrial electrical products. WESCO offers more than 200,000 products. The company sports five distribution centers in the U.S. and Canada; 350 branches in the U.S., Canada, Mexico, Nigeria, Singapore, the U.K., and Venezuela offer walk-in service.

MATERIALS-FIBERGLASS & CARBON COMPOSITES

Check here for sources of lightweight fiberglass and carbon-composite materials. Carbon composites are routinely sold in rod, tube, and sheet form and are among the strongest yet lightest materials commonly available. The same material is used in some fishing poles and tent poles, and hence providers of these items are also good sources for carbon composites.

While not as light, fiberglass is amazingly strong and can be used for both the structural and body components of your robot. Fiberglass can be molded and shaped and in raw form is composed of a fiberglass matting cloth and a two-part resin. The resin chemicals are mixed, and the matting cloth is immersed in this liquid. The resin cures after several hours of exposure to air.

SEE ALSO:

MATERIALS-PAPER & PLASTIC LAMINATES: More lightweight construction materials

MATERIALS-PLASTIC: Consider lightweight plastics, such as foamed PVC, as well

RADIO CONTROL: Carbon composites are often used in lightweight flying models

SUPPLIES-CASTING & MOLD MAKING: Check out medical suppliers for easy-to-use fiberglassing kits for making orthopedic casts and splints.

Aerospace Composite Products　　　203558

14210 Doolittle Dr.
San Leandro, CA 94577
USA

📞 (510) 352-2022
📠 (510) 352-2021
🚫 (800) 811-2009
✉️ Info@acp-composites.com
🌐 http://www.acp-composites.com/

Aerospace Composite Products is a supplier of carbon composites and fiberglass. Also sells Spyderfoam, a semicompressible foam available in thicknesses of 1.5 to 2 inches. It is intended as a substitute for "blue foam," commonly used in building construction as an insulator, but also employed as a substrate for laminates.

Air Dynamics　　　　203574

Fantasia Hobby Shop
171-69 46th Ave.
Fresh Meadows, NY 11358
USA

📞 (718) 396-4765
✉️ nyblimp@rcn.com
🌐 http://www.airdyn.com/

Carbon rods and tubes, carbon fiber fittings, and other lightweight materials. Air Dynamics also offers a number of unusual materials for ultralightweight model aviation (including blimps).

Notable additional products carried by Air Dynamics include:

- JiffyPins-gold-plated male and female connectors that can handle currents up to 20 amps.
- CS (Clearly Superior) Heat-Shrink-Thin, lightweight, and clear heat shrink for larger items; 45% shrinkage.
- 3M Blenderm adhesive bandage tape-1/2- or 1-inch-wide tape that sticks to most any surface, but does not leave a residue.
- Tesa transfer adhesive tape-Sticky stuff that's on a tape roll, but there's no tape.
- HiViz tape-Metallic foil diffraction grating tape.

Carbon rods. Photo Gabe Baltaian, Air Dynamics.

Carbon fittings for carbon rods. Photo Gabe Baltaian, Air Dynamics.

Art's Hobby 203295

P.O. Box 871564
Canton, MI 48187-6564
USA

✆ (734) 455-1927
📠 (413) 618-8961

✉ support@arts-hobby.com
🌐 http://www.arts-hobby.com/

Art's Hobby sells several unusual R/C construction components, including carbon laminate sheets and strips and carbon push rods. Carbon composites are useful because they are very lightweight, yet extremely strong.

🌐 💻

Composite Store, The 203557

P.O. Box 622
Tehachapi, CA 93581
USA

✆ (661) 822-4162
📠 (661) 822-4121
🚫 (800) 338-1278
✉ info@cstsales.com
🌐 http://www.cstsales.com/

Carbon composites, fiberglass, carbon laminates, plates, and foam. According to the site, the company stocks "a full inventory of lightweight composite materials available in small quantities for model builders, prototypes, university projects, movie special effects and fiberglass sculptures. Production quantity discounts are also available."

Web site includes how-to articles. Sold mail order and by dealers in the U.S. and worldwide.

🌐 💻

Dave Brown Products 202133

4560 Layhigh Rd.
Hamilton, OH 45013
USA

✆ (513) 738-1576
📠 (513) 738-0152
✉ sitemstr@dbproducts.com
🌐 http://www.dbproducts.com/

Dave Brown makes a long list of items for radio-control models. Products are available in hobby stores or online. The lines ideally suited to robotics include:

- Adhesives/additives
- Carbon-fiber building materials
- Covering materials
- Lite Flite wheels

The Lite Flite wheels are a common staple of robot builders (I must be on my twentieth pair by now). The wheels are light and fairly inexpensive, and they can be attached to a variety of axles and motor shafts. By gluing a servo horn to the rim of the wheel, you can attach it to a servo that has been modified for continuous rotation.

Carbon strip. Courtesy Dave Brown Products.

Dream Catcher Hobby, Inc.   203556

P.O. Box 77
Bristol, IN 46507
USA

☎ (219) 523-1938
✉ webmaster@dchobby.com
🌍 http://www.dchobby.com/

R/C components, such as carbon-fiber push rods. Also:

- Miniature planetary gearbox
- High-speed 400-size motor
- Battery chargers
- Bell cranks

Fibraplex   203554

1200 El Terraza Dr.
La Habra Heights, CA 90631
USA

☎ (714) 392-0720
📠 (562) 691-4667
🚫 (866) 327-2759
✉ sales@fibraplex.com
🌍 http://www.fibraplex.com/

Manufacturer and distributor of carbon-fiber rods and other materials. Sells items like carbon-composite tent poles on the site at reasonable prices.

Fibre Glast Developments Corporation   203588

95 Mosier Pkwy.
Brookville, OH 45309
USA

📠 (937) 833-6555.
🚫 (800) 330-6368
🌍 http://www.fibreglast.com/

Manufacturer of fiberglass and laminates. According to the company, "Fibre Glast is the world's leading supplier of composite materials for industrial, commercial, and hobbyist applications."

Online catalog or printable catalog.

Infinity Composites   203309

P.O. Box 176
Ashtabula, OH 44005-0716
USA

☎ (440) 992-2331
📠 (440) 992-0631
🚫 (866) 284-1173
✉ info@infinitycomposites.com
🌍 http://www.infinitycomposites.com/

Fiberglass composite materials. Be sure to check out their "b-grade and milled seconds" selection for scrap and odd-size pieces sold at a discount.

Kinetic Composites   203393

2520 Jason Ct.
Oceanside, CA 92056
USA

☎ (760) 945-4470
📠 (760) 945-9118
🚫 (800) 375-7043
✉ info@kcinc.com
🌍 http://www.kcinc.com/

Maker of carbon-composite materials, including panels.

Lightweight Backpacker, The   203555

http://www.backpacking.net/

Information and ideas on lightweight backpacking, including light but strong carbon-composite materials for the backpack frame.

SEE ALSO:

http://www.litebackpacker.com/

Michigan Fiberglass   203489

19795 E. 9 Mile Rd.
St. Clair Shores, MI 48080
USA

((586) 777-2032

(586) 777-0350

(800) 589-4444

 http://www.michiganfiberglass.com/

Online fiberglass products seller. Offers woven fabric, resins, adhesives, fiberglass-working tools, urethane foam, and other components common in the fiberglass trade.

NetComposites   203298

http://www.netcomposites.com/

All about composite materials. Check out the Guide to Composites section.

He Ain't Heavy; He's My Robot!

Batteries, motors, and frame contribute the most to a robot's weight. It's hard to control the weight of the first two, especially for robots built on a budget, where your choices are limited to what you have on hand or what is inexpensively available. Your choice of frame material for your robot, however, can greatly influence its final tally on the scales.

The frame of the robot can add a surprising amount of weight. An 18-inch-square, 2-foot-high robot constructed from extruded aluminum and plastic panels might weigh in excess of 20 or 30 pounds, even before you add motors and batteries. The same robot in wood (of sufficient strength and quality) could weigh even more.

There are ways to lighten your robot without sacrificing strength. This is done by selecting a different construction material and/or using different construction techniques. Instead of building the base of your robot using solid aluminum sheet, consider an aluminum frame, with crossbar members for added stability. Use thin plastic sheets as "skin" over the frame. The plastic is strong enough for mounting circuit boards, sensors, and other lightweight components.

Ordinary acrylic plastic is rather dense, and therefore fairly heavy, considering its size. Lighter-weight plastics are available, but aren't as easy to find—unless you know where to look. Foamed PVC sheets weigh less than half as much as an acrylic sheet of the same size, yet is about as strong. Foamed PVC goes under the trade names of Sintra, Komatex, Komacel, and others. It's available at most plastics specialty retailers, as well as sign-makers' shops.

Robots that require multiple decks—layers of components stacked up over the base—can be lightened using foamed PVC for the decking material and Class 125 for the "risers." Avoid the use of Schedule 40 or Schedule 80 water pipe, as these are thicker, and therefore heavier. You can attach the decks together using 6/32 all-thread, placing the PVC pipe over the all-thread. You need only a washer and a nut to keep the all-thread tight.

For the lightest robots possible, use fiberglass or carbon composites. Hollow fiberglass and carbon-composite tubes are very lightweight, yet are extremely strong. Both can withstand considerable flexing before breakage (many carbon composites will not break unless under very high tension). While cost for fiberglass and carbon composites is higher than for most other materials, the extra can be offset by the ability to use smaller and less expensive batteries and motors.

Olmec Advanced Materials Ltd.   204076

126 Dobcroft Road
Millhouses, Sheffield
S7 2LU
UK

☎ +44 (0) 1142 361606
📠 +44 (0) 1142 621202
✉ olmec@zoo.co.uk
🌐 http://www.olmec.co.uk/

Resellers and fabricators of specialty materials, including carbon-composite materials.

U.S. Composites, Inc.   203240

5101 Georgia Ave.
West Palm Beach, FL 33405
USA

☎ (561) 588-1001
📠 (561) 585-8583
✉ info@uscomposites.com
🌐 http://www.shopmaninc.com/

U.S. Composites sells fiberglass, casting materials, and lightweight composite materials. Useful products for robotics include:

- Fiberglass cloth and fiberglass mat
- Carbon fiber/graphite
- Clear casting resin
- Urethane foam
- Silicone rubber
- Latex

🏭 MATERIALS-FOAM

Who'd a thunk you could make robots out of foam? Well, to be truthful, not the kind of foam used to make beds or pillows, but heavier cellular foams commonly used for such tasks as sign making, floor underlayment, even wall insulation. Some foams are stiff and need no additional reinforcement, but others are best suited for lamination with a rigid material. Combine a hunk of foam, using contact cement, with a thin rigid piece of plastic or metal, and you make yourself a very strong-yet lightweight-substrate for a robot body or shell.

One side benefit of foams is that they are carvable, using either a regular knife, hot knife, or saw (in a pinch, you can also use an electric carving knife).

SEE ALSO:

MATERIALS-FIBERGLASS & CARBON COMPOSITES: Additional lightweight materials
MATERIALS-METAL AND MATERIALS-PLASTIC: Ideas for lamination substrates
SUPPLIES-CASTING & MOLD MAKING: Some casting resins make expandable, carvable foams

Advanced Plastics, Inc.   202676

7360 Cockrill Bend Blvd.
Nashville, TN 37209
USA

☎ (615) 350-6500
🚫 (800) 321-0365
✉ webmaster@advanced-plastics.com
🌐 http://www.advanced-plastics.com/

Offers Sign Foam, a heavier-bodied rigid polyurethane foam that can be drilled, carved, cut, and routed. According to the site: "Sign Foam is a lightweight, high density polyurethane board that possesses remarkable strength and durability. It is unlike any other foam you may have used . . . in strength, hardness, chemical resistance, structural durability, screw holding capabilities, and thermal expansion and contraction rates."

Local outlets in Tennessee, Alabama, and St. Louis, Mo.

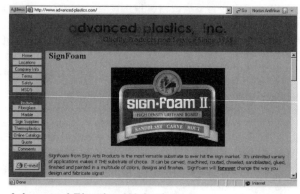

Advanced Plastics Web site.

Aerospace Composite Products   203558

14210 Doolittle Dr.
San Leandro, CA 94577
USA

☎ (510) 352-2022
📠 (510) 352-2021
🚫 (800) 811-2009
✉ Info@acp-composites.com
🌐 http://www.acp-composites.com/

Sells Spyderfoam, a semicompressible foam available in thicknesses of 1.5 to 2 inches. It is intended as a substitute for "blue foam," commonly used in building construction as an insulator, but also employed as a substrate for laminates.

Burman Industries, Inc.   203413

14141 Covello St.
Ste. 10-C
Van Nuys, CA 91405
USA

    (818) 782-9833

   (818) 782-2863

   info@burmanfoam.com

   http://www.burmanfoam.com/

See listing under Materials-Metal.

Composite Store, The   203557

P.O. Box 622
Tehachapi, CA 93581
USA

   (661) 822-4162

   (661) 822-4121

   (800) 338-1278

   info@cstsales.com

   http://www.cstsales.com/

See listing under Materials-Fiberglass & Carbon Composites.

CPE, Inc.   203535

541 Buffalo West Springs Hwy.
Union, SC 29379
USA

   (864) 427-7900

   mail@cpe-felt.com

   http://www.cpe-felt.com/

Manufacturer of felt products, including "stiffened felt," under the name Easy Felt. Available for sale at craft and art supply stores.

Douglas and Sturgess, Inc.   203178

730 Bryant St.
San Francisco, CA 94107-1015
USA

   (415) 896-6283

   (415) 896-6379

   Sturgess@ix.netcom.com

   http://www.artstuf.com/

See listing under Supplies-Casting & Mold Making.

Blue Foam, Pink Foam

Blue foam is the common term for a family of expanded polystyrene foams that are sold at building supply stores. Builders use the stuff for exterior insulation. It's more rigid than fiberglass insulation and has a higher R factor (R factor is the rating they give to insulative materials; the higher the factor, the more insulation the material provides).

Blue foam is from 1 to 3 inches thick, though specialty versions, such as floor sound-deadening foam, are available in thinner sheets. This latter material can be found at flooring stores.

Blue foam is useful as a substrate or "rigidizer" for your robot. It weighs very little for its size and bulk, yet offers remarkable rigidity. It's best used when physically cemented to a carrier, such as 1/8-inch (or even thinner) plywood or plastic. The two materials together provide a strong yet lightweight building platform for your robot. Being foam, it's easy to saw or drill, but it does break off into pieces the way Styrofoam and similar materials do.

Though called "blue foam," its color may be either blue or pink. The foam is available in different densities, with many of the pink variety foams having the lowest densities. You may find the heavier blue foams easier to work with because they're not as floppy.

Jim Allred Taxidermy Supply   203561

216 Sugarloaf Rd.
Hendersonville, NC 28792
USA

((828) 692-5846

✉   jim@jimallred.com

🌐   http://www.jimallred.com/

Taxidermy supplies: eyes (for humanoid or animal robots) and foam.

McKenzie Taxidermy Supply   203563

P.O. Box 480
Granite Quarry, NC 28072
USA

((704) 279-7985

📠   (704) 279-8958

🚫   (800) 279-7985

✉   taxidermy@mckenziesp.com

🌐   http://www.mckenziesp.com/

Sign Makers for Substrates

Sign makers use "substrates" as something to print on. There are many types of substrates for signage, including wood, plastic, metal, foam, and glass. Sign maker suppliers sell these substrates in convenient sizes and in many colors. Many of the substrates can be used to fashion all kinds of robot parts, including bases, frames, and bodies.

Here's a sampling of sign-making substrates that can also be used in robots.

- *MDO plywood*—MDO stands for "medium-density overlay," a lighter-weight plywood than stuff at the lumber yard, yet still very strong
- *Foam board*—Foam laminated with plastic, wood, or metal
- *Alumalite*—Aluminum over corrugated plastic; thicknesses from 1/8 inch; very strong stuff!
- *Clad-tex aluminum sign blanks*—Variations on an aluminum theme, with plastic, vinyl, and other coatings over aluminum sheet
- *Celtec PVC board*—Expanded (foamed) PVC; this is the basic stuff you shouldn't do without (other brands include Sintra, Komacel, and Komatex)
- *Coroplast*—Inexpensive foam board laminate
- *Corrugated plastic*—Goes by trade names like Gatorboard, Gatorplast, and other Gators, looks like the liner in cardboard but is made of plastic
- *DiBond*—Sheet metal over expanded (foamed) PVC; economical alternative to Alumalite
- *Econolite*—Like Alumalite, but with aluminum on one side only
- *Fiber-Brite substrates*—Fiberglass panels; very light but very strong
- *PolyCarve substrates*—Extruded polyethylene that can be carved into 3D shapes
- *Reflective tapes*—Metal or glass tapes that reflect lots of light
- *Holographic tapes and films*—Add rainbow colors and designs with various patterns
- *LusterBoard*—Aluminum on the outside, lightweight wood on the inside; 1/4- and 1/2-inch widths common

Note that prices tend to be very competitive, because sign makers get paid by the job, and they want to reduce their materials costs as much as possible. On the other hand, online retailers of sign-making supplies may require a minimum order of $25 or $50, and some will not cut material to more convenient sizes (typical sign substrate sheets are 4 by 8 feet). Not only do you get a big sheet of material to contend with, it must be shipped via motor freight and not UPS. Because of added expenses of trucking, look for sign maker suppliers that sell cut pieces.

Stuff for when you bag that ol' robot. McKenzie sells taxidermy supplies, of which things like eyes, casting materials, and carvable foam are useful to the robot constructionist.

MIT: Sketch Modeling (lectures)   203559

http://me.mit.edu/lectures/sketch-modelling/1-goals.html

Very helpful informational site on using various foams and substrates to make models. Includes information on using:

- Blue foam
- Cardboard
- Foam core
- Honeycomb board

Tips are provided for cutting, making curves, fastening, and more.

Plaster Master Industries   203188

4308 Shankweiler Rd.
Orefield, PA 18069
USA

- (610) 391-9277
- (610) 391-0340
- http://www.plastermaster.com/

Casting and mold-making supplies:

- Gypsum plaster
- Resin and casting materials
- Mold and model making
- Foam supplies

Public Missiles, Ltd.   204138

25140 Terra Industrial Dr.
Chesterfield Twp., MI 48043
USA

- (586) 421-1422
- (586) 421-1419
- (888) 782-5426
- PMLHighpowerSales@compuserve.com
- http://www.publicmissiles.com/

Public Missiles sells parts and electronics for model rocketry. They offer some nice polymer tubing (strong but light), wrapped phenolic tubing, electronic altimeters, two-part expanding foam, and other odds and ends that an earth-based robot ought to find useful.

R & J Sign Supply   202678

4931 Daggett Ave.
St. Louis, MO 63110
USA

- (314) 664-8100
- (314) 664-1305
- (800) 234-7446
- rich@rjsign.com
- http://www.rjsign.com/

R & J Sign specializes in materials for sign makers. Of particular interest to us robot constructors is Sign Foam; they also sell several other lightweight-yet strong-substrates that can be used to build machine bodies and other parts. Sign Foam is a rigid, high-density ure-thane material available in several different weights. It's easy to cut and can be "sculpted" to various shapes. Also sells Alumalite (aluminum over foam), corrugated plastic, and PVC foam board.

Sign Foam   204025

34700 Pacific Coast Hwy.
Ste. 207
Dana Point, CA 92624
USA

- (949) 489-9890
- (949) 489-1891
- (800) 338-4030
- info@signfoam.com
- http://www.signfoam.com/

Home of Sign Foam

Sign Foam makes and sells (typically through resellers that specialize in the sign-making specialty) a product that bears its name: Sign Foam. The product, used extensively to make raised and 3D signs, is a lightweight, high-density polyurethane board. It's very strong yet light and is easy to cut. You can use it to make robot bases, and because it is thick, it can be routed or carved to make three-dimensional contours.

Taxidermy.net   203562

http://www.taxidermy.net/

Links, forums, and information on taxidermy, includes list of supplies (for things like eyes and carving foam).

ULINE   203490

2200 S. Lakeside Dr.
Waukegan, IL 60085
USA

- (847) 473-3000
- (800) 958-5463
- Customer.service@uline.com

 http://www.uline.com/

Think "out of the box" on this one-literally, out of the shipping box. ULINE caters to people who ship things by mail or freight. That means cardboard boxes . . . and lots of other interesting things, like foam padding, tubes, shrink wrap, and tons of other stuff. ULINE is one of the largest mail room supply companies, and they'll send you a full-color catalog in a heartbeat.

Some interesting products you might want to consider, for robot construction, for shipping, or for the workshop:

- Antistatic packing materials (foam, tubing, shielding bags, and bubble pack)
- Foam for shipping boxes
- Bubble bags and sheets
- Clear and colored mailing tubes, with and without end caps
- Little clear plastic boxes (great for building small modules)

Van Dykes Taxidermy   203560

P.O. Box 278
39771 S.D. Hwy. 34
Woonsocket, SD 57385
USA

Sign Maker Sources

Here are some additional sources of sign maker supplies you can check out.

Custom Cut Aluminum
http://www.customcut.com/

Fiber-Brite
http://www.fiberbrite.com/

Harbor Sales
(very nice selection)
http://www.harborsales.net/

Laminator's Inc.
http://www.signboards.com/

Google.com search phrases you can try:
"sign making" substrates
"sign making substrates"

(605) 796-4425

 (605) 796-4085

(800) 787-3355

taxidermy@cabelas.com

http://www.vandykestaxidermy.com/

No, Van Dykes's taxidermy supplies aren't for hanging up a prize robot after you've bagged it. Instead, you can use Van Dykes for its carving foam, various kinds of glass eyes, and assorted unusual materials. Their foam block is easy to work with and can be shaped with simple tools.

Web page for Van Dykes Taxidermy

MATERIALS-LIGHTING

Not a construction material, per se; lighting elements add style and decoration to your robot. The sources in this section provide (or explain) various high-tech lighting products, including electroluminescent wire, very high brightness LEDs, and ultraviolet lamps.

With some creative thought, some of these lighting components could be put to good use in various robotic subsystems. For instance, electroluminescent wire could be attached to the floor and used to prevent a robot from wandering outside its perimeter. Or high-brightness flashing LEDs could be used in multiple-player robot soccer. Ultraviolet lamps (and ultraviolet detectors, typically used for testing paper money) could provide an optical proximity system based on the fluorescence of materials.

AS&C CooLight   203801

P.O. Box 783054
Winter Garden, FL 34778-3054
USA

 (407) 654-2660

 (413) 669-6842

info@coolight.com

http://www.coolight.com/

Electroluminescent wire, inverters, and sequencers. Products can be purchased from Web site.

Black Feather Electronics   202944

4400 S. Robinson Ave.
Oklahoma City, OK 73109
USA

 (405) 616-0374

 (405) 616-9603

blkfea@juno.com

http://www.blkfeather.com/

Something of a cornucopia of unique electronics items: connectors and cords, audio, gadgets, laser pointers, minicams, soldering, parts and switches, power, test equipment, kits, tools, video, high-brightness LEDs, electroluminescent wire.

Cool Neon/Funhouse Productions   203805

P.O. Box 4672
Berkeley, CA 94704
USA

 (510) 547-5878

 (510) 597-0996

info@coolneon.com

http://www.coolneon.com/

Cool Neon sells electroluminescent (EL) wire, a wire that glows when subjected to high-frequency supply voltages. Select colors and thicknesses, add a driver/inverter, and you're all set to go. Web site includes some details on soldering EL wire.

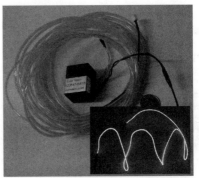

Electrolumniescent wire. Photo Rane Halloran, Funhouse Productions.

Sequencing electroluminescent wire driver. Photo Rane Halloran, Funhouse Productions.

Don Klipstein's Site 203265

http://misty.com/people/don/

Everything you ever wanted to know about lamps, LEDs, lasers, and strobe lights.

ELAM Electroluminescent Industries Ltd. 204203

Har Hotzvim, P.O.B. 45071
8 Hamarpeh Street
Jerusalem 91450
Israel

  972 2 5328888
  972 2 5328889
 http://www.elamusainc.com/

ELAM is the manufacturer of many of the electroluminescent wires sold under varying trade names (such as Neon Trim, Cool Wire, Live Wire, and Cool Neon). ELAM's name for the stuff is LyTec. The company provides technical details and specifications sheets on the wire and inverter/driver products.

USA subsidiary address:
ELAM USA Inc.
2 Seaview Blvd. Ste. # 101
Port Washington, NY 11050

Gilway Technical Lamp 202814

55 Commerce Way
Woburn, WA 01801-1005
USA

  (781) 935-4442
  (781) 938-5867

 http://www.gilway.com/

LEDs and lamps. Specialty products include super-bright LEDs in all colors, from ultraviolet to infrared.

Glowire 203802

10772 E SR 205
LaOtto, IN 46763
USA

  (219) 693-0772
  (775) 418-8613
  sales@glowire.com
  http://www.glowire.com/

Glowire sells electroluminescent wire, in different thicknesses and colors, as well as the necessary DC inverters used to drive the wire. They also provide "laser LEDs," which are really very bright colored LEDs. Useful if your robot needs bright headlights.

J. A. LeClaire 204202

3080 E. Outer Dr.
Ste. 10
Detroit, MI 48234
USA

  (248) 545-6660
  (248) 927-0338
  http://www.neontrim.com/

Sellers of Neon Trim electroluminescent wire, inverters, programmable sequencers.

Olmec Advanced Materials Ltd. 204076

126 Dobcroft Road
Millhouses, Sheffield
S7 2LU
UK

  +44 (0) 1142 361606
  +44 (0) 1142 621202
  olmec@zoo.co.uk
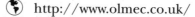 http://www.olmec.co.uk/

Resellers and fabricators of specialty materials, including Surelight electroluminescent wire.

Electroluminescent Wire

Imagine a flexible neon sign. That's what electroluminescent (EL) wire is. EL wire looks a lot like small plastic tubing, but when electricity is applied to it, it glows in a rainbow of colors. How does it work? At the center is a solid copper conductor. This conductor is coated with an electroluminescent phosphor. To excite the phosphor, two very fine wires are wrapped around the center conductor. Over this whole arrangement is a clear plastic sheath, which also protects everything inside.

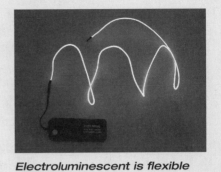

Electroluminescent is flexible "glow wire" that has both functional and decorative uses in robotics.

Apply current to the wires, and the phosphor lights up. Colors are produced by varying the chemical makeup of the phosphor and also by altering the tinting of the plastic protective sheath, varying the voltage, and/or varying the frequency of the current driving the wire. The end result is a brightly colored glowing wire.

EL wire has several uses in robotics. Here are just some of them:

- First and foremost, it looks cool! Wrap some EL wire of different colors around the periphery of your robot to give it some pizzazz.

- Small strips of EL wire, of a certain color, can be used to identify robots in a competition. If the robots are equipped with filtered light sensors, they can even differentiate friend from foe on the battlefield.

- EL wire can provide illumination for the robot for use in object detection. When used in conjunction with cadmium sulfide (CdS) cells, the reflected glow of the EL wire can be detected and used for proximity sensing. (In addition, many red-colored phosphors will emit a certain amount of near-infrared light, which is detectable with ordinary infrared phototransistors.)

- A strip of EL wire on the floor can be used for a line-tracking robot. On the underside of the robot, affix sensors to detect the glow of the wire.

- Strips of EL wire can be placed around the periphery of a room or along the floor to serve as a kind of electronic fence. Sensors on the underside of the robot detect the light from the EL wire. A bonus: Unlike a painted line, the EL wire can be switched on and off, thereby allowing the robot to exit the fenced area, should that be necessary.

Powering EL Wire

EL wire is driven by a high-voltage alternating current (AC). But it need not be plugged into a wall outlet. Rather, the wire uses small self-contained inverters that produce the required voltage from a small DC source (usually 3 to 12 volts; AA batteries are sufficient).

Inverters are rated by their output capacity, which in turn determines the length of EL wire that can be driven. Small inverters can drive from 3 to 6 feet of EL wire; higher-capacity units drive 20 to 30 feet (and more) of wire. Models designed for commercial lighting can drive several hundred feet.

Inverters are not terribly expensive—consumer models retail for $7 to $12. You'll have good results if you add more inverters to drive additional strands of EL wire rather than try to do it all from one unit. Additionally, you can opt for an inverter that blinks the EL wire at specific intervals or keeps it on continuously. Specialty inverters are available with built-in sequencers that selectively activate several strands of EL wire in turn.

Note that inverters are available at different operating frequencies—from 400 Hz to over 12,000 Hz. The brightest outputs are provided at the higher frequencies. The color of some of phosphors can be altered by changing the frequency of the AC power signal. For example, the "blue" phosphors can be changed from green to blue by varying the frequency between 400 and about 6,000 Hz. Changing the input voltage of the inverter, which in turn changes the output voltage, also alters the color of some EL wire.

Available Colors

Color choice varies by manufacturer, but most offer the following, in diameters from 1.3mm (called "angle hair") to 5.0mm:

- Aqua (blue/green)
- Deep red
- Green
- Indigo (deep blue)
- Lime green
- Orange
- Pink
- Purple
- Red
- White
- Yellow

The blues and greens tend to be the most vibrant colors.

One caveat when working with EL wire: Exposed (cut) ends of the wire can let in moisture, which can ruin the phosphor coating. To prevent this, always apply heat-shrink tubing over the ends of the wire. Attach the heat-shrink tubing so that it forms a closed seal.

RepairFAQ: Sam's Laser FAQ   202867

http://www.repairfaq.org/sam/lasersam.htm

From Sam Goldwasser's RepairFAQ: using and abusing (how not to) lasers. Includes a good section on safety.

Spencer Gifts, Inc.   204204

6826 Black Horse Pike
Egg Harbor Township, NJ 08234-4197
USA

☎ (609) 645-3300

🚫 (866) 469-2259

🌐 http://www.spencergifts.com/

Spencer Gifts is a retailer of the unusual, including gag cards, joke gifts, adult novelties, and unusual lighting effects (such as UV, fiber optics, and electroluminescent). The latter group is useful in robotics. Or maybe the rest, too. . . . I'll leave that up to you.

Web page for Spencer Gifts novelty store.

You can order online or visit a store near you. Most Spencer Gifts stores are in shopping malls.

Surelight   203807

☎ +44 (0) 1142 361606

📠 +44 (0) 1142 621202

info@surelight.com

 http://www.surelight.com/

Surelight sells electroluminescent (EL) wire in various colors and thicknesses, as well as EL drivers, light sticks, and other specialty lighting goods.

That's Cool Wire/Solution Industries   203806

P.O. Box 1692
Cypress, TX 77410-1692
USA

((281) 304-7400

((281) 304-7300

((800) 643-3267

info@thatscoolwire.com

http://www.thatscoolwire.com/

That's Cool Wire sells electroluminescent wire and driver modules.

Xenoline   203808

P.O. Box 2111
Pompano Beach, FL 33061-2111
USA

((561) 289-9500

Info@Xenoline.com

http://www.xenoline.com/

Xenoline sells electroluminescent and high-tech "glow-in-the-dark" products. This stuff is useful to dress up an otherwise boring robot, to provide a guide path or fence, or to provide illumination of a specific color for a robot with vision. Key products include:

- Xenopaks-Kits of different-colored electroluminescent wire and driver circuits
- ZLine-Flat electroluminescent light strips (several colors in the orange to blue spectrum), 1/4-inch wide by 28 inches long
- Gamma Rays-Ultra-bright-colored (blue, red, green, yellow, and white) LEDs intended as light sources
- Krill Lamps-Compact self-contained electroluminescent "lanterns" in a variety of colors
- Laser pointers-Hack 'em to make any pinpoint light source for your robot

All of the above are battery-powered, but do note that electroluminescent wire and strips require a driver consisting of a high-voltage, high-frequency inverter. These are generally inexpensive (under $15 for most models) and can be run from 3 to 12 volts.

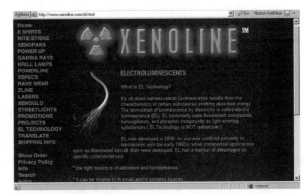

Xenoline Web page.

Electroluminescent Wire Suppliers

Most electroluminescent wire is manufactured by one company, ELAM (based in Israel), but it is available from a number of suppliers. Here are some that sell it mail order over the Internet:

Cool Neon—*http://www.coolneon.com/*
Coolight—*http://www.coolight.com/*
Glowire—*http://www.glowire.com/*
Lightgod.com—*http://www.lightgod.com/*
Surelight—*http://www.surelight.com/*
That's Cool Wire—*http://www.thatscoolwire.com/*
Xenoline—*http://www.xenoline.com/*

Bulb Direct

http://www.bulbdirect.com/

Light bulbs of all sizes, shapes, colors

CircuitToys, Inc.

http://www.circuittoys.com/

"Club" lighting novelties: LED glowsticks; laser pens; "Light Cubes" (plastic ice cubes that light up in colors)

Lightgod.com

http://www.lightgod.com/

Electroluminescent wire, and other rad lighting

LightLizard.com

http://www.lightlizard.com/

Distributors of specialty lighting gear: electroluminescent wire, flat lighting, fiber optics

MATERIALS-METAL

Some lament the end of "Detroit Iron," the days when American car manufacturers used tons of tin and chrome for the bulbous cars of the 1940s and 1950s. Don't let the same thing happen to robots! For a heavy-duty robot, you need to consider metal, if not for the frame or other structural components, then for the body. Most of the winning combat robots are made of aluminum or steel (some even stainless steel) and could dent a '55 Ford if the two ran into one another. Now that's a robot!

Seriously, in this section are sources for and about metals suitable for use in robotics. Several online metal retailers are included, as well as some local establishments that have Web sites you can look at. Buying metal through the mail can be expensive if you're purchasing large pieces of steel, as shipping costs mount. Therefore, depending on the material you are purchasing, you may wish to look for a local resource before turning to mail order.

On the other hand, lightweight aluminum and brass are readily shippable, especially if you purchase them in the sizes closest to your end application. That also limits waste and helps reduce cutting metal to sizes.

SEE ALSO:

FASTENERS: To hold metal together
MACHINE FRAMING: Extruded aluminum (and plastic) for building machine bodies

MATERIALS-PLASTIC: Substitute for metal
RETAIL-HARDWARE & HOME IMPROVEMENT: Handy local resource for metal stock (selection may be limited)

Admiral Metals
203161

11 Forbes Rd.
Woburn, MA 01801
USA

☏ (781) 932-0482
📠 (781) 932-4265
🌐 http://www.admiralmetals.com

Online (and local outlet store, in the Woburn, Mass., area) metals supplier.

Sells full lengths and cut pieces. Rod, bar, tube, sheet, and plate in:

- Aluminum
- Brass
- Copper
- Bronze
- Stainless steel

Alcoa
203183

Alcoa Corporate Center
201 Isabella St. at 7th St. Bridge
Pittsburgh, PA 15212-5858
USA

☏ (412) 553-4545
🌐 http://www.alcoa.com/

Read about Reynolds Wrap and other aluminum products, plus a gaggle of various manufacturing materials. Technical documents are fairly hidden (when I last looked at the site); do a Web search on the Alcoa.com site-minus the business articles-to look for the technical briefs.

Thanks to the Alcoa site, I learned this tidbit: "Making aluminum from recycled scrap takes only 5% of the energy it would take to make new metal from ore." How about that!

Airparts, Inc.
203153

2400 Merriam Ln.
Kansas City, KS 66106
USA

((913) 831-1780

 (913) 831-6797

⊘ (800) 800-3229

✉ airparts@airpartsinc.com

🌐 http://www.airpartsinc.com/

Metal supplies: aluminum and steel sheets, rods, tubes, etc.; fasteners and hardware. An excellent source for aluminum, chromolly, and other homebuilt aircraft construction supplies. You don't need to build an airplane; a robot will do.

Large selection of aluminum extruded pieces, including rods, angles, bars, and tubing. For many of the products, the aluminum alloy (such as 6063/T52 or 6061/T6) is specified.

All Metals Supply, Inc.　　202823

600 Ophir Rd.
Oroville, CA 95966
USA

((530) 533-3445

 (530) 533-3453

⊘ (888) 668-2220

✉ sales@allmetalssupply.com

🌐 http://www.allmetalssupply.com/

All Metals is a distributor of ferrous and nonferrous metals, fasteners, and industrial hardware. Metal products include aluminum, copper, zinc, and others in bar, tube, extruded shapes, sheets, and plate. Fastener products include machine screws and bolts of all sizes in stainless, zinc, and nylon. The Web site is for reference only; order directly from company.

American FlagStore.com　　204017

7714 Roger St.
Orange, TX 77632
USA

 (509)-275-7370

⊘ (888) 317-4594

✉ sales@americanflagstore.com

🌐 http://www.americanflagstore.com/

---

8 Tips for Working with Metal

With the right tools, working with metal is only slightly harder than working with wood or plastic. Here are some tips to streamline your metalworking.

1. Always use sharp, well-made drill bits, saws, and files. Dull, bargain-basement tools aren't worth the trouble.

2. If you're using a power drill or power saw, use the slower speed settings. High speed settings are for wood; slow for metal.

3. For sawing metal, select a fine-tooth blade, on the order of 24 or 32 teeth per inch. Coping saws, keyhole saws, and other handsaws are generally engineered for wood cutting, and their blades aren't fine enough for metalwork.

4. When cutting rod, bar, and channel stock by hand, use a miter box. The hardened plastic and metal boxes are the best buys. Be sure to get a miter box that lets you cut at 45 degrees both vertically and horizontally.

5. Use a punch to ensure accurate drilling. When cutting metal, the bit will skate over the surface until the hole is started. The punch creates a small "dimple" in the metal that reduces skating.

6. Whenever possible, use a drill press to cut metal pieces.

7. Always use a proper vise when working with a drill press. *Never* hold the work with your hands, or serious injury could result. If you can't place the work in the vise, use a pair of Vice Grip pliers or other suitable locking pliers.

8. Cutting and drilling leaves rough edges (called burrs and flashing) in the metal. These must be filed down using a medium- or fine-pitch metal file.

Show your patriotism and put a flag on your robot! Or, snoop around American FlagStore.com for such things as fiberglass poles, lightweight aluminum poles, and brass and aluminum metal balls-the balls sit atop the flagpole and can be used to build all sorts of unusual robot shapes. Their flagpole balls come in sizes from 3 to 12 inches.

ASAP Source   203154

LTEK Industries, Inc.
2284 S. Industrial Hwy.
Ann Arbor, MI 48104
USA

((734) 747-6105

℉ (734) 747-7139

⊘ (877) 668-0676

⊕ http://www.asapsource.com/

Offers aluminum, brass, and steel rods, bars, sheets, plates, and extruded forms. Sold in various sizes and lengths. Available in special cut lengths, for an additional charge. Also: threaded rod, welding rod, acetal and acrylic plastic pipes, rods, sheets, and plate.

Web site for ASAP Source

Atlas Metal Sales   203247

1401 Umatilla St.
Denver, CO 80204
USA

((303) 623-0143

℉ (303) 623-3034

⊘ (800) 662-0143

✉ jsimms@atlasmetal.com

⊕ http://www.atlasmetal.com/

Suppliers of casting ingots (aluminum, bronze, pewter). Billed as "Your Art Sculpture Specialists." Also

sells silicon bronze, rods, sheets, plates, and tubes; copper sheet and coil.

Burman Industries, Inc.   203413

14141 Covello St.
Ste. 10-C
Van Nuys, CA 91405
USA

((818) 782-9833

℉ (818) 782-2863

✉ info@burmanfoam.com

⊕ http://www.burmanfoam.com/

Burman is in the business of monsters. Making them, that is. They sell foam, latex, and other materials for special effects makeup, mold/model-making, puppetry, and animatronics professionals. Their adhesives, clays, latex, mold-making products, tubing, armature wire, aluminum pieces, and other products can be used to make robots, monstrous or not.

Burman also sponsors workshops on various model-making techniques on the first Saturday of each month in beautiful downtown Van Nuys, Calif.

Cal Plastics and Metals   202840

2540 Main St.
Chula Vista, CA 91911
USA

((619) 575-4633

℉ (619) 575-4561

✉ sales@calplasticsandmetals.com

⊕ http://www.calplasticsandmetals.com/

Plastics and metals. In metals (in various shapes, including sheets, bars, and tubes):

- Brass and bronze
- Copper
- Monel
- Aluminum
- Stainless steel

Du-Mor Service & Supply Co.   202635

10693 Civic Center Dr.
Rancho Cucamonga, CA 91730
USA

(909) 483-3330

(909) 483-3123

info@du-mor.com

http://www.du-mor.com/

Fasteners (including stainless steel), steel stock, shop tools, ferrous and nonferrous pipe and fittings, cutting tools and abrasives, hardware (swivels, brackets, etc.), and equipment.

EL-COM   203827

12691 Monarch St.
Garden Grove, CA 92841
USA

(714) 230-6200

(714) 230-6222

(800) 228-9122

http://www.elcomhardware.com/

Fasteners and hardware, mainly for cabinetry. Also casters, aluminum extrusions (squares, channels, bars), plastic laminates, and foam products.

FAQ: Metal Construction Set   202411

http://www.robotics.com/erector.txt

A FAQ on Meccano sets. Maintained by author and editor Jeff Duntemann.

FlagandBanner.com   204018

800 W. Ninth St.
Little Rock, AR 72201
USA

FlagandBanner.com sells metal flag poles and balls.

(501) 375-7633

(501) 375-7638

(800) 445-0653

sales@flagandbanner.com

http://www.flag-banner.com/

Flagpole stuff, including metal flagpole balls (from 2 1/2 to 12 inches; use 'em to make round robots or unusual wheel designs or anything else); metal and fiberglass poles.

Flagpole Components, Inc.   204019

P.O. Box 277
Addison, TX 75001
USA

(972) 250-0893

(972) 380-5143

(800) 634-2926

fci@flagpoles.com

http://www.flagpoles.com/

Flagpole poles (metal, fiberglass), different sizes and types of flagpole balls (from 3 to 12 nches).

Goodfellow   203156

Ermine Business Park
Huntingdon
PE29 6WR
UK

+44 (0) 1480 424800

+44 (0) 1480 424900

enq@goodfellow.com

http://www.goodfellow.com/

Goodfellow supplies metals, polymers, ceramics, and other materials for research, development, and specialist production. Products include sheet, ball, bar, and rod, in all common metals, plus exotic metals such as tungsten and titanium.

Web site in English, French, German, and Spanish.

High Performance Alloys   203387

444 Wilson St.
Tipton, IN 46072
USA

(765) 675-8871

(765) 675-7051

(800) 472-5569

Fast@HPAlloy.com

http://www.hpalloy.com/

Online retailer of specialty alloys for specialty projects, such as Hastelloy, Inconel, Monel, Nitronic, cobalt based, and commercially pure nickel. Stock includes bar, wire, sheet, plate, and fasteners.

Note that many of these alloys are much tougher than aluminum and even steel. If you plan on working with these metals, be sure you have the proper tools. For example, many desktop machine lathes and mills are not powerful enough for the tougher alloys; you need full-size machinery to do the job right.

Industrial Metal Supply Co. 203162

8300 San Fernando Blvd.
Sun Valley, CA 91352-3222
USA

(818) 450-3333

(818) 450-3334

http://www.industrialmetalsupply.com/

Metals: aluminum, brass, cast iron, chrommoly, copper, hobbyist "miniature" shapes and extrusions, silicon bronze, stainless, and steel.

Locations in Sun Valley, Irvine, and San Diego, Calif., and Phoenix, Ariz.

K&S Engineering 202138

6917 W. 59th St.
Chicago, IL 60636
USA

(773) 586-8503

(773) 586-8556

http://www.ksmetals.com/

K&S Engineering makes all those metal rods, bars, and other pieces you see in hobby stores across North America. They sell strictly to retailers (including some home improvement centers), though you can find their wares through online sources as well. Product line consists of steel, aluminum, and brass, in bar, tube, square, sheet, foils, and shim. Lengths from 12 inches to 48 inches.

SEE ALSO:

http://www.specialshapes.com/

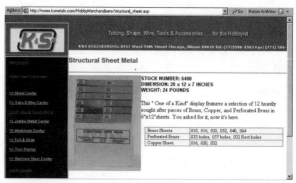

K&S Engineering sells metal products through retailers, but you can view their line on their Web site.

McNichols Company 203157

5505 West Gray St.
Tampa, FL 33609-1007
USA

(813) 282-3828 Ext. 2100

(813) 287-1066

corp@mcnichols.com

http://www.mcnichols.com/

McNichols sells various forms of unusual extruded metals, including perforated, metal bar grating, wire cloth, expanded metal (even some fiberglass thrown in for good measure). Available in sheets and plates, the product is strong yet lightweight.

Metal Supermarkets International 203158

170 Wilkinson Rd.
Unit 18
Brampton, ON
L6T 4Z5
Canada

(905) 459-0466

(905) 459-3690

(800) 807-8755

http://www.metalsupermarkets.com/

Metals; local stores. See the Web site for a store locator.

Metal Suppliers Online 204225

http://www.suppliersonline.com

Metal Stock at the Corner Hardware Store

Have a hankering to construct the next Terminator robot? You'll need about $300,000 in titanium to do it. A better idea: Make another robot and settle for commonly available metal stock, such as aluminum plate, channel, and rod. Your local hardware or home improvement store is the best place to begin. Here's what you'll find at the better-stocked stores.

Extruded Aluminum and Steel

Extruded metal stock is made by pushing molten metal out of a shaped orifice. Extruded aluminum and steel come in 2-, 3-, 4-, 6-, or 8-foot sections; some stores will let you buy cut pieces. Aluminum is lighter and easier to work with, but steel is stronger. Use steel when you need the strength; but otherwise, opt for aluminum.

Extruded aluminum and steel is available in more than two dozen common styles, from thin bars to pipes to square posts. Although you can use any of it as you see fit, a couple of standard sizes may prove to be particularly beneficial in your robot-building endeavors.

Extruded aluminum (and steel) comes in common bar, angle, and channel shapes.

- 1 by 1 by 1/16-inch angle
- 57/64 by 9/16 by 1/16-inch channel
- 41/64 by 1/2 by 1/16-inch channel
- Bar stock, widths from 1 to 3 inches; thicknesses 1/16-inch to 1/4-inch

Mending Plates

The typical wood-frame home uses galvanized mending plates, joist hangers, and other metal pieces to join lumber together. Much of it is weird shapes, but flat plates are available in a number of widths and lengths. You can use the plates as is or cut to size (the material is galvanized steel and is hard to cut; be sure to use a hacksaw with a fresh blade). The plates have numerous predrilled holes in them to facilitate hammering with nails, but you can drill new holes where you need them.

Mending plates are available in lengths of about 4, 6, and 12 inches by 4 or 6 inches wide, and also in 2-inch-wide T shapes. You can usually find mending plates, angles, and other steel framing hardware in the nail and fastener section of the home improvement store.

Iron Angle Brackets

You need a way to connect all the metal pieces together. The easiest way is to use galvanized iron brackets, located in the hardware section of the store. Angle brackets come in a variety of sizes and shapes and have predrilled holes to facilitate construction. The 3/8-inch-wide brackets fit easily into the two sizes of channel stock mentioned previously. You need only to drill a corresponding hole in the channel stock and attach the pieces together with nuts and bolts. The result is a very sturdy and clean-looking frame. You'll find the flat corner angle iron, corner angle ("L"), and flat mending iron to be particular useful.

Extensive listings of metal suppliers and fabricators, with addresses and (for most) brochures or catalogs.

MetalMart.com   202822

W229 N2464 Joseph Rd.
Waukesha, WI 53186
USA

((262) 547-3606

ℂ (262) 547-3860

🌐 http://www.metalmart.com/

Metals (and some plastic) by mail. Walk-in stores, primarily in Midwest and Great Lakes areas of the U.S., as well as Texas and North Carolina.

The following major product types are available:

- Aluminum
- Alloy and steel "cold rolled"; steel "hot rolled"
- Brass, bronze, copper
- Cast iron
- Drill rod
- Nickel silver
- Plastic
- Sheet and plate, tubing
- Stainless, tool steel

Web site for MetalMart.com

MetalsDepot   203163

4200 Revilo Rd.
Winchester, KY 40391
USA

((859) 745-2650

ℂ (859) 745-0887

✉ mdsales@metalsdepot.com

🌐 http://www.metalsdepot.com/

Metals-rod, sheet, and all the rest. Materials include steel, aluminum, stainless, brass, copper, bronze, cold-finish steel, alloy steel, tool steel, spring steel, cast iron, and titanium.

Metalworking.com   203703

http://www.metalworking.com/

Online resources for the rec.crafts.metalworking community of hobbyists and small businesses.

🗣

Nolan Supply Corporation   204242

111-115 Leo Ave.
P.O. Box 6289
Syracuse, NY 13217
USA

((315) 463.6241

ℂ (315) 463.0316

🚫 (800) 736.2204

✉ sales@nolansupply.com

🌐 http://www.nolansupply.com/

Nolan sells industrial supplies, tools, and metals. They have a separate 100-page metal shop catalog, where you can purchase a variety of types, sizes, and alloys, and with no minimums. They have alloy steel, copper, brass, aluminum, cast iron, and many other types in rod, tube, sheet, and plate form. Their tool section includes abrasives, air tools, cutting tools, electric tools, hand tools, machinery and machine tool accessories, precision tools, and vices.

Online Metals   202851

366 West Nickerson
Lower Level
Seattle, WA 98119
USA

((206) 285-8603

ℂ (206) 285-7836

🚫 (800) 704-2157

🌐 http://www.onlinemetals.com/

Mail-order metal supply: aluminum, stainless, copper, brass, steel. Available in plates, sheets, rod, tube, etc. Seems to be pretty good pricing.

OrderMetals.com 204086

Joseph Fazzio, Inc.
2760 Glassboro-CrossKeys Rd.
Glassboro, NJ 08028
USA

☎ (856) 881-3185
📠 (856) 881-0214
✉ products@josephfazzioinc.com
🌐 http://ordermetal.com/

A 26,000-ton inventory of steel and metal. Stock includes aluminum, stainless, copper, and brass in extruded shapes, plates, sheets, and expanded sheets. They will cut and shape to order.

Purity Casting Alloys Ltd. 203248

18503-97th Ave.
Surrey, BC
V4N 3N9
Canada

☎ (604) 888-0181
📠 (604) 888-8318
✉ dougb@purityalloys.com
🌐 http://www.purityalloys.com/

Specializes in custom nonferrous alloys (mostly aluminum); accepts small orders.

🌐 🏭

Sam Schwartz Inc. 202677

932-40 Hunting Park Ave.
Philadelphia, PA 19124
USA

☎ (215) 744-9996
🚫 (800) 440-9996
🌐 http://www.samschwartzinc.com/

Retailer of metals, plastics, and substrates: aluminum and other metals, Showcard, Sign Foam, Foamcore, Gatorfoam, Coroplast, Sintra (PVC), Signboards, Styrene, and others.

🌐

Slice of Stainless, Inc. 203159

6566 State Route 48
Goshen, OH 45122
USA

☎ (513) 722-1290
📠 (513) 722-1291
✉ TBRslice@aol.com
🌐 http://www.sliceofstainless.com/

Stainless steel sheets. For the hard-bodied bots.

T N Lawrence & Son Ltd. 202738

208 Portland Road
Hove
BN3 5QT
UK

☎ +44 (0) 1273 260260
📠 +44 (0) 1273 260270
✉ artbox@lawrence.co.uk
🌐 http://www.lawrence.co.uk/

Intaglio printing plates involve copper or zinc sheet metal. These sheets can be used in the construction or embellishment of robot bodies. Available in different thicknesses (1 to 2mm is average) and sizes, and the smaller sheets are quite affordable.

TM Technologies 203152

P.O. Box 429
North San Juan, CA 95960
USA

☎ (530) 292-3506
📠 (530) 292-3533
✉ kent@tinmantech.com
🌐 http://www.tinmantech.com/

TM Technologies provides tools and methods for metalworkers, such as that in body shop repair. Products include fairing angle curving dies, silver brazing rod, and fasteners.

🌐 🖥

Wicks Aircraft Supply 203160

410 Pine St.
Highland, IL 62249
USA

☎ (618) 654-7447
📠 (618) 654-6253
🚫 (800) 221-9425
✉ info@wicksaircaft.com
🌐 http://www.wicksaircaft.com/

Small aircraft parts; specialty fasteners. Note: This is not little model aircraft stuff, but stuff for small aircraft-ones people can climb into. Products of particular interest to robobuilders are:

- Composite materials (epoxy, foam, cloth)
- Steel, aluminum, plastic
- Hardware (bolts, nuts, washers, etc.)
- Control system accessories
- Wheels, brakes, tires

You can order most products online either by browsing or by index search.

XPress Metals

203155

Midland Aluminum Corp.
4635 W. 160th St.
Cleveland, OH 44135
USA

- (216) 267-7893
- (216) 267-7899
- (800) 321-1820
- midlandxpress@midlandaluminum.com
- http://www.midlandxpressmetals.com/

Aluminum, brass, copper, steel, stainless steel, and other metals in tube, bar, pipe, square, sheet, plate, and other shapes. The Web site lists 1,000 items for online sale; the company warehouses 10,000 items, so be sure to call if you need something special. Local retail stores in Cleveland, Ohio, and Clearwater, Fla.

MATERIALS-OTHER

Herein is the catch-all section for materials that don't fit neatly elsewhere. Includes wood (including specialty stocks for wood turning-handy stuff for superstrong robot bodies), wire, various hunks of specialty hardware, specialty adhesive tapes, and jewelry findings.

What are jewelry findings? They are small plastic or metal pieces used to make jewelry. Several common types are useful in the construction of smaller robots. For example, earring posts and earring backs can be used to make miniature linkages. Small beads can be used as spacers and even bearings; jump rings can be used to attach small parts (they're bendable with pliers); and spacer bars can be used as tiny linkages.

SEE ALSO:

RETAIL-ARTS & CRAFTS: More sources for specialty materials, including miniature jewelry stuff

Belden Inc.

204038

7701 Forsyth Blvd.
Ste. 800
St. Louis, MO 63105
USA

- (314) 854-8000
- (314) 854-8001
- info@belden.com
- http://www.belden.com/

Belden makes wire. Lots of it. You probably won't buy any directly from Belden, but check out the site anyway for their copious technical details on wire and fiber optics, including tables and formulas. Belden's products can be viewed online or in a printed catalog.

Constantines Wood Center

203216

1040 E. Oakland Park Blvd.
Ft. Lauderdale, FL 33334
USA

- (954) 561 1716
- (954) 565 8149
- info@constantines.com
- http://www.constantines.com/

Constantines is an established mail-order retailer specializing in wood products for the craftsperson. These include special veneers, woodworking jigs, exotic hardwoods, and tools. For robots, you can use their hard-

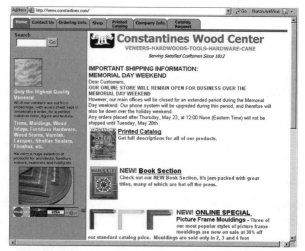

Constantines Wood Center Web site.

wood plywoods for robot bases. Precision routing accessories are useful for making special cuts in wood and plastics.

Craft Supplies USA 203932

1287 E. 1120 S.
Provo, UT 84606
USA

☎ (801) 373-0917
✆ (801) 377-7742
⊘ (800) 551-8876
✉ cust@woodturnerscatalog.com
🌐 http://www.woodturnerscatalog.com/

Supplies and tools for the turning (lathe) woodworker. Among their product line:

- Adhesives-Insta-Bond, Hot Stuff, Excel Glue, epoxy
- Alternative materials-ivory(fake), Tagua Nuts, Acrylester
- Finishing-buffing, finishes, waxes, polishes, dyes
- Sanding-power sanding, Velcro discs, Power-Lock, paper
- Shop accessorieswork lamps, face shields, airugs
- Wood blanks-bowl blanks, turning squares, exotics
- Wood lathes
- Wood-turning tools

Of particular interest are the wood blanks. Most of these are fairly exotic woods, but they're not all that expensive. A benefit of these exotics is that many are extremely hard and/or dense (like carob) and are a lot stronger than a piece of pine or some other common wood. You might want to consider one of these exotics if you are building a robot with wood and need strength for the base, risers, or some other component.

Smaller variations of these blanks can be found in the Pen Blanks category and include both wood, celluloid, and something called Environ, a manufactured material that looks like granite or wood, but is actually newsprint and soybean by-products. Pen blanks are usually about 3/4 by 5 inches.

Darice, Inc. 203520

13000 Darice Pkwy.
Park 82
Strongsville, OH 44149
USA

> **Electric Highways, Inc.**
> http://www.evparts.com/
> Parts for electric vehicles: motor parts, batteries, speed controllers, hardware
>
> **ForceField / Wonder Magnet**
> http://www.wondermagnet.com/
> Surplus high power magnets
>
> **Golf Cart Trader**
> http://www.golfcarttrader.com/
> Buy and sell golf cart parts, including clutches, electric motors, batteries, and rechargers
>
> **Systems Material Handling Company**
> http://www.smhco.com/
> Materials handling replacement and repair parts; large catalog

✆ (440) 238-1680
⊘ (800) 321-1494
✉ webmaster@darice.com
🌐 http://www.darice.com/

Manufacturer of jewelry findings and arts/crafts products. Catalog (450+ pages) costs $20. Darice sells their product though arts and crafts stores or online.

Eberhard Faber GmbH 203534

EFA-Strasse 1
92318 Neumarkt
Germany

☎ +49 (0) 9181 4300
✆ +49 (0) 9181 430222
✉ info@eberhardfaber.de
🌐 http://www.eberhardfaber.de/

Among many product lines, manufacturer of FIMO modeling clay; available at most craft, hobby, or art supply stores. Other modeling products include efaplast, CERAMOFIX, and Holzy. Web site is in German and English.

Fire Mountain Gems 203202

One Fire Mountain Way
Grants Pass, OR 37526-2373
USA

 (800) 355-2137

questions@firemtn.com

http://www.firemountaingems.com/

Small precision tools; jewelry supplies. Think out of the box on this one.

Globe Electronic Hardware, Inc.   203619

34-24 56th St.
Woodside, NY 11377
USA

(718) 457-0303

(718) 457-7493

(800) 221-1505

help@globelectronics.com

http://www.globelectronics.com/

Manufacturer of electronics hardware-things like case handles, spacers, thumbscrews, retainers, and fasteners.

Golf Car Catalog, The   202317

Mountaintop Golf Cars, Inc.
9647 Hwy. 105 South
Banner Elk, NC 28604
USA

(828) 963-6775

(828) 963-8312

(800) 328-1953

FindIt@GolfCarCatalog.com

http://www.golfcarcatalog.com/

All replacement parts for golf cars, including motors and batteries.

HUT Products   203933

15361 Hopper Rd.
Sturgeon, MO 65284
USA

(573) 443-6747

(800) 547-5461

HUTpfw@aol.com

http://www.hutproducts.com/

Wood, tools, and supplies for the precision wood-turning crafter. They offer:

- Wood and synthetic pen blanks-small blocks of wood (usually about 3/4 by 5 inches) intended for making pens using a lathe. You can use the blanks for anything. Pen-blank material is typically strong and dense and is ideal when you require structural strength in some part of your robot.

- Wood-turning and metal lathes-resellers of the Sherline lathes, as well as the VEGA Mini Lathe Duplicator.

- Acrylic rod-clear and colored.

- Dyed plywood-Small sheets (11 by 12 inches) of Baltic birch plywood; 0.350-inch thick.

J. W. Winco   202717

P.O. Box 510035
New Berlin, WI 53151-0035
USA

(262) 786-8227

(262) 786-8524

(800) 877-8351

http://www.jwwinco.com/

Industrial components (casters, handles, cranks, handwheels, knobs, O rings), available in both metric and inch sizes. Check out the technical section for helpful engineering tidbits.

Jewelry Supply   203530

503 Giuseppe Ct.
Ste. 4
Roseville, CA 95678
USA

Jewelry findings at Jewelry Supply.

Hardware for Hard Robot Bodies

Small, lightweight robots don't need heavy-duty construction. A little scrap of wood or plastic, some glue, maybe a piece of Velcro, and you're done. But bigger robots need a stronger means of holding things together. Enter construction hardware—things like nuts, bolts, angles, and extruded aluminum.

Here are some common items you'll want to have around your shop. Much of it can be found at the corner hardware or home improvement store. Online and mail-order sources offer greater variety. If you're looking to buy a lot of some given hardware, you can often purchase it in quantity at a significant discount.

Fasteners

The most basic hardware is fasteners. The following descriptions pertain to SAE-size threads, but similar sizes are available in the metric world.

- *Machine screws.* Available in different diameters, amateur robots use mainly number 4, 6, 8, and 10 machine screws (4/40, 6/32, 8/32, and 10/24, respectively). Common lengths are from 1/4 inch to 2 inches; you'll probably use the 1/2-inch and 3/4-inch lengths the most.
- *Nuts.* Hex nuts are the most common and cheap, but also consider *T-nuts* (also called blind nuts), which can be used with wood and soft plastics. Also look for locking nuts, which are like standard hex nuts but with a nylon insert; the nylon helps prevent the screw from working itself loose.
- *Washers.* Flat washers act to spread out the compression force of a fastener. They're available in sizes to complement the size of machine screw or bolt you are using. Tooth (internal or external) and split lock washers help prevent the nut from coming loose.
- *All-thread rod.* This is sold in 1- to 3-foot lengths. It comes in the same standard thread sizes and pitches. All-thread is good for shafts and linear motion actuators and to make bolts of any given length. Buy it when you need it.

The cheap zinc-plated screws, nuts, and washers are fine, but if you want a more hi-tech look, try the black anodized variety. When weight is a problem, opt for nylon hardware. It's a little more expensive than metal, and not quite a strong, but considerably lighter.

When buying machine screws, you have a choice of a variety of heads and drivers. Common machine screws come in round-head, flat-head, pan-head, or Fillister-head varieties; round-head or pan-head is adequate for most uses, and they are cheaper when bought in quantity. (Okay, okay, there are about three dozen additional machine screw head geometries, but I'm not going to list them all. Most of the specialty varieties are not generally available except in large quantities.)

Most machine screws available at the hardware store are slotted for flat-bladed screwdrivers. You may instead wish to use Phillips, Torx, square drive, Pozidrive, or hex head (Allen wrench) screws, which require the proper screwdriver. Slotted screws are cheaper to make, so they cost less. But there's a risk of stripping out the slot if overtightened. Specialty drive screws can be tightened and loosened without as much risk of stripping out the head.

Extruded Aluminum

Robot frames and other pieces can be constructed out of extruded aluminum. Designed for such applications as show trim, picture frames, and other handy applications, extruded aluminum comes in various sizes, thicknesses, and configurations. Length is usually 8 or 12 feet, but if you need less, most hardware stores will cut to order.

Useful sizes:

- 41/6 by 1/2 by 1/16-inch U-channel
- 57/64 by 9/16 by 1/16-inch U-channel
- 1 by 1 by 1/16-inch angle
- 1/2- and 3/4-inch-wide bar

Use extruded aluminum to make robot frames, bodies, arms, and other components.

Zinc-Plated Steel Angle Brackets

Standard hardware angle brackets are ideal for general robotics construction. You can use the brackets to build the frame of a robot constructed with aluminum extrusion stock. Common sizes are:

- 1 1/2-inch by 3/8-inch flat corner brackets—used when joining pieces cut at 45-degree angles to make a frame.

- 1-inch by 3/8-inch, and 1 1/2-inch by 3/8-inch corner angle brackets—used when attaching the stock to base plates and when securing various components (like motors) to the robot.

Keep in mind that angle brackets are heavy, and if you use a lot of them, they can add considerably to the weight or your robot. If you must keep weight down, consider substituting angle brackets for other mounting techniques, including gluing, brazing (for metal), or screws fastened directly into the frame or base material of your robot.

Angle brackets are used in general robot construction.

((916) 780-9610
📞 (916) 780-9617
✉ sales@jewelrysupply.com
🌐 http://www.jewelrysupply.com/

Online jewelry-making and craft supplies, miniature tools. Check the section on jewelry findings. Great source for teeny-tiny parts and tools.

K-Surplus Sales Inc.   202644

1403 Cleveland Ave.
National City, CA 91950
USA

((619) 474-6177
📞 (619) 474-3521
✉ kplus@pacbell.net
🌐 http://www.ksurplus.com/

Surplus fasteners and hardware.

$ 📠

Midwest Products Co., Inc.   203524

400 S. Indiana St.
Hobart, IN 46342
USA
📞 (219) 942-5703

(800) 348-3497

 customerservice@midwestproducts.com

 http://www.midwestproducts.com/

Balsa wood and basswood and kits. Sold in hobby stores or by mail Other woods available, including thin plywood, cherry, and spruce.

Musical Instrument Technicians Association   204104

http://www.mitatechs.com/

MITA stands for Association of Worldwide Professional Technicians. Their Web site is really a gold mine of links, resources, and catalogs of companies that specialize in parts for musical instruments. These parts include all kinds of little worm gears (guitar tuners), clamps, valves, and lots more. A robot built from musical instrument parts may not be able to carry a tune, but it can be cheaper to make.

Notions Marketing   203533

 contact@notions-marketing.com

http://www.notions-marketing.com/

Distributor of a broad line of notions to sewing and craft stores. Publishes a huge printed catalog. There is a $250 minimum opening order, so your best bet is to find their wares at local fabric and craft stores.

$

Stockade Wood & Craft Supply   203217

785 Imperial Rd. North
Guelph, ON
N1K 1X4
Canada

((519) 763-1050

(519) 763-1981

(800) 463-0920

info@stockade.ca

http://www.stockade-supply.com/

Wood parts. Lots of 'em. Includes wheels, pins, biscuits (oval-shaped thin wood for parts and shims), cubes, dowels, and lots more. Most of the product is used to make small wooden toys or as miniatures for doll houses. Obvious uses as robobits.

Tower Fasteners Co., Inc.   202639

1690 North Ocean Ave.
Holtsville, NY 11742
USA

((631) 289-8800

(631) 289-8810

(800) 688-6937

http://www.towerfast.com/

Master distributor, with online ordering, for several fastener brands, 3M adhesives, hardware for electronics, clamps and couplers, and power transmission. Distribution centers located along the East Coast.

Westrim Crafts   203503

9667 Canoga Ave.
Chatsworth, CA 91311
USA

((818) 998-8550

(818) 709-0928

(800) 727-2727

customerservice@westrimcrafts.com

http://www.westrimcrafts.com/

Westrim Crafts is a distributor of craft-making products, including things like jewelry findings. These can be used for construction of small parts. Westrim sells exclusively to retailers; most any local or online craft store will carry their product.

Woodcraft Supply Corp.   203218

P.O. Box 1686
Parkersburg, WV 26102-1686
USA

((304) 428-4866

(304) 428-8271

(800) 225-1153

custserv@woodcraft.com

http://www.woodcraft.com/

Woodworker tools and supplies. Be sure to check out their extensive line of plywoods (if you're building a robot base using wood). Of course, they offer the regular hand tools, like drills, saws, and planes, for working with wood. Order online or visit one of their 61 retail locations.

Gummy Transfer Tape

Here's an old trick: Suppose you want some "stickum," but can't use tape. One way is to apply a coating of rubber cement over some waxed paper. Wait half a minute for the cement to congeal, then gently "roll up" the residue into small clumps. You can now stick the stickup where you need it.

One disadvantage of this technique is that rubber cement tends to dry out, and its stickum doesn't stay sticky for long.

Transfer tape is an option. This stuff, which is also called *unsupported adhesive tape*, looks like double-sided tape, but it's engineered to leave its sticky residue, and no tape. It's meant for such jobs as electronics production, where workers apply small dabs of sticky substances to hold down wires and components. Transfer tape, such as the stuff made by tesa AG in Germany, is applied by first placing the material as you would any other tape. You then peel off the waxed paper, and what's behind is a layer of very sticky goo. Unlike rubber cement, transfer tape stickum remains tacky and semiflexible.

Transfer tape is available from art supply stores, adhesive specialists, and electronics production supply outfits. You can find it online with these Google.com searches (don't look just for "transfer tape"; that won't cut it):

"transfer tape" "tesa 4900"

+"transfer tape" +"waxed backing"

"unsupported adhesive" tape

Here are some online sellers of adhesive transfer tape (note that some tapes work best when used with a specially made dispenser):

Curry's Art Store Limited
http://www.currys.com/

Hillas Packaging
http://www.hillas.com/

Jerry's Artarama
http://www.jerryscatalog.com/

ULINE
http://www.uline.com/

Blenderm: Sticky Stuff without Gummy Residue

3M's Blenderm is a waterproof adhesive bandage that is intended to be used on people, but is also pretty handy with robots. The semiclear tape is available in different widths up to 2 inches, and it sticks to practically anything. Yet it doesn't leave much gummy junk behind, and it stays flexible for an eternity. Possible ideas: Use it to construct hinges or movable flaps.

You can sometimes find Blenderm at local drug stores, but be wary of the cheaper brands. You want the real stuff, so make sure it says *3M* and *Blenderm*.

If you can't find it locally, try on the Web, using the following Google search:

blenderm

Typical online sellers of Blenderm are medical supply outfits. Here are some to get you started:

Elite Medical
http://www.elitemedical.com/

Medical Supply Company
http://www.medsupplyco.com/

Global Drugs
http://www.globaldrugs.com/

Wrights 203502

85 South St.
P.O. Box 398
West Warren, MA 01092
USA

⊘ (877) 597-4448

✉ help@wrights.com

🌐 http://www.wrights.com/

Manufacturer of sewing and craft notions, including fusing tape (heat it up and it congeals with a stickiness). The products are available at fabric and craft stores.

🌐 🏭

MATERIALS-PAPER & PLASTIC LAMINATES

A laminate is a sheet composed of two materials glued together, almost like a sandwich.

Laminates save money by using only a small amount of the expensive outside material, relying on the inside material to supply most of the bulk. Because two kinds of materials are cemented together, they tend to reinforce one another.

A commonly available laminate is foam core, available at most craft and art supply stores. The material is composed of compressed Styrofoam (or similar), lined on both sides with heavy colored paper. You can cut foam core with a knife or a small hobby saw, and ordinary paper glue can be used to hold the cut pieces together.

Foam core is the most common paper laminate, but it's by far not the only one. There is a wide variety of paper and even plastic laminate sheets available, and many brands are listed here. Some manufacturers of paper and plastic sell directly, but most want you to purchase their goods from arts and craft retailers. For those makers that don't sell directly, you can see their Web sites for handy application notes and material-handling datasheets.

SEE ALSO:

MATERIALS-FOAM AND MATERIALS-PLASTIC:
Make your own foam core

RETAIL-ARTS & CRAFTS: Sources for paper laminate materials

Advantage Distribution 202679

112 E. Railroad Ave.
Monrovia, CA 91016
USA

📞 (626) 359-0778

📠 (626) 359-2778

⊘ (800) 601-1169

✉ JP@advantagedistribution.com

🌐 http://www.advantagedistribution.com/

Plastic products, including acrylic sheet, expanded foamed PVC (Sintra), corrugated plastic. Also sells Bienfang Foamboard, Pillocore Foamcore, and Ultra Board (polystyrene foam core).

Ships by UPS or orders available for will-call pickup. Retail store room in Monrovia, Calif.

$ 💻 🛍

ASW-Art Supply Warehouse 202741

5325 Departure Dr.
Raleigh, NC 27616-1835
USA

📞 (919) 878-5077

📠 (919) 878-5075

⊘ (800) 995-6778

✉ aswexpress@aol.com

🌐 http://aswexpress.com/

Artist accessories, brushes, foam boards, and paints.

📄 🌐

Douglas and Sturgess, Inc. 203178

730 Bryant St.
San Francisico, CA 94107-1015
USA

📞 (415) 896-6283

📠 (415) 896-6379

✉ Sturgess@ix.netcom.com

🌐 http://www.artstuf.com/

See listing under Supplies-Casting & Mold Making.

🌐 💻

GoldenWest Manufacturing 203237

P.O. Box 1148
Cedar Ridge, CA 95924
USA

📞 (530) 272-1133

📠 (530) 272-1070

🌐 http://www.goldenwestmfg.com/

Manufacturer of casting resins (both rigid and flexible), machinable plastic, foam board, and tools for cast and mold making. Their machinable plastic includes a product known as Butter-Board, a lightweight plastic block that is nonabrasive and very easy to machine or work with hand tools.

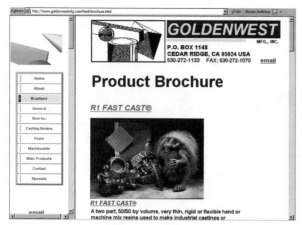

GoldenWest Manufacturing Web site.

International Paper

202673

P.O. Box 1839
Statesville, NC 28687-1839
USA

📟 (704) 878-2708
🚫 (800) 438-1701
🌐 http://www.gatorfoam.com/

In addition to paper products used for just about everything, International Paper makes Gatorfoam, GatorFlex, and Foam-Cor products. These can be purchased at local art supply stores. Some materials information is provided on the Web site.

Public Missiles, Ltd.

204138

25140 Terra Industrial Dr.
Chesterfield Twp., MI 48043
USA

☎ (586) 421-1422
📟 (586) 421-1419
🚫 (888) 782-5426
✉ PMLHighpowerSales@compuserve.com
🌐 http://www.publicmissiles.com/

Public Missiles sells parts and electronics for model rocketry. They offer some nice polymer tubing (strong but light), wrapped phenolic tubing, electronic altimeters, two-part expanding foam, and other odds and ends that a earth-based robot ought to find useful.

Cardboard Substrates

When I was a little kid, I sent away for a fiberboard submarine. I didn't know what "fiberboard" meant, and I hoped they would send me a real metal submarine. It turned out that fiberboard is just another name for cardboard, the stuff they make shipping boxes out of. In fact, my submarine was just a square box with a submarine shape (two colors, no less!) printed on it.

Cardboard may not make for good submarines, but—surprise!—it can be used as an effective construction material for robotics. Not just any cardboard, though. Common cardboard consists of two thin sheets (liner boards) of paper sandwiched to a corrugated inner core, or medium. The thickness and material used for the inner corrugated medium determines the overall rigidity of the cardboard. Cardboard for shipping boxes is too lightweight for most robot projects.

A better kind is called *honeycomb board*. This is a special kind of cardboard that uses a thick honeycomb medium. Thickness can vary from 1/2 inch to several inches. Honeycomb board is available at many art supply stores and is used in model making. For robotics, it can be sandwiched to a thin carrier, such as plywood or plastic, or even metal. The carrier need not be thick; the strength of the sandwich material will come from the cardboard.

Compared to both cardboard and foam core, honeycomb board (that's a generic term for it, by the way) is quite expensive. However, it has good structural characteristics. When used properly, honeycomb board can hold over 100 pounds.

R & J Sign Supply 202678

4931 Daggett Ave.
St. Louis, MO 63110
USA

☎ (314) 664-8100
🖷 (314) 664-1305
🚫 (800) 234-7446
✉ rich@rjsign.com
🌐 http://www.rjsign.com/

R & J Sign specializes in materials for sign makers. Of particular interest to us robot constructors is Sign Foam; they also sell several other lightweight—yet strong—substrates that can be used to build machine bodies and other parts. Sign Foam is a rigid high-density urethane material available in several different weights. It's easy to cut and can be "sculpted" to various shapes. Also sells Alumalite (aluminum over foam), corrugated plastic, and PVC foam board.

Home page for R & J Supply.

ULINE ☷ 203490

2200 S. Lakeside Dr.
Waukegan, IL 60085
USA

☎ (847) 473-3000
🚫 (800) 958-5463
✉ Customer.service@uline.com
🌐 http://www.uline.com/

Think out of the box on this one-literally, out of the shipping box. ULINE caters to people who ship things by mail or freight. That means cardboard boxes . . . and

ULINE

At first glance, the ULINE catalog is just mailroom supplies—boxes, packing tape, Styrofoam "peanuts," and "Packing List Enclosed" labels.

Ah, but look closer, and you might find a treasure trove of robot-building goodies:

- Inexpensive cardboard bins (well under a dollar each), ideal for organizing all the goodies of ongoing projects. Available in sizes from 6 to 12 inches deep.

- Gusseted polyethylene bags, in various sizes, for storing small parts. The bags are clear, so you can see what's in them. Their squared bottoms allow them to stand upright. Good (and cheap!) way to manage screws, nuts, washers, and other fasteners. You can write on the plastic with a Sharpie.

- Special thick-walled cardboard boxes for shipping heavy items. Though made of paper, this kind of cardboard is so sturdy it can be used to make robot bases and structures, especially if it's laminated with thin (1/16 inch) plastic sheets.

- Antistatic bags and foam, for storing your electrical items and for making simple touch and pressure sensors.

- Nylon cable ties, including the "releasable" kind that can be reused. Handy for works-in-progress. Also, low-cost cable tie tools.

- Steel, nylon, and polyester strapping tape (and associated tools), which can be used for securing robot parts and as drive belts for rotary and linear drives and lots more.

- A product called Instapak Quick, self-expanding foam contained in a thin plastic sheath. To activate the contents, you apply some heat and mix them together; the foam expands and hardens in a few seconds. The foam conforms to any shape. One idea: Use for building sturdy but very lightweight robot shells.

- Specialty tapes, including Kapton (can withstand high temperatures and still holds), glue dots (use for holding down wires and small components), Teflon tape (for any job where a self-lubricated surface is needed), 3M Dual Lock (like Velcro, but for very heavy-duty applications), adhesive transfer tape (tape that leaves the gummed adhesive but no tape carrier), and double-sided masking tape (like double-sided foam tape, but not as thick, and very, very sticky).

Many of ULINE's products are sold in bulk—they don't want to sell just one 35-cent cardboard box—but single quantities of specialty items are available. The company offers a printed catalog, or you can order online.

The Web site for ULINE is:

http://www.uline.com/

Foam Core Substrates

Foam core is part foam, part paper. It is available from art supply stores, in thicknesses from 1/8 to 1 inch. Construction is simple: An inner foam sheet is sandwiched on both sides with high-quality paper. Foam core is cheap, easy to cut with a hobby knife or small scroll saw, and can be readily glued.

You can provide extra rigidity to foam core by combining it with another thick substrate, such as cardboard, thin metal, or plastic. However, depending on thickness and weight loads, foam core can be used by itself.

lots of other interesting things, like foam padding, tubes, shrink wrap, and tons of other stuff. ULINE is one of the largest mail room supply companies, and they'll send you a full-color catalog in a heartbeat.

Some interesting products you might want to consider for robot construction, for shipping, or for the workshop:

- Antistatic packing materials (foam, tubing, shielding bags, and bubble pack)
- Foam for shipping boxes
- Bubble bags and sheets
- Clear and colored mailing tubes, with and without end caps
- Little clear plastic boxes (great for building small modules)

MATERIALS-PLASTICS

Most everyone knows the famous line in the movie The Graduate, where, at Dustin Hoffman's graduation party, a man comes up to him and tells him to get into the one business of the future, "plastics." What audiences worldwide don't know is that the man was really talking about plastics used in amateur robots. Both plastics and robots are the wave of the future, so it makes sense they are connected at the hip.

This section details resources for affordable plastics that can be used in the construction of robot frames, bases, and shells. Many of the listings are for online retailers, though some are for local establishments. Plastics specialty outlets and sign makers are among the best sources of plastics for robot bodies.

In a pinch, you can purchase a small sheet of acrylic or polycarbonate plastic at the local hardware store. But these plastics are harder—and in the case of acrylics,

prone to cracking—than such specialty plastics as PVC, ABS, Delrin, styrene, and nylon.

Note that if you purchase sheet plastics from a specialty retailer, you may have to buy an entire sheet, which is 4 by 8 feet. Most will cut it for you. If you don't need that much material, most retailers have a "junk bin" of excess plastic pieces; scrounge through it to see what you can find.

One possible problem with this method: Unless you know your plastics, you may end up buying material that is more trouble than it's worth. There are some resin plastics that are difficult to cut and drill and which will not respond to solvent cements intended for the more common PVC, ABS, and styrene plastics. If you're not sure what kind of plastic you've found, be sure to ask.

SEE ALSO:

MATERIALS-METAL: Stronger than plastic; use when strength is important

RETAIL-ARTS & CRAFTS: Another potential source of plastics

SUPPLIES-GLUES & ADHESIVES: Cements to glue plastics together

---

Advanced Plastics, Inc.   202676

7360 Cockrill Bend Blvd.
Nashville, TN 37209
USA

- (615) 350-6500
- (800) 321-0365
- webmaster@advanced-plastics.com
- http://www.advanced-plastics.com/

Advanced Plastics is a wholesale distributor of a variety of fiberglass-reinforced plastic products, as well as cast polymers, thermoplastics, and sign products/supplies. Product offerings include expanded foam PVC in a variety of colors.

Also offers Sign Foam, a heavier-bodied rigid polyurethane foam that can be drilled, carved, cut, and routed. According to the site: "Sign Foam is a light-weight, high density polyurethane board that possesses remarkable strength and durability. It is unlike any other foam you may have used . . . in strength, hardness, chemical resistance, structural durability, screw holding capabilities, and thermal expansion and contraction rates."

Local outlets in Tennessee, Alabama, and St. Louis, Mo.

Advantage Distribution   202679

112 E. Railroad Ave.
Monrovia, CA 91016
USA

((626) 359-0778
℡ (626) 359-2778
⊘ (800) 601-1169
✉ JP@advantagedistribution.com
🌐 http://www.advantagedistribution.com/

Plastic products, including acrylic sheet, expanded foamed PVC (Sintra), corrugated plastic. Also sells Bienfang Foamboard, Pillocore Foamcore, and Ultra Board (polystyrene foam core).

Ships by UPS or orders available for will-call pickup. Retail store room in Monrovia, Calif.

$ 💻 🛒

American Plastics   202675

7451 Dogwood Park
Fort Worth, TX 76118
USA

⊘ (800) 869-4061
✉ sales@americanplastics.net
🌐 http://americanplastics.net/

American is a wholesale supplier of plastic sheet and related products. Products include:

- ABS sheet
- PBV sheet and rod
- Acrylic/PVC sheet
- Delrin sheet and rod
- Komatex foam PVC
- Polypropylene rod
- Styrene

- Welding rod (thermoplastic, ABS)
- Solvents and cements

ASAP Source   203154

LTEK Industries, Inc.
2284 S. Industrial Hwy.
Ann Arbor, MI 48104
USA

((734) 747-6105
℡ (734) 747-7139
⊘ (877) 668-0676
🌐 http://www.asapsource.com/

Offers aluminum, brass, and steel rods, bars, sheets, plates, and extruded forms. Sold in various sizes and lengths. Available in special cut lengths for an additional charge. Also: threaded rod, welding rod, acetal and acrylic plastic pipes, rods, sheets, and plate.

🌐 💻

Aspects, Inc.   203811

P.O. Box 408
245 Child St.
Warren, RI 02885
USA

((401) 247-1854
℡ (401) 247-1820
⊘ (888) 277-3287
🌐 http://www.aspectsinc.com/

Aspects is a birdfeeder supply company. They offer acrylic dome hemispheres (clear half-round plastic spheres) of several different sizes. The domes are intended to keep nasty squirrels from eating the birdfeed, but you can use them for building robots instead.

🌐 💻

Bay Plastics Ltd.   203317

Unit H1, High Flatworth
Tyne Tunnel Trading Estate
North Shields, Tyne & Wear
NE29 7UZ
UK

(+44 (0) 1912 580777
℡ +44 (0) 1912 581010
✉ enquiries@bayplastics.co.uk
🌐 http://www.bayplastics.co.uk/

Types of Plastics

There are literally thousands of types of plastics. Here is a listing of the more common varieties you will encounter as you build robots.

- *ABS* (acrylonitrile butadiene styrene). The most common application for ABS is sewer and waste-water plumbing systems. ABS is the large black pipes and fittings you see in the hardware store. Despite its shiny black appearance in plumbing material, ABS is really a glossy, translucent plastic that can take on just about any color and texture. It is tough and hard and yet relatively easy to cut and drill. Besides plumbing fittings, ABS comes in rods, sheets, and pipes and is the plastic used to construct LEGO pieces.

- *Acrylic.* Acrylic is clear and strong, the mainstay of the decorative plastics industry. It can be easily scratched, but if the scratches aren't too deep, they can be rubbed out. Acrylic is somewhat tough to cut without cracking and requires careful drilling. The material comes mostly in sheets, but is also available in extruded tubing, rods, and is the coating in pour-on plastic laminate.

- *Cellulosics.* Lightweight, flimsy, but surprisingly resilient, cellulosic plastics are often used as sheet coverings. They have minor uses in robotics. One useful application, however: The material softens at low heat, and can be slowly formed around an object. Comes in sheet or film form.

- *Epoxy.* Very durable clear plastic, often used as the binder in fiberglass. Epoxies most often come in liquid form, for pouring over something or onto a fiberglass base. The dried material can be cut, drilled, and sanded.

- *Nylon.* Tough, slippery, self-lubricating stuff most often used as a substitute for twine. Nylon also comes in rods and sheets from plastics distributors. Nylon is flexible, which makes it moderately hard to cut.

- *Phenolic.* An original plastic, phenolics are usually black or brown in color, easy to cut and drill, and smell terrible when heated. The material is usually reinforced with wood or cotton bits or is laminated with paper or cloth. Even with these additives, phenolic plastics are not unbreakable. Comes in rods, sheets, and pour-on coatings. Minor application in robotics except as circuit board material.

- *Polycarbonate.* Polycarbonate plastic is a close cousin to acrylic but more durable and more resistant to breakage. Polycarbonate plastics are slightly cloudy in appearance and easy to mar and scratch. Comes in rods, sheets, and tubing. A common inexpensive window-glazing material, polycarbonates are hard to cut and drill.

- *Polyethylene.* Polyethylene is lightweight and translucent and is often used to make flexible tubing. It also comes in rod, film, sheet, and pipe form. The material can be reformed with application of low heat, and when in tube form, can be cut with a knife.

- *Polypropylene.* Like polyethylene, but harder and more resistant to heat.

- *Polystyrene.* A mainstay in the toy industry. This plastic is hard, clear (can be colored with dyes), and cheap. Although often labeled as "high-impact" plastic, polystyrene is brittle and susceptible to damage by low heat and sunlight. Available in rods, sheets, and foam board. Moderately hard to cut and drill without cracking and breaking.

- *Polyurethane.* These days, polyurethane is most often used as insulation material, but it's also available in rod and sheet forms. The plastic is durable, flexible, and relatively easy to cut and drill.

- *PVC* (polyvinyl chloride). PVC is an extremely versatile plastic best known as the material used in freshwater plumbing and outdoor plastic patio furniture. Usually processed with white pigment, PVC is actually clear and softens in relatively low heat. PVC is extremely easy to cut and drill and almost impervious to breakage. Beside plumbing fixtures and pipes, PVC is supplied in film, sheet, rod, tube, even nut and bolt form.

- *Silicone.* A large family of plastics all in its own right. Because of their elasticity, silicone plastics are most often used in molding compounds. Silicone is slippery to the touch and comes in resin form for pouring.

Local plastic retailer open to the trade and public. Check out the plastics application guides on their Web site, as well as the "DIY Plastic Materials" section.

BioPlastics 203673

34655 Mills Rd.
North Ridge, OH 44039
USA

☎ (440) 327-0485

📠 (440) 327-3666

🚫 (800) 487-2358

✉ info@bioplastics.com

🌐 http://www.bioplastics.com/

Manufacturer of plastic belts. In the words of the Web site: "BioPlastics manufactures BioThane belting, a strapping made of plastic coated webbing which is superior to nylon and leather in many ways. We use polyurethane or vinyl to coat narrow webbing resulting in a cleanable and strong belt with a consistent tensile strength."

Birdfeeding.com 203810

http://www.birdfeeding.com/

Forget the birds. Look for the large acrylic hemispheres. Use 'em to build dome bodies for your robots. See the listing for Aspects, Inc. (this section) for company address and phone numbers.

Cal Plastics and Metals 202840

2540 Main St.
Chula Vista, CA 91911
USA

☎ (619) 575-4633

📠 (619) 575-4561

✉ sales@calplasticsandmetals.com

🌐 http://www.calplasticsandmetals.com/

Plastics and metals. Among the products in the plastics category:

- ABS, Cycolac
- Acetal, Delrin
- Acrylic, Plexiglas
- Tubing
- Coroplast
- Nylon
- Phenolic, Micarta, Textolite
- Polycarbonate, Lexan
- Polypropylene, polyurethane
- PVC-CPVC, vinyl, polyvinyl chloride
- Weld-on adhesive

Web site for Cal Plastics and Metals.

EL-COM 203827

12691 Monarch St.
Garden Grove, CA 92841
USA

☎ (714) 230-6200

📠 (714) 230-6222

🚫 (800) 228-9122

🌐 http://www.elcomhardware.com/

Fasteners and hardware, mainly for cabinetry. Also casters, aluminum extrusions (squares, channels, bars), plastic laminates, and foam products.

GoldenWest Manufacturing 203237

P.O. Box 1148
Cedar Ridge, CA 95924
USA

☎ (530) 272-1133

📠 (530) 272-1070

🌐 http://www.goldenwestmfg.com/

Manufacturer of casting resins (both rigid and flexible), machinable plastic, foam board, and tools for cast and mold making. Their machinable plastic includes a product known as Butter-Board, a lightweight plastic block that is nonabrasive and very easy to machine or work with hand tools.

Laird Plastics

202485

1400 Centrepark, Ste. 500
West Palm Beach, FL 33401
USA

📞 (561) 684-7000
📠 (561) 684-7088
🚫 (800) 610-1016
✉️ feedback@lairdplastics.com
🌐 http://www.lairdplastics.com/

Sells plastics (sheets, rods, tubes, films) and related plastic products.

Multi-Craft Plastics, Inc.

202457

240 N. Broadway
Portland, OR 97227-1874
USA

📞 (503) 288-5131
📠 (503) 282-5696
🚫 (800) 488-9030
✉️ sales@multicraftplastics.com
🌐 http://www.multicraftplastics.com/

Plastics of all kinds, including sheet, tube, rod, and profiles (profiles are fancy extruded shapes). The company also makes plastic hemispheres of most any size. Retail stores in Portland and Eugene, Ore.

Plastic Products, Inc.

203319

P.O. Box 188
Bessemer City, NC 28016
USA

📞 (704) 739-7463
📠 (704) 739-5566
🚫 (800) 752-7770
🌐 http://www.plastic-products.com/

Plastic Products offers sheets, rods, tubes, profiles, shapes, slabs, and "massive blocks," thick hunks of plastic suitable for machining or making doorstops. Their product line is intended for plastic molding, machin-

ing, and fabrication and is well suited for the robotics trade. The company also offers a full line of stock materials, such as foamed PVC rod and conveyor components (sprockets, cams, raceways, and other goodies). Check the Steals and Deals page.

Plastic Products Web site.

Plastic Specialties Inc.

202322

10630 Marina Dr.
Olive Branch, MS 38654-3712
USA

📞 (662) 895-8777
📠 (662) 895-8796
🚫 (866) 638-7926
✉️ info@psilighting.com
🌐 http://www.psilighting.com/

Plastic domes and other shapes. Their plastic (white) domes are available in three basic configurations: flanged, flangeless, and step domes. Sold in packs.

Plastic World

203176

1140 Sheppard Ave. West, Unit 8
Downsview, ON
M3K 2A6
Canada

📞 (416) 630-6745
📠 (416) 630-9272
🌐 http://www.plasticworld.ca/

Plastics and plastic casting. Offers cut-to-order plastic, fiberglass cloth and mat, Devcon products (putties), and other casting supplies.

PlasticsNet 203320

http://www.plasticsnet.com/

Portal for plastics. Mostly for industry. According to the Web site, "Find products and suppliers, read current headlines, and keep up with the latest information in your professional community through these featured resources."

Plasticsusa.com 204088

http://www.plasticsusa.com/

Links, classified ads, sources, user-to-user forums, and technical information about plastics and plastic-making processes.

See also http://www.polymerweb.com/

Decoding Plastic ID Symbols

You've seen them on the bottom of bottles and canisters. They're the funny circular arrow thingamajigs with numbers inside. You know this already, but they're ID symbols, which are designed to help consumers and recyclers tell the difference between various plastics.

Well, if you're building robots out of junk you find around the house, it can be darned useful to know what kind of plastic you are using, in case you want to join two or more pieces with some solvent cement. Why is this important? Because different plastics require different kinds of solvent cements. A solvent for one plastic may do absolutely nothing for another.

Here's what the ID symbols mean.

PETE—polyethylene terephthalate

HDPE—high-density polyethylene

PVC—polyvinyl chloride

LDPE—low-density polyethylene

PP—polypropylene

PS—polystyrene

A combination of plastics, or none of the above

Plasti-kote   203169

1000 Lake Rd.
Medina, OH 44256
USA

📞   (330) 725-4511
📠   (330) 723-3674
🚫   (800) 431-5928
🌐   http://www.plasti-kote.com/

Plasti-kote makes a variety of spray coatings like paints, but more-such as Color Fleck, Epoxy Appliance Paint, Classic Metals, and Cracklin. These coatings can be used to add unusual finishes to your robotic creations. Available in paint departments of home improvement stores.

Plastruct, Inc. 🎖   202416

1020 South Wallace Pl.
City of Industry, CA 91748
USA

📞   (626) 912-7016
📠   (626) 965-2036
🚫   (800) 666-7015
✉   Plastruct@Plastruct.com
🌐   http://www.plastruct.com/

Plastruct is the world's leading supplier of plastic scale model parts. Of their product, their structural shapes, tubing, sheet, and patterned sheet materials are of keen interest to robot builders.

Polymorf, Inc.   204096

11500 NE 76th St.
#A-3, #309
Vancouver, WA 98682
USA

📞   (360) 449-3024
✉   morfun@polymorf.net
🌐   http://www.polymorf.net/

See listing under Toys-Construction.

Port Plastics   203316

16750 Chestnut St.
City of Industry, CA 91747
USA

📞   (626) 333-7678
📠   (626) 336-3780
🚫   (800) 800-0039
🌐   http://www.portplastics.com/

Distributor of engineering, high-performance, and commodity plastics. Lots of technical info.

PTG/Patios To Go   203315

307 N. Highway 27
Clermont, FL 34711
USA

📞   (352) 243-3220
📠   (352) 243-3221
✉   jsoakes@mindspring.com
🌐   http://www.patiostogo.com/

PVC pipe and fittings for furniture; use for robot frames. Also provides hands-on (at their local store) instruction on using PVC to build furniture. Sells in quantity.

PVC Store, The   203487

P.O. Box 924
Huntsville, AL 35804
USA

📞   (256) 859-4957
📠   (256) 851-7723
✉   Info@ThePVCStore.com
🌐   http://www.thepvcstore.com/

Distributors of PVC pipe and fittings, including furniture grade. Sells to local hardware stores, but the site provides some background information and product listings. There's also a store locator.

PVC pipe and fitting can be used to construct robot frames and bodies.

The Joys of Rigid Expanded PVC

What (sort of) looks like wood, drills like wood, cuts like wood, and sands like wood—but isn't wood? Expanded polyvinyl chloride (PVC). This material, commonly used for both sign making and construction, is manufactured by mixing an inert gas with the molten plastic. The plastic is then extruded into various shapes: sheets, rods, tubes, bars, and more. The gas (usually nitrogen or carbon dioxide) forms tiny microscopic bubbles in the plastic, which makes the material bulkier when it cools.

Foamed (expanded) PVC comes in a variety of thicknesses and colors.

This means foamed PVC contains less plastic than ordinary PVC materials. This has two advantages:

- Less plastic = less weight. That's important in building robots for which added weight drains the battery faster.
- Less plastic = less density. This makes foamed PVC easier to drill, cut, and mill. If you've ever cut acrylic plastic, you know it chips and breaks easily, and its high density makes using hand tools a real chore. The thinner, foamed PVC materials can be cut using a knife; the thicker stuff, with an ordinary saw blade.

PVC Sheets for Signage

Foamed PVC is often used as a replacement for wood in some outdoor applications, typically where moisture and rot are a problem. The material is formed into familiar wood-like shapes, and some even has a wood grain. As robot builders, we're not too interested in this stuff. Rather, we're after the foamed PVC sheets used to make signs (sign makers refer to this raw material as *substrate*). It's available in a variety of sizes and thicknesses, in a rainbow of colors: blue, red, orange, tan, black, brown, yellow . . . you name it.

Foamed PVC goes by many trade names, such as Sintra, Celtec, Komatex, Trovicel, Versacel, and Komcel, but it's probably easiest if you just ask for it by its generic "foamed PVC" moniker. Sheets are commonly available in any of several "mil" sizes (a *mil* is one millimeter). Of the more common thicknesses:

- 3 mil, or roughly 1/8 inch
- 6 mil, or roughly 1/4 inch
- 10 mil, or roughly 13/32 inch

Not all colors are available in all thicknesses. I've found the thicker pieces are available in basic black or white only, with colors a special-order item (often requiring large volumes).

The same cements and adhesives for PVC irrigation pipes can be used with foamed PVC. I like using a fairly thin cement, such as Weld-On 66. It has a medium consistency and is clear. It can be brushed on or applied using a suitable needle applicator (you can get cement brushes and applicators at the same place you buy the PVC plastic). You can also use a specialty adhesive, such as Foamex contact adhesive, for fast production assembly.

PVC cements bond by partially "melting" the plastic pieces and are recommended over traditional adhesives, including hot glue and epoxies. For best results, rough up the pieces to be cemented by lightly sanding the surfaces. Everything has to be squeaky clean for a good bond. You can clean PVC using ordinary acetone.

If you're assembling large quantities of PVC parts, consider the investment of a thermoplastic welder. These welders apply concentrated heat that melts the plastic pieces together. Get a model that has several temperature settings, as you'll need to experiment with the proper heat level for a strong, consistent weld. Thermoplastic welding kits for everything but heavy-duty assembly are available in the $200 to $500 range; tips wear out over time, and replacements cost $25 to $50.

Though foamed PVC can be "worked" just like wood, if you use power tools, adjust the speed of the drill, saw, or sander so that it's slower than you'd use for wood. This reduces heat from friction; the heat can cause the plastic to melt, and it makes for a messy cut. If the drill or blade is thin, the melted plastic may recongeal, and you'll have to recut it. Almost all of the better saws and drills, and many bench sanders, allow for adjusting speed.

Pioneer Supply Co.

http://www.pioneersupply.com/

Supplies, materials, and tools for sign makers and screen printers

---

Regal Plastics   202674

5265 South Rio Grande St.
Littleton, CO 80120
USA

☎   (303) 794-9800
🖶   (303) 794-0126
⊘   (800) 777-7342
🌐   http://www.regalplastics.com/

Plastics distributor-look particularly at their industrial engineering plastics, which include ABS, PVC, acetal, laminates, acrylic, and many others. Excellent collection of technical documentation on various plastics. Structural plastics include ABS in sheet, rod, and tube;

Web site for Regal Plastics.

PVC in sheet, rod, tube, bar, and rolls (including clear PVC); and Kydex thermoformable sheet. Retail stores in Southwest U.S. and Mexico.

Ridout Plastics   204139

535 Ruffin Rd.
San Diego, CA 92123
USA

☎   (858) 560-1551
🖶   (858) 560-1941
⊘   (800) 474-3688
✉   info@ridoutplastics.com
🌐   http://www.ecomplastics.com/

Industrial plastics distributor and fabricator. Products include acrylics, ABS, and PVC in sheet, rod, and other forms. Located in San Diego, but will ship. Check out their Plexiglas Primer for information about Plexiglas and other plastics.

Sam Schwartz Inc.   202677

932-40 Hunting Park Ave.
Philadelphia, PA 19124
USA

☎   (215) 744-9996
⊘   (800) 440-9996
🌐   http://www.samschwartzinc.com/

Retailer of metals, plastics, and substrates: aluminum and other metals, Showcard, Sign Foam, Foamcore, Gatorfoam, Coroplast, Sintra (PVC), signboards, styrene, and others.

San Diego Plastics, Inc. 202841

2220 McKinley Avenue
National City, CA 91950
USA

((619) 477-4855

(619) 477-4874

(800) 925-4855

http://www.sdplastics.com/

Local (San Diego) plastics supply. Good selection, good prices.

Savko Plastic Pipe & Fittings, Inc. 203486

683 East Lincoln Ave.
Columbus, OH 43229
USA

Foamed PVC—The Devil's in the Details

Foamed (or expanded) PVC is one of the best materials for constructing strong robot bodies. It's relatively cheap, lightweight, and easy to work with. The material is available in sheet form, as well as tubes, squares, rods, and other shapes. Here are some details about working with the stuff.

- While foamed PVC can be sawed and drilled like wood, the feed rate should be slower. This prevents gumming up the drill or saw.
- During cutting or drilling, cool the cut with a blast of compressed air. A flexible hose can be set up on your saw or drill press for this. The air also removes chips that can gum up the saw or drill.
- You can use most any high-speed cutting tools with foamed PVC. Band saws, circular saws, and panel saws are good choices. I use a radial arm saw, equipped with a carbide-tipped combination blade, to cut out pieces of PVC sheets.
- When using a reciprocating saw, blades with a tooth pitch of 0.080-inch to 0.32-inch give best results. For circular saws, blades with a minimum of 80 teeth are recommended.
- Wood, metal, or plastic drills can be used. If using milling tools, ground them the same as you would when using other plastics. Drills should have a 30-degree angle of twist in order to facilitate chip removal (or else the plastic may remelt into the cut).
- For faster cutting, scoring, punching, or perforating, you can warm PVC to 85 to 105 degrees (an incandescent or infrared heat lamp is a good choice).
- Deep groves or notches should be avoided, as this reduces the strength of the material.
- Solvent-based materials containing MEK (methyl ethyl ketone) or THF (tetra hyrdo furan) are often used to join PVC pieces together. Both of these chemicals are highly toxic, however. Treat them with respect. These chemicals are the most common solvent ingredients found in PVC cement sold at hardware and home improvement stores.
- You can either apply the solvent with a narrow artist's brush, or better yet, use a needlepoint applicator. For the latter, you need low-viscosity solvent, such as WELD-ON 2007.
- Clean all surfaces to be joined with isopropyl alcohol. Acetone can be used to remove sticky residue or inks from the plastic, but apply sparingly, as the acetone partially melts the plastic.
- Bear in mind that adhesives cannot fill gaps in the materials. Strive for straight, even cuts.
- Foamed PVC is thermoformable, so it can be heated and bent into shapes. Use a heat gun, heat strip, or other apparatus approved for use with bending plastics. Never, ever use an open flame to heat PVC, or any other plastic. Though foamed PVC is flame retardant, it can catch on fire; if it does, it releases toxic gas, and the molten plastic can drip and cause very serious burns.

(614) 885-4445

(614) 885-4470

(877) 885-4445

http://www.savko.com/

Manufacturers and suppliers of plastic plumbing pipe. Offers furniture-grade PVC pipe fittings, and will do mail order. You can use PCV pipe for robot frames, risers, battle rinks-you name it. You can get it in thicknesses less than standard irrigation PVC pipe, which saves weight.

Specialty Resources Company
203649

1156 W. 7th St.
Ste. B
San Pedro, CA 90731-2930
USA

(310) 514-1581

---

(310) 514-2997

rocky@aliendecor.com

http://www.aliendecor.com/

Weird stuff, like alien mannequins (or maybe they're real aliens sold as fake?), but also clear plastic hemispheres of various sizes. Sizes from 5 inches to 96 inches with colors and hex patterns available.

TAP Plastics
202458

6475 Sierra Ln.
Dublin, CA 94568
USA

(925) 829-6921

(800) 894-0827

info@tapplastics.com

http://www.tapplastics.com/

The Case Against PVC

Plumbers and robot builders love PVC (polyvinyl chloride). According to industry sources, it's the second most commonly used plastic in the world. And to some, it's also a major polluter.

The international Greenpeace organization contends that PVC is one of the world's largest sources of dioxin. The dioxin is released during manufacture of the plastic, but also when the plastic is burned. Unlike many other plastics, most PVC products are not "recycled" (in the traditional sense), and more often than not are simply disposed of in landfills or incinerated. By the way, if you can find a Recycling Plastic ID symbol on the product, you'll know it's made of PVC if it says "3."

Does this mean you should not use PVC in your robotic creations? That's up to you. There are disadvantages to most any substrate or building material you may use. A metal or wood base may be no less environmentally unfriendly; toxins are released during the manufacture of all metals, and the hardwoods in high-quality plywood may come from rain forests that are not being replanted. What's a robot builder to do?

First and foremost, use your materials wisely. Don't waste. Not only does this save you money, it helps reduce scrap sent to the landfill. Design your robot so that you can use standard-size materials with little or no excess. Except for shavings or very small pieces, save all your discards. You can use them for braces and cross pieces.

Second, some kinds of PVC material are more air than PVC. So-called foamed PVC (sometimes called expanded or cellular PVC) is made by injecting air or other inert gas into the slurry, so that the plastic is "bulked up." This reduces the weight of a given piece, and less PVC material is used.

Finally, when possible, raid the scrap and salvage bins at your local plastics store or sign maker store. Just because a sheet of plastic has been previously used doesn't mean it can't be cut up and used to build your next robot. You'll save some money, and you'll know you're doing your part to clean up the earth.

Plastics retailer, with stores in northern California, Oregon, and Washington.

TAP Plastics provides fiberglass, plastic, and signage products. They also carry resins and fiberglass mat, mold-making supplies, casting and sculpting products, fiberglass composite adhesives, sealants, tools, and plastics forming heater strips.

An assortment of colored plastic rod. Photo TAP Plastics.Inc.

Clear plastic pipe. Photo TAP Plastics Inc.

United States Plastic Corp.   202610

1390 Neubrecht Rd.
Lima, OH 45801-3196
USA

  (419) 228-2242
  (419) 228-5034
  (800) 809-4217
  usp@usplastic.com

Celastic—Plastic-Impregnated Fabric

I've always envied prop makers for movies and theater. They get to play with the neatest stuff. Case in point is Celastic, a woven fabric that is impregnated with polymer (plastic). Out of the box, the fabric is a little stiff. It is "activated" by drenching it in acetone, which effectively melts the plastic.

After soaking in acetone, the fabric becomes quite flexible and can be molded to most any shape. As the acetone evaporates, the plastic in the cloth recongeals, and the fabric hardens to a semistiff mat. The fabric retains its shape enough to apply latex, fiberglass, urethane resin, or urethane foam, if you want a more robust structure.

You can use Celastic to create an unusual robot body. If you apply urethane foam to the "dried" Celastic, you can sand down the contours and even paint it. Celastic is sold in 48- to 50-inch-wide strips, usually by the yard. The acetone to activate the material can be purchased in the paint department at any hardware or home improvement store. Do remember to use Celastic with proper ventilation only, and keep the acetone—and the drying Celastic cloth—away from open flames.

Celastic is available from most casting and mold-making supply outlets, taxidermy retailers, and specialty retailers that deal with puppet-making goods. Online retailers of Celastic include:

Douglas and Sturgess, Inc.

http://www.artstuf.com/

Freeman Manufacturing & Supply

http://www.freemansupply.com/

Kindt-Collins Company

http://www.kindt-collins.com/

http://www.usplastic.com/

U.S. Plastics sells-well, plastics. Things like pipe, sheets, jugs, buckets, and tubing. Most of it has only marginal use in robotics, except their line of White Pipe Fittings. This is furniture-grade PVC that can be joined together to make all varieties of shapes. The pipe and fittings are available in sizes from 1 inch to 1 1/2 inch.

The company's printed catalog lists over 14,000 plastic products.

Urethane Supply Company 203819

1128 Kirk Rd.
Rainsville, AL 35986
USA

((256) 638-4103
(256) 638-8490
(800) 633-3047
info@urethanesupply.com
http://www.urethanesupply.com/

Tips for Drilling and Cutting Acrylic

All things considered, acrylic plastic isn't the best material for building robots, but it's relatively cheap and can be found everywhere. You're better off using ABS or PVC plastic (or a mix of the two). They're easier to cut and drill than acrylic, and they're lighter per square inch, too.

But, ABS and PVC plastic sheets can be hard to find. Sometimes, we have to use what we can get, so if you're stuck building a robot out of acrylic plastic, here are 10 tips for drilling and cutting it into the shape you need.

1. First and foremost, *always wear protective goggles or safety glasses*. Acrylic can shatter when cutting or drilling it, and pieces may fly off and lodge into your eye. Use caution—plastic can be nasty!

2. *Avoid using wood drill bits*. Drill bits designed for plastic or glass drilling yield better, safer results. You probably only need one or two plastic bits; 1/8-inch and 3/16-inch sizes should suffice for most work.

3. *Use a power drill* and not a hand drill. You get best results by using a medium drill speed; about 750 to 1,000 rpm. (For wood, you want a fast speed; for plastic a medium speed; and for metal, a slow speed).

4. When drilling, always *back the plastic with a wooden block* or a piece of masking tape. Without the block or tape, the plastic is almost guaranteed to crack

5. Holes larger than 1/4-inch should first be made by drilling a smaller *pilot hole*. Start with a small drill and work your way up several steps.

6. *Use a radial arm or table saw for straight cuts*. Though there are specialty blades for cutting plastics, they are expensive. A regular combination blade is acceptable for all but the thinnest of acrylic sheets.

7. *Use only a sharp blade.*

8. *Avoid fast feed rates, and don't force the material into the blade*. When cutting (and drilling, for that matter), the plastic tends to remelt at the cut, and pushing the work too fast can generate lots of friction.

9. For rounded cuts, *use a saber saw or scroll saw* (the latter is my preference). Or, if you don't have to cut much, a manual coping saw. Use a slow speed to avoid remelting the cut plastic due to friction.

10. Finish the cut ends and sharp corners of the plastic with a *metal file*. If you have the equipment, you can also *burnish the edges* with plastic rouge and a buffing wheel. The experts use flame burnishing to obtain smooth edges, but I don't recommend it if you're not skilled in the art.

Hot and Cold Dipping Plastics

Dipping plastics are like fondue with polymers. Perhaps the best-known application for dipping plastics is applying a kind of rubberized grip to hand tools—just dip the handle of the tool into a can of dip, pull out after a few seconds, and let dry. Most dip can also be applied by brushing, but the idea is the same.

Dipping plastic is available in dip or brush-on cans.

Dipping plastics harden to a tough shell or a rubbery sheath, depending on the chemistry used. There are two basic flavors: cold and hot. Cold dip is not heated before use, and it cures with exposure to air. Hot dip is cured by warming it in an electric oven as per the manufacturer's directions (typically 350 to 375 degrees for 15 or 20 minutes). The plastic completes its curing process when cooled.

For robotics, dipping plastic can be used to add a rubberized soft coating to the legs of walking bots, to frame bumpers to reduce damage, to add traction to wheels, and even to provide extra insulation for wire connectors.

Here are some tips when using dipping plastics:

- Curing times for cold dip can be reduced by gently warming with a hair dryer.

- Apply several thin coats rather than one thick coat.

- Hot dipping is best when the object being dipped is preheated. Obviously, this won't work if you're using plastic or wood. It works well for metal objects, like the aluminum legs of a walking robot.

The most popular dipping material is plastisol, a PVC compound that can be made in most any color, thickness, and hardness (when cured). In addition to being used as a heavy coating, plastisol serves as a semiflexible mold.

Plastic welders and supplies; mainly for automotive and boat repair. Good technical information, regular newsletter in Adobe Acrobat PDF format.

Westward Plastics 203321

Unit 19, Cater Road
Bishopsworth
BS13 7TW
UK

(+44 (0) 1179 358058

📠 +44 (0) 1179 784114

🌐 http://www.westwardplastics.co.uk/

Check out their engineering plastics, which include acetal copolymer, ABS, nylon, PVC, polycarbonate, and many others. Retail stores in Bishopsworth, Wentloog, and Marsh Barton.

World of Plastics, Inc. 203318

110 Taylor Industrial Blvd.
Hendersonville, TN 37075
USA

((615) 244-4949

📠 (615) 822-8062

🚫 (800) 866-8660

✉ sales@worldofplastics.net

🌐 http://www.worldofplastics.net/

Plastics: supplies and fabricator. Retail store in Hendersonville, Tenn.

 MATERIALS-STORE FIXTURES

I know it's weird, but bear with me: Retail store fixtures can be used to build robots. Honest! Store fixtures include:

- 1- to 2-inch diameter lightweight PVC and aluminum tubing, used to create displays and racks. Along with the tubing are various types and styles of connectors: 2-, 3-, 4-, 5-, even 6-way connectors-all of which can be used to build cheap (as in inexpensive) robot frames.

- Ball bearing lazy Susan turntables, intended to make rotating displays, but useful for any rotational movement in a robot. Sizes from 3 to 12 inches in diameter and even larger.

- Slat wall is used to mount peg bars for holding merchandise. This material is plastic with T-shaped slots grooved into it at even intervals. With simple hardware, you can use slat wall as a reconfigurable robot base, no drilling required.

The list goes on. Peruse the online and printed catalogs of these vendors for more ideas.

SEE ALSO:

MACHINE FRAMING: Professional-level aluminum and plastic construction materials

MATERIALS-METAL AND MATERIALS-PLASTIC

Alpha Store Fixtures, Inc. 203858

6808 Oporto Madrid Blvd.
Birmingham, AL 35261
USA

- (205) 833-8700
- (800) 451-6127
- CustomerService@storefixtures2000.com
- http://www.storefixtures2000.com/

Store fixtures: slat wall, body forms, acrylic plastic bins and trays.

Display Warehouse 203833

8820 Kenamar Dr.
San Diego, CA 92121
USA

- (858) 271-0492
- (858) 271-1999
- (800) 842-5501
- design@displaywarehouse.com
- http://www.displaywarehouse.com/

Store fixture components: pipe, clamps, slat board, etc.

Tebo Store Fixtures 203859

5771 Logan St.
Denver, CO 80216
USA

- (303) 292-2446
- (303) 292-2410
- (800) 525-2646
- http://tebostorefixtures.com/

A place to buy store fixtures. What you're looking for here are things like tubing, slat wall, and other building materials for your robots.

WR Display & Packaging 203824

30 Plymouth St.
Winnipeg, Manitoba
R2X 2V7
Canada

- (204) 925-7900
- (204) 925-7910
- (800) 665-8447
- info@wrdisplay.ca
- http://www.wrdisplay.ca/

Store supplies and display fixtures. Some interesting product include:

- Jewelry boxes-they make great holders for small parts
- Slat wall and accessories-possible use in making robot bases
- Display hardware-for making robot frames

▣ MATERIALS-TRANSFER FILM

Listings in this section are for various laser and ink-jet printer transfer films. These films are used to make color decals, printed circuit board layouts, signs, control panels, and more.

SEE ALSO:

RETAIL-ARTS & CRAFTS: Additional transfer film sources

AutoGraphics of California 202129

4609 #3 New Horizon Blvd.
Bakersfield, CA 93313
USA

(661) 836-2886

(661) 836-0938

autogfx1@aol.com

http://www.autographics-decals.com/

AutoGraphics of California makes decals for hobby-related products. Decals include numbers, signs, shapes, and designs (flames, cartoon characters, racing stripes, lightning bolts, stars, and more). Custom decals are also available.

Bel Inc.   203962

6080 NW 84th Ave.
Miami, FL 33166
USA

(305) 593-0911

(305) 593-1011

beldecal@bellsouth.net

http://www.beldecal.com/

Bel markets an extensive line of custom decals, including ink-jet decals for doing it yourself. The ink-jet (also works with many laser printers) decal film is the water slip type; after printing, place in water for a short period of time, then transfer to a flat, clean surface-wood (sanded, closed-grained are best), plastic, or metal. They even sell a "tattoo paper" that lets you make temporary tattoos of your rad robots.

Web site in English, German, French, Portuguese, and Spanish. See also:

http://www.waterslide-decals.com/

Web page for Bel Inc.

Dyna Art   203789

1947 Sandalwood Pl.
Clearwater, FL 33760-1713
USA

(727) 524-1500

(727) 524-1225

mail@dynaart.com

http://www.dynaart.com/

Makes and sells a PCB transfer system; uses laminator-like machine to fuse artwork (prepared by laser printer) onto copper clad.

HPS Papilio   203985

P.O. Box 855
Rhome, TX 76078-0855
USA

817-489-5249

Sales@papilio.com

http://www.papilio.com/

Manufacturers of water-slide decals for laser and ink-jet printers, as well as specialty laser and ink-jet papers and coatings, including adhesive-backed ink-jet vinyl, printable magnetic media, temporary tattoo papers, and self-adhesive window film.

Image Solutions   203959

108 North Harris St.
Sandersville, GA 31082
USA

(478) 553-0134

(478) 552-4747

(877) 544-5270

Info@BestImageSolutions.com

http://bestimagesolutions.com/

Image Solutions distributes heat-transfer (or digital-transfer) supplies, vinyl media, water-slide decal media, and bumper sticker media. Can be used to customize robots, create labels and control panels, etc.

Lazertran Limited   203957

8 Alban Square
Aberaeron, Ceredigion
SA46 OAD
UK

+44 (0) 1545 571149

+44 (0) 1545 571187

✉ mic@lazertran.com

🌐 http://www.lazertran.com/

Lazertan is a thin paper that is used to transfer full-color images to flat surfaces, such as wood, tile, metal, or plastic. You first print the image onto the paper using a color photocopier. The paper is then left to soak in water, like a decal, until its emulsion is soft. The emulsion is then transferred to the surface you want to cover. You can use Lazertran in lieu of painting or as labeling for a control panel.

Available by mail order (including local distribution offices in the U.S., U.K., Canada, and Australia) or through many arts and crafts retailers.

McGonigal Paper & Graphics

203961

P.O. Box 134
Spinnerstown, PA 18968-0134
USA

☎ (215) 679-8163

📠 (215) 679-8163

✉ help@mcgpaper.com

🌐 http://www.mcgpaper.com/

Online retailers of specialty papers for arts and crafts, including:

- Water-slide decals
- Glow-in-the-dark transfer film
- Super Color Shrink (shrinks when heated)
- Backlight film (fluoresces under ultraviolet light)
- Window-cling decals (clings to glass and other very smooth surfaces)

SuperCal Decals

203960

Micro Format Inc.
830-3 Seton Ct.
Wheeling, IL 60090
USA

☎ (847) 520-4699

📠 (847) 520-0197

🚫 (800) 333-0549

🌐 http://www.supercaldecals.com/

SuperCal makes and markets a line of water-transfer decal sheets that can be printed on an ink-jet printer. Print your design, soak in water, and transfer the design to plastic, metal, wood, and many other nonporous surfaces. Online sales available, but the transfer sheets are available at many hobby stores.

Lazertran Water-Slide Color Transfer Decals

With Lazertran, you can copy your ideas from your computer directly to your robot. Lazertran is a water-slide color decal transfer paper, designed for art school students but used by most anyone with an interest of transferring a color-copied image to all kinds of surfaces. Full-color art can be reproduced onto paper, canvas, metal, plastic, and wood.

You can use Lazertran transfer papers to emboss a design on your robot's base or create the lettering and artwork for its control panels. Though intended to create large pieces of art, you are certainly free to use the material to add small spots of color to your robot. How about your name, school, or team mascot in the corner of your battling bot?

Most transfer papers require the use of heat, pressure, or chemicals. Lazertran uses water as a transfer medium. After printing using one of the recommended copiers, the Lazertran sheet is immersed in a shallow pan of warm water.

Lazertran water-slide decal film.

The sheet is then positioned and the backing paper is removed, leaving just the color image. After drying, the surface must be "fixed" using a spray protective coating. Spray coatings with satin or gloss finishes are available at art supply stores.

According to the makers of Lazertran, the material has been used successfully with the following makes of photocopiers: Minolta, Canon, Ricoh, and Xerox. Hewlett-Packard printers, and in fact all laser printers (despite the name of the product), are not recommended.

Learn more about **Lazertran** at:

http://www.lazertran.com/

You can purchase Lazertran transfer sheets directly from the manufacturer or from online retailers. Be sure to read the detailed instructions on using the material to save you time and expense.

Mirror Image Multisurface Transfer Paper

You can print on most anything with Mirror Image transfer paper. Use a color copier or printer to create the artwork. Then, using a heat press or a clothes iron (the heat press is better), apply the transfer paper to the surface you want to use—it can be plastic, metal, wood, even glass. The image is transferred, in brilliant color, to the surface. A clear gloss or satin spray coating is recommended to "fix" the image so it won't smear or rub off.

Mirror Image Multisurface transfer paper is sold by Visual Communications at:

http://www.visual-color.com/

The transfer papers are available in sizes ranging from 8 1/2 by 11 inches to 11 by 17 inches. Papers are available for application to fabric or many kinds of solid surfaces. Cost is reasonable, starting at about 45 cents per sheet. Other products by the manufacturer include professional heat presses and water slide-off color decals.

SEE ALSO:

http://www.paper-paper.com/

T N Lawrence & Son Ltd. 202738

208 Portland Road
Hove
BN3 5QT
UK

(+44 (0) 1273 260260
℘ +44 (0) 1273 260270
✉ artbox@lawrence.co.uk
🌐 http://www.lawrence.co.uk/

Lawrence Art Materials offers online sales of arts and crafts supplies. See full listing in Retail-Arts & Crafts.

Visual Communications 203958

95 Morton St.
6th Fl.
New York, NY 10014
USA

((212) 741-5700
℘ (212) 741-5701
⊘ (800) 624-4210
✉ info@visual-color.com

Fun with Transfer Films

Thanks to a wide variety of transfer films, expressing your creativity and showing it to the world has never been easier. With just a laser printer, ink-jet printer, or copier, you can transfer images to many kinds of surfaces, including fabric, wood, metal, and plastic.

Among the most common transfer films are iron-on transfer for fabric, for making custom T-shirts. Elsewhere in this section you can learn about Lazertran, a water-slide decal transfer film that works with color copiers. Once printed, you can transfer the Lazertran image to a variety of hard surfaces (there is also a version for printing on fabric, to make framable art).

And yet there are still others, all driven by a robust appetite for such custom-made items as mugs, mouse pads, the aforementioned T-shirts, amateur sports trophies—you name it, there's probably a transfer film for it. Some require a special kind of printer, such as the Alps and Citizen Printiva, which use a dry-resin thermal transfer ink for the most vibrant colors. Others work with ordinary ink-jet and laser printers. The transfer method is usually either heat—the toner/ink transfers from film to surface using an iron or heat press—or by immersion in liquid, usually water.

Here are some additional resources for transfer films. Be sure to check out their offerings to see what they have available.

1 Source Mouse Pads
http://www.pilgrim-co.com/

ACP Technologies
http://www.acp.com/

Award Line
http://www.awardline.com/

Beacon Graphic Systems
(also sells aluminum sign blanks and other interesting substrates)
http://www.beacongraphics.com/

Google.com search phrases you can try:

sublimation transfer film

ink-jet "transfer film"

"thermal transfer" "transfer films"

http://www.visual-color.com/

Makers of Mirror Image iron-on transfer papers. Available for copiers, laser printers, and ink-jet printers. For color or black and white. Suitable substrates include many woods, metal, plastic, and fabric. Also sells presses, heat press stand, and transfer tools.

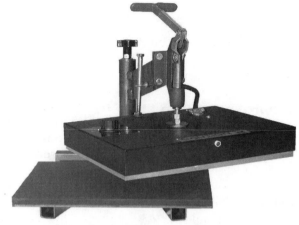

Compact heat press. Photo Visual Communications.

Making special license plates (adaptable to robotics) with heat transfer films. Photo Visual Communications.

Color Anything with Water-Slide Decals

If you've ever assembled a plastic model kit you know about water-slide decals: Dip the decal into warm water, wait a few seconds, then slide the clear decal emulsion off the backing paper and onto your model.

Modern marvels being what they are, you can print your own water-slide decals using a laser or ink-jet printer. The decal "paper" can accept black-and-white or color images. While you must print a full sheet at a time (though some printers will accept smaller sizes down to about 4 by 6 inches), you can cut out individual decals and apply the smallest to various parts of your robot.

You can use water-slide decals for making control panels or labels, and, of course, to dress up your robot with color or design. The decal will stick to most any smooth surface and should be "fixed" into place with a gloss or matte overspray. These chemicals are available at any hobby or art supply store and are expressly designed for use with decals.

Water-slide decal paper is but one subgroup of a larger family of printer transfer films, which also includes clear and translucent "sticky-back"—used primarily for overlays—and adhesive-backed opaque vinyl, often used for bumper stickers and signs. These materials share the common trait of compatibility with color copiers and laser and/or ink-jet printers, allowing you to prepare your own designs quickly and relatively cheaply.

Some of the suppliers of water-slide decal paper and similar products are:

SuperCal Decals
http://www.supercaldecals.com/

Bel Inc.
http://www.beldecal.com/

Lazertran
http://www.lazertran.com/

Papilio Supplies
http://www.papilio.com/

MICROCONTROLLERS

Microcontrollers helped bring amateur robotics into the mainstream. These single-chip computers are expressly designed to be used in so-called embedded applications, where control of some external device is the main goal. Microcontrollers can be connected to sensors and motors-input and output. Based on the robot's input, the internal programming in the microcontroller can command the motors in such a way that the robot exhibits quasi-intelligent functions. For instance, the 'bot might move toward a light or away from a barking dog. The possibilities are endless.

The following sections list numerous resources for microcontrollers. Included are manufacturers of complete board-level products with built-in robotic functions, coding examples, and software to aid in programming microcontrollers. The sections are:

- Hardware: Lists resources for microcontroller hardware, particularly ready-made boards that incorporate the microcontroller chip, voltage regulator, extra memory (when needed), and interface elements.
- Programming: Examples of programming microcontrollers, with an emphasis on code useful in robotics.
- Software: Add-in software for writing your own microcontroller programs.

SEE ALSO:

COMPUTERS-SINGLE BOARD COMPUTERS: Like microcontrollers, but typically more powerful

DISTRIBUTOR/WHOLESALER-INDUSTRIAL ELECTRONICS: Sources for microcontrollers, particularly in quantity

MANUFACTURERS-SEMICONDUCTORS: Lists several microcontroller chip makers

PROGRAMMING-EXAMPLES: Additional programming examples you may be interested in

RETAIL-GENERAL ELECTRONICS: More sources for buying microcontrollers

RETAIL-ROBOTICS SPECIALTY: See the listings for robotics-specific sensors that are readily interfaced to popular microcontrollers

MICROCONTROLLERS-HARDWARE

See the main Microcontrollers section for a description of the listings in this section.

Antratek Electronics   203452

Postbus 356, 2900
AJ Capelle A/D IJSSEL
The Netherlands
☎ 010 4504949
✂ 010 4514955
✉ info@antratek.nl
🌐 http://www.antratek.nl/

Distributor of microcontroller hardware and software: Basic Stamp, compilers, DSP starter kits, emulators, EPROM-emulators, I2C adapters, I2C bus monitors, IC testers, LCD/VFD modules, modem modules, programmers, serial displays, and starter kits.

The Web site is predominately in Dutch with some English.

SEE ALSO:

http://www.robotwinkel.nl/ for robot specialty products

ASA Micros   202339

21 Edmunds Fold
Littleborough
Lancs
OL15 9LS
UK

Brad Mock's Technical Works

http://www.technicalworks.com/

MockUp embeddable development boards for the Microchip, Ubicom (Scenix), and Parallax Basic Stamp

eMicros

http://www.emicros.com

Interface adapters for Controller Area Networks (CAN) and I2C

MVS

http://www.star.net/people/~mvs/

Embedded systems, single board computers, LCDs

RoboMinds

http://www.robominds.com/

Motorola-based microcontrollers for robotics

+44 (0) 1706 371695

+44 (0) 1706 375896

info@asamicros.com

http://www.asamicros.com/

Manufacturers and resellers of microcontrollers, including SimmStick, microEngineering Labs products, and PIC C compiler.

Athena Microsystem Solutions   202522

10624 Rockley Rd.
Houston, TX 77099
USA

(281) 418-5631

(281) 256-3851

info@athenamicrosystems.com

http://www.athenamicro.com/

Athena sells microcontrollers and single-board computer peripherals. Product highlights include:

- AMS-HE/DE-Precision Hall-effect DC current sensing module
- AMS876-SIMMStick plug-in module based on Microchip's PIC16F876 Flash memory microcontroller
- SLI-OEM-Serial LCD controller
- AMS-900PA/232-Spread spectrum wireless transceiver

AVR-based Robot Hardware   204115

http://homepage.ntlworld.com/seanellis/
    avrrobot_hw.htm

Circuit and construction details on:

- Processor module
- Eyes
- Motor driver
- RS232 buffer
- Programming
- Laying out strip board with Eagle

Axiom Manufacturing, Inc.   202891

2813 Industrial Ln.
Garland, TX 75041
USA

(972) 926-9303

(972) 926-6063

sales@axman.com

http://www.axman.com/

In the words of the Web site: "Axiom Manufacturing is a diverse microcontroller company specializing in single board computers, embedded controllers, custom design, and manufacturing solutions."

Products include single board computers based on the Motorola 68HC11 and 68HC12 microcontrollers, 80CXX microprocessor, MPC555 PowerPC, and MMC2001 Mcore microcontroller.

Basic Micro, Inc.   202087

34391 Plymouth Rd.
Livonia, MI 48152
USA

(734) 425-1744

(734) 425-1722

Info@basicmicro.com

http://www.basicmicro.com/

Basic Micro produces the MBasic line of compilers for PICmicro microcontrollers. Here's a listing of some of their products:

- Development boards
- Getting-started kits
- ISP-PRO programmer
- Atom and OEM Atom microcontroller
- MBasic compilers
- Education-Microcontrollers in the classroom
- Solderless development boards
- Prototyping boards
- ISP-PRO programmer

Let's examine the Atom-IC Module (available in 24-pin and other packages). This product looks a lot like a Basic Stamp, but has a number of hardware features that make it ideal for use in robots:

- SpMotor-Control stepper motors
- Servo-Easily control servo motors
- HPWM-Use the hardware PWM of a PICmicro
- Interrupts-Use interrupts with Basic
- ADin-Easily use internal A/D on PICmicro
- OWIN/OWOU-Use Dallas 1-wire
- XIN/XOU-Use X-10 devices

The 24-pin version of the Atom is pin-compatible with the Basic Stamp II, and the programming language used with the Atom is a superset of that found on the Basic Stamp II.

The MBasic compilers let you program a PIC microcontroller in Basic, compile the program to the PIC's native assembly language, and download that program to chip. Available in two versions, Basic and Pro; I recommend the Pro version, though it costs a little more.

The Basic Micro Atom family. Photo Basic Micro Inc.

BasicX ⚇   203981

10940 N. Stallard Pl.
Tucson, AZ 85737
USA

☎   (520) 544-4567
📠   (520) 544-0800
✉   sales@basicx.com

🌐 http://www.basicx.com/

The BasicX is a general-purpose microcontroller with a built-in programming language. You write programs on the PC using a Basic-like syntax, then download them, via a cable, to the BasicX. Unplug the cable, and the program now resident on the BasicX runs.

There are several flavors of the BasicX:

The BasicX-1 is a 40-pin chip that requires an external crystal and capacitors and voltage regulator (if one is not provided with your other circuitry). The BasicX-24 is a 24-pin chip that has everything on-board to run from a 6- or 9-volt battery. The pinout of the BX-24 is the same as the Basic Stamp II from Parallax.

Several advantages of the BasicX include its fast processing time, plus its ability to directly access most of the hardware registers on the microcontroller (both chips use Atmel AVR controllers). The company sells development boards to make it easier to prototype your projects, as well as a serial LCD module.

The product is available through distributors or directly from the Web site.

See also the following Web sites run by NetMedia; the company that produces the BasicX:

Microcontrollers Are Self-Contained Computers

A key benefit of microcontrollers is that they combine a microprocessor component with various input/output typically needed to interface with the real world. The typical microcontroller sports these features:

- Central processing unit (CPU)
- Hardware interrupts
- Built-in timer/counter
- Programmable full-duplex serial port
- Multiple I/O lines (typically from 8 to 32)
- RAM and ROM/EPROM in some models

Some microcontrollers will have greater or fewer input/output (I/O) lines, and not all have hardware interrupt inputs (an interrupt is a way for an external device to "get the attention" of the controller); some will have special-purpose I/O for such things as voltage comparison or analog-to-digital conversion.

Just as there is no one car that's perfect for everyone, the design of each microcontroller makes it more suitable for a given application than another. When selecting the microcontroller for your robot, first analyze the hardware requirements. Most microcontrollers are part of a family or line, with siblings that have fewer or more features. Check out the specifications of another controller in the same line to see if a variation provides just the right hardware for your robo needs.

The Basic Stamp Microcontroller

The Basic Stamp microcontroller.

Since its inception, the Basic Stamp has provided the "onboard brains" for countless robotics projects. This thumbprint-sized microcontroller uses Basic language commands for instructions and is popular among robot enthusiasts, electronics and computer science instructors, and even design engineers looking for an inexpensive alternative to microprocessor-based systems.

The Basic Stamp, which is manufactured by Parallax is really an off-the-shelf Microchip PICmicro microcontroller. Embedded in this PIC is a proprietary Basic-like language interpreter, called *PBasic*. The chip stores commands downloaded from a PC or other development environment. When the program is run, the language interpreter built inside the Stamp converts the instructions into code the chip can use.

As a result, the Basic Stamp acts like a programmable electronic circuit, with the added benefit of intelligent control—but *without* the complexity and circuitry overhead of a dedicated microprocessor. Instead of building a logic circuit out of numerous inverters, AND gates, flip-flops, and other hardware; you can use just the Basic Stamp module to provide the same functionality, doing everything in software.

The Basic Stamp uses two kinds of memory: PROM (programmable read-only memory) and RAM. The PROM memory is used to store the PBASIC interpreter; the RAM is used to store data while a PBASIC program is running. Housed in a separate chip (but still part of the Basic Stamp itself; see the description of the BSII module, later) is EEPROM memory for the programs you download from your computer.

In operation, your PBASIC program is written on a PC, then downloaded—via a serial connection—to the Basic Stamp. The program is stored in EEPROM.

The Basic Stamp is available directly from its manufacturer or from a variety of dealers world over. Prices from most sources are about the same. The Basic Stamp is available in several versions, including the older BSI, the ever-popular BSII, and new BSII-SX, 2p, and 2e. The Stamp comes stand-alone or is part of several kits:

- *BSII Module.* The Basic Stamp module contains the actual microcontroller chip, as well as other support circuitry. All are mounted on a small printed circuit board that is the same general shape as a 24-pin IC. In fact, the BSII is made to plug into a 24-pin IC socket. The BSII module contains the microcontroller, which holds the PBASIC interpreter, a 5-volt regulator, a resonator (required for the microcontroller), and a serial EEPROM chip.

- *BSII Starter Kit.* The starter kit is ideal for those just, well, starting out. It includes a BSII module, a carrier board, programming cable, power adapter, and software on CD-ROM. The carrier board has a 24-pin socket for the BSII module, a connector for the programming cable, power adapter jack, and a prototype area for designing your own interface circuitry.

- *Basic Stamp Activity Board.* The Activity Board, which is typically sold without a BSII module, offers a convenient way of experimenting with the Basic Stamp. It contains four LEDs, four switches, a modular jack for experimenting with X-10 remote-control modules, a speaker, and two sockets for easy interfacing of such things as serial analog-to-digital converters (ADCs).

- *Growbot and BOE Bot.* The Growbot and BOE Bot products are small mobile robot kits that are designed to use the Basic Stamp microcontroller. The robots are similar, with the BOE Bot a little larger and heavier and able to accommodate more experiments. A BSII module is generally not included with either robot kit.

- *Basic Stamp Bug II*. Another robot kit, the Basic Stamp Bug II is a six-legged walking robot. The Bug is meant to be controlled with a BSI microcontroller, though it could be refitted to use the BSII. The Basic Stamp module is extra.

The PBASIC language supports several dozen special functions, many of which are ideal for robotics. You'll want to study these statements more fully in the Basic Stamp manual, which is available for free download from Parallax and is also included in the Starter Kit as a printed book.

- *button*—The *button* statement momentarily checks the value of an input and then branches to another part of the program should the button be in a LOW (0) or HIGH (1) state. The *button* statement lets you choose the I/O pin to examine, the "target state" that you are looking for (either 0 or 1), and delay and rate parameters that can be used for such things as switch debouncing.
- *debug*—The Basic Stamp Editor (which runs on your PC) has a built-in terminal that displays the result of bytes sent from the Basic Stamp back to the PC. The *debug* statement "echos" numbers or text to the screen and is highly useful during testing.
- *freqout*—The *freqout* statement is used to generate tones primarily intended for audio reproduction. You can set the I/O pin, duration, and frequency (in Hertz) using the *freqout* statement. An interesting feature of *freqout* is that you can apply a second frequency, which intermixes with the first.
- *input*—The *input* statement makes the specified I/O pin an input. As an input, the value of the pin can be read in the program. Many of the special function statements, such as *button* and *pulsin*, automatically set an I/O pin as an input, so the *input* statement is not needed for these.
- *pause*—The *pause* statement is used to delay execution by a set amount of time. To use *pause* you specify the number of milliseconds (thousandths of a second) to wait. For example, *pause 1000* pauses for one second.
- *pulsin*—The *pulsin* statement measures the width of a single pulse, with a resolution of two microseconds (2 µs) on the Basic Stamp II.
- *pulsout*—Pulsout is the inverse of *pulsin*: with *pulsout* you can create a finely measured pulse, with a duration of between 2 µs (on the Basic Stamp II) and 131 milliseconds. The *pulsout* statement is ideal when you need to provide highly accurate waveforms.
- *rctime*—The *rctime* statement measures the time it takes for an RC (resistor/capacitor) network to discharge to an opposite logical state. The *rctime* statement is often used to indirectly measure the capacitance or resistance of a circuit, or simply as a kind of simplified analog-to-digital circuit.
- *serin* and *serout*—Serin and *serout* are used to send and receive asynchronous serial communications. It is one method of communicating with other devices, even other Basic Stamps, all connected together. You use *serout* to send commands and text to the LCD.
- *shiftin* and *shiftout*—The *serin* and *serout* statements (see previous) are used in one-wire asynchronous serial communications. The *shiftin* and *shiftout* statements are used in two- or three-wire synchronous serial communications. The main difference is that with *shiftin/shiftout* a separate pin is used for clocking the data between its source and destination.

Additional information on the Basic Stamp can be found at the following Parallax Web sites:

http://www.parallaxinc.com/

http://www.stampsinclass.com/

Also try these Google.com search phrases to locate program examples and circuit diagrams using the Basic Stamp:

"basic stamp" program examples

"basic stamp" programming examples

"basic stamp"—site:www.parallaxinc.com

http://www.siteplayer.com/

http://www.netmedia.com/

http://www.web-hobbies.com/

The BasicX-35 microcontroller and development board, from NetMedia. Photo NetMedia Inc.

Bill Ruehl   202061

http://www.robotdude.com/

Microcontroller info and projects:

- Hardware hacks
- Robot-building info
- Links

Chuck Hellebuyck Electronics   203057

Electronic Products
1775 Medler
Commerce, MI 48382

((248) 515-4264

((413) 825-0377

 http://www.elproducts.com/

Chuck's resells the Atom from Basic Micro, MBasic compiler software, ePic boards, and bootloader packages.

Crownhill Associates Ltd.   203650

32 Broad St.
Ely
Cambridgeshire
CB7 4PW
UK

(+44 (0) 1353 666709

(+44 (0) 1353 666710

(http://www.crownhill.co.uk

Resellers of PC microcontrollers, development boards, PIC compilers, and related hardware/software.

CSMicro Systems   203450

213 Sage St.
Ste. #3
Carson City, NV 89706
USA

((775) 887-0505

((775) 887-8973

((888) 820-9570

(CSMicroSystems@CSMicroSystems.com

 http://www.csmicrosystems.com/

Embedded systems reseller (Basic Stamp, BasicX, PIC micros). Also carries the CodeVision editor for PIC Basic, BasicX, and others and the EPIC PIC programmer modules.

Dontronics, Inc.   202041

P.O. Box 595
Tullamarine, 3043
Australia

☎ +61 3 9338 6286

✉ don@dontronics.com

🌐 http://www.dontronics.com

Dontronics specializes in microcontrollers, as well as the SimmStick prototyping development board system. Highlight products:

- DT007 Micro Motherboard
- DT104 Atmel Micro on a SimmStick
- DT107 SimmStick for 8051, 8252, AVR 8515, and AVR 28-pin Micros
- DT108 SimmStick Video
- DT205 Relay Board
- SIMM100 SimmStick compatible for the AT90S8535
- Gigatechnologies USB

There's lots more; Don's Web site is jammed with useful trinkets for robot builders and electronics experimenters. He ships worldwide.

SimmStick DT003, from Dontronics.

Dontronics: PIC List   204117

http://www.dontronics.com/piclinks.html

Don McKenzie's listing of useful PIC Web sites.

Elan Microelectronics Corp.   202027

No. 12, Innovation Rd. I
Science-Based Industrial Park
Hsinchu City
Taiwan

🌐 http://www.emc.com.tw/

Elan makes a number of microcontroller, interface, and specialty ICs. These include:

- 4-bit microcontroller, general purpose
- 4-bit microcontroller, with DTMF
- 4-bit microcontroller, for LCD
- 8-bit microcontroller, general purpose
- Mouse controller
- Keyboard encoder
- Analog-to-digital converter high-speed ADC

Data sheets and application notes provided on the Web site.

Elektronikladen Mikrocomputer GmbH   202326

Wilhelm-Mellies-Str
88, D-32758
DETMOLD
Germany

☎ +49 (0) 5232 8171

📠 +49 (0) 5232 86197

✉ detmold@elektronikladen.de

🌐 http://www.elektronikladen.de/

Microcontroller development boards, including:

- USB08 Starter Kit-MC68HC908 Evaluation Board and USB Reference Design
- HC08 Welcome Kit-Low-cost MC68HC908 Evaluation Board
- HC12 Welcome Kit-A Starter Kit with Motorola's 68HC812A4

Web site is in German and English.

Embedded Acquisition Systems   203059

c/o Kin Fong
2517 Cobden St.
Sterling Heights, MI 48310
USA

📠 (240) 266-4252

✉ sales@embeddedtronics.com

🌐 http://embeddedtronics.com/

Makers of MiniDaq, a small data acquisition module for the PC. Also offers the EAS Finger Board II; scaled-down Handy Board. The Web site includes pics of prototype robots the company has made using their products.

Embedded Systems Design Website
202264

http://www.microcontroller.com/

News, product announcements, tutorials, references, selection guides, and more for the serious microcontroller developer. You can search for information by microcontroller brand to help you zero in on the data you want.

Embedded Systems, Inc.
202164

11931 Hwy. 65 NE
Minneapolis, MN 55434
USA

📞 (763) 767-2748
📠 (763) 767-2817
🌐 http://www.embedsys.com/

Makers of low-cost development systems and add-ons for (among other things) Atmel AVR microcontrollers:

- AVR Sprint 2313 development system
- AVR Sprint 2313 Basic starter kit
- AVR Sprint 2313 microprocessor module
- LCD Display
- Sprint Basic (private label version of BASCOM AVR)

Eric's PIC Page
203067

http://www.brouhaha.com/~eric/pic/

Links, products, and information on using Microchip PICmicro microcontrollers.

Gleason Research
202648

P.O. Box 1494
Concord, MA 01742-1464
USA

📞 (636) 536-7179
📠 (978) 287-4170
🚫 (800) 265-7727
🌐 http://www.gleasonresearch.com/

Sellers of the MIT Handy Board and Handy Cricket single board computers. The Handy Board is a favorite at MIT and for many university and college robotics courses.

Handy Board, The
202936

http://www.handyboard.com/

The Handy Board uses a Motorola 68HC11 microcontroller to build a sophisticated robotics central brain. The Handy Board is used in many college and university robotics courses (it was originally developed at MIT) and is suitable for education, hobby, and industrial purposes. As the Web site says, "People use the Handy Board to run robot design courses and competitions at the university and high school level, build robots for fun, and control industrial devices."

The features of the Handy Board are:

- Motorola 68HC11 microcontroller
- 2-line LCD
- Integrated 700mAh ni-cad rechargeable battery (not included in some versions)
- 8 analog inputs
- 9 digital inputs
- Infrared output and input
- Start and stop buttons
- Piezo buzzer
- 32K battery-backed memory to store programs
- 4 1.1 amp H-bridge motor drivers (not included in some versions)
- Serial (RS-232) and SPI interfaces

The HandyBoard.com Web site is a facilitator of the Handy Board; MIT allows the board to be reproduced by anyone for noncommercial purposes. However, you may find it easier to get one already made, either assembled or in form. Vendors are listed on the Web site and include Gleason Research, The Robot Store in Hong Kong, and Acroname (all are listed in this book).

A variety of software choices are available for the Handy Board, but the most common recommended is Interactive C, a multitasking C language that allows for compiling programs, as well as line-at-a-time command execution. A free version of Interactive C is available for

Main informational Web site for the Handy Board.

the PC (running in a DOS window), Mac, and Unix-based computers.

A great deal of documentation, user-supplies programs, and other material exists to support the Handy Board. But one of the best is a book by the Handy Board's creator, Fred Martin. Check out *Robotic Explorations: A Hands-on Introduction to Engineering* (ISBN 0130895687). It is not an inexpensive book, but it does an excellent job of teaching robotic concepts. Though "controller agnostic," most of the examples revolve around the Handy Board.

High-TechGarage.com   202650

2615-1/2 Taylor Ave.
Racine, WI 53403
USA

☏ (413) 714-4523

✉ support@High-TechGarage.com

🌐 http://www.high-techgarage.com/

Basic Stamp enhancement products (timer, coprocessor). The company sells what they describe as "innovative enhancements for your BS2, BX-24 and ATOM." Datasheets (in Adobe Acrobat PDF format) available for all products.

HTH/High Tech Horizon   203066

Asbogatan 29 C
S-262 51 Angelholm
Sweden

☏ +46 (0) 431 410 088

☏ +46 (0) 431 410 088

✉ info@hth.com

🌐 http://www.hth.com/

Resellers of Basic Stamp, Atmel AVR, and Basic-X microcontrollers, as well as the BASCOM programming software. Web site is in Swedish and English.

iButton   204068

4401 South Beltwood Pkwy.
Dallas, TX 75244
USA

☏ (972) 371-4448

☏ (972) 371-6600

🌐 http://www.ibutton.com/

The iButton is a miniature special-function microcontroller made by Maxim/Dallas Semiconductor. Application includes nonvolatile memory, time and temperature, and personal code key (think "magic decoder ring"). Can be programmed in Java. Development samples can be ordered directly at the Web site.

Intec Automation, Inc.   203733

2751 Arbutus Rd.
Victoria, British Columbia
V8N 5X7
Canada

☏ (250) 721-5150

☏ (250) 721-4191

✉ info@steroidmicros.com

🌐 http://www.steroidmicros.com/

High-end microcontroller and microcontroller boards. Includes the SS555 ("Steroid Stomp"), a $500 microcontroller with an "obscenity of features." Software options include ImageCraft C and Dunfield C.

JStamp

See the listing for Systronix (this section).

Kanda Systems Ltd.   203311

Units 17-18 Glanyrafon Enterprise Park
Aberystwyth, Ceredigion
SY23 3JQ
UK

☏ +44 (0) 1970 621030

☏ +44 (0) 1970 621040

✉ info@kanda.com

🌐 http://www.kanda.com/

Programmers for microcontrollers for the following microcontrollers and subsystems:

- 8051
- Atmel AVR
- CAN
- Internet/Ethernet
- Scenix
- ST7
- Xicor

The BasicX Microcontroller

Microcontrollers are fast becoming a favorite method of endowing a robot with smarts. Offering both speed and ability is the BasicX by NetMedia, a company that previously devoted itself to home automation and small Web cams. The BasicX-24 is actually a member of a family of microcontrollers from NetMedia, which also includes the less expensive (but network-capable) BasicX-1, as well as the BasicX-35. We'll concentrate just on the BasicX-24 (or BX-24) from here on out.

The BasicX-24 (BX-24) micro-controller, on a development board.

A selling point of the BX-24 is that it is pin-for-pin compatible with Parallax's Basic Stamp II. It's important to note that the BX-24 is not a Stamp "clone." The two microcontrollers don't share the same programming languages, so programs written for one will not work on the other.

The BX-24 directly supports 16 input/output (I/O) lines. For each I/O line, or pin, you can change the direction from an input or an output. When an I/O line is an output, you can individually control the value of the pin, either 0 (logic LOW) or 1 (logic HIGH). When an I/O line is an input, you can read a digital or analog value of a TTL-compatible device connected to the BX-24. Eight of the 16 I/O lines can be used for analog connections. The BX-24 incorporates its own built-in 10-bit analog-to-digital converter (ADC). Under software control, you can indicate which of the eight input lines is to be read.

Three of the plated-through holes of the BX-24 serve as optional I/O and are programmatically referred to as pins 25, 26, and 27, making a total of 19 input/output pins. (The remaining plated-through holes provide a way to connect to the chip's serial peripheral interface, or SPI, lines. Connecting to these lines is not recommended unless you're familiar with SPI interfaces, especially as the BX-24's EEPROM is controlled by these same I/O lines.)

A nice touch on the BX-24 is its two LEDs: one red and one green. The green LED is normally used to indicate power-on for the chip, but you can individually control both LEDs from your own programs. You might use the LEDs as status indicators, for example. The LEDs share two of the additional plated-through hole connectors on the BX-24.

The BX-24 board comes with its own 5-volt voltage regulator, which provides enough operating current for all the components on the board, plus several LEDs or logic ICs. If you plan on using the BX-24 to operate a robot, you'll want to provide a separate power supply of adequate current rating to the other components of the robot. You should not rely on the BX-24's onboard regulator for this task.

In order to program the BX-24 you need to purchase the BasicX-24 developer's kit, which contains one BX-24, a programming cable, a power supply, a "carrier board," and programming software on CD-ROM. Cost as of this writing is $99 for the developer's kit, with additional BX-24s at $49.95 each. You plug the BX-24 into the carrier board, which has a 24-pin socket and empty solder pads that you can use to add your own circuitry. The programming cable connects between the carrier board and a serial port on your PC. The power supply is the "wall wart" variety and provides about 12 to 16 vdc.

The BX-24 uses a proprietary programming environment, which consists of an editor and a download console, which also serves double-duty as a terminal for data sent from the microcontroller. The program editor supports the BasicX language, which is a subset of Microsoft Visual Basic. Don't expect all Visual Basic commands to be available in BasicX, however. BasicX supports the same general syntax as Visual Basic and many of the same data types (bytes, integers, strings, and so forth).

If you're familiar with Visual Basic, then you should feel right at home with BasicX. The BasicX language supports the usual control structures, such as If/End If, While/Wend,

For/Next, and Select/Case. Your BasicX programs can be subroutines, and you can call those subroutines from anywhere in the program.

The BX-24 is a general-purpose microcontroller, so many of its built-in features are geared toward any typical personal or commercial microcontroller application. Still, a number of features of the BasicX programming language lend themselves to use in robotics. These features are implemented as functions added to the BasicX language. To use a feature, you merely include it in your program, along with any necessary command parameters.

- *GetADC and PutDAC.* Recall from previously that the BX-24 has its own eight-channel, 10-bit ADC. With the GetADC function, you can read a voltage level on any of eight I/O pins and correlate that voltage level with a binary number (from 0 to 1,023). Conversely, you can use the PutDAC function to output a pulse train that will mimic a variable voltage.

- *ShiftIn and ShiftOut.* With ShiftIn you can receive a series of bits on a single I/O pin and convert them to a single byte in a variable. ShiftOut does the inverse, where you can convert a byte into a series of bits. Both functions allow you to specify an I/O pin for use as the data source and another I/O pin for the clock. The BasicX software automatically triggers the clock pin for each bit received or sent.

- *OpenCom.* The BX-24 supports as many serial ports as you have available I/O pins. With OpenCom you can establish serial communications with other BX-24 chips or any other device that supports serial data transfer. One common use for OpenCom is to establish a link from the BX-24 chip back to the download window of your PC; this window can serve as a terminal for debugging and other monitoring tasks.

- *PulseIn and PulseOut.* The PulseIn function waits for the level at a given I/O pin to change state. One practical application of this feature is to watch for a critical button press to activate some function on your robot. PulseOut sends a pulse of a certain duration (in 1.085-microsecond units) out a given I/O pin. PulseOut is one of the most commonly used functions, used to blink LEDs, trigger sonar pings, and command servomotors to move to a new location.

- *InputCapture.* Somewhat akin to PulseIn, InputCapture watches for signal transition on a specific I/O pin of the BX-24. InputCapture can time the duration of these transitions, thereby giving you a "snapshot" of a digital pulse train, including how long each pulse lasted.

- *PlaySound.* The PlaySound function outputs a waveform that, when connected to an amplifier via a decoupling capacitor, allows you to play previously sampled sound that has been stored in the EEPROM. You can play back sounds at various sampling rates and control the number of times the sound is repeated.

Find out more about the BasicX-24 and other members of the BasicX family at the company's Web site:

http://www.basicx.com/

Try these Google.com search phrases to locate program examples and circuit diagrams using the BasicX:

basicx program examples

basicx programming examples

basicx—site:www.basicx.com—site:basicx.com

Support for the Atmel AVR line is a specialty. Also sells starter kits, microcontroller chips and development boards, project boards, compilers and programming software (for both Basic and C), books, and PC interfaces. Additional offices in the U.S.

STK300, for programming Atmel AVR microcontrollers.

Kevin Ross   203983

P.O. Box 1714
Duvall, WA 98019
USA

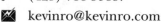

(425) 788-5985.

kevinro@kevinro.com

http://www.kevinro.com/

Kevin Ross sells a whole mess of BotBoard Plus microcontroller boards and BotBoard interface products. Many of the boards are available in parts kit or assembled form.

For those who are new to the BotBoard Plus, it uses a Motorola 68HC11-based microcontroller and provides various connectors to attach robotic parts to it. According to Kevin, "The BotBoard Plus is widely used by Universities and hobbyist for learning and experimentation. The members of the Seattle Robotics Society have been using the BotBoard design for several years."

Additional boards using other models of Motorola microcontrollers, such as the MC68HC912D60 and MC68HC812A4, are available as well. All are professionally produced, with green solder mask and plated-through holes.

Kevin is also the editor of Encoder, the official publication of the Seattle Robotics Society; see:

http://www.seattlerobotics.org/

LART Pages, The   202037

http://www.lart.tudelft.nl/

According to the Web site: "The LART is a small yet powerful embedded computer capable of running Linux. Its performance is around 250 MIPS while consuming less than one watt of power. In a standard configuration it holds 32MB DRAM and 4MB Flash ROM, which is sufficient for a Linux kernel and a sizeable ramdisk image."

LAWICEL   204207

Klubbgatan 3
S-282 32 Tyringe
Sweden

+46 (0) 451 598 77

+46 (0) 451 598 78

info@lawicel.com

http://www.lawicel.com/

LAWICEL makes and markets a line of inexpensive microcontroller board products, many with Controller Area Network (CAN). An example product is the StAVeR24-4433, a 24-pin microcontroller based on the Atmel AVR AT90S4433, which uses the same pinout as the Basic Stamp.

Web site is in English and Swedish.

SEE ALSO:

http://www.candip.com/

Lugnet: Handy Board Mail List Archive   202937

http://news.lugnet.com/robotics/handyboard/

Lugnet's Handy Board group is focused on the MIT-developed Handy Board. Topics include "obtaining, debugging, or using the Handy Board design; troubleshooting problems; exchanging ideas and techniques; sharing code; etc."

M2L Electronics 202359

250 CR 218
Durango, CO 81303
USA

✆ (970) 259-0555

📠 (970) 259-0777

✉ sales@m2l.com

🌐 http://www.m2l.com/

Device programmers for EPROM, 8751 devices, Microchip PICmicro controllers.

Micro Engineering Labs 203254

Box 60039
Colorado Springs, CO 80960
USA

✆ (719) 520-5323

📠 (719) 520-1867

✉ support@melabs.com

🌐 http://microengineeringlabs.com/

Micro Engineering Labs makes and sells development tools for the Microchip PICmicro microcontrollers. Their products in review:

- PicBasic Compiler-Compatible with the Basic Stamp I, adds I2C support, instructions to access external serial EEPROMs, serial speeds to 9600 baud, in-line assembler code.

- PicBasic Pro Compiler-Compatible with the Basic Stamp II, adds I2C support, direct and library routine access to any pin or register, automatic page boundary handling past 2K, real If..Then...Else..Endif structures, built-in LCD support, to access more external devices including serial EEPROMs.

- EPIC Plus PICmicro Programmer-For Windows and DOS compatible, capable of in-circuit serial programming, parallel port interface, and works with most PICmicro microcontrollers.

- LAB-X1 experimenter board-Includes its own 2x20 LCD, 16-button keypad, serial port with 9-pin D connector, programmable oscillator, speaker, and more. Also offered is the LAB-X2 with less built-in hardware.

Additional support products include:

- 2x16 LCD
- Serial graphics LCD module
- Coprocessor modules (analog input, pulse out, PWM)

- PICProto boards
- Loader

Microchip Technology 202371

2355 W. Chandler Blvd.
Chandler, AZ 85224
USA

✆ (480) 792-7200

📠 (480) 899-9210

🌐 http://www.microchip.com/

Microchip makes a broad line of semiconductors, including the venerable PICmicro microcontrollers. Their Web site contains many datasheets and application notes on using these controllers, and you should be sure to download and save them for study.

The company is also involved with radio frequency identification (RFID), selling readers and tags, as well as developer's kits.

National Control Devices 202140

P.O. Box 455
Osceola, MO 64776
USA

✆ (417) 646-5644

📠 (417) 646-8302

✉ ryan@controlanything.com

🌐 http://www.controlanything.com/

NCD wants you to control anything and everything. They offer some microcontroller-enabled products to help, including:

- A/D converters
- Character displays
- Graphic displays
- Input/output devices
- IO expansion modules
- Microcontrollers
- Motor controllers
- Quick-start kits
- Relay controllers
- Serial interface services
- Video switchers

Many of the products are connected to a host (microcontroller, PC, etc.) via an addressable serial line, meaning you connect several of them on a single pair of wires and talk to each one using its unique identification number. Up to 256 such devices can share a single serial port.

The owner of NCD wrote several articles for Nuts & Volts magazine in 1998/99 describing practical uses for the products; the articles are reprinted for your edification at the Web site.

The URL http://www.controleverything.com/ gets you to the same Web site.

NetMedia Inc./BasicX   202150

10940 N. Stallard Pl.
Tucson, AZ 85737
USA

☎   (520) 544-4567

✆   (520) 544-0800

✉   sales@basicx.com

🌐   http://www.basicx.com/

NetMedia's BasicX family of rapid development microcontrollers includes the BasicX-1, BasicX-24, and the BasixX-35, plus various development boards and serial LCD modules. The BasicX sports a Windows-based Basic language development platform, and the BX-24 product has the same form factor and pinout as the Parallax Basic Stamp microcontroller.

    🏭

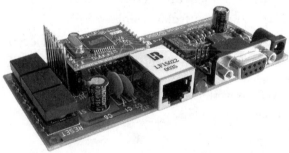

SitePlayer embedded Internet Web server microcontroller. Photo NetMedia Inc.

New Micros, Inc.   202007

1601 Chalk Hill Rd.
Dallas, TX 75212
USA

☎   (214) 339-2204

✆   (214) 339-1585

🌐   http://www.newmicros.com

New Micros is a leading manufacturer of single board computers (SBC), peripherals, and support electronics. The company specializes in embedded systems and particularly the Motorola processor line.

Robotics is singled out as an ideal application for the company's line of DSP-based microcontrollers. As noted on their Web site: "The DSP803-Mini is the perfect board for robotic applications. Small in size, offering many features. NMIN-0803 Mini Features: Memory 32K x 16, program Flash 512 x 16, program RAM 2K x 16, data RAM 4K x 16, data Flash 2K x 16, boot Flash, 10 I/O(s), 8-channel 12-bit A/D 6 PWMs, one quadrature decoder, two timers, two external interrupts, one serial communication interface, CAN 2.0, A/B JTAG, input power: 7 to 9V DC (3.3V and 5V regulators on board). With LCD interfacing and two LED indicators."

The IsoPod 16-bit based microcontroller, from New Micros. Photo New Micros, Inc.

The New Micros DSP803-Mini microcontroller. Photo New Micros, Inc.

The OOPic Microcontroller

Taking an unusual approach to embedded hardware is the OOPic, from Savage Innovations. This chip, which comes on its own carrier/developer board, uses *object-oriented* programming rather than the "procedural" programming found in the Basic Stamp and similar products. The OOPic—which is an acronym for "Object-Oriented Programmable Integrated Circuit"—is said to be the first programmable microcontroller that uses an object-oriented language. The language used by the OOPic is modeled after Microsoft's popular Visual Basic. You don't need Visual Basic on your computer to use the OOPic; the OOPic programming environment is completely stand-alone and available free.

The OOPic microcontroller.

The OOPic supports 31 input/output (I/O) lines, and with few exceptions, any of the lines can serve as any kind of hardware interface—using what the OOPic documentation calls "hardware objects," which are digital I/O lines that can be addressed individually or by nibble (4 bits), by byte (8 bits), or by word (16 bits). The OOPic also supports prede-fined objects that serve as analog-to-digital conversion inputs, serial inputs/outputs, pulse-width modulation outputs, timers/counters, radio-control (R/C) servo controllers, and 4x4 matrix keypad inputs. The device can even be networked with other OOPics, as well other components that support the Philips I2C network interface.

The OOPic comes with a 4K EEPROM for program storage, but memory can be expanded to 32K, which will hold some 32,000 instructions. The EEPROM is "hot swap-pable," meaning that you can change EEPRPOM chips even while the OOPic is on and running. When a new EEPROM is inserted into the socket, the program stored in it is immediately started.

Additional connectors are provided on the OOPic for add-ins such as floating point math, precision data acquisition, DTMF/modem/musical-tone generator, digital ther-mometer, and even a voice synthesizer, currently under development. The OOPic's hard-ware interface is an open system; the I2C interface is published by Philips, allowing any IC that uses the I2C interface to "talk" to the OOPic.

While the hardware capabilities of the OOPic are attractive, its main benefit is what it offers robot hackers: Much of the core functionality required for robot control is already embedded in the chip. This will save you time in writing and testing your robot control programs. Instead of several dozen lines of code to set up and operate an R/C servo, you need about four lines when programming the OOPic.

A second important benefit is that the OOPic's various hardware objects are multitask-ing, which means they run independently of and concurrently with one another. You might command a servo in your robot to go to a particular location.

Though the OOPic is meant as a general-purpose microcontroller, many of its objects are ideally suited for use with robotics. Of the built-in objects of the OOPic, the oA2D, oDio*x*, oKeypad, oPWM, oSerial, and oServo objects are probably the most useful for robotics work.

- *Analog-to-Digital Conversion.* The oA2D object converts a voltage present on an I/O line and compares it to a reference voltage. It then generates a digital value, which rep-resents the percentage of the voltage in relation to the reference voltage. There are four physical analog-to-digital circuits implemented within the OOPic, available on I/O lines 1 through 4.

- *Digital I/O.* Several digital I/O objects are provided in 1-bit, 4-bit, 8-bit, or 16-bit blocks. In the case of the 1-bit I/O object (named oDio1), the Value property of the object repre-sents the electrical state of a single I/O line. In the case of the remaining digital I/O objects, the Value property presents the binary value of all the lines of the group (4, 8, or 16, depending on the object used). There are 31 physical 1-bit I/O lines implemented within the OOPic. The OOPic offers six physical 4-bit I/O groups; three 8-bit groups, and one 16-bit group.

- *R/C Servo Control.* The oServo object outputs a servo control pulse on any IO line. The servo control pulse is tailored to control a standard radio-controlled (R/C) servo and is capable of generating a logical high-going pulse from 0 to 3 ms in duration in 1/36-ms increments. A typical servo requires a 5-volt pulse in the range of 1 to 2 ms in duration. This allows for a rotational range of 180 degrees.

- *Keypad Input.* The oKeypad object splits two sets of four I/O lines in order to read a standard 4x4 keypad matrix. The four row lines are individually and sequentially set low (0 volts), while the four column lines are used to read which switch is pressed within that row. If any switch is pressed, the Value property of the oKeypad object is updated with the value of the switch; a Received property is used to indicate that at least one button of the keypad is pressed. Once all keys are released, the Received property is cleared to 0.

- *Pulse Width Modulation.* The oPWM object provides a convenient pulse-width-modulated (PWM) output, suitable for driving motors (through an appropriate external transistor output stage, of course). The oPWM object lets you specify the I/O line to use—up to two at a time for PWM output—the cycle frequency, and the pulse width.

- *Asynchronous Serial Port.* The oSerial object transmits and receives data at a baud rate specified by the Baud property. The baud rate can be either 1,200, 2,400, or 9,600 baud. The oSerial object is used to communicate with other serial devices, such as a PC or a serial LCD display.

Several versions of the OOpic are available: the original OOPic (or OOPic I) and the OOPic II. The OOPic II adds several features useful for robotics, including built-in support for many popular sensors and motor controllers. These include:

- oCompassDN—Reads a Dinsmore compass (*http://www.dinsmoresensors.com/*)
- oLCDSE(T)—Set of objects that control Scott Edwards serial LCDs (*http://www.seetron.com/*)
- oSonarPL—Reads a Polaroid 6500 sonar ranging module (*http://www.polaroid-oem.com/*)
- oSonarDV—Reads a Devantech (*http://www.robot-electronics.co.uk/*) SRF04 ultrasonic range finder
- oTracker—Tracks the position of a line
- oUVTronHM—Reads a Hamamatsu UVTron flame detector (*http://www.hamamatsu.com/*)
  More information on the OOPic can be found here:

http://www.oopic.com/

http://www.robotprojects.com/ (Web site sponsored by Savage Innovations)

Also check the following Google.com searches for programming examples and project tips:

oopic example robot OR robotics

oopic motor control—site:www.oopic.com

oopic OSC robot—site:www.oopic.com

Olimex Ltd. 202226

89 Slavjanska St.
P.O. Box 237
Plovdiv, 4000
Bulgaria

☎ +35 9-32-626259
📠 +35 9-32-621270
✉ info@olimex.com
🌐 http://www.olimex.com/

Olimex produces printed circuit boards and also sells ready-made boards. Ready-made board products include:

- PIC-Programmers, ICD, prototype boards
- AVR-Programmers, prototype boards
- MSP430-Flash emulation tool, prototype boards

Oricom Technologies 202284

P.O. Box 68
Boulder, CO 80306
USA

☎ (303) 449-6428
✉ support@oricomtech.com
🌐 http://www.oricomtech.com/

Oricom develops "Bot-CoPs"-coprocessors for off-loading computation-intensive real-time tasks from main controllers in small robotic systems. Web site includes experimental project info, links, and articles.

🌐 🗣

Parallax, Inc. 👤 202149

599 Menlo Dr.
Ste. 100
Rocklin, CA 95765
USA

☎ (916) 624-8333
📠 (916) 624-8003
🚫 (888) 512-1024
✉ info@parallaxinc.com
🌐 http://www.parallaxinc.com/

The Basic Stamp revolutionized amateur robotics, yet the concept is simple: Take an 8-bit microcontroller, normally intended to be programmed in assembly language. Instead of requiring folks to learn assembly, embed within the microcontroller a language interpreter, so that it can be programmed in a simpler language, namely Basic. The Basic Stamp is a PICmicro microcontroller with such a language interpreter. It also includes additional basic components so that it is completely self-contained, and runs just by applying power to it. A voltage regulator, crystal, and additional memory are mounted on the Basic Stamp chip, which is the same size as a "fat" 24-pin integrated circuit.

While the Basic Stamp is a main product for Parallax, they recognize that robotics is a central area of interest, so they also offer a number of robotcentric items, including robot kits (GrowBot and BOE-Bot), various sensor packages (line following, compass, etc.), and development boards. They also team up with third-party companies to offer integrated products, such as RF modules, video cameras, LCD panels, and sound modules.

Some additional URLs of interest related to the Basic Stamp:

Stamps in Class-http://www.stampsinclass.com/
SX Tech-http://www.parallaxinc.com/
Spanish S.I.C.-http://www.stampsenclase.com/

🌐 🗋 🖥

The venerable Basic Stamp II microcontroller. Photo Parallax, Inc.

Stamp carrier, from Parallax.

Peter H. Anderson　　　　202028

915 Holland Rd.
Bel Air, MD 21014
USA

☏　(410) 836-8526

✉　pha@phanderson.com

🌐　http://www.phanderson.com/

Professor Anderson provides products and help when he's not teaching class at Morgan State University. You'll find sales and tutorials on PC parallel port interfacing, BasicX microcontroller, BASIC Stamp microcontroller, 68HC11 microcontroller, PIC, JKMicro Flashlite V25 single board computer, and others. Prices are reasonable, and there's a ton of information at the Web site.

PMB Electronics　　　　202972

1A Beth St.
Trentham
Wellington
New Zealand

☏　64 4-970-7268

☏　64 4-970-7269

✉　info@pmb.co.nz

🌐　http://www.pmb.co.nz/

Microcontroller mail-order retail, including Motorola HC11 microcontroller boards, PICmicro project boards, and LCD modules.

Protean Logic　　　　204040

11170 Flatiron Dr.
Lafayette, CO 80026
USA

☏　(303) 828-9156

☏　(303) 828-9316

🌐　http://www.protean-logic.com/

Makers of single board computers and microcontroller boards, many using the TICKit interpreter engine, said to offer faster processing than the Basic Stamp. Products include:

- TICKit 63 processor IC
- TICKit 63 computer module
- RSB509b serial data buffer IC
- TICKit 63 single board computer

Rabbit Semiconductor　　　　202006

2932 Spafford St.
Davis, CA 95616
USA

☏　(530) 757-8400

☏　(530) 757-8402

✉　sales@rabbitsemiconductor.com

🌐　http://www.rabbitsemiconductor.com/

Rabbit makes a popular 8-bit microcontroller and associated developer kits; the Rabbit system is known for its speed (minimum CPU speed is 20 MHz, compared to 1, 4, or 8 MHz of most other microcontrollers). In addition to bare controllers, the company also sells "core modules" such as the RabbitCore RCM2200 with Ethernet connectivity built in.

Technical documentation and other support documents (most in Adobe Acrobat PDF format) available at the Web site.

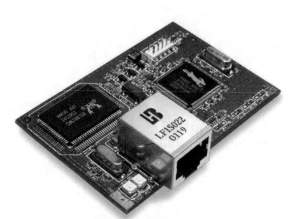

RCM3000 RabbitCore Ethernet-based microcontroller. Photo Ryan Fardo, Z-World.

Reynolds Electronics 🎖　　　　202009

3101 Eastridge Ln.
Canon City, CO 81212
USA

☏　(719) 269-3469

☏　(719) 276-2853

✉　support@rentron.com

🌐　http://www.rentron.com/

Rentron offers kits and ready-made products for the electronics enthusiast and robotmeister, including PicBasic and PicBasic Pro compilers, Basic Stamp, Microchip PICmicro, Intel 8051 microcontrollers, remote controls, tutorials, projects, RF components, RF remote-control kits, and infrared kits and components.

Rho Enterprises 202148

Box 33
4100 W. Colfax Ave.
Denver, CO 80204-1405
USA

((720) 359-1467

🌐 http://www.rhoent.com/

Kits and modules for microcontrollers, including serial
LCD interface, SX28 proto board, and SX28 proto
board starter kit.

Roboblock System Co., Ltd. 204072

137-070, #707 ilkwang Bldg.
1656-2, Seocho-Dong
Seocho-Gu
Seoul
South Korea

(+82 2-597-8224

📠 +82 2-597-9441

✉ info@roboblock.co.kr

🌐 http://www.roboblock.com/

Sells Roboblock kits, mobile robot kits, toy robot kits,
microcontrollers, AVR C and Basic programming software, parts, and other robot/electronics products. Web
page is a mix of English and Korean.

Robotics Building Blocks 202981

2639 West Canyon
San Diego, CA 92123 9212
USA

((858) 715-6695

📠 (650) 592-2503

🌐 http://www.rdk2001.com/

RDK sells Atmel AVR microcontrollers and robotics kits
(minibot, Micro Mouse). The Web site provides useful
examples of using the AVR with the BASCOM AVR
Basic language compiler.

Sage Telecommunications Pty Ltd. 202341

P.O. Box 2171 Warwick
Western Australia 6024
Australia

(+61 8 9344 8474

📱 +61 8 9345 4975

✉ sales@sages.com.au

🌐 http://www.sages.com.au/

Embedded systems: Dontronics SimmStick, electronic
components, PIC microprocessors, Atmel AVR microprocessors, LCDs, and RF data modules.

Savage Innovations/OOPic 203982

http://www.oopic.com/

Manufacturer of the OOPic and OOPic2 microcontrollers that offer multitasking and built-in "objects"
that simplify programming. Many of the objects are
directly suitable for robotics. Sold by distributors.

SEE ALSO:

http://www.oopic2.com/
http://www.robotprojects.com/

Small PC Board for the 68HC812A4 202253

http://www.rdrop.com/users/marvin/other/other-
prj.htm

Circuit schematic and PCB layout for a 68HC812A4-
based microcontroller board.

Systronix 203654

555 South 300 East
Salt Lake City, UT 84111
USA

((801) 534-1017

📱 (801) 534-1019

✉ info@systronix.com

🌐 http://www.systronix.com/

Embedded control hardware, software, enclosures,
components, etc. Java and non-Java systems (such as
JStamp), high-speed 8051s.

TECEL 202519

2508 Spruce SE
Albuquerque, NM 87106
USA

✆ (505) 239-8483

📠 (505) 243-7514

✉ tecel@tecel.com

🌐 http://www.tecel.com/

Microcontroller boards using 80C251, 80C552, 8051, and 68HC11 controllers. Compiler, assembler and loader software included upon purchasing any of the microcontroller boards. Also:

- High-power motor driver board
- Sensors
- Encoder disks
- IC components
- Diodes
- Wire and connectors
- Misc: capacitors and bare boards

Technological Arts 202004

Technological Arts
819-B Yonge St.
Toronto, ON
M4W 2G9
Canada

✆ (416) 963-8996

📠 (416) 963-9179

🚫 (877) 963-8996

✉ sales@technologicalarts.com

🌐 http://www.technologicalarts.com

Technological Arts produces postage stamp-sized single board computers using the Motorola 68HC1x microcontrollers. A number of special-purpose application boards are also offered, and many are suitable for robotics. These boards include:

Adapt 912B32 controller. Photo Carl Barnes of Technological Arts

- Voice Record/Play
- Display/keyboard
- X-Y-Z stepper
- X-Y stepper
- Data acquisition
- Quad 12-bit DAC
- 8-channel differential amplifier
- CAN interface
- Quad motor driver

Vikon Technologies 202338

6 Way Rd.
Middlefield, CT 06455
USA

✆ (860) 349-7055

📠 (860) 349-7088

✉ info@vikon.com

🌐 http://www.vikon.com/

Vicon makes and sells embedded systems, including single board computers, Atmel AVR programming development boards, PIC development kits and boot loaders, PIC prototyping boards, and SimmStick bus compatible products.

Web Ring: AVR Microcontroller 202060

http://r.webring.com/hub?ring=avr

Atmel AVR microcontroller-related Web sites.

Web Ring: Embedded 202347

http://s.webring.com/hub?ring=embedded

Microcontrollers and embedded RTOS (real-time operating systems)-related Web sites.

Wilke Technology GmbH 204041

Krefelder Str. 147
52070 Aachen
Germany

✆ +49 (0) 2419 18900

📠 +49 (0) 2419 189044

info@wilke-technology.com

🌍 http://www.wilke-technology.com/

Single-tasking and multitasking single board computers. Products include BASIC-Tiger professional software

🏭

Zorin Microcontroller Products 202649

1633 4th Ave. West
Seattle, WA 98119
USA

☎ (206) 282-6061

📠 (206) 282-9579

 info@ZORINco.com

🌍 http://zorinco.com/

Makers and sellers of HC11-based embedded microcontroller systems. Products include:

- ModCon Microcontroller-Modular controller
- Digital Input and Event Processing System-Take actions based on sensors, push buttons, and other inputs
- SPI-X10 Controller-Control lights and household appliances
- Audio Record/Playback-Record and play back up to 90 seconds of audio
- MIDI Gizmo—An easy way to interface the HC11 to MIDI

🌐 💻

 # MICROCONTROLLERS-PROGRAMMING

See the main Microcontrollers section for a description of the listings in this section.

Anvil 203384

http://www.focalpoint.freeserve.co.uk/

Anvil combat robot. With building diary and pictures. Also how-to pages using the OOPic microcontroller.

🗣

AWC/Al Williams 202361

310 Ivy Glen Ct.
League City, TX 77573-5953
USA

☎ (281) 334-4341

📠 (281) 754-4462

 alw@al-williams.com

🌍 http://www.al-williams.com

Home page of author Al Williams. From the main page you can select one of several "subbusinesses":

- AWC-Windows and Web development, consulting, training, books, and Al Williams's Java@Work columns.
- AWCE-Microcontroller development, consulting, our solderless breadboard products, and PAK coprocessors. Look here for Stamp-related info, including a BS2 to PS/2 keyboard interface.
- WD5GNR-Ham radio or hobby electronics

🌐 💻

Beginners' PIC and AVR page 202390

http://homepage.ntlworld.com/
matthew.rowe/micros/

A page for beginners: how to use the Microchip PICmicro and the Atmel AVR microcontrollers.

🗣

Beyond Logic 202085

http://www.beyondlogic.org/

Broad and detailed information on microcontroller and computer topics with an emphasis on interfacing. Articles include:

- uClinux-Linux for microcontrollers
- Universal serial bus
- AT keyboards

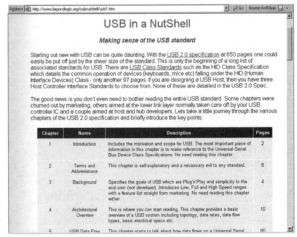

One of a number of informational pages at Beyond Logic.

- Windows device drivers (Windows NT/2000/XP and 98/ME)
- CMOS image sensors
- Parallel port interfacing
- RS-232 interfacing
- Miscellaneous

See, for example:

http://www.beyondlogic.org/keyboard/keybrd.htm

Brunning Software   203855

138 The Street
Little Clacton
Clacton-on-sea
Esssex
CO16 9LS
UK

℡ +44 (0) 1255 862308

✉ sales@brunningsoftware.co.uk

🌐 http://www.brunningsoftware.co.uk/

Programming training and development labs for PICmicro microcontrollers; software, breadboards, and parts kits.

Byron Jeff's PIC Page   203729

http://www.finitesite.com/d3jsys/

Among the topics on Byron's page is the "Trivial 16F87X LVP programmer."

C-AVR   203082

http://www.spjsystems.com/cavr.htm

A C compiler for the Atmel AVR family of microcontrollers. Free demo.

Connecting a PC Keyboard
to the BS2   203069

http://www.emf-design.com/bs2/reader.htm

Instructional article on how to connect a PS/2 keyboard to a Basic Stamp. Includes programming code.

Control a Serial LCD from a PIC   202147

http://www.mastincrosbie.com/mark/
    electronics/pic/lcd.html

Sample code for interfacing a PIC to an LCD, via a serial line. Requires a serial LCD, such as those sold by Scott Edwards Electronics.

Custom Computer Services, Inc   202187

P.O. Box 2452
Brookfield, WI 53008
USA

℡ (262) 797-0455

📠 (262) 797-0459

✉ ccs@ccsinfo.com

🌐 http://www.ccsinfo.com/

Sells a C compiler for Microchip PIC controllers for Windows and Linux.

EASY START: PIC microcontroller
prototype boards   202260

89 Slavjanska St., P.O.Box 237, Plovdiv 4000
Bulgaria

℡ +35 932 626259

📠 +35 932 621270

🌐 http://www.olimex.com/easystart.html

EASY START PIC microcontroller prototype boards:
    PIC16_84 EASY START development board
    PIC12Cxxx prototype boards (MINI, MIDI, MAXI)

Available assembled or bare board only.

FSMLabs, Inc.   202672

P.O. Box 1822
Socorro, NM 87801
USA

℡ (505) 838-9109

✉ business@fsmlabs.com

🌐 http://www.fsmlabs.com/

Publishers of RTLinux, a real-time version of Linux for industrial and embedded applications.

Generating Sony Remote Control Signals with a BASIC Stamp II   203068

http://www.whimsy.demon.co.uk/sircs/index.html

As the title says.

L.O.A.A.-List of AVR Applications   204194

http://www.hth.com/loaa/

Programming examples for the Atmel AVR line of microcontrollers. Most examples are in either C or assembler, with a few in BASCOM Basic.

L.O.S.A.-List of Stamp Applications   204193

http://www.hth.com/losa/

Lots and lots of programming examples using the Basic Stamp microcontroller.

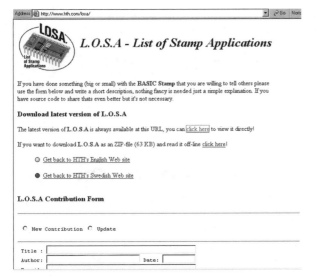

L.O.S.A. - List of Stamp Applications

Lud's Open Source Corner   204056

http://drolez.com/hardware/

How-to articles with code and programming examples on:

- Atmel AVROpen Source Software: PWM/Servo controllers with GPIO and serial interface
- Palm Cybot-Control a Cybot from your PalmOS-based device and SmallBASIC

Myke Predko, Author   203734

http://www.myke.com/

Web site of author Myke Predko. Plenty of goodies for the PIC microcontroller.

Robot Projects   202066

http://www.robotprojects.com/

This Web site provides hands-on examples of a variety of interesting robotics projects, most of which revolve around using the OOPic microcontroller (the Web site is maintained by Scott Savage, the developer of the OOPic). Projects include:

- Racing Rover-Collision avoidance sensors on a high-speed robot.
- Big-OTrak-Retrofitting a Milton Bradley Big Trak with an OOPic.
- WilbyWalker-CADD drawings and source code for a six-legged walker.
- Contactless Angular Measurement-Measure the angle your robot is to a wall.
- Recycling the sonar unit from a Polaroid Camera.
- Experiments with the SP0256 speech synthesizer.
- Controlling 21 servos from your PC.

Robotics Information and Articles   204120

http://www.leang.com/robotics/

Example 'bots and online articles, good ones on such subjects as:

- Controlling servo motors with various microcontrollers
- RF serial communication for the MIT Handy Board
- H-bridge motor driver circuit
- Infrared proximity sensor

For Kam Leang's past and current robot projects, see also:

http://www.leang.com/robotics/

Serial LCD interface using AVR 90s2313   202155

http://members.tripod.com/Stelios_Cellar/AVR/

SerialLCD/serial_lcd_interface_using_avr.htm

Tutorial and code examples (for the Atmel AVR controller) on interfacing the chip to an LCD panel using a Hitachi HD44780A LCD controller.

Serial Servocontroller with AT90S1200   203010

http://mariabonita.hn.org/servo/en.htm

How to program a serial R/C servo controller using an Atmel AVR AT90S1200 microcontroller. Stop paying $20 or $30 for a serial servo controller when you can program your own for less than $4.

Steve Curtis: Robotics Experiments   202254

http://www.freenetpages.co.uk/hp
      /SteveGC/index.htm

Steve Curtis shares his robot designs, including schematics and programming examples.

TechTools   202492

P.O. Box 462101
Garland, TX 75046-2101
USA

((972) 272-9392
⌣ (972) 494-5814
✉ support@tech-tools.com
🌐 http://www.tech-tools.com/

PIC, EEPROM programmers, ClearView Mathias (in-circuit emulator), compilers, and debuggers.

Web Ring: Basic Stamp SX   202105

http://v.webring.com/hub?ring=stamp

Basic Stamp computers, PIC microcontrollers, and related embedded technologies.

Web Ring: HC11   202104

http://n.webring.com/hub?ring=hc11

Motorola 68HCxx microcontroller family and related embedded technologies.

Web Ring: PICMicro   202107

http://o.webring.com/hub?ring=picmicro

Web sites and pages dedicated to the Microchip PIC microcontroller and related projects.

Web Ring: Zilog   202106

http://v.webring.com/hub?ring=zilog

Resources for Zilog microprocessors and compatibles (e.g., Hitachi and Rabbit), and microcontrollers and related embedded technologies.

Zoomkat: Web Based Control Panel   202229

http://www.geocities.com/zoomkat/

Web-based control of servos (via a Mini SSC II servo controller). Details and programming examples for using servo control with a Web cam.

〰 MICROCONTROLLERS-SOFTWARE

See the main Microcontrollers section for a description of the listings in this section.

Antratek Electronics   203452

Postbus 356, 2900
AJ Capelle A/D IJSSEL
The Netherlands

(010 4504949
⌣ 010 4514955
✉ info@antratek.nl
🌐 http://www.antratek.nl/

Distributor of microcontroller hardware and software: Basic Stamp, compilers, DSP starter kits, emulators, EPROM emulators, I2C adapters, I2C bus monitors, IC testers, LCD/VFD modules, modem modules, programmers, serial displays, and starter kits.

Web site is predominately in Dutch with some English.

For robot specialty products, see also:

http://www.robotwinkel.nl/

AVR + GameBoy(tm) Camera = Fun   202277

http://pages.zoom.co.uk/andyc/camera.htm

Detailed information, circuits, and sample programming (for the Atmel AVR microcontroller) for using the GameBoy camera for crude machine vision.

AvrX Real Time Kernel   204114

http://www.barello.net/avrx/

From the Web site: "AvrX is a Real Time Multitasking Kernel written for the Atmel AVR series of micro controllers." Distributed free. See the author's Cherry Blossom mini sumo robot for details on how AvrX is used in robotics. Thanks Larry Barello.

Basic Micro, Inc.   202087

34391 Plymouth Rd.
Livonia, MI 48152
USA

((734) 425-1744
((734) 425-1722
Info@basicmicro.com
http://www.basicmicro.com/

Crownhill Associates Limited

http://www.letbasic.com/

PICBasic compilers for the Microchip PICmicro line of microcontrollers

Hitex Development Tools

http://www.hitex.com/

Development tools and programs for microcontrollers: 8051, ARM, Pentium, Motorola HC1x, and others

---

Basic Micro produces the MBasic line of compilers for PICmicro microcontrollers. See the full listing under Microcontrollers-Hardware.

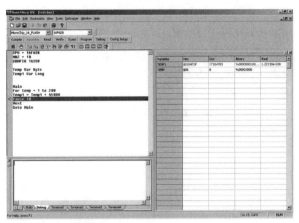

Debug Screen for the MBasic compiler. Courtesy Basic Micro Inc.

Chuck Hellebuyck Electronics   203057

Electronic Products
1775 Medler
48382, MI Commerce
USA

((248) 515-4264
((413) 825-0377
http://www.elproducts.com/

Chuck's resells the Atom from Basic Micro, MBasic compiler software, ePic boards, and bootloader packages.

CodeVisionAVR   203464

HP InfoTech S.R.L.
Str. Liviu Rebreanu 13A
Bucharest 746311
Romania

(+40 9346 9754
(+40 9346 9754
office@hpinfotech.ro
http://www.hpinfotech.ro/

Sellers of a high-performance C compiler, integrated development environment, automatic program generator, and in-system programmer for the Atmel AVR family of microcontrollers

Free evaluation and paid full version (paid version in Standard and Lite editions).

Crownhill Associates Ltd. 203650

32 Broad Street
Ely
Cambridgeshire
CB7 4PW
UK

✆ +44 (0) 1353 666709
📠 +44 (0) 1353 666710
🌐 http://www.crownhill.co.uk

Resellers of PC microcontrollers, development boards, PIC compilers, and related hardware/software.

CSMicro Systems 203450

213 Sage St.
Ste. #3
Carson City, NV 89706
USA

✆ (775) 887-0505
📠 (775) 887-8973
⊘ (888) 820-9570
✉ CSMicroSystems@CSMicroSystems.com
🌐 http://www.csmicrosystems.com/

Embedded systems reseller (Basic Stamp, BasicX, PICmicros). Also carries the CodeVision editor for PIC Basic, BasicX, and others and the EPIC PIC programmer modules.

Dontronics, Inc. ⚕ 202041

P.O. Box 595
Tullamarine, 3043
Australia

✆ +61 3 9338 6286
✉ don@dontronics.com
🌐 http://www.dontronics.com

Dontronics specializes in microcontrollers, as well as the SimmStick prototyping development board system. See the full listing under Microcontrollers-Hardware.

Embedded Systems, Inc. 202164

11931 Hwy. 65 NE
Minneapolis, MN 55434
USA

✆ (763) 767-2748

📠 (763) 767-2817
🌐 http://www.embedsys.com/

Makers of low-cost development systems and add-ons. See listing under Microcontrollers-Hardware.

HTH/High Tech Horizon 203066

Asbogatan 29 C
S-262 51 Angelholm
Sweden

✆ +46 (0) 431 410 088
📠 +46 (0) 431 410 088
✉ info@hth.com
🌐 http://www.hth.com/

Resellers of Basic Stamp, Atmel AVR, and BasicX microcontrollers, as well as the BASCOM programming software. Web site is in Swedish and English.

Imagecraft Software 202373

706 Colorado Ave. #10-88
Palo Alto, CA 94303
USA

✆ (650) 493-9326
📠 (650) 493-9329
✉ info@imagecraft.com
🌐 http://www.imagecraft.com/

C compiler for embedded applications. The company's compilers work with Atmel AVR, MegaAVR, tinyAVR; Cypress MicroSystems PSoC; Motorola's HC08, HC11, HC12, HC16, and Texas Instruments MSP430 Microcontrollers.

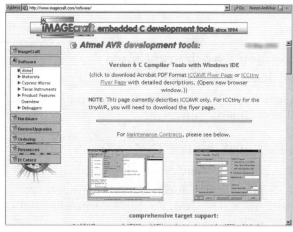

Web page for Imagecraft Software.

Kanda Systems Ltd.   203311

Units 17-18 Glanyrafon Enterprise Park
Aberystwyth, Ceredigion
SY23 3JQ
UK

✆ +44 (0) 1970 621030

📠 +44 (0) 1970 621040

✉ info@kanda.com

🌍 http://www.kanda.com/

Programmers for microcontrollers. See the full listing under Microcontrollers-Hardware.

LART Pages, The   202037

http://www.lart.tudelft.nl/

See listing under Microcontrollers-Hardware.

MCS Electronics   202340

G. Brautigamstraat 11
1506WL
Zaandam
Holland

✆ +31 75 6148799

📠 +31 75 6144189

✉ info@mcselec.com

🌍 http://www.mcselec.com/

Publishers of BASCOM AVR and BASCOM-8051. Also provides embedded system development and support. Sells SimmStick, AM RF modules, programs, and simulators.

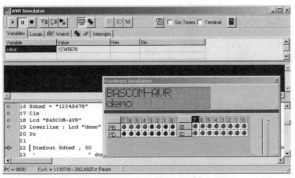

BASCOM AVR compiler and simulator.
Courtesy Mark Alberts.

Micro Engineering Labs   203254

Box 60039
Colorado Springs, CO 80960
USA

✆ (719) 520-5323

📠 (719) 520-1867

✉ support@melabs.com

🌍 http://microengineeringlabs.com/

Micro Engineering Labs makes and sells development tools for the Microchip PICmicro microcontrollers. See product listing under Microcontrollers-Hardware.

Roboblock System Co., Ltd   204072

137-070, #707 ilkwang Building
1656-2, Seocho-Dong
Seocho-Gu
Seoul
South Korea

✆ +82 2-597-8224

📠 +82 2-597-9441

✉ info@roboblock.co.kr

🌍 http://www.roboblock.com/

Sells Roboblock kits, mobile robot kits, toy robot kits, microcontrollers, AVR C and Basic programming software, parts, and other robot/electronics products. Web page is in English and Korean.

Wilke Technology GmbH   204041

Krefelder Str. 147
52070 Aachen
Germany

✆ +49 (0) 2419 18900

📠 +49 (0) 2419 189044

✉ info@wilke-technology.com

🌍 http://www.wilke-technology.com/

Single-tasking and multitasking single board computers. Products include BASIC-Tiger professional software.

MOTOR CONTROL

In this section, you'll find hardware for controlling DC, stepper, and servomotors (controllers for R/C servo motors can be found under Radio Control-Servo Control). Many of the listings in this section are for industrial motion control and are therefore priced accordingly. However, a company's product information can be useful for general education, and many motor control electronics makers also publish free online application notes about the art and science of controlling motors.

Of course, there are also plenty of listings for affordable motor control products that are sold directly to individuals. This section also includes Web sites that provide handy information on motor control techniques.

SEE ALSO:

ACTUATORS: Additional motor control resources

ACTUATORS-MOTORS: Motors to control

COMPUTERS-I/O: Look for interface electronics; also includes motor control functions

MICROCONTROLLERS-Inexpensive hardware for operating a motor

Chapp.com

http://www.chapp.com/

Servo controller, stepper controller, video cam, programmable R/C controller

DMachine

http://www.dmachine.tv/

Controllers and other parts for combat robots; UK based

Magnevation

http://www.magnevation.com/

Motor controllers for the OOPic and other microcontrollers

Motion Online

http://www.motiononline.com/

Industrial servo motors and controllers

4QD 202651

30 Reach Road
Burwell
Cambridgeshire
CB5 0AH
UK

☏ +44 (0) 1638 744080
✉ sales@4qd.co.uk
🌐 http://www.4qd.co.uk/

4QD manufactures speed controllers for battery-operated electric motors.

The Web site is also a good source of information on speed control and H-bridges for DC motors. You can find a number of schematics for advanced H-bridge designs. Be sure to read the "Electronics Circuits Reference Archive PWM speed control" at:

http://www.4qdtec.com/pwm-01.html

Baldor Electric Company. 203222

5711 R.S. Boreham, Jr. St.
P.O. Box 2400
Fort Smith, AR 72901
USA

☏ (501) 646-4711
☏ (501) 648-5792
🚫 (800) 828-4920
🌐 http://www.baldor.com/

Baldor is a manufacturer of industrial motors. Product suitable for robotics include DC, DC gearhead, and servomotors. The Web site sports a product cross-reference style of product selection. Prices are provided. You won't be using these puppies to build a small tabletop bot, but they're good candidates for your next combat robot. Products are sold through local distributors.

Basic Stamp: Stepper Motor Driver 202046

http://www.mastincrosbie.com/mark/stamp/stepper.html

Sample code for Basic Stamp II to drive a unipolar four-coil stepper motor. The code also provides for two switches to increase and decrease the speed of the motor and an LCD output to display the motor speed.

Bodine Electric Company 203880

2500 West Bradley Pl.
Chicago, IL 60618-4798
USA

- ((773) 478-3515
- ✆ (773) 478-3232
- ⊘ (800) 726-3463
- 🌐 http://www.bodine-electric.com/

Bodine is a well-known and respected manufacturer of motors of all shapes, sizes, and kinds. Of Bodine's line of most interest to robotics folk are their DC motors and gearmotors, and brushless DC motors and gearmotors. The Bodine product catalog is on CD-ROM, and the Web site contains numerous technical articles that apply to Bodine products, as well as general information about motors from many other companies. Their white paper "Brushless DC Motors Explained" is particularly interesting.

Brookshire Software 202250

113 Maywood Ln.
Charlottesville, VA 22904
USA

- ((703) 850-0470
- ✆ (661) 288-1867
- ⊘ (800) 999-2734
- ✉ Info@BrookshireSoftware.com
- 🌐 http://www.BrookshireSoftware.com/

Servo control software. According to the company, "Visual Servo Automation (VSA) is the first visual solution to servo control and automation. Using the Mini SSC, SV203, and SMI [these are popular third-party servo controller boards], VSA breaks away from traditional solutions and realizes that complex animatronics, robotics, and other servo systems demand a sophisticated solution."

Or, put another way, the company's VSA software lets you program or record motions for R/C servos to replicate. The software uses a graphical user interface and operates in Windows.

Carlo Gavazzi Holding AG 203992

Sumpfstrasse 32
CH-6312 Steinhausen
Switzerland

- (+41 41 747 4525
- ✆ +41 41 740 4560
- ✉ gavazzi@carlogavazzi.ch
- 🌐 http://www.carlogavazzi.com/

High-end industrial automation components. Sensors (proximity and photoelectric), solid state relays, and motor controllers.

Compumotor Engineering Reference Guides 203927

http://www.compumotor.com/
    catalog%5Feng%5Fref.htm

The technical white papers (all in Adobe Acrobat PDF format) available at this Web site would easily be the equivalent of a hundred-dollar textbook. Download these to your computer, and refer to them often. Note particularly the last documents on calculations; useful for the engineering inclined. These documents are from an older Compumotor catalog; the information is so good they kept it around for free download.

- Motor Applications
- Step Motor Technology
- Linear Step Motor Technology
- Common Questions Regarding Step Motors
- DC Brush Motor Technology
- Brushless Motor Technology
- Hybrid Servo Technology
- Direct Drive Motor Technology
- Step Motor Drive Technology
- Microstepping Drive Technology
- Analog and Digital Servo Drives
- Brushless Servo Drive Technology
- Servo Tuning
- Feedback Devices
- Machine Control
- Control System Overview
- Serial & Parallel Communications
- Electrical Noise Symptoms & Solutions
- System Selection Considerations
- Motor Sizing and Selection Software
- System Calculations-Move Profiles
- System Calculations-Leadscrew Drives
- System Calculations-Direct Drives
- System Calculations-Gear Drives
- System Calculations-Tangential Drives

- System Calculations-Linear Motors
- Glossary of Terms

SEE ALSO:

http://www.compumotor.com/

Control of Stepping Motors, a tutorial

203007

http://www.cs.uiowa.edu/~jones/step/

Dr. Douglas Jones (of the University of Iowa Department of Computer Science) explains all about stepper motors. Moderately to fairly technical.

Control Technology Corp.

203274

25 South St.
Hopkinton, MA 01748
USA

- (508) 435-9595
- (508) 435-2373
- (800) 282-5008
- ctcwebmaster@ctc-control.com
- http://www.ctc-control.com/

CTC offers industrial motion control as well as "Web-enabled automation" products. Much of the product CTC provides is a bit over the top for the average amateur robot builder, but the Web site also contains a number of popular tutorials on the inner workings of servo systems. Among the topics are:

In Review: DC Motor Bridge Techniques

Technique	Pros	Cons
Relays	• Inexpensive • Easy to implement	• Maximum amperage of about 10–20 amps • Heavy-duty relays can be large and noisy
Solid-state relays	• Act like relays but use solid-state components inside • Some models able to switch currents up to 50–100 amps	• Can be frightfully expensive • High control voltages (20–30 volts) often required • Limited switching arrangements; most are SPST only
Bipolar transistors	• Cheap • Components easy to find	• Requires careful selection of components to suit motor, load • High currents may require heavy heat sinking • Motor doesn't get full voltage because of voltage drop through transistors
MOSFET transistors	• High efficiency • Minimal voltage drop through transistor, so motor gets nearly full voltage	• Not effective for motors under about 6–9 volts • High currents may require heavy heat sinking • Careless handling or high currents readily blow out MOSFETs
H-bridge IC	• Best "universal" approach to motor bridge	• Harder to find • Some H-bridge packages use "offset" pins that can't be easily breadboarded • ICs more difficult to heat sink when high currents are required

- Adjusting PID gains
- The Bode diagram
- PID and servos
- Servo types

Controller Design and System Modelling: Tuning a Controller · 202269

http://www.chemeng.ed.ac.uk/ecosse/
    control/sample/system.html

Technical details on using a PID algorithm to control servos.

CTC Control: Literature and Resources · 203652

http://www.ctc-control.com/litres/

Compendium of white papers and technical notes on automation, servos, and control algorithms.

Diverse Electronics Services · 202282

Carl A. Kollar
1202 Gemini St.
Nanticoke, PA 18634-3306
USA

((570) 735-5053

✉ carl@diverseelectronicservices.com

🌍 http://www.diverseelectronicservices.com/

PIC-based motor controllers, radio-controlled device controllers, and transmitter/receiver sets.

MC7 motor controller. Photo Diverse Electronic Services.

EA Electronics · 204090

8 Maple St.
Ajax, ON
L1S 1V6
Canada

((905) 619-1813

✉ email@eaelec.com

🌍 http://www.eaelec.com/

Purveyors of motors and motor control electronics for remote-control model ships. But what's a robot if not a ship with wheels instead of a rudder? In other words, just about everything for R/C model ships will work in a robot, too. Their speed-control boards are designed for some high-current DC motors and will handle 10 to 20 amps.

Among the products offered:

- Multifrequency speed control
- Optically isolated smart control
- DC PM motors (from Johnson, Coleman, and Pittman, the big names in small DC motors)

Electronics are sold though distributors; motors directly.

Easy Step'n, An Introduction to Stepper Motors for the Experimenter · 202180

http://www.stepperstuff.com

Easy Step'n, from Square 1 Electronics, is a how-to book on stepper motors. According to the Web site, "The book provides the experimenter with the information needed to use stepper motors. Determine important surplus motor electrical and mechanical specs using simple, easy to build electrical and mechanical test equipment. Design and build microcontroller-based control systems for stepper motor applications. . . ."

See the company's main Web site:

http://www.sq-1.com/

eduRobotics.com · 203391

http://www.eduRobotics.com/

See listing for Innovation First (this section).

EFFECTive ENGINEERING · 202669

9932 Mesa Rim Rd.
Suite B
San Diego, CA 92121
USA

☎ (858) 450-1024

📠 (858) 450-9244

🌐 http://www.effecteng.com/

Produces animatronics products for stage and special effects. Products include animatronics (e.g., Halloween), mechanical effects, and unusual props.

E-Lab Digital Engineering, Inc.   202490

Carefree Industrial Park
1600 N. 291 Hwy. Ste. 330
P.O. Box 520436
Independence, MO 64052-0436
USA

☎ (816) 257-9954

📠 (816) 257-9945

✉ support@elabinc.com

🌐 http://www.elabinc.com/

E-Lab makes a series of "building block" ICs and modules that support various microcontrollers, including the Microchip PIC or Atmel AVR. Their products include:

- Serial text LCD controller IC
- Octal seven-segment LED decoder
- Unipolar stepper motor controller
- Bipolar stepper motor controller
- Serial to parallel-printer IC

FAQ on PID Controller Tuning   202268

http://www.tcnj.edu/~rgraham/PID-tuning.html

Excellent FAQ on PID control of motors and other devices.

FerretTronics   202682

P.O. Box 89304
Tucson, AZ 85752-9304
USA

☎ (520) 572-6824

📠 (240) 526-8985

✉ sales@ferrettronics.com

🌐 http://www.ferrettronics.com/

FerretTronics is a developer of custom chips and software for robotic and electronic control devices. Highlights of their product line include:

- FT609-stepper motor controller. Using just a 2400-baud serial communications link (from a computer or microcontroller), you can control the step, speed, direction, and acceleration of a stepper motor. The chip lacks high-current capacity, so you need to augment it with an H-bridge like the L293D, a UDN2540 quad Darlington power driver or other suitable transistor circuit.

- FT639 -R/C servo controller. Also using a 2400-baud serial line, this chip can control up to five R/C servos simultaneously (of course, you have full independent control of each servo). You can precisely control the position of the servo in 256 steps.

- FT936-Sony infrared remote-control decoder. This chip will decode the infrared signals from a Sony TV remote and output the codes on five I/O lines. You can use the FT936 to add remote-control capability to your robot without having to write the decoding software yourself. (And, by having the decoding done off-board, your robot's microcontroller can be busy doing other important things.) The unit comes with an infrared receiver module.

You may want to consider the company's "sampler pack," a grab bag of several of their products.

GE Industrial Systems   203947

41 Woodford Ave.
Plainville, CT 06062
USA

☎ (860) 747-7111

📠 (860) 747-7393

🌐 http://www.geindustrial.com/

GE Industrial Systems supplies a wide range of products for residential, commercial, industrial, institutional, and utility applications. Mostly high-end industrial components; their product is also a common find on the surplus market. The GE Industrial Web site provides downloaded catalogs and technical references for you to study. Will sell industrial product direct, limited to territory.

Among some of the cogent GE products for larger robots are:

- Contactors
- Controllers and I/O
- Drives
- Embedded computers
- Motion control
- Motor control centers
- Motors
- Sensing solutions

- Sensors, solenoids, limit switches

Georgia Tech: Motors 202288

http://srl.marc.gatech.edu/education/
ME3110/primer/motors.htm

Tutorial on motors. Part of Georgia Tech's curriculum resources.

IFI Robotics

See listing for Innovation First (this section).

Innovation First 203389

9701 Wesley St.
Ste. 203
Greenville, TX 75402
USA

((903) 454-1978

✉ info@innovationfirst.com

🌐 http://www.innovationfirst.com/

Innovation First manufactures and sells a small but potent line of robot power control systems. Their Isaac32 robot controller and operator interface are common finds in the combat robot circuit. The company also offers an interesting "Robot Prototyping Kit" that includes standard-size building block components.

Additional products:

- Solid state relay
- 20A H-bridge ("small" motors)
- Radio modems

SEE ALSO:

http://www.eduRobotics.com/.

4 Ways to Control a DC Motor from a Computer

All of the following motor control techniques can be controlled by a computer or microcontroller.

- *Relays.* A single-pole, single-throw (SPST) relay can turn a motor on or off. A double-pole, double-throw (DPDT) relay can reverse the direction of a motor. A variation is the *solid-state relay*, which uses solid-state components (such as power MOSFETs, described later) rather than electromechanical contacts.

- *Bipolar transistors.* A single bipolar transistor can be used to switch a motor on or off. Four transistors, in an H-bridge arrangement, can be used to control the direction and power to a motor. Two transistors in the bridge should be NPN types; the remaining two, PNP types. Power transistors are needed for larger currents. *Transistor arrays* (ICs that contain several transistors) can be used in place of discrete components.

- *MOSFET transistors.* Like bipolar transistors, MOSFETs can be used to either switch a motor on or off or to control its direction. Four transistors are used in an H-bridge arrangement to control direction and power to a motor. Two transistors should be the N-channel type; the remaining two the P-channel type.

- *H-bridge IC.* H-bridge chips contain bipolar, MOSFET, or other types of transistors. There are either four transistors in the chip that can be externally connected to form an H-bridge, or the transistors are already internally connected in the H-bridge arrangement. H-bridge ICs can control direction and power to a motor.

In each case, components must be selected to properly handle the current demand from the motor. Damage to the component—as well as the computer or microcontroller connected to it—can result if the motor draws excessive current. Heat sinks, which help dissipate heat due to high current loads, are generally used with the preceding components (except electromechanical relays).

J R Kerr 203854

990 Varian St.
San Carlos, CA 94070
USA

✉ sales@jrkerr.com

🌐 http://www.jrkerr.com/

J R Kerr manufactures and distributes DC servo modules and drivers. They offer complete boards as well as preprogrammed component chips for DC servo control. They sell mainly through distributors, including Jameco. The company's main product is the PIC-SERVO line of low-cost, easy-to-use motion controllers.

Lab Electronics 202062

253 James St.
Mt. Ephraim, NJ 08059
USA

☎ (609) 933-0351

📠 (609) 933-0351

✉ sales@lab-elec.com

🌐 http://lab-elec.com/

Lab Electronics makes and sells motor controllers and related electronics. Their products include:

- MC-003 DC motor controller board
- DR-002 decoder counter for incremental encoder
- MD-18201 H-bridge driver board
- SD-001 stepper motor driver

Lud's Open Source Corner 204056

http://drolez.com/hardware/

How-to articles with code and programming examples on:

- Atmel AVR-Open Source Software: PWM/Servo controllers with GPIO and serial interface
- Palm Cybot -Control a Cybot from your PalmOS-based device and SmallBASIC

Microcontroller Based Motor Speed Control 202232

http://www.webelectricmagazine.com/01/2/speed.htm

In multiple parts: how to construct a motor control circuit, complete with power H-bridge using MOSFET transistors. The microcontroller in the example is an 89C2051.

Micromech 202189

5-8 Chilford Court
Braintree, Essex
CM7 2QS
UK

☎ +44 (0) 1376 333333

📠 +44 (0) 1376 551849

🌐 http://www.micromech.co.uk/

Purveyors of stepper and servomotors and their associated control circuitry. Also: DC gearhead motors, gearheads, and X-Y translation tables. Online sales, including clearance items of old, discontinued, and demo product.

Mobot Building Info Pages 202214

http://www.mobots.com/makingMobots/

From Mobots.com; a small handful of useful tutorials on motor drive and motion control topics:

- PWM (Pulse-Width Modulation) DC Motor Speed Control with 555 Timer
- Stepper Motor Tester Using the AVR AT90S1200
- MCUs or Controller Boards?
- Quadrature Decoding Demo

Motion Control Buyer's Guide 🏅 204051

http://www.motioncontrolmall.com/

Motion Control Buyer's Guide also calls itself "Motion Control Mall." This is one mall Kristen and Todd and all their buddies from high school won't be hanging

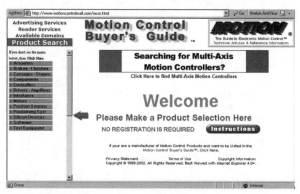

Motion Control Buyer's Guide Web page.

out at. No, this mall is for locating industrial motion control product, including actuators and motors, drivers, position sensors, and linear stages.

Ad based, but not in your face. The design of the Web site is quite good, and there are tons of listings.

MotionShop.com 204016

http://www.motionshop.com/

Everything on motion control: catalogs, links, listings, manufacturer summaries, and news stories.

MotorControl.com 203139

http://www.motorcontrol.com/

Portal for motor control engineering community.

Open Source Motor Controller 202516

http://www.dmillard.com/osmc/

The OSMC is an open source project to build (and improve) a 50-amp motor controller. See the mailing list at:

http://groups.yahoo.com/group/osmc/

PID and Servos 203651

http://www.ctc-control.com/litres/tutorials/pid.asp

L293D Push Pull 4-Channel Driver

The L293D is a 16-pin integrated circuit (the surface mount version is 20-pin) designed to drive loads such as motors, relays, and solenoids. It sports four separate drivers and can be used as an H-bridge for two motors (each motor uses two of the chip's outputs). The "D" suffix means that the IC has built-in clamping diodes on its outputs. These clamping diodes help prevent electrical damage when driving an inductive load.

The L293D has two supply voltages: one for the logic portion of the chip, and one for the driver output portion. The logic voltage and driver output voltages can be 4.5 to 36 volts, though the driver output voltage cannot be any less than the logic voltage (that is, if the logic voltage is 5 volts, that's the minimum for the driver output as well).

With a small heat sink, the L293D can handle loads of up to 600 milliamps per channel, with a peak draw of 1.2 amps. This isn't enough for large motors on combat robots, of course, but it's sufficient for the typical small carpet-roving 'bot.

Higher current carrying is possible merely by stacking two L293D chips on top of one another—just be sure to provide a good heat sink. Another option is to use the L293 (no D suffix) or the L293B, both of which can carry up to 1 amp per channel. If you use either of these versions, however, you will need to add your own clamping diodes on each output. The datasheets for the L293 and L293B provide an example schematic on how to do this.

The L293D is manufactured by several companies. ST Microelectronics and Texas Instruments both make compatible devices, and TI offers an upgraded version, the SN75441, which sports clamping diodes and a 1-amp-per-channel current capacity. The SN75441 is pin-for-pin compatible with the L293D, so no change in your circuit is required to use it.

Datasheets for the L293, L293D, and Texas Instruments SN75441 can be found at these locations (you need Adobe Acrobat Reader to view these files):

Chip	Datasheet
ST Micro.—L293D	http://us.st.com/stonline/books/pdf/docs/1330.pdf
TI—L293D	http://www-s.ti.com/sc/ds/l293d.pdf
ST Micro.—L293B	http://us.st.com/stonline/books/pdf/docs/1681.pdf
TI—L293	http://www-s.ti.com/sc/ds/l293.pdf
TI—SN75441	http://www-s.ti.com/sc/ds/sn754410.pdf

Informational page on using PID for servomotor control.

PID Without the Math   203081

http://members.aol.com/pidcontrol/booklet.html

Informational page about PID Without the Math, a semitechnical overview of controlling servomotors.

Precision MicroDynamics, Inc.,   204074

#3-512 Frances Ave.
Victoria, British Columbia
V8Z 1A1
Canada

((250) 382-7249

((250) 382-1830

✉   mailto:sales@pmdi.com

🌐   http://www.pmdi.com/

Motion control hardware and software for Windows and QNX operating systems, DC amplifiers, and data acquisition cards.

Robot Electronics   202242

Unit 2B Gilray Road
Diss
Norfolk
P22 4EU
UK

(+44 (0) 1379 640450

(+44 (0) 1379 650482

✉   sales@robot-electronics.co.uk

🌐   http://www.robot-electronics.co.uk/

Robot Electronics (sometimes referred to as Devantech) manufactures unique and affordable robotic components, including miniature ultrasonic sensors, electronic compass, and a 50-amp H-bridge for motor control.

The company's SRF08 high-performance ultrasonic rangefinder module can be connected to most any computer or microcontroller and provides real-time continuous distance measurements using ultrasonics. The measurement values are sent as digital signals and are selectable between microseconds, millimeters, or inches.

50 amp motor controller from Robot Electronics. Photo Robot Electronics.

Robot Power   203978

31808 8th Ave. S.
Roy, WA 98580
USA

((253) 843-2504

✉   sales@robot-power.com

🌐   http://www.robot-power.com/

Develops motor control products derived from the Open Source Motor Control (OSMC) project. The company's main product is a 160-amp, 50V forced-air-cooled H-bridge driver for permanent magnet motors.

Robot Powertrain Calculator   203399

http://www.killerhurtz.co.uk/howto/calculator.htm

Calculate torque, efficiency, and other important motor data using this handy JavaScript calculator. See also the Java applets and JavaScript programs at:

http://www.johnreid.demon.co.uk/howto/

RobotLogic   300008

Greg Hjelstrom
10416 Snowdon Flat Ct.
Las Vegas, NV 89129
USA

✉   support@robotlogic.com

🌐   http://www.robotlogic.com/

RobotLogic makes and sells a special controller for operating three-wheeled robots that use omnidirectional wheels (also called omniwheels, or multidirectional wheels). The company's OMX-3 Omni-Directional R/C Mixer is designed for three-wheeled omnidirectional robots (these include the PalmPilot

Robot Kit, sold by Acroname and others). According to the Web site, the controller "mixes three input R/C channels: x, y, and spin, to generate the appropriate signals for motors controlling three omni-directional wheels mounted 120 degrees from each other."

Other products sold by the company include a mixer controller for four-wheeled vehicles that use omnidirectional wheels, as well as mixers for standard differentially steered (two-wheeled) robots, a serial servo controller, and an R/C PWM controller designed to permit PWM control of DC motors using R/C pulse-duration-type signals.

The owners of the Web site are combat robot builders. . . be sure to check out their bots!

Rockwell Automation   203871

Firstar Building
777 East Wisconsin Ave.
Suite 1400
Milwaukee, WI 53202
USA

((414) 212-5200
🖰 (414) 212-5201

 http://www.rockwellautomation.com/

Rockwell Automation manufactures a broad line of automation electronics and components. See the listing under Actuators-Motion Products.

Sensoray   203201

7337 S.W. Tech Center Dr.
Tigard, OR 97223
USA

((503) 684 8005
🖰 (503) 684 8164
🌐 http://www.sensoray.com/

Sensoray's products are used in a wide range of applications, from motor control to digital video security.

Simple Step LLC   203740

400 Morris Ave.
Suite #273
Denville, NJ 07834
USA

Motor Bridge ICs

A motor bridge is an electronic circuit capable of controlling the operation and direction of a motor. You can build a motor bridge out of relays or transistors or purchase a specially made integrated circuit that does it all for you. Though motor bridge ICs used to be quite expensive, they are routinely available in the $2 to $20 range, from a variety of online and mail-order outlets.

Motor control bridges have two or more pins on them for connection to control electronics. Typical functions for the pins are:

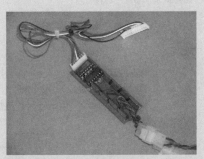

Bridges using discrete components, like this one, tend to make much larger modules than motor bridge ICs.

- *Motor enable.* When enabled, the motor turns on. Some bridges let the motor "float" when disabled, which causes the motor to stop. On other bridges disabling the motor causes a full or partial short across the motor terminals, which acts as a brake to stop the motor very quickly. By pulsing the enable pin using pulse-width modulation (PWM), the H-bridge is able to change the speed of the motor.

- *Direction.* Setting the direction pin changes the direction of the motor.

- *Brake.* On bridges that allow the motor to float when the enable pin is disengaged, a separate brake input is used to specifically control the braking action of the motor.

Better motor control bridges incorporate overcurrent protection circuitry to avoid damage to themelves if the motor pulls too much current and overheats the chip.

(973) 423-2395

 (253) 981-9343

general@simplestep.com

http://www.simplestep.com/

Simple Step makes products for automation, instrumentation, and robotics. Their Simple Step Motion Control system control stepper motors with single-voltage operation and plain ASCII commands sent via RS-232. Versions for microstepping and quadrature encoder feedback are available.

Solutions Cubed   204042

256 East First St.
Chico, CA 95928
USA

(530) 891-8045

(530) 891-1643

solcubed@solutions-cubed.com

http://www.solutions-cubed.com/

Solutions Cubed manufactures a number of motor control and other "modules" for common microcontroller applications. These products include:

- Motor Mind B-DC motor control via serial line; controls DC motor direction and speed (up to 30VDC motors, 3.5A peak)

- Motor Mind C-Like MMB, but with optional active cooling and with maximum continuous current of 4.0A

- ICON H-Bridge-DC motor control interface, can control brushed DC motors up to 40V and 12.0A

- ICON Serial Adapter Board-For RS-232 connections from a PC

- Easy Roller Robot Wheel Kit-Motor, wheel, and nice aluminum mount in one simple unit

ICON Interface Module, from Solutions Cubed.

All boards are expertly engineered and produced. Products are available through distributors or directly from the company.

ICON Adapter, from Solutions Cubed.

Stepbots   203008

http://www.voti.nl/stepbots/index.html

Robots made with stepper motors. Web site includes a brief explanation of how stepper motors work, plus schematics and programming stepper motors with PIC microcontrollers.

Stepper Motor Archive, The   203196

http://www.wirz.com/stepper/

The Stepper Motor Archive "is an online database of stepper motor data including coil-winding diagrams and electrical and mechanical data." A handy reference.

Team Delta Engineering   202174

1035 North Armando St., Unit D
Anaheim, CA 92806
USA

(208) 692-4502

dan@teamdelta.com

http://www.teamdelta.com/

Home of TeamDelta robot combat group. Also sells components and other goods in support of combat robotics. See the listing under Actuators-Motors.

The Basics of Relays

Relays in your robot? Some robot builders blanch at the idea, considering relays to be old-fashioned. Better to use all-electronic solutions, such as H-bridges or power transistors. While modern electronics certainly has its place, so do older components. Circuits can be built without resistors and capacitors, yet these components don't suffer the same ridicule as relays.

The truth is, in certain situations, the electromechanical relay may be the best and cheapest solution. They're not always the right solution, so like everything else in robots, you need to weigh your choices. But, if you value your time, energy, and pocketbook, you won't automatically discount relays simply because they are old-fashioned.

How Relays Work

A relay is composed of two sections: an electromagnetic coil and a set of contacts. These sections are electrically isolated, so that the coil is energized with one circuit, and the contacts are used in another circuit.

When electricity (AC or DC) is applied to the coil, the contacts close or open, depending on their design. In most relays, when the electricity is removed from the coil, the contacts return to their original state, because they are spring loaded. In some relays, the contacts "latch," and their position must be explicitly changed. Latching relays aren't used extensively in robotics, but they are in security systems and other applications where the contacts must remain open even if power is removed from the coils.

The contacts of a relay are like a mechanical switch. There can be one set of contacts or many. Each set of contacts is referred to as a pole. In addition, each pole can have one or two positions. A single-position pole is either open or closed, but a double-position pole has contacts for both open *and* closed.

The pole and position nomenclature for relays is the same as it is for switches:

SPST	Single-pole, single-throw	The relay contacts are either open or closed when energized. The normal condition (when not energized) is indicated: NO, for normally open, and NC, for normally closed.
SPDT	Single-pole, double-throw	The relay has a common and both NO and NC contacts. This type of relay can be used for NO or NC operation or both.
DPST	Double-pole, single-throw	Same as SPST, except a separate set of poles is added for an additional circuit. For example, the relay can control both a 5-volt and a 12-volt circuit.
DPDT	Double-pole, double-throw	Two sets of poles, with NO and NC contacts for each.

Specialty relays have additional poles—3, 4, and even more poles—but these are seldom used in amateur robotics. This type of relay tends to be physically large, as well.

Coil Voltage and Current Draw

The coil of the relay is energized by a control circuit. Coils are rated for a specific voltage. Relay coils for 5 volts are ideal for use on robotics, because 5 volts is the common operating voltage of many computer circuits. This makes it easy to interface the relay to a microcontroller or computer.

When energized, relay coils consume a certain amount of current. It is important to note this current, not only because any current draw will discharge the batteries of your robot, but also some relays may draw too much current from the control electronics connected to it. The average 5-volt SPST relay may draw from 20 to 30 milliamps (mA). This is no more than a brightly lit LED, but some microcontroller or computer I/O ports may not be able to provide this much current. In such cases, you need to add a buffer, such as a TTL gate or a small signal transistor, to increase the current-handling capacity.

The basic approach for connectiing an I/O port to a relay through a transistor is available in most any book on beginning electronics. Common in most such circuits is a resistor, from 1.0K to 4.7K at the base of the transistor. Start with a high value and work down; you've picked the right resistor value when the relay engages when you activate the I/O port. A diode is also used across the terminals of the coil. This is important because when the relay turns off, a rush of current will flow back through the coil and possibly into the transistor. The diode prevents damage to your electronics.

Contact Amperage

The contacts of the relay are rated for the current they can carry. The rating is typically specified as amps at a particular voltage, usually 250 VAC volts or 30 VDC. Forgetting the voltage specification, you need to ensure that the contacts will handle the current you wish to pass through the relay. If you are using the relay to control motors that draw upward of 3 amps each, then the relay contacts must be able to handle no less than three amps.

There is generally no need to overspecify the amperage-handling capacity of the relay, unless you notice the contacts are "sticking" (they don't release when you deenergize the coil), if there is a nasty spark when the contacts are disengaged, or if the relay becomes noticably hot.

Relay Case Styles

Relays designed for heavy-duty use, controlling several amps of current, are more likely to be the big, gompy, and ugly things your mother warned you against. Many are enclosed in clear plastic so you can see the innards of the relay in all its glory. There's no reason to you can't cover or paint the plastic to "dress up" the relay, but just don't use any material that's flammable.

Relays for smaller loads are smaller—some no bigger than an integrated circuit mounted on a socket. These types of relays are popular for controlling lighter loads, like small motors or LED lamp clusters. The small relay variety may be electromechanical—equipped with a coil and spring-loaded contacts—or it may be the reed type. Reed relays use a much simpler construction: A coil is wrapped around two pieces of metal enclosed in a glass ampoule. When the coil is energized, the metal pieces are drawn toward (or away, in the NC style) from one another.

Torque Speed Applet   204075

http://www.pmdi.com/calculator/tsp/tspApplet.html

Java applet for calculating torque/speed and power/speed of a DC motor. You need to input several values (which you obtain from the specification sheet for the motor). Requires Java.

From Precision MicroDynamics, Inc.— http://www .pmdi.com/

Using a PID-based Technique for Competitive Odometry and Dead Reckoning   202239

http://www.seattlerobotics.org/encoder/200108/using_a_pid.html

Lengthy and detailed article, written by G. W. Lucas, on using PID motor control algorithms to develop a system of dead-reckoning navigation for competition robots.

Pros and Cons of Relays

First, let's look at some advantages of using relays.

- Relays can offer a size advantage on systems that require heavy heat sinking for power electronics. A relay with 10-amp contacts may be only 3/4 inch square, but a heat sink for a 10-amp power MOSFET can be many times that.

- Relays don't require adapting to ensure proper operation in the circuit. Many transistor-based power circuits require you to determine the characteristics.

- Relays are relatively inexpensive, so if you wreck one while experimenting, replacing it won't set you back as much. Replacing an all-electronic control can be quite costly. Now for a few of the more salient disadvantages:

- Relays for very high current (25 or 50 amps and above) can be big, loud, and heavy. All-electronic controls—the solid-state relay or the power MOSFET—become the better choice for robotic applications.

- Relays wear out over time, though the number of switches is in the tens or hundreds of thousands. When used within their design limits, electronic controls never wear out and can be expected to last years, trouble free.

- Some applications, like pulse-width modulation of motors, cannot be accomplished with relays. An electronic motor control solution is required in these instances.

OUTSIDE-OF-THE-BOX

Many products have obvious uses: a machine screw is meant to fasten two or more pieces together. But sometimes, looks can be deceiving, and items that appear to be useful for a narrow application can be adapted to serve other purposes. This is called "outside-of-the-box" thinking, where you take something meant for one job and give it a completely different job to do.

The sources in this section are merely representative of outside-of-the-box products that can be applied to robot building. Most offer a free catalog, printed or online, and it costs nothing to scan their offerings to see what piques your interest. The more you study catalogs, the more you'll be able to find "just the right thing" for your next robot.

Archie McFee & Co.   202590

P.O. Box 30852
Seattle, WA 98103
USA

((425) 349-3009
 (425) 349-5188
 mcphee@mcphee.com
 http://www.mcphee.com/

Novelties and some surplus. Everyone needs a rubber chicken. Retail store in Seattle, Wash.

Archie McFee sells rubber chickens, and other novelties. Be brave and use them with your robots!

Carol Wright Gifts   202591

P.O. Box 7823
Edison, NJ 08818-7823
USA

((732) 287-8811
 (732) 572-2118
 cwcustserv@CarolWrightGifts.com

 http://www.carolwrightgifts.com/

General low-cost gifts and trinkets. How about hacking the motorized nose- and ear-hair remover to make a miniature robot?

Discount Package Supply, Inc.   203083

2415 S. Roosevelt
Ste. 101
Tempe, AZ 85282-2015
USA

((480) 921-7429
⊘ (800) 373-7713
info@discountpackage.com
http://www.dispac.com/

Packaging materials. Good source for clean cardboard, bubble pack, and cushioning foams.

EZ Flow Nail Systems   203308

10561 Dale St.
Stanton, CA 90680
USA

((714) 236-1188
(714) 236-1185
⊘ (800) 552-1477
ezflow@ezflow.com
http://www.ezflow.com/

Stay with me here. Consider: metallic power, acrylic paint, fiberglass filler, tiny drill, and shaping bits. Fingernail system, ha! Sounds like robot-building stuff to me.

Lifestyle Fascination Inc.
http://www.shoplifestyle.com/

Gifts for those into high-tech and "neato" stuff; include some high-end robotics, as well as robotic toys

Brainwaves
http://www.brainwaves.co.uk/
Novelties, high-power LEDs

Fun Express, Inc.
http://www.funexpress.com/
Mainly a distributor for retail stores and amusement prizes

Fun For All! Toys 202762

P.O. Box 985
Rehoboth Beach, DE 19971
USA

☎ (302) 227-1015
📠 (630) 839-7252
🚫 (877) 332-7697
✉ info@ffat.com
🌐 http://www.funforalltoys.com/

Novelties and toys.

Heartland America 202594

8085 Century Blvd.
Chaska, MN 55318-3056
USA

🚫 (800) 943-4096
🌐 http://www.8002292901.com/

General (and mostly inexpensive) gifts and novelties. Stuff you see *As Advertised on TV*. Now, what kind of robot can you build with that slicer-n-dicer?

Johnson-Smith Co. 🎖 202596

4514 19th St. Ct. E
Bradenton, FL 34203
USA

🚫 (800) 551-4406
✉ custservjs@jsls.com
🌐 http://www.johnson-smith.com/

Well-known novelty toy company and creator of the phrase, "Things You Never Knew Existed." Give your robot a pair of X-Ray Vision glasses.

Or how about a remote-control flatulence transmitter? Your robot doesn't need gas to benefit from this; what you end up with is a hackable RF transmitter and receiver with a hundred-foot range for less than you'd probably pay elsewhere for the same technology.

Lightweight Backpacker, The 203555

http://www.backpacking.net/

Information and ideas on lightweight backpacking, including light but strong carbon-composite materials for the backpack frame.

SEE ALSO:

http://www.litebackpacker.com/

Oriental Trading Company 202763

4206 South 108th St.
Omaha, NE 681370-1215
USA

☎ (402) 331-6800
📠 (402) 596-2364
🚫 (800) 875-8480
🌐 http://www.oriental.com/

Oriental Trading Company sells all those little trinkets and toys you get in exchange for tickets at the kids' zone amusements. Well, you can buy them for yourself, usually in boxes of dozens, hundreds, and even thousands. The good thing about these products is that they're cheap and many can serve as materials for constructing robots. The company's catalog also includes craft supplies.

Oriental Trading Company sells bulk items of toys and novelties, some of which can be used for robot parts.

Silly Universe 202779

2612 Needmore Rd.
Ste. 108
Dayton, OH 45414
USA

☎ (937) 277-2000
📠 (937) 277-2508
✉ customerservice@sillyuniverse.com
🌐 http://www.sillyuniverse.com/

Wacky products . . . for wacky wobots? How about a robot with a realistic (and life-size) mutant hand? Or rolling wheels made of feet?

 **PORTAL**

The listings in the following three sections are portals, all-purpose Web sites that offer a mix of news, links, articles, and possibly user-to-user forums. MSN.com and Yahoo.com are good examples of generic portals; the ones here are strictly related to robotics, electronics, and associated endeavors. You'll find portal listings for:

- Other. Portals that don't fit the other two categories.
- Programming. Gathering grounds for those who like to program in various languages and on various platforms.
- Robotics. Specialty Web sites just for and about robotics. Most are aimed at amateur robotics.

SEE ALSO:

INTERNET-BULLETIN BOARD/MAILING LIST: More user-to-user forums

INTERNET-SEARCH: Search for robotics and related topics

INTERNET-USENET NEWSGROUPS: Even more user-to-user forums

USER GROUPS: Share ideas with others, either locally or via the Internet

 **PORTAL-OTHER**

Control.com

203881

http://www.control.com/

In the words of the Web site: "Welcome to Control.com, the global online community of automation professionals. Check out the technical articles, scan the industry news, and participate in the technical discussions."

Controlled.com

202488

http://www.controlled.com/

Links to control—related products for various bus architectures, for "data acquisition, testing, automation, motion control, robotics, communications and more."

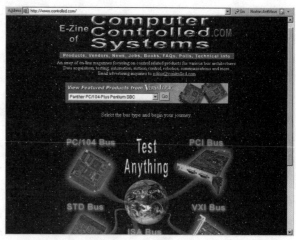

Controlled.com Web page.

eg3.com

202933

http://www.eg3.com/

The eg3.com Web site is a portal that indexes sites, services, and informational Web pages on embedded systems, DSP, real-time/RTOS, board-level computing, and related topics.

MotionNET.com

202570

http://www.motionnet.com/

A directory for mechanical engineers, especially those involved in motion mechanicals.

MotionShop.com

204016

http://www.motionshop.com/

Everything on motion control: catalogs, links, listings, manufacturer summaries, and news stories.

MotorControl.com

203139

http://www.motorcontrol.com/

Portal for motor control engineering community.

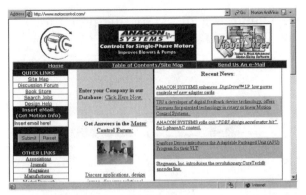

MotorControl.com Web page.

PaintballWatch.com 203864

http://www.paintballwatch.com/

Comparison shopping guide for paintball air-pressure system components. Some of these components can be used to construct pneumatic actuators for robots. Or, to make robots that spit paint at one another. Your choice.

Plasticsusa.com 204088

http://www.plasticsusa.com/

Links, classified ads, sources, user-to-user forums, and technical information about plastics and plastic-making processes. See also:

http://www.polymerweb.com/

Rocketry Online 203020

http://www.rocketryonline.com/

Links, how-tos, resources, sources, classifieds, and more on model rockets. Some rocketry materials have uses in robotics. These include altimeters, accelerometers, and construction materials, such as foam, tubes, and plastic sheets.

Wearables Central-Wearable Computing 203660

http://wearables.blu.org

Portal; links and information on computers you wear.

Be like Seven of Nine, every guy's favorite Borg. Wear your next computer. Or how about a "wearable robot"? -a robot you wear facilitates your daily life. It might hold your cell phone while you're driving and talking to Mom;

it might carry the third soda at the movie theater; it might nudge you awake if you doze off while at work. (Okay, forget the last idea. The first two sound good, though.)

Woodworking Pro 202995

http://www.woodworkingpro.com/

A portal for the woodworking professional. News, links, buyer's guide, and a woodworkers magazine (free to qualified readers).

World Tube Audio Portal 202999

http://www.worldtubeaudio.com/

Specialty source for electronic tubes (audio and power), but also provides useful links to general electronics sources, such as soldering and tools. Look in the Directory section.

 PORTAL-PROGRAMMING

AVR Forum 203310

http://www.avr-forum.com/

Stomping grounds for geeks who are into the Atmel AVR line of 8-bit microcontrollers. Includes links, sample code, user-to-user forums, and an AVR FAQ.

AVRFreaks 203022

http://www.avrfreaks.org/

User-to-user forums, code examples, resources, application notes, articles, and links for those involved with programming the Atmel AVR line of 8-bit microcontrollers. Be sure to check out the free code library (requires free registration).

Future AI 202234

http://www.futureai.com/

"Your page for artificial intelligence." Tutorials, files, articles, news, and user-to-user forum.

Pekee.fr

http://www.pekee.fr/

Robot news, forum, newsletter; in French and English

TINI Resources 204067

http://www.tiniresources.com/

Resources for TINI and iButton projects. TINI and iButton were originally developed by Dallas Semiconductor, now owned by Maxim. Additional information can be found at:

http://www.ibutton.com/TINI/

AceUpLink 203373

http://www.aceuplink.com/

Robot Arena forum; also robotics portal: links, forums, news. Fiction and nonfiction robots.

AIBO-Life 203715

http://www.aibo-life.com/

The AIBO-Life Web site provides forums, chat, and news catering to the owners of the Sony AIBO and other robotic pets.

BEAM Online 202380

http://www.beam-online.com/

MOVERS AND SHAKERS

Hans Moravec

http://www.frc.ri.cmu.edu/~hpm/

Dr. Moravec has been designing and building robots since 1963. That's even before *Star Trek*! He is now Principal Research Scientist in the Robotics Institute of Carnegie Mellon University and the author of several books on robotics and artificial intelligence.

BEAM Online is a BEAM-specific portal for robots. Includes news, links, and a gallery of various BEAM robots.

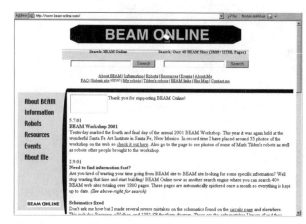

Web page for BEAM Online.

Botic: The World of Robotics 202051

http://www.botic.com/

News, links, and forums about robots.

GoRobotics.net 202056

http://www.gorobotics.net/

GoRobotics is an all-purpose robotics portal, with articles, reviews, links, resources, polls, project repository, and news.

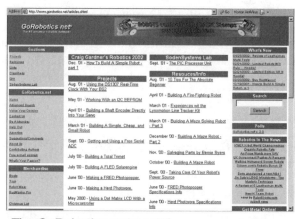

The GoRobotics.net portal.

Home-Robot.com 202518

http://www.home-robot.com/

General information Web site on home and hobby robotics.

hwatkins.com 203741

http://hwatkins.com:81/

Robotics portal, with links, news, project descriptions, and user-to-user forums.

Jonny555.co.uk 203708

http://www.jonny555.co.uk/

A robot gallery (industrial, commercial, movie, etc.) and a portal. Says the Web site: "Welcome to the best robotic site on the net. We pride ourselves on bringing you the most up to date robotic information available. You will discover facts about all kinds of robots, from simple robotic toys to the most advanced robots on earth and beyond."

MachineBrain.com 202819

http://www.machinebrain.com/

Robot news and links.

Netsurfer Robotics 204083

http://www.netsurf.com/nsr/

News, views, and book reviews, and more about our favorite subject.

OnRobo.com 203434

http://www.onrobo.com/

Home and entertainment robotics. Includes news, reviews, and a user-to-user forum.

Reconn's World 202979

http://www.reconnsworld.com/

Andrew, a.k.a. "Reconn," provides a message board on robots, some circuit examples, along with examples of his creations.

How Fast Can Your Robot Move?

If you know how fast the wheels of your robot turn, you can calculate the speed, in inches per second, that your robot will move. Making the calculation is easy:

1. Divide the speed of the robot, in revolutions per minute (rpm) by 60. The result is the revolutions of the motor per second (rps). A 100-rpm motor runs at 1.66 rps.
2. Multiply the diameter of the drive wheel by *pi*, which is approximately 3.14. This yields the circumference of the wheel. A 7-inch wheel has a circumference of about 21.98 inches.
3. Multiply the speed of the motor (in rps) by the circumference of the wheel. The result is the number of linear inches covered by the wheel in one second.

With a 100-rpm motor and 7-inch wheel, the robot will travel at a top speed of 35.168 inches per second, or just under 3 feet per second.

You can adjust the traveling speed of your robot by altering the diameter of its wheels. By reducing the diameter of the wheel by half, you reduce the traveling speed of the robot by half.

Bear in mind that the actual travel speed, once the robot is all put together, may be lower than this. The heavier the robot, the larger the load on the motors, so the slower they will turn.

Robot Channel, The 202421

http://www.therobotchannel.com/

Editorials and links about autonomous robots.

RobotCombat.com 202978

http://www.robotcombat.com/

RobotCombat.com (operated by ro-battler Jim Smentowski) is a portal for robotics, where the specialty of the house is machines that bash up each other.

A very useful feature is their continually updated links pages is:

http://www.robotcombat.com/links.html

Nightmare, a robot built by Jim Smentowski, host of the RobotCombat.com portal. Photo Jim Smentowski.

RobotGeeks.com 203462

http://www.robotgeeks.com/

News about robots. See also GoRobotics.net (this section).

Robotics Online 202710

http://www.roboticsonline.com/

Robotic Industries Association online Web site.

Robots.net 202409

http://www.robots.net/

Robots.net provides news on personal and industrial robotics, robot competitions, robotic sensors and com-

ponents, and more. Check out the Projects page, which lists dozens of robot creations from contributors.

Solarbotics.net 202030

http://www.solarbotics.net/

Solarbotics.net is a BEAM robotics community server, sponsored by the folks at Solarbotics (see http://www.solarbotics.com/). Here, you'll find user-to-user forums, columns by BEAM fans, links to Web sites, design tips and pictures, videos of BEAM robots walking, rolling, or hopping about, and more.

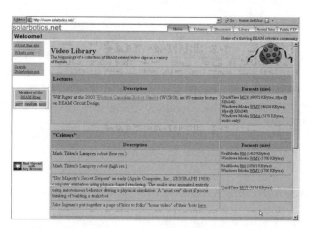

Solarbotics.net portal page, for BEAM robot enthusiasts.

Tech Geek 203454

http://www.techgeek.com/

A specialty portal all about "Building a robotic community."

Technopia: Robotics 203453

http://www.technocopia.com/robots.html

News and views about robots and automation.

TrueForce 203721

http://trueforce.com/

"The ultimate source of insight into robotics and automation."

POWER TRANSMISSION

"Power transmission" is anything that transfers power or movement from one device to another. This category is quite large and encompasses gears, belts, chains, bearings, bushing, shafts, collars, retaining clips, ball bearings, and other components used in moving mechanisms.

Many sources for power transmission components also sell other materials of practical application for robots. Be sure to review all of the offerings of a given source as they may also provide something else you've been looking for.

It's not uncommon for sellers of power transmission components to provide you with lots of technical detail but no prices! This is common in the industrial supply business, where finding just the right part is more critical than pricing. As amateur roboticists are typically long on dreams and short on cash, you'll want to specifically ask for a pricing sheet so you can compare costs. The outfits that require you to speak with a sales representative for every price you want are probably not worth the effort. Skip to the next source.

SEE ALSO:

ACTUATORS (VARIOUS): Additional motion products, including motors

FASTENERS: Used in machine construction, but also low-end linear actuators and other mechanical devices

MACHINE FRAMING: Build frames and rails for robots

MATERIALS-METAL AND MATERIALS-PLASTIC: Substitute less-expensive general metal and plastic for some power transmission products

RETAIL-SURPLUS MECHANICAL: An alternative source for overstocked and used components

SUPPLIERS-CASTING & MOLD MAKING: Cast your own gears and sprockets in plastic . . . yes, it can be done!

WHEELS AND CASTERS: So your robot can move around

Allied Devices   202122

325 Duffy Ave.
Hicksville, NY 11801
USA

☎   (516) 935-1300

📠   (516) 937-2499

✉   info@allieddevices.com

🌐   http://www.allieddevices.com/

Manufacturers and distributors of high-precision motion products and mechanical components. Offerings include:

• Rotary motion assemblies (gearheads, speed reducers, differentials)

• Rotary motion components (shafts, couplings, shaft adapters)

• Gears (including metric)

• Linear motion assemblies (racks, pinions, linear slides, ACME screws and leadnuts)

• Assembly hardware (screws, nuts, hangars, set screws, springs)

Bayside Automation Systems and Components   202478

27 Seaview Blvd.
Port Washington, NY 11050
USA

☎   (516) 484-5353

🚫   (800) 305-4555

🌐   http://www.baysidemotion.com/

Precision motion products: bearings, linear-positioning slides, gear reducers, and servomotor amplifiers.

Bearing Belt Chain   203509

3501 Aldebaran St.
Las Vegas, CA 89102
USA

☎   (702) 876-4225

📠   (702) 364-0842

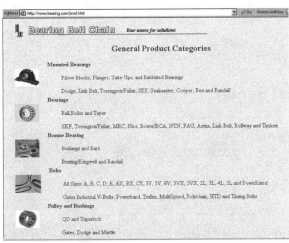

Bearing Belt Chain

sales@bearing.com

http://www.bearing.com/

Local and online retailer of bearings (linear, roller, taper, pillow, etc.), belts (including V and timing), sprockets, and chains. Large inventory.

Bearing Headquarters Co. 204186

P.O. Box 6267
Broadview, IL 60155
USA

(708) 681-4400

(708) 681-4462

http://bearingheadquarters.com/

Industrial bearings (all types), couplings, clutches, belt drives and rollers, gears, conveyor rolls and chain, sprockets, and chains.

See also Headco Industries:

http://www.headco.com/

Belt Corporation of America 202286

Dept L-12
3455 Hutchinson Rd.
Cumming, GA 30040
USA

(770) 887-4138

(800) 235-0947

sales@beltcorp.com

http://beltcorp.com/

Belts, and not the kind you wear. BCA offers timing belts, woven endless belts (can be useful to construct robot tank treads), natural rubber and neoprene stretch belts, and countless round belts.

Boca Bearing 204101

1500 S.W. 30th Ave #3
Boynton Beach, FL 33426
USA

(561) 998-0004

(561) 998-0119

(800) 332-3256

bearing@gate.net

http://www.bocabearings.com/

Boca specializes in small and miniature bearings for such applications as radio-control vehicles, inline skates, power tools, small appliances, fishing reels, and, of course, robotics. Bearings are listed by size, type, and general application.

Be sure to check out their engineering section, which has some one dozen helpful technical backgrounders on using bearings.

Boston Gear 202124

14 Hayward St.
Quincy, MA 02171
USA

(617) 328-3300

(617) 479-6238

(888) 999-9860

info@bosgear.com

http://www.bostgear.com/

Gears, yes, but also bearings, transmissions, clutches, pneumatics, and assorted other power transmission and actuation products. Boston Gear also offers free literature, maintenance manuals, and operating instructions for their products.

Selection of gears. Courtesy Boston Gear.

Helical gears. Courtesy Boston Gear.

Commonly Used Power Transmission Components

There are literally thousands of power transmission components, but the following comprise the most commonly used and the most critical.

Gears

Gears are a principle component of power transmission and are primarily used in robotics to reduce the speed and increase the torque of the wheel drive motors. Because of the mechanical precision required to properly mesh gears, most amateur robot builders do not construct their own gear assemblies. Gears are more fully detailed elsewhere in these sections.

Spur gears.

Timing Belts

Also called synchronization belts. Typical timing belts for small mechanisms range from 1/8 to 5/18 inch in width and sizes from just a few inches to several feet in diameter. Material is usually neoprene, with metal or fiberglass reinforcement. Belts are rated by the pitch between "nubs" or "cogs," which are located on the inside of the belt. Timing belts are used with matching timing belt pulleys, which come with either ball bearing shafts (used for idler wheels) or with press-on or set screw shafts for attaching to motors and other devices.

Timing belts (or synchronization belts) use cogged belts and sprockets.

V-belts

V-belts have a tapered V shape and are used to transfer motion and power from a motor to an output when synchronization of that motion is not critical (because the belt could slip). V-belts, which are often made with metal- or fiberglass-reinforced rubber, are used with V-grooved pulleys. By changing the diameter of the pulleys, it's possible to alter the speed and torque of the output shaft in relation to the drive shaft. The same physics that apply to gears and gear sizes apply to V-belt pulleys as well.

Endless Round Belts

Endless round belts are used to transfer low-torque motion. The belts look like overgrown O-rings and, in fact, are often manufactured in the same manner. Other endless round belts are made by fusing the ends of rounded rubber (usually neoprene). Some belt makers provide splicing kits so you can make custom belts of any length. Grooved pulleys are used with round belts; as with V-belt pulleys, the diameter of the round belt pulley can be altered to change the torque and speed of the output.

Ladder Chain

Ladder chain resembles the links of a ladder and is used for fairly low torque and slow speed operations. Movement of a robotic arm or shoulder is a good application for ladder chain. With most chain, links can be removed and added using a pair of pliers. Special toothed sprockets, engineered to match the pitch (distance from link to link) of the chain is used.

Roller Chain

Roller chain is exactly the same kind as for bicycles, except for most small-scale machinery, the chain isn't as big. Roller chain is available in miniature sizes, down to 0.1227-inch pitch (distance between links). More common is the #25 roller chain, which has 0.250-inch pitch. For reference, most bicycle chain is #50, or 0.50-inch pitch. Sprockets with matching pitches are used on the drive and driven compo-

Roller chain and sprocket.

nents. Roller chain comes in metal or plastic; plastic chain is easier to work with and links can be added or removed. Many types of metal chains are prefabricated using hydraulic presses and require the use of "master links" to make a loop.

Idlers

Idlers (also called idler pulleys or idler wheels) take up slack in belt- and chain-driven mechanisms. The idler is placed along the length of the belt or chain and is positioned so that any slack is pulled away from the belt or chain loop. Not only does this allow more latitude in design, it also quiets the mechanism. The bores of the idlers are fitted with appropriate bearings or bushings.

Couplers

Couplers come in two styles

- Rigid
- Flexible

Rigid coupler.

Flexible coupler.

Couplers are used to directly connect two shafts together, thus obviating the need for any kind of gear or belt. Rigid and flexible couplers are detailed more fully in these sections.

Bearings

Bearings are used to reduce the friction of a spinning component, such as a wheel or idler, around a shaft. Several bearing constructions exist, with ball bearings being the most common. The bearing is composed of two concentric rings; between each ring is a row of ball bearings. The

Ball bearing.

Bearing in a pillow block.

rings—and the ball bearings—are held in place by a mechanical flange of some type. Bearings can be mounted directly to a device, which requires precision machining and a press to securely insert the bearing into place.

Another form of bearing uses narrow pieces of metal rod, called needles, and works in a similar manner.

Pillow blocks are available that allow bearings to be readily mounted on any frame or device.

Bushings

Bushings and bearings serve the same general purpose, except a bushing has no moving parts. (Note: Some people also call these bearings or dry bearings, but I prefer to use the term *bushing* in order to differentiate them.) The bushing is made of metal or plastic and is engineered to be "self-lubricating."

An example is Oilite, a self-lubricated bronze metal commonly found in industrial bushings. Several kinds of plastics, including Teflon, exhibit a self-lubricating property. Bushings are used instead of bearings to reduce cost, size, and weight and are adequate when friction between the moving parts can be kept relatively low. Bushings, and not the more expensive bearings, are used in the output gear of the less expensive R/C servos, for example.

A self-lubricating bronze bushing.

BRECOflex Co., LLC 203621

P.O. Box 829
Eatontown, NJ 07724
USA

☎ (732) 460-9500
📠 (732) 542-6725
🚫 (888) 463-1400
✉ info@brecoflex.com
🌎 http://www.brecoflex.com/

Manufacturer of belts: timing belts, profiled belts, flat belts, pulleys, belt tensioners, and slider beds.

Canadian Bearings Ltd. 203512

1401 Courtneypark Dr. E,
Mississauga, Ontario
L5T 2E4
Canada

☎ (905) 670-6700
📠 (905) 670-2632
🌎 http://www.canadianbearings.com/

Motion mechanicals: bearings, brakes, bushings, casters, clutches, motors, couplers, reducers, gears, sprockets, and conveyor belts. Locations across Canada.

Cruel Robots 202533

32547 Shawn Dr.
Warren, MI 48088
USA

✉ Dan@cruelrobots.com
🌎 http://www.cruelrobots.com/

Performance materials and products for combat robots. Colson wheels (some are "combat ready" with heavy-duty hub already attached), axles, reducers, hubs, sprocket and chain, weapons, and casters.

Danaher Motion MC 203507

45 Hazelwood Dr.
Amherst, NY 14228
USA

☎ (716) 691-9100
📠 (716) 691-9181
🚫 (800) 566-5274
🌎 http://www.danahermcg.com/

U.S.-based manufacturer and distributor of several top-quality motion product brands. See listing under Actuators.

David Price: On Bearings 202820

http://www.soton.ac.uk/~cds/bearings/
    bearings.htm

David gives us information about various kinds of mechanical bearings (plain bearings, ball bearings, roller bearings, thrust bearings) and how they are used in design.

Drives, Incorporated 204195

901 19th Ave.
Fulton, IL 61252
USA

 (815) 589-4420

⊘ (800) 435-0782

✉ custserv@drivesinc.com

🌐 http://www.drivesinc.com/

Drives, Incorporated makes and sells roller chain and "attachment products," as well as chain for conveyors. The chain is available in sizes from #35 (slightly smaller than bicycle chain) up to A2060, which has a pitch of 1 1/2 inches. So-called attachments include mechanical clips that seat into the chain-ideal for making heavy-duty tracked robots.

🏭

Dura-Belt, Inc. 204185

2909 Scioto-Darby Exec. Ct.
Hilliard, OH 43026-8990
USA

📞 (614) 777-9448

⊘ (800) 770-2358

✉ durabelt@iwaynet.net

🌐 http://www.durabelt.com/

Makers and sellers of round urethane endless belts (O-rings), quick-disconnected twisted belts, flat belts (in different thicknesses and widths), groove sleeves for round belts, idlers, and belt-splicing kits.

Dyna-Veyor 203768

10 Hudson St.
Newark, NJ 07103-2804
USA

📞 (973) 484-1119

📠 (973) 484-7790

⊘ (888) 484-1119

✉ DynaVeyor@aol.com

🌐 http://www.dyna-veyor.com/

Maker of plastic conveyor belt chain, sprockets, idlers, and related conveyor components. Intended mainly for the food industry, the components can also be used in the design of tracked robots.

Electronics Parts Center 202902

1019 S. San Gabriel Blvd.
San Gabriel, CA 91776
USA

📞 (626)-286-3571

📠 (775) 257-1375

⊘ (800) 501-9888

🌐 http://www.electronicsic.com/

Specializes in replacement/service parts for electronics products (power supplies, monitors, and TVs-you name it). Look up parts by part number or function. Includes mechanical VCR parts, such as rollers, gears, and belts. This is one way to get mechanical components for cheap, though your engineering selection is somewhat limited.

Emerson Power Transmission Manufacturing 204030

8000 W. Florissant Ave.
St. Louis, MO 63136
USA

📞 (314) 553-2000

📠 (314) 553-3527

🌐 http://www.emerson-ept.com/

Mondo major manufacturer of power transmission and motion products. Brands include:

- Browning-world leader in V-belt drives
- Morse-roller chain drives
- SealMaster-bearings, rod ends
- US Gearmotors-fractional-horsepower AC and DC gearmotors
- Rollway-2,000 types of bearings
- Kop-Flex-industrial shaft couplings

Gates Rubber Co. 203038

900 South Broadway
Denver, CO 80217-5887
USA

📞 (303) 744-1911

🌐 http://www.gates.com/

Gates is a major supplier of belts or timing and power transmission, for both industry and automotive applications. Among the most useful belts (for robotics) in their line are:

- Synchronous belts-or "timing" belts, they keep parts of a mechanism working together
- Vectra and V-belts-belts with trapezoidal shapes that work in V-shaped pulleys

The company's products are available through distributors, though for low-end amateur robotics, you may find your best bet is buying them at surplus. In exchange for spending a lot less money, you must give up the ability to specify an exact size of belt you need. However, in many cases, standard-sized belts available on the surplus market can be made to work in your robot projects simply by rearranging the position of the parts.

Synchronized belt and pulley. Courtesy Gates Rubber Company.

Georgia Tech: Gears and Gear Trains 202287

http://srl.marc.gatech.edu/education/
   ME3110/primer/geartit.htm

Tutorial on gears and gear trains. Part of Georgia Tech's curriculum resources.

Go Kart Supply 204230

12784 Mansfield Rd.
Keithville, LA 71047
USA

((318) 925-2224

 gokarts@gokartsupply.com

 http://www.gokartsupply.com/

Parts for go-karts and mini bikes (and therefore for the plus-size robots out there), including bearings, drive sprockets and chain, axles, wheels, replacement tires, control cables, and clutches.

Helical Products Co., Inc. 202479

901 W. McCoy Ln.
P.O. Box 1069
Santa Maria, CA 93456-1069
USA

((805) 928-3851

(805) 928-2369

 sales@heli-cal.com

 http://www.heli-cal.com/

Helical flexible couplings. Many different sizes, styles, and materials.

Huco Engineering Industries Ltd 204199

Merchant Drive
Hertford
Hertfordshire
SG13 7BL
UK

(+44 (0) 1992 501900

+44 (0) 1992 500035

hei.sales@huco.com

 http://www.huco.com/

Manufacturer of flexible couplers. Products include three-part couplers with replaceable wear elements, one-piece couplers, and plastic universal joints.

igus GMBH 203444

Spicher Straße 1 a
D-51147 Köln
Germany

(+49 (0) 2203 96490

+49 (0) 2203 9649222

 http://www.igus.de/

Makers of polymer (plastic) bearings, chain, linear slides, and other mechanicals. Web site is in many languages, including English and German.

Industrial Links Ltd 203670

19 Ventura Place
Upton Industrial Estate
Upton Poole
BH16 5SW
UK

A spur gear, with setscrew hub.

✆ +44 (0) 1202 632996
🖷 +44 (0) 1202 632997
✉ sales@industrial-links.com

🌐 http://www.industrial-links.com/

Industrial mechanical-bearings, belts, seals, fasteners, chain, sprockets, etc.

Invensys Plc 204159

Carlisle Place
London
SW1P 1BX
UK

✆ +44 (0) 2078 343848
🖷 +44 (0) 2078 343879
🌐 http://www.invensys.com/

Getting Geared Up

Of course, you can always buy gears from Gears R Us. (Okay, most go by far more mundane names like Boston Gear, Small Parts, W. M. Berg, and Stock Drive.) You'll get just what you're looking for from these sources, but it'll cost you. The average machined 1-inch-diameter aluminum gear can cost between $20 and $30.

As long as your requirements aren't too unusual, you may be able to locate the gears you want from other products and sources.

- *Toy construction sets.* Don't laugh! Toys like LEGO, Erector, and Inventa come with gears you can use in your robotics projects. Most are on the large size and are made of plastic.

- *Hobby and specialty retailers.* Next time you're at the hobby store look for replacement gear sets for servos and drive motors for R/C cars and airplanes. Some are plastic; others are metal (usually either aluminum or brass). Typically, you'll have to buy the whole set of replacement gears for whatever motor or servo the set is for, but in other cases you can purchase just one gear at a time. Some online retailers, such as ServoCity.com and Jameco.com, sell gears specifically for hobby applications (like robots). The price is reasonable.

- *Surplus catalogs.* New gears can be expensive; surplus gears can be quite affordable. You can often find new gears, plastic or metal, for about 10 cents on the dollar, compared to the cost of the same gear new. The only problem: Selection can be limited, and it can be hard to match gear sizes and pitches even when buying gears from the same outlet.

- *Rechargeable electric screwdrivers.* Inside are numerous gears, typically in a "planetary" configuration, used to produce their very high speed reductions. Before raiding the screwdriver for just its gears, consider using the motor, too. The motor and gearing system of a typical electric screwdriver makes for a fine robot drive system.

- *Hacked toys.* Discarded and discounted toys make for good gear sources. These include friction and battery-powered toy cars, "'dozer" toys, even some action figures. Tear the toy apart for the treasure inside. These gears tend to be small and made of plastic.

- *Old kitchen appliances.* Go to thrift stores and garage sales and look for old food mixers, electric knives, even electric can openers. Unlike toys, kitchen appliances commonly use metal gears—or, at the least, very strong plastic gears.

Invensys is a large corporate parent of many motion control and automation brands. See the listing under Actuators.

JJC & Associates
202480

1386 Bello Mar Dr.
Encinitas, CA 92056
USA

- (760) 635-9183
- (760) 635-9184
- ⊘ (800) 576-1035
- john@jjcassociates.com
- 🌐 http://www.jjcassociates.com/

Custom and standard drive components. Belts, timing belts, pulleys, gears, plastic power drive components, rollers, collars, and clamps,

Karting Distributors Inc.
204232

700 N. University Dr.
Fort Worth, TX 76114
USA

- (817) 625-2562
- karting@flash.net
- 🌐 http://www.kartingdistributors.com/

Though intended for go-karts, the company's bearings, axles, sprockets, chain, and other mechanical components are useful on larger robots, especially those intended for mortal combat.

Lovejoy Inc.
204196

2655 Wisconsin Ave.
Downers Grove, IL 60515
USA

- (630) 852-0500
- (630) 852-2120
- feedback@lovejoy-inc.com
- 🌐 http://www.lovejoy-inc.com/

Lovejoy manufactures a line of affordable flexible couplers. These are designed to connect a motor drive with some driven device, like a pump or a wheel. Because they are flexible, the coupler allows the shafts of the driver and the drivee to be slightly out of whack from one another, and yet they won't tear each other apart.

One of the more common Lovejoy connectors in use for robotics is the jaw coupling, which consists of two machine halves that fit together with "fingers." A flexible material, called the spider, is sandwiched between. There are two benefits of jaw connectors: Halves in the same series can be mixed and matched, so the shaft sizes can be different for each side. The spider material is available from soft to fairly rigid, to better suit the application.

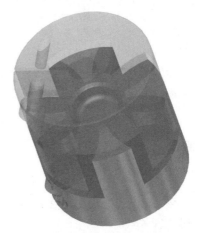

Three-piece jaw flexible coupler.

Manufacturer's Supply Inc
202638

P.O. Box 167
Dorchester, WI 54425
USA

- ⊘ (800) 826-8563
- sales@mfgsupply.com
- 🌐 http://www.mfgsupply.com/

Chainsaw, motorcycle, and engine parts. Includes wheels, chain, bearings, axles, snowmobile treads, and more. Check out the Go-Kart page:

http://www.GoKartParts.com/

Maryland Metrics
204105

P.O. Box 26
Owings Mills, MD 21117-0261
USA

- (410) 358-3130
- (410) 358-3142
- ⊘ (800) 638-1830
- sales@mdmetric.com
- 🌐 http://www.mdmetric.com/

Something of a one-stop shop, Maryland Metrics carries bearings, linear bearings, fasteners, rods, gears, pneumatic and hydraulic fittings, and a variety of power transmission items. Good assortment of technical info.

Minarik Corporation 203510

905 East Thompson Ave.
Glendale, CA 91201
USA

((818) 637-7500
(818) 637-7509
⊘ (800) 427-2757
 http://www.minarikcorp.com/

Full-line mechanical (bearings, shafts, gears, chain, etc.); electronics (PWM drives, sensors); online ordering plus many local warehouses throughout the U.S.

Miniature Bearings Australia Pty. 204015

Unit 4, 224 Wishart Rd.
Wishart, Queensland 4122
Australia

(+61 7 3349 1400
+61 7 3349 3801
✉ sales@minibearings.com.au
 http://www.minibearings.com.au/

Miniature bearings and other mechanicals. Offers wares to both industry and the hobbyist.

A bellows coupler.

Motion Industries 203508

1605 Alton Rd.
Birmingham, AL 35201-1477
USA

((205) 956-1122
⊘ (800) 526-9328
 http://www.motion-industries.com/

Full-line distributor/retailer of industrial products: bearings (linear and other), sprockets, actuators, chain, many others.

MSC Industrial Direct Co., Inc. 202826

75 Maxess Rd.
Melville, NY 11747-3151
USA

((516) 812-2000
⊘ (800) 645-7270
 http://www.mscdirect.com/

See listing under Materials.

Applied Controls
http://www.appliedc.com/
NJ-based distributor of motion products and sensors

Belt Corporation of America
http://www.timing-belt.com/
Industrial timing and synchronous belts

Bunting Bearings
http://www.buntingbearings.com/
Bearings, all types and materials

Gopher Bearing Co.
http://gopherbearing.com/
Bearings, belts, couplings, sprockets, and many other mechanicals

MRC Bearings
http://www.mrcbearings.com/
Industrial bearings

RBC Bearings
http://www.rbcbearings.com/
Bearings, ball screws, roller bearings

NAPSCO (North American Parts Search Company)　204098

4411 East Amberwood Dr.
Phoenix, AZ 480 759 820
USA

　(480) 759-8072

　(480) 759-8205

　napsco@itol.com

　http://www.napsco.com/

NAPSCO is a distributor of multiple lines of power transmission and industrial parts.

Nordex, Inc.　202331

426 Federal Rd.
Brookfield, CT 06804
USA

　(203) 775-4877

　(203) 775-6552

　(800) 243-0986

　info@nordex.com

　http://www.nordex.com/

Gears, miniature instrument bearings, shafts, Geneva mechanisms, fasteners, ball (linear and rotary) slides, brakes, clutches, couplings, assemblies, enclosed geartrains, and many other related precision components.

Northern Tool & Equipment Co.　202606

2800 Southcross Dr. West
Burnsville, MN 55306
USA

　(952) 894-9510

　(952) 894-1020

　(800) 221-0516

　http://www.northerntool.com/

See listing under Tools.

NSK　202922

3861 Research Park Dr.
P.O. Box 1507
Ann Arbor, MI 48106-1507
USA

　(734) 761-9500

　(734) 668-7888

　comm@nsk-corp.com

　http://www.nsk.com/

Power transmission -bearings, bushings, gears, sprockets, and more. Extensive technical details provided on the site, including online engineering calculators.

Universal joint coupler.

PIC Design　202483

86 Benson Rd.
P.O. Box 1004
Middlebury, CT 06762
USA

　(203) 758-8272

　(203) 758-8271

　(800) 243-6125

　sales@pic-design.com

　http://www.pic-design.com/

Precision mechanical components, motion control mechanicals, X-Y translation tables, leadscrews, belts, pulleys, and gear products.

Plastic Products, Inc.　203319

P.O. Box 188
Bessemer City, NC 28016
USA

　(704) 739-7463

　(704) 739-5566

　(800) 752-7770

　http://www.plastic-products.com/

Plastic Products offers sheets, rods, tubes, profiles, shapes, slabs, and "massive blocks," thick hunks of plastic suitable for machining or making doorstops. Their product line is intended for plastic molding, machining, and fabrication and is well suited for the robotics trade. The company also offers a full line of stock materials, such as foamed PVC rod and conveyor components (sprockets, cams, raceways, and other goodies). Check the Steals and Deals page.

PowerTransmission.com   204002

http://www.powertransmission.com/

PowerTransmission.com helps you find suppliers of gears, motors, bearings, clutches, couplings, speed reducers, and other components that transmit mechanical power. Most suppliers have Web sites, where you can compare products (but usually not prices, as you have to call or write for those). You'd be amazed just how many outfits are out there involved in motion products. Advertiser-based.

Buyers' guides are provided for these and other topics:
- Actuators
- Adjustable speed drives
- Bearings
- Belt and chain drives

Understanding Gears

Gears are used for two purposes:

- To transfer power or motion from one mechanism to another.
- To reduce or increase the speed of the motion between two linked mechanisms.

The simplest gear systems use just two gears: a drive gear and a driven (or output) gear. More sophisticated gear systems, referred to as *gear trains*, *gearboxes*, or *transmissions*, may contain dozens or even hundreds of gears. Motors with attached gearboxes are said to be *gearbox motors*.

Gear Teeth

Gears are specified not only by their physical size, but also by the number of teeth around the circumference. *Spur* gears are most common and are used when the drive and driven shafts are parallel. *Bevel* gears have teeth on the surface of the circle, rather than the edge. They are used to transmit power to perpendicular shafts. *Miter* gears serve a similar function but are designed so that no reduction takes place.

Spur, bevel, and miter gears are reversible—the gear train can be turned from either the drive or the driven end. Conversely, w*orm* and *leadscrew* gears transmit power perpendicularly and are not usually reversible The leadscrew resembles a threaded rod.

Rack gears are like spur gears unrolled into a flat rod. They are primarily intended to transmit rotational motion to linear motion.

The teeth of a gear provides mechanical traction.

Gear Reduction = Torque Increase

When gears are used to reduce the output speed of a mechanism—say, a motor—the torque at the output is increased. Gears are basically a form of lever; power can be increased by changing the ratio of the lever over the fulcrum. Substituting the fulcrum in a gear system is the number of teeth on each gear.

Gear reduction is accomplished by changing the ratio of teeth of mating gears: A two-gear system with a 100-tooth gear and a 50-tooth gear is said to have a 2:1 reduction. With such a system, output speed is reduced by 50 percent, and torque is roughly doubled.

- Brakes
- Clutches
- Controls
- Couplings and U-joints
- Gears and gear drives
- Linear motion
- Motors
- Sensors

Putnam Precision Molding, Inc.   204198

11 Danco Rd.
Putnam, CT 06260
USA

☎ (860) 928-2229
⊘ (877) 477-6462
✉ ccampbell@putnamprecisionmolding.com
🌐 http://www.putnamprecisionmolding.com/

Manufactures and sells the Plastock line of mechanical drive components. Products include:

- Timing belt pulleys
- Timing belts
- Chain sprockets
- Roller chain
- Spur gears

Quality Transmission Components   203516

2101 Jericho Tnpk.
Box 5416
New Hyde Park, NY 11042-5416
USA

☎ (516) 437-6700
☎ (516) 326-8827
🌐 http://www.qtcgears.com/

Medium and coarse metric pitch small gears and other power transmission goodies. Part of Stock Drive Products (see http://www.sdp-si.com/). Offered:

- Spur gears
- Helical gears
- Ring gears (internal and external)
- Racks (straight, helical, and flexible)
- Miter gears (straight and spiral)
- Bevel gears (straight and spiral)
- Worms and worm wheels (standard and duplex)

- Screw gears
- Involute splines (internal and external)
- Ratchets and pawls
- Gear couplings

Reid Tool Supply Co.   203820

2265 Black Creek Rd.
Muskegon, MI 49444
USA

⊘ (800) 253.0421
✉ mail@reidtool.com
🌐 http://www.reidtool.com/

Reid is an all-purpose industrial supply resource. They carry tens of thousands of items, including bearings, gears, linear shafts, leadscrews and nuts, ballscrews and ball nuts, multidirectional rollers (omniwheels), ball transfers and ball casters, light- to heavy-duty casters, machine framing (reseller of 80/20 Inc.), fasteners of all kinds, and much more.

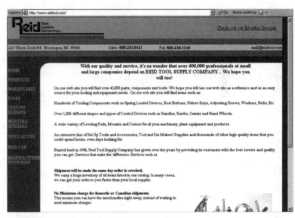

Web site for Reed Tool Supply.

RG Speed Control Devices Ltd.   203513

40A Courtland Ave.
Concord, Ontario
L4K 5B3
Canada

☎ (905) 761-7893
☎ (905) 761-6242
⊘ (800) 265-5304
🌐 http://www.rgspeed.com/

Gears, bearings, bushings, actuators, stepper motors, and motor controllers.

RMB Roulements Miniatures SA 204011

Eckweg 8
Box 6121
CH-2500 Biel-Bienne 6
Switzerland

(+41 32 344 4300
🖐 +41 32 344 4301
✉ info@rmb-group.com
🌐 http://www.rmb-ch.com/

Makers of incredible small miniature bearings and Smoovy motors. For the latter, see:

http://www.smoovy.com/

RS Components Ltd 203551

Birchington Road
Corby
Northants
NN17 9RS
UK

(+44 (0) 1536 201234
🖐 +44 (0) 1536 405678
🌐 http://www.rs-components.co.uk/

Multinational (based in the U.K.) retailer of various electronic and mechanical products. Mechanicals include pneumatics, gears, fasteners, bearings, and bushings.

SEE ALSO:

http://rswww.com/

Secs, Inc. 202123

520 Homestead Ave.
Mt. Vernon, NY 10550
USA

((914) 667-5600
🖐 (914) 699-0377
🌐 http://www.prosecs.com/

Gears and such. See the listing under Actuators-Motion Products.

Seitz Corp. 202484

212 Industrial Ln.
Torrington, CT 06790-1398
USA

🖐 (860) 496-1949
🚫 (800) 261-2011
🌐 http://www.seitzcorp.com/

Plastic gears, gears, and motion control mechanicals.

Common Gear Specifications

Here are some common gear specifications to keep you warm at night.

- *Pitch.* The size of gear teeth is expressed as pitch, which is roughly calculated by counting the number of teeth on the gear and dividing it by the diameter of the gear. Common pitches are 12 (large), 24, 32, 48, and 64. Odd-size pitches exist, of course, as do metric sizes.

- *Pressure angle.* The degree of slope of the face of each tooth is called the pressure angle. The most common pressure angle is 20 degrees, although some gears, particularly high-quality worms and racks, have a 14 1/2-degree pressure angle.

- *Tooth geometry.* The orientation of the teeth on the gear can differ. The teeth on most spur gears are perpendicular to the edges of the gear. But the teeth can also be angled, in which case it is called a helical gear. There are a number of other unusual tooth geometries in use, including double-teeth and herringbone.

Serv-o-Link 202064

5356 West Vickery
Fort Worth, TX 76107
USA

☎ (817) 732-4327

✉ servolink@home.com

🌐 http://servolink.com/

Serv-o-Link is the source for power transmissions with precision plastic gears, chain, and sprocket drives. The products are injection molded, so they're less expensive than machined gears in Delrin or metal, yet they are precise enough for many robotic applications. The snap-lock chain link design allows any pitch length; master links are not necessary, nor are special tools. The chain can be operated at relatively high speeds-1,000 feet per minute.

Miniature sprocket and chain, from Serv-O-Link. Photo Serv-O-Link, Corp.

Small Parts Inc. 202120

13980 N.W. 58th Ct.
P.O. Box 4650
Miami Lakes, FL 33014-0650
USA

☎ (305) 557-7955

📠 (305) 558-0509

🚫 (800) 220-4242

✉ parts@smallparts.com

🌐 http://www.smallparts.com/

Small Parts is a premier source for—get this!—small parts. All jocularity aside, Small Parts is a robot builder's dream, selling most every conceivable power transmission part, from gears to sprockets, chains to belts, and bearings to bushings. Product is available in a variety of materials, including brass, steel, and aluminum, as well as nylon and Delrin. Rounding out the mix is a full selection of raw materials: metal rod, sheets, tubes, and assorted pieces, as well as a huge assortment of fasteners.

Now about prices. Small Parts is for the serious builder, both amateur and pro. A little brass gear might cost $6, but what you pay for (apart from the precision, of course), is the ability to find just about everything you need.

Get their printed catalog, or you can browse through their online catalog at:

http://www.engineeringfindings.com/

Bevel gears from Small Parts. Photo Small Parts Inc.

Stock Drive Products 202486

2101 Jericho Tnpk.
Box 5416
New Hyde Park, NY 11042-5416
USA

☎ (516) 328-3300

📠 (516) 326-8827

✉ webmaster@sdp-si.com

🌐 http://www.sdp-si.com/

If Stock Drive doesn't have it, it probably doesn't exist. SDP is a manufacturer and seller of power transmission products-gears, bearings, bushings, shafts, sprockets, chain, and dozens of other categories. They specialize in the smaller-scale stuff that is most useful in amateur robotics.

Stock Drive's Handbook of Drive Components details thousands of power transmission parts. Courtesy Stock Drive Products-Sterling Instrument.

The company sells their products online, but because of the sheer number of items they stock, it's probably easier to get their printed catalogs. They also, at no cost, provide extremely useful technical design books and a "Designer Companion" CD-ROM, featuring "the largest selection of mechanical drive components," as well as automated design function programs.

For technical tidbits from Stock Drive's reference material, be sure to check out:

http://www.sdp-si.com/Sdptech_lib1.htm

And, for specialty CNC and linear motion products from SDP, see also:

http://www.techno-isel.com/

Tower Fasteners Co. Inc.   202639

1690 North Ocean Ave.
Holtsville, NY 11742
USA

- (631) 289-8800
- (631) 289-8810
- (800) 688-6937
- http://www.towerfast.com/

Master distributor, with online ordering, for several fastener brands, 3M adhesives, hardware for electronics, clamps and couplers, and power transmission. Distribution centers located along the East Coast.

TS Racing, Inc.   204231

123 West Seminole Ave.
Bushnell, FL 33513
USA

- (352) 793-9600
- (352) 793-4027
- (800) 962-4108
- Info@TSRacing.com
- http://www.tsracing.com/

Go-kart and motorbike racing supplies, also suitable for use for robots (racing or non): bearings, axles, sprockets, sprocket hubs, wheels (5- and 6-inch diameter), and assorted mechanical parts.

Vaughn Belting   203622

200 Northeast Dr.
Spartanburg, SC 29304
USA

- (864) 574-0234
- (864) 574-4258

Flexible Linkages

Flexible linkages allow mechanical power or movement to be transferred from one place to another using some form of bendable material. Examples are as follows:

- *Pulleys and belts.* Pulleys are like wheels, and the belts ride over the wheels. Most pulleys incorporate a sleeve or rim to keep the belt in place.
- *Sprockets and chain.* Sprockets are also wheels, but incorporate teeth around their circumference in order to mesh with a chain.
- *Cable.* A flexible cable, made of plastic or metal, transfers power/movement by spinning within some protective sheath. The speedometer cable on older-model cars is a good example of how these work.

Except for cable, flexible linkages can function in a similar manner to gears, including reducing or increasing speed and torque. This is accomplished by using different sprocket or pulley diameters.

A benefit of using pulleys/belts or sprockets/chain is you needn't be so concerned with absolute alignment of the mechanical parts of your robot. When using gears it is necessary to mount them with high precision.

⊘ (800) 533-9086

✉ Vaughnbe@bellsouth.net

🌐 http://www.vaughnbelting.com/

Local distributor of rubber, nylon, steel, and plastic timing belts, conveyor belts, and other belts used in industry.

🌐

W. M. Berg, Inc./Invensys 202487

499 Ocean Ave.
East Rockaway, NY 11518
USA

☏ (516) 599-5010

📠 (516) 599-3274

⊘ (800) 232-2374

✉ wmbergsales@wmberg.com

🌐 http://www.wmberg.com/

W. M. Berg manufactures and distributes precision mechanical components, including gears, rotary bearings, pulleys. belts, hardware and fasteners, linear bearings and slides, couplings, flexible ladder chain (useful as miniature robot tracks), and roller chain (both plastic and metal).

An online catalog is available at the Web site, but you'll want the printed one so you can readily refer to it while

---

Rigid and Flexible Couplers

Couplers are used to connect two drive shafts. A common application is to use a coupler to connect the drive shaft of a motor with the axle of a wheel. Connectors can be rigid or flexible. Rigid couplers are best used when the torque of the motor is low, as it would be in a small tabletop robot. Flexible couplers are advised for higher torque applications, as they are more "forgiving" of errors in alignment.

Rigid couplers can be made using metal or plastic tubing, selected for its inside diameter. You can purchase suitable tubing at a hobby or hardware store. Cut the tubing to length, then drill two small holes at both ends for setscrews. Use a tap to thread the hole for the size of setscrews you wish to use—4/40 is a good all-around size for most applications.

Steel tubing provides the most strength, but is harder to cut, drill, and tap. If the thickness of the tubing is sufficient, aluminum will work well for most low-torque applications. Brass and bronze should be avoided because these metals are too soft. For very low-torque jobs, plastic or even rubber tubing will work. Select the rubber tubing so that it is just slightly smaller than the motor shaft and axle you are using, and press it on for a good fit.

There are many types of commercially available rigid and flexible couplers, and cost varies from under a dollar to well over $50, depending on materials and sizes. Common flexible couplers include helical, universal joint (similar to the U-joint in the drive shafts in older cars), and three-piece jaw. The couplers attach to the shafts either with a press fit, by a clamping action, by setscrews, or by a keyway. Press fit and clamp are common on smaller couplers for low-torque applications; setscrews and keyways are used on larger couplers.

Three-piece jaw couplers, like those made by Lovejoy, consist of two metal or plastic pieces that fit over the shafts. These are the "jaws." A third piece, the spider, fits between the jaws and acts as a flexible cushion. One advantage of three-piece couplers is that, because each piece of the jaw is sold separately, you can readily "mix and match" shaft sizes. For example, you can purchase one jaw for a 1/4-inch shaft and another for a 3/8-inch shaft. Both jaws must have the same outside diameter.

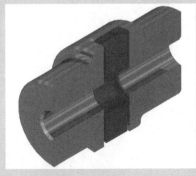

A three-piece jaw coupler.

you're building your robot. Now, the stuff W. M. Berg sells isn't cheap. You pay for the quality and for getting exactly what you want (instead of trying to make do with something you found surplus). Product is available through local distributors.

For additional mechanical components and motion products, see:

Rexnord—http://www.rexnord.com/

Invensys Control Systems— http://www.invensyscontrols.com/

Wholesale Bearing & Drive Supply   204099

P.O. Drawer 370690
Miami, FL 33137
USA

☎ (305) 573-7678
📠 (305) 573-2410
✉ sales@wbds.com
🌐 http://www.wbds.com/

Online sales of bearings and other power transmission components. Their slogan: "On-Line Purchasing of Quality Industrial Products at Discount Prices."

PROFESSIONAL SOCIETIES

This section contains information on various professional societies aligned with robotics, electronics, and engineering.

See also User Groups for others who share your robotics passion.

American Society of Mechanical Engineers ♀   202420

http://www.asme.org/

In the words of the Web site, "Founded in 1880 as the American Society of Mechanical Engineers, today ASME International is a nonprofit educational and technical organization serving a worldwide membership of 125,000."

ASME provides technical publications, conferences, standards, and an online magazine.

ARRL: Home Page   203550

http://www.arrl.org/

This is the main Web page for the Amateur Radio and Relay League (ARRL), an international organization promoting amateur radio.

Association for Unmanned Vehicle Systems International (AUVSI)   203876

http://www.auvsi.org/

In the words of the Web site, "The Association for Unmanned Vehicle Systems International (AUVSI) is the world's largest nonprofit organization devoted exclusively to advancing the unmanned systems community. AUVSI, with members from government organizations, industry and academia, is committed to fostering, developing, and promoting unmanned systems and related technologies."

BARA/British Automation and Robot Association   203954

http://www.bara.org.uk/

BARA is an lobbying group in the U.K. for matters pertaining to industrial robotics. The Web site maintains a useful "Information Encyclopedia" of robotics and automation topics.

Hobby Industry Association   203885

http://www.hobby.org/

The Hobby Industry Association serves the interests of over 4,000 member companies engaged in the manufacturing and merchandising of craft and hobby products. The Web site is useful because it contains information on upcoming hobby trade shows, many of which are the staging grounds for the announcement or introduction of new products. These shows are not ordinarily open to the general public, but press announcements, news, and exhibitor lists are often posted.

Web site for the Hobby Industry Association.

IEE/Institution of Electrical Engineers   203955

http://www.iee.org/

The IIE is the largest professional engineering society in Europe, with a worldwide membership of close to 140,000. The Web site provides news, journals and books (for a fee), and a calendar of important upcoming events.

IEEE Robotics and Automation Society   204128

http://www.ncsu.edu/IEEE-RAS/

IEEE-RAS is the robotics and automation specialty group with the Institute of Electrical and Electronics Engineers. Membership is open to both professionals and students.

International Conference on Field and Service Robotics   202249

http://www.automaatioseura.fi/fsr2001/

Annual conference. "The conference will provide a forum at which manufacturers, users and researchers in the field can exchange their views and receive the very latest information on on-going research and development." Consult the Web page for the date and place of the next conference.

International Conference on Mechatronics
202935

http://eyrie.shef.ac.uk/mech2k2/

In the words of the Web site, "The aim of this symposium is to bring together engineers and scientists who are concerned with modeling, analysis, measurement and control of sound and vibration. It is hoped that this conference will stimulate cross-fertilization of different disciplines involved in acoustics, noise and vibration aspects of systems and to promote the fundamentals as well as industrial applications."

International Technology Education Association
203937

http://www.iteawww.org/

ITEA is a professional association for technology education teachers who teach a curriculum called "technology education." This curriculum involves problem-based learning utilizing math, science, and technology principles.

JEDEC
203266

http://www.jedec.org/

JEDEC is a solid-state technology association and is the semiconductor engineering standardization body of the Electronic Industries Alliance (EIA).

Robotics Industries Association
202557

http://www.robotics.org/

Professional trade group for industrial and commercial robotics. (If you sign up be careful about giving your permission to give out your personal information to third parties. I got numerous unsolicited phone calls from companies I'd never heard of, plus a ton of e-mail spam.)

Robotics Industries Association Web site.

Robotics Society of Japan, The
203862

http://www.rsj.or.jp/

Home page for the Robotics Society of Japan, an independent group serving researchers concerned with the opportunity to exchange information and to present their papers on robotics. Web site is in Japanese and English.

Society of Robotic Combat (SORC)
202412

http://www.sorc.ws/

The WWF of battling bots. Not as much sweat, but the same amount of swearing.

Society of Robotic Combat (SORC).

PROGRAMMING

The sections that follow detail programming resources and products. The listings are divided into several sub-groups to help you locate the field of programming most closely aligned with your interests. The links in these sections are geared toward embedded systems, single board computers, and microcontrollers.

- Examples: Others share with you their programming code

- Languages: Programming languages, compilers, and linkers for embedded systems and microcontrollers

- Platforms & Software: Operating systems, applications developer interfaces (such as speech), and artificial intelligence

- Robotic Simulations: Play with robots on your computer; some are games, but others are real-life simulations of mechanisms

- Telerobotics: Programming to operate a robot via long-distance remote control, typically the World Wide Web

- Tutorial & How-to: Hands-on examples and articles on programming

SEE ALSO:

ELECTRONICS-CIRCUIT EXAMPLES: Web pages often also contain programming examples

INTERNET-INFORMATIONAL: More code examples, some with circuits

MICROCONTROLLERS-SOFTWARE AND MICROCONTROLLERS-PROGRAMMING: Software and programming examples expressly for microcontrollers

PROGRAMMING-EXAMPLES

See the main Programming section for a description of the following listings.

All Basic Code Archives 204155

http://www.allbasiccode.com/

User-to-user forums and some 2,600+ examples of various programs written in Basic. Most examples are in Microsoft QuickBasic and/or QBasic, and others are in PowerSoft PowerBasic.

CodeHound 203636

http://www.codehound.com/

CodeHound is an online search engine that contains programming and software examples for programmers and software developers. The Web site includes coverage of:

- Java
- VB/VB.NET
- Delphi
- C/C++/C#
- SQL
- XML
- Perl
- PHP

The Web site also sponsors a regular newsletter of interest to the developer community and lists useful third-party resources, including books and upcoming conferences.

CodeHound Web site.

CPAN: Comprehensive Perl Archive Network 203643

http://www.cpan.org/

A zillion (or so it seems) examples, libraries, modules, and other goodies for the Perl programming language. You can browse by subject of interest, or search through the code examples.

DevX 203628

http://www.devx.com

Programming portal for many popular languages, include Visual Basic, C++, and Java. Some content is by subscription only. The Web site is maintained by

Fawcette Technical Publications, a San Francisco publisher of magazines catering to Microsoft programming solutions. You'll find code examples, articles, how-tos, tutorials, and newsletters.

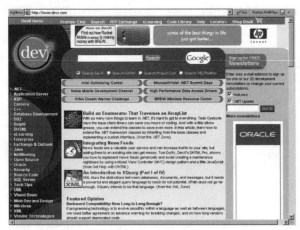

Web site for DevX.

Microsoft Developer's Network　　203629

http://msdn.microsoft.com/

A Big Daddy do-everything site for programming languages and developer's support for current (and some past) Microsoft product, including Visual Basic and Windows programming.

Serial Servocontroller with AT90S1200　　203010

http://mariabonita.hn.org/servo/en.htm

How to program a serial R/C servo controller using an Atmel AVR AT90S1200 microcontroller. Stop paying $20 or $30 for a serial servo controller when you can program your own for less than $4.

WGM Consulting　　203019

http://www.wgmarshall.freeserve.co.uk/

See the Robotics link. Several construction plans and programming code for MicroMouse class robots.

 PROGRAMMING-LANGUAGES

See the main Programming section for a description of the following listings.

All Basic Code Archives　　204155

http://www.allbasiccode.com/

User-to-user forums and some 2,600+ examples of various programs written in Basic. Most examples are in Microsoft QuickBasic and/or QBasic, and others are in PowerSoft PowerBasic.

Borland Software Corporation　　203634

100 Enterprise Way
Scotts Valley, CA　95066-3249
USA

[　(831) 431-1000

✆　(831) 431-4122

⊘　(800) 457-9527

✉　customer-service@borland.com

🌐　http://www.borland.com/

Publisher of a variety of development tools, including Delphi and Borland C++.

Bytecraft Ltd.　　203863

4@listing-address:21 King Street North
Waterloo, ON
N2J 4E4
Canada

[　(519) 888-6911

✆　(519) 746-6751

✉　info@bytecraft.com

🌐　http://www.bytecraft.com/

C cross compilers for microcontrollers:

- Motorola 68HC05 and 68HC08

MOVERS AND SHAKERS

Joseph F. Engelberger

Joseph Engelberger is widely considered the "father of modern robotics," having begun (with inventor George Devol) the first robot company (Unimation) in 1960. Just a few years later, Unimation's "Unimate" robots were installed at General Motors and were busy churning out parts for cars. Today, robots are found in just about every car manufacturing factory in the world. Without robots, your car would be more expensive and less reliable.

- 8051
- Ubicom SX
- Microchip PIC 12Cxxx, 14000, and 16/17Cxx families
- National COP8C
- Zilog Z8

Available through distributors or directly from Byte Craft.

Bywater BASIC 203730

https://sourceforge.net/projects/bwbasic/

Open source Basic interpreter for Linux.

Chipmunk Basic 202221

http://www.rahul.net/rhn/basic/

Chipmunk Basic for MacOS.

In the words of the Web site, "Chipmunk Basic is an old fashioned Basic interpreter which runs on almost all Macs, and is accelerated for PowerMacs. Supported features include AppleScript, Drag&Drop, graphics, sprites, sound, speech and OOP (object oriented programming). Runs on systems from System 6.0.7 thru MacOS 9.2 and OS X 10.1, old Mac 512Ke's thru the latest Apple G3/G4 iMacs and PowerMacs."

FORTH, Inc. 203637

5155 W. Rosecrans Ave.
Ste. 1018
Hawthorne, CA 90250
USA

   (310) 491-3356

   (310) 978-9454

   (800) 553-6784

   sales@forth.com

   http://www.forth.com/

Programming solutions in the FORTH language.

Free Software Foundation, Inc 203640

http://www.gnu.org/

Web site for the GNU project, dedicated to free and open source software. Among the projects are GNU GCC compiler for Linux (and various Unix variants) and Windows. GCC supports C, C++, Objective C, Chill, Fortran, and Java.

LEGO Mindstorm with Linux Mini-HOWTO 203125

http://www.linuxdoc.org/HOWTO/mini/Lego/

How to get Linux to work with the LEGO Mindstorms, including its infrared programming tower.

Liberty BASIC 203624

http://www.libertybasic.com/

QuickBasic compatible compiler. Available for download (free and paid versions available).

Linux Assembly 203632

http://linuxassembly.org/

Portal site for writing assembly language programs for Linux and Linux-like operating systems. According to their Web site: "If you are looking for information on assembly programming under UNIX-like operating systems (Linux/BSD/BeOS/etc), this is the right place to be. Here you can find various resources, ranging from tutorials and documentation, to actual programs written in assembly language."

Metrowerks, Inc./Codewarrior 203639

9801 Metric Blvd.
Austin, TX 78758
USA

   (512) 997-4700.

   (512) 997-4901

   (800) 377-5416

   info@metrowerks.com

   http://www.metrowerks.com/

Developers of the CodeWarrior C++ compiler, ideally suited for use in embedded controller applications.

Opensource.org
203631

http://www.opensource.org/

Informational Web site on open source software, such as Linux, Apache Web Server, Python, Zope, and Perl.

Home page for Opensource.org.

O'Reilly Perl.com
203641

http://www.perl.com/

The main train stop for the Perl programming language. O'Reilly publishes many of the authoritative texts on Perl programming, including books by Larry Wall, the creator of the Perl language.

Perl Mongers/Perl.org
203642

http://www.perl.org

Support Web site for the advocacy or all things Perl.

For Perl user groups, see also:

http://www.pm.org/

PowerBASIC, Inc.
203626

316 Mid Valley Center
Carmel, CA 93923
USA

((831) 659-8000

✆ (831) 659-8008

⊘ (800) 780-7707

✉ sales@powerbasic.com

🌐 http://www.powerbasic.com/

PowerBasic Basic compiler and programming tools. This is one of the last commercially available versions of Basic.

Production Basic
203731

http://probasic.sourceforge.net/

Open source Basic for Linux. Distributed under the GNU Public license.

Project JEDI
203635

http://www.delphi-jedi.org/

Support Web site for Borland Delphi and Kylix.

Says the Web site: "Project JEDI is an international community of Delphi developers with a mission to exploit our pooled efforts, experiences and resources to make Delphi and Kylix—the greatest Windows and Linux application development tools-even greater."

Python
203644

http://www.python.org/

Official home page of the Python programming language (irreverently named after the Monty Python comedy troupe, and not the reptile). Python is available for a variety of computing platforms, including the IBM PC, Macintosh, and Linux.

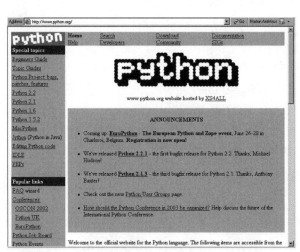

Official home for the Python programming language.

QB45/Future Software 203627

http://qbasic.qb45.com/

Support Web site for the (now-obsolete) Microsoft QuickBasic/QBasic programming language.

QB4all.com 203625

http://qb4all.cjb.net/

QB4All is a QuickBasic-centric portal, links, and forum. "The Home of the QBasic Community on the net!"

QBasic.com 202842

http://www.qbasic.com/

Microsoft QBasic lives-here at this Web site. QBasic.com provides programming examples and tutorials on using the QBasic and QuickBasic programming languages.

QBasicNews.com 203623

http://www.qbasicnews.com/i

Informational Web site on Microsoft QBasic and QuickBasic.

SmallBASIC 203732

http://smallbasic.sourceforge.net/

SmallBASIC is a simplified open source version of the Basic language for PalmOS, Linux, Win32 and DOS. As noted on the Web site, "SmallBASIC is not a developer tool for professional programmers (there are no data-types, no binary output, no QB compatibility, no low-level access, etc)." Still looks useful, though.

Smalltalk.org 203630

http://www.smalltalk.org/

Portal site for the Smalltalk programming language. Smalltalk is perhaps the definitive object-oriented programming language and is favored by many robot researchers at universities.

Sun Java 203645

http://java.sun.com/

Main support Web site for Java, the wildly successful programming language by Sun Microsystems. The Web site acts as a clearinghouse for ideas, articles, links, and how-tos on programming in Java. The basic Java software developer's kit is available for download, at no charge, from this Web site. Choose the operating system platform you're interested in, either Windows, Linux, or Solaris.

You will also want to look over the available optional modules for Java, including the Communications API. This set of programming interfaces allows you to write Java applications that can communicate through the computer's serial and parallel ports.

Terrapin Software 202424

10 Holworthy St.
Cambridge, MA 02138
USA

((617) 547-5646
♨ (617) 492-4610
⊘ (800) 774-5646
✉ info@terrapinlogo.com
🌍 http://www.terrapinlogo.com/

Developers of the Terrapin Logo programming language for home computers and LEGO Mindstorms RCX. The company also sells some LEGO and Fischertechnik kits.

UltraTechnology 203726

2512 10th St.
Berkeley, CA 94710
USA

🌍 http://www.ultratechnology.com/

Information on the Forth programming language, as well as details regarding a number of integrated circuits with embedded Forth. Forth-related products for sale.

VLL Library for NQC 203104

http://www.mi-ra-i.com/JinSato/
        MindStorms/DDK/vll-e.html

Information on using VLL with the Not Quite C programming language.

PROGRAMMING-PLATFORMS & SOFTWARE

See the main Programming section for a description of the following listings.

Apple Speech Recognition 203759

http://www.apple.com/macos/speech/

Developer's Web page for speech recognition technology for the Apple Macintosh.

In the words of the Web site, "Apple Speech Recognition lets your Macintosh understand what you say, giving you a new dimension for interacting with and controlling your computer by voice. You don't even have to train it to understand your voice, because it already understands you, from your very first word."

EGg0 Educational Robotics 203147

http://eggo.sourceforge.net/software2.php

According to the Web site, "The EGg0 robotics system will provide software tools as a suite of Python modules called The EGg0 Toaster. These modules are being designed specifically to help programmers access all of the functionality of the EGg0 computer system, without requiring them to learn the intricacies of input/output and other low-level programming ideas."

FreeDOS 202989

http://www.freedos.org/

Open source DOS alternative.

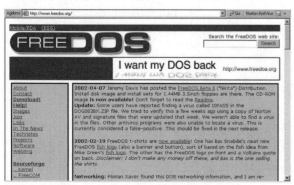

Web site of the FreeDOS movement.

Generation 5 202261

http://www.generation5.org/

Resource for artificial intelligence; member-posted articles.

OpenDOS 204154

http://www.deltasoft.com/opendos.htm

Caldera OpenDOS is an open source version of the infamous DOS operating system. Many users report good compatibility. Includes a FAQ, downloadable distributions, and documentation.

PalmOpenSource.com 204170

http://www.palmopensource.com

Billing itself as the "PalmOS Open Source Portal," the Web site is a magnet for those interested in extending the Palm Pilot and PalmOS operating system beyond canned applications. Articles and user-to-user forums provide lots of reading. The Palm is a common platform for robotics, and there are always a few robot-centric articles posted.

Rossum Project-Open-Source Robotics Software 202467

http://rossum.sourceforge.net/

Says the Web site, "Building a robot isn't easy. Robotics presents a challenging intersection of hardware and software. It reaches across disciplines. . . . The Rossum

Open source robotics at the Rossom Project

ActiveState Corporation

http://www.activestate.com/

Programming tools for Perl, Python, PHP, and others

Project is an attempt to help. Our goal is to collect, develop, and distribute software for robotics applications. Over the last few years, open-source and free software initiatives have given computer users a remarkable collection of tools and capabilities."

Uniform Robotic Control Protocol   203589

http://www.urcp.com/

In the words of the Web site, "The Uniform Robotic Control Protocol (URCP) is a protocol that is the result of developing control systems for several unrelated devices that use a common numbering system. The common numbering system allows the devices to communicate with and/or control other devices which use the same numbering system without the need for conversion."

PROGRAMMING-ROBOTIC SIMULATIONS

See the main Programming section for a description of the following listings.

Bug Brain   203750

http://www.australianwildlife.com.au/bugbrain/

In BugBrain, you build brains to run a bug. The game features rendered graphics, and puzzles, and along the way teaching you about neural networks.

Co-Evolutionary Robot Soccer Show   202258

http://www.legolab.daimi.au.dk/cerss/

Here's what the Web page has to say, "The Co-Evolutionary Robot Soccer Show is a game that allows you to develop robot soccer players by using the concept of co-evolution. You can develop the robot soccer

players in the software provided for free at this Web site. You evolve different robot soccer players by changing the parameters for the evolution (the population parameters and the fitness formula parameters).

"When you have evolved a good robot soccer player, you can send the player to our server. Each night (European time), the server will play 50,000 matches between the uploaded players and generate a new Highscore List every morning. So you can keep track on how your player(s) is/are doing by going to this Web site every morning. At the end of the competition, the first players on the Highscore List will win the sponsored prizes. These prizes include a LEGO Mindstorms Robotic Invention System."

CogniToy LLC/MindRover   203093

236 Central St.
Acton, MA 01720
USA

  (978) 264-3945

  (978) 264-3946

  (888) 788-1792

  support@cognitoy.com

  http://www.mindrover.com/

MindRover is a simulation game where you develop robots to complete a task. Free demo.

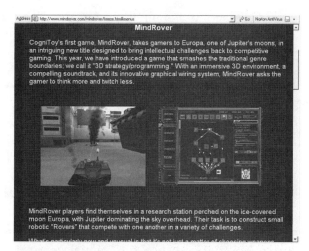

Play with robots at Mindrover.com

Colobot   203376

http://www.epsitec.ch/colobot/index.htm

Game: Colonize space with your virtual robots.

Cyberbotics 203723

Cyberbotics S.r.l.
Ch. de Vuasset
CP 98
1028 Prverenges
Switzerland

☏ +41 21 693 8624
✆ +41 21 693 8624
✉ info@cyberbotics.com
www.cyberbotics.com
🌐 http://www.cyberbotics.com/

Publishers of the Webots 3D robot simulator: simulates any robot using two-wheel differential steering.

Enigma Industries 204214

P.O. Box 27522
Anaheim, CA 92809-0117
USA

✉ info@EnigmaIndustries.com
🌐 http://www.enigmaindustries.com/

Enigma sells a unique real-world simulation program for developing drive trains for coaxially driven (wheel on either side) robots, especially larger robots for combat. You enter the technical details of the motor and the details of your robot design, including gear ratios, tire diameters, motor voltage and capacity, and weight. The program then simulates how the robot will perform given the motor you've selected.

The technical specs of common motors are already in the program's database, and you can edit those figures if neededor create new entries. The program simulates

---

RobotWorks

http://www.compucraftltd.com/rbw.html

Robotics Interface and path generator for SolidWorks

Webots

http://www.cyberbotics.com/webots/

Webots 3.0 is a 3D robot simulator which simulates any robot using two-wheel differential steering.

New River Kinematics RobotAssist

http://www.kinematics.com/robot/

RobotAssist is a full-featured robot modeling and control package designed specifically for the Windows environments.

---

battery voltage, state of charge, and internal resistance, as well as over- and under-volting of the motor (e.g., the performance of motor driven at 12 volts, but whose faceplate rating is 19 volts).

Another unique aspect is the database includes the approximate price and sources for the motors.

GNU Robots 204189

http://www.gnu.org/software/robots/

Open source and free game software for controlling and interacting with robots, written (currently) for Unix machines.

MATHWorks, Inc. 203646

3 Apple Hill Dr.
Natick, MA 01760-2098
USA

☏ (508) 647-7000
✆ (508) 647-7001
✉ info@mathworks.com
🌐 http://www.mathworks.com/

Developers of MATLAB, Simulink, and Stateflow. "The MathWorks offers a set of integrated products for data analysis, visualization, application development, simulation, design, and code generation."

MazeBots 203095

http://www.innovatus.com/Products/
Education/MazeBots/mazebots.html

In the words of the Web site: "MazeBots is a simply 'a-maze-ing' way to learn programming in a safe and fun environment. The student programs robots to solve mazes. While this is exciting for the student, this is not a game but a tool for education and introducing the student to programming."

Robocode 202065

http://robocode.alphaworks.ibm.com/
home/home.html

From the Web site, "Build the best. Destroy the rest. In Robocode, you'll program a robotic battle tank in Java for a fight to the finish. The game is designed to help you learn Java, and have fun doing it . . . from a simple 10 line robot to a very sophisticated, intelligent robot that destroys the competition!"

RoboForge 202166

http://www.roboforge.net/

RoboForge is a game played on the PC. The object: build robot gladiators, train them for combat, then submit them to online tournaments. "Win fame and prizes," says the Web site.

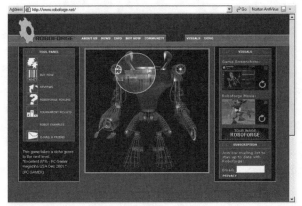

Roboforge game Web site.

Robot Arena 203767

http://www.robotarena.com

Robot Arena is simulation game published by Infogrames. The aim is to build bots, then pit them against one another. Game play is similar to the real-life robot "death match" competitions.

Robot Battle 202469

http://www.robotbattle.com/

The Web site says it best: "Robot Battle is an evolving programming game. The game allows players to design and build intelligent robots using a simple scripting language. Robot Battle is both an exciting game and an excellent way to learn programming. Most modern computer systems are based on event driven techniques very similar to those you will learn playing Robot Battle. Several schools are currently using the game as part of their introductory programming classes. Robot Battle was created by Brad Schick."

Robotics Toolbox for MATLAB 203647

http://www.cat.csiro.au/cmst/staff/pic/robot/

From the Web site: "The Robotics Toolbox provides many functions that are useful in robotics such as kinematics, dynamics, and trajectory generation. The Toolbox is useful for simulation as well as analyzing results from experiments with real robots." Requires MATLAB (http://www.matlab.com/).

PROGRAMMING-TELEROBOTICS

See the main Programming section for a description of the following listings.

ActivMedia: Mobile Robots 203477

http://www.mobilerobots.com/

Show-off Web site for demonstrating the telerobotics capability of an ActivMedia intelligent mobile robot. You need the company's WorldPass software to use the demonstrator.

See also the main ActivMedia Robotics Web page:

http://www.activmedia.com/

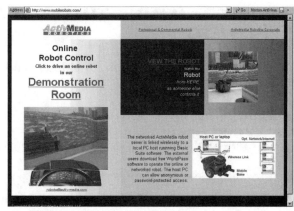

ActivMedia: Mobile Robots.

Australia Telerobot 203418

http://telerobot.mech.uwa.edu.au/

Control a robot in Western Australia from your Web browser! Just remember robots spin backward Down Under.

Autonomous versus Teleoperated Robots

Autonomous robots are those that act upon their own control. They consist of the following systems:

- Sensors to monitor the environment
- Some form of processor or computer
- Actuators to move wheels, arms, or other mechanisms

A teleoperated robot built by Michael Owings and operated by anyone with a Web browser.

The processor or computer runs a program whose aim is some goal for the robot. That goal might be waxing the open areas of the floor in a large warehouse, for instance. The program also includes alternative actions for the robot to perform should an obstacle stand in the way of completing the goal. Such obstacles include objects in the robot's way, low battery power, or mechanical breakdown.

A teleoperated robot is one that is commanded by a human and operated by remote control. The robot may be in physical sight of its human controller, like the combat robots in BattleBots and similar competitions, or it may be linked via camera, such as police robots that are used to defuse bombs. Of the three systems in an autonomous robot, the telerobot lacks the first two, as sensing and computing are handled by the human operator.

Some telerobots, like the world famous Mars Pathfinder Sojourner, the first interplanetary dune buggy, are actually half remote controlled and half autonomous. A growing number of telerobots fit into this category. The low-level functions of the robot are handled by a microprocessor on the machine. The robot is able to carry out basic instructions on its own, freeing the human operator from having to control every small aspect of the machine's behavior.

PROGRAMMING-TUTORIAL & HOW-TO

See the main Programming section for a description of the following listings.

Code Corner: Accurate timing in Win32   202663

http://www.bitbanksoftware.com/code4.htm

Accurate timing under 32-bit Windows, especially when using Visual Basic, requires some extra effort. This Web page describes ways to overcome the problem.

Code Corner: How to efficiently emulate polyphonic waveform generation   202664

http://www.bitbanksoftware.com/code15.htm

Program example (in C) for emulating polyphonic waveforms; useful for adding sound effects (à la R2-D2) to your robot.

Code Corner: How to emulate DAC-based arcade sounds   202665

http://www.bitbanksoftware.com/code7.htm

Basic information for using a digital-to-analog (DAC) converter to produce arcade-type sound effects.

Device Control via the PC   202301

http://www.elabinc.com/note1.pdf

Application note (in Adobe Acrobat PDF format) on connecting stuff to the PC parallel port, using the E-Lab EDE300.

SEE ALSO:

http://www.elabinc.com/

DevX   203628

http://www.devx.com

Programming portal for many popular languages, including Visual Basic, C++, and Java. Some content is by subscription only. The Web site is maintained by Fawcette Technical Publications, a San Francisco publisher of magazines catering to Microsoft programming solutions. You'll find code examples, articles, how-tos, tutorials, and newsletters.

John McCarthy: LISP Designer   203633

http://www-formal.stanford.edu/jmc/

Home Web page of the creator of the LISP programming language.

Microsoft Developer's Network   203629

http://msdn.microsoft.com/

A Big Daddy do-everything Web site for programming languages and developer's support for current (and some past) Microsoft product, including Visual Basic and Windows programming.

 **RADIO CONTROL**

Though not primarily intended for amateur robotics, radio control products are nevertheless a real boon to the art and science of robot building. There are many more radio control (model car, boat, airplane) enthusiasts than there are robot makers, and the population of buyers for radio control products helps keep the prices down.

The listings that follow in this section pertain to generic resources for radio control and how-to products. Additional subsections include listings for specific radio control product categories that are especially useful in robotics.

SEE ALSO:

RADIO CONTROL (VARIOUS SUBCATEGORIES): Specific categories of products you might be interested in

RETAIL-ARTS & CRAFTS: Good sources for construction and repair goods

RETAIL-TRAIN & HOBBY: Where to buy radio-control products

Airtronics Inc. 202127

1185 Stanford Ct.
Anaheim, CA 92805
USA
☎ (714) 978-1895
📠 (714) 978-1540
✉ info@airtronics.net
🌐 http://www.airtronics.net/

Airtronics is a maker of radio control electronics for flight models. Their product line is extensive; for robotics, we're mostly interested in their servos (and servo accessories, including replacement gears, servo trays, and servo control horns), as well as their piezo-rate gyro.

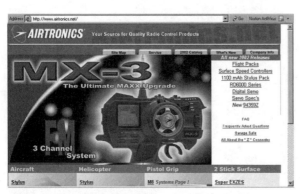

Web home of Airtronics.

Sales are through a worldwide chain of retailers and distributors. The Airtronics Web site provides some technical information, an updated catalog of products, and suggested retail prices.

BestRC / Hobbico, Inc. 202526

http://www.bestrc.com/

Portal site to Hobbico R/C product companies, which includes Futaba, Great Planes, and DuraTrax. The product lines are sold through dealers.

SEE ALSO:

http://www.hobbico.com/

Cermark 202729

107 Edward Ave.
Fullerton, CA 92833
USA
☎ (714) 680-5888
📠 (714) 680-5880
🚫 (800) 704-6229
✉ customerservice@cermark.com
🌐 http://www.cermark.com/

R/C: servos, batteries, electric motors, and gearboxes.

Draganfly Innovations 204238

2108A St. George Ave.
Saskatoon, Saskatchewan S7M-0K7
Canada
☎ (306) 955-9907
📠 (306) 955-9906
🚫 (800) 979-9794
🌐 http://www.rctoys.com/

R/C blimps, gyro-stabilized helicopters, wireless microcam

ElectroDynamics 203296

31091 Schoolcraft Rd.
GAZ Commerce Center
Livonia, MI 48150
USA

((734) 422-5420

(734) 422-5338

(800) 337-1638

support@electrodynam.com

http://www.electrodynam.com/

R/C electronics: batteries and R/C connectors, servo testers, battery monitors, and digital servos.

Electronic Gadgets for Radio Control 202502

http://www.uoguelph.ca/~antoon/
  gadgets/gadgets.htm

Circuits and information on: servo testers, ni-cad battery sensors, ni-cad battery rechargers, flashers, electronics tutorials (555 timer IC, 741 op-amp IC, and others).

Electronic Model Systems 203304

22605 East La Palma Ave.
Ste. 516
Yorba Linda, CA 92887
USA

((714) 692-1393

(714) 692-1330

(800) 845-8978

mark@emsjomar.com

http://emsjomar.com/

EMS sells electronic products and other accessories for R/C aircraft and boats: battery packs, speed controls, and rechargers.

Futaba 202142

Great Planes Model Distributors
P.O. Box 9021
Champaign, IL 61826-9021
USA

((217) 398-6300

(800) 637-7660

gpinfo@gpmd.com

http://www.futaba-rc.com/

Futaba makes R/C transmitters, receivers, and servos. Their Web site provides some handy technical reference on their product, including the latest digital servos.

Note that the address and phone number provided is for the Futaba North American distributor.

Futaba R/C Web site.

Hitec/RCD 202144

12115 Paine St.
Poway, CA 92064
USA

((858) 748-8440

http://www.hitecrcd.com/

Hitec/RCD makes transmitters, receivers, and servos for hobby R/C (airplanes and cars). Hitec servos are my favorite as they tend to be a little less expensive than the others, yet are still well made and durable. Product is available from hobby dealers everywhere. The Web site provides a smattering of technical documents.

Hobbees.com 204029

c/o What Sounds Good Inc.
306 West State St.
Olean, NY 14760
USA

(716) 372-9428

(877) 663-4447

wsg@eznet.net

http://www.hobbees.com/

Online R/C and hobby products at a discount. Very wide selection (includes servos, wheels, and hardware), most major brands. You can search by category, manufacturer, part number, or product description.

Hobbico, Inc. 202132

http://www.hobbico.com

Hobbico is a master distributor in North America for several popular brands of radio control products, including Futaba and Great Planes. Consumers can't buy from them directly, but this Web site provides links to the brands, a vendor lookup list (by state or zip code), and various product overviews and reviews.

Maxx Products International, Inc. 202735

815 Oakwood Rd.
Unit D
Lake Zurich, IL 60047
USA

((847) 438-2233
(847) 438-2898
⊘ (800) 416-6299
🌐 http://www.maxxprod.com/

High-performance R/C; sells complete line of NiMH batteries, as well as full line of servos.

Model Builders Supply 203814

40 Engelhard Drive, Unit 11
Aurora, Ontario
L4G 6X6
Canada

((905)-841-8392
(905)- 841-8399
🌐 http://www.modelbuilderssupply.com/

MBS manufactures an assortment of tools, materials, and supplies for building miniatures. Much of their product line is geared toward making dollhouses and architectural models, but the same materials can be used to construct small robots. The line includes basswood sticks and sheets (basswood is a light but resilient wood, similar to balsa but stronger), gears, motors, pulleys, and chains, casting materials (rubber, Vinamold hot melt, and Por-a-Kast), and precision tools.

Model Rectifier Corporation 203493

80 Newfield Ave.
P.O. Box 6312
Edison, NJ 08837
USA

((732) 225-2100
🌐 http://www.modelrec.com/

Train and R/C electronics. Their train gear includes DCC transmitters and receivers.

Ohmark Electronics 203171

P.O. Box 45
Leeston
New Zealand

(+64 3 324 4463
✉ sales@ohmark.co.nz
🌐 http://www.ohmark.co.nz/

Specialty R/C components like elevon mixers and servo reversers. Some robotics electronics, as well, based around radio control servos. These include a digital servo controller and a mini R/C relay driver. Direct sales or through dealers; check the Web site for dealer locations.

Plantraco Ltd. 204229

1105 8th St. East
Saskatoon, Saskatchewan
S7H 0S3
Canada

((306) 955-1836
(306) 931-0055
✉ ufoman@plantraco.com
🌐 http://www.plantraco.com/

Sellers of upscale radio controlled toys, including blimps and little tracked vehicles. Their Desktop Rover tracked vehicle can be controlled via a hand-held remote or by software running on your computer. The company also sells a miniature wireless camera for use on its R/C products.

RC Electronics Projects of Ken Hewitt 203392

http://www.welwyn.demon.co.uk/

Reprints from Ken's articles in the U.K. magazine Radio Control Models and Electronics, such as a motor speed controller V-tail mixer. Parts for some projects available for sale.

RC Yellow Pages 204023

http://www.rcyellowpages.com/

Links to over 1,300 companies providing products and services to the worldwide radio control community.

Model Railroading

At a time when amateur robot builders were constructing their automatons with tubs and relays, model railroaders were experimenting with electronic speed controls, touch sensors, miniature motorized armatures, and much more. Even if you're not into model railroading, our track-centric brethren can show the robot builder a thing or two about electromechanical construction and control. They even have plenty of specialized goodies that just beg to be used in robots.

Modern model railroading is centered around the DCC—Digital Command Control—a worldwide standard that permits multiple locomotive engines and other devices to be individually controlled via a signal passed through the metal tracks of the model. In the "old days," each engine was controlled simply by varying the voltage on the track, so there could only be one engine per track (or track segment) at one time. With DCC, the track uses a centralized command center, and each locomotive is equipped with a DCC decoder. Each decoder "listens" to the command signals meant for it and ignores any others.

What's more, the power for the locomotive engine also comes from the track, just as it always has for most kinds of model railroads. The difference is that instead of controlling the motor in the locomotive directly by varying the voltage to the track, the DCC decoder controls the motor if, and only if, it receives a digital signal from the command center to do so.

It doesn't take a rocket scientist to figure out that DCC is also handy for robotics. A DCC command center and decoders could be used to not only transmit power (12-14 volts AC) to different parts of a robot, but to control it as well. Or, if you outfit a robot arena with a metal floor and metal mesh on top, you could construct an army of semiautonomous robots that don't carry their own power supply—the whole thing would work along the same lines as bumper cars at carnivals.

DCC setups can be a tad expensive, though there are still bargains; so be on the lookout for them. The basic unit controls just the on/off and direction of a motor. That's perfectly fine for most robotic application. Advanced DCC command centers can control a half dozen or more locomotives, along with "extras" such as lamps and whistles.

Higher-end setups even offer a wireless command center, so you can control your robot from anywhere in the room, and others can be controlled by the RS-232 port of a personal computer. Because many of the basic functions of a DCC decoder can be preprogrammed—such as motor speed and acceleration rates— when you operate your robot through the command center (or the PC), you need not manipulate every little nuance of the robot's operation.

Of course, there's more to model railroading than just DCC, and a surprising amount of it can be applied to robotics, as well:

- Construction materials, including lightweight plastics, rubber and plastic castings, and spray foam.

- Drive motors, often in unusual shapes and arrangements, for locomotives. These are very high quality with great efficiency in a miniature scale. Many come with gearboxes attached or can be readily attached to a gearbox.

- Mechanicals, trellis, crossing arms, etc.

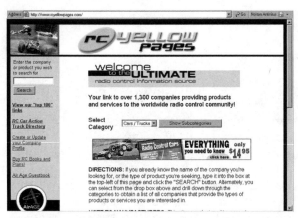

RC Yellow Pages.

RepairFAQ: RC (Radio Control)   202866

http://www.repairfaq.org/filipg/RC/

From Sam Goldwasser's RepairFAQ: links, FAQs, and tips about the radio control hobby.

Subtech   202273

501 Albert Ave.
Lakewood, NJ 08701
USA

📞 (732) 363-7426

📠 (732) 363-7427

✉ skip@rcboats.com

🌐 http://www.rcboats.com/

SubTech makes electronics and hardware for aquatic-style radio controlled vehicles, including submarines. Hardware products include shaft seals, important if you are designing an autonomous submarine with electric motors and batteries!

Vantec   202178

460 Casa Real Plz.
Nipomo, CA 93444
USA

📠 (805) 929-5056

🚫 (888) 929-5055

✉ input@vantec.com

🌐 http://www.vantec.com/

Vantec is a major manufacturer of remote radio control systems for mobile battery-powered robots, as well as motor controllers, servos, and servo amplifiers. Their products, which sport an R/C or computer interface,

are commonly used in combat robotics. Product lines include:

- DC PM motor speed controls
- Radio-control systems
- Servo components

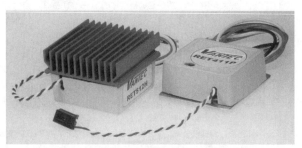

Vantec reversing electronic throttle controllers. Photo VANTEC.

DuraTrax

http://www.duratrax.com/

R/C vehicles, tires, NiCd batteries, spur gears

Great Planes

http://www.greatplanes.com/

Distributor of kits, hardware, and accessories for radio control models

Hobby Express

http://www.hobbyx.com/

Distributor for R/C products

Kyosho

http://www.kyosho.com/

Model R/C airplanes and cars; sells replacement tracks for their Nitro Blizzard racer

MotoCalc

http://www.motocalc.com/

Software for calculating electric flight performance

Novak

http://www.teamnovak.com/

Brushless motors, motor speed controls, battery chargers

Central Hobbies

http://www.centralhobbies.com

Full line of RC parts; carbon fiber pushrods

Radio-Control Model Hobby Parts and Components

When you think about it, a small mobile robot is not much different from a radio control model car, boat, or airplane—except that its activity is largely controlled by a computer, rather than remotely by a human being. The mechanical aspects of the robot share many of the same parts and components as radio controlled (R/C) models. Because R/C modeling is such a popular hobby, practiced all over the world, a vast array of products are widely available at affordable prices.

Transmitter and Receiver

The centerpiece of the typical R/C model is the radio control transmitter. In a fully autonomous robot, the transmitter is not used because the automaton is controlled by an electronic circuit of some type (exception: when the robot is equipped with a wired or wireless remote control, so it can be operated by a human).

Transmitters are complemented by receivers in the vehicles (or in our case, the robots). The transmitter and receiver operate over radio frequency waves and are on a very specific frequency. Most hobby R/C models use a transmitter/receiver capable of at least two functions, or channels. In the case of a model car, for example, one channel might be for speed and another for steering. Additional channels are used for other functions, such as direction.

For hobby R/C, there are two general types of transmitters: air and land. Depending on the country of use, there are various restrictions for both kinds, and the transmitters operate at different frequencies to avoid interference. Because robots are land-based vehicles, you should use only a transmitter meant for land applications. This helps avoid potential interference, where your robot experiments may cause the crash of a nearby model airplane.

The transmitter/receiver pair is crystal controlled and operates on a specific frequency (often referred to as a "channel," but this kind of channel should not be confused with the various function channels—that is, speed or steering—supported by the transmitter). If you plan on using your robot in a competition, with other remotely controlled machines, you may need to change the operating frequency of the transmitter and receiver to avoid conflicts with other entrants.

You can often save some money by purchasing a starter kit that includes a transmitter and matching receiver, as well as one or more servos. You can then add to it with additional servos or alternative frequency crystals and other goodies.

A few final words about R/C transmitters and receivers: Don't be lulled into thinking you need lots and lots of channels (that is, channels for functions like speed and steering). A three- or four-channel transmitter is probably more than sufficient.

Among the highest priced transmitters are those that employ FM-PCM (pulse code modulation) circuitry. For the most part, it's money wasted on a mobile robot; PCM is ideal for fine-control applications for model airplanes and helicopters. Land-based robotics can make do with the "older-fashioned," but still quite capable, FM-PPM (pulse position modulation) or the even less expensive AM circuitry.

R/C Servos

An R/C servo is a motor that always knows "where it is." The output of the motor is meant to connect some part of the model, like the steering wheels of a car or the rudder of

Model R/C servos are great for motorizing robots.

an airplane. Obviously, the positioning of these parts needs to be precise, and electronics inside the servo provide for that.

More about servos throughout this section.

Rechargeable Battery Packs

R/C applications are power hungry, and rechargeable battery packs are the norm. The battery packs are available at several voltages and current capacities. Common voltages are as follows:

4.8-volt

7.2-volt

9.6-volt

The 7.2-volt packs are perhaps the most useful for robotics. Note the "unusual" voltages; these are the result of the 1.2-volts-per-cell batteries used in the packs. Current capacities range from about 350 milliamp-hours (mAh) to over 1,500 mAh. The higher the current capacity, the longer the battery can provide juice to your robot. Unfortunately, higher capacity batteries also tend to be larger and heavier. You should always pick the current capacity based on the estimated needs of your robot, rather than just selecting the biggest brute of a battery that you can find.

There are two general types of batteries used in rechargeable packs: nickel-cadmium (ni-cad) or nickel metal hydride (NiMH). Both can be recharged many times, but of the two, ni-cad batteries are the least expensive because they've been around the longest. NiMH batteries provide for high current capacities, with ratings of 600 to 3,000 mAh and over.

There are other advantages to NiMH batteries. For years, users have complained about the "memory effect" of ni-cad cells (though ni-cad battery makers say this problem has long been corrected). The memory effect is simply this: If a ni-cad battery is recharged before being completely discharged, it may "remember" this shortened current capacity. The next time the battery is used, it may not last as long as it should before needing a recharge.

Additionally, ni-cad batteries contain cadmium, a highly toxic metal. Ni-cad batteries can be bad for the environment, because the cadmium in them can leach out and filter down to the water table and into underground streams or rivers. Exposure to cadmium is known to cause everything from flu-like symptoms to kidney failure to cancers of the lung and prostate. As a result, ni-cad batteries should never just be thrown away in the trash, but properly recycled or disposed in a hazardous waste facility.

Both ni-cad and NiMH battery packs require rechargers specially designed for them. The better battery rechargers work with a variety of pack voltages.

Wheels and Tires

The local hobby store provides a gamut of wheels and tires of all sizes for your robot, as well as mounting hardware.

Drive Wheels

Wheels for model radio controlled planes are ideally suited for the main drive of mobile robots, as they are both strong and lightweight. The wheels, which typically have a metal or plastic hub drilled out for a 1/8-inch-diameter axle shaft, are available in sizes from 1.5 to over 6 inches in diameter.

You can chose between rubber and foam tires; rubber tires come with or without treads and tend to be heavier and sturdier. They are ideal for robots weighing over one pound. Foam wheels are lighter and in many cases cheaper, but heavier robots may cause the wheels to deform if used on a heavy robot.

Tires are used on model cars and are generally smaller than the wheels for R/C airplanes. This doesn't mean they are less expensive; on the contrary, tires for R/C cars can cost $30 to $50 each, though most are under $15 per pair. Hubs are plastic or metal. The main benefit of R/C car tires is the traction they can provide. A wide rubber tire affords considerable traction on all kinds of surfaces. Robots destined for use outdoors, or in sumo-style robot competitions, benefit from this extra traction.

Foam wheels are both inexpensive and lightweight.

Tailwheel

Tailwheels are the "third wheel" for when a plane is on the ground. Because tailwheels spin and rotate around a center column, they are perfectly suited for use as support casters in coaxial-drive robots. Tailwheels are made to match certain scales of model airplanes, and the rubber wheels come in sizes from 3/4 to 2 inches. The wheel is mounted on a prebent metal post, and the post fits into a holder which attaches to the robot (see the following figure). The height of the tailwheel can be adjusted in its holder to accommodate a variety of robot designs.

Dura-Collars

Though used for many applications, Dura-Collars are typically employed to keep R/C wheels on their axle shafts. The collar is made of plated metal and is drilled for a setscrew. You use a hex wrench to tighten the collar around the axle shaft. Dura-Collars come in a variety of sizes; you match collar to the diameter of the axle shaft. Common sizes are 1/16 inch (1.5mm) to 3/16 inch (4.7mm).

Pushrods, Cables, and Linkages

Less used in amateur robotics, but still very useful, is the pushrod, which is basically a piece of heavy, thick metal. More often than not, the pushrod is threaded on one end, so that it firmly connects into a clevis, swivel ball, or other linkage fitting. The other end of the pushrod is bent to make a hook and is attached to the servo by way of a servo horn (see the following section for more information on servo horns).

You can use pushrods in robotic designs that require you to transfer linear motion from one point to another, such as an arm or finger grippers. This allows you to place the heavy and bulky servo in the base of the arm.

You must match the size of the pushrod and the clevis. You can chose from plastic or metal clevises; the metal variety can be either screw-on or solderable. The screw-on type is definitely easier to work with, but requires threaded pushrod, and this adds to the expense.

Flexible cables, with and without an outer plastic sheathing, are used in much the same way as pushrods, but are useful when the linkage cannot be rigid.

Servo Horns and Bellcranks

Servo horns fit onto the shafts of R/C servos and are most often used to convert the rotational movement of the servo to a linear movement. In a model airplane, for example, this linear movement might be to move the ailerons up and down. Servo horns come in a variety of shapes and sizes; it's best to simply take a look at what's available and chose the kind you think will work best. You need to get servo horns for your make and model of servo, because the mounting holes vary in size.

By the way, there's also plain ol' control horns, which are not for attaching to the output shafts of servos. Control horns are most often used on the other end of the mechanical linkage from the servo—you'd use a control horn on the surface of an aileron.

Bellcranks are similar to control horns and serve as levers. They are often used to enlarge or reduce the amount of linear movement.

Miscellaneous R/C Hardware

It's not possible to describe every piece of hardware available for R/C modeling, but here's a quick rundown of some of the more useful components.

- *Control hinges.* Plastic or metal hinges with mounting holes. In a variety of sizes. For robotics, go for the best you can get, so the hinge doesn't fall apart from extra wear.

- *Bolt/nut/screw hardware.* In a variety of sizes, from tiny 2/56 threads to standard 6/32 and 8/32 threads. Look for blind nuts, which let you mount things like servos and motors flush to the body of the robot. Blind nuts need soft plastic, wood, or similar materials (but not metal or "hard" plastic like acrylics). The "fins" of the nut must be able to dig into the material to provide a secure fit.

- *Threaded inserts.* Also used with softer materials, they screw into the material to create a standard-size threaded hole.

- *Servo tape.* Wide and supersticky tape for holding servos, batteries, and other objects to the frame of a robot.

- *Hex-socket-head screws.* Precision machined with threads from 2/56 through 6/32. Has hex-socket head, sometimes with a knurled knob to allow for easier manual tightening.

- *Threaded couplers.* Extend the length of threaded rods by allowing them to be connected together.

- *Replacement servo gears.* Though intended to repair broken or worn-out servos, these gearsets can also be used for any other purpose you choose. Replacement gears may be plastic or metal, depending on the make and model of servo they are for.

Walther's Model Railroad Mall   202415

5601 W. Florist Ave.
Milwaukee, WI 53218-1622
USA

((414) 461-1050

⊘ (800) 487-2467

✉ custserv@walthers.com

🌐 http://www.walthers.com/

Manufacturers and distributors of model railroad products. Carries 85,000 model railroad items from over 300 manufacturers; the products are supplied to hobby shops around the world. Direct orders are also welcome.

Wangrow   202398

P.O. Box 98
Park Ridge, IL 60068-0098
USA

✉ systemone@wangrow.com

🌐 http://www.wangrow.com/

Manufacturer of high-end digital model-train controllers. Available through dealers.

Woody's Servo Page   203306

http://www.rc-soar.com/woodys_servos.htm

Basic reference specifications on dozens of R/C servos.

RADIO CONTROL-ACCESSORIES

This section contains specialty accessories for radio-control models. Included are paints, special foils and materials, and gears and sprockets for R/C servos. See also **Radio Control-Hardware** for fasteners, wheels, and other R/C hardware.

Airline Hobby Supplies   203165

P.O. Box 2128
Chandler, AZ 85244-2128
USA

☎ (480) 792 9589
📠 (480) 792 9587
✉ rmbrown@ican.net
🌐 http://www.airline-hobby.com/

Small assortment of R/C paints, bare-metal foil, adhesives, and accessories. Also publishes Airline Modeler Magazine.

Hobby Stuff Inc.   203314

11239 E. Nine Mile Rd.
Warren, MI 48089
USA

☎ (586) 754-6412
📠 (586) 754-7402
✉ info@hobbystuffinc.com
🌐 http://hobbystuffinc.com/

Hobby Stuff deals with the in-ordinary . . . stuff few others carry. They sell unusual hardware pieces for R/C airplanes that might be used in a robot, along with vacuum formers (you supply the vacuum and the heat source) for making your own formed parts using thin plastic sheets. The prices won't break you.

R.C. Scrapyard   203712

66 Cross Lane
Stocksbridge
Sheffield
S36 1AY
UK

✉ Sales@RCScrapyard.bizland.com
🌐 http://rcscrapyard.bizland.com/

U.K.-based resource for R/C enthusiasts. The company sells new, surplus, and used R/C parts, with an emphasis on R/C car racing.

Servo City   202079

BTR
620 Industrial Blvd.
Winfield, KS 67156
USA

☎ (620) 221-7071
📠 (620) 221-0858
🚫 (877) 221-7071
🌐 http://www.servocity.com/

Servo City sells R/C transmitters, receivers, servos, and servo components. The latter two are of keen interest to robot builders, and Servo City offers a wide assortment at terrific prices. They carry the full lines of Futaba and Hitec servos. Be sure to check out the Servo Accessories and Mechanical Parts pages for such things as:

- Universal servo mounts-mount servos on flat surfaces (no more tape or glue!)
- Servo-to-servo X-Y axis mount-small mechanical dohicky for two degrees of freedom
- Servo sprockets and chain-attach sprockets to servos and power with strong plastic chain
- Plain bore and hub-mounted 48- and 32-pitch gears-build power transmissions with gears

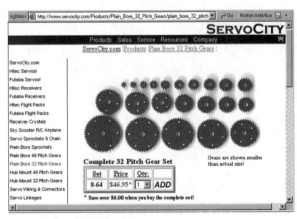

ServoCity.com Web site.

Sullivan Products   204217

1 North Haven St.
P.O. Box 5166
Baltimore, MD 21224
USA

✆ (410) 732-3500

📠 (410) 327-7443

✉ sales@sullivanproducts.com

🌐 http://www.sullivanproducts.com/

Sullivan manufactures accessories for performance R/C aircraft. They offer a number of hardware-related items like pushrods and landing gear, but also of interest is their electric starter for gasoline-powered airplane engines. Starter motors exhibit high torque and are ideal for use in robot drives, and their prices tend to be more affordable when compared to many other motors within the same class

RADIO CONTROL-HARDWARE

This section contains specialty hardware for radio control models. Included are fasteners, carbon composites for strong yet light frames and bodies, small wheels, shaft collars, linkages, and specialty connectors. See also **Radio Control-Accessories**.

Art's Hobby   203295

P.O. Box 871564
Canton, MI 48187-6564
USA

✆ (734) 455-1927

📠 (413) 618-8961

✉ support@arts-hobby.com

🌐 http://www.arts-hobby.com/

Art's Hobby sells several unusual R/C construction components, including carbon laminate sheets and strips and carbon push rods. Carbon composites are useful because they are very lightweight, yet extremely strong.

Du-Bro Products, Inc.   202134

P.O. Box 815, 480 Bonner Rd.
Wauconda, IL 60084
USA

🌐 http://www.dubro.com/rcproducts.html

Du-Bro is a leading manufacturer of hardware, wheels, and accessories for radio controlled models. If you build a robot, odds are it'll have at least one Du-Bro product in it. Things you might be interested in:

- Adjustable motor mount
- Ball links
- Building supplies and finishing
- Control hook-ups
- Control horns
- Fasteners
- Landing gear
- Motor mounts
- R/C boat accessories
- R/C car accessories
- Tools
- Wheels

Also check out their "teeny-weeny hardware" line, such as 1-inch mini lite wheels, micro push-pull system, and micro control horn. Useful for midget robots.

Most product is available in small-quantity packaging.

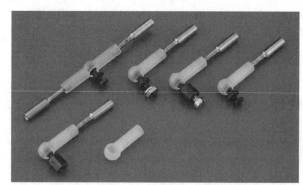

Du-Bro ball links. Photo Du-Bro Products, Inc.

Hammad Ghuman   202143

6 Tower Heights
Albany, NY 12211
USA

✆ (518) 782-9255

🌐 http://www.1hg.com/

Precision parts for competition R/C model cars. Each part is CAD/CAM designed and precision CNC machined. Products include metal-alloy wheels, shock pistons, pinion gears, titanium ball ends, and shafts. Materials are aluminum, stainless, titanium, and Delrin.

Parts are listed by type and also under the model of R/C car they are for. For robotics, you'll want to select the part based on its specifications—number and pitch of the gear, or the length of the shaft, for example.

Custom made R/C racing gear from Hammad Ghuman. Photo HG / Hammad Ghuman Inc.

R/C racing hub from Hammad Ghuman. Photo HG / Hammad Ghuman Inc.

RADIO CONTROL-SERVO CONTROL

The following listings are for companies that specialize in making and selling electronics for operating radio-control servos from devices other than R/C receivers. Most servo controllers can operate multiple servos at a time and are meant to be commanded by a microcontroller or computer.

Medonis Engineering 202452

3126 SW 153rd Dr.
Beaverton, OR 97006
USA

((503) 860-1980

((503) 605-1471

✉ info@medonis.com

🌐 http://www.medonis.com/

Servo controllers, animatronic head kit.

Says the Web site, "My name is Mark Medonis, and after building robots for nine years, I created Medonis

Engineering in 1999 to market some of my robot products for sumo, animatronics, and hobby robot building in general. Because I love building robots and I think more people should build them."

How can you disagree with that!

MIDI Robotics 202980

http://www.midivid.com/robasic.html

According to the Web site, "Midivid has the ability to control up to 512 R/C servo motors or devices with the use of inexpensive serial microcontroller interfaces called the mini SSC II." The product uses the MIDI interface of a PC or other computer.

Mister Computer 203500

P.O. Box 600824
San Diego, CA 92160
USA

((619) 281-2091

((619) 281-2073

✉ info@mister-computer.com

🌐 http://www.mister-computer.com/

Mister Computer develops and sells animation and animatronics software that support the Scott Edwards SSC (serial servo controller). Their Animatronics Kit provides hardware (serial controller, two servos) and Windows-based software for recording, storing, and playing back animations.

National Control Devices 202140

P.O. Box 455
Osceola, MO 64776
USA

((417) 646-5644

((417) 646-8302

✉ ryan@controlanything.com

🌐 http://www.controlanything.com/

See listing under **Microcontrollers-Hardware**.

Review of Seven Serial Servo Controllers

Servo controllers operate multiple radio control servos from a single serial line. The benefit of servo controllers in robotics is obvious: Rather than devote the majority (or all) of a robot's controller to servo functions, this task is handed off to a "coprocessor" that does all the work.

In operation, the robot's computer or microcontroller sends a short set of instructions to the serial controller, telling it which servos to operate and where to move them. These instructions are sent through a simple one- or two-wire serial connection. In the typical two-wire connection, one line is used for the actual data, and the other line is used as a synchronizing clock. Even those microcontrollers that lack true serial communications capability can use the two-wire approach by applying what's known as "bit banging"— sending data to a pin one bit at a time.

There are a number of serial servo controllers on the market. Here's a quick rundown on several of the most popular ones.

Mini-SSC (Scott Edwards Electronics)

http://www.seetron.com/

The Mini-SSC set the stage for the other servo controllers that followed, and as a result, many products are functional duplicates of this one. The Mini-SSC connects to a serial communications port at 9,600 bps or 2,400 bps and controls up to eight standard hobby servos at one time. It's possible to link several Mini-SSC boards in parallel and therefore control even more servos.

The Mini-SSC servo controller from Scott Edwards Electronics.

SV203 (Pontech)

http://www.pontech.com/

The Pontech SV203 series of serial servo controllers operates up to eight standard servos at speeds of 2,400, 4,800, 9,600, or 19,200 bps. Like the Mini-SSC, each board can be assigned a different "address," and therefore you can control multiple banks of eight servos. The SV203B/C boards also feature infrared, digital I/O, and onboard program memory, allowing them to be used in stand-alone mode, with a controller or computer.

The SC203 servo controller from Pontech.

Servo 8T (Web-Hobbies.com/NetMedia)

http://www.web-hobbies.com/

The Servo 8T controls up to eight servos, and eight units can be daisy-chained to control up to 256 servos. The Servo 8T supports serial speeds up to 19,200 bps. What sets this controller apart from the others is that it incorporates force feedback for each servo. This information is communicated back to the computer or microcontroller and can be used to moderate the position of the servo.

The 8T servo controller from Web Hobbies (NetMedia).

FT639 (FerretTronic)

http://www.ferrettronics.com/

The FT639 is the smallest servo controller chip; everything is contained in an 8-pin integrated circuit. The chip can control up to five servos and supports 2,400 bps communications. The FT639 cannot be paralleled to control additional servos; however, the company offers another product, the FT649, that can control five FT639 servo controllers (for a total of 25 servos, which is usually more than enough for anybody).

ASC16 (Medonis Engineering)

http://www.medonis.com/

The ASC16 operates up to 16 R/C servos. To help separate it from the rest of the pack, the product also sports eight high-current digital outputs and eight inputs configurable as either analog or digital inputs.

Pololu Servo Controller (Polulu)

http://www.pololu.com/

Provided in kit form (you must solder it together), this controller operates up to 16 servos, at data rates from 1,200 to 19,200 bps.

Ohmark Digital Servo Controller (Ohmark)

http://www.ohmark.co.nz/

The Ohmark DSC controls up to eight servos and can be daisy-chained to operate up to 32 servos using the stock product. In addition to being able to set the position of the servo, the DSC can control the "rate" or speed of the servo movement (this is a unique and great feature) and can disable individual servos so that they do not consume power.

NetMedia Inc./Web-Hobbies 202152

10940 N. Stallard Pl.
Tucson, AZ 85737
USA

☏ (520) 544-4567

📠 (520) 544-0800

🌐 http://www.web-hobbies.com/

NetMedia's Web Hobbies makes the "worlds only serial servo controller with torque feedback." As stated on the Web site, "Unlike standard serial servo controllers the Servo 8T can also provides live torque/load information for each connected servo. This information can then be used to ascertain a wide variety information about the connected servos."

Pontech 202837

9978 Langston St.
Rancho Cucamonga, CA 91730
USA

☏ (413) 235-1651

🚫 (877) 985-9286

📧 info@pontech.com

🌐 http://www.pontech.com/

Pontech produces low-cost servo and stepper motor controllers. See listing under **Actuators-Motors**.

RobotLogic 300008

Greg Hjelstrom
10416 Snowdon Flat Ct.
Las Vegas, NV 89129
USA

Roll Your Own Serial Servo Controller

In addition to commercial serial servo control products, a number of Web sites illustrate how to make your own, typically with a PICmicro controller and assembly language or Basic programming. Here are two:

16-Channel Serial Servo Controller
http://www.seattlerobotics.org/encoder/200106/16csscnt.htm

5-Channel Serial Servo Controller
http://www.frii.com/~dlc/robotics/projects/botproj.htm

---

support@robotlogic.com

http://www.robotlogic.com/

See listing under **Motor Control**.

Scott Edwards Electronics, Inc.   202179

1939 S. Frontage Rd. #F
Sierra Vista, AZ 85635
USA

((520) 459-4802

(520) 459-0623

info@seetron.com

http://www.seetron.com/

Scott Edwards Electronics (otherwise known as Seetron) manufactures and sells serial LCD and VFD displays that easily interface to a computer or microcontroller. The company also offers the Mini SSC II interface to control up to eight R/C servos from a single serial connection.

 RADIO CONTROL-SERVOS

Refer to these listings for makers and sellers of radio control servos. All sizes and styles are represented here. Note that most R/C servo makers do not sell directly to the public, but their Web sites provide useful technical information and datasheets. Refer to a hobby retailer if you wish to purchase R/C servos. See also the main **Radio Control** section for additional sources for servos.

Airtronics Inc.   202127

1185 Stanford Ct.
Anaheim, CA 92805
USA

((714) 978-1895

(714) 978-1540

info@airtronics.net

http://www.airtronics.net/

Airtronics is a maker of radio control electronics for flight models. Their product line is extensive and includes many servos and servo accessories; also sells replacement gears, servo trays, and servo control horns.

Futaba   202142

Great Planes Model Distributors
P.O. Box 9021
Champaign, IL 61826-9021
USA

((217) 398-6300

○ (800) 637-7660

gpinfo@gpmd.com

http://www.futaba-rc.com/

Futaba makes R/C transmitters, receivers, and servos. Their Web site provides some handy technical reference on their product, including the latest digital servos.

Global Hobby Distributors   203494

18480 Bandilier Cir.
Fountain Valley, CA 92708
USA

Inside an R/C Servo

The typical R/C servo comprises three major parts:

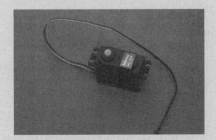

An R/C servo motor.

- *Motor.* The motor is a small DC-operated permanent magnet unit, capable of reversing direction.

- *Reduction gear.* The high-speed output of the motor is reduced by a gearing system. Many revolutions of the motor equal one revolution of the output shaft of the servo. In most servos, the output shaft turns no more than 90 degrees in either direction.

- *Control circuit.* The output shaft is connected to a potentiometer, a common electronic device very similar to the volume control on a radio. The potentiometer (or "pot") connects with a control circuit, and the position of the pot naturally indicates the position of the output shaft.

In a radio-control application, the receiver—mounted someplace in the vehicle—both powers and controls the servo. The control signal for the servo is in the form of a series of pulses. The duration of the pulses is what determines the desired position of the servo.

Specifically, the servo is set at its center point if the duration of the control pulse is 1.5 millisecond (one-thousandth of a second). Durations longer or shorter command the servo to turn in one direction or the other. A duration of 1.0 millisecond (ms) causes the servo to turn all the way in one direction; a duration of 2.0 ms causes the servo to turn all the way in the other direction.

Note that the pulse-width variance of 1.0 to 2.0 ms is average for most R/C servos, and that the full rotation of a servo is typically 130 to 180 degrees, depending on its mechanical design.

---

  (714) 963-0133

  (714) 962-6452

✉ info@globalhobby.net

🌐 http://www.globalhobby.com/

Global Hobbies is a national distributor of hobby products specializing in radio controlled models, including Cirrus servo motors. They are also a distributor for a wide selection of general hobby product brands.

🌐 🏭

Hitec/RCD   202144

12115 Paine St.
Poway, CA 92064
USA

☎ (858) 748-8440

🌐 http://www.hitecrcd.com/

Hitec/RCD makes transmitters, receivers, and servos for hobby R/C (airplanes and cars). Hitec servos are my favorite as they tend to be a little less expensive than the others, yet are still well made and durable. Product

is available from hobby dealers everywhere. The Web site provides a smattering of technical documents.

Hitec model HS-625MG.
Photo Hitec RCD USA, Inc.

Hobby Horse Wisconsin, Inc. 202730

1769 Thierer Rd.
Madison, WI 53704
USA

 (608) 241-1313

(800) 604-6229

hobbyhorse@hobbyhorse.com

http://www.hobbyhorse.com/

R/C products including a full line of Hitec servos. Discounts on most items.

Horizon Hobby 203495

4105 Fieldstone Rd.
Champaign, IL 61822
USA

(217) 355-1552

(800) 338-4639

websales@horizonhobby.com

 http://www.horizonhobby.com/

Distributor of land, air, and water R/C products, with online sales and distribution in stores across North America. Among product line most useful to robot builders: JR servos, adhesives, and tools.

Robots.net: Servo Hacking Tutorials 203714

http://robots.net/article/208.html

Links and info on hacking R/C servos for continuous rotation.

Servo City 202079BTR

620 Industrial Blvd.
Winfield, KS 67156
USA

Modifying Servos for Continuous Rotation

R/C servos have built-in mechanical stops that prevent them from turning more than about 180 degrees (exception: servos meant for specialty applications, like landing gear retraction, which may turn 360 to 720 degrees). By removing the mechanical stops, and making a change in the electrical connections inside, it's possible for the output of the servo to turn continuously in either direction. Modified servos are often used for the drive wheels of small robots because their use simplifies the connection between the control electronics of a robot and the motor drive. The servo package includes motor, reduction gearing, and power drive electronics and can be directly connected to a microcontroller, computer port, or other digital interface.

The Internet and World Wide Web provide a number of sources for information on how to modify popular standard-size servos. Check out the following:

Seattle Robotics Society

http://www.seattlerobotics.org/guide/servohack.html

Al's Robotics

http://alsrobotics.botic.com/

Robot Store

http://www.robotstore.com/download/Servo_Mod_Notes_1.0.pdf

(requires Adobe Acrobat Reader)

Representative model R/C servo.

 (620) 221-7071

 (620) 221-0858

 (877) 221-7071

🌐 http://www.servocity.com/

Servo City sells R/C transmitters, receivers, servos, and servo components. See listing under. **Radio Control-Accessories**.

Digital Proportional versus Digital Servos

Standard servos for radio controlled models use digital signals to control their position. This is called digital proportional control. The position of the output of the servo is proportional to the width of the on/off (digital) signals being fed to it.

But internally, these servos are old-fashioned analog devices. The latest twist in R/C is the *digital servo*; these use onboard digital circuitry to enhance the functionality of the motor. Inside the motor case is a microprocessor that is controlled by commands you send from the transmitter. In addition to the standard digital pulses used for positioning the servo, additional programming commands can be sent for such functions as controlling the speed of the servo motor and for readjusting the center point of the servo.

Digital servos can be operated from the same R/C receivers or robotic servo controllers as standard servos. However, the programmability feature (for those digital servos that support it) requires a specialized programmer module.

Key benefits of digital servos include higher resolution of position and faster response time. An example digital servo is the Hitec HS-5925MG. Its specifications are impressive:

- 128 oz. of torque
- 0.08-sec. speed (60 degrees)
- 384 oz. of holding power, at 6 volts

Digital servos cost several times more than their older-fashioned analog cousins, but at least part of the higher cost is in mechanical improvements. The typical digital servo uses high-speed ball bearing construction, more powerful motors, and stronger gears. They are designed as *the choice* for performance R/C applications, such as model helicopters.

If you opt to use digital servos in your robotics project, you must ensure that your battery pack can deliver the extra current they draw. The average digital servo is designed to drive heavier loads at higher speeds, and this requires more current from your battery pack. High-capacity (1,600 mAh or higher) ni-cad batteries, NiMH batteries, or sealed lead-acid batteries are recommended.

Finally, digital servos tend to be much louder than standard servos, both when turning and when "idling" (they can buzz while remaining still). Don't use them if you're building a spy robot!

Futaba provides some technical details on digital servos on their Web site (some pages may be in PDF format, requiring you to install and use the Adobe Acrobat Reader):

http://www.futaba-rc.com/servos/

 **RETAIL**

The retail establishments in the following sections are open for business to the general public. Online, traditional mail order, and so-called bricks-and-mortar retailers are represented here.

SEE ALSO:

DISTRIBUTOR/WHOLESALER (various)

ELECTRONICS (various)

SUPPLIES (various)

How to Buy Mail Order

It may seem daft to "explain" how to buy mail order, but every year thousands of people get cheated out of millions of dollars with mail order (that is, millions in total, not every person!). So, for reminder's sake, here's a list and dos and don'ts when conducting business by mail.

Do

- Understand exactly what you are buying, when delivery will be made, and how much you're paying *before* sending any money. Sounds simple enough, but it's easy to forget the small stuff when you're excited about finding goodies for your robot.
- Favor those companies that provide a mailing address and a working phone number for voice contact (not just fax). Sellers without one or the other aren't necessarily crooks, but lack of contact information just makes it harder to get a hold of someone should there be a problem.
- Be wary of companies that advertise by sending unsolicited "spam" via e-mail. Spam is basically free to send, so everyone can do it—including the scamsters.
- Verify shipping charges, handling charges, and service fees before finalizing the order. These costs can significantly add to the price, especially for small orders.
- Check out the company before sending them a significant order ("significant" is up to you; it might be anything over $500, or it might be anything over $35). Check for a poor rating with the Better Business Bureau (or similar institution for those outside the United States) in the company's home town; in the appropriate newsgroups; or in online chat or bulletin boards.
- Determine added costs for duty, taxes, and shipping when buying internationally.
- Carefully examine your credit card monthly statement for improper charges.

Don't

- Give your credit card number via e-mail, or on a Web page order form, unless you know the communications link is secure.
- Buy from a source unless you feel very comfortable you can trust your money with them.
- Use a credit card to pay for goods from a company you have not yet dealt with if sending a check or money order is just as easy. This limits the exposure of your credit card accounts to possible Internet fraud.
- Send money to foreign companies unless you're positive they are safe bets. While you're checking them out, be sure they will ship to your country.

Should you have trouble with a mail-order merchant (and you or the business is in the United States), the following two organizations might be able to help you resolve the matter.

Better Business Bureau System

http://www.bbb.org/

National Mail Order Association

http://www.nmoa.org/

Money Changing and Currency Conversions

Robotics is truly a global endeavor, with people all over the world buying and selling goods, services, and software. Web pages in other countries typically show prices in the local currency, such as yen in Japan or pounds or Euros in Great Britain. You can use a currency converter service—of which there are many freely available on the Web—to calculate the going exchange rate of currencies.

One of the easiest (and least marred by pop-up ads) is XE.com:

http://www.xe.com/

To use, specify the amount to exchange and the "from" and "to" currencies. For example, if a Web site shows a price in Japanese yen, and you want dollars, you select *JPN Japan Yen* in the From pull-down list, and *USD United States Dollars* in the To pull-down list. The rates are updated daily, but are not guaranteed to be the ones in effect when you actually make your purchase. Use for ballpark estimates only.

Calculating Shipping Costs via UPS

The United Parcel Service, or UPS, delivers packages to North America and worldwide. For those in the United States, UPS is a common mode of delivery of packages, even over the Post Office. UPS has several service "grades" depending on how soon you want your goodies, from next-day air to "ground"—ground shipping goes by truck or air and is given the lowest priority.

The costs to ship via UPS are readily calculated using the UPS Web site:

http://www.ups.com/

From this centralized location, choose the Rates link, then fill in the following information to calculate shipping costs: shipper location, shippee (that's you) location, weight, size, service type, and whether the package is being dropped off at a UPS service station or if it's being picked up.

You probably won't know some specifics about the package being sent to you, such as its size or whether it's being picked up by a driver, but you can enter nominal values here to arrive at an approximate shipping cost. Most online retailers not operated out of someone's home use the "daily pickup service," where a driver comes by once a day to collect the boxes. Use that option if you're not sure. Unless you know the box will be overly large, use 8-inch-square dimensions for the packaging size.

Note that many mail-order retailers also charge a "handling" or packaging fee, in addition to any direct shipping costs. Remember to tack these fees onto the price you pay.

RETAIL-ARMATURES & DOLL PARTS

Retail stores that specialize in doll and teddy bear-making parts are listed in this section. This includes something known as armatures, which are used in dolls and puppets to replicate bone joints. Basic armatures for teddy bears are made of plastic and act like universal joints, allowing movement in all directions. More sophisticated armatures are made of metal and allow movement in one plane only. They may be on puppets and more lifelike dolls.

Other doll parts you can use in your robots include eyes, noses, and "body foam" (soft and pliable, but thick).

SEE ALSO:

RETAIL-ARTS & CRAFTS: Some art supply and arts and craft stores also carry doll armatures and other doll-making parts

SUPPLIES-CASTING & MOLD MAKING: Body foams and articulated links

Armaverse Armatures   204129

906 E. Walnut St.
Lebanon, IN 46052
USA

 armaverse@in-motion.net

http://www.armaverse.com/

Ball-and-socket doll armatures. Sells individual pieces and kits of parts.

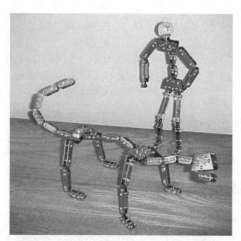

Armatures from Armaverse.
Photograph courtesy of Armaverse
Armatures - www.armaverse.com

Bear Ingredients   202742

588 Edward Avenue, Unit 52
Richmond Hill, ON
L4C 9Y6
Canada

(905) 770-3173

(905) 770-6811

 info@bearingredients.com

http://www.bearingredients.com/

Teddy bear supplies-eyes and joints.

Chatsco Distributions   203186

3998 Malpass Rd.
London, ON
N6P-1E8
Canada

(519) 652-1555

(519) 652-2094

 chatsco@rogers.com

http://www.chatsco.com/

Teddy bear and doll body parts: eyes, noses, squeakers, whiskers, voice modules.

Disco Joints and Teddies   202745

2 Ridgewood Place
Box 468
St. Clements, ON
N0B 2M0
Canada

(519) 699-5762

(519) 699-4525

http://www.discojoints.on.ca/

Teddy bear supplies, including disk joints, screw joints, eyes, and noses.

EZ Pose Flexible Doll Bodies   204132

P.O. Box 97
Crestone, CO 81131
USA

(719) 256-4235

(719) 256-4209

 sandi@ezpose.com

http://www.ezpose.com/

Professional doll artist Sandi Patterson sells a line of posable doll bodies that, with a bit of work, servos, and batteries, could make a nightmare straight out of The Twilight Zone. The bodies are soft fabric and foam covering bendable armature wire. Attach a head, and you've got your own Talking Tina!

Grandma T's 204131

15753 Jolly Rd.
Marysville, OH 43040
USA

((937) 642-7032
✆ (937) 642-7032
🌐 http://www.grandmats.com/

This grandma doesn't bake cookies-or maybe she does, but they aren't available at the site. Rather, Grandma T sells arts and craft supplies, including doll armatures and acrylic craft boxes. Here's a short sampling of material useful in robots:

- Acrylic craft boxes (use for robot parts, bodies; in different colors)
- Acrylic shapes (half rounds, ovals)
- Coiling Gizmo (makes springy coils from wire; use for touch sensors)
- Darice craft supplies (various odds and ends; useful for small parts)
- Doll parts (including armatures)

Animation Supplies

http:// www.animationsupplies.net

Armatures for stop-motion animation (or robots)

Gryphyn Armatures

http://easyweb.easynet.co.uk/edawe/

Fine machined armatures in stainless and aluminum

Jointworks

http://www.jointworks.com/

Armatures for stop-motion animation

Taylor Designs

http://www.taylordesign.org/

Armatures for puppetry and animation

- Foamies (foamed rubber in various thicknesses and colors)

Jim Allred Taxidermy Supply 203561

216 Sugarloaf Rd.
Hendersonville, NC 28792
USA

((828) 692-5846
✉ jim@jimallred.com
🌐 http://www.jimallred.com/

Taxidermy supplies: eyes (for humanoid or animal robots) and foam.

McKenzie Taxidermy Supply 203563

P.O. Box 480
Granite Quarry, NC 28072
USA

((704) 279-7985
✆ (704) 279-8958
⊘ (800) 279-7985
✉ taxidermy@mckenziesp.com
🌐 http://www.mckenziesp.com/

Stuff for when you bag that ol' robot. McKenzie sells taxidermy supplies, of which things like eyes, casting materials, and carvable foam are relevant to the robot constructionist.

Sculpture House Casting 203241

155 W. 26th St.
New York, NY 10001
USA

✆ (212) 645-3717
⊘ (888) 374-8665
✉ info@sculptshop.com
🌐 http://www.sculptshop.com/

"Sculpture House Casting is one of the cradles of fertile ideas and artistic expression." Or, put another way, they sell casting and mold-making supplies and tools, armatures, and modeling accessories. Use their stuff to create fertile and artistic robots.

Spare Bear Parts

202743

792 E. Hwy. 66
Tijeras, NM 87059
USA

 (505) 286-5005

 (505) 286-5018

⊘ (866) 999-2327

 sales@sparebear.com

 http://www.sparebear.com/

Teddy bear parts, including eyes.

Teddy Bear Stuff

202744

JumpStart Marketing
907 Queen's Blvd.
Kitchener, ON
N2M 1B1
Canada

((519) 742-3325

 (519) 742-7348

🌐 http://www.teddybearstuff.com/

Teddy bear supplies, such as eyes and foam.

Web site for Teddy Bear Stuff.

Van Dykes Taxidermy

203560

P.O. Box 278
39771 S.D. Hwy. 34.
Woonsocket, SD 57385
USA

((605) 796-4425

 (605) 796-4085

⊘ (800) 787-3355

 taxidermy@cabelas.com

 http://www.vandykestaxidermy.com/

No, Van Dykes taxidermy supplies aren't for hanging up a prize robot after you've bagged it. Instead, you can use Van Dykes for its carving foam, various kinds of glass eyes, and assorted unusual materials. Their foam block is easy to work with and can be shaped with simple tools.

RETAIL-ARTS & CRAFTS

Here, you'll find retail stores (online, mail-order, and walk-in) that specialize in arts and crafts, as well as professional and amateur artist supplies. Products run the gamut from glue guns and liquid adhesives, to small parts for dollhouses and jewelry making, paints, paintbrushes and airbrushes, fixatives (use for stabilizing water-slide decals), foam board and other substrates, small balsa and metal pieces (e.g., J&S Engineering) for construction, sewing notions (fusible tape, plastic needlepoint cloth, much more), small plastic display boxes (use them for parts containers or cheap electronics housings), polymer clay, casting materials, and a whole lot more!

SEE ALSO:

RETAIL-ARMATURES & DOLL PARTS: Arm and leg joints, eyes for dolls, teddy bears, and small robots

SUPPLIES (VARIOUS): Additional craft supplies, including cast and mold-making products, adhesives, and paints

Activa Products, Inc.

203179

8242B Main St.
Mokelumne Hill, CA 95245
USA

 (209) 286-9603

⊘ (800) 577-1421

🌐 http://www.activaproducts.com/

Arts and crafts supplies. Includes casting and mold-making supplies.

ArtSuppliesOnline.com

202521

718 Washington Ave. North
Minneapolis, MN 55401
USA

((612) 333-3330

(612) 333-0200

(800) 967-7367

info@artsuppliesonline.com

http://www.artsuppliesonline.com/

Art supplies: craft boards (such as foam board), plastics, and adhesives, and lots more. Their online store lets you browse by category or search for specific products by name or brand. Check out their "Imaginative Manikins," small articulated wood models in the shapes of people and animals.

Local store in Minneapolis, Minn.

ASW-Art Supply Warehouse 202741

5325 Departure Dr.
Raleigh, NC 27616-1835
USA

((919) 878-5077

(919) 878-5075

(800) 995-6778

aswexpress@aol.com

http://aswexpress.com/

Artists accessories, brushes, foam boards, and paints.

Clotilde, Inc. 203528

B3000
Louisiana, MO 63353-3000
USA

(800) 545-4002

Clotilde Web home.

Where to Get Stuff: Arts and Craft Stores

Arts and craft stores are veritable gold mines of handy robotic materials, with the added advantage of low cost—few people want to spend lots of money on their leisure projects. Keep an eye out for the following goodies at your neighborhood arts and craft outlet.

- *Foam rubber sheets* in various colors and thicknesses.
- *Foam board,* constructed of foam sandwiched between two heavy sheets of paper.
- *Electronic light and sound buttons* to make Christmas ornaments and custom greeting cards are also well suited for robots.
- *Jewelry findings,* in plastic and metal, for making your own jewelry, are handy for constructing miniature robots and other parts. Findings include metal pushpins and locking fasteners, used to make earrings.
- *Parts for dolls and teddy bears* can often be used in robots. Fancier dolls use "articulations"—movable and adjustable joints—which can be used in your robot creations. Look also for linkages, bendable posing wire, and eyes (great for building robots with personality!).
- *Plastic crafts construction material* can be used in lieu of more expensive building kits.
- *Model-building supplies,* including plastic and metal parts, glues and adhesives, and hand tools.

http://www.clotilde.com/

Online and mail-order (printed catalog) sewing and quilting supplies. Look over things like fusing tape (partially melts when heated), small tools, rotary cutters (useful for foam core and other lightweight laminates), and elastics.

Crafter's Market 203504

Great Bridge Shopping Center
237 S. Battlefield Blvd., 14-B
Chesapeake, VA 23322
USA

(757) 546-8811

(757) 547-8424

CCSstitch@aol.com

http://craftersmarket.net/

Crafts, including knitting and needlepoint. Local and online stores.

Dal-Craft, Inc. 203527

P.O. Box 61
Tucker, GA 30085
USA

(800) 521-7311

sales@lorancrafts.com

http://www.lorancrafts.com/

Manufacturer of 6-inch-long half-cylinder magnifier, used in needlecrafts. But think of the uses in robotics: The magnifier can also be used as a magnifier for line-following and related tasks. Products available in sewing and crafts stores.

Dick Blick Art Materials 202737

P.O. Box 1267
Galesburg, IL 61402-1267
USA

(309) 343-6181

(800) 828-4548

info@dickblick.com

http://www.dickblick.com/

Complete line of craft materials and art supplies. Also local stores (predominately in the Midwest and Great Lakes areas of the United States).

Dick Blick Art Materials Web site.

Discount Art Supplies 202740

P.O. Box 1169
Conway, NH 03818
USA

(603) 447-6612

(603) 447-3488

Shrink Art Plastic

Ask most kids and they know what Shrinky Dink is: It's a plastic that shrinks to one-quarter of its size when heated in an oven at about 225 degrees. Generically, it's called *shrink art plastic sheets* and also goes by such names as Super Color Shrink and PolyShrink. You can draw or print a shape on the plastic, and when "reduced," the printing stays vivid and clear. You can use it to create nameplates and labels, encoder discs, body designs, and more.

Cost is about a dollar per sheet, and you can cut out just the portion you need.

You can find Shrinky Dink at Wal-Mart and other discounters; the generic (and therefore less expensive) material can be found at arts and craft stores.

(800) 547-3264

sales@discountart.com

http://www.discountart.com/

Paint and paint supplies; airbrushes.

Dixie Art & Airbrush Supplies 203204

2612 Jefferson Hwy.
New Orleans, LA 70121
USA

(800) 783-2612

http://www.dixieart.com/

Airbrushes and compressors; general art supplies.

Fastech of Jacksonville, Inc. 203828

P.O. Box 11838
Jacksonville, FL 32239
USA

(904) 721-6761

(800) 940-6934

info@hookandloop.com

http://www.hookandloop.com/

Velcro distributor. Online sales.

FLAX Art & Design 202603

240 Valley Dr.
Brisbane, CA 94005-1206
USA

(888) 352-9278

askus@flaxart.com

http://www.flaxart.com/

Art supplies: adhesives, paints, substrates, craft, drafting.

Focus Group Ltd. 203904

Gawsworth House
Westmere Dr.
Crewe, Cheshire
CW1 6XB
UK

+44 (0) 1270 501555

+44 (0) 1270 250501

info@focusdoitall.co.uk

http://www.focusdiy.co.uk/

Focus is a chain of "do-it-yourself" and craft stores in the U.K. See the Web site for a store locator. Online catalog of several popular product lines.

Grandma T's 204131

15753 Jolly Rd.
Marysville, OH 43040
USA

(937) 642-7032

(937) 642-7032

http://www.grandmats.com/

This grandma doesn't bake cookies-or maybe she does, but they aren't available at the site. Rather, Grandma T sells arts and craft supplies, including doll armatures and acrylic craft boxes. Here's a short sampling of material useful in robots:

- Acrylic craft boxes (use for robot parts, bodies; in different colors)

- Acrylic shapes (half rounds, ovals)

- Coiling Gizmo (makes springy coils from wire; use for touch sensors)

- Darice craft supplies (various odds and ends; useful for small parts)

- Doll parts (including armatures)

- Foamies (foamed rubber in various thicknesses and colors)

Joann.com
http://www.joann.com/
Online and local retailer of fabric and notions

Lacy & Company
http://www.lacytools.com/
Jewelry supplies and tools (Canada)

Rex Art
http://www.rexart.com/
Art supplies

Hobby Lobby   202714

7707 SW 44th St.
Oklahoma City, OK 73179
USA

📞   (405) 745-1100
📠   (405) 745-1636
🌐   http://www.hobbylobby.com/

Hobby Lobby is not to be confused with Hobby Lobby International: the former is a chain of arts and crafts stores; the latter is a mail-order retailer of hobby R/C components. Hobby Lobby locations are throughout the central U.S.

HR Meininger Company   203521

499 Broadway
Denver, CO 80203
USA

📞   (303) 698-3838
📠   (303)-871-8676
🚫   (800) 950-2787
✉️   supplies@meininger.com
🌐   http://www.meininger.com/

Online and local (in Colorado) arts/crafts stores. Mostly for paint artists, but some of the kids project kits have useful material for robot building.

Hygloss Products, Inc.   202319

45 Hathaway St.
Wallington, NJ 07057
USA

📠   (973) 458-1745
🚫   (800) 444-9456
✉️   info@hygloss.com
🌐   http://www.hygloss.com/

Manufacturer of children's arts and crafts supplies. These include specialty paper, precut Styrofoam pieces, and foam sheets.

Ken Bromley Art Supplies   202739

Curzon House, Curzon Road
Bolton
BL1 4RW
UK

📞   +44 (0) 1204 381900
📠   +44 (0) 1204 381123
✉️   kenbromley@artsupplies.co.uk
🌐   http://www.artsupplies.co.uk/

Online seller of art supplies; mostly for paint artists.

McGonigal Paper & Graphics   203961

P.O. Box 134
Spinnerstown, PA 18968-0134
USA

📞   (215) 679-8163
📠   (215) 679-8163
✉️   help@mcgpaper.com
🌐   http://www.mcgpaper.com/

Online retailers of specialty papers for arts and crafts, including:

- Water-slide decals
- Glow-in-the-dark transfer film
- Super Color Shrink (shrinks when heated)
- Backlight film (fluoresces under ultraviolet light)
- Window cling decals (cling to glass and other very smooth surfaces)

Michaels Stores, Inc.   203497

850 North Lake Dr.
Ste. 500
Coppell, TX 75019
USA

🚫   (800) 642-4235
✉️   custhelp@michaels.com
🌐   http://www.michaels.com/

Michaels sells arts and crafts, both online and in some stores across North America and Puerto Rico-they're the largest such retailer, in fact.

MisterArt   203203

913 Willard St.
Houston, TX 77006
USA

📠   (713) 332-0222
🚫   (866) 672-7811
✉️   customerservice@misterart.com

http://www.misterart.com/

The site bills itself as the "world's largest online discount art supply store." Looks to be true-they have a lot of stuff! Quite a bit of supplies for paint artists, as well as foam core, precision tools, and adhesives.

Web site for MisterArt.

MJ Designs   203888

9001 Sterling St.
#120
Irving, TX 76063
USA

((972) 621-8585
 (972) 621-8877
 http://www.mjdesigns.com/

Arts and crafts, Texas style. Locations across the Lone Star State.

Nancy's Notions   203529

333 Beichl Ave.
Beaver Dam, WI 53916
USA

⊘ (800) 833-0690
 http://www.nancysnotions.com/

Online and mail order (printed catalog available) sewing supplies.

Nasco   202566

901 Janesville Avenue
P.O. Box 901
Fort Atkinson, WI 53538-0901
USA

((920) 563-2446
 (920) 563-8296
⊘ (800) 558-9595
 custserv@eNASCO.com
 http://www.enasco.com/

eNasco is the online component of NASCO, a mail-order catalog offering some 60,000 educational supplies. The company also sells farm and ranch supplies (no kidding), construction toys, and books. Outlet stores in Fort Atkinson, Wisc., and Modesto, Calif.

Pearl Paint   202736

1033 East Oakland Pk. Blvd.
Fort Lauderdale, FL 33334
USA

((954) 564-5700
⊘ (800) 221-6845
 http://www.pearlpaint.com/

Arts, crafts, and graphics at a discount. Paints, paper, and boards.

Polymer Clay Express   203804

at TheArtWay Studio
13017 Wisteria Dr.
Box 275
Germantown, MD 20874
USA

((301) 482-0435
 (301) 482-0610
 polymer@erols.com
 http://www.polymerclayexpress.com/

Polymer Clay Express is an online retailer of unusual and useful art supply materials, including plastic shrink art, WireForm wire mesh, casting and modeling clay, Lazertran transfer sheets, and holographic metal foils (who says robots should look dull!).

QUINCY   204027

122 Quincy Cir.
Seaside, FL 32459-4748
USA

((850) 231-0874
 (850) 231-0876

⊘ (800) 299-4242

✉ quincy@gnt.net

🌐 http://www.quincyshop.com/

Art supplies, craft kits, tin toys (Futurama), miscellaneous toys.

🌐 💻 🛍

Reuel's   203501

370 South West Temple
Salt Lake City, UT 84101
USA

☏ (801) 355-1713

⊘ (888) 355-1713

✉ info@reuels.com

🌐 http://www.reuels.com/

Art and framing supply. Local store and online store. Products include adhesives, airbrushes, craft supplies, sculpting supplies, and foam board.

🌐 💻 🛍

Sax Art's & Crafts   203940

2725 S. Moorland Rd.
New Berlin, WI 53151
USA

☏ (262) 784-6880

⊘ (800) 558-6696

✉ info@saxarts.com

🌐 http://www.artsupplies.com/

Online and mail-order retailer of arts and crafts supplies, with emphasis on schools and educators. They charge a small refundable fee for their catalog for noneducators.

Sax's is a unit of School Specialty, Inc. Refer to its listing for additional catalog mail companies that cater to educational supplies.

📄 🌐 💻

T N Lawrence & Son Ltd.   202738

208 Portland Road
Hove
BN3 5QT
UK

☏ +44 (0) 1273 260260

✆ +44 (0) 1273 260270

✉ artbox@lawrence.co.uk

🌐 http://www.lawrence.co.uk/

Lawrence Art Materials offers online buying of arts and crafts supplies. Their product line emphasizes painting and reproduction supplies (acrylics, brushes, easels, and whatnot), but they also offer a number of potentially useful artifacts for bot building. It's all in how you look at things. Here are some ideas:

- Intaglio printing plates involve copper or zinc sheet metal. These sheets can be used in the construction or embellishment of robot bodies. Available in different thicknesses (1-2mm is average) and sizes, and the smaller sheets are quite affordable.

- Fluorescent acrylics can be used to paint your robot in vivid colors. Bright colors can be used when building robot teams for competitions (depending on the vision system, of course).

- Lazertran transfer paper lets you print in full color and transfer the image to most any surface. Use it for your robot's paint job, to label a control panel, and more. See http://www.lazertran.com/ for more info.

- Student-quality rollers, which can be used to press on glued laminates (thin metal over wood or plastic), rub-on lettering, and any other job where gentle but firm pressure must be applied.

- Engraving plastic is a thin semiflexible plastic sheet that can be used for building laminates.

🌐 💻 🛍

Web home for T N Lawrence & Son.

Wandix Zippy Foam

Wandix Zippy Foam is a nontoxic flexible foam material used in arts and crafts projects. It's ideal for making cutouts, for 3D projects, and for pads. The sheets come in 12-18 inches size and are available in over a dozen colors. Available at craft stores.

Manufactured by:

Wandix International, Inc.

17 Di Carolis Court

Hackensack, NJ 07601

Phone: 800-385-6855

 **RETAIL-AUCTIONS**

Auctions used to be reserved for special sales, like entire estates, expensive artwork, or antiques. With the coming of the Internet, auctions have moved mainstream, where people are selling everything-literally, everything.

This section lists a number of auctioneers that provide item descriptions or even complete details online. Listings include regular auctions held by the U.S. government, the world's largest purchaser of goods and, besides online giant eBay, the world's largest auctioneer.

For robotics, auctions can provide inexpensive test equipment, tools, and materials. But be careful: Items at most auctions are sold as-is, and one person's high bid may be another person's junk they're trying to unload. A word of advice: Do not bid on any auction unless you know exactly what you're buying, no matter how attractive the price. This applies particularly to government auctions, where online details are scarce and the condition of the item only marginally described.

DoveBid Inc. 203577

1241 E. Hillsdale Blvd.
Foster City, CA 94404
USA

- ☎ (650) 571-7400
- 📠 (650) 356-6700
- ⊘ (800) 665-1042
- ✉ CustomerService@DoveBid.com
- 🌐 http://www.dovebid.com/

Industrial equipment auctioneers.

DRMS 203541

http://www.drms.com/

This is the "master federal directory" of U.S. military and related government auctions. Straight from the horse's whatever. It's the place to obtain original U.S. Government surplus property.

See also, which contains a list of DRMS sites across the country:

http://www.drms.dla.mil/

eBay 203542

http://www.ebay.com/

Online auctions. You can search for what you want, or browse by topics. Here are some topics where robot parts and information can often be found (additional "specialty" categories can be found under most of these, as well):

- Books:Textbooks, Education:Engineering
- Books:Nonfiction:Instructional
- Business, Office & Industrial:Industrial Supply, MRO
- Business, Office & Industrial:Electronic Components

- Toys & Hobbies:Hobbies & Crafts
- Toys & Hobbies:Hobbies & Crafts:Radio Control

An eBay auction search for robotics goodies.

FedSales.gov 203571

http://www.fedsales.gov/

FedSales.gov is the official U.S. federal government portal for all government asset sales worldwide. The site includes links to surplus, art, books, even NASA surplus (sorry, no used Space Shuttles available at this time). A searchable database is provided at the site to aid in locating the kinds of products you're interested in. But don't expect eBay here. . . . Uncle Sam designed this site, and it can be difficult to find what you want.

Hint: Look for the By Agency search link, and then choose from among these most likely candidates for robotics goodies (Note: Names can change to protect the guilty parties, so be prepared to do a little sleuthing):

DRMS

DRMS stands for "Defense Reutilization and Marketing Service," an agency of the U.S. government (at least for now) that sells stuff back to the taxpayers, who already paid for it to begin with. The government calls it "surplus," but it can be new or used, in perfect or in junk condition. Prices are typically pennies on the dollar, and goods are typically offered as sealed bids.

(By the way, you'll also see the notation DRMO. That stands for "Defense Reutilization and Marketing Offices," the name the government gives to each local DRMS field office.)

What can you get government surplus? If you're lucky, you might find a Jeep for $22, but more likely, you'll find reasonably good deals on motors, test equipment, tools, computer gear, and other electronics. You won't be able to purchase entire Titan missiles from Uncle Sam, but you might get at auction some of its nonclassified subsystems. (They destroy the classified stuff.)

In the old days, government surplus was sold by the pallet-full at depot stations around the country—often, but not always, at military bases. Now, the auctions may take place on base or more often than not, over the Internet or at a private salvager's lot. With the latter, the government has handed over the details of the auction to a private company; the company handles all the financial and material transactions, taking a service fee for doing so.

An outfit known as Government Liquidation, LLC, handles most (if not all) of the surplus sales for the U.S. government. Their Web site should be your first stop for any government surplus auction you may wish to participate in:

http://www.govliquidation.com/

The process is fairly simple: First, you register with the site, providing your e-mail name and physical address. You then cruise the listings. Like all auctions, be sure you're serious about buying before you bid. And unlike eBay and most online auctions, the typical government auction is "closed," and you don't get to see what others are bidding for on the stuff you want. This can be a disadvantage or an advantage, depending on how you bid.

One approach: Consult the purchase price amount, which is usually indicated in the auction listing. Then base your bid on some percentage of this amount, say 10 percent. This assumes the item you're bidding on is in good working order; if it's damaged or nonworking, even if listed as repairable, you'll want to offer less than your usual.

- DRMS-Public Sales
- GSA Federal Supply Service (FSS)
- Federal Surplus Property Acquisitions
- NASA Surplus and Sales
- NASA-LaRC Surplus Property Program
- Parcel's, DoD Base Closing Property Development Site

Government Liquidation LLC   203540

15051 North Kierland Blvd.
3rd Fl.
Scottsdale, AZ 85254
USA

☏ (480) 367-1300
🖷 (480) 367-1450

Hobbies and Crafts on eBay

Buy used. Buy new. Buy junk. It's all available on eBay, the world's largest online auction site. What you get all depends on how careful a buyer you are.

eBay categorizes their auctions; hobby and craft products are listed under:

http://pages.ebay.com/catindex/hobbies.html

Some of the main and subcategories of interest are as follows:

- *Model RR, Trains*—Model trains and accessories
- *Arts & Crafts*—Painting, scrapbooking, ceramics, handcrafted arts, more
- *Models*—Plastic, metal, and wood kits, model-making supplies
- *Radio Control Vehicles*—Radio-controlled cars, boats, and aircraft
- *Supplies*—Art supplies and more

If you've used eBay before, then you know all the details. If you haven't used eBay, keep the following in mind:

- You must sign up first before you can bid on any auction, and you must be 18 years or over to sign up.
- Don't bid unless you are willing to spend the money. eBay lets sellers write negative comments about customers who don't pay up. (As a buyer, you can also write things about the sellers you've dealt with. This ensures everyone plays fair.)
- Read the item description carefully before bidding. Know exactly what you are getting before you bid.
- If you have any questions, e-mail the seller. If the seller does not respond, do not bid!
- Double-check shipping and handling charges. Some sellers on eBay charge excessive handling and make much of their money that way. Don't be a sucker.
- Compare the going price for the item you're bidding on against other current auctions, if any. You can also compare the final high-bid price using the Completed Item search option.
- Use proxy bidding to set your "best and final" high price. Decide the maximum amount you want to pay for an item, then *stick to it*. Set this as your high price; eBay will automatically ratchet up your bid, up to your maximum, if others bid against you. Refrain from getting into "bidding wars" at the last minute and going over your maximum.
- It's quite common to be "sniped" by last-minute high bidders. They wait until the auction is about to close, then place their bids, hoping no one outbids them in the final seconds. Don't get mad and vow to do the same yourself next time. You're bound to overbid.

info@govliquidation.com

🌐 http://www.govliquidation.com/

U.S. government surplus auctioneers.

🗣

GSA Auctions 203572

https://www.gsaauctions.gov/

Online U.S. government surplus sales.

From the Web site: "GSAAuctions.gov offers Federal personal property assets ranging from commonplace items (such as office equipment and furniture) to more select products like scientific equipment, heavy machinery, airplanes, vessels and vehicles. GSAAuctions.gov's online capabilities allow GSA to offer assets located across the country to any interested buyer, regardless of location."

Internet Auction List 203539

http://www.internetauctionlist.com/

Link lists for local and online auctions.

🗣

🚗 RETAIL-AUTOMOTIVE SUPPLIES

Even if you're not building a robotic car you will find automotive supply stores handy resources for tools and parts. Granted, much of the components are geared toward cars and trucks, and specific models at that, but the general-purpose merchandise such as switches, wires, electrical connectors, batteries, battery chargers, 12-volt vacuum cleaners, rubber gasket-making and RTV materials, and trim can be useful to robot builders.

Not included here, but nevertheless useful sources, are automotive recyclers, better known as junkyards. These are local in nature, so look in the Yellow Pages or other business phone listings. Junkyards are good sources for things like used windshield or power window motors. Some require you to remove the part from the car, and others have done the dismantling for you. Obviously, you'll save money with the former.

SEE ALSO:

FASTENERS: Bolts, nuts, and washers, large and small

POWER TRANSMISSION: Wide selection of gears, bearings, and associated items

TOOLS-HAND: A bevy of hand tools for robot building

Advance Auto Parts, Inc. 203893

5673 Airport Rd. Northwest
Roanoke, VA 24012
USA

☎ (540) 362-4911

📠 (540) 561-1448

🌐 http://www.advance-auto.com

Auto parts store; retail stores in the U.S. (most on East Coast and in Midwest), under the names Advance Auto Parts Stores, Western Auto, and Discount Auto Parts Stores. Also sells online. Check the Web site for store locator.

See also:

http://www.partsamerica.com/

CARQUEST Corporation 203894

2635 Millbrook Rd.
Raleigh, NC 27604
USA

☎ (919) 573-2500

📠 (919) 573-2501

🚫 (800) 492-7278

🌐 http://www.carquest.com/

CARQUEST is an auto supply retail chain with over 4,000 stores in North America. No online sales; check the Web site for the location of a retail store near you.

CSK Auto Inc. 202315

645 E. Missouri Ave.
Ste. 400
Phoenix, AZ 85012
USA

☎ (602) 265-9200

AutoZone, Inc.
http://www.autozone.com/
Chain of auto parts stores in the United States

(602) 631-7321

(877) 808-0698

customerservice@partsamerica.com.

http://www.cskauto.com/

Chain of auto parts stores: Checker Auto Parts, Schuck's Auto Supply, and Kragen Auto Parts. Use the Web site to find local stores.

For online sales, see also:

http://www.partsamerica.com/

Web home for Checker, Schucks, and Kragen automotive supply retailers.

J. C. Whitney, Inc. 202589

1 JC Whitney Way
LaSalle, IL 61301
USA

(800) 529-4486

http://www.jcwhitney.com/

Even before I had a driver's license I studied the J. C. Whitney catalog of automotive parts. I never got into fixing cars (building robots is enough!), but I still pore over the J. C. Whitney catalog for interesting automotive accessories, tools, batteries, and more. Useful items include:

- Windshield wiper motors (replacement, upgrade)
- Windshield washer pumps (can pump any noncaustic watery liquid)
- Electric trunk release kits (mechanical and electric)
- Electric window motors (replacement, upgrade)
- High-wattage 12-volt amplifiers (for robots with big mouths!)
- Polyurethane end link bushings (set of four or eight; use as tires)

- Rubber gasket material for door frames (bumpers and padding)
- Chrome trim (for dressing up your robot, of course!)

SEE ALSO:

http://www.CarParts.com/

Pep Boys 202314

3111 W. Allegheny Ave.
Philadelphia, PA 19132
USA

(215) 430-9000

http://www.pepboys.com/

Chain of auto parts stores (in most states of the U.S.). Use the site to locate a store in your area.

Pep Boys Web site.

RETAIL-DISCOUNT & DEPARTMENT

I just love it when I can walk into a general-purpose store and find the perfect part for my robot projects. One reason for this joy is the cost savings in finding merchandise meant for one thing, but readily adaptable to my robotics hobby.

This section lists a variety of national, even international, discount and department stores. The product carried by these stores is varied, and the listings are provided here more to get your creative juices flowing. You have to actually walk through the aisles and pick things up to determine which stuff is useful in your robot-building endeavors.

Several of the big discount chains, such as Kmart and Target, are also ideal sources for robotic toys. Watch for regular clearance sales, which occur in the months before Christmas and immediately after the holidays. Additional clearance sales may be held in late January or February, after the annual toy convention held in New York.

Do consider there are many more discount and department stores than those listed here. In your area may be a number of smaller chains, even independent stores. Don't forget to include them your regular "parts hunts."

SEE ALSO:

> **RETAIL** (various): Additional product-specific retailers
>
> **TOYS:** Toy-specific retailers

---

99 Cents Only Stores
203889

4000 Union Pacific Ave.
Commerce, CA 90023
USA

- ☎ (323) 980-8145
- 📠 (323) 980-8160
- ⊘ (888) 582-5999
- ✉ contact@99only.com
- 🌐 http://www.99only.com

Operating mostly in the western U.S., 99 Cents Only Stores sell everything at 99 cents. Much of it is common staples and food, like candy bars, soda pop, toothpaste, and soap. But you'll also find (primarily name brand) batteries, tools, even some discontinued toys that can be hacked for different robot projects. The company is owned by a wholesale importer that caters to other retailers, so they aren't at risk from running out of things to put in their stores.

You may not have a 99 Cents Only Store near you, but odds are, something similar will be in your town. The idea is the same.

Big Lots
203890

300 Phillipi Rd.
Columbus, OH 43228-0512
USA

- ☎ (614) 278-6800

- 📠 (614) 278-6676
- 🌐 http://www.cnstore.com

You may know them as Odd Lots, Big Lots, Pic 'N' Save, and Mac Frugal's. They're all owned by the same outfit, and they're all pretty much the same kind of store: closeouts and "surplus" goods, including mostly name brand household items. Stuff for robot builders include batteries, hand tools, and toys (okay, and the occasional Milky Way candy bar).

As I write this, all the stores (about 1,300 in total) are in the process of being renamed to Big Lots, but their merchandise will remain the same. Refer to the Web site for a store locator.

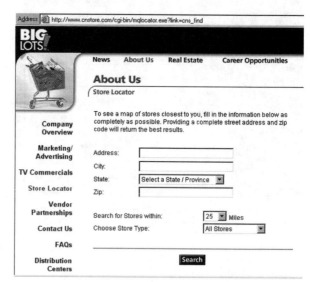

Find-a-store search for the Big Lots discount chain.

BJ's Wholesale Club, Inc.
203891

1 Mercer Rd.
Natick, MA 01760
USA

- ☎ (508) 651-7400
- 📠 (508) 651-6114
- 🌐 http://www.bjswholesale.com/

BJ's is a membership (gotta pay to shop) retailer, mostly in the eastern U.S. (Costco and Sam's Club have a lock everywhere else). The attraction of membership stores is that you buy in bulk. Don't pay $5 a pack for a set of four AA alkalines; they're a lot cheaper in packs of 24. You'll also find office supplies, books, and even toys. Except for office supplies, these aren't in bulk, but you still save. Check the Web site for a store locator.

Costco   202783

999 Lake Dr.
Issaquah, WA 98027
USA

(425) 313-8100
(425) 313-8103
(800) 774-2678
http://www.costco.com/

Costco is the Mondo Burger of membership warehouses, with some 330 stores worldwide (U.S., Canada, Japan, Mexico, South Korea, Taiwan, and the U.K.). Here you buy 30 rolls of toilet paper at a time, or you don't buy any at all. When you don't need toilet paper, there's always huge packs of alkaline batteries at an equally huge discount,as well as computer books, software, office products, and toys-including the occasional LEGO set, R/C car, or whatever else is hot at the time.

You can order products at the Costco Web site, but they're generally not the same as the ones for sale in the stores. Check the Web site for store locations near you.

Helpful tip: Costco labels all their product with large pricing signs, usually stuck to the shelves with magnets or hung overhead on a wire. If you notice a big asterisk (star) on the pricing sign, it means it's a "seasonal" and/or temporary item, and when the current stock is gone, it's gone. It probably won't be reordered. Keep this in mind should you want to stock up on some product that they are discontinuing.

Costco online.

Kmart   202784

3100 W. Big Beaver Rd.
Troy, MI 48084
USA

(248) 463-1000
(248) 463-5636
http://www.bluelight.com/

Now considered the "granddaddy" of discount department stores in the U.S. and North America, Kmart stocks a healthy number of toys, including robotic pets, LEGO, and motorized vehicles.

While their Web site provides online buying, you really should browse the aisles of your nearest Kmart, as a lot of the really good finds will be on the shelves and not on your browser.

As I write this, Kmart is considering not using their www.bluelight.com Web address (Kmart shoppers know this refers to their "blue light specials," temporary markdowns indicated in the store by a flashing blue police light). So, if bluelight.com doesn't work, you can always try http://www.kmart.com/.

Target Stores   202782

777 Nicollet Mall
Minneapolis, MN 55402-2055
USA

(612) 370-6948
(612) 370-5502
http://www.target.com/

Once known for its abundant use of orange Formica, Target is now one of the leading retailers in North America, with stores from Bangor, Maine, to Honolulu, Hawaii-and about 1,000 more in between. Target stores come in a variety of flavors that fit the local neighborhood. For lots of choices, nothing beats Target Greatland and Super Target, which boast an extended product mix (and, in my observation, slightly lower prices). The "ordinary" Target stores offer a few less items, but I found they are more likely to stuff the end caps (the ends of shelves) with clearance items. I once bought a *new* LEGO Mindstorms set, on clearance, at my local "ordinary" Target store for under $100-about half off.

My usual procedure is to make an afternoon of visiting the several Target stores in my area. Pickin's are particularly good after the holidays, as you can imagine.

Visit the Target.com Web site for a store locator. You can also purchase many items, including toys, online.

Like their bricks-and-mortar stores, Target.com offers product clearances, so check it often, before I do. (My wife will thank you for it, as maybe things will be sold out by the time I get there.)

The Target company, formerly known as Dayton Hudson, also owns the Mervyn's discount apparel store chain, the upscale Marshall Field department store, and catalog retailer Rivertown Trading. Few robot things sold in these stores, though.

Toy section on Target.com.

Wal-Mart   202786

702 SW Eighth St.
Bentonville, AR 72716
USA

((501) 273-4000

((501) 273-1917

⊘ (800) 925-6278

🌐 http://www.walmart.com/

Wal-Mart stores and Sam's Club stores. Sam's Club stores are "membership" warehouses, where you buy larger quantities or in bulk and save a bundle.

In 2001 Wal-Mart made over $200 billion in revenue, with over 4,100 stores-most of them in North America, but others across the globe. (There are three-three, count 'em!-Wal-Mart's near me within a 10-minute drive.) I know a few bucks of that $200 billion came from me.

Woolworths PLC   203892

Woolworth House, 242-246 Marylebone Road
London
NW1 6JL
UK

(+44 (0) 8456 081102

(+44 (0) 1293 744040

🌐 http://www.woolworths.co.uk/

Woolworths was born in the U.S. and is now gone from the USA. But it's alive and well in the U.K., and local stores sell batteries, tools, arts and craft supplies, and other sundries useful in robot building. Their online store sells a small portion of their stock; check the site for store locations.

RETAIL-EDUCATIONAL SUPPLY

Educational supply stores are more popular than ever, thanks to the recent growth in home education. Yet these stores have existed for a long time, serving primarily schoolteachers and day care providers. Much of the items sold by educational supply stores will have minimal use in your robots, but you'll want to look at their stock of "manipulatives" (small plastic and wood pieces used by preschool children), construction sets, and science kits.

SEE ALSO:

RETAIL-SCIENCE: Retailers that specialize in science-oriented kits and products

TOYS-CONSTRUCTION: Toy sets that develop mechanical construction skills

Childcraft Education Corp.   203212

P.O. Box 3239
Lancaster, PA 17604
USA

⊘ (800) 631-5652

✉ service@childcrafteducation.com

🌐 http://www.childcraft.com/

Toys and manipulatives. What's a manipulative, and why is it a cool thing for robots? *Manipulatives* are small shapes used to teach counting, patterns, and motor skills. They may be little bits of plastic, wood, or foam, but because they are small and precut, they make for good robot-building materials. Also sells gears, K'NEX construction sets.

ClassroomDirect.com   202608

P.O. Box 830677
Birmingham, AL 35283-0677
USA

⊘ (800) 599-3040

🌐 http://www.classroomdirect.com/

School supplies; arts and crafts supplies.

Discovery Mart.com   203140

16200 Bear Valley Rd., #110
Victorville, CA 92392
USA

☎ (760) 843-0030

📠 (760) 843-6830

⊘ (888) 638-8004

🌐 http://www.discoverymart.com/

Learning toys and educational materials; science kits, fun gadgets.

Edmund Scientific ⚕   203054

60 Pearce Ave.
Tonawanda, NY 14150-6711
USA

☎ (716) 874-9091

⊘ (800) 728-6999

🌐 http://www.scientificsonline.com/

See listing under Retail-Science.

Educational Experience   203800

P.O. Box 860
Newcastle NSW 2300
Australia

☎ +61 2 4923 8222

📠 +61 2 4942 1991

✉ hotline@edex.com.au

🌐 http://www.edex.com.au/

Early and middle educational supplies:

- LEGO Dacta
- LEGO (e.g., Early Simple Machines)
- Science and technology
- Construction toys
- Arts and crafts

Educational Insights, Inc.   203942

18730 S. Wilmington Ave.
Rancho Dominguez, CA 90220
USA

☎ (310) 884-2000

⊘ (800) 995-4436

✉ service@edin.com

🌐 http://www.educationalinsights.com

Over 900 items (including Amazing Live Sea-Monkeys), for home and school. Primary emphasis is on prekindergarten to eighth grade. Products are offered internationally. Highlight: Educational Insights resells Capsela construction toys.

Hope Education   203799

Hyde Buildings
Ashton Road
Hyde, Cheshire
SK14 4SH
UK

☎ +44 (0) 8702 412308

📠 +44 (0) 0800 929139

✉ enquiries@hope-education.co.uk

🌐 http://www.hope-education.co.uk/

Early and middle educational supplies, including LEGO Dacta.

J. L. Hammett Co.   203796

P.O. Box 859057
Braintree, MA 02185-9057
USA

⊘ (800) 955-2200

🌐 http://www.hammett.com/

Early education materials, including manipulatives (buttons, beads, blocks, etc.; use the material as robot pieces).

Lakeshore Learning Materials   203797

2695 E. Dominguez St.
Carson, CA 90810
USA

☎ (310) 537-8600

📠 (310) 537-5403

⊘ (800) 421-5354

lakeshore@lakeshorelearning.com

http://www.lakeshorelearning.com/

Educational materials. Includes some younger-age manipulative sets, with possible uses as parts for robots.

Web site for Lakeshore Learning Materials.

Learning Resources, Inc.   202601

380 N. Fairway Dr.
Vernon Hills, IL 60061
USA

(847) 573-8400

(847) 573-8425

info@learningresources.com

http://www.learningresources.com/

Educational materials: K-12, with a decided leaning toward the younger of this group. Check out their Gearbotics products.

Logiblocs Ltd.   203936

P.O. Box 375, St.
Albans
AL1 3GA
UK

+44 (0) 1727 763700

+44 (0) 1727 763700

feedback@logiblocs.com

http://www.logiblocs.com/

Logiblocs are electronic building blocks for elementary and junior high school students. They plug together to make complete and working circuits.

Nasco   202566

901 Janesville Ave.
P.O. Box 901
Fort Atkinson, WI 53538-0901
USA

(920) 563-2446

(920) 563-8296

(800) 558-9595

custserv@eNASCO.com

http://www.enasco.com/

eNasco is the online component of NASCO, a mail-order catalog offering some 60,000 educational supplies. The company also sells farm and ranch supplies (no kidding), construction toys, and books. Outlet stores in Fort Atkinson, Wisc., and Modesto, Calif.

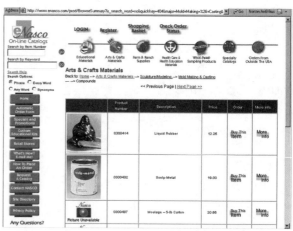

Nasco online.

Creative Kids Stuff

http://www.creativekidsstuff.com/

Educational materials and suplies; OWI robot kits

Discount School Supply

http://www.earlychildhood.com/

School and educational supplies

Imagination Village

http://www.imaginationvillage.com/

Arts & Crafts; Manipulatives (can be used to construct parts); Science

SmarterKids

http://www.smarterkids.com/

Construction and other toys for younger children

Sax Art's & Crafts   203940

2725 S. Moorland Rd.
New Berlin, WI 53151
USA

📞   (262) 784-6880
🚫   (800) 558-6696
✉   info@saxarts.com
🌐   http://www.artsupplies.com/

Online and mail-order retailer of arts and crafts supplies, with emphasis on schools and educators. They charge a small refundable fee for their catalog for noneducators.

Sax's is a unit of School Specialty, Inc. Refer to its listing for additional catalog mail companies that cater to educational supplies.

School Specialty, Inc.   203941

http://www.schoolspecialty.com/i

Corporate owners of a number of catalog mail-order firms that cater to educational supplies.

  School Specialty—http://www.schoolspecialty.com/
  Sax's Arts & Crafts—http://www.artsupplies.com/
  Childcraft—http://www.childcraft.com/
  Brodhead Garrett—
      http://www.brodheadgarrett.com/
  Classroom Direct—
      http://www.classroomdirect.com/
  Beckley Cardy—http://www.beckleycardy.com/

Spectrum Educational Supplies   203798

125 Mary St.
Aurora
Ontario
L4G 1G3
Canada

📞   (905) 841-0600
📠   (905) 727-6265
🚫   (800) 668-0600
✉   customerservice@spectrumed.com
🌐   http://www.spectrumed.com/

Huge selection of educational supplies and kits, from trinkets to LEGO Robolab robots. Source for hard-to-locate replacement parts for LEGO Mindstorms, including additional infrared towers.

Young Explorers   202602

P.O. Box 2257
Loveland, CO 80538
USA

🚫   (800) 239-7577
✉   customerservice@youngexplorers.com
🌐   http://www.youngexplorers.com/

Catalog of unusual science toys, educational materials and supplies. Sample offerings include the Arctimorph Transforming Speeder and the Robo-Dog.

∿ RETAIL-GENERAL ELECTRONICS

Found here are retailers that supply general electronics parts, such as capacitors, resistors, transistors, ICs, fuses, switches, connectors, soldering irons and other electronics tools, wire, cable, and printed circuit board makings. These stores are your front line of defense for collecting the electronic components necessary for your robotics projects.

Some of those listed here are local only; most sell at least a portion of their product line by mail order. You will also want to check the Yellow Pages or other business directory for local electronics stores in your area. Check under **Electronics-Retail** for starters.

SEE ALSO:

  DISTRIBUTOR/WHOLESALER-INDUSTRIAL ELECTRONICS: Additional resources for electronics components

  RETAIL-SURPLUS ELECTRONICS: Where to buy used and overstock electronics

A-1 Electronics   203062

718 Kipling Ave.
Toronto, ON
M8Z 5G5
Canada

📞   (416) 255-0343
📠   (416) 255-4617
✉   email@a1parts.com
🌐   http://www.a1parts.com/

Kits: electronics, radio, educational lab. Also soldering equipment, surplus, pinhole cameras, tubes, and technical books.

Local store in Toronto, Canada.

AbleTronics   203610

9155 Archibald Ave.
Unit E
Rancho Cucamonga, CA 91730
USA

((909) 987-7606

℘ (909) 945-9449

✉ ablesales@abletronics.com

🌎 http://www.abletronics.com/

Local retail store in California.

General electronics: passive and active components, tools, connectors, chemicals, relays, switches, enclosures, test equipment, etc. A small selection is available for online ordering.

ABRA Electronics Corp.   203060

1320 Route 9
Champlain, NY 12919
USA

🚫 (800) 717-2272

✉ sales@abra-electronics.com

🌎 http://www.abra-electronics.com/

ABRA offers an extensive line of electronics components, as well as custom kits and school curriculum kits. Prices are quite reasonable.

ACK Electronics   203594

554 Deering Rd. NW
Atlanta, GA 30309
USA

((404) 351-6340

℘ (404) 351-1879

Where to Get Stuff: Mail Order

There is no limit to what you can buy mail order (at one time, they even sold houses through the mail!). So, rather than tell you what to look for in mail order, it makes more sense to remind you of the different kinds of mail order. (And yes, I know that these days, less and less "mail order" is going through the mails—it's a term of convenience.)

- *Retail catalog sales.* Some retail stores also offer goods via a separate mail-order branch. If the store has a printed catalog, obtain a copy and look through it for special products not carried in the bricks-and-mortar store.

- *Mail-order catalogs.* This is the typical form of mail order, where a company sends you a catalog or brochure, and you order from it. More and more outfits are conducting this kind of business from the Internet, saving the cost of printing catalogs.

- *Internet.* Thanks to the Internet you can now find the most elusive part for your robot. The Internet is an extension of catalog sales, where Web pages take the place of printed pages. A disadvantage is that even the best e-commerce shopping cart is not as handy as a nice color catalog that you can read at your leisure, in any part of the house.

- *Online auctions.* Typified by eBay, online auctions provide a means for individual buyers and sellers to trade. Goods are shipped via the mail.

When shopping mail order, always compare prices of similar items offered by various companies before buying. Consider all the variables, such as the added cost of insurance, postage and handling, and COD fees. Be sure that the mail-order firm has a lenient return policy. You should always be able to return unsuitable goods if they are not satisfactory.

(800) 282-7954

 http://www.acksupply.com/

Mail-order industrial and commercial parts supply, since 1946. Searchable database of parts, by part number, description, and manufacturer.

Also store in Birmingham, Ala.

Action Electronic Wholesale Company   202815

1300 E. Edinger Ave.
Santa Ana, CA 92705
USA

 (714) 547-5169

 (714) 547-3291

 (800) 563-9405

 sales@action-electronics.com

http://www.action-electronics.com/

New and surplus electronics, including test equipment, active and passive components, relays, switches, soldering stations and supplies, connectors, wire and cable, electronics kits, and more.

Active Electronics

See listing under **Distributor/Wholesaler-Industrial Electronics**.

Active Electronics Components Depot   203861

http://www.activestores.com/

Retail general electronics stores in Canada and the U.S. See the Web site for locations.

RK Distributing
http://www.r-k.com/
Electronic components; books

RNJ Electronics
http://www.rnjelect.com/
Video, security, wiring, and cable

Active Surplus   203031

345 Queen St. West
Toronto, ON
M5V 2A4
Canada

(416) 593-0909

(416) 593-0057

(800) 465-5487

info@activesurplus.com

http://www.activesurplus.com/

Though by name an electronics surplus retailer, Active Surplus also sells new and used medical instruments (for tools, like hemostats), electronics kits, components, new semiconductors, hand tools, and hardware for electronics project construction.

Addison Elctronique Ltd.   203063

8018, 20 ime Ave.
Montreal, Quebec
H1Z 3Z7
Canada

(514) 376-1740

(514) 376-9792

http://www.addison-electronique.com/

New and surplus electronics, including components, mechanical parts for VCRs and other consumer products, motors, and relays.

Web site is in French and English. Local retail store in Montreal, Quebec, Canada.

Advanced Component Electronics (ACE)   202906

1534 Berger Dr.
San Jose, CA 95112
USA

(408) 297-1383

(408) 297-5617

prince@acecomponents.com

http://www.acecomponents.com/

Local store in San Jose, Calif. Inventory list is available as a downloadable file. According to the site: "We have millions of ICs, transistors, diodes, capacitors, relays and more."

AE Associates, Inc.

7733 Densmore Ave. #5
Van Nuys, CA 91406
USA

((818) 997-3838

203426

((818) 997-0136

info@ae4electronicparts.com

http://www.ae4electronicparts.com/

New and used electronics, including switches, connectors, electronic components (resistors, capacitors,

Stocking Up on Everyday Electronic Components

There are thousands of different electronic components, but most circuits use just a handful of them. If you do any amount of electronic circuit building, you'll want to stock up on the following standard components. Keeping spares handy prevents you from making repeat trips to the electronics store.

- *Resistors.* Get a good assortment of 1/8- and 1/4-watt resistors. Make sure the assortment includes a variety of common values, and that there are several of each value. Supplement the assortment with individual purchases of the following resistor values: 270 ohm, 330 ohm, 1K ohm, 3.3K ohm, 10K ohm, and 100K ohm. The 270- and 330-ohm values are often used with light-emitting diodes (LEDs), and the remaining values are common to TTL and CMOS digital circuits.

- *Variable resistors.* Variable resistors, or potentiometers (pots), are relatively cheap and are a boon when designing and troubleshooting circuits. Buy an assortment of the small PC-mount pots (about 80 cents each, retail) in the 2.5K, 5K, 10K, 50K, 100K, 500K, 250K, and 1-megohm values. You'll find 500K and 1-megohm pots often used in op-amp circuits, so buy a couple extra of these.

- *Capacitors.* For a well-stocked shop, get a dozen or so each of the following inexpensive ceramic capacitors: 0.1, 0.01, and 0.001 µF (microfarad). Many circuits use in-between values of 0.47, 0.047, and 0.022 µF. A small stock of 0.1-µF tantalum capacitors is useful; these capacitors are used for power supply decoupling. Larger electrolytic and specialty capacitors can be purchased on an as-needed basis, though a small assortment of values from 1 to 100 µF will be a time-saver.

- *Transistors.* There are thousands of transistors available, but a few "generic" ones are typical in many circuits. Common NPN signal transistors are the 2N2222 and the 2N3904 (some transistors are marked with an "MPS" prefix instead of the "2N" prefix; they are the same). Common PNP signal transistors are the 2N3906 and the 2N2907. The NPN TIP31 and TIP41 are familiar to most anyone who has dealt with power switching or amplification of up to about 1 amp. PNP counterparts are the TIP32 and TIP42. These transistors come in the T0-220-style package. A common larger-capacity NPN transistor that can switch 10 amps or more is the 2N3055. It comes in the T0-3- and T0-220-style packages.

- *Diodes.* Common diodes are the 1N914, for light-duty signal-switching applications, and the 1N4000 series (1N4001, 1N4002, 1N4003, and 1N4004). Get several of each and use the proper size to handle the current in the circuit. A special kind of diode is the zener, which is typically used for regulating voltage. Zener diodes are available in a variety of voltages and wattages. Purchase these as needed.

- *LEDs.* All semiconductors emit light, but light-emitting diodes (LEDs) are especially designed for the task. Small red LEDs are among the most common, but other colors and sizes are available, as well. White and blue LEDs are the most expensive, so you'll want to save these for special purposes. Green and red are far less expensive, and you can stock up on a number of these. Infrared LEDs, which emit no light, are used to build several kinds of robotic sensors.

diodes, transistors, etc.), and test equipment. Searchable database. Also sells a small number of compact B&W and color video cameras. Local store in Van Nuys, Calif.

Al Lasher's Electronics 203449

1734 University Ave.
Berkeley, CA 94703
USA

((510) 843-5915

((510) 843-9475

info@allashers.com

http://www.allashers.com/

General electronics; CCD cameras. Local store in Berkeley, Calif.

All Electronics Corp. 202160

P.O. Box 567
Van Nuys, CA 91408-0567
USA

((818) 997-1806

((818) -781-2653

(888) 826-5432

allcorp@allcorp.com

http://www.allcorp.com/

All Electronics is one of the primary sources in the U.S. for new and used robotics components. Prices and selection are good. Walk-in stores in the Los Angeles area are located at:

Los Angeles store: 905 S. Vermont Ave., Los Angeles, CA; (213) 380-8000

Van Nuys store: 14928 Oxnard St., Van Nuys, CA; (818) 997-1806

Product line includes motors, switches, discrete components, semiconductors, LEDs, infrared and CdS sensors, batteries, LCDs, kits, and much more. Specifications sheet for many products are available at the Web site.

Same as http://www.allelectronics.com.

Alltronics 202352

P.O. Box 730
Morgan Hill, CA 95038-0730
USA

((408) 847-0033

((408) 847-0133

ejohnson@alltronics.com

http://www.alltronics.com/

Not to be confused with All Electronics in southern California, this northern California electronics retailer is known for a good assortment and reasonable prices. New and surplus merchandise. Online catalog and sales via the Internet; the company used to provide a walk-in store in San Jose, but this has closed. The company provides mail order service only. A will-call window is available in Gilroy; check the company's Web page for details. Some product is also available for auction on eBay.

Web page for All Electronics Corp.

Alltronics.com home page.

A printed catalog costs $3, or you can download it free from the site (you need Adobe Acrobat Reader to view it).

Among their product line useful in robotics is:

- Motors (DC and stepper)

- Stepper motor controllers

- Power MOSFETs

- H-bridge ICs (including the oft-cited L293D, L297, and L298)

- Atmel AVR microcontrollers

- Small CCD video cameras

- Tools

- Solenoids and relays

Product datasheets (in PDF format) are available for download for many of the specialty semiconductor products.

Stocking Up—Integrated Circuits

Integrated circuits let you construct fairly complex circuits from just a couple of components. Although there are literally thousands of different ICs, some with exotic applications, a small handful crops up again and again in hobby projects. You should keep the following common ICs in ready stock:

- *555 timer.* This is, by far, the most popular integrated circuit for hobby electronics. With just a couple of resistors and capacitors, the 555 can be made to act as a pulser, a timer, a time delay, a missing-pulse detector, and more.

- *LM741 op-amp.* The 741 can be used for signal amplification, differentiation, integration, sample-and-hold, and a host of other useful applications. The 741 is available in a dual version, the 1458. Note that there are numerous other op-amps available, and some have design advantages over the 741. These include the LM324, LF356, and the TL082.

- *LM339 quad comparator.* Comparators are used to compare two voltages. The output of the comparator changes depending on the voltage levels at its two inputs. The comparator is similar to the op-amp, except that it does not use an external feedback resistor. You can use an op-amp as a comparator, but a better approach is to use something like the 339 chip, which contains four comparators in one package.

- *LM386 audio amplifier.* This common and versatile single-package audio amplifier can drive an ordinary 8-ohm speaker with up to a quarter of a watt. Only a couple of capacitors are needed to complete the circuit.

- *ULN2003 transistor array.* A transistor array is an integrated circuit containing multiple transistors. The device is commonly used to increase the current capacity of microcontrollers and other devices that cannot provide enough juice to run large LED panels or relays. Other transistor array ICs are available, with some able to drive loads of up to 500 mA per channel.

- *TTL logic chips.* TTL ICs are common in computer circuits and other digital applications. There are many types of TTL packages, but you won't use more than 10 to 15 of them unless you're heavily into electronics experimentation. Specifically, the most common and most useful TTL ICs are the 7400, 7402, 7404, 7407, 7408, 7414, 7420, 7430, 7432, 7473, 7474, 74123, 74154, 74174, 74175, 74193, 74240, 74244, 74273, and 74274.

- *CMOS logic chips.* Because CMOS ICs require less power to operate than the TTL variety, you'll often find them specified for use with low-power robotic and remote-control applications. Like TTL, there is a relatively small number of common packages: 4001, 4011, 4013, 4016, 4017, 4024, 4027, 4040, 4049, 4066, 4069, 4071, 4081, 4093, 4511, 4543, and 4584.

Altronics Distributors Pty. Ltd. 203667

174 Roe Street
Perth
Western Australia 6000
Australia

☎ +61 8 9328 1599

📠 +61 8 9328 3487

✉ info@altronics.com.au

🌏 http://www.altronics.com.au

Full-line electronics online retailer. Mail order and several stores across Oz.

Amazon Electronics/Elecronics123 202506

14172 Eureka Rd.
P.O. Box 21
Columbiana, OH 44408-0021
USA

☎ (330) 549 3726

📠 (603) 994 4964

🚫 (888) 549-3749

✉ amazon@electronics123.com

🌏 http://www.electronics123.com/

See listing under Kits-Electronic.

Ametron Electronic Supply, Inc. 203608

1546 N. Argyle Ave.
Hollywood, CA 90028-6410
USA

☎ (323) 464-1144

📠 (323) 871-0127

✉ fredr@ametron.com

🌏 http://www.ametron.com/

Passive general electronics (switches, wire, connectors). No online sales. Local store in Hollywood, Calif.

Arcade Electronics, Inc. 204144

5655-F General Washington Dr.
Alexandria, VA 22312
USA

☎ (703) 256-4610

✉ sales@arcade.va.net

🌏 http://www.arcade-electronics.com/

Arcade sells batteries and chargers, cables and connectors, electronics kits, active components (resistors, capacitors, transistors, ICs, etc.), passive components, tools, test equipment, and other general electronics products.

B. G. Micro 202210

555 N. 5th St.
Ste. #125
Garland, TX 75040
USA

☎ (972) 205-9447

📠 (972) 205-9417

🚫 (800) 276-2206

✉ bgmicro@bgmicro.com

🌏 http://www.bgmicro.com/

B. G. Micro is a haven for the electronics tinkerer and robotics enthusiast. Much of the stock is surplus, so it comes and goes, but while it's being offered, it has a good price attached to it. Get it while you can, because someone else surely will.

The company also offers a fairly complete line of passive and active general electronics components, including IC sockets, resistors, capacitors, linear and digital ICs, switches, relays, solenoids, pin headers, transistors, diodes, voltage regulators, LEDs, infrared phototransistors, batteries and battery holders, cables, wire, hand tools, connectors, optics, standoffs, and other plastic and metal hardware.

Web catalog page for B.G. Micro.

Specific products of interest to the robot builder include LCD displays (including serial LCD), motors, and the OOPic microcontroller.

A word of caution! Don't skim the B. G. Micro catalog too quickly. Its design is fairly "compact" and it's easy to miss things. They sell a lot of parts that are best described as miscellaneous, and there can be some real gems hidden between product listings. Some super finds are just a single line of text in the 30-odd-page catalog!

Baynesville Electronics   203411

1631 E. Joppa Rd.
Baltimore, MD 21286
USA

☎ (410) 823-0082

✉ bayelec@ix.netcom.com

🌐 http://www.baynesvilleelectronics.com/

Baynesville is an old-fashioned retail electronics store located in Baltimore, Md. Look to the Web site for special promotions and even how-to information.

Black Feather Electronics   202944

4400 S. Robinson Ave.
Oklahoma City, OK 73109
USA

☎ (405) 616-0374

📠 (405) 616-9603

✉ blkfea@juno.com

🌐 http://www.blkfeather.com/

Something of a cornucopia of unique electronics items: connectors and cords; audio; gadgets; laser pointers; minicams; soldering; parts and switches; power; test equipment; kits; tools; video; high-brightness LEDs; electroluminescent wire.

Brigar Electronics 🏆   202455

7-9 Alice St.
Binghampton, NY 13904
USA

☎ (607) 723-3111

📠 (607) 723-5202

✉ BRIGAR2@aol.com

🌐 http://brigarelectronics.com/

Electronic components and parts; new and surplus. Excellent selection of electronics and mechanical components for robobuilders everywhere:

- Ballscrews (prices are pretty good for what you get!)
- Bearings, linear slides
- Cameras (CCD)
- Capacitors, resistors, potentiometers, semiconductor
- Cords and cables
- Hardware, solder, terminals, wire, tubing
- Lamps, LEDs, LCDs
- Materials handling
- Meters
- Motors (DC, stepper, gear, motor control circuits)
- Pneumatic (cylinders, 24-volt solenoid valves)
- Power supply
- Relay, switches
- Sensors
- Tools

Very good prices for most items.

Bull Electrical   203287

Unit D
Henfield Business Park
Shoreham Road (A2037)
Henfield
Sussex
BN5 9SL
UK

☎ +44 (0) 1273 491490

📠 +44 (0) 1273 491813

✉ sales@bull-electrical.com

🌐 http://www.bullnet.co.uk/shops/test/

Full line of electronics parts, motors, optics, digital CCD cameras, lasers, active and passive components, relays, tools, and much more. New and surplus.

See also:

http://www.veronica-kits.co.uk/ (RF kits)

http://www.lockpicks.co.uk/ (video cameras, transmitters)

http://www.home-cctv.co.uk/ (cameras)

Central Utah Electronics Supply 203604

735 South State St.
Provo, UT 84606
USA

 (801) 373-7522
(801) 373-7736
(800) 805-7522
 http://www.utahelectronics.com/

General electronics, plus robot kits; local store in Provo, Utah.

CID Inc. 202573

4758 Hammermill Rd.
Ste. 307
Tucker, GA 30084
USA

(770) 908-9883
(800) 205-9471
http://www.cidonline.com/

Electronics tools, materials, components, and books.

Circuit Specialists, Inc. 202364

220 S Country Club Dr. #2
Mesa, AZ 85210
USA

(480) 464-2485
(480) 464-5824
(800) 528-1417
jr@cir.com
http://www.web-tronics.com/

Sellers of ICs, active and passive components, test equipment, tools, microcontrollers and programmers, switches, relays, kits, lab trainers, chemicals, and more.

Additionally, they sell several robotics specialty products, including motion control cards, Arrick Robotics robotic workcells, and positioning tables.

Computronics Corporation Ltd. 203662

Locked Bag 20
Bentley
Western Australia 6983
Australia

+61 8 9470 1177
+61 8 9470 2844
 kdare@computronics.com.au
 http://www.computronics.com.au

Industrial electronics: electronic displays, tools, and components. Includes soldering stations, high-brightness LEDs, chemicals, RF transmitter and receiver modules.

Dan's Small Parts and Kits 202898

Box 3634
Missoula, MT 59806-3634
USA

(406) 258-2782
http://www.fix.net/dans.html

Grab bag of electronics components. The emphasis is on amateur radio, but the company also sells general electronics components (such as trimmer pots, diodes, MOSFETs, and the usual others). A good source for specialty parts on the cheap.

Debco Electronics, Inc. 202054

4025 Edwards Rd.
Cincinnati, OH 45209
USA

(513) 531-4499
(513) 531-4455
800) 423-4499
 debc@debco.com
http://www.debco.com/

Components (active, passive, cables, connectors, fasteners, etc.), kits, and hand tools.

Dick Smith Electronics 202011

P.O. Box 500
Regents Park DC
NSW, 2143
Australia

+61 2 9642 9100
+61 2 9642 9111
postmaster@dse.com.au
http://www.dse.com.au/

Australia's biggest electronics retailer, well known for Mr. Smith's head pasted over everything-though he is no longer connected with the company. For a time, DSE also had stores in the U.S., but they exited the market during the downturn of the late 1980s and early 1990s. Stores across Oz and New Zealand, and they ship worldwide.

Digi-Key   202358

701 Brooks Ave. South
Thief River Falls, MN 56701
USA

☎ (218) 681-6674
✆ (218) 681-3380
⊘ (800) 344-4539
🌐 http://www.digikey.com/

Digi-Key is one of the largest mail-order retailers/distributors of electronic components in North America. Their printed catalog has everything-and is so jammed, they had to make the type very small, you'll probably need a magnifying glass to read it! Fortunately, they also offer a very fast and efficient online ordering system, complete with links to datasheets (when available). Digi-Key regularly lists prices for product in single quantity and volume, for OEMs. If you use their online ordering system, they'll even tell you how many of an item they have on hand, in case you're thinking of breaking the bank.

Display Electronics 203809

29 / 35 Osborne Road
Thornton Heath
Surrey
CR7 8PD
UK

☎ +44 (0) 2086 533333
✆ +44 (0) 2086 538888
🌐 http://www.distel.co.uk/

Surplus, general electronics, test equipment (new and used), components, electromechanical (relays, motors, solenoids).

🌐 🖥

Edlie Electronics 202945

2700 Hempstead Tpke.
Levittown, NY 11756
USA

☎ (516) 735-3330
✆ (516) 731-5125

All in the TTL Chip Family

There are several families of TTL chips. The family is denoted by letter identifiers, such as 74HC193. They differ in the amount of current they consume, their speed, and their manufacture. The most commonly used letter identifiers are as follows:

A—Advanced

C—CMOS

F—Fast

H—High speed

L—Low power

S—Schottky

T—TTL compatible

It is common to combine letter identifiers, such as LS (low-power Schottkey), HC (high-speed CMOS), and HCT (high-speed CMOS TTL-compatible).

Which one(s) to use depends on the application. If a circuit specifies a certain TTL family, be sure to use it, as the designer of the circuit may have included it for a reason. If there is no specific family mentioned, an S, LS, or HCT version will *usually* (note the emphasis on *usually*) work.

✉ ELieblinng@aol.com

🌐 http://www.edlieelectronics.com/

Edlie provides an extensive supply of electronic educational kits, component kits, tools, and test equipment.

Electro Sonic Inc.   203850

Suite 110
4020 Viking Way
Richmond, BC
V6V 2N2
Canada

☎ (604) 273-2911

📠 (604) 273-7360

🌐 http://www.e-sonic.com/

Canadian-based general electronics online/mail-order retailer. Retail stores across Canada.

Electronic Goldmine 🎖   202652

P.O. Box 5408
Scottsdale, AZ 85261
USA

☎ (480) 451-7454

📠 (480) 661-8259

🚫 (800) 445-0697

✉ goldmine-elec@goldmine-elec.com

🌐 http://www.goldmine-elec.com/

Electronic Goldmine sells new and used electronic components, robot items, electronics project kits, and more.

- General electronic components: capacitors, crystals, displays, fuses, heat sinks, ICs, infrared items, LEDs, potentiometers, resistors, semiconductors (misc.), thermal devices/thermistors, transistors, transformers, voltage regulators.

- Passive and electromechanical components: batteries, cables and wire, circuit boards, fans, hardware, keychain lights, knobs, sockets, power supplies, relays, solenoids, switches, test equipment, tools, voltage converters.

- Sensor components: bar code scanners, microphones, piezos, sensors, video cameras.

- Kits and specialty: alarms and sirens, electronics project kits, motors, optics, robot items, strobe, ultrasonic items.

Catalog in PDF format available; printed catalog sent to U.S. addresses only. Be sure to check out the interesting and unusual (and low-cost) robotics kits.

Electronic Parts Outlet   203605

3753-B Fondren
Houston, TX 77063
USA

☎ (713) 784-0140

📠 (713) 268-1044

🚫 (800) 403-3741

✉ webmaster@epo-houston.com

🌐 http://www.epo-houston.com/

General electronics components, supplies, and tools. Used and surplus, stores in the Houston, Tex., area.

Electronics Plus   203595

10302 Southard Dr.
Beltsville, MD 20705
USA

☎ (301) 937-9009

📠 (301) 937-5092

🚫 (888) 591-9009

✉ mail@electronics-plus.com

🌐 http://www.electronics-plus.com/

Online surplus retailer. Electronics kits. Local store near Washington, D.C.

Product lines include batteries and battery accessories; cable and wire; capacitors and resistors; chemicals; circuit boards and accessories; coils, chokes, inductors, and more; connectors; crystals and oscillators; hardware and mechanical parts; heat-shrink tubing; integrated circuits; kits ("huge selection"); lamps, lights, and indicators; LEDs and accessories; motors and accessories; potentiometers, trimmers, etc.; relays, solenoids, switches; test equipment; tools.

Electronix Express   202854

365 Blair Rd.
Avenel, NJ 07001
USA

☎ (732) 381-8020

📠 (732) 381-1006

⊘ (800) 972-2225

✉ electron@elexp.com

🌐 http://www.elexp.com/

Electronics parts, supplies, components, hardware, switches, relays, test gear, tools. New and used; large inventory.

🌐 ▯ 🖳

ESR Electronic Components   203853

Station Road
Cullercoats
Tyne & Wear
NE30 4PQ
UK

📞 +44 (0) 1912 514363

📠 +44 (0) 1912 522296

✉ sales@esr.co.uk

🌐 http://www.esr.co.uk/

Silicon components (transistors, diodes, thyristors, etc.) for industry, hobby. Also general electronics: connectors, kits, test equipment, tools.

🌐 🖳

Farnell   202399

Castleton Road
Maybrook Industrial Estate
Leeds
West Yorkshire
LS12 2EN
UK

📞 +44 (0) 8701 200296

📠 +44 (0) 8701 200297

✉ enquiries@farnell.com

🌐 http://www.farnell.com/

U.K.-based electronics mail order. Will ship most anywhere. Offers a complete line of electronics and mechanical.

According to the company, "Farnell is a leading high-service, low-volume distributor of electronic, electrical and industrial products. Headquartered in Leeds, England, it services more than 100,000 active customers through 15 business units across Europe, Australia, Asia Pacific and South America."

 🖳

Frigid North Company   203613

3309 Spenard Rd.
Anchorage, AK 99503
USA

📞 (907) 561-4633

📠 (907) 563-0836

⊘ (800) 478-4633

✉ frigid@gci.net

🌐 http://www.frigidn.com/

Active and passive components, connectors, switches, relays, and other general electronics wares. Local store in Anchorage, Alaska, with some sales online.

🖳   🛒

Fry's Electronics   202882

600 E. Brokaw Rd.
San Jose, CA 95112
USA

📞 (408) 487-4500

📠 (408) 487-4700

🌐 http://www.frys.com/

Fry's is an electronics superstore chain operating primarily on the West Coast. Fry's has an odd mix, with prices ranging from deep discount to manufacturer's suggested list price. Among the things you'll find at Fry's are:

- Robots (like the Parallax BOE-Bot)
- Basic Stamp starter kits
- Tamiya Educational motor sets
- OWI/MOVITS robots
- Integrated circuits, transistors, diodes, resistors, capacitors, connectors
- Soldering irons, PCB supplies, breadboard supplies
- Books (more than you knew were ever published)
- Computer software
- Soda pop and candy
- Video games and video game consoles
- X-rated movies
- Washing machines

I left out about 1,000 other product lines, but you get the idea. Weekdays are the best time to shop; the aisles get really cluttered on the weekends. Watch for previously opened (and/or returned) merchandise, the latter of which is usually marked. If the box can be opened-that is, it's not shrink wrapped-check its contents before buying.

If there isn't a Fry's near you, they offer online shopping (limited merchandise selection) at:

http://www.outpost.com/

FYI Department:

- The store is named after its founders, the Fry brothers (John, Randy, and Dave).
- The person who guides you to the next available numbered register when you're waiting in line at the checkout is called a Vanna White (of Jeopardy fame).

Future Electronics 202567

237 Hymus Blvd.
Pointe-Claire, Quebec
H9R 5C7
Canada

 (514) 694-7710
 (514) 695-3707
 http://www.futureelectronics.com/

See listing under Distributor/Wholesaler-Industrial Electronics.

Greenweld Ltd. 203590

Unit 24
West Horndon Industrial Park
West Horndon
Brentwood
Essex
CM13 3XD
UK

+44 (0) 1277 811042
+44 (0) 1277 812419
service@greenweld.co.uk.
http://www.greenweld.co.uk/

New and surplus electronics; some electronics kits, including OWI robots. They also sell small mechanical parts, such as threaded rod, gear shafts, spacers, books, power supplies, and pneumatic cylinders. A little of everything.

Hosfelt Electronics 202883

2700 Sunset Blvd.
Steubenville, OH 43952
USA

(614) 264-6464
(614) 264-5414
(800) 524-6464
order@hosfelt.com
http://www.hosfelt.com/

General electronics. New and surplus. Great prices.

- LEDs, especially superbrights, combo-color, and blue
- Lasers, pointers, and key-ring holders
- Miniature video cameras
- Tools (hand, solder)
- Chemicals
- Components, passive and active
- Velleman electronics kits

HSC Electronic Supply/Halted 202163

3500 Ryder St.
Santa Clara, CA 95051
USA

(408) 732-1573
(408) 732-6428
sales@halted.com
http://www.halted.com/

Mail order, with walk-in retail stores: Santa Clara, Sacramento, and Santa Rosa.

Says the Web site: "For over 30 years HSC has been the Bay Area's source for a unique mix of day-to-day electronic components, test equipment, computers and parts, as well as a mind-boggling array of Silicon Valley exotica!"

Jameco Electronics 202874

1355 Shoreway Rd.
Belmont, CA 94002
USA

 (650) 592-8097
 (650) 592-2503
(800) 831-4242
info@jameco.com
http://www.jameco.com/

Full-service general electronics mail order. Jameco carries just about everything you need and often at a price less than the other guys. I regularly check the Jameco catalog to make sure I'm not overpaying.

Of course, they have resistors, capacitors, ICs, transistors, diodes, and other active components; connectors, plugs, jacks, headers, and other passive components; batteries and battery chargers; microcontrollers and microcontroller development boards; books; kits; electronics hardware; LEDs and LCDs; test and measurement tools (hand-held and bench); tools; and plenty more.

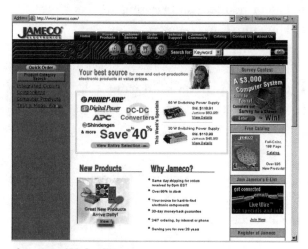

Jameco Web site.

Jaycar Electronics 203668

P.O. Box 6424
Silverwater. NSW 1811
Australia

- +61 2 9741 8555
- +61 2 9741 8500
- techstore@jaycar.com.au
- http://www.jaycar.com.au

Jaycar sells a bunch of stuff, including personal electronics, video cameras, test gear, electronics kits, passive and active components, hardware and fasteners, batteries and chargers, and more.

JDR Microdevices 202158

1850 South 10th St.
San Jose, CA 95112-4108
USA

- (408) 494-1400
- (408) 494-1420
- (800) 538-5000
- sales@jdr.com
- http://www.jdr.com/

JDR is a direct marketer of electronic components to hobbyists and the technical engineering communities. Their catalog is skewed toward PC components, but they carry plenty of general electronics (active and passive components, tools, wire and cable, etc.) as well.

JK Electronics 202141

6395 Westminster Blvd.
Westminster, CA 92683
USA

- (714) 890-4001
- (714) 892-6175
- http://www.jkelectronics.com/

Complete line of resistors, capacitors, relays, switches, semiconductors, electromechanical, batteries, adhesives, tools, and lots more. The company also offers belts, pulleys, and other replacement parts for VCRs, which can also be used in robots. Just don't expect your bot to play back the rerun of Baywatch you recorded last week.

J-Tron Inc. 202505

P.O. Box 378
324 Gilbert Ave.
Elmwood Park, NJ 07407
USA

- (201) 398-0500
- (201) 398-1010
- (888) 595-8766
- J-Tron@erols.com
- http://www.j-tron.com/

Datak kits, active and passive electronics components, switches, elecromechanical, wire and cable, fasteners, etc.

LNL Distributing Corp. 203600

235 Robbins Ln.
Syosset, NY 11791
USA

- (516) 681-7270
- info@lnl.com
- http://www.lnl.com/

General electronics; vacuum tubes; online and local retail store.

Maplin Electronics   202871

National Distribution Centre
Valley Road, Wombwell
Barnsley, South Yorks
S73 0BS
UK

☎   +44 (0) 1226 751155
🖷   +44 (0) 1226 272499
✉   sales@maplin.co.uk
🌐   http://www.maplin.co.uk/

Maplin Electronics is a got-it, have-that electronic components superstore, based in the U.K. Of course, they offer the traditional passive electronics components, relays, connectors, and other prime parts. But they also deal in things like video wireless transmitters and receivers, GPS modules, optical components, and test tools.

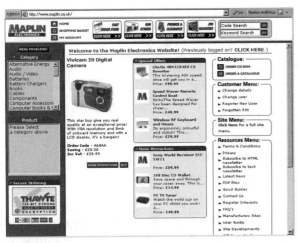

Maplin Electronics online.

Marlin P. Jones & Assoc. Inc. 🔱   202209

P.O. Box 530400
Lake Park, FL 33403
USA

☎   (561) 848-8236
⊘   (800) 652-6733
✉   mpja@mpja.com
🌐   http://www.mpja.com/

First, a story. Years ago I was interested in exploring radiation and radioactive substances (nothing dangerous, mind you, just naturally occurring materials). I wanted to build a Geiger counter, which is basically a specially designed gas tube (called a Geiger-Mueller tube) and some very simple electronics. Though I searched everywhere, few sources were willing to sell just the G-M tube . . . except Marlin P. Jones & Assoc.

(MPJA). I got a beautiful large tube in an aluminum housing for 10 bucks, and they got a customer for life.

MPJA sells both new and surplus electronic and mechanical products. Their assortment of such items as motors is fairly small, but they make up for it with a wide selection of other common (and some not-so-common) products, like LEDs, DC-DC converters, power supplies, sugar-cube-sized video cameras, tools, LCD displays (including a nice serial interface board), ultrasonic transducers, microphones, and tons of other stuff. Prices are very good.

Get their color catalog, but make it a point to regularly use the Web site for closeouts, specials, and new items. Be sure to study it well so you don't miss the interesting items. For example, an addition that almost slipped by me is an affordable (under $3) ultraviolet LED. These are useful for a number of reasons. UV light causes certain materials to fluoresce, which can be used (with the proper sensor) for robotic vision.

Online Web site for Marlin P. Jones.

MarVac Electronics   203609

2001 Harbor Blvd.
Costa Mesa, CA 92627
USA

☎   (949) 650-2001
🖷   (949) 642-0148
⊘   (800) 606-2782
✉   sales@marvac.com
🌐   http://www.marvac.com/

General electronics; retail stores in southern California.

MCM Electronics   202873

650 Congress Park Dr.
Centerville, OH 45459
USA

⊘ (800) 543-4330

✉ mcmtalk@mcmelectronics.com

🌐 http://www.mcmelectronics.com/

MCM is a mail-order retailer of general electronics and repair parts for consumer electronics. They offer active and passive components (stuff like capacitors, resistors, and semiconductors), connectors, wire and cable, OEM replacement parts (VCR belts, motors, gears), electronics kits, tools, chemicals (cleaners, lubricants, etc.), test and measurement gear, batteries, and chargers.

Of particular interest is the replacement parts for VCRs, cassette players, and other consumer electronics. These can be a good source of relatively inexpensive belts, pulleys, and even motors. To be useful, you need to know the part number to understand exactly what you're getting; you can get part numbers from online schematics and service notes from various manufacturer Web pages.

MECI-Mendelson Electronics Company, Inc. ♀   202886

340 E. First St.
Dayton, OH 45402
USA

☎ (937) 461-3525

📠 (937) 461-3395

⊘ (800) 344-4465

✉ meci@meci.com

🌐 http://www.meci.com/

Well, MECI sells just about everything you could want or need in electronics components. If you can think of

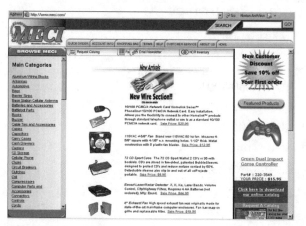

MECI Web page.

it, they probably have it. They also provide a "BattleBots Parts" page on their Web site, with some new and surplus components suitable for combat robots. These include heavy-duty motors, batteries, and more.

Milo Associates, Inc.   203602

5736 N. Michigan Rd.
Indianapolis, IN 46228
USA

☎ (317) 257-6811

📠 (317) 257-1590

✉ mai@iquest.net

🌐 http://www.websitea.com/mai/

General electronics; new and surplus, particularly Allen-Bradley.

Mouser Electronics ♀   202357

1000 North Main St.
Mansfield, TX 76063-1511
USA

☎ (817) 804-3888

📠 (817) 804-3899

⊘ (800) 346-6873

✉ sales@mouser.com

🌐 http://www.mouser.com/

Mouser Electronics is a mail-order components distributor, providing a full line of products for industry and hobbyists. All general electronics are carried, including active and passive components, wire and cable, hardware, relays, switches, fans, heat sinks, batteries, component kits, chemicals, and tools.

M-Tronics Inc.   203599

1175 Post Rd.
Warwick, RI 02888
USA

☎ (401) 941-7400

📠 (401) 941-1222

⊘ (877) 687-6642

✉ sales@mtronicsinc.com

🌐 http://www.mtronicsinc.com/

General electronics.

Local store in greater Providence, R.I., area.

N R Bardwell, Ltd./Bardwells 202879

288 Abbeydale Road
Sheffield
South Yorkshire
S7 1FL
UK

☎ +44 (0) 1142 552886

📠 +44 (0) 1142 555039

✉ Sales@Bardwells.co.uk

🌐 http://www.bardwells.co.uk/

General electronics.

Oatley Electronics 202255

P.O. Box 89
Oatley
NSW 2223
Australia

☎ +61 2 9584 3563

📠 +61 2 9584 3561

✉ sales@oatleyelectronics.com

🌐 http://www.oatleyelectronics.com/

Oatley sells test equipment, electronics kits, mechanical components (switches, motors, etc.), active and passive components, lasers and LEDs, RF remote control, FM transmitter kits, and video cameras.

Parts Express 202612

725 Pleasant Valley Dr.
Springboro, OH 45066-1158
USA

☎ (937) 743-3000

📠 (937) 743-1677

🚫 (800) 338-0531

✉ sales@partsexpress.com

🌐 http://www.partsexpress.com/

Parts Express is an all-around electronics retailer, selling everything from sound systems to test equipment, and from stage lighting to electronic components. Of their products, those most closely aligned with amateur robotics are:

- Batteries

- Chemical products (includes cleaners and adhesives)
- Connectors
- Electronic parts
- Hobbyist/prototyping
- Power/electrical (includes miniature high-efficiency inverters; run your PC from a 12-volt battery)
- Security products (such as wireless color cameras)
- Test equipment
- Tools
- Wire/cable

Web site for Parts Express.

Parts on Sale/Solatron Technologies Inc. 204102

19059 Valley Blvd.
Ste. 219
Bloomington, CA 92316-2219
USA

☎ 909-877-8981

🚫 877-744-3325

✉ sales@partsonsale.com

🌐 http://www.partsonsale.com/

TTL, CMOS, and linear ICs, microprocessors, crystals and clock oscillators, voltage regulators, transistors, diodes, optoelectronics, capacitors, resistors, speakers and sounders, transformers, power supplies, wire, switches, connectors, tools, robotics parts (motors, solenoids, sonar modules, more).

Philcap Electronic Suppliers 203601

275 East Market St.
Akron, OH 44308
USA

(330) 253-2109

(330 253-2618

philcap@philcap.com

http://www.philcap.com/

General electronics.

Quality Kits/QKits   202200

49 McMichael St.
Kingston, ON
ON K7M 1M8
Canada

(613) 544 6333

(613) 544 4944

(888) 464 5487

tech@qkits.com

http://www.qkits.com/

Reseller of Velleman, Minikits, and Kits R Us electronics kits.

Radar, Inc.   203616

168 Western Ave. West
Seattle, WA 98119
USA

(206) 282-2511

(206) 282-2511

(206) 282-1598

Radar is a general electronics distributor and retailer, with bricks-and-mortar stores throughout the Pacific Northwest (including Seattle, Spokane, and Boise).

Radio Shack   202580

100 Throckmorton St.
Ste. 1800
Fort Worth, TX 76102
USA

(817) 415-3700

http://www.radioshack.com/

Radio Shack is the world's largest electronics chain, with some 8,000 company-owned and franchise stores worldwide. These days, The Shack has fewer components and other electronics items for sale at each store,

but they do carry the basics-common value resistors, capacitors, switches, solder, electronics construction tools, that sort of thing. Additional items can be ordered through the Radio Shack online store.

Radio Shack online.

RF Parts Company   202581

435 So. Pacific St.
San Marcos, CA 92069
USA

(760) 744-0700

(760) 744-1943

(800) 737-2787

http://www.rfparts.com/

Specializes in RF tubes, transistors, and other components for amateur radio, but also offers a wide variety of electronics parts (capacitors, connectors, power supplies, relays, etc.).

Rockby Electronic Components   203669

P.O. Box 1189 Huntingdale
Melbourne, Victoria 3166
Australia

+61 3 9562 8559

+61 3 9562 8772

sales@rockby.com.au

http://www.rockby.com.au/

Electronics, new and surplus.

RS Components Ltd.   203551

Birchington Road
Corby
Northants
NN17 9RS
UK

(+44 (0) 1536 201234

(+44 (0) 1536 405678

   http://www.rs-components.co.uk/

Multinational (based in the U.K.) retailer of various electronic and mechanical products. Mechanicals include pneumatics, gears, fasteners, bearings, bushings.

See also:

http://rswww.com/

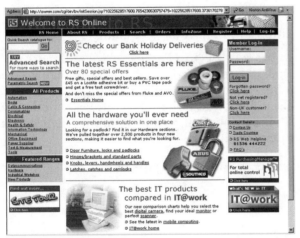

Web site for RS Components.

Stark Electronic   203552

444 Franklin St.
Worcester, MA 01604
USA

((508) 756-7136

((508) 756-5752

   starkel@ma.ultranet.com

   http://www.starkelectronic.com/

Local and mail-order electronics: security cameras, test equipment, tools, general components.

Supertronix Inc.   203612

16550 W. Valley Hwy.
Tukwila, WA 98118
USA

((425) 251-8484

((425) 251-5402

   info@supertronix.com

   http://www.supertronix.com/

General electronics; local store in the Seattle, Wash., area.

Surplustronics Trading Ltd.   203293

P.O. Box 90439
A.M.S.C
Auckland
New Zealand

((09) 302 0688

((09) 302 0686

   inquire@surplustronics.co.nz

   http://www.surplustronics.co.nz/

New and surplus general electronics, including soldering stations, relays, motors, LEDs, battery holders.

SWS Electronics   203607

153 S. Plumer Ave.
Tucson, AZ 85719
USA

((520) 628-1613

((520) 628-1921

⊘   (800) 279-7970

   info@sws-electronics.com

   http://www.shopsws.com/

General electronics and computer components: active/passive components, electromechanical, wire and cable, tools, etc.

Tanner Electronics, Inc.   202285

1100 Valwood Pkwy.
Ste. #100
Carrollton, TX 75006
USA

((972) 242-8702

((972) 245-7415

   http://www.tannerelectronics.com/

Dallas/Fort Worth-area electronic components store. New and surplus.

Tech America

See RadioShack (this section).

Unicorn Electronics  202351

1142 State Route 18
Aliquippa, PA 15001
USA

☏ (724) 495-7882
⊘ (800) 824-3432
✉ unielect@aol.com
🌐 http://www.unicornelectronics.com/

General components, EEPROMS, laser diodes, books, and a lot more.

URS Electronics Inc. 203615

123 N.E Seventh Ave.
Portland, OR 97293-0040
USA

☏ (503) 233-5341
☏ (503) 232-3373
⊘ (800) 955-4877
✉ sales@ursele.com
🌐 http://www.ursele.com/

General industrial electronics; wire and cable; video cameras; test equipment. Local store in Portland, Ore.

Wacky Willy's 203431

2900 S.W. 219th Ave.
Hillsboro, OR 97123
USA

☏ (503) 642-5111
☏ (503) 642-9120
✉ wacky@wackywillys.com
🌐 http://www.wackywillys.com/

Surplus electronics and test gear. Local stores in Oregon and Hawaii; mail order of limited product.

Walker Electronic Supply Co. 203606

3347 Columbia NE
Albuquerque, NM 87107
USA

☏ (505) 883-2992
☏ (505) 883-1889
⊘ (800) 824-9064
✉ wemail@earthlink.net
🌐 http://www.walkerradio.com/

Full-line general electronics: tools, components, connectors, etc. Local store in Albuquerque, N.M.

Wiltronics 202857

P.O. Box 43
Alfredton, 3350
Australia

☏ +61 3 5334 2513
☏ +61 3 5334 1845
✉ sales@wiltronics.com.au
🌐 http://www.wiltronics.com.au/

Wiltronics sells electronics kits, active and passive components, test equipment, tools, and the regular lineup of general electronics. Plus motors, gears, plastic wheels, pulleys, and other mechanicals for robot building, and the RoboBall, a functional robot enclosed in a clear plastic ball.

Web site is in English and Japanese.

Wirz Electronics 202157

P.O. Box 457
Littleton, MA 01460-0457
USA

☏ (978) 448-0196
✉ sales@wirz.com
🌐 http://www.wirz.com/

In the words of the Web site: "Wirz Electronics offers a wide variety of electronic components and modules to both hobbyist and OEMs. In addition to our in-house products, we carry items from other manufactures such as microcontroller development tools and electronic modules."

You-Do-It Electronics Center 203598

40 Franklin St.
Needham, MA 02494
USA

☏ (781) 449-1005

(781) 449-1009

✉ sales@youdoitelectronics.com

🌐 http://www.youdoitelectronics.com/

Local general electronics retailer in the greater Boston area, catering to the electronics repair shop (with replacement parts, belts, etc.) and to the electronics experimenter. The Web site offers limited online selling.

RETAIL-HARDWARE & HOME IMPROVEMENT

Everyone else may spend their Saturdays fixing a clogged kitchen drain, but we robot builders are busy in our workshops, hammering together our next walking 'bot. Still, both the robot makers and the drain fixer-uppers go to the local hardware and home improvement stores. We may actually buy much of the same things, but the end use is completely different.

Hardware stores (and their close but larger cousin, the home improvement outlet) are excellent sources for fasteners, basic hardware, tools, tool accessories (saw blades, bits, etc.), paint, glues and adhesives, casters, wheels, plastic sheeting, metal stock, and hundreds of other items.

The resources listed here either have an online presence or are national chains. In the U.S., many hardware stores belong to one of several cooperative "chains," such as True Value and Ace. Other independent hardware stores may be located near you, and you should not hesitate to visit them. Keep a notebook of what each area hardware store offers, so the next time you need a particular item, you'll know just where to go to get it.

SEE ALSO:

MATERIALS-METAL AND MATERIALS-PLASTIC: Specialty sources for plastic and metal

RETAIL-AUTOMOTIVE SUPPLIES: Also stocks tools, fasteners, and some hardware-related items

RETAIL-SURPLUS MECHANICAL: Wheels, casters, and other hardware items at a discount

RETAIL-TRAIN & HOBBY: Source of metal stock, usually in smaller sizes than available at the hardware store

TOOLS (VARIOUS): Tools for building robot bodies

WHEELS AND CASTERS: Specialty sources for these; larger selection than at most hardware stores

Ace Hardware 203898

2200 Kensington Ct.
Oak Brook, IL 60523
USA

📞 (630) 990-6600

📠 (630) 990-6838

🌐 http://www.acehardware.com/

Independently owned chain of hardware stores. In my experience, some Ace Hardware stores I frequent have a number of products not carried by the "Big Guys" (Lowe's and Home Depot), such as unusual fasteners and hardware. Don't overlook the small stores in your area for unique components for your robots.

Ace has store locations across America and in 70 other countries. Check the Web site for a store locator. See also:

Ace Commercial & Industrial Supply— http://www.acecisupply.com/

Ace Contractor Center—http://www.acelbm.com/

Ace Hardware Club—http://www.helpfulhardware- club.com/

Aubuchon Hardware 203832

W.E. Aubuchon Co., Inc.
95 Aubuchon Dr.
Westminster, MA 01473
USA

🚫 (800) 282-4393 Ext. 2000

✉ mailbox@aubuchon.com

🌐 http://www.aubuchon.com/

Online and local stores hardware across the Northeast. Online catalog boasts over 70,000 items. Products include hand and power tools, fasteners, hardware, plumbing, and electrical.

B&Q 202327

Portswood House
1 Hampshire Corporate Park
Chandlers Ford
Eastleigh
Hants
SO53 3YX
UK

🌐 http://www.diy.com/

Hardware and home improvement products; online sales and local stores in the U.K.

B&Q is a large D.I.Y. chain, with lots of goodies. Check the Web site for store locations and phone numbers.

B&Q online.

CornerHardware.com   202666

P.O. Box 3964
Bellevue, WA 98004
USA

 (425) 455-7642

 (425) 467-6655

 (800) 361-1787

 info@cornerhardware.com

 http://www.cornerhardware.com/

Now you can go to the hardware store using only your computer. CornerHardware.com is like the neighborhood hardware store, except it's open on Sunday. Products potentially useful in robotics are:

- Cabinet/furniture hardware (for swivels, brackets, etc.)
- Power tools
- Hand tools
- Hardware (fasteners, wheels and casters, etc.)

DoItYourself.com Inc.

http://doityourself.com/

Hardware, home improvement, and craft supplies

Focus Group Limited

http://www.focusdiy.co.uk/

Focus is a chain of "Do-it-yourself" and craft stores in the UK; see the Web site for a store locator; online catalog of several popular product lines

---

CWIKship.com   203823

3222 Winona Way #201
North Highlands, CA 95660
USA

 (916) 331-5934 Ext. 154

 (916) 331-8732

 Help@CWIKship.com

 http://www.cwikship.com/

"Contractor's warehouse"-sells to the public. Hardware, fasteners, power tools, hand tools, and other usual hardware store items. Company operates online store, retail outlets, and clearance outlet.

Do It Best/FixitCity.com   203860

P.O. Box 868
Fort Wayne, IN 46801
USA

(260) 748-7175

(260) 748-5664

helpline@doitbest.com

http://fixitcity.com/

Independent chain of hardware stores in the U.S.and worldwide. Some online sales of smaller, shippable items.

Focus Group Ltd.   203904

Gawsworth House
Westmere Drive
Crewe
Cheshire
CW1 6XB
UK

+44 (0) 1270 501555

+44 (0) 1270 250501

info@focusdoitall.co.uk

http://www.focusdiy.co.uk/

Focus is a chain of "do-it-yourself" and craft stores in the U.K. See the Web site for a store locator. Online catalog of several popular product lines.

HadwareShop   202643

ABN 46 091 084 909
1018-1022 Old Princes Highway
Engadine NSW 2233
Australia

☎ +61 2 9548 2746

📠 +61 2 9520 3977

✉ info@hardwareshop.com.au

🌐 http://www.hardwareshop.com.au/

HardwareShop bills itself as "Australia's premier online hardware and home improvement store." The online store boasts hand and power tools, fasteners, hardware, automotive supplies, and electrical (but not electronic) parts.

Home Depot-Maintenance Warehouse
202632

10641 Scripps Summit Ct.
San Diego, CA 92131
USA

⊘ (800) 431-3000

🌐 http://www.mwh.com/

Home Depot has a printed catalog of maintenance and repair supplies. A big one, with lots of pictures, illustrations, and specifications-ideal for figuring out exactly

Don't Forget the Corner Hardware Store

In the age of the Internet, where everything seems but a click away, it's easy to forget that the best place to buy things is in your own neighborhood. Buying local helps you handle the merchandise yourself; a picture on a Web page may say a thousand words, but being able to hoist some artifact in your hand is worth a lot more. And, of course, returns and exchanges are easier when you buy locally, because you don't have to hassle with packing things in a box and standing in line at the post office or shipper's.

Any self-respecting robot hobbyist should know every location of every hobby store and hardware store in a 5- or 10-mile radius. This is especially true of hardware stores, which are more plentiful and varied than hobby stores. Whether you live in an area, as I do, with four or five home improvement superstores nearby, or have only a single small-town retailer to rely on, hardware stores are *the* source for common items, including metal stock, fasteners, tools, saws and bits, adhesives, and paints.

I've memorized every aisle of the stores near me and have taken notes about pricing, so that if I need some widget, I know where I saw it and about how much it cost.

Humongous home improvement outlets aren't necessarily better than the small, independent retailer. In fact, in my area, the local Ace Hardware store carries a better variety of casters and specialty fasteners (including metric and nylon) than the big Lowe's and Home Depot.

Of course, not everyone lives within a reasonable distance of any hardware or home improvement store, and here is where the Internet can help. In the last few years, hardware-specialty retailers have set up shop on the Internet, either in direct competition with local stores or in support of their own chain of bricks-and-mortar outlets. For example,

- B&Q has many stores in the U.K., and they also sell many products over the Internet.

http://www.diy.com/

- Home Depot in North America augments their huge warehouses with Maintenance Warehouse, a catalog and online retailer of hardware goods intended for home and building repair/maintenance.

http://www.mwh.com/

- CornerHardware.com offers a virtual hardware store experience, offering hand and power tools, electrical items, and most everything else you'd find in a real hardware store.

http://www.cornerhardware.com/

what you need for your 'bot. Depending on where you live, same-day or next-day shipping may be available to you; otherwise, you'll wait two or three days to get your stuff.

Use the locator at the Web site to find a warehouse near you (they're in major cities across the U.S.).

See also:

http://www.homedepot.com/

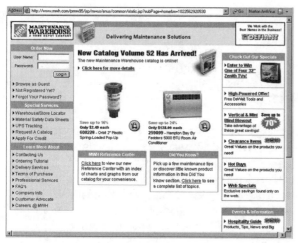

Home Depot, on the Web.

Home Depot, The   203899

2455 Paces Ferry Rd.
Atlanta, GA 30339
USA

✆   (770) 433-8211
✆   (770) 384-2337
⊘   (800) 430-3376
🌐   http://www.homedepot.com/

The world's largest home improvement chain, with 1,300 stores in the U.S., as well as others in Canada, Puerto Rico, and Mexico. Use the store locator to find a Home Depot near you.

Be sure to check out Home Depot's maintenance supply catalog/online sales at:

http://www.mwh.com/

Home Hardware Stores Ltd.   203903

34 Henry Street West
St. Jacobs, ON
N0B 2N0
Canada

✆   (519) 664-2252

✆   (519) 664-2865
🌐   http://www.homehardware.com/

Hardware and home improvement, Canadian style. Local stores across Canada.

Homebase   203902

Beddington House
Railway Approach
Wallington
SM6 0HB
UK

✆   +44 (0) 8709 008098
✆   +44 (0) 2087 847690
✉   enquiries@homebase.co.uk
🌐   http://www.homebase.co.uk/

Homebase is a hardware and home improvement store in the U.K. (stores in the U.S. with the name Homebase recently closed their doors). The Web site provides a store locator and online shopping.

Lowe's Companies, Inc.   203900

1605 Curtis Bridge Rd.
Wilkesboro, NC 28697
USA

✆   (336) 658-4000
✆   (336) 658-4766
⊘   (800) 445-6937
🌐   http://www.lowes.com/

An alternative to Home Depot, with a selection of fasteners, hardware, and much more. Retail stores and

Lowe's online.

online sales. Check the Web site for a store locator. Lowe's has 600+ superstores in some 40 states.

The site includes a "how-to library" on home repair and remodeling. I looked . . . there's nothing about building robots. But some of the articles might be useful to learn about materials and tools and the best way to use them.

Menard, Inc. 203901

4777 Menard Dr.
Eau Claire, WI 54703
USA

((715) 876-5911
(♛ (715) 876-2868
(🌍) http://www.menards.com/

Hardware stores in the Great Lakes and central states. Check the Web site for a store locator.

Orchard Supply Hardware ⚕ 203950

6450 Via Del Oro
San Jose, CA 95119
USA

((408) 281-3500
⊘ (888) 746-7674
(🌍) http://www.osh.com/

Orchard Supply Hardware (OSH) is a California-based hardware and home improvement store known for its wide selection. I've built entire robots with the stuff I've bought at OSH. Currently does not sell online.

Rockler Woodworking and Hardware 203215

4365 Willow Dr.
Medina, MN 55340
USA

((763) 478-8201
(♛ (763) 478-8395
⊘ (800) 279-4441
✉ support@rockler.com
(🌍) http://www.rockler.com/

Rockler carries hand and power woodworking tools, hardware, and wood stock (including precut hardwood plywoods). Among important hardware items are

medium-sized casters, drop-front supports (possible use in bumpers or joints in robots), and drawer slides.

📄 🌐 💻 🛍

TruServ Corporation 203897

8600 W. Bryn Mawr Ave.
Chicago, IL 60631-3505
USA

((773) 695-5000
(♛ (773) 695-6516
✉ email@truserv.com
(🌍) http://www.truserv.com/

TruServ is the corporate parent of a number of hardware stores, home improvement centers, and industrial supply outlets.

TrueValue-Major hardware store chain in the U.S.:

http://www.truevalue.com/

Induserve Supply and Commercial Supplies-210,000 items for small to large businesses:

http://www.induserve.com/
http://www.commercialsupplies.com/

🛍

 RETAIL-OFFICE SUPPLIES

What do office supplies and robots have in common? Depends on how you look at the products carried by office supply companies. Consider alternative uses for such items as packing tape, nonpermanent adhesives, ink-jet transfer film, plastic for spiral-binding machines (both the plastic covers and the combs), reinforced strapping tape, and lots more.

And, of course, there's always the mundane, but still useful, label makers, labels, parts bin organizers, batteries, and other products useful for maintaining a well-ordered robot-building shop.

The chain store listings here are representative of office supply stores worldwide. These resources offer online sales, and you can browse their Web catalogs for ideas.

SEE ALSO:

MATERIALS-TRANSFER FILM: Heat or water-slide transfer films for use in copiers, ink-jet printers, and laser printers

RETAIL-ARTS & CRAFTS: Good source for glues, adhesives, unusual parts

SUPPLIES-GLUES & ADHESIVES: All kinds of glues and adhesives

Office Depot   202963

2200 Old Germantown Rd.
Delray Beach, FL 33445
USA

- ☎ (561) 438-4800
- 🖂 (561) 438-4001
- ⊘ (888) 463-3768
- 🌐 http://www.officedepot.com/

Office supplies is the name of the game at Office Depot, the largest office supply biz in the world. Shop both their online store and their many bricks-and-mortal retail stores across the U.S.

Office Depot also owns Viking Office Products, which caters to both U.S. and non-U.S. paper clip buyers. See:

http://www.vikingop.com/

Office Max   202964

3605 Warrensville Center Rd.
Shaker Heights, OH 44122
USA

- ☎ (216) 921-6900
- 🖂 (216) 491-4040
- ⊘ (800) 283-7674
- 🌐 http://www.officemax.com/

Office supplies. Examples of the kind of stuff you need: batteries; drafting and design; labels and label makers; storage and organization; tape, and adhesives and glues. Local stores and online mail order.

Staples, Inc.   202965

500 Staples Dr.
Framingham, MA 01702
USA

- ☎ (508) 253-5000
- 🖂 (508) 253-8989
- ⊘ (800) 378-2753
- 🌐 http://www.staples.com/

Staples operates a couple (okay, some 1,300) office supply stores in the U.S., Canada, Germany, the U.K., the Netherlands, and Portugal. They also sell most of their shippable products online from their Web site and through direct catalog sales. Refer to the Web site for a store locator.

Web site for Staples office supplies.

Viking Office Products   202966

950 West 190th St.
Torrance, CA 90502
USA

- ☎ (310) 225-4500
- 🖂 (310) 327-2376
- ⊘ (800) 711-4242
- ✉ support@vikingop.com
- 🌐 http://www.vikingop.com/

Viking is an international office retailer. They offer mail-order sales by phone or Internet, throughout the world.

⚙ RETAIL-OPTICALS AND LASERS

Optical components, including lasers, are used in robotic vision and object detection. The resources in this section specialize in optical components, such as lenses, filters, prisms, diffraction gratings, beam splitters, and polarizers. Also included here are sources for low-powered lasers (both gas and solid state), which can be employed in vision and distance measurement systems.

SEE ALSO:

RETAIL-SURPLUS MECHANICAL: New and surplus optical components, especially military pullouts

SENSORS-OPTICAL: Ready-made optical sensors (see how they work)

American Science & Surplus 202881

3605 Howard St.
Skokie, IL 60076
USA

((847) 982-0870

✉ info@sciplus.com

🌐 http://www.sciplus.com/

See listing under Retail-Surplus Mechanical.

Black Feather Electronics 202944

4400 S. Robinson Ave.
Oklahoma City, OK 73109
USA

((405) 616-0374

ℂ (405) 616-9603

✉ blkfea@juno.com

🌐 http://www.blkfeather.com/

See listing under Retail-General Electronics.

BMI Surplus 203582

P.O. Box 652
Hanover, MA 02339
USA

((781) 871-8868

ℂ (781) 871-7412

🚫 (800) 287-8868

🌐 http://www.bmius.com/

Electronic surplus, much of it high-end industrial or scientific; opticals, laser.

DiscountLasers.com 202764

ilium 2001
552 Grass Valley St.
Simi Valley, CA 93065
USA

✉ ilium@hollywoodfactory.com

🌐 http://discountlasers.com/

Killer ray guns for the financially challenged. Or, if you don't believe that, then pen lasers, ultra-high-brightness LEDs, and tools (like levelers) with lasers.

Edmund Industrial Optics 203053

101 East Gloucester Pk.
Barrington, NJ 08007-1380
USA

((856) 573-6250

ℂ (856) 573-6295

🚫 (800) 363-1992

✉ sales@edmundoptics.com

🌐 http://www.edmundoptics.com/

Lasers, optics, optical mounts, coatings, filters, polarizers, lenses, beam splitters.

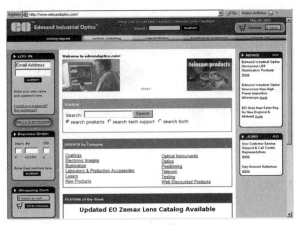

Web site for Edmund Industrial Optics.

Edmund Scientific 203054

60 Pearce Ave.
Tonawanda, NY 14150-6711
USA

((716) 874-9091

🚫 (800) 728-6999

🌐 http://www.scientificsonline.com/

See listing under Retail-Science.

Hosfelt Electronics 202883

2700 Sunset Blvd.
Steubenville, OH 43952
USA

((614) 264-6464

ℂ (614) 264-5414

🚫 (800) 524-6464

✉ order@hosfelt.com

🌐 http://www.hosfelt.com/

General electronics. New and surplus. See listing under Retail-General Electronics.

Lasermotion   203578

LMDC 3101 Whipple Rd.
Union City, CA 94587
USA

☎   (510) 429-1060

✉   office@lasermotion.com

🌐   http://www.lasermotion.com/

Industrial surplus. Offers include:

- Air bearings-spindles, slides, stages, systems, and kits
- Electronics-components, test equipment, power supplies
- Optics and lasers -accessories, lenses, tables, mounts
- Robotics-commercial robots, sensors
- Stages-linear, rotary
- Servo components-amplifiers, controllers, motors, encoders
- Step motor-software, controllers, drivers
- Video-cameras, crosshair generators, displays, printers

Meredith Instruments ♀   202354

P.O. Box 1724
5420 W. Camelback Rd., #4
Glendale, AZ 85301
USA

☎   (623) 934-9387

📠   (623) 934-9482

🚫   (800) 722-0392

✉   info@mi-lasers.com

🌐   http://www.mi-lasers.com/

Lasers and optics at Meredith Instruments.

Meredith Instruments is one of the oldest and most respected sellers of new and surplus lasers and optical components. They deal in gas and solid state lasers, various types of optics, parts for laser light shows, and laser safety products.

Midwest Laser Products   203584

P.O. Box 262
Frankfort, IL 60423
USA

☎   (815) 464-0085

📠   (815) 464-0767

🌐   http://www.midwest-laser.com/

New and surplus lasers. Good selection.

 🖥

MWK Laser Products Inc.   203258

455 West La Cadena 14
Riverside, CA 92501
USA

📠   (909) 784-4890

🚫   (800) 356-7714

✉   support@mwkindustries.com

🌐   http://www.mwkindustries.com/

Lasers, including gas and solid state. Also optics, power supplies, and assorted goodies.

📄 🖥

New Method Lasers   203084

10530-72nd St. N., #706
Largo, FL 33777
USA

☎   (727) 545-0376

📠   (727) 547-1760

✉   NML@laser-light-show.com

🌐   http://www.laser-light-show.com/

Laser light shows, materials, products.

Resources Un-Ltd.   202499

300 Bedford St.
Manchester, NH 03101
USA

☎   (603) 668-2499

(603) 644-7825

(800) 810-4070

info@resunltd4u.com

http://www.resunltd4u.com/

New and surplus video, infrared illuminators, laser, optics, stepper motors, linear translation tables, and assorted odds and ends.

Science Kit & Boreal Laboratories   203055

777 E Park Dr.
P.O. Box 5003
Tonawanda, NY 14150
USA

(716) 874-6020

(716) 874-9572

(800) 828-7777

http://www.sciencekit.com/

See listing under Retail-Science.

SmartTechToys.com   204095

301 Newbury St.
Dept. 131
Danvers, MA 01923
USA

(800) 658-5959

CustomerService@smarttechtoys.com

http://smarttechtoys.com/

Robots and more robots (LEGO Mindstorms and K'NEX); toys to hack (R/C cars, Rokenbok vehicles); lasers and laser pointers.

Surplus Shed   203425

407 U.S. Route 222
Blandon, PA 19510
USA

(610) 926-9226

(877) 778-7758

surplushed@aol.com

http://www.surplusshed.com/

Surplus Shed has surplus and used optical and electronic items. Useful stuff for robotics include optics (lenses, mirrors, prisms, beam splitters, filters) and electronic test equipment.

 **RETAIL-OTHER**

In this section you'll find miscellaneous retailers who carry some parts useful in robot building. Scan their product offerings for ideas.

Apogee Components, Inc.   203170

630 Elkton Dr.
Colorado Spings, CO 80907-3514
USA

(719) 535-9335

(719) 534-9050

tvm@apogeerockets.com

http://www.apogeerockets.com/

Model rocketry stuff. For building supplies: epoxy clay, cardboard tubes, and parachutes (in case you build a hang-gliding robot). Full catalog available on CD-ROM.

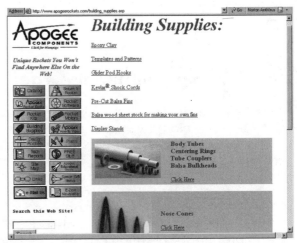

Apogee Components Web page.

Big 5 Sporting Goods   203564

2525 E. El Segundo Blvd.
El Segundo, CA 90245
USA

(310) 536-0611

(310) 297-7585

http://www.big5sportinggoods.com/

Sporting goods chain store. Use the Store Locator, or search for product information.

Darby Drug Co., Inc.

http://www.darbydrug.com/

Sells fiberglass casting tape, used to construct light-weight bodies and structures

Resco/Cresco

http://www.restaurantequipment.com/

Local (Northern California, Colorado, Nevada) and online restaurant supply retailer (look for stainless, aluminum, and plastics)

Shoplet.com

http://www.shoplet.com/

General store on the Internet; includes office supplies

Canadian Tire

203949

P.O. Box 770, Stn. K
Toronto, ON
M4P 2V8
Canada

🚫 (800) 387-8803

 feedback@canadiantire.ca

🌐 http://www.canadiantire.ca/

Canadian Tire is a local (in Canada) and chain retailer of automotive, household, workshop, and garden products. Useful products include batteries, battery chargers, hand tools, power tools, and welding and soldering tools. Web site is in English and French.

Cynthia's Bar & Restaurant

204197

Under London Bridge
4 Tooley Street
London
SE1 2SY
UK

📞 +44 (0) 2074 036777

📠 +44 (0) 2073 781918

✉ cynthia@cynbar.co.uk

🌐 http://www.cynbar.co.uk/

The world's first robotic bar and restaurant! Where robots serve drinks. The chef makes a mean braised lamb knuckle. Yum.

Researching with Mail-Order Catalogs

I once saw a documentary about how actor and martial arts expert Jackie Chan made it a habit to save old newspaper and magazine clippings as ideas for his films. He'd save some for months, even years, and hang them up on a wall for easy reference. The process must work because Jackie Chan movies are always inventive and original, even though many of the sequences in them may have been inspired by real-life events!

The moral of this story is really about finding inspiration and how you can take ordinary things, find some new slant about them, and end up with an original idea. Same thing with products found in mail-order catalogs. Most catalogs describe products for the use they were intended. Yet many ordinary, everyday products have many more applications.

Make it a point to go through every mail-order catalog that comes into the house or classroom. Just because the catalog may not sell "Robot Parts" doesn't mean it doesn't have robot parts to sell. Obviously, apparel catalogs have less relevance to robot building than, say, a household tools catalog or even the monthly Home Depot flyer.

Instead of throwing away catalogs when they get old, go through them once more and tear out products you think might have a use beyond the obvious application. Maybe that fiberglass replacement shovel handle could be used to build a super lightweight robot frame, for far less than the same amount of material in wood or metal. Or maybe the dog-training collar in the latest pet supply catalog could be adapted to signal a robot when it's about to venture outside your yard.

Even if you can't think of a good application now, save the item, and put in a scrapbook. Periodically, when the mood strikes you, pull out the scrap book to review your special finds.

FARMTEK

202611

1395 John Fitch Blvd.
South Windsor, CT 06074
USA

 (800) 327-6835

sales@farmtek.com

http://www.farmtek.com/

Dispel the images of a John Deere robot . . . this catalog is useful for the nontractor goodies it offers, such as fasteners, springs, clamps, tubing, PVC, conduit, tools, and lots more.

Mending Shed, The

202335

1735 South State St.
Orem, UT 84097-8008
USA

(801) 225-8012

(801) 225-9297

(800) 339-9297

info@mendingshed.net

http://www.mendingshed.com/

Appliance parts, including replacement parts for Fisher Price Power Wheels.

Small Parts, for Robots of Any Size

Large mechanical components are relatively plentiful, and easy to buy. Bike stores, lawnmower repair shops, auto parts marts, auto recyclers (junk yards), and similar resellers offer a wide variety of heavy duty bearings, axles, sprockets, chains, gears, and other power transmission components.

Too bad it's not nearly as simple to find the small stuff. And when you can find it, miniature power transmission components tend to be just as expensive as their bigger cousins. Salvaging components from old VCRs, compact disc players, even wall clocks is an alternative, but you never know what you'll get, and the parts are often manufactured for a specific narrow application, and may not adapt to the role you want it to play in your robot creation.

This doesn't mean small power transmission parts don't exist; you just have to go to the right sources. Premier mail order sources of small power transmission components—gears, sprockets, roller chain, bearings, shafts, etc.—are Serv-o-link, Small Parts, Stock Drive Products, W. M. Berg, Didel, Hammad Ghuman, Kelvin, and Miniature Bearings. There are others, of course, but this is a good well-rounded group that represents a variety of components, all the way from inexpensive plastic molded parts, to hand-machined goods in titanium.

When searching for small parts, you'll want to be mindful of the intended application, and then look for the cheapest material that will do the job. In most cases, small parts are not used for high torque applications, so molded plastic parts may suffice. If this is the case for your application, you'll find that by selecting plastic over metal you can save considerably.

Didel (based in Swizterland)
http://www.didel.com/

Hammad Ghuman
http://www.1hg.com/

Kelvin
http://www.kelvin.com/

Miniature Bearings Australia Pty.
http://www.minibearings.com.au/

Serv-o-link
http://servolink.com/

Small Parts Inc.
http://www.smallparts.com/

Stock Drive Products
http://www.sdp-si.com/

W. M. Berg
http://www.wmberg.com/

Nelson Appliance Repair, Inc.   203773

1220 E. Fillmore St.
Colorado Springs, CO 80907
USA

📞 (719) 630-1577
🚫 (800) 291-8010
📧 information@nelsonappliance.com
🌐 http://www.nelsonappliance.com/

Yes, replacement parts for appliances. Like food mixers, bread makers, espresso coffee makers, and more. But most valuable to us robot builders is their selection of Fisher Price Power Wheels parts: motors, gearboxes, switches, and other parts.

🌐 💻

Sports Authority, Inc.   203895

3383 N. State Rd. 7
Fort Lauderdale, FL 33319

📞 (954) 735-1701
📞 (954) 484-0837
🚫 (888) 801-9164
🌐 http://www.thesportsauthority.com/

America's number 1 sporting goods chain. Retail stores; online shopping also available.

🌐 💻 🛒

RETAIL-OTHER ELECTRONICS

The following listings are sources for specialty or unusual electronics items, such as tubes and other parts for repairing antique radios, high-end gadgetry, and alarm system components. The products sold by these resources represent the hackable side of robotics.

SEE ALSO:

DISTRIBUTOR/WHOLESALER-OTHER COMPONENTS: Miscellaneous industrial electronic components

RETAIL-GENERAL ELECTRONICS: General electronics outlets

RETAIL-SURPLUS ELECTRONICS: Also offer unusual parts, usually at a discount because of the used/overstock nature of the merchandise

Antique Electronic Supply   202885

6221 S Maple Ave.
Tempe, AZ 85283
USA

📞 (480) 820-5411
📞 (480) 820-4643
📧 info@tubesandmore.com
🌐 http://www.tubesandmore.com/

Parts and supplies for restoring old radios and other antique electronics, but some interesting items for robotics.

📄 🌐 🔋 💻

Chip-Sources.com   204168

7860 W. McLellan Rd.
Glendale, AZ 85303
USA

📞 (623) 521-6580
📞 (623) 623-2860
📧 riccf@earthlink.net
🌐 http://www.chip-sources.com/

Reseller of specialty electronics:

- SimmStick boards
- Sensors (digital compass, infrared distance)
- LCDs (serial controller and panels)
- BasicX microcontroller
- Epic PIC programmers; PIC Basic compilers
- BOB video overlay

🌐 💻

Clever Gear   202592

4514 19th St. Court East
P.O. Box 25600
Bradenton, FL 34206-5600
USA

🚫 (800) 829-2685
🌐 http://www.clevergear.com/

General gifts and novelties. Some hackables for robot building.

📄 🌐 💻

Hammacher Schlemmer & Company, Inc.   202593

303 W. Erie St.
IL 60610
USA

(312) 664-8170

(312) 664-8618

 customerservice@hammacher.com

 http://www.hammacher.com/

High-end gifts; offers some robotic toys and gadgets. Definitely not for the cash-challenged.

The company published a famed catalog in print form, which is a reading experience all in itself. Retail stores are located in New York and Chicago.

Web page for specialty retailer Hammacher Schlemmer.

Hawkes Electronics   202372

469 Parker Branch Rd.
Barnesville, GA 30204
USA

(770) 358-0623

 http://www.hawkeselectronics.com/

Hawkes sells "products and information for the student and hobbyist in electronics and robotics." Wares include:

- 8051 project board
- LCDs
- A/D relay modules

HMC Electronics   202569

33 Springdale Ave.
Canton, MA 02021
USA

(781) 821-1870

(781) 821-4133

(800) 482-4440

 sales@hmcelectronics.com

 http://www.hmcelectronics.com/

HMC offers a full line of tools and test equipment, with an emphasis on production floor needs. Here, you'll find pro-level soldering irons, static control gear, and work area products.

Hub Material Company

See the listing for HMC Electronics (this section).

Resources Un-Ltd.   202499

300 Bedford St.
Manchester, NH 03101
USA

(603) 668-2499

(603) 644-7825

(800) 810-4070

info@resunltd4u.com

http://www.resunltd4u.com/

New and surplus video, infrared illuminators, laser, optics, stepper motors, linear translation tables, and assorted odds and ends.

Sharper Image, The   202595

http://www.sharperimage.com/

Gifts for yuppies. Includes some robotic toys.

Wm. B. Allen Supply Company, Inc.   202887

301 N. Rampart St.
New Orleans, LA 70112-3105
USA

(504) 525-8222

(504) 525-6361

(800) 535-9593

info@wmballen.com

http://www.wmballen.com/

Wm. B. Allen supplies 35,000 items from over 150 manufacturers. For the robot builder, their electronics, video, and tools sections will be of most interest. The company is primarily in the alarm, security, and controlled-access business.

  ## RETAIL-OTHER MATERIALS

Look here for unusual retail merchandise you can use for robot-building parts. These include bits and pieces of plastic and metal from restaurant supply stores (many of which provide online sales) and specialty 3M tapes and adhesives.

SEE ALSO:

MATERIALS-METAL AND MATERIALS-PLASTIC: Prime source for metal and plastic pieces

RETAIL-SURPLUS MECHANICAL: Used and surplus materials

SUPPLIES-GLUES & ADHESIVES: More sources for specialty tapes and adhesives

Ace Mart Restaurant Supply 203565

P.O. Box 18100
San Antonio, TX 78218
USA

☎ (210) 323-4400
🖷 (210) 323-4404
⊘ (888) 223-6278
✉ corporate@acemart.com
🌐 http://www.acemart.com/

Online store, and chain of local restaurant supply stores in Texas. Look for inexpensively priced metal bowls, mixing utensils, and interesting kitchen gadgets that can be hacked to make things for your robot.

BigTray, Inc. 203570

1200 7th St.
San Francisco, CA 94107
USA

Robots Don't Eat!

You may be wondering why I listed several online and local restaurant supply outlets. No, I'm not suggesting you build a mechanical Emeril, or that amateur robots make for good chefs—personally, I get enough robotic service at the local fast-food joint.

Rather, restaurant supply outlets tend to be excellent sources for materials at low cost. Typical restaurant supply outlets carry such items as:

- Stainless steel spun bowls (miniature dipping and finger-bowl size, up to gargantuan 100-pound mixing bowls);

- Aluminum and steel baking sheets, with and without antistick coatings (can be a cheap alternative to aluminum or metal sheet purchased at hardware and hobby stores);

- Kitchen and baking utensils, in plastic, metal, or wood—cut them to various shapes to make parts for your robots;

- Miscellaneous unusual items, including strainers, individual bread and cake pans (good as the "mother mold" for making small castings), silverware containers that make for great vertical parts bins, and plastic tumblers (you and I call them "glasses"), which can be sawed to various sizes and shapes and used for protective robotic covers, shells, and other applications.

When shopping the restaurant supply store, you must think outside the box. Don't look at a miniature plastic saltshaker as a dispenser of sodium chloride. Instead, look at it as a housing for a robotic sensor or the leg tips to a medium-sized walking robot. Be creative!

Final note: Steer away from the retailers that specialize in "gourmet kitchens." Their products tend to be more expensive. The typical restaurant supply outlet sells cheap, because eateries are like any other business—they're always looking for ways to trim costs. You get to enjoy those cost savings, too.

(415) 863-3614

(800) 244-8729

help@bigtray.com

http://www.bigtray.com/

Online restaurant supply retailer. Cheap source of stainless steel: steel bowls, stirrers, utensils, etc. Check the "Smallwares" section.

BigTray restaurant supplies.

Brown's Home Kitchen Center & Restaurant Supply   203567

2551 Capital Cir. NE
Tallahassee, FL 32308
USA

(850) 385-5665

(850) 385-5670

staff@brsupply.com

http://www.homekitchencenter.com/

Local (Tallahassee, Fla.) retail restaurant supply store.

Chef's Depot   203569

2509 W. Washington Blvd.
Los Angeles, CA 90018
USA

(323) 730-8987

(800) 979-7253

info@restaurant-supply.com

http://www.restaurant-supply.com/

Chef's Depot is an online restaurant supply retailer. Don't think cooking 'bots; think cheap stainless steel hemispheres (mixing bowls), aluminum and steel sheets (baking pans), plastic, wood, and metal in vari-

ous shapes and sizes (utensils), and food molds (round, square, rectangular bread/cake making) you can use to cast specialty shapes for your robot projects.

G & G Restaurant Supply   203566

126 VanGuysling Ave.
Schenectady, NY 12305
USA

(518) 393-2183

(518) 393-2328

(800) 206-0777

http://www.gandgequipment.com/

Online retailer of restaurant supplies. You can try cooking for your robot, but using the inexpensive stainless steel bowls and other products may be a more workable idea.

Micro Parts & Supplies, Inc.   204235

308 Cary Point Rd.
Cary, IL 60013
USA

(847) 516-0191

(847) 516-0491

(800) 336-2131

ebrown@mpsupplies.com

http://www.mpsupplies.com/

Mail-order resellers of numerous 3M industrial products (in single/small quantities), including conductive tape and adhesives, connectors, and static protection devices.

Premium Supply Company   203568

960 Grand Blvd.
Deer Park, NY 11729
USA

(631) 586-3639

(800) 894-3480

info@premiumsupply.com

http://www.premiumsupply.com/

Online restaurant supply retailer.

Where to Get Stuff: Retail

There aren't many robot-building stores at the mall, but you can do quite well at the following retail establishments around town.

- *Hobby and model stores*. Ideal sources for small parts, including lightweight plastic, brass rod, servo motors for radio control (R/C) cars and airplanes, gears, and construction hardware.

- *Arts and craft stores*. Sell supplies for home crafts and arts. As a robot builder, you'll be interested in just a few of the aisles at most craft stores, but what's in those aisles will be a godsend!

- *Hardware stores and builder's supply outlets*. Fasteners (nuts, bolts), heavy-gauge galvanized metal, hand and motorized tools.

- *Electronic specialty stores*. There's Radio Shack, of course, which continues to support electronics experimenters. Check the local Yellow Pages under *Electronics-Retail* for a list of electronics parts shops near you.

- *Surplus*. Electronics surplus stores specialize in new and used mechanical and electronic parts (these are not to be confused with surplus clothing, camping, and government equipment stores). Finding them is not always easy. Start by looking in the phone book Yellow Pages under *Electronics*, and also under *Surplus*.

- *Sewing machine repair shops*. Ideal for small gears, cams, levers, and other precision parts. Some shops will sell broken machines to you. Tear the machine to shreds and use the parts for your robot.

- *Auto parts stores*. The independent stores tend to stock more goodies than the national chains, but both kinds offer surprises in every aisle. Keep an eye out for things likes hoses, pumps, and automotive gadgets.

- *Auto repair garages*. More and more used parts from cars are being recycled or sent back to a manufacturer for proper disposal, so the pickins at the neighborhood mechanics garage aren't as robust as they used to be. But if you ask nicely enough, many will give you various used parts, like timing belts, or offer them for sale at low cost.

- *Junkyards*. Old cars are good sources for powerful DC motors, used for windshield wipers, electric windows, and automatic adjustable seats (though take note: such motors tend to be terribly inefficient for a battery-based 'bot). Or how about the hydraulic brake system on a junked 1969 Ford Falcon? Bring tools to salvage the parts you want. And maybe bring the Falcon home with you, too.

- *Lawn mower sales/service shops*. Lawn mowers use all sorts of nifty control cables, wheel bearings, and assorted odds and ends. Pick up new or used parts for a current project or for your own stock.

- *Bicycle sales/service shops*. Not the department store that sells bikes, but a *real* professional bicycle shop. Items of interest: control cables, chains, brake calipers, wheels, sprockets, brake linings, and more.

- *Industrial parts outlets*. Some places sell gears, bearings, shafts, motors, and other industrial hardware on a one-piece-at-a-time basis.

RETAIL-ROBOTICS SPECIALTY

When I started robotics as a hobby there was no such thing as a "robot store." The first I had ever seen was started in the mid-1980s by Timothy Knight, a young author who wrote a popular book on personal robots. His store, in the San Francisco Bay area, offered for sale toys and robot kits; it closed a few years later, during the "robot doldrums," a sad period when amateur robotics was on the decline because of lack of decent hardware.

Now, with sophisticated microcontrollers costing less than a night at the movies for a family of four, there are plenty of interesting products to sell, and robotics specialty stores have cropped up all over the world. A few are local retail stores only, but most offer mail-order sales through catalogs or the Internet.

Robotics specialty stores stock a variety of items, from robot toys to kits, sensors, motors, and microcontrollers. The sources listed in this section expressly concentrate on amateur robots and all will sell mail order (many internationally).

The stock at many robotics specialty outlets is much the same, allowing you to compare prices. A few, like Acroname, offer exclusive product not available elsewhere.

SEE ALSO:

KITS-ROBOTIC: Lower-end robotic kits

ROBOTS-HOBBY & KIT: Amateur robots and higher-end kits

TOYS-CONSTRUCTION: Construction toys, like LEGO, include robot kits

TOYS-ROBOTICS: Robotic toys

Acroname Inc. 🐛 203371

4894 Sterling Dr.
Boulder, CO 80301-2350
USA

☎ (720) 564-0373
📠 (720) 564-0376
✉ info@acroname.com
🌍 http://www.acroname.com/

Acroname is an online retailer specializing in robotics. They carry numerous kits, including:

- CMU Camera Kit-low-cost robotics vision
- LEGO Mindstorms-the latest versions
- Handy Board Kit-microcontroller designed expressly for robotics

Acroname, Inc.

I don't know what the name Acroname means, but I know what the company does: sells reasonably priced robot kits and components. Acroname is a licensed reseller for a number of popular products, including the OOPic microcontroller, LEGO Mindstorms, Sony AIBO, PalmPilot Robot Kit (PPRK), and Rug Warrior.

They round out their main product lines with specialty sensors, including the Polaroid 6500 ultrasonic transducer and board, the Devantech (RobotElectronics.com) ultrasonic sensor, various Sharp infrared proximity and distance sensors, Hamamatsu UVTron sensor, Pontech SV203 serial servo controller, Magnevation motor driver kit, omnidirectional wheels of various sizes, the popular CMUcam digital camera kit, and a number of robotics books.

The company also developed and sells their own controller that they call BrainStem. The BrainStem is a combination of hardware and software modules that make it easier to interconnect the various parts of a robot. The General Purpose Module provides coprocessor power for the robot's main controller and integrates the following hardware:

- 5 10-bit-resolution A/D ports
- 5 digital I/O ports
- 1 Sharp GP2D02 ranger port
- 4 high-resolution servo ports
- 1 serial port
- 1 1MBit IIC port (master and slave)

A nice feature of the Acroname Web site is their "code store" of example programming for common microcontrollers. The main description page for many of their products, such as sensors, also include links to source code for such controllers as the OOPic, Basic Stamp, and BasicX. The Web site also sponsors a gallery of other people's robots, articles of interest (such as "Using an Omni-directional Wheel as a Caster" and "Using the OOpic oPWM Object to Generate Sounds," as well as a user-to-user forum.

Acroname's BrainStem provides a means to interconnect parts of a robot together.

Acroname is located in Boulder, Colo., and can be found here on the Web:

http://www.acroname.com/

- PalmIII PPRK-build a bot with your PalmPilot
- Rug Warrior-Popular intermediate robot from the book Mobile Robots
- Omniwheels-Used in the PPRK and as casters

The company also sells numerous (and some hard-to-find) sensors: ultrasonic, infrared (including the Sharp infrared distance and proximity), and flame, as well as the OOPic and Basic Stamp microcontrollers.

An extensive array of online documentation, gallery of customers' robots, and datasheets round out this excellent resource.

The Sony AIBO robot. Courtesy Sony Corp and Acroname, Inc.

Android World-Anthropomorphic Robots & Animatronics
202225

http://www.androidworld.com/

Android World is dedicated to the art and science of building android (human-like) robots. Includes some custom products for sale, news, links, and user-to-user forums.

Blue Point Engineering
202050

213 Pikes Peak Pl.
Longmont, CO 80501-3033
USA

((303) 651-3794

 http://www.bpesolutions.com/

Says the Web site: "One of the main goals of Blue Point Engineering is to provide a product line of low cost, high quality software, hardware, electronics, and supplies used in animatronics, robotics, haunted industry, and technology education, for.hobbyist and professional designers, imagineers and dreamers creating their own forms of animatronic and robotic life."

Budget Robotics
204255

P.O. Box 5821
Oceanside, CA 92056
USA

((760) 941-6632

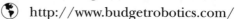 info@budgetrobotics.com

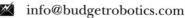 http://www.budgetrobotics.com/

Disclosure alert! This is my company, and I won't hide that fact. I didn't start out with any notion of selling goods to the public. But after seeing many gaps in the product offerings of companies and the high prices some specialty retailers ask, I decided to offer an alternative.

Budget Robotics sells a narrow range of product, with an emphasis on structural elements, such as robot bodies and frames. Much of it is custom-made and is explicitly designed for affordable amateur robotics. We also buy some hard-to-find items in bulk and provide them in smaller quantities for your convenience.

Products include:

- Low-cost metal-framing components to build strong robots at a minimal cost
- Precision-cut robot bases, in different colors and shapes, to make unusual robot bodies

Pololu
http://www.pololu.com/
Various robotics parts: Tamiya Educational products, infrared beacon, motor controller, servo controller

Robotic Education Products
http://www.robotics.com.au/
Online robotics retailer; based in Australia

Robotics World
http://www.roboworld.com.sg/
Reseller of robotics kits, microcontrollers, and other parts; based in Singapore

- Omnidirectional wheels, in a variety of sizes and styles
- Affordable new (not surplus) gearmotors
- Strain gauge sensors, at prices that won't strain your wallet
- Specialty sensor material such as Kynar piezo-electrics
- Robot Parts Play kits-assortment of useful small robots for robot building
- Spherical casters (these aren't ball transfers and are quite unique)
- Specialty fasteners (lots of nylon pieces), hardware (like hex spacers), and more

Future-Bot Components   202653

106 Commerce Way
Ste. A8
Jupiter, FL 33458
USA

☎ (561) 575-1487

📠 (561) 575-1487

✉ fuboco@bellsouth.net

🌐 http://www.futurebots.com/

Sells numerous robot electronic and mechanical parts:

- Motorola 68MC11 microcontroller "brain"
- Gearmotors
- Plastic domes
- Tilt switches
- ICs (various)
- Recycled computer parts

Future-Bot Components on the Web.

HVW Technologies Inc. ♂   202146

3907-3A St. N.E. Unit 218
Calgary, AB
T2E 6S7
Canada

☎ (403) 730-8603

📠 (403) 730-8903

✉ Info@HVWTech.com

🌐 http://www.hvwtech.com/

HVW Tech is a leading online Canadian retailer of robotics, microcontrollers, and related products. The product mix is varied and includes:

- Books
- Development tools-PIC programmers, prototyping boards
- Displays-LCD serial interface, video-text overlay
- Educational tools-BOE-Bot robotics kits, Stamps in Class
- Microcontrollers-Basic Stamp, OOPic, PICmicro, 68HC11
- Miscellaneous components-High-current driver chips, serial EEPROMs
- Robotics-Cybug robotic kits, motor controllers, motors and wheels, motorized gearsets (Tamiya), servos and controllers
- Sensors-Infrared, temperature
- Software-Compilers for PICmicros (BASIC, C)
- Radio frequency-2.4 GHz Wireless Video Transceiver
- Static control

Web page of HVW Technologies.

- Sharp infrared distance and proximity sensors

Ships internationally.

Images SI Inc.   202153

39 Seneca Loop
Staten Island, NY 10314
USA

- (718) 698-8305
- (718) 982-6145
- images@imagesco.com
- http://www.imagesco.com/

See listing under Retail-Science.

Kronos Robotics 202503

P.O. Box 4441
Leesburg, VA 20175
USA

- (703) 779-9752
- msimpson@kronosrobotics.com
- http://www.kronosrobotics.com/

Kronos sells parts and kits for amateur robotics. They are a reseller for Dontronics and Basic Micro products, and they carry the Tamiya Educational gearmotor kits. The site provides numerous how-to articles, with circuit diagrams and construction details.

As an aside, the name Kronos comes from the 1957 movie of the same name, where a robot from outer space comes to earth to suck up energy for its power-depleted civilization, which apparently has not yet dis-

The Kronos Crawler, from Kronos Robotics. Photo Michael Simpson, Kronos Robotics.

covered the concept of turning the light off at night. The robot is destroyed, of course, but is it the end?

Lynxmotion, Inc. 202034

P.O. Box 818
Pekin, IL 61555-0818
USA

- (309) 382-1816
- (309) 382-1254
- sales@lynxmotion.com
- http://www.lynxmotion.com/

Lynxmotion sells high-quality kits and parts for mobile robots-both wheeled and legged-as well as robotic arm trainers. Their distinctive yellow plastic robot kits include rad four- and six-legged walkers, which come complete with servo motors, linkages, and other hardware. Most kits are available with or without processor, and you can add sensors as you wish.

See also: http://www.lynxmotion.com/smodh1.htm

See also: http://www.lynxmotion.com/smodh2.htm for Hitec HS-422 servo mod.

Lynxmotion Carpet Rover 3. Photo Jim Frye, Lynxmotion, Inc.

Milford Instruments Ltd. 202271

Milford House
120 High Street
South Milford
Leeds
LS25 5AQ
UK

- +44 (0) 1977 683665
- +44 (0) 1977 681465
- sales@milinst.com
- http://www.milinst.com/

Milford sells robot and mechatronics goodies through the mail. Their products include Basic Stamps and accessories, LCDs, robotics, animatronics, and PIC stuff.

The Stamp Bug II walker. Photo Milford Instruments.

Mondo-tronics, Inc. 202048

4286 Redwood Hwy. PMB-N
San Rafael, CA 94903
USA

((415) 491-4600
  415-491-4600
  (415) 491-4600

✆ (415) 491-4696

⃠ (800) 374-5764

✉ info@RobotStore.com

🌐 http://www.robotstore.com/

The Robot Store sells all kinds of robotics goodies, from kits to books to individual parts (like motors, sensors, and wheels). Carries LEGO Mindstorms, Parallax Basic Stamp and BOE-Bot, Sony AIBO Robot Dog, servos, servo controllers and motor drivers, batteries and chargers, speech recognition modules, and more.

📄 🌐 💻

Mr. Robot 202428

1615 Chatsworth Ave.
Richmond, VA 23235
USA

((804) 272-5752

🌐 http://www.mrrobot.com/

Mr. Robot specializes in robots, parts, and supplies (sorry, there is no Mrs. Robot). Products include microcontroller kits, robot kits, servos, DC gearhead motors, infrared and ultrasonic sensors, wireless mini color

cameras, Fischertechnik construction kits, omnidirectional wheels (omniwheels).

🌐 💻

Robologic 203575

202 St. Margarets Rd.
Bradford
West Yorkshire
BD7 2BU
UK

✉ sales@robologic.co.uk

🌐 http://www.robologic.co.uk/

U.K.-based retailer of robotics equipment. They offer microcontrollers (including the BasicX from NetMedia), infrared sensors, and motors. They also provide tutorials on robot building.

🌐 💻

Robot Store 202049

🌐 http://www.robotstore.com/

See Mondo-tronics, Inc. (this section).

📄 🌐 💻

Robot Store (HK) 202847

7th Floor, Fok Wa Mansion
No.19 Kin Wah St.
North Point
Hong Kong

(+85 2 9752 0677

✆ +85 2 2887 2519

✉ info@RobotStoreHK.com

🌐 http://www.robotstorehk.com/

Robot Store (HK) is a robotics mail-order retailer based in Hong Kong. Not to be confused with RobotStore.com, based in the U.S.

Robot Store (HK) is an authorized distributor of MIT Handy Board microcontroller and expansion board developed by Dr. Fred Martin of Massachusetts Institute of Technology (MIT) Media Lab. They also offer the licensed Interactive-C development tool developed by Newton Research Labs. The company sells assembled and unassembled MIT Handy Board systems, robot bases, sensors, gearbox kits (from the Tamiya Educational line), servos, DC motors, and various mechanical parts.

🌐 💻

RobotStore.com.

Robot Zone 202984

BTR
620 Industrial Blvd.
Winfield, KS 67156
USA

📞 (620) 221-7071
📠 (620) 221-0421
🚫 (877) 221-7071
🌐 http://www.robotzone.com/

RobotZone (same corporate owner as Servo City) sells robot-specific parts and assemblies.

See also:

http://www.servocity.com/

RoboToys 202865

12025 Ventura Blvd.
Studio City, CA 91604
USA

📞 (818) 769-5563
🌐 http://www.RoboToys.com/

RoboToys Web home.

According to the Web site: "We carry a variety of robot toys, educational scientific kits, classic robots, robot construction kits, Gundam and anime model kits, tin windup robots, Giant Robot anime videos and more!"

Site provides a user-to-user forum for robonuts to chat about whatever is on their minds.

RobotOz 204212

P.O. Box 635
Inglewood, WA, 6932
Australia

📞 +61 8 9370 3456
📠 +61 8 9370 2323
📧 kits@www.robotoz.com.au
🌐 http://www.robotoz.com.au/

Robotics, Aussie style. RobotOz is a reseller of:

- Lynxmotion products
- Parallax Inc. Basic Stamp microcontrollers
- Muscle Wire shape memory alloy

They also sell various BEAM kits, as well as the Handy Board microcontroller from RobotStoreHK.com. On-site tutorials include "Make your own BEAM Photovore," with step-by-step directions and pictures. You'll also find user-to-use forums and robotics links to fill out your day.

ScienceKits.com, Inc. 202855

785F Rockville Pk., #515
Rockville, MD 20852
USA

📞 (301) 294-9729
📧 info@sciencekits.com
🌐 http://www.sciencekits.com/

Science Kit Center sells robot (OWI) kits, Fischertechnik, Logiblocs (electronic building blocks), Electronics and Electricity Science Kits.

Selectronic 203451

B.P. 513
59022 Lille Cedex
France

📞 +33 (0) 328 550328
📠 +33 (0) 328 550329

http://www.selectronic.fr/

Emporium of fun electronic and robotic stuff:

- Fischertechnik robots
- Flexinal shape memory alloy
- Speech chips
- Gearmotors
- Servo motors

. . . and more. Web site is in French and English.

Tech-supplies.co.uk 203036

Revolution Education Ltd.
4 Old Dairy Business Centre
Melcombe Rd.
Bath
BA2 3LR
UK

✆ +44 (0) 1225 340563
📠 +44 (0) 1225 340564
✉ info@tech-supplies.co.uk
🌐 http://www.tech-supplies.co.uk/

Says the Web site: "The new online shop for all your electronic, robotic and educational technology projects. The site provides a comprehensive one-stop shop for students and hobbyists looking for resources for electronic and robotic projects."

Products include:

- Robot kits: OWI robots, LEGO Mindstorms
- PIC and Basic Stamp microcontrollers, programmers
- Modeling materials: corrugated plastic, metal sheets
- Motors and servos

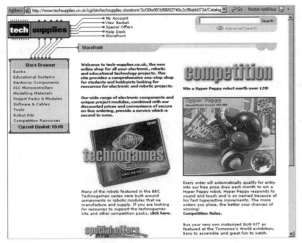

Web page for Tech-supplies.

- Electronic components: LEDs, light sensors, H-bridge motor driver chips
- Tools

Total Robots Ltd. 202417

49 Church Road
Epsom
Surrey
KT17 4DN
UK

✆ +44 (0) 1372 741954
📠 +44 (0) 1372 729595
✉ enquiry@totalrobots.com
🌐 http://www.totalrobots.com/

Reseller of the OOPic microcontroller, several BEAM robot kits, and the ARobot from Arrick Robotics.

Wizard Devices, Inc. 203499

7731 Tuckerman Ln., #186
Potomac, MD 20854
USA

✆ (301) 309-6825
📠 (301) 309-6825
✉ wizard1@wizard-devices.com
🌐 http://www.wizard-devices.com/

Wizard Devices sells products primarily designed for the special effects and stage pyrotechnic industries, but they also carry (as a reseller) several robotics products. They carry the Milford Instruments animatronic and robotic kits, as well as a variety of servo control electronics, including IR Decoder Module, Animate Card (up to 40 seconds of recorded servo control movement), and the Wizard Card (up to five minutes of recorded servo control movement, for four servos).

World of Robotics Online 202860

c/o Robotic Education Products (Aust.) P/LTD
ABN 800 772 10934
110 Mt. Pleasant Road Belmont.
Geelong
VIC 3216
Australia

✆ +61 3 5241 9581
📠 +61 3 5241 9089

⊘ 1 800 000 745

✉ wworldof@bigpond.net.au

🌏 http://www.robotics.com.au/

Online robot store Down Under. In the words of the Web site: "The World of Robotics has a huge range of kits and purpose built robots. We have the big brands including, MOVIT, OWIKIT, LEGO Mindstorms, Fischertechnik, Robotix by Learning Curve, Logiblocs, Lynxmotion, Unimat, Wonderborg, Basic Stamp, BEAM, Cybug, Angelus Research, CYE the Home Office Robot, RB5X by General Robotics, Micro Bot, Micro Mouse, OOpic, Hotchips, Capsela and Silverlit."

 RETAIL-SCIENCE

In this section you will find retail outlets for amateur scientific kits, demonstrators, and components. Some of the resources listed here supply science and lab equipment to schools, and there may be restrictions on selling directly to the general public. Check the ordering terms for any purchasing limitations.

Some sources also sell robots, in complete or kit form.

SEE ALSO:

KITS-ELECTRONIC AND KITS-ROBOTIC: Kits for building electronic circuits and robots

---

Future-toys.com

http://www.future-toys.com/

Sellers of kits: robots (mostly the OWI line), electronics projects, soldering kits, learning toys, and hobby tools

Invention Factory

http://www.inventionfactory.com/

Science kits; magnets, motors

OMSI Science Store

http://www.omsi.edu/store/

Science kits, including construction toys (T-rex robots, other)

Patent Cafe

http://www.patentcafe.com/

Links, resources, and products for educational and construction materials

---

RETAIL-EDUCATIONAL SUPPLY: Another source for science parts

Analytical Scientific, Ltd. 202863

11049 Bandera Rd.
San Antonio, TX 78250
USA

☏ (210) 684-7373

🖥 (210) 520-3344

⊘ (800) 364-4848

✉ asltd@intersatx.net

🌏 http://www.analyticalsci.com/

Scientific goodies that would make even Mr. Wizard flush with joy. Analytical Scientific carries chemistry, astronomy, anatomy, and biology kits and supplies. They offer the full line of OWI robot kits and other nicknacks of interest to automaton builders.

Arbor Scientific 202588

P.O. Box 2750
Ann Arbor, MI 48106
USA

☏ (734) 913-6200

⊘ (800) 367-6695

✉ mail@arborsci.com

🌏 http://arborsci.com/

Science and educational materials, for teachers and hobbyists. Product categories include Force & Motion; Light & Color; Electricity; Chemistry; Astronomy; Measurement; Lasers; Holography; Sound & Waves; Magnetism; Fiber Optics; Books & Videos; Science Toys; Science Software; Computer DataLoggers.

Among Arbor Scientific's more notable products useful in robotics are:

- Color Filters Kit-Large sheets of colored gel filters; primary and secondary colors
- Color Filter Swatch Book-Small pieces of dozens of colored gels
- Holographic Glasses-Cheap source for diffraction gratings
- Spring Scales-For measuring motor force
- Laser Pointers-Hackable laser diodes
- Helical Spring "Snaky"-Long acoustic spring, for possible use in bumper detection
- Sound Pipes-Basically medium-diameter flexible tubes, for use in construction or looks

- Magnets-Rare earth, ceramic, and others; for use in sensors, construction, etc.

Barkingside Co. 202862

6417 Lyndale Ave. South
Richfield, MN 55423-1405
USA

((612) 869-4445
☏ (612) 869-4445
⊘ (800) 917-2275
✉ sales@barkingside.com
🌐 http://www.barkingside.com/

Barkingside is one of the more eclectic robotics sources out there. In addition to selling OWI robot and hobby kits, Tamiya Educational, and Technicraft kits (including pulleys, motors, and wheels), they also offer leaf tea, beer-brewing equipment, and KitchenAid appliances!

They do provide a printed catalog, but the hobby and robotic kits don't appear to be included.

The company sells mail order, and a walk-in retail store is in Richfield, Minn.

Carolina Science & Math 202565

2700 York Rd.
Burlington, NC 27215-3398
USA

((336) 584-0381
⊘ (800) 334-5551
🌐 http://www.carolina.com/

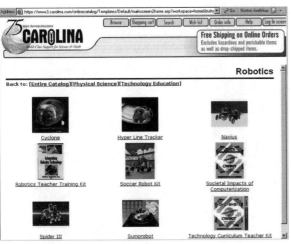

Web site for Carolina Science.

Carolina sells a massive amount of educational supplies and materials for schools and researchers. Their (online and print) catalog contains a number of kits and products in the fields of physics, technology ed (including electronics and robotics), and lab equipment/supplies. The company's robotic offerings are the OWI robot kits.

Cole-Parmer Instrument Company 204021

625 East Bunker Ct.
Vernon Hills, IL 60061-1844
USA

((847) 549-7600
☏ (847) 247-2929
⊘ (800) 323-4340
✉ sales@coleparmer.com
🌐 http://www.coleparmer.com/

Institutional lab supplies, including motors, safety equipment, sensors, pumps, syringes, and tools. Site provides technical info about many of the products.

Edmund Scientific 🎖 203054

60 Pearce Ave.
Tonawanda, NY 14150-6711
USA

((716) 874-9091
⊘ (800) 728-6999
🌐 http://www.scientificsonline.com/

Edmund Scientific is dead. Long live Edmund Scientific!

Actually, the company didn't die, but it was sold by its original owner and is now in the able hands of Science Kit and Boreal Laboratories. (Note: If you're looking for optical parts, that portion of the business was retained by the Edmund family and can be found at Edmund Industrial Optics: http://www.edmundoptics.com/.)

The new Edmund is very much like the old one, selling science-oriented products for home, school, and small business. Of their product lines, the following are of keen value to robot builders:

- Laser pointers-Hack these for various projects, or make a Borg Bot
- Metal detectors-Treasure-hunting robot, or maybe a robot that is guided by metal

- Optics-Every robot needs glasses for eyes; plus diffraction gratings, filters
- Unique lighting-Includes live wire (electroeluminescent wire)
- Magnets-Rare earth and not-so-rare-earth (Alnico and ceramic)
- General science-Motors, gears, robots, Fischertechnik, K'NEX Ultra
- Tools-Mostly hand tools

Edmund Scientific on the Web.

Educational Innovations, Inc.   203737

362 Main Ave.
Norwalk, CT 06851
USA

((203) 229-0730
℧ (203) 229-0740
✉ info@teachersource.com
🌐 http://www.teachersource.com/

Science supplies, kits, and demonstrators. Among their products useful for robotics are:

- CLIMBaTRON window-climbing robots
- Polarizing filters
- Refracting, diffracting, and reflecting light
- Magnets
- Ferrofluids
- Nitinol memory metal/Muscle Wire

And these power sources, for a rad robot from the future, or one trapped in the past:

- Fuel cell car kit
- Jensen steam engine

Efston Science   202853

3350 Dufferin Street
Toronto, ON
M6A 3A4
Canada

((416) 787 4581
℧ (416) 787 5140
🚫 (888) 777-5255
✉ info@e-sci.com
🌐 http://www.e-sci.com/

Science kits, supplies; includes mechanical and physical science, astronomy, kits for kids, science fair projects, Jensen tools.

Exploratorium   202324

3601 Lyon St.
San Francisco, CA 94123
USA

((415) 561-0360
✉ shipping@exploratorium.edu
🌐 http://www.exploratorium.edu/

The Explaratorium is a museum in San Francisco, Calif.; the museum gift shop sells various science kits and other trinkets. Product comes and goes, but invariably there's a robot or robot kit, plus other interesting mechanical devices.

Flinn Scientific, Inc.   203736

P.O. Box 219
Batavia, IL 60510
USA

🚫 (800) 452-1261
✉ flinn@flinnsci.com
🌐 http://www.flinnsci.com/

K-12 educational materials. Science supplies. Mostly biology and chemistry, though.

IDEA ELETTRONICA   203739

Via XXV
Aprile n°76
21044 Cavaria con Premezzo
Varese
Italy

☎ +39 (0) 331 215081

✉ idele@tiscalinet.it

🌐 http://www.ideaelettronica.it/

Science projects, science kits, electronics kits. Web site is in Italian.

Images SI Inc. ⚥   202153

39 Seneca Loop
Staten Island, NY 10314
USA

☎ (718) 698-8305

📠 (718) 982-6145

✉ images@imagesco.com

🌐 http://www.imagesco.com/

Images Co. (operated by book author John Iovine) offers a wide range of high-tech goodies well suited to robotics. John has an eye for the special, and he's often one of the first retailers to offer a new technology. Among the products carried by Images Co. are:

- Air muscles
- Nitinol shape memory alloy
- OWI 007 arm
- Robot hardware, wheels (including omnidirectional), motors, gearboxes
- Plastic domes
- Aluminum sheet and bar stock
- Flex and pressure sensors
- Compasses, tilt switches
- Piezoelectric film
- PIC programmers
- Assorted semiconductors and parts

A series of articles on the Web site provide background information for using many of the products.

Imaginarium.com   202582

🌐 http://www.imaginarium.com/

Famous museum in San Francisco. They also sell their museum products by mail order. The Web site is now under the auspices of Amazon.com.

Indigo Instruments   203261

169 Lexington Court, Unit I
Waterloo, ON
N2J 4R9
Canada

☎ (519) 746-4761

📠 (519) 747-5636

🚫 (877) 746-4764

✉ info@indigo.com

🌐 http://www.indigo.com/

Science kits; science supplies (test tubes, etc.). Specializes in organic chemistry parts and kits. Some products, like the 3mm-diameter rare earth Neodymium magnets, are useful in robotics sensors.

Informal Education Products   202584

Museum Tour Catalog
2525 SE Stubb St.
Milwaukie, OR 97222
USA

☎ (503) 496-1258

📠 (503) 794-7111

🚫 (800) 360-9116.

✉ emilieb@museumtour.com

🌐 http://www.museumtour.com/

Online retailer of products common in science and educational museum gift shops. Includes a number of robotic toys and mechanical and electronic learning sets. Construction toys include Erector Set; robotics include Capsela and OWI robots.

Kelvin   202877

280 Adams Blvd.
Farmingdale, NY 11735
USA

☎ (631) 756-1750

📠 (631) 756-1763

🚫 (800) 535-8469

✉ kelvin@kelvin.com

🌐 http://www.kelvin.com/

Kelvin sells educational kits and materials for a high-tech teaching world. Their Technology series includes a number of very useful products for robot building:

- K'NEX
- LEGO Dacta
- Robotics kits (BOE-Bot, MAZER, OWI robots, many others)
- PIC programmers
- Fischertechnik

- Tamiya Educational kits
- Plastics vacuum former
- Plastics injection molder
- Science and chemistry lab components

They also offer project materials in metal, plastic, and wood; magnets; various sizes and types of gearboxes (and motors with and without gearboxes); motor holders; linear actuator motors; wheels; gears; sprockets and sprocket chain; and hundreds of additional products.

Electronics include trainers, board-level solder kits, electronics construction tools, test gear, components (active and passive), and others.

Kelvin's sales are intended for educational institutions. While they will sell to individuals, they say some products may cost more and that some products are only available to schools and teachers. If you're ordering for a school, they accept school POs. A printed catalog is available to teachers and schools only.

Kelvin Web site.

Out of This World 203738

P.O. Box 1010
Mendocino, CA 95460
USA

((707) 937-3324
[(707) 937-1303
⊘ (800) 485-6884
✉ orders@DiscountTelescopes.com
🌍 http://www.discounttelescopes.com/

Check out their Science Fun link: They offer electronics kits, robot kits, and Muscle Wire projects.

Pitsco 203448

P.O. Box 1708
Pittsburg, KS 66762
USA

⊘ (800) 835-0686
✉ orders@pitsco.com
🌍 http://www.pitsco.com/

Online shopping through their e-commerce portal, www.shop-pitsco.com.

- Brutus robotic arm
- Pitsco Blinky robot kit
- Sensor kits and software
- S-Cargo and other OWI robots
- Space Wings electronics kit (shape memory alloy)
- R/C servos
- Servo power transmission (hub mounts, sprockets, sprocket chain, gears, wheels)
- Plastic injection-molding tools and supplies
- Plastic vacuum-forming tools and supplies
- Aircraft birch plywood

Though expensive, they have foam wheels with servo mounts so they can be directly attached to an R/C servo without any extra hassle.

See also:

http://www.pitsco-legodacta.com/

School-Tech Inc. 203206

745 State Cir.
Box 1941
Ann Arbor, MI 48106
USA

((734) 761-5072
[(734) 761-8711
✉ service@school-tech.com
🌍 http://www.school-tech.com/

Science kits-the physical science line of kits includes magnetism, electricity, and robotics (the latter, the OWI robot kits).

Science & Hobby 203979

http://www.sciencehobby.com/

Ways to Steer Your Robot

There are a variety of methods used to steer wheeled robots. Here are the most common approaches.

Differential

For wheeled and tracked robots, *differential steering* is the most common method of getting the machine to go in a different direction. The technique is exactly the same as steering a military tank: One side of wheels or treads stops or reverses direction while the other side keeps going. The result is that the robot turns in the direction of the stopped or reversed wheel/tread. Because of friction effects, differential steering is most practical with two-wheel drive systems. Additional sets of wheels can increase friction during steering.

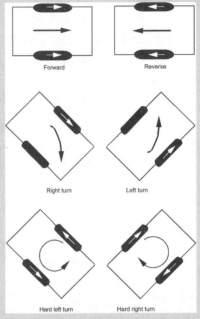

Differential steering allows a
robot to spin in place.

Car-Type

Pivoting the wheels in the front is yet another method of steering a robot. Robots with *car-type steering* are not as maneuverable as differentially steered robots, but they are better suited for outdoor use, especially over rough terrain. Somewhat better traction and steering accuracy are obtained if the wheel on the inside of the turn pivots to a greater extent than the wheel on the outside. This technique is called *Ackerman steering* and is found on most cars, but not as many robots.

Three-Wheel Tricycle

Car-type steering, described previously, is one method that avoids the problem of "crabbing" due to differences in motor speed (simply because the robot is driven by just one motor). But car-type steering makes for fairly cumbersome indoor mobile robots; a better approach is to use a single drive motor powering two rear wheels and a single steering wheel in the front; the arrangement is just like a child's tricycle. The robot can be steered in a circle just slightly larger than the width of the machine. Be careful of the wheel base of the robot (distance from the back wheels to the front steering wheel). A short base will cause instability in turns, causing the robot to tip over in the direction of the turn.

Tricycle-steered robots require a very accurate steering motor in the front. The motor must be able to position the front wheel with subdegree accuracy. Otherwise, there is no guarantee the robot will be able to travel a straight line. Most often, the steering wheel is controlled by a servomotor; servomotors used a "closed-loop feedback" system that provides a high degree of positional accuracy.

Three-Wheeled Omnidirectional

Three drive motors, placed 120 degrees apart (basically at the points of an equilateral triangle), can be used to drive a robot in any direction. In order for the design to work, special multidirectional wheels are required. These wheels have rollers around their circumference; they provide traction at angles other than perpendicular to the hub of the wheel. The robot moves "forward" by activating any two motors; it turns by adjusting the speed and/or direction of any and all three of the motors.

As you can imagine, this system requires three drive motors instead of just two. An alternative design uses four wheels, with either two motors (two wheels per motor) or four motors. The wheels are mounted in traditional car fashion. The robot is differentially steered, as explained previously.

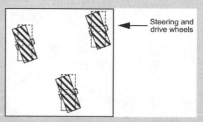

Steering and drive wheels

Three-wheeled omnidirectional drive provides movement in all directions, without requiring a support caster.

Multidrive Omnidirectional

For the highest-tech of all robots, omnidirectional drive uses multiple steerable drive wheels, usually at least three such drive wheels, but sometimes four, five, and even six. The wheels are operated by two motors: one for locomotion, and one for steering. In the usual arrangement, the drive/steering wheels are "ganged" together using gears, rollers, chains, or pulleys. Multidrive omnidirectional robots exhibit excellent maneuverability and steering accuracy, but they are technically more difficult to construct.

Science & Hobby is an alternative Web interface for the products sold at Tower Hobbies:

http://www.towerhobbies.com/

For additional educator's supplies, see also:

http://www.hearlihy.com/

Science City   203260

3009 Redstone Dr.
Arlington, TX 76001
USA

☎ (817) 465-1566

✉ info@science-city.com

🌐 http://www.science-city.com/

Educational science kits: electronics, robotics, mechanics.

Science Experience   202861

Hearlihy & Company
P.O. Box 929
Springfield, OH 45501-929
USA

🚫 (800) 622-1000

✉ info@scienceexperience.com

🌐 http://www.scienceexperience.com/

ScienceExperience.com is intended for educators, but is open to all buyers. They carry K'NEX, robot kits (mostly OWI but also some Robotix), and various electronics kits.

Science Kit & Boreal Laboratories   203055

777 E Park Dr.
P.O. Box 5003
Tonawanda, NY 14150
USA

☎ (716) 874-6020

🖰 (716) 874-9572

🚫 (800) 828-7777

🌐 http://www.sciencekit.com/

Web page for Science Kit & Boreal Laboratories.

Selling some 15,000 items, Science Kit specializes in products for education, and many of their offerings are packaged for demonstration and multistudent exploration. This can be a good thing: May of their products are "samplers" with a little bit of a lot of things. This can save you from buying larger quantities of individual parts when you need only a little bit yourself.

The Science Kit Web site is broken down into a main section, with multiple subsections. You have to do a lot of clicking to find what you want, but here's a quick overview of the cogent products for robot enthusiasts:

- Robotics (Cybug, Arrick ARobot, OWI kits)
- Construction (K'NEX)
- Laser pointers
- Air toys (hack these, like the Air Hog, for their pneumatic rotary pistons)
- Science fun (miscellaneous, including mold-making material, Hoberman Spheres)
- Components (piezo discs, LEDs, Sorbothane shock-absorbing rubber)
- Motors and gears

Science Source, The   204085

P.O. Box 727
Waldoboro, ME 04572
USA

 (207) 832-7281

(800) 299-5469

info@thesciencesource.com

http://www.thesciencesource.com/

The Science Source is designed for the upper-grade science teacher, and if that's what you are, you probably know about this place already. But if you're not a science teacher, you'll want to know about it anyway, as they have truly unique products (many are intended for classroom demonstrations or group study) that have definite applications in amateur robotics. Some items that should pique your interest:

- Liquid-filled accelerometer (part #10-100)
- Spring and pass accelerometer (part #10160)
- Super Slinky (part #15815)
- Color filters (part #33250)

Thinker Toys   202856

P.O. Box 6297
7th and San Carlos Ave.
Carmel, CA 93921
USA

(831) 624-0441

(831) 624-0551

info@thinkertoys.com

http://www.thinkertoys.com/

Wide variety of toys, including LEGO Mindstorms, robotics kits and sets (mostly OWI), building logs, Zoob, etc., arts and crafts. Local stores in Carmel, Monterey, and Morgan Hill, Calif.

See also Thinker Toys's "sister" store:

http://www.walnutgrovetoys.com/

Zany Brainy   202585

2520 Renaissance Blvd.
King of Prussia, PA 19406
USA

(610) 278-7800

(610) 278-7805

(888) 548-8531

http://www.zanybrainy.com/

Zany Brainy is a retailer of educational toys and books. They sell through a national chain of retail stores, as well as mail order. Their product offering includes LEGO and K'NEX sets, along with some robotic toys. Stores are located throughout the U.S.; check the Web site for a store finder.

Zany Brainy online.

RETAIL-SURPLUS ELECTRONICS

Surplus doesn't mean junk; it simply means someone doesn't need it anymore and is selling their excess stock. In the case of electronics, surplus seldom means "used" as it may for other surplus components, such as motors or mechanical devices. The listings in this section are for local and mail-order retailers of surplus electronics. Note also that many such retailers also sell new components. As such, there is a blurred line between what it surplus and what is new.

One benefit of shopping the surplus electronics retailer is cost: Even for new components, prices are generally lower than from general electronics retailers. On the downside, selection may be limited to whatever components the store was able to purchase. Don't expect every value and size of resistor or capacitor to be available.

SEE ALSO:

RETAIL-GENERAL ELECTRONICS: New electronic components

RETAIL-SURPLUS MECHANICAL: Surplus mechanisms to go with your surplus electronics

---

A-2-Z Solutions, Inc.   203579

P.O. Box 740756
Boynton Beach, FL 33474-0756
USA

✆ (561) 967-4646
📠 (561) 967-2524
✉ sales@a2z-solutions.com
🌐 http://a2z-solutions.com/

New and surplus electronics. Mostly computer equipment (PCs, monitors, scanners, and so forth).

Active Surplus   203031

345 Queen Street West
Toronto, ON
M5V 2A4
Canada

✆ (416) 593-0909
📠 (416) 593-0057
🚫 (800) 465-5487
✉ info@activesurplus.com
🌐 http://www.activesurplus.com/

See listing under **Retail-General Electronics**.

AE Associates, Inc.   203426

7733 Densmore Ave. #5
Van Nuys, CA 91406
USA

✆ (818) 997-3838
📠 (818) 997-0136
✉ info@ae4electronicparts.com
🌐 http://www.ae4electronicparts.com/

New and used electronics, including switches, connectors, electronic components (resistors, capacitors, diodes, transistors, etc.), and test equipment. Searchable database. Also sells a small number of compact B&W and color video cameras. Local store in Van Nuys, Calif.

All Electronics Corp. ⚡   202160

P.O. Box 567
Van Nuys, CA 91408-0567
USA

Electronic Surplus Inc.
http://www.electronicsurplus.com/
Surplus components, test equipment. Local store in Cleveland, OH.

Hoffman Industries
http://www.hoffind.com/
Surplus electronics; active and passive components, including ICs, resistors, capacitors, diodes, switches

San Mateo Electronic Supply
http://www.smelectronics.com/
General electronics; wholesale and surplus; local store.

Surplus Traders
http://www.73.com/
Surplus; specializes in lot sales, as well as wall warts (plug-in transformers for providing DC power to some electronic product); Also offers motors, computer equipment, telephone, and electronic components

((818) 997-1806

☏ (818) -781-2653

⊘ (888) 826-5432

✉ allcorp@allcorp.com

🌐 http://www.allcorp.com/

All Electronics is one of the primary sources in the U.S. for new and used robotics components. Prices and selection are good. Walk-in stores in the Los Angeles area are located at:

Los Angeles Store: 905 S. Vermont Avenue, Los Angeles, CA; (213) 380-8000

Van Nuys Store: 14928 Oxnard Street, Van Nuys, CA; (818) 997-1806

Product line includes motors, switches, discrete components, semiconductors, LEDs, infrared and CdS sensors, batteries, LCDs, kits, and much more. Specifications sheets for many products are available at the Web site.

Same as http://www.allelectronics.com.

Alltronics   202352

P.O. Box 730
Morgan Hill, CA 95038-0730
USA

((408) 847-0033

☏ (408) 847-0133

✉ ejohnson@alltronics.com

🌐 http://www.alltronics.com/

Not to be confused with All Electronics in southern California, this northern California electronics retailer is known for their good assortment and reasonable prices. New and surplus merchandise. Online catalog and sales via the Internet; the company used to provide a walk-in store in San Jose, but this has closed. The company provides mail order service only. A will-call window is available in Gilroy; check the company's Web page for details. Some product is also available for auction on eBay.

A printed catalog costs $3, or you can download it free from the Web site (you need Adobe Acrobat Reader to view it).

Among their product line useful in robotics are:

- Motors (DC and stepper)
- Stepper motor controllers
- Power MOSFETs
- H-bridge ICs (including the oft-cited L293D, L297, and L298)

- Atmel AVR microcontrollers
- Small CCD video cameras
- Tools
- Solenoids and relays

Product datasheets (in PDF format) are available for download for many of the specialty semiconductor products.

Apex Jr.   203580

3045 Orange Ave.
La Crescenta, CA 91214
USA

((818) 248-0416

☏ (818) 248-0490

✉ steve.apexjr@prodigy.net

🌐 http://www.apexjr.com/

Surplus electronics and mechanicals. General electronics, transformers, and "movie props."

Ax-Man Surplus   203596

1639 University Ave.
St. Paul, MN 55104
USA

((651) 646-8653

☏ (651) 646-1819

✉ axmansurplus@cs.com

🌐 http://www.ax-man.com/

Local (St. Paul, Fridley, and St. Louis Park, Minn.) electronics and mechanical surplus.

B. G. Micro   202210

555 N. 5th St.
Ste. #125
Garland, TX 75040
USA

((972) 205-9447

☏ (972) 205-9417

⊘ (800) 276-2206

✉ bgmicro@bgmicro.com

🌐 http://www.bgmicro.com/

See listing under Retail-General Electronics.

BCD Electro Inc. 203581

2525 West Commerce
Dallas, TX 75212
USA

((214) 630-4298

℘ (214) 267-1127

✉ company@bcdelectro.com

🌐 http://www.bcdelectro.com/

Surplus electronics: active and passive electronics, motors, relays, switches, etc.

Web site for BCD Electro.

BMI Surplus 203582

P.O. Box 652
Hanover, MA 02339
USA

((781) 871-8868

℘ (781) 871-7412

🚫 (800) 287-8868

🌐 http://www.bmius.com/

Electronics surplus, much of it high-end industrial or scientific; opticals, laser.

Boeing Surplus Store 203617

20651 84th Ave S.
Kent, WA
USA

((425) 393-4065

🌐 http://www.boeing.com/assocproducts/surplus/

All sorts of surplus, from small plastic parts to large machine tools-but no aircraft parts. My guess is that Boeing buys this stuff, puts it in a warehouse somewhere for a few years, then sells it at their surplus store at great prices!

Local only; Seattle, Wash.

C & H Sales 202190

2176 E. Colorado Blvd.
Pasadena, CA 91107
USA

((626) 796-2628

℘ (626) 796-4875

🚫 (800) 325-9465

✉ candhsales@earthlink.net

🌐 http://www.candhsales.com/

See listing under Retail-Surplus Mechanical.

CTR Surplus 203288

202 West Livingston Ave.
Crestline, OH 44827
USA

((419) 683-3535

℘ (419) 683-3230

✉ buy@ctrsurplus.com

🌐 http://www.ctrsurplus.com/

Surplus electrical:
- Computer
- Electrical
- Fans/blowers
- Motors/gearboxes
- RF equipment
- Test equipment
- Power supplies
- Optics
- Generators

Dexis Corporation 203597

9749 Hamilton Rd.
Eden Prairie, MN 55344
USA

((952) 944-7670

☎ (952) 942-9712

✉ info@dexis.com

🌐 http://www.dexis.com/

Electronics surplus. Test and measurement. Local store in Minnesota.

EIO.com   203195

P.O. Box 3148
Redondo Beach, CA 90277
USA

☎ (310) 217-8021

☏ (310) 217-0950

🚫 (800) 543-0540

✉ ecsc@eio.com

🌐 http://eio.com/

A Web site with lots of information and some sales, too. Surplus stuff, plus lots of links and resources on a number of surplus electronics topics:

- Batteries
- Capacitors, resistors
- CCDs, video
- Electro optics, fiber optics, lasers
- LCD, LEDs
- Microcontrollers
- Power supplies
- Prototyping
- Relays, stepper motors
- Robotics
- Solar cells
- Transformers

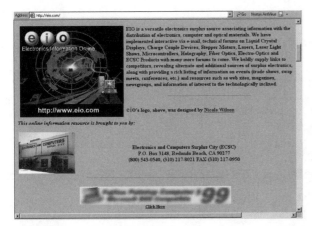

EIO.com.

Electro Mavin   202194

2985 E. Harcourt St.
Compton, CA 90221
USA

☎ (310) 632-9867

☏ (310) 632-3557

🚫 (800) 421-2442

✉ john@mavin.com

🌐 http://mavin.com/

Electronic components, motors, batteries, optics, and test equipment.

Electronic Dimensions   203424

424 Puyallup Ave.
Tacoma, WA 98421
USA

☎ (253) 272-1061

☏ (253) 383-2088

✉ eldim@worldnet.att.net

🌐 http://www.el-dim.com/

Military and industrial surplus, electronics, radio receivers, transmitters and parts, electron tubes, test equipment, and ham gear.

Electronic Goldmine   202652

P.O. Box 5408
Scottsdale, AZ 85261
USA

☎ (480) 451-7454

☏ (480) 661-8259

🚫 (800) 445-0697

✉ goldmine-elec@goldmine-elec.com

🌐 http://www.goldmine-elec.com/

See listing under Retail-General Electronics.

Electronic Surplus Co.   203422

9012 Central Ave. SE
Albuquerque, NM 87123
USA

☎ (505) 296-6389

☏ (505) 296-3922

elecsurp@nmia.com

http://www.surplus-electronics.com/

Surplus audio, ham, computer, and general electronics (stuff like ICs, motors, capacitors, and cables).

Electronics Plus 203595

10302 Southard Dr.
Beltsville, MD 20705
USA

(301) 937-9009

(301) 937-5092

(888) 591-9009

mail@electronics-plus.com

http://www.electronics-plus.com/

See listing under Retail-General Electronics.

EOL Surplus 202670

P.O. Box 7348
Laguna Niguel, CA 92607-7348
USA

(949) 388-1282

(949) 203-8652

http://eolsurplus.com/

See listing under Retail-Surplus Mechanical.

Excess Solutions 202907

430 E. Brokaw Rd.
San Jose, CA 95112
USA

(408) 573-7045

(408) 573-7046

info@excess-solutions.com

http://www.excess-solutions.com/

Surplus computer parts and general electronics, including stuff from companies in the Silicon Valley area that are no longer on any map (I mean, Silicon Valley is still on the map; the companies aren't because they're out of business . . . oh, never mind). Components include ICs, fans, heat sinks, connectors, resistors, capacitors, motors, inductors, lamps, LCDs and LEDs, speakers, tools, and lots more.

Local store in San Jose-and that's still on the map, so you're sure to know the way.

Fair Radio Sales 202186

1016 E. Eureka
P.O. Box 1105
Lima, OH 45802
USA

(419) 227-6573

(419) 227-1313

fairradio@fairradio.com

http://www.fairradio.com/

Fair Radio Sales primarily caters to ham operators, with their radio sets and old gear. But they have plenty of test equipment and general surplus electronics to tide anyone over. Prices are always reasonable. I've bought from them for over three decades.

The company provides a yearly catalog, with updates.

Fair Radio on the Web.

Forest City Surplus 203058

1712 Dundas St.
London, ON
N5W3E1
Canada

(519) 451-0246

(519) 451-9341

(877) 393-0056 Ext. 55

http://www.fcsurplus.ca/

Everything from electronic gizmos to camping surplus stuff. Electronics include power supplies, test equipment, batteries and battery chargers, and switches.

Gateway Electronics, Inc. 202185

8123 Page Blvd.
St. Louis, MO 63130
USA

(314) 427-6116
(314) 427-3147
(800) 669-5810
http://www.gatewayelex.com/

Gateway is a general electronics mail-order supplier and retailer. Among their product are passive and active components, motors, electronics kits, gadgets, books, and tools. Some of their goods are new; others are surplus.

They operate local stores in St. Louis, Mo.; San Diego, Calif.; and Denver, Colo. See store info at:

http://www.gatewayelex.com/storeinfo.htm

Keystronics 203432

88 Hadham Rd.
Bishops Stortford
Hertfordshire
CM23 2QT
UK

+44 (0) 1279 505543
+44 (0) 1279 757656
http://www.keytronics-uk.co.uk/

Surplus ICs, transistors, crystal oscillators, crystals, passive components, optoelectric, and photoelectric.

KRP Electronic Supermarket 203618

219 West Sunrise Hwy.
Freeport, NY 11520
USA

(516) 623-3343
(516) 623-3391
info@krp.com
http://www.krp.com/

Surplus stuff. Local store in Freeport, N.Y.

LabX 203282

P.O. Box 216
478 Bay Street
Midland, ON
L4R 1K9
Canada

(705) 528-6888

(705) 528-0270
(888) 781-0328
help@labx.com
http://www.labx.com/

Says the Web site: "LabX.com is the largest independent marketplace for pre-owned and surplus scientific equipment. LabX is not a dealer, or a broker-we don't buy or sell equipment, and do not get involved with negotiations or transactions. You deal directly with the buyer, seller, or sellers agent."

Mark Hannah Surplus Electronics 203427

822 NW Murray Blvd.
PMB #250
Portland, OR 97229
USA

(503) 591-7391
(503) 591-8391
mhannah1@nwlink.com
http://www.markhannahsurplus.com/

Surplus electronics, tools. Good selection.

Parts for Industry 203583

http://www.partsforindustry.com/

Though categorized in the retail section, Parts for Industry doesn't actually do any selling. Instead, they act as facilitators between buyers and sellers of surplus ("excess inventory") mechanical and electrical components, such as bearings, sprockets, linear actuators, motors, etc.

The site makes its money charging a small listing fee to sellers. Buyers can browse for free.

Skycraft Parts & Surplus Inc. 203421

P.O. Box 536186
Orlando, FL 32853-6186
USA

(407) 628-5634
(407) 647-4831
Info@Skycraftsurplus.com
http://www.skycraftsurplus.com/

Skycraft is a veritable surplus mall of accessories, power supplies, transistors, relays, ICs, wire, cable, heat shrink, transformers, motors, fiber optics, test equipment, resistors, diodes, and other goodies.

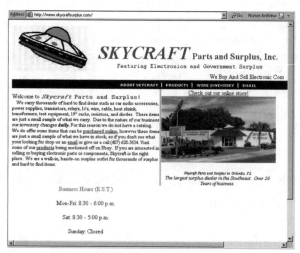

Web site for Skycraft surplus.

Surplus Sales of Nebraska

202884

1502 Jones St.
Omaha, NE 68102
USA

  (402) 346-4750

  (402) 346-2939

  http://www.surplussales.com/

Surplus electronic parts, including connectors, passive and active components, electronics hardware, relays and solenoids, chemicals, and antique radio parts (to build antique robots, of course).

Surplustronics Trading Ltd.

203293

P.O. Box 90439
A.M.S.C
Auckland
New Zealand

  (09) 302 0688

  (09) 302 0686

  inquire@surplustronics.co.nz

  http://www.surplustronics.co.nz/

New and surplus general electronics, including soldering stations, relays, motors, LEDs, battery holders, etc.

Timeline Inc.

202184

2539 West 237 St.
Building F
Torrance, CA 90505
USA

  (310) 784-5488

  (310) 784-7590

  mraa@earthlink.net

  http://www.digisys.net/timeline/

Surplus of all kinds: electronics, computer peripheral, laser, motors, LCDs, and more. (Though on a "free" Web page, they are included in this book because of their long years in business.)

Transtronics, Inc.

203790

3209 W.9th St.
Lawrence, KS 66049
USA

  (785) 841-3089

  (785) 841-0434

  index@xtronics.com

  http://www.xtronics.com/

Electronics kits, electronic programmers, surplus, and other goodies. Plus lots of useful information and tidbits. Read the explanation of the Jack and Jill rhyme.

Weird Stuff Warehouse

202910

384 West Caribbean Dr.
Sunnyvale, CA 94089
USA

  (408) 743-5650 Ext. 324

  (408) 743-5655

  http://www.weirdstuff.com/

Weird Stuff Warehouse sells surplus, including electronics. Not really "weird" stuff-just the ordinary fodder for robot building. Retail store in Sunnyvale, Calif.

W. J. Ford Surplus Enterprises

203423

4 Wellington St.
P.O. Box 606
Smith's Falls, ON
K7A 4T6
Canada

(613) 283-5195

(613) 283-0637

e-mail: testequipment@falls.igs.net

http://www.testequipmentcanada.com/

Surplus electronics: transmitters, receivers, test gear, lab equipment, and components. Local stores and mail order.

⚙ RETAIL-SURPLUS MECHANICAL

This section lists sources for surplus mechanical parts, all selected for their relevancy to amateur robotics. Surplus mechanical parts may be either used (called RFE, for "removed from equipment"), or new. New parts may be from a manufacturer who made too many of them, unused goods from the government, or are repair/replacement parts for a product no longer in broad use.

No matter what their source, surplus mechanical components can represent a real savings over buying the same goods new. Selection can be severely limited, but you never know when you'll find exactly what your robot needs. For this reason, collect as many surplus mechanical catalogs as you can, and consult each issue for the best components.

SEE ALSO:

ACTUATORS-MOTION PRODUCTS: New mechanical components

POWER TRANSMISSION: Sources for new gears, belts, and other power parts

RETAIL-GENERAL ELECTRONICS: New electronic components

RETAIL-SURPLUS ELECTRONICS: Retailers of new and overstocked electronics

---

American Science & Surplus 🛢   202881

3605 Howard St.
Skokie, IL 60076
USA

(847) 982-0870

info@sciplus.com

http://www.sciplus.com/

Some time ago, a fellow named Jerry used to sell odds-and-ends surplus stuff, having taken over the business from his parents. That in itself wasn't unusual, but the humor he brought to the "write-ups" for the products in his printed catalog made each edition worth waiting for. Every robot builder who knew how to use a screwdriver received-and-studied-the Jerryco catalog.

Alas, Jerry passed away a few years back, but his spirit is still alive in the "new" Jerryco, American Science & Surplus. From old optics to new lasers, to books, gears, pulleys, tools, pumps, magnets, electronics, batteries, arts and crafts, and everything in between, AS&S offers a wide range of product at good prices. They still send out a printed catalog, but all of their wares are available for viewing online, where you can order as well.

Realizing that robot building is an important aspect of their business, AS&S dedicates a special section of the Web site to robot parts. Find the Robot Parts link in the table of contents area, and you'll find the latest offerings. When I last looked, they had ball transfers (great for robot support caster wheels), large heavy-duty wheels, pneumatic cylinders, roller chain, and more.

Retail stores are in the Chicago and Geneva, Ill., areas and in Milwaukee, Wisc.

Web site for American Science & Surplus

American Surplus Inc.   203593

1 Noyes Ave.
East Providence, RI 02916
USA

(401) 434-4355

(401) 434-7414

(800) 989-7176

info@American-Surplus.com

http://www.american-surplus.com/

New and used industrial surplus, specializing in materials handling and conveyors. Among the product lines carried are:

- Light- and medium-duty casters

- Thermoplastic wheels (3- to 8-inch diameter)

- Polyurethane wheels (with metal hubs); diameters 4 to 12 inches

- Phenolic canvas wheels

- Moldon rubber wheels (with metal hubs); diameters 4 to 14 inches

- Industrial steel shelving (use for robot frames)

APEX Electronics   203592

8909 San Fernando Rd.
Sun Valley, CA 91352
USA

✆ (818) 767-7202

📠 (818) 767-1341

✉ mrybak@jps.net

🌐 http://www.apexelectronic.com/

Military and industrial surplus. Huge selection, but the store is disorganized, and virtually nothing is priced (you have to ask, and don't be afraid to haggle). Plan to spend several hours there. Special emphasis on copper wiring. The store is not in the best part of town, but you're going there for the surplus not the sights, right?

Some items can be ordered online.

Ax-Man Surplus   203596

1639 University Ave.
St. Paul, MN 55104
USA

✆ (651) 646-8653

📠 (651) 646-1819

✉ axmansurplus@cs.com

🌐 http://www.ax-man.com/

Local (St. Paul, Fridley, and St. Louis Park, Minn.) electronic and mechanical surplus.

Boeing Surplus Store   203617

20651 84th Ave S.
Kent, WA
USA

✆ (425) 393-4065

🌐 http://www.boeing.com/assocproducts/surplus/

All sorts of surplus, from small plastic parts to large machine tools-but no aircraft parts. My guess is that Boeing buys this stuff, puts it in a warehouse somewhere for a few years, then sells it at their surplus store at great prices!

Local only; Seattle, Wash.

Bill Boeing's excess goodies sold here.

Brigar Electronics   202455

7-9 Alice St.
Binghampton, NY 13904
USA

✆ (607) 723-3111

📠 (607) 723-5202

✉ BRIGAR2@aol.com

🌐 http://brigarelectronics.com/

See listing under Retail-General Electronics.

---

equip.recycle.net

http://equip.recycle.net/

Links and lists of sources for surplus and recycling

KW Surplus

http://www.kwsurplus.com/

Mechanical, electronic, kitchenware, and more

TechMax

http://www.techmax.com/

Optics; motors; miniature bearings; other surplus

Burden's Surplus Center   203987

1015 West O St.
P.O. Box 82209
Lincoln, NE 68501
USA

   (402) 474-5198

   800-228-3407

   (none specified)

Major catalog retailer of mechanical items, from tiny gears to huge generators.

C & H Sales   202190

2176 E. Colorado Blvd.
Pasadena, CA 91107
USA

(626) 796-2628

(626) 796-4875

(800) 325-9465

candhsales@earthlink.net

http://www.candhsales.com/

C & H sells motors, gears, pneumatics, pumps, solenoids, relays, and lots of odds and ends. Their catalog (both printed and online) regularly contain dozens of quality surplus DC (geared and non) and stepper motors.

Their store on Colorado Blvd. in Pasadena is small, but packed with all kinds of goodies. If you're in the Los Angeles area, be sure to make an afternoon trip to the C & H retail store.

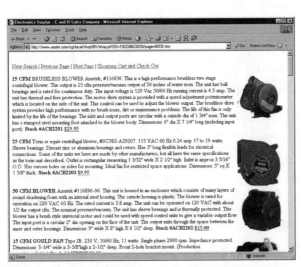

Plain but serviceable C & H Sales Web catalog.

Electro Mavin   202194

2985 E. Harcourt St.
Compton, CA 90221
USA

(310) 632-9867

(310) 632-3557

(800) 421-2442

john@mavin.com

http://mavin.com/

Electronic components, motors, batteries, optics, and test equipment.

EOL Surplus   202670

P.O. Box 7348
Laguna Niguel, CA 92607-7348
USA

(949) 388-1282

(949) 203-8652

http://eolsurplus.com/

EOL offers a good assortment of CNC hardware, including linear bearings, leadscrews, stepper motors. Some of their categories useful in robotics include:

- Lab lasers
- Optical
- Motion control
- Test equipment
- Tools
- Hardware

Stock comes and goes, so check often.

Forest City Surplus   203058

1712 Dundas St.
London, ON
N5W3E1
Canada

(519) 451-0246

(519) 451-9341

(877) 393-0056 Ext. 55

http://www.fcsurplus.ca/

Everything from electronic gizmos to camping surplus stuff. Electronics include power supplies, test equipment, batteries and battery chargers, and switches.

Gateway Electronics, Inc. 202185

8123 Page Blvd
St. Louis, MO 63130
USA

 (314) 427-6116

 (314) 427-3147

Ø (800) 669-5810

🌐 http://www.gatewayelex.com/

See listing under Retail-Surplus Electronics.

H&R Company, Inc. (Herbach and Rademan) 202878

353 Crider Ave.
Moorestown, NJ 08057
USA

 (856) 802-0422

 (856) 802-0465

Ø (800) 848-8001

✉ sales@herbach.com

🌐 http://www.herbach.com/

Surplus mechanicals: motors, relays, gears, optics, and lots, lots more. I've built lots of robots using parts I've purchased from H&R.

(Interesting story: H&R was one of the first mail-order surplus catalogs to carry my book, *Robot Builder's Bonanza*. They still carry it, and in their description of the book they remark, "Many of the components shown are available from H&R's vast inventory." That's because I bought them from H&R!)

Web side for Herbach & Rademan.

HGR Industrial Surplus 203587

20001 Euclid Ave.
Euclid, OH 44117-1480
USA

 (216) 486-4567

 (216) 486-4779

✉ sales@hgrindustrialsurplus.com

🌐 http://www.hgrindustrialsurplus.com/

Industrial surplus: electronics, pumps, everything. Many large machines, including used industrial machinery-screw machines, turret lathes, you name it.

Marlin P. Jones & Assoc. Inc. 202209

P.O. Box 530400
Lake Park, FL 33403
USA

((561) 848-8236

Ø (800) 652-6733

✉ mpja@mpja.com

🌐 http://www.mpja.com/

See listing under Retail-General Electronics.

Murphy's Electronic & Industrial Surplus Warehouse 203591

401 N. Johnson Ave.
El Cajon, CA
USA

((619) 444-7717

 (619) 444 6750

✉ murphy@cts.com

🌐 http://maxpages.com/murphyjunk/

Industrial and military surplus of all types. Local store (El Cajon, Calif.), but also sells via eBay. Some products listed on the site.

Okay, they're on a free page, but I'm making an exception to list them because of my personal experience with the store.

I Love Catalogs!

Ever since I was a kid I collected—and studied—mail-order catalogs. No, I'm not talking about the Frederick's of Hollywood catalogs, either. I'm talking about science, electronics, and surplus catalogs, like Edmund, Fair Radio, Layfayette Electronics (remember them?), Allied, and of course Radio Shack.

Never mind that I couldn't afford anything in these catalogs. I still studied them like they were baseball statistics. I learned quite a bit from these catalogs, and today I continue the habit of poring over the better component and equipment catalogs I receive. I even buy something once in a while!

The Internet makes mail-order buying a cinch, but in my opinion, a printed catalog is still the best way to really discover what's available. Web pages can be slow and cranky, and they're like trying to see the world through tiny portholes on the side of a ship.

Also consider: Printed catalogs can be taken with you. A highlighter and a stack of Post-It notes let you flag items you might be interested in. Be sure to read the descriptions for the catalog items; it's a great way to learn what things are and how they might be used. The better catalogs even suggest ways how you might adapt a product to a certain application.

For those mail-order retailers that still offer printed catalogs, be sure to order a copy (most are free, or cost a buck or two). Some catalogs come with a date stamped on them; for those that don't, be sure to write down the date you received it. That way, you will know when it's time to get a new one. If the company sends out regular catalogs—say, every month or so—odds are you'll be dropped from the mailing list if you don't order regularly. That's okay; keep signing up for more catalogs.

Parts for Industry   203583

 http://www.partsforindustry.com/

See listing under Retail-Surplus Electronics.

RobotPartz.com   202564

http://www.robotpartz.com/

RobotPartz.com is a special "redirect" page to the to robotcentric catalog pages of the online merchant American Science & Surplus (see their listing in this section).

Servo Systems Co.   202599

115 Main Rd.
P.O. Box 97
Montville, NJ 07045-0097
USA

📞 (973) 335-1007
📠 (973) 335-1661
🚫 (800) 922-1103

✉ info@servosystems.com

 http://www.servosystems.com/

Servo Systems Co. is a full-service motion control distributor and robotic systems integrator. The Web site contains copious descriptions and technical data on their industrial components. The company also sells motion mechanicals, such as linear stages. Be sure to check out their "surplus bargains" pages for affordably priced servos and other gear.

✈ RETAIL-TRAIN & HOBBY

This section represents a fairly broad collection of retailers involved with motorized hobbies, particularly trains and radio-control cars, planes, and boats. Trains and hobby R/C are combined here because so many retail establishments do the same. While some model train components can be pressed into service for robotics, the prime focus is on R/C models, particularly servo motors and related hardware, batteries and battery chargers, miniature construction hardware and fasteners, lightweight wheels, gears and gear sets, and high-powered motors.

This section primarily lists online and mail-order sources. Don't forget the local train and hobby retailer. Prices may be comparable to what you can find mail order, with the added benefit of being able to buy and use all in the same day!

SEE ALSO:

POWER TRANSMISSION: Gears, bearings, belts, sprockets, chain, and complementary components

RADIO CONTROL (VARIOUS): Additional sources for hobby parts

SUPPLIES-GLUES & ADHESIVES: Special bonding cements

Ace Hardware Hobbies   204022

1854 Magnolia Ave
Burlingame, CA 94010
USA

✆ (650) 697-3383
📠 (650) 697-6801
🚫 (800) 383-2657
🌐 http://www.ace-hobbies.com/

R/C hobbies, with an emphasis on racing cars.

A2Z Toys.com
http://www.a2ztoys.com/
Hackables: Hobby Zone R/C Forklift; Hobbico chargers; Duratraxx batteries

Hobby Wholesale
http://www.hobbywholesale.com/
Kits, models, wood (balsa and ply), accessories

Hobby Works, Inc
http://www.hobbyworks.com/
Trains, R/C science and rockets

Lenz Agency
http://www.lenz.com/
Model railroad. DCC control; based in Germany

America's Hobby Center   202728

8300 Tonnelle Ave.
North Bergen, NJ 07047
USA

✆ (201) 662-2800
📠 (201) 662-1450
🚫 (800) 242-1931
✉ questions@ahc1931.com
🌐 http://www.ahc1931.com/

Full-line radio-control retailer for R/C (aircraft and racer) and model railroad. Also sells airship models that can be used to create a pilotless air-bot. Stores in the N.Y.-N.J. area.

Balsa Products   204191

122 Jansen Ave.
Iselin, NJ 08830
USA

✆ (732) 634-6131
📠 (732) 634-2777
🌐 http://www.balsapr.com/

Don't let the name fool you; Balsa Products sells more than just balsa wood. They also sell Hitec and GWS servos, batteries and battery chargers, adhesives, electronic speed controllers, and more.

Balsa Products Web site.

CVP Products   202396

P.O. Box 835772
Richardson, TX 75083-5772
USA

✆ (972) 422-2169
📠 (972) 516-9527

kggcvp@aol.com

http://www.cvpusa.com/

Model railroad: digital command control (DCC) systems. Check out their wireless throttle. Useful tutorials and how-tos on the site.

Digitrax   202397

450 Cemetery St., #206
Norcross, GA 30071
USA

(770) 441-7992

(770) 441-0759

 sales@digitrax.com

http://www.digitrax.com/

Makers of model railroad digital command control (DCC) systems.

Discount Train and Hobby   203968

211 E. Oakland Park Blvd.
Ft. Lauderdale, FL 33334
USA

(954) 561-0403

(954) 564-5367

(888) 62-2948

sales@discount-train.com

http://www.discount-train.com/

Hobby stuff: trains; R/C, paints, glues, and tools. Site claims over 20,000 items in stock. Also carries the Tamiya educational, robotics, and construction lines. Prices much lower than average.

DJ Hobby   204035

96 San Tomas Aquino Rd.
Campbell, CA 95008
USA

(408) 379-1696

http://www.djhobby.com/

OWI robot kits, Erector set, arts and crafts supplies, R/C servos, paints, adhesives, precision hand tools, Tamiya gearmotor kits.

DK Models, Inc   202824

St Georges Road
Boston
Lincolnshire
PE21 8RU
UK

+44 (0) 1205 367652

+44 (0) 1205 369949

 sales@dkmodels.co.uk

http://www.dkmodels.co.uk/

Mostly plastic model, but DK also sells Tamiya Educational kits, including gearmotors.

Dream Catcher Hobby, Inc.   203556

P. O. Box 77
Bristol, IN 46507
USA

(219) 523-1938

 webmaster@dchobby.com

http://www.dchobby.com/

R/C components, such as carbon-fiber push rods. Also:
- Miniature planetary gearbox
- High-speed 400-size motor
- Battery chargers
- Bell cranks

eHobbies   204028

http://www.ehobbies.com/

Over 50,000 items in 10 categories. Servos and accessories, gyros, tools and supplies, glues and adhesives, metal and plastic construction sheets, construction hardware (servo links, fasteners, etc.), foam and rubber wheels. All major brands. Search by product type or manufacturer.

eHobbyland   202776

1810 E. 12th. St.
Ste. C
Mishawaka, IN 46544
USA

(219) 256-1364

(219) 256-1213

⊘ (800) 225-6509

✉ sales@hobbylandinc.com

🌐 http://e-hobbyland.com/

Hobbies, toys, plastic models. Model trains.

🌐 💻 🛍

FMA Direct   203772

9607 Dr. Perry Rd.
Unit 109
Ijamsville, MD 21754
USA

☎ (301) 831-8980

📠 (301) 831-8987

⊘ (800) 343-2934

✉ sales@fmadirect.com

🌐 http://www.fmadirect.com/

R/C electronics, including servos (micro to 1/4 scale); batteries.

🌐 💻

Frontline Hobbies   203666

255 Hunter St.
Newcastle, NSW 2300
Australia

☎ +61 2 4929 1140

🌐 http://www.frontlinehobbies.com/

Hobbies: R/C, tools, materials (wood, metal, plastics), Meccano (kits and spares). There's also a user-to-user forum for asking questions.

🌐 💻

Hobby Barn, The   202844

P.O. Box 17856
Tucson, AZ 85731
USA

☎ (520) 747-3792

📠 (520) 747-3792

✉ info@hobbybarn.com

🌐 http://www.hobbybarn.com/

Hobby Barn specializes in R/C aircraft; for the roboticist you'll find servos, drive motors (with and without gearboxes), batteries and chargers, and piezo gyros.

🌐 💻 🛍

Hobby Club   204032

P.O. Box 6004
San Clemente, CA 92674
USA

☎ (949) 425-1362

📠 (949) 349-0829

⊘ (866) 739-5026

✉ hobbyclub@earthlink.net

🌐 http://www.hobbyclub.com/

R/C airplane motors (including Astro Flight), batteries, rechargers, Minicraft tools, servos (Futaba, Airtronics, Cirrus, and others).

🌐 💻

Hobby Lobby International, Inc.   202274

5614 Franklin Pike Cir.
Brentwood, TN 37027
USA

☎ (615) 373-1444

📠 (615) 377-6948

🌐 http://www.hobby-lobby.com/

Hobby Lobby offers a broad line of R/C products, including servos, CO_2 motors (for an unusual power plant for your robot), wheels, collars, push rods, and other hardware, motors, and gear drives.

Be sure to also check out their bulk order page for great quantity discount deals:

http://www.hobby-lobby.com/bulkone.htm

📄 🌐 💻

Web home of Hobby Lobby International.

Hobby Maker
202780

1424F Airport Fwy.
Bedford, TX 76022
USA

☎ (817) 267-0991
📠 (817) 685-9272
🚫 (800) 274-8076
📧 info@hobbymaker.com
🌐 http://www.hobbymaker.com/

Building supplies (paints, adhesives, etc.), hobby tools (Minicraft, Dremel, X-Acto), and entry-level science kits (including electronics theory and projects).

Hobby People
202731

18480 Bandilier Cir.
Fountain Valley, CA 92708
USA

☎ (714) 963-9881
📠 (714) 962-6452
🚫 (800) 854-8471
📧 service@hobbypeople.net
🌐 http://www.hobbypeople.net/

Hobby People is an online store with retail outlets in California and Nevada. They carry an extensive line of R/C parts, including servos, Dave Brown and Du-Bro wheels, Du-Bro hardware (such as blind nuts, collars, and machine screws), battery packs, chargers, and other accessories.

Hobby Shack

See Hobby People (this section).

Hobby Stores on the Net
204036

http://www.hobbystores.com/

Listing of local hobby stores that have Internet Web sites.

Hobby Stuff Inc.
203314

11239 E. Nine Mile Rd.
Warren, MI 48089
USA

☎ (586) 754-6412
📠 (586) 754-7402
📧 info@hobbystuffinc.com
🌐 http://hobbystuffinc.com/

Hobby Stuff deals with the in-ordinary . . . stuff few others carry. They sell unusual hardware pieces for R/C airplanes that might be used in a robot, along with vacuum formers (you supply the vacuum and the heat source) for making your own formed parts using thin plastic sheets. The prices won't make you gag.

Hobbybox
203166

P.O Box 60
Braeside
Victoria 3195
Australia

📠 +61 3 9580 9295
📧 info@hobbybox.com.au
🌐 http://www.hobbybox.com.au/

Modelers tools and supplies, including miniature tools (vice, screwdrivers, saws, etc.) and small hardware.

Hobbyco Pty. Ltd.
203665

Reply Paid Q99
Queen Victoria Building
NSW 1230
Australia

☎ +61 2 9221 0666
📠 +61 2 9221 0710
📧 info@hobbyco.com.au
🌐 http://www.hobbyco.com.au/

Online sellers of hobby stuff. Offers R/C products, plastic kits, tools and supplies, craft supplies, educational (including Meccano [sets and spares], LEGO, and Tamiya Educational). Also components: gear kits, materials (plastic, balsa, metal).

Hobbyhoo-Hobby Site Search Engine
300006

http://www.hobbyhoo.com/

Links to various hobby sites on the Web. Divided into categories.

Hobbylinc.com 202715

76 Bay Creek Rd.
Ste. P
Loganville, GA 30052
USA

☎ (770) 466-2667
📠 (770) 466-0650
🚫 (888) 327-9673
✉️ hobbylinc@hobbylinc.com
🌐 http://www.hobbylinc.com/

Hobbylinc sells a wide range of hobby products. See listing under Kits-Electronic.

Hobby's 203675

Knight's Hill Square
London
SE27 0HH
UK

☎ +44 (0) 2087 614244
📠 +44 (0) 2087 618796
✉️ hobby@hobby.uk.com
🌐 http://www.hobby.uk.com/

Model hobby wholesalers. Parts, kits, fasteners, glues, metal and plastic sheet, tools (including Unimat Basic and Classic miniature lathes).

Little Shop of Hobbies 203965

2309 North Duck Lake Rd.
Ste. #2
Highland, MI 48356
USA

☎ (248) 889-0420
📠 (248) 889-0420
✉️ info@littleshopofhobbies.com
🌐 http://www.littleshopofhobbies.com/

Hobby R/C, trains, and science sets. Sells some Tamiya Educational products, too.

Main Hobby Center Inc., The 203967

1011 Commerce Blvd.
Park Center Plaza
Dickson City, PA 18519
USA

☎ (570) 489-8857
📠 (570) 383-9517
✉️ mainhoby@epix.net
🌐 http://mainhobby.com/

Model tools and supplies; paints and adhesive; Tamiya Educational products (motors, mechanicals), and Robotix construction sets.

Major Hobby 204034

1520 B Corona Dr.
Lake Havasu City, AZ 86403
USA

☎ (928) 855-7901
🚫 (800) 625-6772
✉️ majorhobby@majorhobby.com
🌐 http://www.majorhobby.com/

Online retailer; attractive prices on servos (Hitec, Futaba), rate gyro, NiMH and ni-cad batteries, battery chargers, electronic speed controllers. Monthly specials may include heavily discounted servos.

National Hobby Supply 202733

1975 South Cobb Dr.
Marietta, GA 30060
USA

☎ (770) 333-0190
🚫 (800) 437-2736
✉️ lmmiele275@cs.com
🌐 http://www.nationalhobbysupplyinc.com/

Full line of R/C and model train. Local store in Marietta, Ga.

Radio Model Supplies 202967

235 Albany Highway
Victoria Park
Western Australia 6100
Australia

☎ +61 8 9362 2133
📠 +61 8 9362 2054
✉️ info@radiomodels.com.au
🌐 http://www.radiomodels.com.au/

Model R/C parts (servos, hardware, tires). The company also sells Tamiya Educational parts, a line that includes affordable gearmotor kits.

Sheldons Hobbies 202734

2135 Oakland Rd.
San Jose, CA 95131
USA

((408) 943-0220
⊘ (800) 822-1688
🌐 http://www.sheldonshobbies.com/

Full line of R/C parts, including JR servos, Cirrus servos, Airtronics servos and gyros, Du-Bro hardware, and Dave Brown wheels.

Tower Hobbies 202126

P.O. Box 9078
Champaign, IL 61826-9078
USA

((217) 398-3636
⊘ (800) 637-6050
🌐 http://www.towerhobbies.com/

Tower Hobbies is a leader in mail-order R/C (air, ground, water) products and accessories. They offer all major brands, and prices are competitive. However, you'll want to comparison shop after you've ordered at least one item from Tower; then you get their "preferred customer" catalog, along with coupons for ongoing discounts.

Tower operates several informational Web sites for the benefit of consumers:

Web page for Tower Hobbies.

http://www.sciencehobby.com/
http://www.hobbies.net/

TrainTown Hobbies and Crafts 203966

29 North Main St.
Batesville, IN 47006
USA

((812) 933-0274
🌐 http://www.traintownhobbycrafts.com/

Despite the name, TrainTown carries lots more than just trains. Their hobby shop products include science kits, Tamiya Educational products (like the Wall Hugging Mouse and several motor kits), and mechanical parts, including the Tamiya ball caster kit.

Uptown Sales Inc. 204033

1242 Commons Ct.
Clermont, FL 34711-6513
USA

((352) 243-5985
✆ (352) 243-5987
⊘ (800) 548-9941
✉ help@usahobby.com
🌐 http://www.hobbyplace.com/

Stocks metal tubing, rod and others shapes (brass, aluminum, copper), basswood panels, electronics teaching labs, small hand tools, wide array of brass fasteners, paints, and adhesives.

Worth Marine Inc. 203032

6 Barnard St.
Marblehead, MA 01945
USA

((781) 639-1835
✆ (781) 639-0936
🌐 http://www.worthmarine.com/

Model boats and hardware. Check out their hardware section for unusual fasteners, as well as miniature block and tackle.

ROBOTS

You'll find real, working robots in the sections that follow. Some are available in ready-made or kit form, while others are made for research and cost as much as a luxury car. Sections include:

- Low-cost BEAM robots, many of which are made from materials found around the house
- Hobby and kit robots; either is a great way to start learning about robotics
- Personal robots for mowing the lawn or vacuuming the floor
- Walking robots, which step instead of roll around their environment

SEE ALSO:

INTERNET-PLANS & GUIDES: Other robot builders share their designs with you

KITS-ROBOTIC: Kits for building toy and simple robots

TOYS-CONSTRUCTION AND TOYS-ROBOTS: Some robots can be toys, and some toys can be robots

ROBOTS-BEAM

BEAM stands for "Biology Electronics Aesthetics Mechanics," a philosophy of robotics that suggests-when it's all said and done-smaller is better than larger, simpler is better than complex, and cheaper is better than expensive. It's not surprising then that BEAM robotics has many followers. It's a fun, fast, and relatively inexpensive way to get involved in robotics.

While most BEAM 'bots are extremely simplistic and do just one or two core tasks, a challenge of this style of robotics is how to make the machine more robust, without making it more complex, heavy, or expensive. Turns out this is very hard to do.

The sources in this section sell or support BEAM robotics. Listings include Web sites with free BEAM robot plans, as well as retailers who sell kits for building BEAM robots.

Andy Pang's Homepage 202381

http://home.ust.hk/~bcandyp/

Andy is into BEAM robotics. Site includes BEAM robot examples, pictures, and tutorials.

BEAM Beastiary 202379

http://bestiary.solarbotics.net/default.htm

Something of a "zoo" for various BEAM-style robots that have been built. Interesting classification system defines categories of BEAM robots as Sitter, Squirmer, Slider, Crawler, Jumper, Roller, Walker, Swimmer, or Flier.

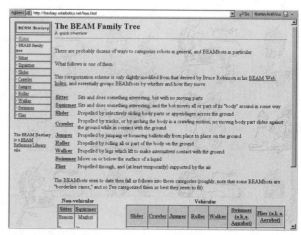

BEAM Beastiary

BEAM Four-Legged Walker 203360

http://www.buildcoolstuff.com/topics/BEAM_Walker.html

BEAM walker instructions.

BEAM Hexapod 203009

http://members.tripod.com/sparkybots/hexapod.htm

A relatively heavy-duty BEAM walking robot. Circuit example and construction pictures.

BEAM Online 202380

http://www.beam-online.com/

BEAM Online is a BEAM-specific portal for robots. Includes news, links, and a gallery of various BEAM robos.

BEAM Robotics 202376

http://www.nis.lanl.gov/projects/robot/

This is the main site to Mark Tilden's LANL (Nonproliferation and International Security) page. Mark is responsible for the BEAM concept, and here we see his philosophy of BEAM-though in fairly typical Tilden-speak: "The idea is to improve robo-genetic stock through stratified competition and have an interesting time in the process." I think he means scientists just want to have fun.

You'll also find links to other Web sites, where to go for more information, and news of upcoming BEAM competitions.

Web home of Mark Tilden, creator of BEAM.

HMBOTS 203433

http://www.hmbots.homestead.com/

BEAM 'bots spoken here. The site provides a number of BEAMish circuits and BEAM robot construction ideas. There's also a very well illustrated guide-with close-up color pictures-to modifying an R/C servo for continuous rotation. One of the better sites I've seen.

MOVERS AND SHAKERS

Mark Tilden
http://www.nis.lanl.gov/projects/robot/

Can a bunch of little robots do the work of one big robot? Mark Tilden, creator of the "BEAM" concept in robotics, thinks so. *BEAM* stands for "Biology Electronics Aesthetics Mechanics," a confluence of attributes that derives locomotion and behavior from nature, with simple electronics and mechanics. Today BEAM robots are popular kits for beginners.

JCM Inventures/JCM Electronic Services 202406

3335 Caribou Dr. NW
Calgary, AB
T2L 0S4
Canada

☎ (403) 284-2876

🌐 http://www.jcminventures.com/

JCM makes and sells the Cybug and BEAM robot kit with a decided buggy appearance. I like the red QueenAnt the best.

Miller Microcore 202378

http://microcore.solarbotics.net/

What's a "microcore"? This Web site answers the question everyone asks.

Robotics and Electronics Links 202382

http://beamlinks.botic.com/

Lots of links to BEAMish robotics sites. Categorized by BEAM robot type, such as photovore and walker, and by content.

Rug-Bug BEAM Photovore 202405

http://www.acesim.com/beam/rugbug.html

Ken Hill's design for a BEAM photovore: "An environmentally aware 'virtually alive' photovore robot bug following the BEAM philosophy. It walks (well, rolls for now), talks-(chirps in response to light level), sees-(photo resistors sense brightness to guide it), feels-(via touch sensing feelers of course!), feeds-(coming soon via solar cells charging rechargeable batteries-once enough light is located, bot will go to 'sleep' to recharge)."

Solarbotics Ltd. 202031

179 Harvest Glen Way N.E.
Calgary, AB
T3K 4J4
Canada

☎ (403) 818-3374

☏ (403) 226-3741

⊘ (866) 276-2687

🌐 http://www.solarbotics.com/

Solarbotics is a primary retailer of BEAM robots, both in kit and ready-made form. BEAM robotics was invented by Mark Tilden in 1989 and follows a "simple is better" approach to design. Though simple, many BEAM robots are actually quite sophisticated in their function. Products include various light-attracted bugs and walking robot kits, motors, solar cells, and electronics.

The Solarbotics site also provides articles on BEAM, a gallery of BEAM robots built by others, and links to other important BEAM sites. Also check out http://www.solarbotics.net/ for more BEAM resources.

Solarbotics Photopopper. Photo Solarbotics Ltd.

Western Canadian Robot Games   202093

http://www.robotgames.com/

The Western Canadian Robot Games is one of the oldest robot competitions, with events that include sumo wresting, something called atomic hockey, a hallway navigation game for walking robots, and a series of challenges specially designed for BEAM robots. The competitions are held annually in Alberta, Canada.

Wilf's E-BEAM   202377

http://wilf.solarbotics.net/

Semiregular online articles about BEAM robotics by Wilf Rigter. Some very nice and detailed discussions of BEAM electronics and design philosophy.

ROBOTS-EDUCATIONAL

Reading books on robot theory is fine, but what better way to learn about robotics than to build or use one? The listings in this section are for robots designed to teach about robotics. Most educational robots are designed (and priced) for classroom study, sometimes by one student at a time, but more commonly, by groups of students. The typical educational robot comprises both a hardware and a software component; the software allows the robot to be programmed for various functions.

ActivMedia: AmigoBots   203719

http://www.amigobot.com/

All about the AmigoBot!, from ActivMedia Robotics. AmigoBot! is a prebuilt (not a kit) advanced personal robot suitable for education and competitions. AmigoBot! isn't cheap by any means and is better suited for a school curriculum, a lottery winner, or Bill Gates's kids.

The robot hardware is supported by a wide range of software:

- Basic Suite (ARBS), a set of plug-n-play robot software modules.
- WorldLink, transforms your PC into a robot server and "chat station."
- WorldPass remote control for telerobotics over the Internet.
- Mapper, for building navigation maps of rooms and floor plans, for use in navigating the robot.
- Simulated Robots, for hands-on PC simulations of common robotic tasks, including navigation.

See also the main ActivMedia Robotics Web site at: http://www.activmedia.com/

Advanced Design, Inc./Robix   202476

6052 N. Oracle Rd.
Tucson, AZ 85704
USA

☏ (520) 575 0703

☏ (520) 544 2390

✉ desk@robix.com

🌐 http://www.robix.com/

Robot lab trainer. The "Rascal robot construction kit" uses servos and modular links to build different kinds of robots. Very creative kit.

Angelus Research 203473

11801 Cardinal Cir.
Unit J
Garden Grove, CA 92843
USA

- (714) 590-7877
- (714) 590-7879
- dgolding@angelusresearch.com
- http://www.angelusresearch.com/

Angelus Research makes and sells a line of robots for industry, education, and hobby. Among their robot family are:

- Whiskers, the intelligent robot (educational)
- Intruder Robot (military/law enforcement)
- Piper, the pipe inspection robot (commercial/industrial)
- Stockboy AGV (commercial/industrial)
- Bugsy AI (low-cost educational)

Arrick Robotics 202558

P.O. Box 1574
Hurst, TX 76053
USA

- (817) 571-4528
- (817) 571-2317
- info@robotics.com
- http://www.robotics.com/

See listing under Actuators-Motion Products.

BBC Robots 202525

http://www.bbc.co.uk/science/robots/

BBC Television robot lab. From the Web site: "Build-A-BotTechlab is ready for action and better than ever! All new robot building assignments, an interactive glossary to help out with those tricky words plus new audio tracks in the music player. Enter the facility now and start building Robot-Walker and Creeper-Bot, the very latest additions to the series."

Sources for purchasing the robot kits provided on the Web site.

EduRobot

- http://www.edurobot.com/

See listing for General Robotics Corporation (this section).

General Robotics Corporation 203033

760 South Youngfield Ct.
Lakewood, CO 80228-2813
USA

- (303) 988-5636
- (800) 422-4265
- sales@generalrobotics.com
- http://www.edurobot.com/

Old robots never die, and some don't even fade away. The RB5X robot was first introduced in the 1980s and is still available today as an education platform. Technical documentation and curricula are provided with the robot; some information can be downloaded for review prior to purchasing.

See also the home page of the parent company at:

http://www.generalrobotics.com/

Other products include the RobotLab & Component Kit and the Robot Arm Module.

Wany Robotics 203467

CEEI Cap Alpha Avenue de l'Europe
Clapiers 34940 Monpellier
Cedex 9
France

- +33 (0) 467 593626
- +33 (0) 467 593010
- contact@wany.org
- http://www.wany.fr/

Robot manufacturer: Pekee robot, Robotic and Educational autonomous platform.

In the words of the Web site: "Developed with the teachers' consultation, the Pekee robot is an educational, evolutional platform. Multi-teaching and multi-levels, the Pekee robot will allow you to fascinate your students during your teachings on electronics, computing, engineering or sciences and technologies."

Web site is in English and French.

ROBOTS-EXPERIMENTAL

Some of the robots that follow are for purchase, others are for show, and still others are for showing off. But all are interesting and offer great insights and ideas.

Compaq Robot Controller Specification
202231

http://www.crl.research.digital.com/projects/
  personalserver/rcc-spec.htm

Specifications of an engineering prototype of a general-purpose robot controller interface. Features include:

- 20 motor drivers
- 5 servo drivers
- 32 analog-to-digital ports
- 6 buttons (4 connected to A2Ds, 2 for resets)
- 3 ports (2 connected to A2Ds, 1 for LCD)
- LCD connector

Dance of the Water Spiders
202039

http://www.remo.net/spiders/

Here's what the Web site has to say: "Eight giant robotic spiders, programmed with artificial instincts (and painted in Beetle colors by Scott Volkswagen), dart about the waterways in a spirited ballroom dance. Pioneer cyber artist, Remo Campopiano, designed and built the spiders with the help of the Robotics Art Club of New England, a group of 10-13 year olds."

Sarcos Inc.
203938

360 Wakara Way
Salt Lake City, UT 84108
USA

✆ (801) 581-0155
📠 (801) 581-1151
✉ sarcosinfo@sarcos.com
🌐 http://www.sarcos.com/

Sarcos is a research and development lab located in Utah that develops industrial (telerobots and medical), educational, and entertainment robots. The latter include animated pneumatic androids and singing toucan birds. Videos and pictures of many of their entertainment robots are provided on the site.

Xerox Palo Alto Research Center (PARC)
202017

http://www.parc.xerox.com/spl/
  projects/modrobots/

Web site of the Xerox PARC Modular Reconfigurable Robotics group. In the words of the site: "Modular Reconfigurable Robotics is an idea about how to build robots for various complex tasks. Instead of designing a new and different mechanical robot for each task, you just build many copies of one simple module. The module can't do much by itself, but when you connect many of them together you get a system that can do complicated things. In fact, a modular robot can even reconfigure itself-change its shape by moving its modules around-to meet the demands of different tasks or different working environments."

ROBOTS-HOBBY & KIT

In this section you'll find robots you can buy, either ready-made or in kit form. The size and style of robots are varied, from miniature bots that fit in the palm of a child's hand, to behemoths intended to rip the batteries out of other robots. Prices vary, too, from just a handful of dollars to several thousands of dollars.

Many of the robots and kits listed here are ready to go, but some may require the addition of a microcontroller or other support computer. Several robots (ready-made and kit) are designed for use with third-party microcontrollers, primarily the Basic Stamp II. To get full use of the robot or kit, you must purchase the microcontroller, which obviously adds to the cost.

SEE ALSO:

KITS-ROBOTIC: Additional robot kits, mostly toy size

TOYS-ROBOT: Robotic toys, either for play or for hacking

BBC Robots
202525

http://www.bbc.co.uk/science/robots/

See listing under **Robots-Educational**.

Blue Bell Designs Inc.
202501

P.O. Box 446
Gwynedd Valley, PA 19437-0446
USA

(215) 643-7012

 harry@bluebelldesign.com

 http://www.bluebelldesign.com/

Preassembled robotic platform, equipped Basic Stamp 2p40. No assembly required.

Photo Blue Bell Design Inc.

Budget Robotics

204255

P.O. Box 5821
Oceanside, CA 92056
USA

 (760) 941-6632

 info@budgetrobotics.com

 http://www.budgetrobotics.com/

Disclosure alert! This is my company, and I won't hide that fact. I didn't start out with any notion of selling goods to the public. But after seeing many gaps in the product offerings of companies and the high prices some specialty retailers ask, I decided to offer an alternative.

Budget Robotics sells a narrow range of product, with an emphasis on structural elements, such as robot bodies and frames. Much of it is custom-made and is explicitly designed for affordable amateur robotics. We also buy some hard-to-find items in bulk and provide them in smaller quantities for your convenience.

Products include:

- Low-cost metal framing components, to build strong robots at a minimal cost
- Precision-cut robot bases, in different colors and shapes, to make unusual robot bodies
- Omnidirectional wheels, in a variety of sizes and styles
- Affordable new (not surplus) gearmotors
- Strain gauge sensors, at prices that won't strain your wallet

- Specialty sensor material such as Kynar piezo-electrics
- Robot Parts Play kits-assortment of useful small robots for robot building
- Spherical casters (these aren't ball transfers and are quite unique)
- Specialty fasteners (lots of nylon pieces), hardware (like hex spacers), and more

Bug'N'Bots

204169

General Delivery
Laurel, ON
L0N 1L0
Canada

 (519) 940-0216

Golem Robotics

http://www.golemrobotics.com/
Robot that play soccer

The Exploratorium

http://www.exploratoriumstore.com/
Museum gift store in San Francisco; science toys, including OWI/Moveit robots

Inception Systems Inc.

http://www.inceptionsystems.com/
PC-based robotics and other kits

Panmanee

http://www.panmanee.com/
Robotic kits; CNC; stepper motors

RoboStuff

http://www.robostuff.com/
BotBones robot kits; heavy-duty metal frames; Pittman motors

Science Electronics

http://www.science-electronics.com/
Robot kits (OWI), Arrick Robots reseller, ActivMedia

Wizard.org

http://www.wizard.org/
Varous new and surplus electronics and robotics

stevejones@bugnbots.com

http://www.bugnbots.com/

Mail-order 'bots; mostly BEAM kits. Also BasicX micro-controller and NetMedia's SRV8-T serial servo board.

Competition-Robotics   203150

P.O. Box 1178
Swindon
SN25 4ZL
UK

  +44 (0) 1793 636119

  +44 (0) 1793 705772

sales@competition-robotics.com

http://www.competition-robotics.com/

Sells robot kits, sensors, microcontrollers for small robots, with an emphasis on sumo and similar competitive robot games.

Creative Learning Systems, Inc.   202427

10966 Via Frontera
San Diego, CA 92127
USA

  (858) 592-7050

  (858) 592-7055

  (800) 458-2880

info@clsinc.com

http://www.clsinc.com/

Sells many Fischertechnik kits and à la carte parts.

Also, Capsela and MOVITS kits. Other science and technical educations kits, books, and products.

Didel   203137

Belmont/Lausanne
CH-1092
Switzerland

  +41 21 728 6156

  +41 21 728 6157

info@didel.com

http://www.didel.com/

Maker of various microrobots, including the Swibot-SST (yes, the company is Swiss); MSwibot-Stamp (Basic Stamp based) Microkit, which contains various small

mechanical parts; and a PIC learning system. Individual parts sales of gears, shafts, clips, little motors (including an ultra-mini stepper originally designed for automobile instrumentation), and optical encoder wheels. Prices stated in Swiss francs, Euros, and U.S. dollars.

Web site is in English and French.

Diversified Enterprises   204245

158 Aero Camino Rd.
Santa Barbara, CA 93117
USA

  (805) 740-1852

http://www.robotalive.com/

Diversified makes and sells the Descartes tabletop robot, which incorporates touch and light sensors, coaxial (two-wheel) drive, and is powered by a Basic Stamp II microcontroller (other microcontrollers can also be used). The company also sells RF communica-

The Descartes robot kit. Photo Diversified Enterprises.

The Pherobot robot kit. Photo Diversified Enterprises.

tion modules as well as the Pocket-Bot miniature educational robot.

EasyBot   202534

Michael Berta Ent
P.O. Box 235
Gustine, CA 95322
USA

✉ easybot@earthlink.net

🌐 http://www.easybot.net/

Prebuilt robot chassis. Base, wheels, motors, battery. Stackable levels. Skids under fore and aft battery compartments (leaves only a tad for floor clearance).

EasyBot kit. Photo Michael Berta

Fred Barton Productions, Inc.   202683

P.O. Box 1701
Beverly Hills, CA 90213-1701
USA

☎ (310) 234-2956

📠 (310) 234-0956

✉ tobor1701@earthlink.net

🌐 http://www.the-robotman.com/

See listing under Internet-Reference.

Hyperbot   202425

905 South Springer Rd.
Los Altos, CA 94024-4833
USA

☎ (415) 949-2566

🚫 (800) 865-7631

✉ hyperbot@hyperbot.com

🌐 http://www.hyperbot.com/

Sells its own line of robotic hardware and software products and selected compatible robotic kits and software from other companies. Custom products include CHiP, a programmable technically oriented self-contained robot.

Also sells many Fischertechnik kits.

Johuco Ltd.   203474

Box 385
Vernon, CT 06066
USA

✉ thefolks@johuco.com

🌐 http://www.johuco.com/

Reasonably priced mobile educational robots. Products include Phoenix II, Muramator, G.R.A.K, and Photovore. Also: some interesting papers on robot control and behavioral robotics.

The Phoenix II. Photo Johuco Ltd.

Joker Robotics   202563

ABN 60 354 369 197
1 Warreen Place
City Beach
Western Australia 6015
Australia

✉ info@joker-robotics.com

🌐 http://www.joker-robotics.com/

Assembled and kit robotics; full-featured. Products include:

- EyeCon
- SoccerBot
- Android
- EyeWalker
- Rug Warrior
- Mobotix Cam

International distributors also.

Kadtronics 202220

 (321) 757-9280

 http://www.kadtronix.com/

Robotics platforms (made from plastic chassis boxes), Robot Design Manual (book).

Lynxmotion, Inc. 202034

PO Box 818
Pekin, IL 61555-0818
USA

(309) 382-1816

(309) 382-1254

sales@lynxmotion.com

http://www.lynxmotion.com/

Lynxmotion sells high-quality kits and parts for mobile robots—both wheeled and legged. See listing under Retail-Robotics Specialty.

Mecarobo: Educational Robots 203132

Somerset House
40-49 Price St.
Birmingham
B46LZ
UK

+(44) (0) 2074 139583

What to Look for in a Robotics Kit

Robot kits offer an ideal way to learn about the science of robot building. Rather than having to gather all the bits and pieces yourself, a kit lets you concentrate on the building and programming aspects. You don't need to take numerous trips to the hardware store, and—depending on the kit—you don't even need to pick up a saw, sander, or drill. (Of course, there are many robot tinkerers who enjoy this aspect the most. It's all in your perspective.)

If you're looking to explore the world of amateur robotics with a kit, here are some pointers to keep in mind to help you decide which one is best for you.

Obviously, cost is a consideration, and most kits fall into three categories.

- At the low end, costing from $20 to $100, is the kit for a basic nonprogrammable or simple programmable robot. Examples in this category are BEAM robots (see **Robots - BEAM**). These make for good starter kits, especially for younger robobuilders.

- The middle ground is the $100 to $300 kit. The Parallax BOE-Bot (*BOE* stands for "Board of Education")usually comes with a more sophisticated means of programming. These kits are perfect for junior high and high school robotics studies.

- The high end comprises specialty kits, such as the walking robots from Lynxmotion or the heavy-duty platforms from Zagros. Prices may vary from a low of $300 to several thousand dollars. They are intended for the serious robotics hobbyist or for educational purposes.

Once you've decided on the price range that suits you, the next step is comparing features. You can judge features based on what the kit comes with (two motors, two wheels, etc.), or what the completed robot does. If you are looking to learn behavior-based programming in robotics, a robot kit that does not permit you to change its built-in programs will not be of much use to you, no matter what hardware it comes with.

Finally, if possible, ask for an electronic copy of the instruction manual that comes with the kit. Many that I've seen are poorly written, and difficult to understand. It can be frustrating enough troubleshooting a belligerent robot; you don't need poor Yoda-ish documentation worsening the situation.

info@mecarobot.com

http://www.mecarobot.com/

Robot kits and toys; Web site is in French and English.

Mekatronix, Inc. 202970

316 NW 17th St.
Suite A
Gainesville, FL 32603
USA

tech@mekatronix.com

http://www.mekatronix.com/

From the Web site: "Mekatronix is a manufacturer of autonomous mobile robots, robot kits, microcontroller kits and robot accessories, as well as educational materials related to science and robotics. Our robots and microcontrollers provide students with valuable hands-on experience in programming and engineering concepts."

Products are available through a few dealers.

Microbtica, S.L. 204078

c/ General Moscard, 7 1B
Madrid, 28020
Spain

info@microbotica.es

http://www.microbotica.es/

Microbotica develops and sells single board microcontrollers for small mobile robots, as well as LEGO-based kits. Their CT6811 board uses a Motorola 68HC11 microcontroller; memory expansion and servo control daughter boards are also available.

The company's Tritt and Tritton kits use LEGO Technic pieces for construction; the kits include the controller, LEGO parts, servomotors, and other hardware.

The Web site is in Spanish and English. A Spanish-language Yahoo! eGroups list for the products is available at:

http://www.egroups.com/group/microbotica/

Microrobot NA Inc. 203939

P.O. Box 310
451 Main Street
Middleton, NS
0S 1P0
Canada

Milford Instruments

Located in Leed, England, Milford Instruments caters to hobbyists and engineer tinkerers. In addition to reselling the Basic Stamp and various Scott Edwards LCD panels, the company offers several unique robotics kits, including several hexapod walking robots, an animated talking head, and a fully jointed stationary arm.

StampBug. The StampBug uses a Basic Stamp 1 (a Basic Stamp 2 version is also available) for control and sensing. Feelers on the robot detect obstacles; when something is in the way, the robot changes course and continues moving. Flashing LEDs act as eyes. The Stamp comes preloaded with a standard roaming program and, of course, can be reprogrammed. See also Hextor, a hexapod with 2 DOF (two servos per leg).

Alex-Animated Head. Alex moves "his" eyes and mouth in response to a 20-second sound clip. All movements are recordable and are activated by ordinary R/C servos (included with the kit). The head turns and nods, eyes swivel, and the lips move.

One of Milford's robotic creations.

TecArm6. The TecArm6 is a complete robotic arm system. Large-scale servos are used for the waist, shoulder, and elbow joints; additional (smaller) servos are used for the gripper and wrist actions. In all, the arm sports five degrees of freedom, not including the opening/closing of the gripper.

Also available:

- LCD displays
- Stepper motor drivers
- Servo drivers
- Ultrasonic ranging modules
- Speech synthesis and recognition modules
- PIC development tools

Milford Instruments can be found online at:

http://www.milinst.com/

---

(902) 825-1726

(902) 825-4906

(866) 209-5327

info@microrobotna.com

http://www.microrobotna.com/

Robot kits, microcontroller boards, and parts for sumo, soccer, and line-following competition 'bots. For example, the company's Robo-Lefter is a maze-solving MicroMouse. (Its name is derived from the left-turn maze-solving algorithm it uses.) The products are available through distributors or directly from Microbot NA.

Norland Research   203145

 (702) 498-5799

 (702) 435-9437

✉ see3peoh@ix.netcom.com

🌐 http://www.smallrobot.com/

Norland offers a small plastic (1/8-inch ABS) robot chassis named S.A.M. reminiscent of the Topo and Bob robots from the old Androbot company of the mid-1980s.

S.A.M. from Norland Research. Photo Rebbecca Rowland.

Parallax, Inc.   202149

599 Menlo Dr.
Ste. 100
Rocklin, CA 95765
USA

((916) 624-8333

↪ (916) 624-8003

🚫 (888) 512-1024

✉ info@parallaxinc.com

🌐 http://www.parallaxinc.com/

See listing under Microcontrollers-Hardware.

The Parallax Toddler walking robot. Photo Parallax, Inc.

Real Robots (Cybot)   202529

http://www.realrobots.co.uk/

Real Robots is a twice-monthly ("every fortnight") magazine that combines full-color how-to construction, along with the parts to built a 'bot.

See also http://www.cybotbuilder.com/, an independent Web page supporting the Real Robots product.

Roboblock System Co., Ltd.   204072

137-070, #707 ilkwang Building
1656-2, Seocho-Dong
Seocho-Gu
Seoul
South Korea

(+82 2-597-8224

↪ +82 2-597-9441

✉ info@roboblock.co.kr

🌐 http://www.roboblock.com/

Sells Roboblock kits, mobile robot kits, toy robot kits, microcontrollers, AVR C and Basic programming software, parts, and other robot/electronics products. Web site is a mix of English and Korean.

Robodyssey Systems LLC   202977

20 Quimby Ave.
Trenton, NJ 08610
USA

(609) 585-8535

(609) 585-8535

info@robodyssey.com

http://www.robodyssey.com/

Makers of walking and rolling robotic platforms, intended primarily as educational platforms. Interesting designs.

Robotikits Direct 202477

17141 Kingsview Ave.
Carson, CA 90746
USA

(310) 515-6800

(310) 515-0927

robotikitsdirect@pacbell.net

http://robotikitsdirect.com/

Robotikits Direct is the U.S. importer and distributor for the OWIKit and MOVIT robot kits. The company sells the product online as well. Kits are divided into skill level: Beginner, Intermediate, and Advanced, and those with/without soldering.

Robotix 203400

Learning Curve International
314 West Superior St.
6th Fl.
Chicago, IL 60610-3537
USA

(312) 981-7000

(312) 981-7500

(800) 704-8697

cs@learningcurve.com

http://www.robotix.net/

To call Robotix a "construction toy" would be insulting, but in fact, the Robotix series construction sets are sold as toys. Originally developed by gamemeisters Milton-Bradley, Robotix is now within the fold of Learning Curve, a company heavily involved in the mechanical construction toys market (other Learning Curve brands include Lionel trains).

Robotix consists of snap-together plastic beams, where electric motors can serve as joints or rotating wheels. The beams are very sturdy, and it's possible to construct a walking Robotix robot that measures over 18 inches tall. The motors are operated either by a wired controller or by remote control. The motors are quite powerful as they have their own gear reduction built-in. The Discovery Channel series of Robotics sets offer a number of unique designs, including insect-style walkers.

I have some Robotix sets going back to the mid-1980s, and I cherish them. Every once in a while, I'll see incomplete sets sold at thrift stores, garage sales, even online at eBay. Though incomplete, the parts from the sets are interchangeable, allowing for the construction of even bigger and better robots.

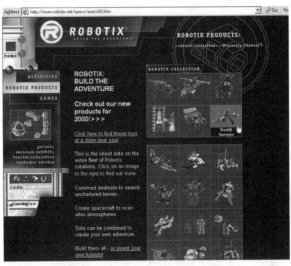

Robotix online.

Robots Wanted: Dead or Alive, Whole or Parts 202280

http://www.robotswanted.com/

Recycle those robots. Possible museum pieces include Heathkit HERO, Androbots (Topo, BOB), RB5X, Omnibots, Rhino Robots, MAXX STEELE, Marvin Mark I, Hearoid, and Turtles.

Seattle Robotics 204208

19336 133rd Ave. SE
Renton, WA 98058
USA

(253) 630-9836

(253) 630-9914

info@seattlerobotics.com

http://www.seattlerobotics.com/

Not to be confused with the Seattle Robotics Society (no relation), Seattle Robotics markets hobby and research robots and cutting-edge robotics add-ons, such as the CMUcam vision sensor.

Tab Electronics Build Your Own Robot Kit

300020

http://www.tabrobotkit.com/

This is the support page for the Tab Build Your Own Robot Kit, a smartly designed robot kit you can buy from many book retailers, including Barnes & Noble. Created by electronics mavens Myke Predko and Ben Wirz, the kit provides an out-of-the-box experience in building and playing with a small two-motor robot. The kit is designed so that you can add an optional Basic Stamp microcontroller for custom programming.

Tab Electronics Build Your Own Robot Kit Support site for the Tab Electronics Build Your Own Robot Kit.

Tamiya America, Inc. -Educational

203143

Attn: Customer Service
2 Orion
Aliso Viejo, CA 92656-4200
USA

((949) 362-6852
Ø (800) 826-4922
🌐 http://www.tamiyausa.com/
   product/educational/

See listing under Actuators-Motors.

Thinker Toys

202856

P.O. Box 6297
7th and San Carlos Ave.
Carmel, CA 93921
USA

((831) 624-0441
✆ (831) 624-0551
✉ info@thinkertoys.com
🌐 http://www.thinkertoys.com/

Wide variety of toys; see listing under Retail-Science.

Tin10.Com

204073

P.O. Box 7711
Pittsburgh, PA 15215
USA

✉ support@tin10.com
🌐 http://tin10.com/

"Robots as educational tools."

Zagros Robotics

202362

P.O. Box 460342
St. Louis, MO 63146-7342
USA

((314) 768-1328
✉ info@zagrosrobotics.com
🌐 http://www.zagrosrobotics.com/

The main product line at Zagros is robotics bases, which include the plastic base plate of the robot, and two motors. They also offer electronic components:

* Sensors-Polaroid 6500 ultrasonic ranger; Dinsmore digital compass; Sharp GP2D02 infrared
* Computers-Eyebot Controller; HC11 MC processor with backplane; 386 Single Board Computer

Robot base from Zagros Robotics. Photo AJ Neal, Zagros Robotics.

- Miscellaneous-H-bridge ICs; 300 MHz RF remote control; DC gearmotors

ROBOTS-INDUSTRIAL/RESEARCH

The robots in this section are commercially available automatons intended primarily for either industrial applications (e.g., sentry robots) or for robotics research. These robots aren't cheap; even if you're not in the market for a high-end robot, the listings that follow are good resources for both inspiration and technical information.

SEE ALSO:

> **INTERNET-EDU/GOVERNMENT LAB:** Additional robots used in research

ActivMedia Robotics, LLC   202859

44-46 Concord St.
Peterborough, NH 03458
USA

((603) 924-9100

⌧ (603) 924-2184

⊘ (800) 639-9481

✉ robots@activmedia.com

🌐 http://www.activmedia.com/

From the Web site: "ActivMedia Robotics designs, integrates and manufactures intelligent mobile robots and their navigation, control, sensing and response systems. Our robots are three-time winners of the World RoboCup Soccer Championship and American Association for Artificial Intelligence contests, as seen on Scientific American Frontiers. Research funding includes grants from DARPA and NIH. We collaborate with SRI's Artificial Intelligence Center and other leading researchers on control software."

ALSTOM Schilling Robotics   202656

201 Cousteau Pl.
Davis, CA 95616-5412
USA

((530) 753-6718

⌧ (530) 753-8092

✉ sales.schilling@powerconv.alstom.com

 http://www.schilling.com/

ALSTOM Schilling Robotics supplies telerobotic manipulator systems for remotely operated vehicles (ROVs) and cable trenching machines used in offshore oil, telecommunications, scientific, and military operations.

Angelus Research   203473

11801 Cardinal Cir.
Unit J
Garden Grove, CA 92843
USA

Applied AI Systems
http://www.aai.ca/
Research robots

Automated Mining Systems Inc.
http://www.robominer.com/
Robotic mining equipment

Cybermotion, Inc.
http://www.cybermotion.com/
CyberGuard security/sentry robot

Hylands Underwater Vehicles
http://www.huv.com/
Manufacturer of micro autonomous underwater vehicles

K-Team
http://www.k-team.com/
K-Team develops, manufactures and markets a family of mobile robotic platforms for use in education and research

MTI Research Corp.
http://www.mtir.com/
Small mobile industrial robots

Robot Entertainment
http://www.entertainmentrobots.com/
Promotional robot rental for special events

Techno-Sommer Automatic
http://www.techno-sommer.com/
High-end industrial robotic end effectors

(714) 590-7877

(714) 590-7879

dgolding@angelusresearch.com

http://www.angelusresearch.com/

See listing under Robots-Educational.

Automated Mining Systems, Inc.   202646

3-16 Mary Street
Aurora, ON
L4G 1G2
Canada

(905) 713-3700

(905) 713-3708

info@robominer

http://www.robominer.com/

Automated Mining Systems makes and markets robots
and other machinery for underground mining operations. Site contains some interesting technical papers
regarding the challenges of mining and communication while underground.

Evolution Robotics, Inc.   203151

130 W. Union St.
Pasadena, CA 91103
USA

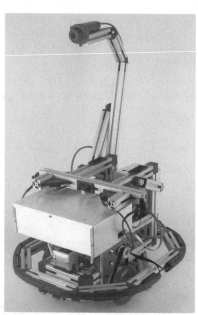

*The Robotics Development Kit from
Evolution Robotics. Photo Evolution
Robotics, Inc.*

(626) 535-2814

(626) 535-2777

info@evolution.com

http://www.evolution.com/

Evolution Robotics wants to build the "open robotics
platform" of the future. They offer a developer's kit
which consists of hardware, computer, and software.
The software is the key in the Evolution plan; it provides standard libraries for developing such things as
mapping, navigation, and behaviors. The hardware-
consisting of aluminum profile extrusions (machine
framing), motors, Razor scooter wheels-and other
pieces make for a well-turned-out robot in its own right.

The price of the developer's kit makes this one for serious players, and it will be interesting to see if Evolution
is able to foster the open platform they envision.

GeckoSystems, Inc.   202430

1820 Hwy. 20
Ste. 132-309
Conyers, GA 30013-2077
USA

(678) 413-9236

(678) 413-9247

(678) 413-9236

info@geckosystems.com

http://www.geckosystems.com/

Manufacturers of the CareBot "care" robot-nursing aid,
fire and safety alarm.

MOVERS AND SHAKERS

Rodney A. Brooks

http://www.ai.mit.edu/people/brooks/

Dr. Rodney A. Brooks is director of the MIT Artificial
Intelligence Laboratory and one of the most famous
roboticists living today. He is the creator of subsump-
tion architecture, used in simplifying artificial intelligence in machines, both large and small. He has
written several books, the latest of which, *Flesh and
Machines: How Robots Will Change Us,* is an international bestseller.

Dr. Brooks is also chairman and chief technical offi-
cer of iRobot Corporation, a major manufacturer of
robots for research and industry.

Haynes Enterprise 203378

1277 Linda Mar Center
Pacifica, CA 94044
USA

(650) 355-2732

(650) 355-4069

info@haynes-world.com

http://www.haynes-world.com/

Robotic "blanks." These full-size (57 inches tall) automated marionettes are programmed to move and talk. You dress them with facial features, headgear, and, of course, clothes (and, no, they are not anatomically correct). Typical applications are conventions and special events, as attention getters.

Photos and video clips show you how these work.

Honda: Humanoid Robot Web Site 202252

http://www.honda.co.jp/robot/

Honda has made a walking robot. Gets 32 miles to the gallon and goes from 0 to 1 miles per hour in six seconds. In all seriousness, the Honda walking robots are pretty cool and fun to watch.

International Robotics 204240

611 Broadway
Ste. 422
New York, NY 10012
USA

(212) 982-8001

(212) 982-8000

info@internationalrobotics.com

http://www.internationalrobotics.com/

International Robotics makes and sells a line of "show bots"—robots used for demonstrations, public appearances, demonstrations, and other interactive events. The robots are available for purchase or rent; the designs are inspiring.

The company also offers such exotica as video projection mannequins, animatronic video busts, and a video statue where the head is a 3D LCD screen fashioned in the likeness of Bert Parks, the legendary Miss America host. Wild.

iRobot 202040

Twin City Office Center
22 McGrath Hwy., Ste. 6
Somerville, MA 02143
USA

(617) 629-0055

(617) 629-0126

info@irobot.com

http://www.irobot.com/

iRobot is a leading developer of mobile robots for commerce and industry. An example product is the CoWorker, a "wireless, mobile, remote telepresence platform that provides control of video, audio and movement through any Internet browser without additional hardware or software." The company is also a think tank for consumer products; they developed MY REAL BABY, an interactive robotic doll that is endowed with artificial intelligence.

The iRobot Co-Worker. Photo iRobot Corporation.

Open PINO Platform 203934

http://www.openpino.org/

Open PINO is an "open" platform for the development of a humanoid robot, and is the brainchild of the

Kitano Symbiotic Systems Project. The Open PINO platform is based on the concept of "copyleft." All the source code is provided under GPL GNU general public license.

The Web site is in Japanese and English.

RoboProbe Technologies, Inc.
203410

P.O. Box 1037
Palatine, IL 60078
USA

((847) 934-5567
℡ (847) 934-9434
✉ solutions@roboprobe.com
🌎 http://www.roboprobe.com/

RoboProbe specializes in remotely operated robots and machines for hard to reach areas and hazardous applications. Products include video systems, crawlers and traction devices, claws, grippers, and manipulators, and robotic components.

Shadow Robot Company Ltd.
203753

251 Liverpool Road
London
N1 1LX
UK

(+44 (0) 2077 002487
🌎 http://www.shadow.org.uk/

Projects: bipedal robot, six-legged hexapod robot (using air), Biomorphic Arm, design for a "de-miner," a robot designed to traverse a minefield so that mines may be detected and removed.

Also sells products including air muscles that move arms and legs using a bladder of air.

Sine Robotics
202986

P.O. Box 172
Woodinville, WA 98072-0172
USA

((425) 788-0160
✉ bill@sinerobotics.com
🌎 http://www.sinerobotics.com/

Sine Robotics bills itself as a research and development firm specializing in small mobile robotics.

SpringWalker
203089

Applied Motion, Inc.
935 North Indian Hill Blvd.
Claremont, CA 91711
USA

✉ jdick@springwalker.com
🌎 http://www.springwalker.com/

The SpringWalker is known as a "body amplifier"-a bipedal machine that you literally ride in. In the basic model no motors or other assistive force is used except for levers and springs; a new model incorporates electric-powered servo control. Basic engineering details are provided from the company's patent pages.

This thing is, of course, a mechanized suit (à la Gundam Wing), taken from the pages of science fiction and demonstrated in real life. The idea may be old, but the technology to make it work is far harder than it looks.

Viva Robotics, Inc.
202419

21606 Stonetree Ct. #160
Dulles, VA 21066
USA

((703) 444-7300
℡ (703) 444-7840
🌎 http://www.vivarobotics.com/

Viva makes robots for trade shows. Check out their gallery of robots—they have a number of interesting designs worth investigating.

Yobotics
202445

850 Summer St.
Ste. #201
Boston, MA 02127
USA

MOVERS AND SHAKERS

Maja J. Mataric
http://www-robotics.usc.edu/~maja/

Maja Mataric is the director of the Robotics Research Lab at the University of Southern California, where (among other things) they study the interaction of robots and robot teams. Dr. Mataric has written extensively on the subject of behavior-based multiple robot coordination.

((617) 464-2144

((617) 464-2146

 sales@yobotics.com

(http://yobotics.com/

Yobotics makes legs . . . artificial legs to help disabled people walk and the legs of walking robots. See the company's RoboWalker, a powered orthotic brace, designed to augment or replace muscular functions of the lower extremities. Also check out the Yobotics Simulation Construction Set, a software package simulating dynamical systems of legged robots and other biomechanical systems.

Zaytran Automation   203372

P.O. Box 1660
Elyria, OH 44036
USA

((440) 324-2814

((440) 324-3552

 info@zaytran.com

(http://www.grippers.com/

Manufacturers of precision grippers for industrial robots, mainly severe highly controlled environments, such as semiconductor manufacturing or handling deadly bacteria collected from outer space in a top-secret underground research laboratory. High end. Available through distributors—the grippers, that is; you're on your own finding space bacteria.

 ## ROBOTS-PERSONAL

Personal robots are those that function as personal valets, doing the chores you'd otherwise do. A number of commercially available robots, listed in this section, are able to do such tasks as vacuuming the floor or mowing the lawn.

Dyson

http://www.dyson.com/
Robotic vacuum cleaners

RoboScience

http://www.roboscience.com/
High-end robotic pooch; kids can ride this thing

FloorBotics   202465

http://home.swbell.net/fontana/

Informational site on floor-cleaning robot.

Friendly Robotics   202562

8336 Sterling Dr.
Irving, TX 75063
USA

((214) 277-8181

((888) 404-7626

 friendly@friendlyrobotics.com

(http://www.friendlymachines.com/

Makers of a commercial robotic mower, Robomower.

*The Friendly Robotics Robomower.
Photo Friendly Robotics, Inc.*

Omnibot   202462

http://omnibot.forsite.net/omnibot/

The Omnibot was made is the mid-1980s by a company called TOMY. Message board, parts sales, and general how-to.

Probotics, Inc./Cye   203478

Ste. 223
700 River Ave.
Pittsburgh, PA 15212-5907
USA

((412) 322-6005

((412) 3220-3569

((888) 550-7658

 customersupport@personalrobots.com

http://www.personalrobots.com/

Probotics makes Cye, a home/office robot that can be programmed from your computer. One of its jobs is to vacuum. Their Map-N-Zap software is a room-mapping program to allow Cye *a priori* navigation capabilities.

The Cye-sr from Probotics. Photo Probotics, Inc.

Solar Mower   203476

http://www.solarmower.com/

Husqvarna's solar lawnmower robot. It's powered by solar energy and automatically cuts your grass without any help from you. Not quite as cheap as the neighbor kid down the street, but you can't have everything.

ROBOTS-WALKING

Walking robots are able to travel over rough terrain and can more readily operate in environments not specifically designed for autonomous machinery. Walking robots include small four- and six-legged robot kits, as well as sophisticated experimental robots that travel on 1, 2, 4, 8, and even 12 legs.

BARt-UH   202024

http://www.irt.uni-hannover.de/~iped/

BARt-UH is a bipedal autonomous walking robot designed at the University of Hannover in Germany. Web site is in English.

BIP2000 Anthropomorphic Biped Robot   202973

http://www.inrialpes.fr/bip/Bip-2000/

Fancy two-legged walking robot Laboratoire de Mcanique des Solides and INRIA Rhne-Alpes. Web site is in English and French.

BiPed Robot   203035

http://members.chello.at/alex.v/

Alex gone and built himself a bipedal (two-legged) walking robot that exhibits dynamic balance. Operated by aircraft servos. Watch the MPEG movies to see the machine in action.

The Web site provides hardware and software design overview, including a 3D exploded view of the robot's parts. Notice the two servos in the ankles of both legs. This is critical in allowing the robot to keep balance.

Hexplorer 2000   203761

http://real.uwaterloo.ca/~bot/

The Hexplorer is a six-legged walking robot at the University of Waterloo, located in Ontario, Canada. Construction details and programming overview are provided.

See also the main page for the Motion Research Group at:

http://real.uwaterloo.ca/

Legged Robot Builder, The   202259

http://www.joinme.net/robotwise/

This site is a resource for builders of autonomous legged robots. Whether your robot walks, crawls, runs, or hops, these links may inspire you.

Lynmotion

http://ww.lynxmotion.com

Maker of several walking robot kits. See listing in Robots-Hobby & Kit.

MHEX-My Six-Legged Walking Robot 202971

http://www.geocities.com/viasc/mhex/mhex.htm

Very nicely done 12-servo hexapod, created out of machined aluminum.

Petzi, My 4-Legged Walking Wonder 202256

http://www.gel.usherb.ca/caron/petzi/petzi.html

Petzi is a homemade walking dog, using R/C servos. Plenty of pictures and a discussion of walking gaits for quadrapedal robots.

Poly-PEDAL Lab 204158

http://polypedal.berkeley.edu/

The Poly-PEDAL Lab studies motion in animals and insects. The walk (gait) and balance studies often help in designing legged robots.

Ray Van Elst 203379

http://home.hccnet.nl/raymond.van.elst/

Walker plans, robots. Check out the CAD files for the walker robot. Some of the Web site is in Dutch.

RHex 202023

http://ai.eecs.umich.edu/RHex/

RHex is a wicked compliant hexapod robot. It is capable of walking, running, leaping over obstacles, and climbing stairs. From the Artificial Intelligence Laboratory at the University of Michigan (Ann Arbor).

Walking Machine Catalogue 203086

http://www.fzi.de/divisions/ipt/WMC/
    walking_machines_katalog/
    walking_machines_katalog.html

In the words of the Web site: "At the beginning of my work, I wanted to create a catalogue of all the walking machines that were ever built. This collection should represent the state of the art in this interesting research field, as well as the history of walking machines."

Pictures and text of walking robots.

Walking Machine Challenge 202561

http://www.sae.org/students/walking.htm

The Society of Automotive Engineers sponsors a "challenge" in college-level engineering to design, build, and test a walking machine with a self-contained power source. Many of the resulting designs are quite sophisticated.

Walking Robots 203470

http://www.walkingrobots.com/

Presented is a collection of walking robots, most of which were machines using an abrasive waterjet (apparently, this is a kind of machining tool, not a description of a mean boss who spits when yelling at you). Close-up photos but no construction details.

For more information on the manufacturer of specialty parts using the abrasive waterjet, see:

http://www.ormondllc.com/

Wilby Walker 203781

http://members.aol.com/wilbywalker/

Description and construction details of the Wilby Walker, a six-legged hexapod. DXF CAM/CAD files are provided.

SENSORS

A robot without sensors is just a fancy machine. If "clothes make the man" (applies to women, too, of course), then sensors make the robot. In this section, and the several sections that follow, you'll find various types of sensors suitable for use in robotics. Sensors can be quite expensive, and several high-end variations are listed. However, most of these sources are affordable. And, most sensor makers and sellers provide copious datasheets and application notes about their products, which you can study as you learn how the various sensor technologies work.

Note that while some sensor manufacturers will sell directly to the public, those that do often have minimum-order requirements.

Also included for reference are Web resources on sensors and sensor technology, how to build homemade sensors, and how to interface sensors to microcontrollers and computers.

SEE ALSO:

LEGO-MINDSTORMS: Web sites and retailers of sensors made to work with the LEGO Mindstorms

RETAIL-ROBOTICS SPECIALTY: Sources of sensors specifically for amateur robots

---

Baumer Electric Ltd.   202114

122 Spring St., C-6
Southington, CT 06489-1534
USA

☏ (860) 628-6280
⊘ (800) 937-9336
🌐 http://www.baumerelectric.com/

Baumer makes industrial sensors: inductive capacitive; photoelectric, retroreflective, and thru-beam; ultrasonic, proximity, and rotary encoders. This stuff isn't cheap, but if you need quality, this is where you'll find it. Web site is in English and German.

Common Robotic Senses

Robots need to sense the world around them in order to interact with it. Without such senses, they become little more than machines. We have five senses with which to experience our world: sight, hearing, touch, smell, and taste. A robot can be endowed with any and all of these senses as well.

- Sensitivity to *sound* is a common sensory system given to robots. The reason: Sound is easy to detect, and unless you're trying to listen for a specific kind of sound, circuits for sound detection are simple and straightforward.

- Sensitivity to *light* is also common, but the kind of light is usually restricted to a slender band of infrared, for the purpose of sensing the heat of a fire or for navigating through a room by way of an invisible light beam.

- In robotics, the sense of *touch* is most often confined to collision switches mounted around the periphery of the machine. On more sophisticated robots, pressure sensors may be attached to the tips of fingers in the robot's hands. The more the fingers of the hand close in around the object, the greater the pressure.

- The senses of *smell* and *taste* aren't generally implemented in robot systems, though some security robots designed for industrial use are outfitted with a gas sensor that, in effect, smells the presence of dangerous toxic gas.

Not all robotic senses are well developed. Robot eyesight is a good example. While electronic cameras, which can serve as the robot's eyes, are both affordable and easy to connect to a computer, the processing of the information from a camera is a complex task. The visual scene must be electronically rendered and programming must make sense of the object the robot sees.

Carlo Gavazzi Holding AG

Sumpfstrasse 32
CH-6312 Steinhausen
Switzerland

📞   +41 41 747 4525

📠   +41 41 740 4560

203992

✉ gavazzi@carlogavazzi.ch

🌐 http://www.carlogavazzi.com/

High-end industrial automation components. Sensors (proximity and photoelectric), solid-state relays, and motor controllers.

Object Detection: The Close, the Far, and the In-between

For robots to be self-sufficient in the human world, they must be able to determine their environment. They do this by sensing objects, obstacles, and terrain around them. This can include you, the cat, an old sock, the wall, the little hump on the ground between the carpet and the kitchen floor, a rock, another robot, a stair, a table leg, and a million other things. We'll lump it all under "object detection" and move on.

Robots perform object detection using either contact or noncontact means. *Contact* detection is when the robot, or some appendage of the robot, touches the object. Typical examples are leaf switches with pieces of wires, serving as whiskers, connected to them. *Noncontact* detection relies on sensing proximity to an object, without actually touching that object. This can be done using vision, ultrasonics, infrared light, inductance, capacitance, and many other techniques.

Once an object is detected, collision with it is avoided, or avoided as much as possible. In the case of contact sensing, the robot has already touched the object, but ideally not in a way that causes damage to either object or robot. For proximity sensing, the robot may be inches, feet, or yards away from the object. In any case, the robot "sees" the object, and then goes about deciding how best to avoid it.

Near- and Far-Object Detection

Proximity detection can be further broken down into two subgroups: near-object and far-object. The difference is relative and depends on such things as the size and speed of the robot, the size and speed of the object, and the type of object. For a typical carpet-roving bot, anything more than 8 to 10 feet away could be considered "far" and is unlikely to be a major influence on the machine. However, the robot still may need to be aware of the object's presence in order to formulate all of its operating plans.

Conversely, objects that are closer must be dealt with in a more immediate and aggressive manner. Such objects are within the robot's immediate sphere of operation. Detecting them is more critical, because they are the ones the robot will likely bump into or fall over.

Proximity or Distance?

There are two ways to approach near-object detection: proximity and distance.

- *Proximity* sensors care only that some object is within a zone of relevance. That is, if an object is near enough in the physical scene the robot is looking at, the sensor detects it and triggers the appropriate circuit in the robot. Objects beyond the proximal range of a sensor are effectively ignored, because they cannot be detected.

- *Distance measurement* sensors determine the distance between the sensor and whatever object is within range. Distance measurement techniques vary; almost all have notable minimums and maximums. Few yield accurate data if an object is smack-dab next to the robot; likewise, objects just outside range can yield inaccurate results.

Crossbow Technology, Inc.   202272

41 Daggett Dr.
San Jose, CA 95134
USA

✆ (408) 965-3300
📠 (408) 324-4840
✉ info@xbow.com
🌐 http://www.xbow.com/

Crossbow is into industrial sensors. Among their product line:

- Inertial and gyro systems
- Accelerometers
- Wireless sensor networks
- Tilt sensors
- Magnetometers

Davis INOTEK   202604

4701 Mount Hope Dr.
Baltimore, MD 21215
USA

🚫 (800) 492-6767
✉ info@inotek.com
🌐 http://www.inotek.com/

Sensors (Omron proximity and others); test equipment; and RFID

Entran Devices, Inc.   203324

10 Washington Ave.
Fairfield, NJ 07004-3877
USA

Proximity Detection versus Distance Measurement

Robots employ a number of noncontact methods to determine if an object is nearby (*noncontact* meaning nothing on the robot physically touches the object). There are two forms of noncontact detection, proximity detection and distance measurement.

- *Proximity detection.* Proximity detection is concerned only that some object is within a specified zone in front of or around the robot. Proximity simply means "close by"; it does not take into account how close or far the object is from the robot, nor does it concern itself with the size of the object. A common nonrobot example of proximity detection is the automatic security light: Walk in front of the sensor at night, and the light turns on.

- *Distance (or range) measurement.* With distance measurement, detection involves measuring the physical range between the object and the robot. Depending on the type of sensor used to measure the distance, relative size of the object may also be inferred.

Both detection schemes use similar technologies. The most common proximity detection schemes use infrared light or ultrasonic sound. If enough light (or sound) is reflected off the object and received back by the robot, then an object is within proximity.

Distance measurement sensors also use infrared light and ultrasonic sound, but the mechanisms tend to be more sophisticated. A popular group of sensors made by Sharp employs what's know as parallax to measure the distance to an object. In operation, infrared light is directed at an angle at an object. The light bounces from the object at the same angle and reenters the sensor. The displacement of the reflected beam indicates the distance between the sensor and the object. The greater the displacement, the greater the distance.

Ultrasonic sensors (proximity or ranging) measure the time it takes for a burst of high-frequency sound to travel from the sensor, strike an object, and return. Though the speed of sound varies depending on atmospheric conditions (humidity, temperature, and altitude), ultrasonic sensors are surprisingly accurate. The popular Polaroid 6500 sensor and control board boast a +/- 1% accuracy over the effective range of the sensor, which is 6 inches to 35 feet. A 1% error at 35 feet is less than five inches.

(973) 227-1002

 (973) 227-6865

 (888) 836-8726

sales@entran.com

http://www.entran.com/

Manufacturer of strain gauges, load cells, accelerometers, and pressure sensors. Not cheap.

Web site is in English, French, German, and Spanish.

Honeywell International Inc.   203919

101 Columbia Rd.
P.O. Box 4000
Morristown, NJ 07962-2497
USA

((973) 455-2000

(973) 455-4807

(800) 707-4555

http://www.honeywell.com/

Honeywell is a manufacturer of automation and control products. Several of their products are available through distributors such as Digi-Key. The company also sells some products directly.

Measurement Specialties, Inc. ⚬   202119

P.O. Box 799
Valley Forge, PA 19482
USA

(610) 650-1509

(888) 215-1744

http://www.measurementspecialties.com/

Measurement Specialties makes and sells sensors, particularly piezo sensors using Kynar plastic. These sensors can be used for such things as ultrasonic measurement, touch, vibration, and accelerometer. The

A bend sensor made from piezo film.

company provides online buying, but the minimum order is $100. Some of their products are also sold by Digi-Key and other distributors.

Merlin Systems Corp, Ltd.   202086

ITTC Tamar Science Park
1 Davy Road
Derriford
Plymouth, PL6 8BX
UK

(+44 (0) 1752 764205

+44 (0) 1752 772227

info@merlinsystemscorp.co.uk

http://www.merlinsystemscorp.co.uk/

Makers of the Humaniform Muscle, a lightweight actuator technology ideal for robotics. Other products include:

- MIABOTS—Intelligent autonomous microrobots
- LEX Sensor—Digital absolute position sensor
- Humaniform robotics and control systems technology
- Stretch sensor

Murata Manufacturing Co.   202472

26-10, Tenjin 2-chome
Nagaokakyo, Kyoto 617-8555
Japan

(+81 75 955 6502

+81 75 955 6526

http://www.murata.com/

Makers of:

- Pyroelectric infrared sensor
- Piezoelectric gyroscope
- Piezoelectric ceramics sensor
- Thermistors
- Magnetic pattern recognition
- Shock sensors
- Piezoelectric sound components

Lots and lots of datasheets. Offices in Japan, North America, and Europe. See also:

http://www.murata-northamerica.com/

http://www.murata-europe.com/

Picard Indistries 202360

4960 Quaker Hill Rd.
Albion, NY 14411
USA

☎ (716) 589-0358

📠 (716) 589-0358

✉ jcamdep4@iinc.com

🌐 http://www.picard-industries.com/

Picard specializes in miniature smart motors and sensors. Their product line includes programmable solenoids, motor control, and sensors.

Robot Electronics 🧲 202242

Unit 2B Gilray Road
Diss
Norfolk
IP22 4EU
UK

☎ +44 (0) 1379 640450

📠 +44 (0) 1379 650482

✉ sales@robot-electronics.co.uk

🌐 http://www.robot-electronics.co.uk/

Robot Electronics (sometimes referred to as Devantech) manufactures unique and affordable robotic components, including miniature ultrasonic sensors, electronic compasses, and 50-amp H-bridges for motor control.

The company's SRF08 high-performance ultrasonic rangefinder module can be connected to most any computer or microcontroller and provides real-time continuous distance measurements using ultrasonics. The measurement values are sent as digital signals and are selectable between microseconds, millimeters, or inches.

*The SRF08, from Robot Electronics.
Photo Robot Electronics.*

Schaevitz 202918

1000 Lucas Way
Virginia, VA 23666
USA

Sensor Beam Spread

Infrared and ultrasonic sensors emit light or sound, respectively, in order to detect nearby objects. With both sensor types, the further the object is from the sensor, the more the light or sound spreads by the time it gets there.

Infrared light is easier to focus into a small beam so that beam spread is minimized. With proper optics, beam spread over a distance of 10 or 15 feet may be only a few inches in diameter. This allows infrared sensors to be more selective in the objects they detect. However, it also requires more stringent alignment and maintenance of the sensor. If the focusing lenses of the sensor are not properly aligned, no object may ever be detected!

Sound waves can also be focused, but in most ultrasonic sensor applications this is not required; by its nature, high-frequency sound disperses (spreads) less readily than low-frequency sound. Even without acoustic focusing, the beam pattern of an ultrasonic sensor such as the Polaroid 6500 is approximately 25 degrees. (The beam pattern also contains nodes to either side of the main beam, but these are at relatively low acoustic power.)

A narrow beam isn't always desirable. Sometimes you want the widest spread possible. In these instances, it's best not to purposely "defocus" the beam, but to incorporate several sensors to create an array. The elements in the array are carefully positioned so that cover is broadened, without excessive overlap.

(757) 766-1500

(757) 766-4297

http://www.schaevitz.com/

Schaevitz manufactures industrial sensors, including LVDTs (linear variable differential transformers), pressure, tilt (clinometers and protractors), accelerometers, and inertial sensors. Many products are for sale on the Web site.

Sensorland.com
204226

http://www.sensorland.com/

Online repository of sensors: overviews, suppliers, technical articles, and new product releases. Check the "How It Works" section for semitechnical articles on how different sensors do what they do.

Sensors, Inc.
203991

3338 Republic Ave.
Miineapolis, MN 55426
USA

(952) 920-0939

(952) 920-9839

(888) 920-0939

custserv@sensorsincorporated.com

http://www.sensorsincorporated.com/

Sensors, what else? Online retailer/distributor for Hohner (encoders), Carlo Gavazzi (proximity), Cutler-Hammer, SICK, and others.

SICK, Inc.
202117

6900 West 110th St.
Bloomington, MN 55438
USA

(952) 941-6780

(952) 941-9287

(800) 314-4071

http://www.sickoptic.com/

SICK is a manufacturer of high-end industrial sensors and electronic measurement systems, including laser proximity scanners, bar coders, and 2D laser radar. Technical white papers are available on the Web site.

Sunx Sensors USA
202919

1207 Maple
West Des Moines, IA 50265
USA

(512) 225-6933

(512) 225-0063

(800) 280-6933

sunx@sunx-ramco.com

http://www.sunx-ramco.com/

Specialty miniature sensors for industrial control applications: photoelectric, fiber optic, inductive proximity, microphoto, laser beam, color and mark detection, ultraviolet, ultrasonic, pressure, and vacuum. Spec sheets are in Adobe Acrobat PDF format.

Vishay Intertechnology, Inc.
203906

63 Lincoln Hwy.
Malvern, PA 19355-2120
USA

(610) 644-1300

(610) 889-9429

http://www.vishay.com/

See listing under **Manufacturer-Semiconductors**.

 ## SENSORS-ENCODERS

This section contains resources for shaft and linear encoders, which includes both optical and mechanical devices. Shaft encoders are used to detect extent of motion. As sensors, encoders tend to be on the expensive end of the spectrum, though several low-cost variations exist. Many of these are low-resolution devices meant as substitutes for mechanical potentiometers. They are not made for use in high-speed or heavy-duty applications, but they should suffice for many jobs in amateur robotics.

Agilent Technologies, Inc.
202010

SPG Technical Response Center
3175 Bowers Ave.
Santa Clara, CA 95054
USA

(408) 654-8675

(408) 654-8575

(800) 235-0312

 SemiconductorSupport@agilent.com

www.semiconductor.agilent.com

Agilent manufactures a wide variety of semiconductor and electronics products.

See listing under **Manufacturer-Semiconductors**.

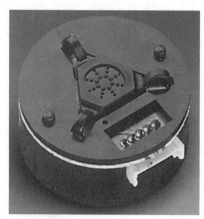

Agilent HEDR54xx series encoder. Photo Agilent Technologies.

Automationdirect.com   202829

3505 Hutchinson Rd.
Cumming, GA 30040
USA

((770) 889-2858

((770) 889-7876

⊘ (800) 633-0405

✉ sales@automationdirect.com

🌐 http://www.automationdirect.com/

Northstar Encoders

http://www.northstarencoders.com/

Encoders and tachometers

Renco Encoders Inc.

http://www.renco.com/

Industrial encoders; check out "Tech Info" technical section

Sumtak Corporation

http://www.sumtak.com

Industrial optical encoders

Online e-commerce and catalog mail-order source for motion control products.

See listing under **Actuators-Motion Products**.

Bourns, Inc.   202830

1200 Columbia Ave.
Riverside, CA 92507-2114
USA

((909) 781-5690

((909) 781-5273

✉ trimcus@bourns.com

🌐 http://www.bourns.com/

Switches, encoders, potentiometers, linear motion potentiometers. See listing under **Manufacturer-Components**.

Clarostat Manufacturing Co.   202833

12055 Rojas Dr.
Ste. K
El Paso, TX 79936
USA

((915) 858-2632

⊘ (800) 872-0042

✉ sensorproducts@clarostat.com

🌐 http://www.clarostat.com/

Encoders and "resistive" products (potentiometers, both rotary and linear). Some of the product line is carried by Digi-Key and other online distributors.

Cole Instrument Corp.   202828

2650 South Croddy Way
P.O. Box 25063
Santa Ana, CA 92799-5063
USA

((714) 556-3100

((714) 241-9061

✉ switch@earthlink.net

🌐 http://www.cole-switches.com/

Makers of small panel-mount optical encoders and mechanical rotary switches. Specification sheets are provided in Adobe Acrobat PDF format.

Odometry: Five Clicks off the Starboard Bow

Odometers measure distance. There's one in your car, so you know how many miles it is to the nearest electronics surplus store. Likewise, robots can have odometers. They are used for two purposes:

- *Dead reckoning.* Most robots lack the ability to determine their location the same way humans can (using familiar places and other landmarks). If a robot knows where it is when it starts, it can use odometry for dead reckoning.
- *Speed control.* By counting the clicks of an odometer over a specific time period, it's possible for a robot to determine its speed. It can then adjust its speed if necessary.

Optical Encoders for Odometry

Perhaps the most common form of robotic odometer is the optical encoder: a small disc is fashioned around the hub of a drive wheel or even the shaft of a drive motor. The disc contains a pattern of slots or dashes and can be used as an optical encoder. With a *reflectance* disc, infrared light strikes the disc and is reflected back to a photodetector. With a *slotted* disc, infrared light is alternatively blocked and passed and is picked up on the other side by a photodetector. With either method, a pulse is generated each time the photodetector senses the light.

An optical encoder can measure distance traveled by a robot.

Magnetic Encoders for Odometry

A magnetic encoder is constructed using a Hall-effect switch (a semiconductor sensitive to magnetic fields) and one or more magnets. A pulse is generated each time a magnet passes by the Hall-effect switch. A variation on the theme uses a metal gear and a special Hall-effect sensor that is sensitive to the variations in the magnetic influence produced by the gear.

How Encoders Work

Encoders (optical or magnetic) produce a series of pulses as they turn. The number of pulses is relative to the distance the robot travels. Suppose a wheel is 3 inches in diameter (that is 9.42 inches in circumference; computed by multiplying the diameter by pi, or roughly 3.14156). Further suppose the encoder wheel has 32 slots. Dividing the number of slots into the circumference of the wheel yields the number of pulses the encoder will produce at each revolution: 0.294 inches of travel (9.42 / 32) for each "tick" of the encoder. If the robot senses 10 pulses, it knows it has moved 2.94 inches.

Most robots use coaxial drive, with two drive wheels on either side and a supporting caster on the front and/or the back. With this common arrangement, encoders are attached to each wheel, so that the robot can determine the distances traveled by both wheels. This is necessary because the drive wheels of a robot are bound to turn at slightly different speeds over time. By integrating the results of both optical encoders, it's possible to determine where the robot really is, as opposed to where it should be.

Collecting the Pulses

On robots outfitted with a microcontroller, odometry measurements are best made if the chip has an onboard *pulse accumulator* or *counter* input. These kinds of inputs independently count the number of pulses received since the last time they were reset and do not require software to be constantly monitoring the activity. This allows the microcontroller to do other tasks simultaneously with encoder monitoring.

Odometry is one of the least expensive methods of robot navigation; but it is not perfect. Odometers are far from accurate: wheels slip, especially if the surface is hard and smooth or in turns. The wheel encoder may register a certain number of pulses, but because of slip the actual distance of travel will be less. Wheel diameters may change due to tire pressure and even heat. Such error is referred to as *unbounded*, because it can grow—getting worse and worse—the more the robot travels.

If you require absolute accuracy in navigation, it is necessary to combine odometry with other navigation techniques, such as active beacons, distance mapping, or landmark recognition.

Donovan Micro-Tek   203980

67 W. Easy St.
Ste. 112
Simi Valley, CA 93065
USA

✆ (805) 584-1893
📠 (805) 584-1892
✉ info@dmicrotek.com
🌐 http://www.dmicrotek.com/

Itty-bitty stepper motors—8, 10, and 15mm. Also carries encoders, drive electronics, and gearboxes. Everything is on the small side.

Encoder Products Group   202703

1601B Highway 2
P.O. Box 1548
Sandpoint, ID 83864-0879
USA

✆ (208) 263-8541
📠 (208) 263-0541
🚫 (800) 366-5412
🌐 http://www.encoderproducts.com/

Makes incremental and absolute encoders (NEMA sizes 15 to 58), including shaft and hollow designs, for industrial and automotive applications. Also sells encoder accessories, such as shaft couplers, mounting brackets, pulse-train converters, and connectors/cables.

Grayhill, Inc.   202698

561 Hillgrove Ave.
LaGrange, IL 60525-5997
USA

✆ (708) 354-1040
📠 (708) 354-2820
🚫 (800) 426-4383
✉ info@grayhill.com
🌐 http://grayhill.com/

Makers of mechanical and optical switches, including rotary encoders. Grayhill's product line is sold through distributors, but some of it is carried by Digi-Key. Mechanical and optical encoders are available, many are low-cost alternatives to industrial quadrature incremental encoders.

Gurley Precision Instruments   203872

514 Fulton St.
Troy, NY 12180
USA

✆ (518) 272-6300
📠 (518) 274-0336
🚫 (800) 759-1844
✉ m.gordinier@gurley.com
🌐 http://www.gpi-encoders.com/

High resolution, high-end linear and rotary encoders. You'll also find numerous technical briefs about optical encoders at the Web site (in Adobe Acrobat PDF format).

Hohner Corp. 203990

5536 Regional Road 81
Beamsville, Ontario
L0R 1B3
Canada

- (905) 563-4924
- (905) 563-7209
- (800) 295-5693
- hohner@hohner.com
- http://www.hohner.com/

Optical incremental and absolute encoders. Sells direct or through distributors. For color detectors, roughness detectors, and fluorescence detectors,

see also:

http://www.surfacesensor.com/

Industrial Encoders Direct 202704

Unit D1, Dutton Road
Redwither Business Park
Wrexham
LL13 9UL
UK

- +44 (0) 1978 664722
- +44 (0) 1978 664733
- sales@industrialencodersdirect.co.uk
- http://www.industrialencodersdirect.co.uk/

Industrial encoders (shaft and hollow), available direct from the manufacturer. Datasheets available.

A shaft encoder.

Servo-Tek Products Co., Inc. 202702

1086 Goffle Rd.
Hawthorne, NJ 07506
USA

- (973) 427-3100

About Pulse and Quadrature Shaft Encoders

Two common forms of shaft encoders are pulse and quadrature.

- With a pulse encoder, a single train of digital pulses (high/low/high/low, etc.) is emitted as the shaft spins. Spin the shaft slowly, the pulses come out slowly. Spin the shaft quickly, the pulses come out rapid fire.

- With a quadrature encoder, two separate trains of pulses are emitted. These outputs are called *channels*. The pulses are timed 90 degrees apart from one another. That is, if you divide a digital pulse into four equal segments, the pulses from the two channels will be offset from one another by one of those segments.

The main use of pulse encoders is to measure speed, and they are sometimes called digital tachometers (analog tachometers output a voltage proportional to the speed of the shaft). Pulses are emitted whether the shaft is turning forward or backward, and there is no way of determining the direction of the shaft.

Quadrature encoders, on the other hand, indicate not only number of pulses per revolution, but direction. This is possible because the pulses from one channel will "lead" the other, depending on the direction the shaft is turning.

Internally, the quadrature encoder is not that much different from a pulse encoder. The quad encoder uses a separate sensing element, typically a light-emitting diode and phototransistor in the optical variety. The LEDs and photosensors are mechanically spaced to produce the 90-degree phase difference required for quadrature encoding.

(973) 427-4249

sales@servotek.com

http://www.servotek.com/

Makers of standard and economy rotary encoders, tachometers, and velocity sensors.

Stegmann, Inc. 202701

7496 Webster St.
Dayton, OH 45413-13596
USA

(937) 454-1956

(937) 454-1955

sales@stegmann.com

http://www.stegmann.com/

Stegmann manufactures linear and rotary encoders (both incremental and absolute) and servomotor feedback systems. Tech sheets and information available in Adobe Acrobat PDF format.

US Digital 202699

11100 NE 34th Cir.
Vancouver, WA 98682
USA

(360) 260-2468

(360) 260-2469

(800) 736-0194

info@usdigital.com

http://www.usdigital.com/

US Digital manufactures and sells reasonably priced linear and rotary quadrature and absolute encoders. They have a variety of models to choose from, and you can select such options as the number of counts per resolution or whether an index pulse is generated at each revolution. Encoders are available with standard or heavy-duty bearings, depending on their intended application.

The company also sells modules, discs, and linear strips to construct your own encoder. This is not so much to save money (though it is cheaper than a full encoder), but it allows you greater flexibility in adding an encoder to your existing hardware setup. Linear strips are sold by the inch; discs are sold by diameter and counts per revolution.

Additionally, US Digital supplies counter chips and interface electronics that make it very easy to attach an encoder to a PC or other controller. Among their interface electronics:

- AD6—Quadrature encoder to parallel port adapter
- LS7083—Quadrature encoder to counter (up/down clock) interface chip
- LS7166—Quadrature encoder to microprocessor interface chip
- A2—Absolute encoder interface to RS232 port

Application notes and datasheets accompany the interface electronics products (you can download the application notes and datasheets and look them over before you buy). In some cases, sample software is included. A simple Visual Basic demo accompanies the AD6 parallel port adapter, for example. Anyone with modest experience in Visual Basic can adapt the demo to suit their needs.

Finally, if you don't feel like making your own cables, custom-made cables and connectors can be ordered to match most any configuration.

H1/H3 heavy-duty shaft encoders. Photo US Digital Corp.

AD6 parallel port interface. Photo US Digital Corp.

SENSORS-GPS

Global positioning satellite (GPS) is a system of special communications satellites used to pinpoint locations on the ground. Though once strictly used by the military and select commercial applications, GPS systems are now routinely available for consumer use. Several GPS receivers come ready made for connection directly to a computer, which—with proper software—can interpret positioning signals. GPS receivers can be used with outdoor robots to give them a sense of exactly where they are in the world.

Bike World   203848

5911 Broadway
San Antonio, TX 78209
USA

((210) 828-5558
📠 (210) 828-3299
🚫 800.928.5558

GST-1 GPS Sentence Translator

http://www.byonics.com/gst-1/gst-1.html

Translates Delorme Earthmate data to NMEA-0183 standard

Laipac Technology, Inc.

http://www.laipac.com/

Maker of GPS modules

---

 http://www.bikeworld.com/

Bike parts, including chains and sprockets, bars and bar stock (some of it carbon composite or fiberglass), and control cables, as well as small GPS receivers.

Garmin Ltd.   203544

Harbour Place
5th Fl., 113 S. Church St.
George Town
Grand Cayman
Cayman Islands

((913) 397-8200
🚫 (800) 800-1020
🌐 http://www.garmin.com/

Garmin is a major manufacturer of GPS systems, including OEM modules. A popular GPS unit used in robotics is the eTrex miniature GPS handheld.

Garmin model GPS76. Photo Garmin Ltd.

Hacking a Consumer GPS Module

Some of the least expensive GPS units are commercial models that are intended to interface directly to a PC (often using proprietary data structures) or display position on an LCD panel. Here are some informational Web sites on hacking popular GPS products, in order to obtain the data stream for use in standard computer applications or with microcontrollers.

Hacking the Delorme Earthmate

http://www.hamhud.net/earthmate.htm

Hacking Garmin eTrex GPS Receiver

http://www.nomad.ee/micros/etrex.shtml

Hacking a Rand McNally/Magellan GPS

http://www.radiohound.com/randgps.htm

You can buy accessories (data cables, mounting brackets, etc.) from Garmin, but the GPS units themselves are only sold through resellers. Online resellers include GPS City, GPS Discount, and others.

GPS City   202216

6 Sunset Way, Ste. 108
Henderson, NV 89014
USA

  (702) 990-5603
  (800) 231-7540
✉ sales@gpscity.com
🌐 http://www.gpscity.com/

Sells GPS units for all occasions. Among many products, sells the Garmin GPS 35 OEM Sensor, which can be connected to any PC or microcontroller through an RS-232 serial interface.

GPS Warehouse   203846

Unit 9, The Lion Centre
Hanworth Trading Estate
Hampton Road West
Hanworth, Middlesex
TW13 6DS
UK

📞 +44 (0) 2088 939393
📠 +44 (0) 2088 946347
✉ sales@gpsw.co.uk
🌐 http://www.gpsw.co.uk/

Online GPS retailer. All major brands. Sells throughout Europe.

GPS World Supply   203847

3N060 Powis Rd.
West Chicago, IL 60185
USA

GPS for Robotics

There is no cost to use GPS, except for the price of a receiver. For robotics applications, a GPS receiver can be used to tell the robot where it is in the world—or even your back yard. A GPS-equipped robot could be programmed to know that it's wandered too far from home. Or, outfitted with a compass, such a robot might be able to navigate back home.

While GPS for a small mobile robot can be useful, there are some "gotchas" you need to be aware of.

- By itself GPS will work only outdoors. The signal from the satellite is not strong enough to penetrate walls. One way around this limitation is to use a *pseudolite*, a receiver that is positioned outdoors, and which communicates with specially modified receivers contained inside a building. (Pseudolites can also used a kind of local-area GPS that covers a building or small areas, but these are beyond the discussion here.)

- It takes time—up to several minutes—for a GPS receiver to lock onto the signals from its satellites. Each time the GPS signals are interrupted, for whatever reason, there may be another 2- to 3-minute delay to fully reacquire the satellites.

- The orbits of the GPS satellites tend to favor North America and other countries where the United States has a strategic interest. In such areas, the receiver will almost always pick up signals from at least three satellites, and often four to eight satellites. But in some other parts of the world, only one or two satellite signals may be received, and this is insufficient to provide a fix.

A GPS equipped with a standard serial link can be connected directly to a computer or microcontroller. Photo Garmin Ltd.

((630) 584-3557

🖰 (630) 584-4105

🚫 (800) 906-6600

✉ info@gpsworldsupply.com

🌐 http://store.gpsworldsupply.com/

Online GPS retailer. Specializes in Garmin product.

🌐 💻

Joe Mehaffey and
Jack Yeazel's GPS Information   203430

http://www.gpsinformation.net/

Joe and Jack yack about GPS receivers and how to interface with them using computers.

🗣

Lowrance Electronics, Inc.   203917

12000 E. Skelly Dr.
Tulsa, OK 74128
USA

((918) 437-6881

🖰 (918) 234-1705

🚫 (800) 324-1356

🌐 http://www.lowrance.com

Lowrance is in the business of GPS and sonar devices. Check out their GPS Tutorial.

Magellan / Thales Navigation   203545

471 El Camino Real
Santa Clara, CA 95050-4300
USA

The ABCs of GPS (Global Positioning Satellite)

The global positioning satellite (GPS) system is a means by which someone on the ground can determine *which ground* they're on . . . are they in Kansas City, Kansas, or Kansas City, Missouri? With GPS, you can know your position on the globe within a few meters. The system is used by the military, aviation, private industry, even individuals, such as hikers.

How It Works

GPS is a collection of two dozen satellites that circle the globe in specific orbits. Each satellite, which spins about 12,000 miles from the earth, contains its own highly accurate atomic clock. The time, as reported by this clock, is broadcast from the satellite to the earth. A receiver on the ground picks up the signal from the satellite. To be effective, the receiver must also pick up the signal from at least two other GPS satellites.

Because each satellite will be at different distances from the ground receiver, it takes a slightly different time for each signal to reach the receiver. The difference in distance causes the signal from each satellite to be delayed from one another—like light, radio signals travel at a known speed, about 186,000 miles per second.

The position of the receiver on the earth can be accurately determined by correlating the "timestamps" from each satellite. As each timestamp is received a fraction of a millisecond after the first one, the delayed timestamps indicate how far away the satellites are from the receiver. Included with the timestamp from each satellite is the satellites own position over the earth.

The GPS system is the result of an initiative by the U.S. Department of Defense. When the first GPS satellite was launched in 1978, the system was off limits to civilian use. That changed in the 1980s, and today, GPS is regularly used in commerce. For just a few hundred dollars, you can purchase all-in-one GPS receivers. The receiver will indicate your position in the wilderness to within 10 to 100 meters (usually on the lower side of this range). You might still get lost, but at least you'll know where you are!

(408) 615-5100

(408) 615-5200

http://www.magellangps.com

Manufacturer of GPS systems.

National Marine Electronics Association (NMEA)

203725

http://www.nmea.org

A technical association that helps set standards for marine electronics. One such standard of importance to amateur robot builders is NMEA-0183. This is a voluntary standard followed by many manufacturers of global positioning satellite receivers. It allows the GPS module to interface with other electronics, such as a computer.

Navtech Seminars and GPS Supply

203997

Ste. 400
6121 Lincolnia Rd.
Alexandria, VA 22312-2707
USA

(703) 256-8900

(703) 256-8988

(800) 628-0885

orders@navtechgps.com

http://www.navtechgps.com/

Navtech is a reseller of GPS equipment, including receivers, antennas, differential GPS modules, OEM GPS kits, and books. They also provide seminars on GPS.

Things to Know About GPS

GPS is both pricey and temperamental for use in amateur robots. But if you're seriously considering it, keep these points in mind:

- Perhaps the easiest way to experiment with GPS is to purchase a developer's kit, including a bare-bones GPS module (a module without a display or control buttons). Cost is $50 to over $200. In most cases, the module connects via a serial link to your microcontroller or computer.

- Using a consumer-ready GPS product may require hacking to interface it to your microcontroller or computer.

- GPS modules use a variety of data formats. The NMEA-0183 is an official (but voluntary) standard, but other proprietary and semiproprietary formats are available as well, including Rockwell binary, SiRF binary, and Delorme Earthmate.

- Your choice of which module to use depends on what you're connecting it to and the availability of software that will understand the data format from the module. When available, opt for a module supporting NMEA-0183 or a module capable of a number of data formats.

- Some new GPS receivers incorporate a technology known as Wide Area Augmentation System, or WAAS, which increases accuracy to about 3 to 5 meters, on average.

- Additionally, certain high-priced GPS receivers use Differential GPS (DGPS), a system that correlates signals from the GPS satellites with that of known land-based transmitters. DGPS offers accuracy to under 1 meter, but its coverage is limited, usually to major waterways or aircraft flight patterns.

- GPS satellites for civilian use transmit at 1575.42 MHz, one of two frequencies used by the GPS system. This frequency is sometimes referred to by its designation L1; L2 is for military use.

- GPS is best in wide, open, unobstructed areas. Inaccurate results can occur if the signal bounces off nearby buildings or mountains.

Google.com Search Phrases for GPS

GPS module

GPS module robot OR robotics

GPS robot OR robotics

 "global positioning satellite" module

"global positioning satellite" robotics

intitle:"global positioning satellite"

intitle:gps robot

Starlink Incorporated   203546

500 Center Ridge Dr.
Ste. 600
Austin, TX 78753
USA

((512) 454-5511

📠 (512) 454-5570

🚫 (800) 460-2167

✉ sales@starlinkdgps.com

🌐 http://www.starlinkdgps.com/

Starlink is a manufacturer of differential GPS systems. Differential GPS provides far greater accuracy than regular GPS and, of course, costs a whole lot more.

Synergy Systems, LLP   203751

P.O. Box 262250
San Diego, CA 92196
USA

((858) 566-0666

📠 (858) 566-0768

🚫 (888) 479-6749

✉ info@synergy-gps.com

🌐 http://www.synergy-gps.com/

OEM and board-level GPS systems, using Motorola modules. Sells starter kits for quick prototyping and developing.

---

u-blox ag   203996

Zuercherstrasse 68
8800 Thalwil
Switzerland

(+41 1 722 74 44

📠 +41 1 722 74 47

✉ info@u-blox.com

🌐 http://www.u-blox.ch/

What the Heck Are NMEA Sentences?

Scan through manuals and literature about GPS systems, and you'll likely read about "NMEA sentences." No, this is not some new form of punishment; rather it's a description of the data format used in many GPS systems—NMEA-0183 (which is technically a standard for interfacing marine electronics, not just GPS units).

An NMEA sentence is normal text (alphabetical and numerical characters) sent serially, at a certain speed (baud rate). Products that use the NMEA-0183 standard know how to receive, parse, and construct NMEA sentences, so that other compatible gear will understand the data.

An example NMEA sentence looks like this:

$GPRMC,092204.999,A,4250.5589,S,14718.5084,E,0.00,89.68,211200,,*25

This sentence provides position and time information. The commas are used to parse (separate out) the data, in order to make sense of it.

Google.com Search phrase: NMEA sentence OR sentences

When the Swiss see something, they want to make it smaller. Never fails. They saw a GPS module and figured they could make that smaller, too. U-blox sells exactly that: ultraminiature GPS modules of about 1 inch square and just an eighth of an inch thick. (Their "big" modules are about 1.5 inches square and some 3/8 inch thick, which is still pretty tiny.) Products are available through distributors.

Regional offices of u-blox are in Switzerland, the U.S., Germany, Hong Kong, and the U.K. Web site is in English and German.

 SENSORS-OPTICAL

Optical sensors use light to detect objects. Depending on the sensor technology used, it's possible to use light to not only determine if an object is near (proximity) but also how far away an object is (distance).

The resources in this section specialize in optical sensors, which include infrared, passive infrared (like the kind used in motion detectors), and ultraviolet. Each variation has its own unique applications.

See also the **Video** sections for additional sensors that use light.

Glolab Corporation   203040

307 Pine Ridge Dr.
Wappingers Falls, NY 12590
USA

✆ (845) 297-9772
✉ kits@glolab.com

PIR detector sensor. Photo Frank Montegari, Glolab.com.

 http://www.glolab.com/

Glolab manufactures and sells multichannel wireless transmitters and receivers, encoder and decoder modules (to permit controlling more than one device through a wireless link). They also provide pyroelectric infrared sensors and suitable Fresnel lenses. An amplifier and hookup diagram from the PIR sensor is available on the Web site.

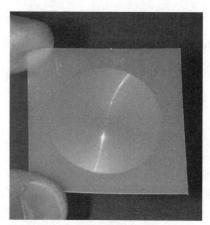

Fresnel lens for PIR sensor. Photo Frank Montegari, Glolab.com.

Hamamatsu Corp.   203742

325-6 Sunayama-cho
Hamamatsu City
Shizuoka Pref.,430-8587
Japan

✆ +81 53 452 2141
📠 +81 53 456 7889
✉ usa@hamamatsu.com
🌐 http://www.hamamatsu.com/

Main Japan office listed; Web site is provided in many languages, and local offices in many countries, including U.S., France, U.K., Germany, and Italy.

Provides photonics detectors, flame detectors, photomultiplier tubes, imaging systems, optical-linear arrays. Product is available in limited sample quantities and is sold through distributors.

Of particular interest are (app notes provided for many sensor types):

- Flame sensors (UV TRON)
- CdS photoconductive cells
- Infrared detectors
- Photo ICs

Measuring Distance with Infrared Sensors

Infrared light sensors use a technique known as parallax to measure distance to an object. Parallax is the angle between the two straight lines that intersect some object (near or far) when viewed at two different points of observation. A good example of parallax is how a person looking at the moon in Los Angeles sees it in a different portion of the sky than a person looking at the exact same time in New York.

The difference in the angles can be computed, and if the distance between observers is known, then the distance from the observers to the object can be calculated.

For infrared sensors, the technique works using a slightly different approach, but the concept is the same. A beam of infrared light illuminates a scene. The beam is aimed at the object at a slight angle. The beam reflects off an object in front of the sensor and bounces back into the sensor. A rule of physics known as *angle of incidence* ensures that the reflected beam is the same angle as the incident beam.

The reflected beam falls onto a linear array of very small photodetectors. The detectors are arranged in a long line; the sensor basically detects where on the line the reflected light strikes. This linear array is connected to internal circuitry that resolves the distance of the object. The circuitry can provide either a digital or an analog output.

A key maker of infrared distance measurement sensors is Japan-based Sharp. The sensors are engineered not for amateur robotics (though they are ideally suited for the tasks), but instead are intended for use in such applications as proximity devices for automobiles, for paper detection in copiers, and similar mundane tasks.

What's Available

Depending on the model, the sensors have a minimum range of about 4 inches (10cm) and a maximum of 31.5 inches (80cm).

One of Sharp's many infrared distance/proximity sensors.

- GP2D02—Digital serial output indicates range as an 8-bit value. Range is from about 4 inches to 30 inches.

- GP2D05—Digital HIGH/LOW output registers whether an object is within a preset range. The range can be adjusted by making an internal adjustment.

- GP2D12—Analog output indicates range as a voltage level.

- GP2D15—Like the GP2D05, the output triggers when the object is within a certain range (10 inches).

- GP2Y0D02YK—Similar to the GP2D05, HIGH/LOW output registers whether an object is within a preset range (31.5 inches).

- GP2Y0A02YK—Extended range version detects objects from 8 to 60 inches.
  Refer to the following Web sites for information on the Sharp infrared sensors:

http://www.sharpsma.com/

http://www.sharp.co.jp/ecg/opto/products/

In all cases, the Sharp infrared sensors share better-than-average immunity to ambient light levels, so you can use them under a variety of lighting conditions (except perhaps very bright light outdoors). The sensors use a modulated—as opposed to continuous—infrared beam that helps reject false triggering and makes the system accurate even if the detected object absorbs or scatters infrared light, such as heavy curtains or dark-colored fabrics.

Proximity Sensing versus Distance Measuring

Several of the Sharp sensors are for proximity detection rather than distance measurement. These have a 1-bit output that is either HIGH or LOW depending on whether an object has been detected within a threshold range.

Your choice of whether to use a proximity sensor or a distance-measuring sensor depends on your application. If you

- are merely concerned that an object is nearby, use a proximity sensor. The HIGH/LOW nature of the output of the sensor makes it very easy to use. The GP2D05 sensor can be modified to alter the proximity threshold distance. The working range is from 4 inches to 30 inches (10cm to 80cm).
- need to detect "far" and "close" proximity, consider adding additional proximity sensors set at different threshold distances. For instance, one sensor can be set at 70cm to 80cm to detect objects that are relatively far from the robot. Another could be set at 10cm to 12cm to detect those objects that are at the risk of being run into.
- must determine relative distances, then a distance-measuring sensor is needed. Two options are available: analog output and digital output. Microcontrollers and computer I/O ports with analog inputs can be used with analog sensors. The sensor provides a varying voltage in relation to the distance of the object. Digital sensors provide the distance information in a simple serial data train. Your microcontroller or computer reads one bit at a time to reconstruct the data. Examples of using both kinds of outputs are provided by many retail sources that offer the Sharp sensors.

Increasing Field of View

The Sharp sensors are restricted to a fairly narrow field of view and are no wider than about 16cm at their widest point. To increase the field of view, you can either use multiple sensors, with slight overlapping coverage, or you can rotate the sensors using an R/C servo.

For the latter, mount the sensor on a plastic control horn, and attach the horn to the servo. Under control of your robot, you can scan the servo up to 90 degrees to either side, thereby taking in an almost 180-degree panoramic view. Avoid taking readings while the sensor is in motion as you could get spurious results. Instead, stop the servo, then take the reading.

Controlling Current Consumption

One disadvantage of the Sharp and other IR sensors is that current consumption is relatively high when the sensor is activated. The GP2D02 and GP2D05 sensors are normally not active; they must be triggered to take a reading by your computer or microcontroller. When inactive, they draw a miniscule 2 μA (microamps) of current. When active, they consume about 25 mA, average for a light-emitting diode at regular intensity. The GP2D12 and GP2D15 sensors take continuous readings, so they are always on.

If current consumption is a concern in your robot, opt for an infrared sensor type that can be turned on and off at will. This reduces unnecessary battery drain. And, since sensors like the Sharp units use modulated light, turning them off when not needed helps ensure the pulsing light doesn't interfere with other infrared systems, including remote control.

Interfacing Jacks and Plugs

One final thought: The Sharp IR proximity and distance-measuring sensors are miniature devices, and they use a special miniature plug-in connector. When purchasing a Sharp sensor, be sure to get the proper connector to go along with it. These connectors, sometimes referred to as Japan Solderless Terminals (JSTs), are not stocked by most electronics retailers, but are available from companies that sell the Sharp units.

Resellers of the Sharp distance and proximity sensors include:

Acroname
http://www.acroname.com/

HVW Technologies Inc.
http://www.hvwtech.com/

Zagros Robotics
http://www.zagrosrobotics.com/

Kodenshi Korea Corp. 203794

513-5
Eoyang-dong
Iksan
Korea

(82 63-839-2111
(82 63-839-2005
✉ kodenshi@kodenshi.com
🌐 http://www.kodenshi.com/

Photosensitive Elements

You have a variety of photosensitive elements to choose from when building a robot. The three most common (and most affordable) are the photoresistor, the phototransistor, and the photodiode.

- *Photoresistors.* Typically made from cadmium sulfide, and therefore referred to generically as Cds cells, photoresistors acts like a light-dependent resistor: The resistance of the cell varies depending on the intensity of the light striking it. While Cds cells are easy to interface to other electronics, they are somewhat slow reacting. This can be a benefit when used indoors with AC-operated lights.

- *Phototransistors.* All semiconductors are sensitive to light, and phototransistors exploit this trait. They are basically regular transistors with their metal or plastic top removed. Unlike Cds cells, phototransistors are very quick acting, able to sense tens of thousands of flashes of light per second. The output of a phototransistor is not "linear"; that is, there is a disproportionate change in the output of a phototransistor as more and more light strikes it.

- *Photodiodes.* A simple form of the transistor, photodiodes are likewise made with a glass or plastic cover to protect the semiconductor material inside them. One common characteristic of most photodiodes is that their output is rather low, even when fully exposed to bright light. This means that to be effective the output of the photodiode must usually be connected to a small amplifier of some type.

Phototransistors (and their cousins photoresistors and photodiodes) are used for simple robotic vision.

Maker of optical sensors, infrared LEDs, and integrated photosensors. Web site is in English, Korean, and Japanese.

Leica Disto   204254

4855 Peachtree Industrial Blvd.
Ste. 235
Norcross, GA 30092
USA

((770) 447-6361

✆ (770) 447-0710

🌐 http://www.disto.com/

Manufacturers of handheld laser range finders. Cost isn't exactly cheap, but reasonable for a high-end 'bot. Part of the worldwide Leica Geosystems group (address provided is for the U.S. office); products are available from distributors or online.

Robobix   203706

http://www.geocities.com/robobix/

Circuits include ultrasonic distance measurement, light-reflection distance measurement, and simple infrared object detection

Feeling Is More Than Touch

Consider the human body: It has many kinds of touch receptors embedded within its skin. Some receptors are sensitive to physical pressure, while others are sensitive to heat. Similarly, you may wish to add these types of sensors to your robots.

- *Heat sensors* can detect changes in heat of objects within grasp. Heat sensors are available in many forms, including thermisters (resistors that change their value depending on temperature) and solid-state diodes that are specifically made to be ultrasensitive to changes in temperature.

- *Air-pressure sensors* can be used to detect physical contact. The sensor is connected to a flexible tube or bladder (like a balloon); pressure on the tube or bladder causes air to push into or out of the sensor, thereby triggering it. To be useful, the sensor should be sensitive down to about 1 pound per square inch, or less.

- *Resistive bend sensors*, originally designed for use with virtual reality gloves, vary their resistance depending on the degree of bending. Mount the sensor in a loop, and you can detect the change in resistance as the loop is deformed by the pressure of contact.

- *Strain gauges* measure the stress on parts of the robot. You can mount strain gauges on the front and back bumpers of a robot, for example, and they will detect when the 'bot has collided with an object. Strain gauges that detect a DC bias (in other words, a constant pressure) can be used to determine if the robot remains pressed against an object. Strain gauges that cannot detect a DC bias (piezoelectric cells are included in this group) can only determine change between contact and noncontact.

- *Microphones and other sound transducers* make effective touch sensors. Microphones, either standard or ultrasonic, can be used to detect sounds that occur when objects touch ("microphonic conduction," for the lack of a better term). Mount the microphone element on the robot. Place a small piece of felt directly under the element, and cement it in place using a household glue that sets up hard. Run the leads of the microphone to a sound trigger circuit. As things move past the sensor, it will pick up the sound.

Sharp Sensor Hack for Analog Distance Measurement
202475

http://www.cs.uwa.edu.au/~afm/robot/
harp-hack.html

Reengineering a Sharp GPIU5 infrared detector module to determine distance.

Thermo Centrovision
203988

2088 Anchor Ct.
Newbury Park, CA 91320
USA

📞 @listing-phonefaxweb:

📠 (805) 499-7770

🚫 (800) 700-2088

✉ info@centrovision.com

🌐 http://www.centrovision.com/

Provides standard and specialty photodiodes: infrared, high-speed infrared, ultraviolet, eye-response sensors, more.

 SENSORS-OTHER

In this section are several makers and sellers of miscellaneous sensor types, such as fence vibration sensors, magnetic sensors, and toxic gas sensors.

Agilent Technologies, Inc.
202010

SPG Technical Response Center
3175 Bowers Ave.
Santa Clara, CA 95054
USA

Robots That Follow Walls

Mice exhibit a curious behavior: When indoors, they tend to run alongside the walls, rather than out in the center of the room. This is for their protection: Mice are prey, and they can be more easily seen in the open.

A common robot experiment is to construct a wall-following "mouse," a mechanical rodent that favors walls and even actively seeks them out before exploring the rest of the room. Most all wall-following robots use a touch or proximity sensor to detect the wall. There are several common ways of accomplishing this:

- *Whisker contact.* The robot uses a mechanical switch, or a stiff wire that is connected to a switch, to sense contact with the wall. This is by far the simplest method, but is prone to mechanical damage after a period of use.

- *Noncontact active sensor.* The robot uses active proximity sensors, such as infrared or ultrasonic, to determine distance from the wall. No physical contact with the wall is needed. In a typical noncontact system, two sensors are used to judge when the robot is parallel to the wall.

- *Noncontact passive sensor.* The robot uses passive sensors, such as linear Hall-effect switches, to judge distance from a specially prepared wall. In the case of Hall-effect switches, the baseboard or wall might be outfitted with an electrical wire, through which a low-voltage alternating current is fed. When in proximity, the sensors will pick up the induced magnetic field provided by the alternating current. Or, if the baseboard is metal, the Hall-effect sensor (when rigged with a small magnet on its opposite side) could detect proximity to a wall.

- *"Soft-contact."* The robot uses mechanical means to detect contact with the wall, but the contact is "softened" with the use of pliable materials. For example, a lightweight foam wheel can be used as a "wall roller." The benefit of soft contact is that mechanical failure is reduced or eliminated, because the contact with the wall is through an elastic or pliable medium.

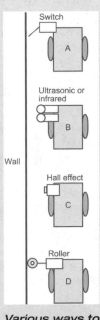

Various ways to follow a wall.

Making a Conductive Foam Pressure Sensor

Besides a microcontroller or computer, sensors tend to be the most expensive component on a robot. On the cheap end of the scale are simple touch sensors, which trigger whenever the robot bumps into something. The typical touch sensor is made from a switch, which is either open or closed. The disadvantage of such switch sensors is that they do not register force. To do that, you need a strain gauge, load cell, or similar technology, all of which can be quite expensive.

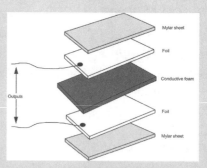

A cross section of a homemade conductive foam pressure sensor.

You can make your own pressure sensor out of a piece of discarded conductive foam, the stuff used to package CMOS integrated circuits. The foam is like a resistor. Attach two pieces of wire to either end of a 1-inch-square hunk and you get a resistance reading on your volt-ohm meter. Press down on the foam and the resistance lowers.

The foam comes in many thicknesses and densities. You'll have best luck with the semistiff foam that bounces back to shape quickly after squeezing it. The more dense the foam, the less compliant it is, and the slower it returns to its original shape.

To make a simple pressure sensor, first use transparent tape to attach the stripped end of a length of 28-30 AWG wire-wrap wire ends to a very thin piece of 1-inch-square unvarnished aluminum sheet. The sheet can be purchased at most any hobby store and should be flexible but not as thin as aluminum foil. Make two of these. Be sure the wires attached to the squares are at least 5 or 6 inches long.

Next, sandwich a piece of conductive foam between the two squares. Wrap up everything in one or two layers of transparent or masking tape. Don't apply the tape too thickly.

The interface circuit can be quite simple. The output of the conductive foam is a resistance, which is lowered when the foam is compressed. A typical RC (resistor-capacitor) circuit is employed, where the foam sensor is the resistor. This circuit is then used with a Basic Stamp or other microcontroller; the time it takes for current to discharge through a capacitor is directly related to the resistance R. The Basic Stamp has a special command for just this, RCtime. A typical RCtime program looks like this:

```
result    var    word        ' establish Word-sized variable named result
high 5                        ' bring I/O pin 5 high
pause 1                       ' for one millisecond
rctime 5, 1, result           ' read value RC charge time
debug ? result                ' display it on the screen
```

The Basic Stamp manual provides more details on how to use RCtime, and you'll want to experiment with different values of the capacitor C in order to derive meaningful results.

Note that conductive foam pressure sensors are not linear (their output does not track the pressure applied), and they are subject to settling times of several seconds to several minutes. During this settling time, after pressure has been removed, the foam is still returning to its original thickness. Therefore, it takes longer for the sensor to reset to normal than it does to register pressure.

(408) 654-8675

(408) 654-8575

(800) 235-0312

SemiconductorSupport@agilent.com

www.semiconductor.agilent.com

Makes and sell unique optical sensors for use in desktop computer mice. See listing under **Manufacturer-Semiconductors**.

Banner Engineering Corp. 202115

9714 Tenth Ave. North
Minneapolis, MN 55441
USA

(763) 544-3164

(763) 544-3213

(800) 345-1629

sensors@bannerengineering.com

http://www.bannerengineering.com/

Manufacturer of industrial photoelectric and fiber-optic sensors.

Dinsmore Instrument Co. 202471

P.O. Box 345
Flint, MI 48501
USA

(810) 744-1330

(810) 744-1790

sensors@dinsmoregroup.com

http://www.dinsmoresensors.com/

Dinsmore manufactures inexpensive digital and analog compass sensors. The popular 1490 outputs eight

The Dinsmore 1490 compass, from the bottom. Photo The Robson Company, Inc.

Fire Detection with Pyroelectric Sensors

You know about *pyroelectric* sensors: They are the "magic eye" used in modern motion-sensitive security lights. The sensor use body heat and motion to detect when a human, large animal, or alien is nearby. Infrared pyroelectric sensors require both heat and movement to work: Movement alone won't trigger them, nor will stationary heat. The sensors are passive—hence the acronym commonly used for them: *PIR*, for "passive infrared"—and emit no light or heat radiation of their own.

PIR units are sensitive to the radiation emitted by most fires. If the flame is large enough, there will be enough movement to trigger the sensor. Small flames, like a cigarette lighter or candle, are not enough to trigger the sensor, unless the flame is very close by. The larger the fire, the more the flame will flicker and move, and therefore the more reliable the detection.

Most PIR sensors are designed for security applications, where you want the widest field of view possible. That field of view is increased with the use of a plastic Fresnel lens, which is mounted in front of the sensor. For firefighting, a wide field of view not only increases the possibility of false alarms, but reduces the overall sensitivity of the sensor. You can increase the sensitivity and decrease the field of view by placing a small plastic lens (about 25mm or less) over the sensing window. Use a positive diopter lens: Plano convex or double convex are suitable. Shield the sensor window from stray light.

With a smaller field of view, you may need to "sweep" the room in order for your robot to see everything around it. You can either move the robot itself, or mount the sensor on a servomotor. The sweeping must stop periodically in order to take a "room reading." Otherwise, the motion of the sensor could trigger false alarms.

Sensing Touch with Piezoelectrics

Pierre and Jacques Curie discovered a new form of electricity just a little more than a century ago. The two scientists placed a weight on a certain crystal. The strain on the crystal produced prodigious amounts of electricity. The Curie brothers coined this new electricity "piezoelectricity"—*piezo* is derived from the Greek word meaning "press."

Later, the Curies discovered that the piezoelectric crystals used in their experiments underwent a physical transformation when voltage was applied to them. Thus, the piezoelectric phenomenon is a two-way street: Depress press the crystals and out comes a voltage; apply a voltage to the crystals and they respond by flexing and contracting.

Piezoelectric materials are commonly used in small speakers and buzzers. The beep-beep-beep of a wristwatch comes from a tiny piezo annunciator. These are made by depositing a piezo ceramic material on a thin piece of meal. Yet piezo activity is not confined to brittle ceramics. PVDF, or polyvinylidene fluoride is a semicrystalline polymer that lends itself to unusual piezoelectric applications. The plastic is pressed into thin, clear sheets and is given precise piezo properties during manufacture.

PVDF piezo film is currently in use in many commercial products, including noninductive guitar pickups, microphones, even solid-state fans for computers and other electrical equipment. One PVDF film you can obtain and experiment with is *Kynar*, available directly from the manufacturer, Measurement Specialties. The company sells a variety of Kynar sensors in single quantities, but (as of this writing) does have a $100 minimum order.

Kynar piezoelectric film is available in a variety of shapes and sizes. The wafers, which are about the same thickness as the paper in this book, have two connection points. Like ceramic discs, these two connection points are used to activate the film with an electrical signal or to relay pressure on the film as an electrical impulse.

When purchasing Kynar piezo sensors, opt for the ones with the lead wires already attached. This makes connecting the sensor to your circuit a lot easier. Measurement Specialties provides application notes and sample schematics you can follow to interface the Kynar material to other circuits.

One application of Kynar is as a bend sensor. Attach a strip of Kynar piezo material to a piece of thin flexible plastic, available at craft stores. Cut the plastic to about 1/2 inch by 6 or 7 inches long. You can use double-sided tape to secure the Kynar to the plastic. Next secure the plastic to the front of your robot to make a kind of bumper.

When the robot bumps up against something, the plastic will flex, and this in turn bends the Kynar sensor. The sensor will output a small voltage, which is detected by your robot's computer or microcontroller.

Note that as with all piezoelectric materials, the output of Kynar is an AC signal. That is, it doesn't produce a voltage and maintain it. Press the bumper, and the sensor outputs a momentary voltage, even if you keep applying pressure. After a fraction of a second, the output of the sensor returns to zero volts. Release the bumper so it returns to its natural shape, and an equal but opposite voltage is produced by the sensor.

PVDF piezo film is available from:

Budget Robotics (reseller)

http://www.budgetrobotics.com/

Images Co. (reseller)

http://www.imagesco.com/

Measurement Specialties, Inc. (manufacturer)

http://www.measurementspecialties.com/

digital compass positions (N-NE-E-SE-S-SW-W-NW). The 1525 sensor outputs a continuous analog sine/cosine signal capable of being decoded to any degree of accuracy.

Figaro USA Inc.   203041

3703 West Lake Ave.
Ste. 203
Glenview, IL 60025
USA

((847) 832-1701
℘ (847) 832-1705
✉ figarousa@figarosensor.com
🌐 http://www.figarosensor.com/

Makers of toxic gas and oxygen sensors

PERIM-ALERT III Fence Sensor   202078

http://www.perim-alert.com/sensor.shtml

Perim-Alert is a sensor sensitive to vibrations of chain link fences. Possible uses in robotics.

PNI Corp. / Precision Navigation   202448

5464 Skylane Blvd.
Ste. A
Santa Rosa, CA 95403
USA

((707) 566-2260
℘ (707) 566-2261
✉ customerservice@pnicorp.com
🌐 http://www.pnicorp.com/

PNI makes compass, radar, magnetometer, and inclinometer sensors. Note: The company's Vector 2X digital compass is no longer offered, though it may still be available from some other retailers.

Watson Industries, Inc.   204014

3041 Melby Rd.
Eau Claire, WI 54703
USA

((715) 839-0628
℘ (715) 839-8248
⊘ (800) 222-4976
✉ inforeq@watson-gyro.com
🌐 http://www.watson-gyro.com/

Electronic Compasses: Where on Earth Is Your Robot?

Just like sea and land explorers of old, robots can use magnetic compasses for navigation. A number of electronic and electromechanical compasses are available for use in hobby robots. The most basic compasses are accurate to about 45 degrees and provide heading information (N, S, E, W, SE, SW, NE, NW) by measuring the earth's magnetic field.

One of the least expensive electronic compasses is the Dinsmore 1490, from Dinsmore Instrument Company. The 1490 detects the earth's magnetic field by using miniature Hall-effect sensors and a rotating compass needle (similar to ordinary compasses). The sensor is said to be internally designed to respond to directional changes similar to a liquid-filled compass, turning to the indicated direction from a 90-degree displacement in approximately 2.5 seconds.

Dinsmore Instrument Co.

http://www.dinsmoresensors.com/

Another option is the Compass Module from Robot-Electronics. With a stated accuracy of ±3–4 degrees, this compass uses a variety of interface methods, including a digital pulse train, or an I2C serial network.

Robot-Electronics.com

http://www.robot-electronics.com/

Hamamatsu UVTron Flame Detector

The UVTron is an ultraviolet detector that is used to indicate the presence, and strength, of ultraviolet light. It is commonly used in robotics as a flame detector.

The sensor makes use of the photoelectric effect of metal combined with the gas multiplication effect. It has a narrow spectral sensitivity of ultraviolet light, specially from 185 to 260 nanometers. The sensor is completely insensitive to visible light. As the earth's atmosphere blocks most ultraviolet light coming from the sun, the sensor can be used outdoors without the need for extra filtering.

The UVTron sensors are small; the smallest is the R2868, popular in robotics applications. Though small, the R2868 has wide angular sensitivity (it can see a wide arc in front of it) and can reliably and quickly detect weak ultraviolet radiation emitted from a flame from a candle or a cigarette lighter. This makes the sensor well suited for robotic firefighting applications, such as the annual competition held at Trinity College.

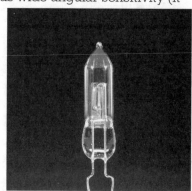
Hamamatsu's R2868 UVTron detector.

A special driver circuit is required to properly use the UVTron sensors. Driver circuits are provided by the manufacturer and can be purchased ready-made. Specialty robotics mail-order retailer Acroname (http://www.acroname.com/) is among a handful of sources that sell both the UVTron sensor and driver board.

For more information on the UVTron, go to:

http://usa.hamamatsu.com/ cmp-detectors/uvtrons/

Fancy sensors for high-end industrial applications:
- Solid-state angular rate sensors
- VSG angular rate gyro
- Attitude and heading reference
- Vertical references
- Inertial measurement unit
- Dynamic measurement system
- Fluxgate magnetometers
- Strapdown heading references

Xilor Inc. 202454

1400 Liberty St.
Knoxville, TN 37909
USA

((865) 546-9863
  (865) 546-8324
∅ (800) 417-6689
✉ info@rfmicrolink.com
🌐 http://www.rfmicrolink.com/

Check out their ZOFLEX ZL Series material, a pressure-activated conductive rubber. According to the Web site, the resistance change with pressure is very drastic. The material is at high resistance (30 Mohms) when pressure is below the actuation pressure. Resistance drops to 0.1 ohms or less when the material is at or above the activation pressure. The pressure required is too much for a "soft-touch" sensor, but other applications are possible.

SENSORS-RFID

Radio frequency identification (RFID) is a kind of sensor that is similar in purpose to bar codes, but it's meant to operate over longer distances and even through other objects. (Emplantable biochips, like the kind used for pets—and now people—are miniature RFID units.)

Applications in robotics are both obvious and numerous: You can use RFID for robot-to-robot identification, robot-to-human identification, navigation, beacon systems, and more. A benefit of RFID is that the sensitivity of the reader electronics can be varied so that you can directly control maximum working distances. In this way, a room could be full of RFID elements, yet your robot will only "see" the one closest to it.

As of yet, there are few RFID systems within affordable reach of most amateur robot builders; still, it's an interesting technology, and it's only a matter of time (perhaps just months) before affordable entry-level solutions become available. If nothing else, you can use the resources in this section to learn more about this technology.

CopyTag Limited
203841

Unit 7 Harold Close
Harlow, Essex
CM19 5TH
UK

(+44 (0) 1279 420438

(+44 (0) 1279 420443

✉ sales@copytag.com

🌐 http://www.copytag.com/

Makers of RFID receivers and tags (transponders).

Microchip Technology
202371

2355 W. Chandler Blvd.
Chandler, AZ 85224
USA

((480) 792-7200

((480) 899-9210

🌐 http://www.microchip.com/

Microchip makes a broad line of semiconductors, including the venerable PICmicro microcontrollers. Their Web site contains many datasheets and application notes on using these controllers, and you should be sure to download and save them for study.

The company is also involved with RFID, selling readers and tags, as well as developer's kits.

OMRON Corporation
203839

14th FL
Gate City Osaki West Tower
1-11-1 Osaki, Shinagawa-ku
Tokyo 141-0032
Japan

(+81 35 435 2016

(+81 35 435 2017

🌐 http://www.omron.com/

Omron is a multitalented company, manufacturing a wide array of sensors and semiconductors, such as RFID tags and readers and machine vision products

RFID—http://www.omron.com/card/rfid/

Machine vision—http://oeiweb.omron.com/oei/Products-VisionSys.htm

A variety of RFID tags. Photo OMRON Corp.

Smart Label. Photo OMRON Corp.

RACO Industries / ID Warehouse
203843

5480 Creek Rd.
Cincinnati, OH 45242
USA

((513) 984-2101

((513) 792-4272

🚫 (800) 446-1991

✉ info@racoindustries.com

🌐 http://www.idwarehouse.com/

Resellers of various bar code and RFID tagging systems.

Active Wave Inc.

http://www.activewaveinc.com/

RFID receivers and transponders

Fractal Antenna Systems, Inc.

http://www.fractenna.com/

Fractal antenna for RFID

Radio Frequency Identification

RFID, or radio frequency identification, uses small devices that radiate a digital signature when exposed to a radio frequency signal. RFID is found in products ranging from toys to employee access cards, to gasoline pump "key fobs" and trucking, to farm animal inventories, automobile manufacturing, and more.

How They Work

RFID tags reradiate a radio frequency signal, embedding a unique data code in it.

A transmitter/receiver, called the *interrogator* (also the *reader* or *host*), radiates a low- or medium-frequency carrier RF signal. If it is within range, a passive (unpowered) or active (powered) detector, called a *tag* or *transponder*, reradiates (or "backscatters") the carrier frequency, along with a digital signature that uniquely identifies the device. RFID systems in use today operate on several common RF bands, including a low-speed 100- to 150-kHz band, typified by the TIRIS line from Texas Instruments, and a higher 13.5-MHz band. The tag is composed of an antenna coil along with an integrated circuit. The radio signal provides power when used with passive tags, using well-known RF field induction principles. Inside the integrated circuit are decoding electronics and a small memory. A variety of data transmission schemes are used, including nonreturn to zero, frequency shift keying, and phase shift keying. Manufacturers of the RFID devices tend to favor one system over another, depending on the intended application. Some data modulation schemes are better at long distances, for example.

Why Use RFID?

How can they be used in robotics? Here are a couple of ideas:

- Place RFID tags on multiple robots that are meant to work together. The robots are able to identify one another as they come in proximity.
- Place RFID tags on people. This would allow a robot, or group of robots, to identify each person.
- Place RFID tags along baseboards of rooms or doorjambs. As a robot passes by a tag, it can determine where in a room, even which room, it is currently located.

It is the last idea that holds much promise for amateur robots. Navigating a robot within a structure or yard is made even more complicated if there are several rooms or areas that the robot must be kept aware of. RFID tags allow the robot to determine which room or area it is currently occupying, without resorting to more complicated mapping or image recognition.

When used for room identification, RFID tags operate as a kind of lighthouse, orienting the robot as it travels. Besides RFID, there are several other ways to provide active navigation signals to a robot: One is infrared beacons placed strategically in a room or area. Surprisingly, the advantage of RFID over infrared beacons is that the coverage of the RF signal is naturally limited. This provides a convenient way to differentiate the areas of a housebound robotic workspace.

Memory and Cost

RFID tags have differing amounts of memory, from just a few bytes to several thousand bytes. Most have 32 to 128 bytes—enough to store a serial number, date, and other limited data. For robotics, this is more than enough to serve as room-by-room or locale-by-locale beacons.

While RFID systems are not complex, cost is not quite in the superaffordable region (demonstration and developer's kits are available from some manufacturers in the $200 to $300 range, and this includes the reader and an assortment of tags). Handheld interrogators cost between $150 and $500; the tags cost under $1 each, but most manufacturers want you to buy them in quantity of 250 or more, making RFID an expensive proposition. However, once implemented RFID is a carefree and long-term solution to helping your robot know where it is.

RFID Components Ltd.   203837

Paragon House
Wolseley Road
Kempston
Bedford
MK42 7UP
UK

☎ +44 (0) 1234 840102
📠 +44 (0) 1234 840707
✉ info@rfid.co.uk
🌐 http://www.rfid.co.uk/

RFID Components is a specialist distributor of products for automatic recognition and identification applications. Their products include the Texas Instruments TI*RFID TIRIS line.

RFID, Inc.   203838

14100 E. Jewell Ave.
Ste. 12
Aurora, CO 80012
USA

☎ (303) 366-1234
🚫 (877) 999-7343
✉ sales@rfidinc.com
🌐 http://www.rfidinc.com/

Makers and sellers of RFID receivers and transponder tags. Offers relatively inexpensive starter kits with sampler tags and receiver.

RFID.org   203836

http://www.rfid.org/

Everything about RFID. Sponsored by AIM, the global trade association for the Automatic Identification and Data Capture industry.

Zebra Technologies Corporation   203842

333 Corporate Woods Pwy.
Vernon Hills, IL 60061-3109
USA

☎ (847) 634-6700
📠 (847) 913-8766
🚫 (800) 423-0442
🌐 http://www.zebra.com/

Zebra manufactures bar code readers (wands, CCD, and laser scanners), RFID readers and tags, and bar code label printers. See listing under **Bar Coding**.

 ## SENSORS-STRAIN GAUGES & LOAD CELLS

Strain gauge sensors—and their close cousin, the load cell—are used to measure a variety of physical attributes, including pressure, torque, tension, and bending. They are routinely used in commercial products, such as bathroom scales and automotive digital torque wrenches.

Though industrial strain gauges and load cells are quite expensive (upward of $500 for even a basic unit), there are a number of sources for low-prevision sensors that are well suited for robotics. These and other sources for strain gauges and load cells are provided in this section. (However, note possible minimum-order requirements.)

Budget Robotics   204255

P.O. Box 5821
Oceanside, CA 92056
USA

☎ (760) 941-6632

✉ info@budgetrobotics.com

🌐 http://www.budgetrobotics.com/

Robotics specialty retailer; components include low-cost strain gauges. See the listing under **Retail-Robotics Specialty**.

Davidson Measurement Pty. Ltd.   204057

1-3 Lakewood Boulevard
Braeside, Victoria, 3195
Australia

☎ +61 3 9580 4366

📠 +61 3 9580 6499

✉ info@davidson.com.au

🌐 http://www.davidson.com.au/

Distributor of industrial precision touch and force sensors. Among their products particularly well-suited to robotics are:

- Force—static and dynamic force sensors
- Strain—strain, stress and photoelastic measurement
- Pressure and Level—sensors, calibrators, indicators, and gauges

  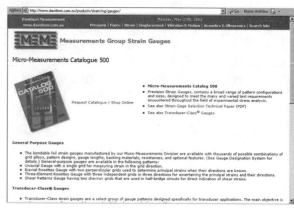

Strain Gauges Web page from Davidson Measurement.

Interlink Electronics, Inc.   203993

546 Flynn Rd.
Camarillo, CA 93012
USA

☎ (805) 484-1331

📠 805) 484-5997

🚫 (800) 340-1331

✉ sales@Interlinkelectronics.com

🌐 http://www.interlinkelec.com/

Strain Gauges

Perhaps the best way to measure physical force is with a strain gauge (and their close cousin, the load cell), used in weight scales, bridge load sensors, and thousands of other applications. Strain gauges are somewhat pricey—about $10 and over in quantity—and they aren't the kind of thing you can find at the corner Radio Shack. The cost may be offset by the increased accuracy the gauges offer. You want a gauge that's as small as possible, preferably one mounted on a flexible membrane.

For robots, strain gauges can be used as collision detectors, as touch pads for grippers and legs, and as position sensors for arms, among other applications.

Sources of stain gauges include:

Interlink Electronics, Inc.
http://www.interlinkelec.com/

OMEGA Engineering, Inc.
http://www.omega.com/

Vishay Intertechnology, Inc.
http://www.vishay.com/

Touch sensors and pads for laptop mice. The touch sensors use strain gauge (they call it a force-sensing resistor) technology. They sell developer's kits online (though they're a bit expensive) and provide free literature on how it all works. The company also manufactures and sells (via their online store) consumer products including keyboards and mice.

Measurement Systems, Inc. 203326

777 Commerce Dr.
Fairfield, CT 06432
USA

((203) 336-4590
🖑 (203) 336-5945
✉ sales@measurementsystemsinc.com
🌍 http://www.measurementsystemsinc.com/

Manufacturer of joysticks and miniature joysticks.

OMEGA Engineering, Inc. 203205

P.O. Box 4047
One Omega Dr.
Stamford, CT 06907-0047
USA

🖑 (203) 359-7811
⊘ (800) 826-6342
🌍 http://www.omega.com/

Omega makes sensors and data acquisition equipment. Of primary importance to robobuilders is their line of low-cost general-purpose strain gauges. These miniature sensors can be used to indicate stress or strain on an object, like the pad of a foot in a walking robot. The sensors are sold in packs of 10, and their per-piece cost is $5 to $8 for many sizes. This is considerably less than

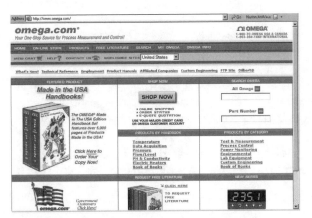

OMEGA Web page.

the average strain gauge, which is designed for super-precise industrial measurements.

The company Web site provides copious amounts of datasheets, application notes, and engineering articles.

Semtech Corporation 203323

200 Flynn Rd.
Camarillo, CA 93012
USA

((805) 498-2111
🖑 (805) 498-3804
✉ sales@semtech.com
🌍 http://www.semtech.com/

Makers of encoders for "pointing stick"–style laptop strain gauge pointing devices. The Web site offers datasheets and application notes. Available in sample quantities and from distributors.

Synaptics, Inc. 203322

2381 Bering Dr.
San Jose, CA 95131
USA

((408) 434-0110
🖑 (408) 434-9819
✉ info@synaptics.com
🌍 http://www.synaptics.com/

Sensors, including strain-gauge type to make pointing devices for handhelds and laptops: touch pad, "ScrollStrip," and "TouchStyk." Downloads for drivers and information. Direct sales may be tough, but you can salvage from older equipment.

 ## SENSORS-TILT & ACCELEROMETER

Tilt sensors and accelerometers can be used to indicate the attitude of a robot (whether it's at an angle—not whether it's having a bad day!). This is especially handy for walking robots, where excessive tilt must be corrected or the machine will fall flat on its nose-or batteries, or whatever.

Accelerometers are further able to determine velocity, sudden jarring movements, and a variety of other conditions. Though these sensors sound complicated and therefore expensive, thanks to the automotive business, which buys these in the millions, these are affordable. This section details a number of sources of tilt sensors and accelerometers, several of which provide sales to individuals directly or through industrial electronics distributors.

Analog Devices, Inc. ☿   202912

One Technology Way
P.O. Box 9106
Norwood, MA 02062-9106
USA

✆ (781) 461-3333

✇ (781) 461-4482

⊘ (800) 262-5643

🌐 http://www.analog.com/

Analog Devices is a key manufacturer of precision linear semiconductors. Among their product line is the low-cost ADXL series of accelerometer/tilt sensors, which can be readily used in amateur robotics. Their Web site contains copious datasheets, white papers, application notes, and other documentation. Samples can be ordered directly from the Web site; a number of industrial electronics distributors, such as Allied Electronics and Newark, carry the Analog line for resale.

The company's main product lines include the following:

- Single- and dual-axis accelerometers
- Analog-to-digital and digital-to-analog converters
- Op-amps

Sensors for Tilt Measurement

A common means of providing a robot with a sense of balance is with a tilt sensor or tilt switch. The sensor or switch measures the relative angle of the robot with respect to the center of the earth. If the robot tips over, the angle of the sensor/switch changes, and this can be detected by electronics in the robot. Tilt sensors and switches come in various forms and packages. Common varieties include:

- *Mercury-filled glass ampoules* that form a simple on/off switch. When the tilt switch is in one position (say, horizontal), the liquid mercury metal touches contacts inside the ampoule, and the switch is closed. But when the switch is rotated to vertical, the mercury no longer touches the contacts, and the switch is open. The major disadvantage to mercury tilt switches is the mercury itself, which is a *highly toxic* metal.

- *Ball-in-cage* all-mechanical switches are popular in pinball machines and other devices were small changes in level are required. The switch is a square or round capsule, with a metal ball inside. Inside the capsule are two or more electrical contacts. The weight of the ball makes it touch the electrical contacts, which forms a switch. The capsule may have multiple contacts, to measure tilt in all directions.

- *Electronic spirit level sensors* use the common fluid bubble, along with some interfacing electronics. A spirit level is the same kind you see in ordinary levels at the hardware store: It's merely a glass tube filled, but not to capacity, with water or some other fluid. A bubble forms at the top of the tube since it isn't completely filled. Because of gravity, titling the tube makes the bubble slosh back and forth. An optical sensor—an infrared LED and detector, for example—can be used to measure the relative size and position of the bubble.

- *Electrolytic tilt sensors* are like mercury switches, but are more complex and a lot more costly. In an electrolytic tilt sensor a glass ampoule is filled with a special electrolyte liquid—a liquid that conducts electricity, but in very measured amounts. As the switch tilts, the electrolyte in the ampoule sloshes around, changing the conductivity between two (or more) metal contacts.

- *Accelerometers*, which can electronically detect the pull of the earth's gravity. The latest accelerometers are available on a single integrated circuit, making them small and relatively inexpensive.

What's an Accelerometer?

An accelerometer is a device that measures *change* in speed. Accelerometers are used in vehicles, rockets, even elevators. Put an accelerometer in a car, for example, and step on the gas. The device will measure the increase or decrease in speed. Most accelerometers only measure acceleration (or deceleration) and not constant speed or velocity. Velocity information can be derived from acceleration data, however.

Although designed to measure changes in speed, many types of accelerometers are also sensitive to the constant pull of the earth's gravity. This type of accelerometer can used to measure the tilt of a robot. This tilt is represented by a change in the gravitational forces acting on the sensor. The output of the accelerometer is either a voltage, or it can be a digital signal that indicates the relative acceleration at any given moment.

Analog Devices ADXL202 Accelerometer

Analog Devices is a semiconductor maker of primarily industrial and military-grade operational amplifiers, digital-to-analog and analog-to-digital converters, and motion control products. One of their key product lines is ADXL line of accelerometers, which uses a patented fabrication process to create a series of near-microscopic mechanical beam. This "micromachining" involves etching material out of a substrate. During acceleration, the beam is distended along its length. This distention changes the capacitance in nearby plates. The change in capacitance correlates as acceleration and deceleration.

The ADXL202 accelerometer mounted on an evaluation board carrier.

In addition to the mechanical portions of the accelerometer, all the basic interface circuitry is part of the device. In fact, looking at one of the ADXL accelerometers, you'd think they were just integrated circuits of some type. Because the basic circuitry is included as part of the accelerometer, only a minimum number of external parts are needed.

Analog Devices makes a lower-cost line of accelerometers specifically designed for consumer products. Their ADXL202 is a dual-axis device with a +/- 2 g sensitivity (if you need more g's, check out the ADXL210, which is rated at +/- 10 g's). The ADXL202 has a simplified output: As acceleration changes, the timing of the pulses at the output of the device changes. This change can be readily determined by a PC or microcontroller that measures the length of the pulse.

The ADXL202 is a surface-mount component, so if you want to use it with a regular prototyping board or solderless breadboard, you'll need to solder it to a carrier, then attach the carrier to the rest of your chip. Analog also sells a handy evaluation board with an ADXL202 and a standard 0.100-inch header for easy attaching to standard breadboards and solder boards.

Analog provides a datasheet and application notes on using the ADXL202 and provides links to resellers of the chip on their Web site (you can also buy some products directly from the company):

http://www.analog.com/

When at the Web site, enter *ADXL202* as the search phrase, and you'll be taken directly to the support pages for the device.

- LCD drivers
- Temperature sensors
- Digital signal processing

Gyration, Inc.   202235

12930 Saratoga Ave.
Bldg.C
Saratoga, CA 95070
USA

((408) 255-3016

℡ (408) 255-9075

✉ sales@gyration.com

🌐 http://www.gyration.com/

Gyration makes the innards of the Gyroscope mouse. They'll sell a gyroscope module developer's kit, but an easier (and perhaps cheaper) way is to buy a Gyropoint mouse at the local computer store.

Says the Web site, "The magic behind the innovative Gyration products is the MicroGyro 100, a revolutionary new gyroscope sensor which can accurately sense the motion of your hand or body. This gives you a new freedom from your desktop, which is enhanced by a robust, long-range radio frequency design for cordless operation."

NEC-Tokin America   203873

32950 Alvarado-Niles Rd.
Ste. 500
Union City, CA 94587
USA

((510) 324-4110

℡ (510) 324-1762

🌐 http://www.nec-tokinamerica.com/

NEC manufactures a variety of industrial sensors:

- Fluxgate-type terrestrial magnetic sensor
- 3D Motion Sensor MDP-A3U7
- Ceramic Gyro

See also their "Flex-Suppressor," a flexible rubberized sheet that acts to reduce RF interference, from 10 MHz to 1 GHz.

PNI Corp. / Precision Navigation   202448

5464 Skylane Blvd.
Ste. A
Santa Rosa, CA 95403
USA

((707) 566-2260

℡ (707) 566-2261

✉ customerservice@pnicorp.com

Piezo Gyros for Inertial Navigation

The same physics that keep a bicycle upright when its wheels are in motion can be used to provide motion data to a robot. Consider a bicycle wheel spinning in front of you while you hold the axle between your hands. Turn sideways, and the wheel tilts. This is the gyroscopic effect in action; the angle of the wheel is directly proportion to the amount and time you are turning. Put a gyroscope in an airplane or ship and even imperceptible changes in movement can be recorded, assuming the use of a precision gyroscope.

Gyros are still used in airplanes today, even with radar, ground controllers, and radios to guide the way. While many modern aircraft have substituted mechanical gyros with completely electronic ones, the concept is the same: During flight, any changes in direction are recorded by the inertial guidance system in the plane (there are three gyros, one for each of the three axes). Any time during the flight the course of the plane can be scrutinized by looking at the output of the gyroscopes.

Inertial guidance systems for planes, ships, missiles, and other such devices are far, far too expensive for robots. However, there are some low-cost gyros that provide modest accuracies. One reasonably affordable model is the Max Products MX-9100 micro piezo gyro, often used in model helicopters. The MX-9100 uses a piezoelectric transducer to sense motion; this motion is converted to a digital signal whose duty cycle changes in proportion to the rate of change in the gyro.

Sources for Piezeo Gyros

Gyroscopes suitable for use in amateur robotics are available from the following companies:

Airtronics Inc.—*http://www.airtronics.net/*
Crossbow Technology, Inc.—*http://www.xbow.com/*
Dragonfly Innovations—*http://www.rctoys.com/*
Gyration, Inc.—*http://www.gyration.com/*
Murata Manufacturing Co.—*http://www.murata.com/*
NEC-Tokin America—*http://www.nec-tokinamerica.com/*

 http://www.pnicorp.com/

PNI makes compass, radar, magnetometer, and inclinometer sensors. Note that the company's Vector 2X digital compass is no longer offered, though it may still be available from some other retailers.

Spectron   203999

595 Old Willets Path
Hauppauge, NY 11788
USA

(631) 582-5600
(631) 582-5671
info@spectronsensors.com
http://www.spectronsensors.com/

Single- and dual-axis tilt sensors, single- and dual-axis inclinometers (digital and analog outputs). Application notes available at the Web site.

 SENSORS-ULTRASONIC

Sound is a convenient and accurate means of measuring distance. Ultrasonic sensors are commonly used in both consumer goods (e.g., cameras) and industrial process control for determining the distance of objects. Ultrasonic sensors run the gamut of the very sophisticated and complex—especially those that measure Doppler shift—to the very inexpensive. This section lists sources within these two extremes. For a hobby robot, no doubt you'll want to concentrate on the low end of this scale. Also included are several Web site articles on how-to build your own ultrasonic sensors.

Automation Sensors   202112

6550 Dumbarton Cir.
Fremont, CA 94555
USA

(435) 753-7300
(435) 753-7490
(888) 525-7300
http://www.automationsensors.com/

Makers of self-contained ultrasonic sensors and pressure products. Check out the technical reference section for a number of application notes, as well as handy white papers (in Adobe Acrobat PDF format) on such things as dielectric constants, bulk densities, engineering unit abbreviations, and thread specifications.

Cheap Ultrasonics   203028

http://www.mindspring.com/~sholmes/
    robotics/ultrasnd.htm

How to build and program an inexpensive ultrasonic ranging system for your robot.

Dissecting a Polaroid Pronto
One Step Sonar camera   203096

http://www.robotprojects.com/sonar/scd.htm

Step-by-step guide to dissecting a Polaroid camera for its ultrasonic sensor. Follow-up article demonstrates using the sensor with an OOPic microcontroller.

Ultrasonic Distance Measurement

In 1947 American test pilot Chuck Yeager broke the sound barrier. The plane he was flying, *Glamorous Glennis* (today on exhibit at the Smithsonian in Washington, D.C.) flew faster than sound waves can travel through the air.

Sound travels at about 1,130 feet per second at sea level. While temperature, humidity, and atmospheric pressure change the speed of sound, for general purposes we can consider it a constant and use it for distance measurement. For years, the fancier Polaroid instant cameras used an ultrasonic focusing system to ensure clear, sharp pictures. Their ultrasonic system worked so well the company was able to make a tidy extra profit on selling the sensors to other manufacturers (Polaroid recently sold the ultrasonic business unit).

Theory of Operation

To measure distance, a short burst of ultrasonic sound—usually 40 kHz for most ultrasonic ranging systems—is sent out through a transducer (essentially a fancy term for a speaker). The sound bounces off an object, and the echo is received by another transducer (this one a specially built ultrasonic microphone). A circuit them computes the time it took between the transmit pulse and the echo and comes up with distance.

Given a speed of 1,130 feet per second (about 344 meters per second), the time it takes for the echo to be received is in microseconds if the object is within a few inches or feet of the robot. Though a few microseconds is a short period of time on the human scale, it's no problem for modern fast-acting CMOS and TTL ICs.

Ultrasonic sensors use the speed of sound to measure distance.

Given a travel time of 13,560 inches per second for sound, it takes 73.7 microseconds (0.0000737 seconds) for sound to travel 1 inch. Or if using centimeters, sound travels 34,442.4 centimeters per second, or 29.03 microseconds per centimeter. In any ultrasonic ranging system, the total transit time between transmit pulse and echo is divided by two, to compensate for the round-trip travel time between the robot and the object. To calculate in inches, the remaining value is divided by 73.7 (for inches) or 29.03 (for centimeters) to determine the distance.

Maximum and Minimum Ranges

Sound waves eventually dissipate, so there is a maximum range you can expect from your ultrasonic distance measurement sensor (the range is reduced even more outdoors, where wind disperses sound waves). The maximum distance of the Polaroid ultrasonic transducer is about 35 feet when used indoors, and a little less when used outdoors, especially on windy days. Other ultrasonic sensors have the same, or even less, maximum range. A range of 3 to 6 meters is usually adequate for robotics, where it is not necessary to detect objects outside the immediate field of interest.

Likewise, ultrasonic systems exhibit minimum working distances. Because the Polaroid ultrasonic transducer is used for both transmitting and receiving, its minimum distance is 6 inches. The reason: The ranging board cannot listen to sound echoes until the transmit phase is complete. The transducer continues to vibrate slightly for a short period after sending a pulse of ultrasonic sound. During this "blanking" period, which is required to eliminate false readings, return echoes are ignored.

Ultrasonic sensors that use separate transmit and receive transducers do not require a blanking period, as long as the transducers are microphonically isolated (that is, mechanical vibrations from the transmitter are not picked up by the receiver). Minimum working distances can be effectively reduced to about an inch.

Fascinating Electronics, Inc. · 204137

31525 Canaan Rd.
Deer Island, OR 97054-9610
USA

☎ (503) 397-1222

📠 (503) 397-1191

🚫 (800) 683-5487

✉ fascinating@columbia-center.org

🌐 http://www.columbia-center.org/
fascinating/

Among other interesting products, Fascinating Electronics markets an affordable ultrasonic sensor and backs it up with experimental software. Attach the sensor to a motor, as shown on the Web site, to build a scanning "radar" unit.

See also their Dual-Wiper Potentiometer and Humidity Sensor.

Felio Parking Sensor · 203023

Lot 5, Jalan Gudang 16/9
Seksyen 16
40200 Shah Alam
Selangor Darul Ehsan
Malaysia

☎ +60 35512 7763

📠 +60 35512 8163

✉ inquiry@feliogroup.com

🌐 http://www.feliogroup.com.my/

Makers of an ultrasonic parking sensor that is attached to the bumper of a car. Interesting idea that can be applied to robotics. Some sample products available for sale directly from the Web site.

Massa Products Corporation · 202473

280 Lincoln St.
Hingham, MA 02043
USA

Ultrasonic Sources

Acroname
http://www.acroname.com/

All Electronics Corp.
http://www.allcorp.com/

B.G. Micro
http://www.bgmicro.com

Hobby-Electronics.com
http://www.robot-electronics.com

Jameco Electronics
http://www.jameco.com/

Marlin P. Jones & Associates, Inc.
http://www.mpja.com/

Polaroid OEM Components Group
http://www.polaroid-oem.com

Robot Store
http://www.robotstore.com/

Robot Store (HK)
http://www.robotstorehk.com

Hacking Polaroid cameras for their ultrasonic sensors:

http://www.robotprojects.com/

http://www.techtoystoday.com/

http://incolor.inetnebr.com/bill_r/robotics.htm

Goolge.com search phrases:
polaroid ultrasonic hack OR hacking
polaroid 6500
ultrasonic robotics

(781) 749-4800

 (781) 740-2045

(800) 962-7543

sales@massa.com

http://www.massa.com/

High-end ultrasonic sensors.

Mekatronix, Inc.   202970

316 NW 17th St.
Ste. A
Gainesville, FL 32603
USA

tech@mekatronix.com

http://www.mekatronix.com/

From the Web site, "Mekatronix is a manufacturer of autonomous mobile robots, robot kits, microcontroller kits and robot accessories, as well as educational materials related to science and robotics. Our robots and microcontrollers provide students with valuable hands-on experience in programming and engineering concepts."

Products are available through a few dealers.

Mekatronix Web page.

Migatron   202116

935 Dieckman St.
Woodstock, IL 60098
USA

 (815) 338-5800

 (815) 338-5803

(888) 644-2876

info@migatron.com

Polaroid 6500 Ultrasonic Ranging Module

Digital photography has all but ended the market for chemical-based instant photography. But for many years, Polaroid had a lock on the technology that allowed people to snap the shutter and within a minute see a clear, full-color picture. The name Polaroid became synonymous with instant photography.

A number of the higher-end Polaroid cameras used another engineering marvel: a high-speed focusing system that used ultrasonic sound waves. The gold-tinted ultrasonic sensor used with these cameras sported a sensing range of under 1 foot to 35 feet. The sensor worked so well for their cameras that Polaroid decided to sell it to other manufacturers, who incorporated it into things like electronic tape-measuring devices.

The fortunes of the Polaroid company have gone through a lot of changes recently, and these days they make few instant cameras. Polaroid recently sold their ultrasonic sensor business unit to SensComp; the latter continues Polaroid's practice of selling the sensors and associated electronics online, and through specialty retailers.

The Polaroid 6500 ultrasonic ranging module.

The Polaroid ultrasonic sensor is a common find in many commercial and university lab robots. The sensor itself is mounted on the front, back, or side of the robot (some robots use multiple sensors, in order to detect obstacles all around). The sensor connects to an interface board, which is located inside the robot. This board is in turn attached to the computer or microcontroller of the robot.

In operation, the robot's computer or microcontroller commands the interface board to "ping" the sensor. The sensor emits a short burst of high-frequency sound (about 50 kHz). The output of the interface board changes state if and when an echo of this sound is detected. Distance is determined by calculating the time between pinging the sensor and receiving its echo. It is up to the robot's computer or microcontroller to perform this timing.

For more information on these ultrasonic products, visit:

http://www.senscomp.com/

http://www.acroname.com/

http://www.migatron.com/

Makers of high-end industrial ultrasonic sensors. Web site contains datasheets in HTML and Adobe Acrobat PDF format.

Paul's Cheap Sonar Range Finder Design

203466

http://www.hamjudo.com/sonar/

How Paul built an inexpensive ultrasonic sonar system using a PIC16F84 microcontroller.

Polaroid Ultrasonic Sensors / SensComp, Inc.

202118

P.O. Box 530790
Livonia, MI
48153-0790
USA

((734) 953-4783

℡ (734) 953-4518

 http://www.senscomp.com/

SensComp, Inc. now wells the Polaroid ultrasonic sensors and developer's kits ("L", "K" & 9000 Series), along with supporting components. You can order products online or from a reseller, such as Acroname.

Sonar Transducers: Buying New or Hacking Old

There are a variety of ways to implement ultrasonic ranging:

- *Purchase a ready-made non-Polaroid sonar ranging system*, such as the SRF08 and SRF04 from Hobby-Electronics.com. The SFR04 requires your robot to calculate transit time of the ultrasonic pulse; the SRF08 handles all calculations for you and returns the result as a digital value.

- *Purchase a ready-made Polaroid 6500 sonar ranging system*. Sources include Acroname.com and RobotStore.com. The 6500 consists of a gold-plated Polaroid ultrasonic sensor and a driving board, as well as engineering notes and application sheets.

- *Purchase transducers and build your own ultrasonic distance board*. Matched sensors are available from a number of sources, including Jameco, Marlin P. Jones, B.G. Micro, and All Electronics. They are relatively inexpensive (usually under $5 for the pair). You need a circuit to provide 40 kHz to the transmitter, and another to amplify and lock onto the 40-kHz signal from the receiver.

- *Salvage a sonar ranging system from a used Polaroid camera*, such as the Sun 660. These use sensor boards similar, but not identical, to the Polaroid 6500. Using the sources provided in this article, you can hack the ranging board and connect it to any of a number of microcontrollers. The ranging board requires its own 6-volt supply.

Acroname—***http://www.acroname.com***
All Electronics—***http://www.allcorp.com/***
B.G. Micro—***http://www.bgmicro.com/***
Jameco Electronics—***http://www.jameco.com/***
Marlin P. Jones—***http://www.mpja.com/***
Robot Electronics—***http://www.robot-electronics.co.uk/***
Robot Store—***http://www.robotstore.com/***
SensComp—***http://www.senscomp.com/***

Robobix 203706

http://www.geocities.com/robobix/

Circuits include: ultrasonic distance measurement, light-reflection distance measurement, and simple infrared object detection.

Senix Corporation 202113

52 Maple St.
Bristol, VT 05443
USA

  (802) 453-5522
  (802) 453-2549
  (800) 677-3649
  http://www.senix.com/

Senix (not the same as semiconductor maker Scenix, which is now Ubicom) is a maker of ultrasonic distance measurement sensors for machinery and industrial automation.

Ultrasonic Imaging (usi) Project 203042

http://www.geocities.com/baja/ravine/
    4301/usi_project/index.htm

Or "Fun with Polaroid 6500 modules and a bit of cunning," written by Jim Whiteside. Jim provides notes on using the Polaroid 6500 ultrasonic board and transducer, writing PIC code for Polaroid sonar modules, and more.

SUPPLIES

Supplies are consumable goods that you use and replace as needed. This section, and the ones that follow, detail consumable supplies commonly used in basic robot construction, including casting and mold-making materials (should you be casting your own parts), chemicals, glues and adhesives, and paints.

Refer also to the resources in this section for companies that provide a wide variety of materials and supplies. These might be considered "general stores" for robot building and carry a broad line of industrial components and consumables.

SEE ALSO:

FASTENERS: To put your robot together

MATERIALS (VARIOUS): Hardware and construction parts

RETAIL-ARTS & CRAFTS: More sources for paints, glues, and adhesives

RETAIL-HARDWARE & HOME IMPROVEMENT: Local and online retailers for a variety of supplies

---

Enco Manufacturing Co. 203672

400 Nevada Pacific Hwy.
Fernley, NV 89498
USA

((770) 732-9099

⊘ (800) 873-3626

✉ info@use-enco.com

🌎 http://www.use-enco.com/

See listing under **Tools**.

Grainger (W.W. Grainger) 202928

100 Grainger Pkwy.
Lake Forest, IL 60045-5201
USA

((847) 535-1000

📠 (847) 535-0878

🌎 http://www.grainger.com/

See listing under **Materials**.

Graybar Electric Company, Inc. 203973

34 N. Meramec Ave.
Clayton, MO 63105
USA

((314) 512-9200

📠 (314) 512-9453

⊘ (800) 472-9227

🌎 http://www.graybar.com/

See listing under **Materials**.

McMaster-Carr Supply Company 202121

P.O. Box 740100
Atlanta, GA 30374-0100
USA

((404) 346-7000

📠 (404) 349-9091

✉ atl.sales@mcmaster.com

🌎 http://www.mcmaster.com/

See listing under **Materials**.

WESCO International, Inc. 203974

4 Station Sq.
Pittsburgh, PA 15219
USA

((412) 454-2200

📠 (412) 454-2505

🌎 http://www.wescodist.com

See listing under **Materials**.

SUPPLIES-CASTING & MOLD MAKING

Casting and mold making aren't skills shared by most robot builders, but they should be. With little training, and only a few dollars in materials, you can produce your own parts for your robo-creations. The process is simple enough: Use plaster or some other material to create a mold from an existing element. One possibility is hubs for wheels. As long as you have one hub for the model, you can make additional ones as needed. The hub can be something you've purchased, retrofitted from some other item, or custom made using wood, metal, or plastic.

After the mold has dried and cured, you can then cast parts from it. Casting can be made using hard or soft rubber, even certain low-temperature metals.

The benefits of casting are twofold: First, you can save money by making your own parts. Buying new sprockets, gears, wheel hubs, and other components can get expensive. Once you make a mold, each casting might cost only 25 cents each, depending on the material-obviously, some casting materials are more expensive than others, but overall, it's pretty inexpensive stuff. Second, you can "mass-produce" unique items, rather than hand-make them one at a time. You need only build one model, then make duplicates with casting.

This section lists resources concerned with casting and mold making, including materials and instruction. In addition to the casting arts, this section also includes sources for vacuum forming, which is a process of making molded shapes with thin plastic sheets.

SEE ALSO:

BOOKS-TECHNICAL: Find books about casting

MATERIALS-PLASTIC: Includes vacuum-formable plastic sheets

RETAIL-ARTS & CRAFTS: Most carry consumer-grade casting and mold-making supplies

---

Abatron, Inc. 203182

5501-95th Ave.
Kenosha, WI 53144
USA

☎ (262) 653-2000

📠 (262) 653-2019

🚫 (800) 445-1754

✉ caporaso@abatron.com

 http://www.abatron.com/

Abatron manufactures and distributes adhesives, sealants, coatings, epoxy, polyurethane, and various other resins. Of particular interest to robot builders is their casting and mold-making supplies.

One interesting product is Woodcast, a two-part light casting material that is machinable and even stainable. You basically mix up the liquid and pour it into a mold. You can shape and sand the material afterward. If the "natural look" isn't right for your robot, there's also Abocast, two-part clear liquid for casting. Abocast is a tooling resin, so you can use it to make complex shapes should you have access to a mill, lathe, or computerized router.

Abatron online.

Alumilite Corporation 203180

315 E. North St.
Kalamazoo, MI 49007
USA

☎ (616) 488-4001

🚫 (800) 447-9344

✉ world@alumilite.com

🌐 http://www.alumilite.com/

Alumilite sells casting resins and mold-making materials. The company sells supplies in small quantities or bulk (up to 100-gallon drums), as well as Super Casting Kit and Mini Super Casting Kit, both designed for the consumer market. Additional products:

- Urethane dyes
- Metallic powder
- Microballoons
- Alumilite 610 foam (expands to 6-10 times the original liquid volume)

The Web site includes numerous how-tos on mold making and casting. Product is available through dealers; a dealer locator map is provided at the site. Dealers are located in North America, Australia, Singapore, and the U.K.

American Art and Clay Co., Inc. (Amaco)

203984

4717 W. 16th St.
Indianapolis, IN 46222
USA

(317) 244-6871

(317) 248-9300

(800) 374-1600

http://www.amaco.com/

Amaco manufacturers and sells (principally through retailers) a line of arts and crafts products which include WireForm wire mesh (can be sculpted to form a shape), Friendly Plastic (softens in warm water), Easy Metal (thin metal foil), and glow-in-the-dark paints.

Armorcast

202045

P.O. Box 14485
Santa Rosa, CA 95402-6485
USA

(707) 576-1619

(707) 576-1619

timdp@armorcast.com

http://www.armorcast.com/

Polyurethane resin molds. Much of Armorcast's product line is "scenery," things like miniature trees and rocks for models. Suitable for robot display or for a playing field for your combat bot.

Bare-Metal Foil Co.

203236

P.O. Box 82
Farmington, MI 48332
USA

(248) 477-0813

Discount Dental Supply

http://207.69.159.104/DDS/Home2.html

Dental plaster

M-PACT Worldwide

http://www.mpactmed.com/

Orthopedic medical supplies: fiberglass casting tape, plaster-of-Paris casting

New Mexico Clay Inc.

http://www.nmclay.com/

Casting plasters, dental plaster

---

(248) 476-3343

contact@bare-metal.com

http://www.bare-metal.com/

Bare-Metal Foil sells bare metal foils, silicone RTV, casting resins, decals, and other products for the model maker. Useful how-to articles.

Barnes Products Pty Ltd.

203174

6 Homedale Road
Bankstown, NSW, 2200
Australia

+61 2 9793 7555

+61 2 9793 7091

info@barnesproducts.com.au

http://www.barnesproducts.com.au/

Barnes is a supplier of silicone rubber, polyurethane, epoxy, polyester resin, foam latex, and other casting and mold-making materials.

Botanical Science

203523

P.O. Box 10909
San Bernardino, CA 92423
USA

(909) 382-0175

(909) 382-0179

leanne@botanicalscience.com

http://www.botanicalscience.com/

Makers of 3-D Gel, a nontoxic, very quick setting molding compound. Also clay and plaster products. Sold at arts and crafts stores.

Burman Industries, Inc. 203413

14141 Covello St.
Ste. 10-C
Van Nuys, CA 91405
USA

((818) 782-9833
📞 (818) 782-2863
📧 info@burmanfoam.com
🌐 http://www.burmanfoam.com/

See listing under **Materials-Metal**.

Castcraft 203181

P.O. Box 17000
Memphis, TN 38187-1000
USA

((901) 682-0961
🌐 http://www.castcraft.com/

Castcraft calls itself "the place for mold making and casting." Along with metal, plastic, and plaster casting, they offer products, plans, and parts to build a home-made vacuum forming machine.

Castcraft Web site.

Compleat Sculptor, Inc., The 203244

90 Vandam St.
New York, NY 10013
USA

((212) 243-6074
📞 (212) 243-6374
🚫 (800) 972-8578
📧 tcs@sculpt.com
🌐 http://www.sculpt.com/

Materials, supplies, tools, and service for sculpting, mold making, and casting.

Composite Store, The 203557

P.O. Box 622
Tehachapi, CA 93581
USA

((661) 822-4162
📞 (661) 822-4121
🚫 (800) 338-1278
📧 info@cstsales.com
🌐 http://www.cstsales.com/

See listing under **Materials-Fiberglass & Carbon Composites**.

Douglas and Sturgess, Inc. 203178

730 Bryant St.
San Francisico, CA 94107-1015
USA

((415) 896-6283
📞 (415) 896-6379
📧 Sturgess@ix.netcom.com
🌐 http://www.artstuf.com/

Casting and mold making for the sculptor: sculpture tools, material, and supplies. Much of it is also useful for most any casting and mold-making project that might involve robots or robot parts. Interesting products include:

- Atomized metal powders
- Coating materials
- Latexes and other flexible mold materials

- Foam board materials (such as pink styrene, polyurethane)

- Fabricform (plastic laminated between pieces of f-abric)

- Altaform (general-use substrate in three different weights)

- FormFast (also known as celastic): plastic-impregnated fabric softens with acetone; can be shaped and left to reharden

DRU Industries, Inc. 203245

3800 Midway Pl. NE
Unit H
Albuquerque, NM 87109
USA

- ((505) 344-0202
- ♆ (505) 345-2900
- ⊘ (800) 600-6653
- ✉ magic@drua-1.com
- 🌐 http://www.wrldcon.com/dru/

Sellers of Smooth-on, clay products, adhesives, tapes, and mold-making supplies.

Eager Plastics, Inc. 203177

3350 W. 48th Pl.
Chicago, IL 60632-3000
USA

- ((773) 927-3484
- ♆ (773) 650-5853
- 🌐 http://www.eagerplastics.com/

Eager Plastics sells materials for plastic casting and mold making. Products include:

- Flexible mold-making silicones

- Flexible mold-making urethanes

- Semirigid and rigid casting urethanes

- Epoxy systems

- Pigments and dyes

- Polyester resins and fiberglassing supplies

- Fiberglass fabric

- Resin fillers and additives

- Release agents

For ideas on what you can do with casting and mold making, check out the finished products photos, such as *Star Trek VI* phasers (made with a smooth casting resin).

Environmental Technology, Inc. 203532

South Bay Depot Rd.
Fields Landing, CA 95537-0365
USA

- ((707) 443-9323
- ♆ (707) 443-7962
- ✉ mail@eti-usa.com
- 🌐 http://www.eti-usa.com/

Manufacturer of casting and molding supplies: epoxies, adhesives, casting compounds, latex rubber, etc. Available in industrial packaging. Contact the company for more information. The Castin'Craft brand, available in smaller quantities, can be purchased at local craft stores.

eWellness 203491

7750 Zionsville Rd.
Ste. 850
Indianapolis, IN 46268-5116
USA

- ⊘ (800) 472-0604
- ✉ info@ewellness.com
- 🌐 http://www.ewellness.com/

Medical supplies, including fiberglass casting supplies—you use it to make stuff for robot bodies, of course, not to fix broken arms and legs. For consumers.

Far West Materials 203238

405 Woodland Ave.
Walla Walla, WA 99362
USA

- ((509) 522-0556

Steps to Casting in Plastic

Casting is a method of creating parts from a master mold. Casting can save you time and money and, if done right, can produce a superior-looking product. Let's say you need to produce 12 sets of servo mounts for your six-legged hexapod robot. You'll need two separate pieces for each servo, making for a total of 24 mounts. You could cut and drill them all, and maybe end up with something respectable. But another approach is to cast them using liquid plastic. Do it right, and you never have to cut or drill anything.

Or, suppose you need two #25 chain sprockets, but only have one in your parts bin. With that one sprocket—the "model"—you can make as many duplicates as you like. Casting can be helpful if you're on a tight budget or can't wait a week or two to order a part.

Casting an object in plastic is not difficult, but it can be time-consuming. The general steps are to create the mold, then create the casting. But of course, there's a bit more to it than that. Here's an overview.

Create the Mold

All castings require a mold of some type. If a mold doesn't yet exist, you will need to make one from a model. The most common molds are made of plaster (also known as plaster of paris), which you can purchase at any arts and craft store. Specialty plasters, such as dental plaster (examples: Hydrocal and Die-Keen), are available from mail-order outlets. The mold is produced by glopping the semiliquid plaster over the model and letting it dry.

Plaster molds are either one-piece of two-piece. A one-piece mold has just one half and is suited for objects where only the front side need be cast because the final object is meant to be placed against a flat panel (examples are decorative trim or some kind of flat hood ornament for your robot). A two-piece mold has a front and back and is meant to cast a stand-alone figure.

Other mold materials exist, of course, and are better suited than plaster for some tasks. Plaster molds can't have any "undercuts" because the plaster dries hard and stiff. The more the model is irregularly shaped, the more you will need to use a flexible mold material, such as latex. Latex molds do not hold their shape for the casting step, however, and need to be reinforced with a plaster "mother mold." In all, making latex and other flexible molds takes extra time.

Create the Casting

A casting is made by pouring a liquid or semiliquid material into the mold and letting the material set up and cure. Some casting materials, like hot melt (very much like hot glue), set up when they cool. These are the fastest to work with. Others, like urethane resin, require a chemical change that can take several hours or even overnight.

Most liquid plastics (like urethane) for casting are composed of two parts: the main plastic resin, and a catalyst. Separately, the materials will never harden to produce a finished plastic piece. When mixed, a chemical reaction causes curing that results in hardening. The catalyst must be added to the plastic in just the right amounts, or the casting will be ruined. The mixture is either by weight or by volume; each manufacturer provides an instruction sheet that you should follow to the letter.

Depending on the materials you use, you may need to apply what's known as *mold release* before pouring in the liquid casting gloop. Petroleum jelly is common for plaster molds, though there are also thinner sprays available that create less mess. It is critical that the mold release be compatible with both the mold and casting materials, or else the casting may never fully cure.

Wait, Wait, Then Wait Some More

The single greatest error made by beginners to the mold-making and casting art is trying to rush things. It can take a day or two between start and finish to complete a casting, especially if a mold is also required. And, depending on the casting materials you use, a plaster mold will need extra time—days or even weeks—for the water in the plaster to dry out.

(However, some moisture is good, as it keeps the mold from becoming too brittle. You may actually have to add water to your molds periodically, especially if you cast with urethane resins, which draw out moisture as they cure.)

Trying to rush things will result in a poor casting that either never cures (it remains soft and sticky), comes out of the mold in pieces, or is too brittle. If you take your time, you'll find your casting results will be up there with the pros.

Safety and Respirator Masks

Materials used for mold making and casting vary from harmless dust producers to downright carcinogens. You must exercise care when working with mold-making and casting materials and remember to wear a respirator mask, safety goggles, and, if necessary, gloves.

Plaster for molds presents little danger, except that you should never use it to make a mold of an arm, leg, or other body part (yours or someone else's). Plaster cures by producing heat, and this heat can burn skin. Urethane and fiberglass resins should always be used in a well-ventilated area. Never try to cast plastic in a closed room, and advise others that they may need to leave for a while.

Smoking is a no-no while working with molds and casting, and keep all ingredients away from open flames. Some casting materials require heating (this is the case with hot melt); use an electric double-boiler or hot plate to heat the material. Don't use an oven or range that uses gas and produces an open flame.

☎ (509) 525-7326

✉ steve@farwestmaterials.com

🌐 http://www.farwestmaterials.com/

Mold-making and casting supplies. The company specializes in the distribution of foundry, sculpture, and molding materials. From their Web site, "It is our aim to supply our customers with low cost products and technical support that will enable the production of art and related products."

🌐 💻

FDJ On Time   203246

1180 Solana Ave.
Winter Park, FL
USA

🚫 (800) 323-6091

✉ info@fdjtool.com

🌐 http://www.fdjtool.com/

See listing under **Tools-Precision & Miniature**.

📄 🌐 💻

Fibre Glast Developments Corporation   203588

95 Mosier Pkwy.
Brookville, OH 45309
USA

☎ @listing-phonefaxweb:☎ (937) 833-6555.

🚫 (800) 330-6368

🌐 http://www.fibreglast.com/

See listing under **Materials-Fiberglass & Carbon Composites**.

🌐 💻

Freeman Manufacturing & Supply Co.   204241

1101 Moore Rd.
Avon, OH 44011
USA

☎ (440) 934-1902

🚫 (800) 321-8511

🌐 http://www.freemansupply.com/

Freeman sells materials for machine tooling, including machineable wax, a kind of hard wax that is used for creating prototypes using a mill or lathe. The wax melts at about 225 degrees and can be "cast" in most any shape. It can then be routed, milled, or lathed into complex machine parts. The resulting wax piece isn't strong enough for a heavy-duty combot, but sometimes can be used for smaller parts on lighter robots. And, of course, you can always use machineable wax for its intended purpose: Create a quick sample in wax, before spending the time and expense doing it in metal.

In addition to machineable wax, Freeman sells high-quality wood laminates (particle board, Armorboard, and so forth), casting urethanes for machining, and machining accessories.

Gibbons Fiberglass   203307

3035 E. Broadway Ave.
Bismarck, ND 58501
USA

- (701) 224-0656
- (800) 424-0656
- gibbons@gcentral.com
- http://www.gibbonsindustries.com/

Source for fiberglass materials and resins. Repair kits provide small amounts of fiberglass cloth and resin.

GoldenWest Manufacturing   203237

P.O. Box 1148
Cedar Ridge, CA 95924
USA

- (530) 272-1133
- (530) 272-1070
- http://www.goldenwestmfg.com/

Manufacturer of casting resins (both rigid and flexible), machineable plastic, foam board, and tools for cast and mold making. Their machineable plastic includes a product known as Butter-Board, a lightweight plastic block that is nonabrasive and very easy to machine or work with hand tools.

Kindt-Collins Company, The   204244

12651 Elmwood Ave.
Cleveland, OH 44111
USA

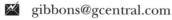

Fiberglass Casting Tape

If you've ever broken an arm or leg, the doctor probably gave you a fiberglass cast. Rather than getting out sheets of fiberglass matting and mixing resin and catalyst in a big jug, the doctor likely just ripped open a package of presoaked fiberglass tape and started wrapping around the broken body part. The material cures in a matter of minutes.

The benefit of using presoaked fiberglass tape is obvious, and ease of use is at the top of the list. The same casting tape is available to mere mortals. You can use it to build robust robot bodies and even to repair ones that get damaged.

3M Scotchcast is a popular casting tape available from medical supply companies. The tape is dipped in water to activate it. It can be cut with scissors (before or after dipping into water) and can be worked into the desired shape over a period of several minutes before it begins to harden. Casting gloves are recommended because the tape gets sticky, and the urethane resin is a skin irritant. The tape is available in widths from 2 to 5 inches.

Scotchcast is made for use with both humans and animals. Veterinary casting supplies are often cheaper. Try the following Google.com search phrases to locate suppliers of casting tape:

veterinary supplies & equipment

casting tape

orthotics supply

((216) 252-4122

📠   (216) 252-5639

🚫   (800) 321-3170

✉   info@kindt-collins.com

🌐   http://www.kindt-collins.com/

See listing under **Materials**.

Kingston Vacuum Works, The   204000

P.O. Box 3301
Kingston, NY 12402
USA

((845) 339-9375

📠   (253) 498-5574

🚫   (877) 560-6398

✉   info@warmplastic.com

🌐   http://www.warmplastic.com/

Vacuum forming tables (quite affordable, actually), supplies, and instruction.

Life-casting.com / Artmolds   202318

18 Bank St.
Summit, NJ 07901
USA

((908) 273-5401

📠   (908) 273-9256

🚫   (866) 278-6653

✉   info@artmolds.com

🌐   http://www.artmolds.com/

Molding and casting supplies. Emphasis is on life casting (heads, bodies, and hands), but the materials can be used for other projects, including almost-alive robots.

Michael Burnett Productions, Inc.   204130

P.O. Box 16627
North Hollywood, CA 91352
USA

((818) 768-6103

📠   (818) 768-6136

✉   mbpfx@aol.com

🌐   http://www.mbpfx.com/

Makeup and special effects products, including casting resins and latex.

Micro-Mark   202609

340 Snyder Ave.
Berkeley Heights, NJ 07922-1595
USA

((908) 464-2984

📠   (908) 665-9383

✉   info@micromark.com

🌐   http://www.micromark.com/

See listing under **Tools-Precision & Miniature**.

Miniature Molds.com   203187

P.O. Box 298
Eastsound, WA 98245
USA

((360) 376-3266

🚫   (800) 346-0567

🌐   http://www.miniaturemolds.com/

Metal casting kits, supplies, instruction. Sells Alumilite, room-temperature vulcanizing (RTV), various types of metals, rubber, and molds. The company specializes in molds and kits to make small-character figures, but of course the same materials can be used for small-character robots.

Moore Medical   203488

P.O. Box 1500
New Britain, CT 06050-1500
USA

((860) 826-3600

🚫   (800) 234-1464

🌐   http://www.mooremedical.com/

Medical supplies; you're interested in fiberglass casting tape, plaster casting materials.

National Gypsum Company   203905

2001 Rexford Rd.
Charlotte, NC 28211
USA

(704) 365-7300

✉ ng@nationalgypsum.com

🌐 http://www.national-gypsum.com/

Plaster-mix products, with spec sheets and technical documentation. The company offers free online courses in gypsum rock, wall board, and plaster technology and methods.

Ortho Cast, Inc. 203175

99 North Main St.
High Bridge, NJ 08829
USA

📞 (908) 638-5610

📠 (908) 638-5663

🌐 http://orthocast.com/

Providers of dental plaster, which is a very fine and quick-setting plaster used in mold making.

Perma-Flex Mold Co., Inc. 203242

1919 East Livingston Ave.
Columbus, OH 43209
USA

📞 (614) 252-8034

📠 (614) 252-8572

🚫 (800) 736-6653

🌐 http://www.perma-flex.com/

Perma-Flex is a supplier of rubber casting and mold-making products. For example, their polysulfide rubber compound creates flexible molds (good for irregularly shaped objects) and can withstand multiple castings. Sold in quantities ranging from one quart to a 600-pound drum. Some basic tutorial and reference material is provided on the site.

Pitsco 203448

P.O. Box 1708
Pittsburg, KS 66762
USA

🚫 (800) 835-0686

✉ orders@pitsco.com

🌐 http://www.pitsco.com/

See listing under **Retail-Science**.

Selecting the Right Plastic Casting Resin

There are plenty of plastic casting resins to choose from. The more you cast your own parts, the more you'll discover which casting resins are best suited for the work you do. (And also such important things as which ones stink less.)

Here's a short rundown of several popular products and manufacturers. Browse each company's Web site to discover the ancillary products that might be of use to you, such as curing additives (these slow down or speed up curing), colorants, additives to change the texture and hardness of the final casting, and much more.

Product	For more info, see:
Alumlite	http://www.alumilite.com/
Armorcast	http://www.armorcast.com/
CR-600	http://www.micromark.com/
Fast Cast	http://www.goldenwestmfg.com/
Foam latex	http://www.burmanfoam.com/
Polyurethane plastics	http://www.polytek.com/
POR-A-KAST	http://www.synair.com/
Urethane casting resins	http://www.bjbenterprises.com/
Urethane casting resins	http://www.smooth-on.com/

Plaster Master Industries 203188

4308 Shankweiler Rd.
Orefield, PA 18069
USA

((610) 391-9277
(610) 391-0340
 http://www.plastermaster.com/

Casting and mold-making supplies.

- Gypsum plaster
- Resin and casting materials
- Mold and model making
- Foam supplies

Plaster Master Industries.

Polytek Development Corp. 203243

55 Hilton St.
Easton, PA 18042
USA

((610) 559-8620
(610) 559-8626
Sales@Polytek.com
http://www.polytek.com/

Manufacturers of flexible molding and resin casting materials, including RTV silicone mold rubber, polyurethane plastics, polyurethane foam, and release agents. Online store.

Schenz Theatrical Supply, Inc. 204243

2959 Colerain Ave.
Cincinnati, OH 45225
USA

((513) 542-6100

(513) 542-0093
info@schenz.com
http://www.schenz.com/

Schenz specializes in products for high school and community theater. Products include premade props (armor, masks, swords, that sort of thing), but they also offer some inexpensive casting and mold-making supplies such as Celastic, liquid latex rubber, and metallic powders.

Sculptor.Org 203472

http://www.sculptor.org/

Resources for sculptors and sculpture on and off the Internet.

Sculpture House Casting 203241

155 W. 26th St.
New York, NY 10001
USA

(212) 645-3717
⊘ (888) 374-8665
info@sculptshop.com
http://www.sculptshop.com/

"Sculpture House Casting is one of the cradles of fertile ideas and artistic expression." Or put another way, they sell casting and mold-making supplies and tools, armatures, and modeling accessories. Use their stuff to create fertile and artistic robots.

Smooth-On 202451

2000 Saint John St.
Easton, PA 18042
USA

((610) 252-5800
(610) 252-6200
⊘ (800) 762-0744
smoothon@smooth-on.com
http://www.smooth-on.com/

Smooth-On is liquid rubber and plastics. With this stuff you can cast or coat items in plastic and rubber by either pouring, spraying, and brushing on. Web site contains products (including samplers), how-tos, and a newsletter. Among the most useful casting materials is rigid urethane resin, which you can use to make your special parts, including sprockets and gears.

Special Effect Supply Co.   203414

164 East Center St. #A
North Salt Lake, UT 84054
USA

☏ (801) 936-9762

✉ spl_efx@xmission.com

🌐 http://www.fxsupply.com/

Where do I start? Okay, how about these goodies for movie and stage special effects, which of course are also useful for robotics:

- Eyeballs
- Hemispheres (clear acrylic domes, diameters to 6 inches)
- Latex rubber
- Makeup prosthetics
- Molding and casting
- Plaster and gypsum
- Pneumatics (solenoid valves, cylinders, hose, and everything but the air)
- Sculpting clay
- Sculpting stuff
- Silicone kits and materials
- Skeletons and corpses (Tales of the Crypt robots?)
- UV and DayGlow makeup, paint, and material
- Vacuum forming

Look well beyond the box for some of these products, and you'll think of some interesting and useful ideas. For example, take their "UV Invisible on White" paint. It looks white in normal light, but glows brightly under ultraviolet light. Combine this paint with a UV LED (available at Hosfelt, Marlin P Jones & Associates, among other sources), and a UV detector, such as those for testing money. With this setup you could create a unique line-following robot that appears to follow an invisible track on a white painted table. Or, use the same idea as a "fence" to prevent a robot from rolling

Casting materials and more (like glass eyes) from Special Effect Supply. Courtesy of Special Effect Supply Corp.

Wormy Plastic

The early plastic catches the worm? In the land of fishing and fishing lures, it can. *Worm plastic* is a make-it-yourself process that uses a special liquid plastic; the plastic cures with the application of 350- to 400-degree heat. The liquid is called worm plastic because it is used to make fishing worm lures, but it has a number of other applications as well.

When cured, the plastic is soft and somewhat rubbery; additives can be used to make it even softer or to harden it up. Worm plastic can be used with simple molds to make shapes, and since it is milky clear, it can be colored with reds, blues, greens, and other hues. Other additives are used for fishing applications to give the lures a scent, but you'll probably want to skip those.

Worm plastic is available at the better fishing bait and lure shops, as well as on the Web. Here are a couple of specialty Web sites that carry or discuss worm plastic, colorants, and various additives:

Lurecraft Industries—*http://www.lurecraft.com/*

Barlow's Tackle-Express—*http://www.barlowstackle.com/*

Tackle Making.com—*http://www.tacklemaking.com/*

Pitsco—*http://www.pitsco.com/* (look in their materials section)

into a forbidden zone (possible use: sumo-class robot wresting).

Most product categories include basic background information.

Syn-Air Corporation   203173

P.O. Box 5269
2003 Amnicola Highway
Chattanooga, TN 37406
USA

✆   (423) 697-0400
📠   (423) 697-0424
🚫   (800) 251-7642
🌐   http://www.synair.com/mold_ind/

Syn-Air makes mold-making products. Offerings include:

- Por-A-Mold—polyurethane rubber, cures at room temperature
- Por-A-Kast—liquid urethane casting compound
- Sil-Mond—two component, cures at room temperature

Also makes polyurethane based in-mold paints, adhesives, binders, and release agents.

TAP Plastics   202458

6475 Sierra Ln.
Dublin, CA 94568
USA

📠   (925) 829-6921
🚫   (800) 894-0827
✉   info@tapplastics.com
🌐   http://www.tapplastics.com/

Plastics retailer, with stores in northern California, Oregon, and Washington. See listing under **Materials-Plastic**.

U.S. Composites, Inc.   203240

5101 Georgia Ave.
West Palm Beach, FL 33405
USA

✆   (561) 588-1001
📠   (561) 585-8583

✉   info@uscomposites.com
🌐   http://www.shopmaninc.com/

See listing under **Materials-Fiberglass & Carbon Composites**.

Urethane Supply Company   203819

1128 Kirk Rd.
Rainsville, AL 35986
USA

✆   (256) 638-4103
📠   (256) 638-8490
🚫   (800) 633-3047
✉   info@urethanesupply.com
🌐   http://www.urethanesupply.com/

Plastic welders and supplies; mainly for automotive and boat repair. Good technical information, regular newsletter in Adobe Acrobat PDF format.

Van Aken International, Inc.   203518

9157 Rochester Court
P.O. Box 1680
Rancho Cucamonga, CA 91729
USA

✆   (909) 980-2001
📠   (909) 980-2333
✉   info@vanaken.com
🌐   http://www.vanaken.com/

Van Aken is a manufacturer of art supplies. Their products include the famous Tempera and Plastelene. The Van Aken Web site has some useful semitechnical backgrounders and application ideas. The product is available through retailers.

W. P. Notcutt Ltd.   203498

25 Church St.
Teddington, Middlesex
TW11 8PF
UK

✆   +44 (0) 2089 772252
📠   +44 (0) 2089 776423
✉   sales@notcutt.co.uk
🌐   http://www.notcutt.co.uk/

W. P. Notcutt sells casting and mold (make that "mould," because they're British)-making supplies, including alginate compounds, hot melts, silicone rubber, and polyurethane rubber. Their Vinamold is a mold-making compound available in three hardnesses, each of which is best suited to given casting material and/or mold master.

 SUPPLIES-CHEMICALS

In this section, you'll find a number of resources for specialty chemicals, including chemicals used in electronics production. Of note are conductive paints and inks, which are used in making and repairing printed circuit boards. These products can also be effectively used to make homemade robotic sensors.

SEE ALSO:

DISTRIBUTOR/WHOLESALER-INDUSTRIAL ELECTRONICS: Good source for electronics chemicals

ELECTRONICS-SOLDERING: Additional chemicals needed for soldering (like flux and cleaners)

RETAIL-GENERAL ELECTRONICS: More electronics chemicals

Alfa Aesar   202587

30 Bond St.
Ward Hill, CA 01835
USA

- (978) 521-6300
- (978) 521-6350
- (800) 343-0660
- info@alfa.com
- http://www.alfa.com/

Spectrum Chemicals and Laboratory Products

http://www.spectrumchemical.com/
Industrial chemicals and lab supply

Tech Spray, L. P.

http://www.techspray.com/
Chemicals

Alfa Aesar is manufacturer and supplier of chemicals, metals, and materials for research, development, and production applications.

CAIG Laboratories, Inc.   202622

12200 Thatcher Ct.
Poway, CA 92064
USA

- (858) 486-8388
- (858) 486-8398
- caig123@caig.com
- http://www.caig.com/

CAIG makes chemicals and production shop accessories for electronics: lubricants, cleaners, solvents, antistatic and shielding compounds, greases, lint-free accessories, and plastic cutting and welding tools. The Web site has extensive tech information on various products and cleaning chemicals.

Fisher Scientific   202597

2000 Park Lane Dr.
Pittsburgh, PA 15275
USA

- (800) 766-7000
- http://www.fishersci.com/

Fisher supplies a full line of chemicals for industrial and educational applications.

See also:

Fisher Educational—*http://www.fisheredu.com/*

Fisher Scientific Worldwide—*http://www.fisherscientific.com/*

GC/Waldom, Inc.   202576

1801 Morgan St.
Rockford, IL 61102-2690
USA

- (815) 968-9661
- (800) 433-7928
- gcw.tech@woodsind.com
- http://www.gcwaldom.com/

GC/Waldom provides a full line of electronic materials, including chemicals, tools, connectors, and switches.

You'll want to review their line of conductive inks and pens, which can be used (among other things) to construct various kinds of sensors. Their Nickel Print and Silver Print products can be brushed on and leave a conductive film—possible uses include both resistive and capacitive sensors, or even crude strain gauges. Silver Print is also useful for EFI shielding.

ITW Chemtronics 202623

8125 Cobb Center Dr.
Kennesaw, GA 30152-4386
USA

((770) 424-4888
(770) 423-0748
(800) 645-5244
AskChemtronics@chemtronics.com
http://www.chemtronics.com/

Chemicals for electronics production and repair applications . . . stuff like solder resin remover, freeze sprays, solder masking agents, cleaners, and lubricators.

You'll also want to check out their CircuitWorks line of printed circuit inks and paints. For instance, their CircuitWorks Conductive Pen applies a bead of conductive silver, which can be used to not only repair printed circuit boards (the intended application of the product), but also to make various types of sensors. You can make primitive touch and pressure sensors with conductive inks and paints. Variations of the Conductive Pen include a paint that is flexible (that should give you plenty of ideas right there), conductive epoxy, and a conductive coating to repair rubber keypads. Again, think sensors.

Conductive ink pen. Photo courtesy of ITW Chemtronics

LPS Laboratories 202624

P.O. Box 105052
4647 Hugh Howell Rd.
Tucker, GA 30085-5052
USA

((770) 243-8800
(770) 243-8899

(800) 241-8334
http://www.lpslabs.com/

Chemicals, including an all-purpose spray lubricant (like WD-40, but not quite as famous).

M.G. Chemicals 203614

9347-193rd St.
Surrey, British Columbia
V4N 4E7
Canada

((604) 888-3084
(604) 888-7754
(800) 201-8822
info@mgchemicals.com
http://www.mgchemicals.com/

Manufacturer of chemicals for electronics production.

WASSCO 202568

12778 Brookprinter Pl.
Poway, CA 92064
USA

((858) 679-8787
(858) 679-8909
(800) 492-7726
sales@wassco.com
http://www.wassco.com/

WASSCO distributes production soldering materials and supplies. Offered are solder, cleaning chemicals, abrasives, soldering tools, static-control products, test and measurement gear, and hand tools.

 ## SUPPLIES-GLUES & ADHESIVES

Not all robots need to be constructed using 1-inch bolts. Some, particularly smaller ones, can be constructed using glues and other adhesives. This section lists a number of resources for glues and adhesives, including specialty extra-sticky tapes, solvent cements, hot glue and hot glue guns, and more. Most of the manufacturers listed in this section do not sell directly to the public, but their products are available from retail stores (such as hardware stores), as well as industrial supply distributors.

The listings in the section are also useful, as company Web sites post materials safety datasheets, which include important safety information. Many glues and adhesives are highly toxic, and you'll want to know the safety warnings before using them.

SEE ALSO

MATERIALS: General retail sources for glues and adhesives

MATERIALS-PLASTICS: Solvent cements for joining plastics

RETAIL-ARTS & CRAFTS: Look for general glues and adhesives, as well as glue sticks

RETAIL-HARDWARE & HOME IMPROVEMENT: Carries many specialty adhesives

SUPPLIES: General retail sources for glues and adhesives

Applied Industrial Technologies   203445

One Applied Plaza
Cleveland, OH 44115-5053
USA

- ((216) 426-4189
- ((216) 426-4820
- ⊘ (877) 279-2799
- ✉ products@apz-applied.com
- 🌐 http://www.appliedindustrial.com/

Industrial bearings, linear slides, gears, pulleys, pneumatics, hydraulics, and other mechanical things. Also hosts Maintenance America, online reseller of industrial maintenance supplies and general industrial supplies (wheels, casters, fasteners, and more), tools, paints, and adhesives.

Devcon
http://www.devcon.com/
Adhesives, for industry and consumer use

Elmer's Products Inc
http://www.elmers.com/
Glue and adhesive manufacturer: white glue, Krazy Glue, All-Purpose Glue Stick, ProBond cements and adhesives

House of Balsa
http://www.mag-web.com/rc-modeler/hobnew/
Zap adhesives

Clean Sweep Supply, Inc.   203817

10424 N. Florida Ave.
Tampa, FL 33612
USA

- ((813) 932-9564
- ((813) 932-6415
- ⊘ (877) 677-7015
- ✉ questions@cleansweepsupply.com
- 🌐 http://www.cleansweepsupply.com/

Office and light industrial supplies; listed here for their hot-melt glue guns, glue sticks, tools.

Cyberbond   202137

401 North Raddant Rd.
Batavia, IL 60510
USA

- ((630) 761-8900
- ((630) 761-8989
- ✉ sales@cyberbond1.com
- 🌐 http://www.cyberbond1.com/

Manufacturer of hobby and automotive adhesives. Technical details about each product is available at the Web site. Available at retail stores; some products are available online.

Hot Melt City   203816

1850 So. Elmhurst Rd.
Mount Prospect, IL 60056
USA

Hot Melt City.

(847) 437-7773

 (800) 323-5158

http://www.hotmelts.com/

Hot-melt glue guns and glues. For both hobby and professional use. Full-line catalogs are in Adobe Acrobat PDF format.

HST Materials, Inc. 202145

777 Dillon Dr.
Wood Dale, IL 60191
USA

(630) 766-3333

(630) 766-6335

info@hstmaterials.com

 http://www.hstmaterials.com/

See listing under **Materials**.

IPS Corporation 203803

455 West Victoria St.
Compton, CA 90220
USA

(310) 898-3300

(310) 898-3392

(800) 421-2677

http://www.ipscorp.com/

IPS makes the WELD-ON brand of solvent cements for PVC and other plastics. Their consumer products are

3M Makes More Than Just Clear Tape

3M is best known for their stickyback Post-It notes and, of course, their Scotch Brand transparent tape. (Years ago, we called this sort of stuff "cellophane" tape, but it's no longer made with cellophane; now we just call it "scotchtape," to the irritation of 3M's trademark lawyers.)

These products may be the company's most famous, but 3M actually makes thousands of other products, far too numerous to mention here. One such product is their XYZ-axis electrically conductive tape (product #9713). Many electrically conductive tapes merely use conductive adhesive. Stick two wires under the tape, press the tape against an insulator, and there will be a signal path between the wires, thanks to the tape adhesive. XYZ-axis tape is conductive within the adhesive, using conductive fibers, but also through the tape.

A variation on the theme is product #9703, an adhesive transfer tape that is electrically conductive in the Z axis only. That is, rather than conduct across the adhesive backing, this stuff is conductive through the adhesive, not across it. You can stick two wires under the adhesive and they will be electrically isolated. Stick two wires on top of one another, with the adhesive in between, and they'll be electrically connected.

One application of these tapes is in the creation of homespun touch sensors and for attaching wires to conductive surface when soldering is not possible. They are also useful in making small and flexible interconnections.

3M describes their electrically conductive tapes on their Web site; use the http://search.3m.com/ search page to locate these, and other, products. You can purchase the tapes from most distributors or industrial electronics suppliers. See the Distributor/Wholesaler - Industrial Electronics section for sources. Digi-Key is one known source of many 3M tape products.

Additional sources for 3M tape products include (minimum orders may apply):

Avnet Production Supplies and Test

http://www.etoolsandtest.com/

Micro Parts & Supplies, Inc.

http://www.mpsupplies.com/

TapeCase Ltd.

http://www.tapecase.com/

available at hardware and home improvement stores; their industrial products, including two-part epoxies and applicators, are available at plastics specialty outlets. Material safety data sheets (MSDSs) are provided online. If you use these products, download these sheets and read them, as solvent cements—as useful as they are—aren't good for your health.

Loctite   203525

1001 Trout Brook Crossing
Rocky Hill, CT 06067
USA

☎ (860) 571-5100

⊘ (800) 562-8483

🌐 http://www.loctite.com/

Using Hot Glue

All glues take time to "set." For most glues, the longer the setting time, the stronger the bond. That's why a 30-minute epoxy is better, all things considered, than a 5-minute epoxy. But the problem with slow-setting glues is that you either need to clamp the pieces together while the joint hardens or sit there holding everything in place. For some jobs, neither is practical.

Enter hot-melt glue. Hot-melt glue is like a sticky wax. It melts at temperatures of 250 to 400 degrees and congeals as it cools. The cooling process doesn't take long—under a minute—and the glued joint is already strong. Additional strength comes as the glue cures, but the pieces are already securely held together. For most tasks, no clamping is necessary.

You need only a hot glue "gun" and glue sticks; both are available at hobby stores, arts and crafts stores, and most home improvement stores. Cost is under $20 for the pair. You can get a regular-size glue gun for bigger household jobs, but for most robotics applications, the small "craft" gun is all you need. They're cheaper, too.

The glue is heated to a viscous state in a glue gun, then spread out over the area to be bonded. Glue sticks are available in "normal" and low-temperature forms; the low-temperature glue is better with most plastics, to avoid any "sagging" or softening of the plastic due to the heat of the glue.

Tip: Before gluing, rough up the plastic surfaces to be bonded. Plastics with a smooth surface will not adhere well when using hot-melt glue, and the joint will be brittle and might break off with only minor pressure.

Finally, using a hot-melt glue gun is easier when the weather is warm. In cold weather, the glue sets very quickly. If it's cold out, move the project indoors. It'll give you a few more seconds to position the pieces together before the glue sets up.

Find More Glues and Adhesives!

Looking for additional sources of glues, adhesives, and cements? Try the following Google.com search phrases. As with the other listings in this section, most glue and adhesive makers do not sell to the public—unless you're purchasing by the 55-gallon drum. But for most companies, you can find their products at local hardware and home improvement stores:

manufacturer glue adhesive

manufacturer glues adhesives

manufacturer glue OR adhesive

manufacturer glues OR adhesives OR "solvent cement"

manufacturer ca glue

Hot Glue Cautions

Though the hot-melt glue is not heated to dangerous temperatures, it's still rather uncomfortable to have a dollop of melting ooze land on your arm or leg. Exercise care when using the glue gun, and keep it away from children. The low-temperature gun and glue sticks are safer (though the glue resets more quickly), but can still burn skin.

Manufacturer of glues and adhesives. Datasheets provided on the Web site.

MSC Industrial Direct Co., Inc.   202826

75 Maxess Rd.
Melville, NY 11747-3151
USA

- (516) 812-2000
- (800) 645-7270
- http://www.mscdirect.com/

See listing under **Materials**.

Uniplast Inc.   203818

616 111th St.
Arlington, TX 76011
USA

- (817) 640-3204
- (817) 649-7095
- (800) 444-9051
- customerservice@uniplastinc.com
- http://www.uniplastinc.com/

Uniplast makes hot glue guns and water-soluble adhesive sticks. They offer industrial and consumer models of glue guns. Glue sticks are available in traditional clear or in colors, as well as glow-in-the-dark.

SUPPLIES-PAINTS

Don't leave that wooden robot the color of dead trees! Paint it! You've got millions of colors to choose from, so feel free to express your creativity in the color schemes you like best. If you're handy with an airbrush—or always

wanted to learn how to use one—you can even dress up your robot with a fancy racing strip or flame motif.

The listings in this section are resources (both makers and sellers) of acrylic and latex paints suitable for use in robotics and other hobby endeavors. Look for specialty paints like ultraviolet and daylight fluorescent.

SEE ALSO:

RETAIL-ARTS & CRAFTS: More sources for paints; most available in small bottles

RETAIL-HARDWARE & HOME IMPROVEMENT: For larger quantities of paints, as well as painting supplies

Airline Hobby Supplies   203165

P.O. Box 2128
Chandler, AZ 85244-2128
USA

- (480) 792 9589
- (480) 792 9587
- rmbrown@ican.net
- http://www.airline-hobby.com/

Small assortment of R/C paints, bare-metal foil, adhesives, and accessories. Also publishes Airline Modeler Magazine.

Badger Air-Brush Co.   202131

9128 W. Belmont Ave.
Franklin Park, IL 60131
USA

- (847) 678-3104
- (847) 671-4352
- (800) 247-2787
- info@badgerairbrush.com
- http://www.badger-airbrush.com/

Manufacturers of a long line of consumer and professional airbrushes, compressors, airbrush-ready paints, canned propellant, and related parts. Airbrushing is the best way to apply paints in a controlled manner.

Model 100G, one of many airbrushes from Badger. Photo Badger Air-Brush Co.

Plaid Enterprises, Inc. 203522

P.O. Box 2835
Norcross, GA 30091-2835
USA

 (800) 842-4197

 http://www.plaidonline.com/

Plaid is a paint and crafts supply manufacturer. Among their products: FolkArt acrylic paints, plaster crafts, and glues. Their wares are available at most arts and crafts stores and online.

Special Effect Supply Co. 203414

164 East Center St. #A
North Salt Lake, UT 84054
USA

(801) 936-9762

spl_efx@xmission.com

Delta Technical Coating Inc.
http://www.deltacrafts.com/
Craft paint (acrylics)

Liquitex
http://www.liquitex.com/
Artist's paints and mediums

Sculptural Arts Coating, Inc.
http://www.sculpturalarts.com/
Specialty paints and coatings for theatrical uses

---

 http://www.fxsupply.com/

UV and daylight fluorescent paints. See listing under Supplies-Casting & Mold Making.

Testors Corporation 203492

620 Buckbee St.
Rockford, IL 61104
USA

(815) 962-7401

(800) 837-8677

 http://www.testors.com/

Testors is a major manufacturer of paint and painting products for the hobby model industry (they also sell plastic models, but that's not of as much interest to us). Testors paints are formulated to adhere well to many kinds of plastics; I've found they are about the best paint for PVC plastic. Once dried and cured (about a day), the paint is not prone to scratching or peeling. Testors also sells tools, painting supplies, and inexpensive airbrushes.

The company sells most of their wares through local hobby retailers but also sells online.

On a personal note, I once wrote a beginner's book for Testors on hobby model building, which they included in some of their paint kits. They no longer sell it, but it talked all about how to assemble and paint plastic models.

Van Aken International, Inc. 203518

9157 Rochester Court
P.O. Box 1680
Rancho Cucamonga, CA 91729
USA

(909) 980-2001

(909) 980-2333

info@vanaken.com

http://www.vanaken.com/

Van Aken is a manufacturer of art supplies. Their products include the famous Tempera and Plastelene. The Van Aken Web site has some useful semitechnical backgrounders and application ideas. The product is available through retailers.

TEST AND MEASUREMENT

In this section, you'll find resources for electronic testing apparatus, such as oscilloscopes, function generators, frequency counters, volt-ohm meters, benchtop power supplies, and logic probes. Many of the listings are for higher-end testing and measuring gear (namely scopes), but are also good sources for high-quality volt-ohm meters and other common testing gear.

SEE ALSO:

KITS-ELECTRONIC: Build your own testing gear with kits for volt-ohm meters, PC oscilloscope interfaces, and logic probes

RETAIL-GENERAL ELECTRONICS: Good source for value-priced test equipment, particularly volt-ohm meters

RETAIL-SURPLUS ELECTRONICS: Used and reconditioned test and measurement equipment

---

Alfa Electronics   202950

P.O. Box 8089
Princeton, NJ 08543-8089
USA

- (609) 897-1135
- (609) 897-0206
- (800) 526-2532

MetersandInstruments.com

http://www.metersandinstruments.com/

Retailer of test gear; sells online

Pomona Electronics

http://www.pomonaelectronics.com/

Connectors, cables, test clips.

Specialized Products Co.

http://www.specialized.net/

Hand tools, test gear

Test Equipment Plus

http://www.testequipmentplus.com/

Used & reconditioned test and measurement equipment

---

 questions@alfaelectronics.com

 http://www.alfaelectronics.com/

Products for use in labs, production lines, repair shops, and classes for engineers, scientists, field service persons, teachers, students, and hobbyists.

BK Precision   202629

22820 Ranch Pkwy.
Yorba Linda, CA 92887
USA

- (714) 237-9220
- (714) 237-9214
- http://www.bkprecision.com/

Test equipment. They have lots of equipment directly suitable for intermediate and advanced robotology: power supplies; function generators; multimeters; oscilloscopes; spectrum analyzers; counters; electrical testers; component testers/logic probes; universal device programmers; battery capacity analyzer. Spec sheets (in Adobe Acrobat PDF format) provided for most products. Downloadable catalog.

C & S Sales   202350

150 W. Carpenter Ave.
Wheeling, IL 60090
USA

- (847) 541-0710
- (847) 541-9904
- (800) 292-7711
- info@cs-sales.com
- http://www.cs-sales.com/

C & S sales deals with test equipment, soldering irons, breadboards, kits (including OWI robot and Elenco electronic), hand tools, and other products. The Web site regularly lists new, sale, and closeout items.

Elenco Electronics   202139

150 W. Carpenter Ave.
Wheeling, IL 60090
USA

(847) 541-3800

 (847) 520-0085

 elenco@elenco.com

 http://www.elenco.com/

From the Web site, "Elenco is a major supplier of electronic test equipment and educational material to many of the nation's schools and hobbyists. We also have a network of distributors selling our products from coast to coast and abroad."

InstrumentWarehouse.com

202755

http://www.instrumentwarehouse.com/

Online source for handheld instruments, meters, and technical tools, often at a good discount. Some hackables to consider, including the Electronic Tape Measure (uses ultrasonics).

Jensen Tools, Inc.

202953

7815 South 46th St.
Phoenix, AZ 85044-5399
USA

(602) 453-3169

(602) 438-1690

(800) 426-1194

http://www.jensentools.com/

Tools for electronics-basically, everything you need, including handheld meters and scopes, precision hand tools, shop supplies, soldering stations, you name it. Wide selection.

Metric Test Equipment

203585

3486 Investment Blvd.
Hayward, CA 94545
USA

(510) 264-0887

(510) 264-0886

(800) 432-3424

 quotes@metricsales.com

 http://www.metricsales.com/

Used test equipment.

Ocean State Electronics

202504

6 Industrial Dr.
Westerly, RI 02891
USA

(401) 596-3080

(401) 596-3590

(800) 866-6626

 ose@oselectronics.com

http://www.oselectronics.com/

Components and kits; mostly radio-oriented. Some unique products that might interest you in your robot-building quests are:

- AM radio on a chip
- Meters, testers, and soldering stations
- Velleman electronics kits

Saelig Co. Inc

202355

1 Cabernet Cir.
Fairport, NY 14450-4613
USA

(716) 425-3753

(716) 425-3835

 saelig@aol.com

 http://www.saelig.com/

PC-based test gear and data acquisition: oscilloscopes, temperature sensing, BITlink industrial control.

Sun Equipment Corp.

202191

P.O. Box 97903
Raleigh, NC 27624
USA

(919) 870-1955

(919) 870-5720

Picking a Good Volt-Ohm Meter

Besides a screwdriver, the most important tool to the robot builder is the volt-ohm meter. This device is also called a *multitester*, and for good reason: It's used to test all sorts of electrical levels and conditions, including voltage, shorts, and open circuits. This moderately priced tool is a basic requirement for working with electronic circuits of any kind.

Digital or Analog

There are two general types of meters available today: digital and analog. The older analog style, which is still preferred by some users, incorporates a mechanical movement with a needle that points to a set of graduated scales. Digital meters display measured values in a digital readout. Analog meters are hard to find, and the digital variety tend to be cheaper, so we'll concentrate on the digital models from here on out.

A volt-ohm meter is a must-have tool for your electronics shop. Photo Elenco Electronics Inc.

Which Functions?

Digital meters vary greatly in the number and type of functions they provide. Features include:

AC volts	Standard feature
DC volts	Standard feature
Milliamps	Standard feature
Ohms	Standard feature
Audible continuity	Common optional feature
Diode check	Common optional feature
Capacitance	Optional; found in some models
Transistor	Optional; found in some models

The maximum ratings of the meter when measuring volts, milliamps, and resistance also vary. For most applications, the following maximum ratings are more than adequate:

DC volts	1,000 volts
AC volts	500 volts
DC current	200 milliamps
Resistance	2 megohms

info@sunequipco.com

http://www.sunequipco.com/

Tools and training kits for electronics. Trainers include modules for microprocessors/controllers, and electronic and electrical control.

Test Equipment Connection Corp.   202954

525 Technology Park
Lake Mary, FL 32746
USA

(407) 804-1780

(407) 804-1277

Do You Need an Oscilloscope?

For serious work in electronics, an oscilloscope is an invaluable tool. Priced at $300 and above, an oscilloscope allows you to perform tests that a volt-ohm meter simply cannot do. Among the many applications of an oscilloscope are:

A dual-trace 60 MHz oscilloscope. Photo BK Precision.

- Test DC or AC voltage levels
- Analyze the waveforms of digital and analog circuits
- Determine the operating frequency of digital, analog, and RF circuits
- Test logic levels
- Visually check the timing of a circuit, to see if things are happening in the correct order and at the prescribed time intervals

A basic, no-nonsense dual-trace (two-channel) model, with a 20- to 25-MHz maximum input frequency, is adequate. Don't settle for the cheap, single-trace units. The two channels let you monitor two lines at once, so you can easily compare the input signal and output signal at the same time. While not absolutely necessary, most expensive scopes come with storage or delayed sweep, and these can be handy features for more complex testing chores.

You can save some money buying used. The desirable Tektronics and Hewlett-Packard brand oscilloscopes are routinely available at eBay and other online auctions. As always, be careful what you buy. Older scopes are expensive to repair, so make sure you get one that is guaranteed to work.

You may also wish to get a PC-based oscilloscope. The better ones tend to be fairly expensive, and, of course, they all require the use of a compatible personal computer.

Oscilloscope Resolution

Resolution refers to the accuracy of the scope. Resolution is measured in bandwidth and vertical sensitivity (there are other factors that contribute to the resolution of an oscilloscope, but they are beyond the scope—no pun intended—of this discussion). The *bandwidth* of the scope is expressed in megahertz (MHz) and is the fastest signal that can be accurately displayed. *Vertical sensitivity* indicates the Y-axis resolution. The low-voltage sensitivity of most average-priced scopes is about 5 mV; higher-priced scopes provide 1- to 2-mV sensitivity. Note that voltage levels lower than this voltage may appear, but they cannot be accurately measured.

(800) 615-8378

 Sales@4testequipment.com

 http://www.testequipmentconnection.com/

Test Equipment Connection deals with reconditioned test and measurement equipment. Rather than just used and of questionable condition, the company cleans, checks out, and if necessary, repairs the equipment before reselling it. Customers are given a "right of refusal period" if the goods they receive don't meet expectations.

Tucker Electronics 202955

1717 Reserve St.
Garland, TX 75042
USA

((214) 348-8800

(214) 348-0367

 (800) 527-4642

http://www.tucker.com/

New and reconditioned test gear. A local favorite of Dallas/Fort Worth–area robot builders.

Wacky Willy's 203431

2900 S.W. 219th Ave.
Hillsboro, OR 97123
USA

((503) 642-5111

(503) 642-9120

 wacky@wackywillys.com

 http://www.wackywillys.com/

Surplus electronics and test gear. Local stores in Oregon and Hawaii; mail order of limited product.

The Value of Good Scope Probes

The probes used with oscilloscopes are not just wires with clips on the end of them. To be effective, the better scope probes use low-capacitance/low-resistance shielded wire and a capacitive-compensated tip—these ensure better accuracy.

Most scope probes are passive, meaning that they employ a simple circuit of capacitors and resistors to compensate for the effects of capacitive and resistive loading. Many passive probes are switchable between 1x and 10x. At the 1x setting, the probe passes the signal without attenuation (weakening). At the 10x setting, the probe reduces the signal strength by 10 times. This allows you to test a signal that might otherwise overload the scope's circuits.

Active probes use operational amplifiers or other powered circuitry to correct for the effects of capacitive and resistive loading and to vary the attenuation of the signal. The following table shows the typical specifications of passive and active oscilloscope probes.

Probe Type	Frequency Range	Resistive Load	Capacitive Load
Passive 1x	dc - 5 MHz	1 megohm	30 pF
Passive 10x	dc - 50 MHz	10 megohms	5 pF
Active	dc - 500 MHz	10 megohms	2 pF

Logic Probes and Logic Pulsers

Logic probes are designed to indicate the logic state of a particular circuit line. One LED on the probe lights up if the logic is 0 (LOW), another LED lights up if the logic is 1 (HIGH). Most probes have a built-in buzzer that has a different tone for the two logic levels.

A third LED or tone may indicate a pulsing signal. A good logic probe can detect that a circuit line is pulsing at speeds of up to 10 MHz. That's fast enough for robot work. The minimum detectable pulse width (the time the pulse remains at one level) is 50 nanoseconds.

A handy troubleshooting accessory when working with digital circuits is the *logic pulser*. This device puts out a timed pulse, letting you see the effect of the pulse on a digital circuit. Normally, you'd use the pulser with a logic probe or an oscilloscope. The pulser is switchable between one pulse and continuous pulsing.

Western Test Systems   202188

2701 Westland Court
Unit B
Cheyenne, WY 82001
USA

((307) 635-2269
((307) 635-2291
⊘ (800) 538-1493

Used instruments and parts (scopes, meters, and analyzers). All come with a 90-day warranty for parts and labor; 10-day return-it-if-you-don't-like-it inspection. Does mail order.

Meter Supplies

Your meter will come with a pair of test leads—one black and one red. The leads are terminated with needle-like metal probes. Quality of the test leads included with the lower-cost meters is usually minimal, so consider purchasing a better set. Coiled leads are handy; they stretch out to several feet yet recoil to a manageable length when not in use.

Standard leads are fine for most routine testing, but some measurements may require the use of a clip lead. These attach to the end of the regular test leads and have a spring-loaded clip on the end. You can clip the lead in place so your hands are free to do other things. The clips are insulated to prevent short circuits.

TOOLS

While it's possible to build a robot with just a screwdriver, the job is made easier with various hand and power tools common to all construction. You don't need a shed full of tools—just the basics will do. The resources in this section, and the ones to follow, provide listings of several dozen hand and power tool makers and retailers. If you choose to augment your main work tools with specialty machinery, such as mills, lathes, and even computer-controlled routers, then that's covered, too.

In this main **Tools** section are resources for general tools, including power and hand tools, as well as tool accessories. The sections that follow break out different types of tools and products and list additional resources you'll want to know about.

SEE ALSO:

DISTRIBUTOR/WHOLESALER-INDUSTRIAL ELECTRONICS: Specialty tools for electronics construction

ELECTRONICS-GENERAL ELECTRONICS: More specialty tools for electronics

RETAIL-HARDWARE & HOME IMPROVEMENT: Great source for tools of all types

TEST AND MEASUREMENT: Electronic testing tools

Applied Industrial Technologies   203445

One Applied Plaza
Cleveland, OH 44115-5053
USA

☎ (216) 426-4189
✆ (216) 426-4820
🚫 (877) 279-2799
✉ products@apz-applied.com
🌐 http://www.appliedindustrial.com/

Industrial bearings, linear slides, gears, pulleys, pneumatics, hydraulics, and other mechanical things. Also hosts Maintenance America, online reseller of industrial maintenance supplies and general industrial supplies (wheels, casters, fasteners, and more), tools, paints, and adhesives.

Enco Manufacturing Co   203672

400 Nevada Pacific Hwy.
Fernley, NV 89498
USA

☎ (770) 732-9099
🚫 (800) 873-3626
✉ info@use-enco.com
🌐 http://www.use-enco.com/

Enco is a premier mail-order source for shop tools, power tools, hand tools, production tools (lathes, mills, hydraulic presses, metal brakes, you name it), bits, saws, casting materials, plastics, hardware (door and cabinet), fasteners, tooling components, ACME rods and nuts, welding equipment and supplies, and lots more. They print a master catalog and send out sales catalogs on a regular basis. The sales catalogs contain some real bargains.

You can also use their shopping cart system. Their online catalog consists basically of scans of their master catalog, in Adobe Acrobat PDF format. The printed catalog is definitely easier to use, and you can "quick order" any part by entering its catalog number in the shopping cart.

Grizzly Industrial, Inc.   202756

P.O. Box 3110
Bellingham, WA 98227-3110
USA

☎ (570) 546-9663
🚫 (800) 523-4777
✉ csr@grizzly.com
🌐 http://www.grizzly.com/

Woodworking and metalworking tools. Large showrooms in Bellingham, Wash., Muncy, Pa., and Springfield, Mo.

Home Lumber Company   203683

499 West Whitewater St.
Whitewater, WI 53190
USA

☎ (262) 473-3538
✆ (262) 473-6908
🚫 (800) 262-5482
🌐 http://www.homelumbercom.com/

Online seller of power and hand tools and accessories (bits, blades, and whatnot). Usually good prices on common items like Dremel and RotoZip.

House of Tools

203829

#100 Mayfield Common NW
Edmonton, Alberta
T5P 4B3
Canada

 (800) 661-3987

 custserv@houseoftools.com

 http://www.houseoftools.com/

Tools of all kinds. Local stores across Canada.

Donegan Optical Company, Inc.

http://www.doneganoptical.com/

Precision eyeglass-style magnifiers for the workroom

Garrett Wade

http://www.garrettwade.com/

Hand and power tools, clamps

Global Tool Supply

http://www.globaltoolsupply.com/

Online tool store offering over 10,000 specialty tools for home, business, and industrial uses. Browse for tools by category or brand

KBC Tools

http://www.kbctools.com/

Power and hand tools; bits; blades; tool accessories; mail order in North America

Pricecutter.com

http://www.pricecutter.com/

Discount tools and woodworking supplies

SEELYE, Inc.

http://www.seelyeinc.com/

Hot air thermoplastic welding kits, heat guns

Surplus Toolmart

http://www.surplustoolmart.com/

Surplus and over-stocked tools

TheToolman.com

http://www.thetoolman.com/

Hand and power tools; tool accessories

International Tool Corporation

203208

2590 Davie Rd.
Davie, FL 33317
USA

(954) 792-4403

(800) 338-3384

 http://www.internationaltool.com/

Online and local retailer of all things tools: hand, power, pneumatic, bits, accessories, and more. Search the Web site by manufacturer (most all recognized brands are represented) or by tool function.

Jensen Tools, Inc.

202953

7815 South 46th St.
Phoenix, AZ 85044-5399
USA

(602) 453-3169

(602) 438-1690

(800) 426-1194

 http://www.jensentools.com/

Tools for electronics—basically, everything you need, including handheld meters and scopes, precision hand tools, shop supplies, soldering stations, you name it. Wide selection.

Kaufman Company

204006

110 Second St.
Cambridge, MA 02141
USA

(617) 491-5500

(617) 491-5526

(800) 338-8023

mailto:postmaster@kaufmanco.com

http://www.kaufmanco.com/

Online retailer of power and hand tools, tool accessories (bits, saws, etc.).

Lee Valley Tools Ltd.

202757

P.O. Box 1780
Ogdensburg, NY 13669-6780
USA

(613) 596-0350

(613) 596-6030

 (800) 267-8735

customerservice@leevalley.com

http://www.leevalley.com/

Wide assortment of woodworking tools.

Malcom Company, Inc.   203815

1676 East Main Rd.
Portsmouth, RI 02871
USA

(401)-683-3199

(401)-682-1904

(888) 807-4030

http://www.malcom.com/

Malcom makes and sells hot-air thermoplastic welding kits, soldering tools, and plastic extruders. Intended for manufacturers and auto-repair, the tools are useful in the robotics lab.

MSC Industrial Direct Co., Inc.   202826

75 Maxess Rd.
Melville, NY 11747-3151
USA

(516) 812-2000

(800) 645-7270

http://www.mscdirect.com/

See listing under **Materials**.

MyToolStore.com   203168

4730 W. Spring Mountain Rd.
Las Vegas, NV 89102
USA

(702) 871-7178

(702) 871-2815

(800) 347-5096

mytoolstore@mytoolstore.com

http://www.sktoolstore.com/

Power and hand tools from major manufacturers. Also drill bits, saw blades, sanding discs, and other accessories.

Nolan Supply Corporation   204242

111-115 Leo Ave.
P.O. Box 6289
Syracuse, NY 13217
USA

(315) 463.6241

(315) 463.0316

(800) 736.2204

sales@nolansupply.com

http://www.nolansupply.com/

See listing under **Materials-Metal**.

Northern Tool & Equipment Co.   202606

2800 Southcross Dr. West
Burnsville, MN 55306
USA

(952) 894-9510

(952) 894-1020

(800) 221-0516

http://www.northerntool.com/

Northern Tool & Equipment Catalog Company is a supplier of products to the DIY crowd, whether they be small businesses, auto shops, or home tinkerers. Northern's main product lines are generators, small engines, pressure washers, and hand-, air-, and power tools. The company is also known for stocking thousands of mechanical parts, including:

- Casters, from small to heavy-duty
- Plastic, metal, and other construction materials
- Shaft couplers, including Lovejoy three-piece jaw couplers

Web page for Northern Tool & Equipment.

- Go-Kart parts (such as centrifugal clutches, chains, sprockets, and wheels)
- Hydraulic cylinders, pumps, and hoses

Penn State Industries   202614

2850 Comly Rd.
Philadelphia, PA 19154
USA

 (215) 676-7603

🚫 (800) 377-7297

✉ psind@pennstateind.com

🌐 http://www.pennstateind.com/

Hand and motorized tools (mostly for wood). The Library section contains information about the tools (in Adobe Acrobat PDF format), as well as plans for home-based projects (no robots that I could find . . .).

Right-Tool.com   203682

6 Ledge Rock Way, #6
Acton, MA 01720
USA

📞 (978) 635-1355

✉ sales@right-tool.com

🌐 http://www.right-tool.com

Internet online retailer of small power tools, including the RotoZip rotary saw (great for cutting out shapes in wood, plastic, and even light metal). Also carries a full line of hand, power, and pneumatic tools, plus accessories.

See also:

http://www.rotozip.net/

Rockler Woodworking and Hardware   203215

4365 Willow Dr.
Medina, MN 55340
USA

📞 (763) 478-8201

📠 (763) 478-8395

🚫 (800) 279-4441

✉ support@rockler.com

🌐 http://www.rockler.com/

Rockler carries hand and power woodworking tools, hardware, and wood stock (including precut hardwood plywoods). Among important hardware items are medium-sized casters, drop-front supports (possible use in bumpers or joints in robots), and drawer slides.

Rockler online.

Satco Supply   203986

2021 West County Rd. C2
Roseville, MN 55113
USA

📞 (651) 604-6602

📠 (651) 604-6606

🚫 (800) 328-4644

✉ satco@mindspring.com

🌐 http://www.tools4schools.com/

Tools for schools. Several catalogs available: general tools, art materials, safety, and industrial machine shop.

Screwfix Direct Ltd.   203857

FREEPOST
Yeovil
Somerset
BA22 8BF
UK

🚫 0500 41 41 41

✉ online@screwfix.com

🌐 http://www.screwfix.com/

E-tailer of fasteners, tools, hardware, and other home improvement items.

Sears, Roebuck & Co. 202785

3333 Beverly Rd.
Hoffman Estates, IL 60179
USA

((847) 286-2500

📠 (847) 286-7829

🌐 http://www.sears.com/

Sears sells lots of stuff, but of prime interest to robot builders are their tools. They used to have a big catalog sales department, but didn't see the Internet coming, and got rid of it. You can buy Sears tools at their local stores, from Bob Vila's TV commercials, and online and mail order. Like the good old days.

Sears also operates specialty hardware and home improvement stores, Sears Hardware and Orchard Supply Hardware. The latter, a real gem, is found primarily in California.

Techni-Tool, Inc. 202605

1547 N. Trooper Rd.
P.O. Box 1117
Worcester, PA 19490-1117
USA

🚫 (800) 832-4866

✉️ sales@techni-tool.com

🌐 http://www.techni-tool.com/

Techni-Tool sells lubricators, testers, hand tools, and soldering stations, among other products.

📄 🌐 🖥️

Tool Peddler 202459

9907 SE 82nd Ave.
Portland, OR 97266
USA

((503) 777-8665

📠 (503) 777-0246

🚫 (800) 344-8469

✉️ tools@toolpeddler.com

🌐 http://www.toolpeddler.com/

Hand and motorized tools of all kinds, tool accessories, and wheels and casters.

Toolsforless.com 203167

527 Danforth St.
P.O. Box 8361
Portland, ME 04104
USA

📠 (207) 772-3173

🚫 (888) 295-4880

✉️ customerservice@toolsforless.com

🌐 http://www.toolsforless.com/

Power tools; hand tools. Stocks some 50,000 power and hand tools, hardware, parts, and accessories.

ToolSource.com 204097

6277 Sugartown Rd.
P.O. Box 950
Ellicottville, NY 14731-0950
USA

📠 (716) 699-8337

🚫 (888) 220-8350

🌐 http://www.toolsource.com/

Hand tools, power tools, specialty tools, with an emphasis on tools for the automotive mechanical and auto body repairer. Browse by category or search for tool name, manufacturer, or part number.

Tools-Plus 202759

53 Meadow St.
Waterbury, CT 06702
USA

((203) 573-0750

📠 (203) 753-9042

🚫 (800) 222-6133

🌐 http://www.tools-plus.com/

Tools-Plus is an online tool retailer. They have everything: hand tools, power tools, metalworking machines and helpers, accessories, and more. Brand names.

TP Tools & Equipment 203210

7075 Route 446
P.O. Box 649
Canfield, OH 44406
USA

((330) 533-3384

📠 (330) 533-2876

(800) 321-9260

http://www.tiptools.com/

Hand and power tools, with an emphasis on heavy-duty stuff for professional contractors and shops. Products include air compressors and tools, plasma cutters and welding machines, and auto body repair tools.

Victor Machinery Exchange, Inc. 202617

251 Centre St.
New York, NY 10013
USA

((212) 226-3494

(212) 941-8465

(800) 723-5359

sales@victornet.com

http://www.victornet.com/

Victor Machinery Exchange was founded in 1918. They provide machine shop supplies, metal-cutting tools, precision measuring instruments (electronic calipers and dial indicators), carbide end mills, drill bits, reamers, taps and dies, lathe chucks, and socket screws.

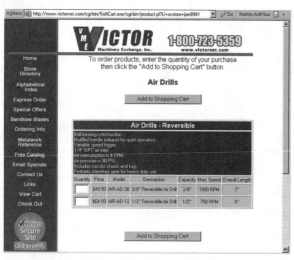

Web site for Victor Machinery Exchange.

Wiha Quality Tools 203211

1348 Dundas Cir.
Oakwood Industrial Park
Monticello, MN 55362
USA

((763) 295-6591

(763) 295-6598

cservice@wihatools.com

http://www.wihatools.com/

Hand tools. Emphasis is on fastener system bits: screwdrivers, screwdriver bits (flat, Philips, Torx, hex, Pozidrive, etc.), Allen wrenches, in metric or standard.

Woodworker's Warehouse 203896

126 Oxford St.
Lynn, MA 01901-1132
USA

((781) 853-0900

(781) 485-2150

(800) 877-7899

http://www.woodworkerswarehouse.com/

Woodworking tools, fasteners, hardware, accessories, and supplies. Local retail stores in the Northeast, and offers online sales.

  TOOLS-ACCESSORIES

Tool accessories include saw blades, drill bits, router bits, sand paper and other abrasives, and similar products. See the **Tools-Hand** and **Tools-Power** for additional sources of tool accessories.

Craft Supplies USA 203932

1287 E. 1120 S.
Provo, UT 84606
USA

((801) 373-0917

(801) 377-7742

(800) 551-8876

cust@woodturnerscatalog.com

http://www.woodturnerscatalog.com/

See listing under **Materials-Other**.

Cutting Tool Mall, The 203831

4 Donna Ct.
Selden, NY 11784
USA

((631) 451-6020

(631) 451-7383

info@cuttingtoolmall.com

ICS Cutting Tools, Inc.

http://www.icscuttingtools.com/

Drill bits; saw blades; tap & die; files

MLCS

http://mlcswoodworking.com/

Router bits and woodworking supplies

 http://www.cuttingtoolmall.com/

Power saws, blades, power tools, and machine tools.

Discount-Tools.com   204100

3001 Redhill Ave.
#6-204
Costa Mesa, CA 92626
USA

((714) 751-3844

☏ (714) 751-3894

🚫 (877) 848-8665

✉ info@discount-tools.com

🌐 http://www.discount-tools.com/

Routers, mills, and bits for drills.

Vermont American Corporation   202307

National City Tower
Ste. 2300
101 South Fifth St.
Louisville, KY 40202
USA

((502) 625-2000

☏ (502) 625-2199

🌐 http://www.vatool.com/

Tool manufacturer: saw blades, screwdriver bits, router products, drill bits, and abrasives. Browsable Web catalog; products are available through retail stores, such as Lowe's and Home Depot.

W. L. Fuller Co., Inc.   203231

P.O. Box 8767
7 Cypress St.
Warwick, RI 02888
USA

((401) 467-2900

☏ (401) 467-2905

✉ info@wlfuller.com

🌐 http://www.wlfuller.com/

Drills and bits: countersinks, counterbores, plug cutters, taper-point drills, brad-point drills, and step drills. Available through distributors and retail stores.

TOOLS-CNC

CNC stands for "computer numeric controlled," a process whereby a computer controls the operation of a machine tool, such as a lathe or mill. Once the domain solely of well-heeled industrial factories, CNC tools are now affordable by even budget-minded hobbyists. The listings in this section are resources for desktop CNC power tools. Many come complete with the necessary software to construct 3D parts. This section also lists resources for CNC software.

For a CNC mill, three stepper or servomotors are used to control the X, Y, and Z axes. (For the uninitiated, a mill is like a drill press, except that the work piece can be moved in two planes.) For a CNC lathe, two stepper or servomotors are used, one for horizontal tool position, and one for tool depth. Additionally, the lathe motor may be computer controlled.

A third type of CNC tool is the router. These serve a similar purpose as the mill, but use a routing tool to carve or cut out pieces in larger sheets of material. Like the mill, the CNC router uses three motors, one each for the X, Y, and Z axes.

Some additional acronyms you should know:

- CAD means "computer-aided design," the software used to create accurate models of 3D objects
- CAM means "computer-aided manufacture," the software that makes a CAD drawing and operates a CNC machine to produce the final part.
- DXF is a popular file format created by CAD programs; many CAD programs can share files in this format.
- G-Code is a popular machine-level output format that instructs the motors of a mill or lathe to move to specific positions.

SEE ALSO:

TOOLS-MACHINERY and **TOOLS-PRECISION & MINIATURE.**

CNC Routers

For the robotics hobbyist with a bit of extra spending cash, a *computer numerically controlled (CNC) router* will have you building all sorts of bots in record time. A CNC router combines a high-speed motorized cutting tool (the router) with computer-controlled movement. This movement is controllable in three planes: X, Y, and Z.

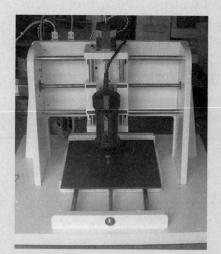

To use, you first secure a piece of wood, plastic, metal, foam, or other material onto the base of the device. You then program the "moves" the router will take over the material. For example, if you're cutting out a shape, the router will first move downward (the Z axis) to drill into the material, then move back and forth (the X and Y axes) to cut out the shape.

Depending on the routing tool used in the machine, you might also be able to mill parts out of softer materials, like plastics and soft woods. (For metals and other hard materi-

Gantry designs are typical of CNC routers.

als, a mill is the better choice, as it has more cutting power.) When milling on a CNC router, the Z axis of the router is varied ever so slightly, in order to produce a 3D cut surface.

While CNC routers are simple in principle, they are not cheap in cost. The typical CNC router package for hobbyist use costs $2,000 and more. Prices go up from there, with $5,000 being typical for an entry-level professional model.

The design of the typical CNC router is the *gantry*. The gantry slides back and forth, and is the X axis. Attached to the gantry is a Z-axis plate, which moves up and down; the router tool is physically attached to this. Depending on the design of the machine, the Y axis is produced by either moving the entire gantry or by moving the work piece itself. The latter is probably more common in the smaller "desktop" CNC machines—the ones most suitable, in size and price, for the amateur robot builder.

The *travel distances* of the three axes define the maxi-mum size of material you can work with. A small CNC router may be limited to a 12 by 12-inch piece of material, and with a maximum thickness of 2 or 3 inches. When comparing CNC routers, check the extents of the X, Y, and Z axes, and be sure they will be adequate for your needs.

You'll want to verify that these are travel extents (the tool actually travels this distance to cut) and not merely the maximum dimensions of the material you can fit into the machine. A given CNC router may be able to accept mate-rial up to 12 by 12 inches, but may only be able to cut out a shape of 8 by 11 inches.

The router cuts and mills material by moving either the tool, the material, or both.

The movement along the three axes of the CNC router is performed by a *stepper motor* (less common is the servomotor, which adds considerably to the cost of the machine). The stepper motor drives the mechanics of the CNC router via an acme screw, ballscrew, trapezoidal leadscrew, rack and pinion gear, belt, or chain. I'll let the manufacturers tout the pros of their specific systems, but in the end, you'll want to ensure your machine has the *repeatable accuracy* you need for your work. Any CNC machine with a repeatable accuracy of less than 0.010 inch is not worth your investment.

Note the term *repeatable* above. Some CNC router vendors list the positioning accu-racy of the stepper (or servo) motor. This is not the same as the repeatability of the cut-ting tool; the latter takes into consideration the flatness of the worktable, the type of drive mechanism, the effects of backlash as the tool moves back and forth, and other variables.

One simple way to test the accuracy of a CNC router is to replace the router tool with a fine-tipped felt pen. Securely tape a piece of paper to the worktable and have the router draw a shape—such as the figure of a dog or person—onto the paper. Do it twice. Carefully examine the drawing: You should not see two "double traces" anywhere. If you do (it'll likely be at the corners or intersections of lines), that router lacks sufficient accuracy.

When purchasing a CNC router, consider the *software* you will use with the machine. Many commercially made CNC routers come with software; others don't. CNC software can cost several hundred, to several thousand, dollars; if your machine lacks software, be sure to add this cost to the final price. (Note: Most CNC routers can be used with software from a variety of vendors, but it's still good to make sure yours doesn't use some proprietary control technique that limits your choices.)

Finally, if the cost of a ready-made CNC router is too rich for your blood, you might want to consider making your own. It's not *quite* as easy as some Web sites and magazine ads make it out to be, but you can save 40 to 60% by going the DIY route.

Stepper motors provide exact movement of the router for precision cuts.

BobCAD CAM, Inc.   203701

1440 Koll Cir. 106
San Jose, CA 95112

((408) 436-7777

((408) 436-7910

⊘ (800) 732-3051

🌍 http://www.bobcadcam.com/

CAD/CAM CNC software for Windows.

Caligari Corporation / trueSpace   203699

1959 Landings Dr.
Mountain View, CA 94043
USA

((650) 390-9755

⊘ (800) 351-7620

✉ sales@caligari.com

🌍 http://www.caligari.com/

Makers of trueSpace 3D modeling software. The trueSpace software outputs in DXF format, which can be used to import into CNC programs.

Carken Co. / Deskam   203702

6404 Curwood Dr.
E. Syracuse, NY 13057
USA

((315) 278-6757

((707) 516-2868

✉ support@deskam.com

🌍 http://www.deskam.com/

Carken publishes CAD/CAM CNC software:

- DesKAM-2 1/2D CAM from DXF or 3D CAM from STL files
- DeskART-Carve or engrave your computer image files.
- Desk Engrave-Turn your True Type fonts into G-Code or DXF.
- DeskNC for DOS or Windows-Run your CNC equipment directly from your PC.
- DeskNCrt-Operate your CNC equipment in closed loop using encoders.

CNC Retro-Fit Links   202213

http://www.mendonet.com/cnclinks/

Links to other Web sites on retrofitting manual lathes and mills for computer control.

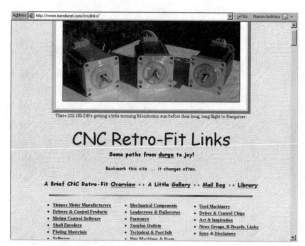

CNC Retro-Fit Links.

CNCez PRO   203698

440 Ridge St.
Lewiston, NY 14092
USA

 (716) 754-2690
(716) 754-2807
(888) 561-7521

moncefm@torcomp.com
http://www.cncezpro.com/

CNC simulation/educational software. Web site is in English, French, and Italian.

cncKITS.com   203074

pentam@cncKITS.com
http://www.cnckits.com/

Kits and parts and plans for building CNC lathes and routers and mills. Good source for leadscrew assemblies, at fair prices. Includes links to other resources on the Internet.

Delcam plc/MillWizard   203700

Small Heath Business Park
Birmingham
B10 0HJ
UK

+44 (0) 1217 665544
+44 (0) 1217 665511
marketing@delcam.com
http://www.millwizard.com/

MillWizard is CAD/CAM CNC software. Produced by Delcam, one of the world's leading developers and suppliers of CAD/CAM software for the 3D design.

See also:

http://www.delcam.com/

Delft Spline Systems / DeskProto   203696

P.O. Box 2071
3500 GB Utrecht
The Netherlands

+31 30 296 5957
+31 30 296 2292
info@spline.nl
http://www.deskproto.com/

3D software for CNC machines.

Desktop Machine Shop.com   203684

RD 2 Box 1982
Mansfield, PA 16933
USA

(570) 549-3044
marty@desktopmachineshop.com
http://www.desktopmachineshop.com/

Reseller of TAIG mills and lathes.

DesktopCNC   203688

http://www.desktopcnc.com/

This is an informational Web site for people wanting to build a desktop CNC machine. I particularly liked the comparison tables about CNC desktop mills, lathes, routers, and software.

Diversi-Tech Inc.   203766

P.O. Box 342
Wyoming, PA 18644
USA

(570) 693-5980
dtech123@diversi-tech.com
http://www.diversi-tech.com/

Plans, videotape, and basic parts for the AutoRout plotter-style router. For small tools like Dremel.

Flashcut CNC 202289

1263 El Camino Real
Ste. W
Menlo Park, CA 94025
USA

☎ (650) 853-1444
📠 (650) 853-1405
🚫 (888) 883-5274
🌎 http://www.flashcutcnc.com/

CNC mini mills and lathes. Based on Sherline products. Complete and retrofit.

HobbyCNC 202645

Dave Rigotti
8502 Mulberry Rd.
Chesterland, OH 44026
USA

✉ DRigotti@AOL.com
🌎 http://www.hobbycnc.com/

Plans and basic starter kits for building your own CNC router. Their "CNC package" includes three stepper motors, stepper motor controller electronics, and assorted hardware (minus the case).

Three-axis CNC kit. Photo Dave Rigotti

Home Build Hobby Plotter / Engraver 202222

http://plotter.luberth.com

Plans for a CNC plotter, hardware, and software. Includes a forum and many other useful resources.

Many CNC System / EasyCut 203689

Rua de S. Miguel
No.3 R/C
Macau

✉ sales@easycut.com
🌎 http://www.easycut.com/

Makers and sellers of CNC routers and 3D engravers, from-12 _ 12 inches to 108 _ 60 inches.

MAXNC Inc. 202834

4122-A West Venus Way
Chandler, AZ
USA

☎ (480) 940-9414
📠 (480) 940-2384
🚫 (888) 327-9371
🌎 http://www.maxnc.com/

Makers of desktop CNC mills and lathes.

MicroKinetics Corporation 203685

2117-A Barrett Park Dr.
Kennesaw, GA 30144
USA

☎ (770) 422-7845
📠 (770) 422-7854
🌎 http://www.microkinetics.com/

Desktop mills and lathes, as well as full-size production machines. Stepper motors, servomotors, and motor controllers for CNC.

MicroProto Systems 203691

12419 E. Nightingale Ln.
Chandler, AZ 85249
USA

☎ (602) 791-0219
📠 (480) 895-9648
✉ info@microproto.com
🌎 http://www.microproto.com/

MicroProto is the "CNC branch" of TAIG Tools, makers of precision desktop lathes and mills. The standard TAIG is manually operated; MicroProto adds stepper motors and control circuits so that you can control your machine from computer. Available with or without the SuperCam software from Super Tech & Associates (http://www.super-tech.com/).

See also TAIG Tools:

http://www.taigtools.com/

Minitech Machinery Corp.   203697

6463 Atlantic Blvd.
Norcross, GA 30071
USA

  (770) 441-8525
  (770) 441-8526
  (800) 662-1760
  sales@minitech.com
  http://www.minitech.com/

Minitech manufactures and sells desktop CNC mills, lathes, and routers. Middle to high end.

Nick Carter's Taig Lathe Pages   203693

http://www.cartertools.com/

Informational site on TAIG lathes. From the site, "Welcome to my pages devoted to the Taig Lathe. Since buying one over five years ago I have become increasingly enthusiastic about the Taig lathe, its economy, capability and over all style. The Taig lathe is especially good if you are a novice to metalworking, and seek to learn the basics without a large investment of money and space. It is my hope that these pages are a useful resource for all Taig users."

Lots of links. Nick is also a dealer for the TAIG.

PCB Milling   203788

http://www.pcbmilling.com/

Informational Web site on PC board fabrication using mechanical etching.

Quantum CNC   203692

Wall End House
Moor Lane
Coleorton
Leicestershire
LE67 8FQ
UK

  +44 (0) 1530 834376
  +44 (0) 1530 834369
  info@quantumcnc.co.uk
  http://www.quantumcnc.co.uk/

European sales, distribution, and support for the TAIG Micro Mill CNC Desktop Machining Systems.

Robo Systems   204224

P.O. Box 290
Newtown, PA 18940-0290
USA

  (800) 221-7626
  robo@robosys.com
  http://www.robosys.com/

Robo Systems makes Accucadd, RoboCAD, and related CAD/CAM software for Windows.

Secrets of CNC   202270

http://www.seanet.com/~dmauch/secrets2.htm

Dan Mauch tells us the "Secrets of CNC, or "How I Learned to Stop Worrying and Love Computer Numerical Control." (Okay, the last part I made up.)

Super Tech & Associates   203690

3313 East Hillery Dr.
P.O. Box 30729
Phoenix, AZ 85046
USA

  (602) 867-1755
  (602) 867-1426
  info@super-tech.com
  http://www.super-tech.com

Super Tech manufactures and sells desktop CNC and mills, as well as low-cost general-purpose CNC software. Their MiniRobo, which I purchased for my own shop, is a compact yet versatile router that uses a Dremel or RotoZip tool for cutting, drilling, and engraving into plastic, wood, and soft metals. Other products include TAIG mills and lathes and the RoboTorch, a large gantry-style CNC plasma cutting rig.

TAIG Tools 203678

12419 E. Nightingale Ln.
Chandler, AZ 85249
USA

 (408) 895-6978

 (480) 895-9648

✉ sales@TaigTools.com

🌐 http://www.taigtools.com/

See listing under **Tools-Precision & Miniature**.

Three-axis mill from TAIG Tools. Photo Susan Daley/Wild West BEST.

CadSoft CAD/CAM

http://www.camsoftcorp.com/

Software for CNC machine: mills, lathes, routers, water jets, lasers, punch presses and EDMs

GriffTek

http://www.grifftek.com/

For CNC: controllers, motor mounts, retrofit kits. Sherline retros a specialty

Ormond, LLC

http://www.ormondllc.com/

Ultra high pressure waterjet CNC machines

Techno-Isel 203515

2101 Jericho Tnpk.
New Hyde, NY 11040
USA

 (516) 328-3970

 (516) 358-2578

 http://www.techno-isel.com/

Techno-Isel is part of catalog retailer Stock Drive, specialists in gears, bearings, and other power transmission products. Techno-Isel specializes in CNC routers. Check out their Mechanical Model Kits.

TOOLS-HAND

This section provides resources for hand tools of various types-screwdrivers, wrenches, pliers, that sort of thing.

My word of advice: Buy the best hand tools you can afford. Don't get cheap with your hand tools. Better hand tools help you do a better job, and they'll last longer, too.

Airparts, Inc. 203153

2400 Merriam Ln.
Kansas City, KS 66106
USA

📞 (913) 831-1780

📠 (913) 831-6797

🚫 (800) 800-3229

✉ airparts@airpartsinc.com

🌐 http://www.airpartsinc.com/

See listing under **Materials-Metal**.

American Tool Companies, Inc. / Vice-Grip 202306

701 Woodlands Pkwy.
Vernon Hills, IL 60061
USA

📞 (847) 478-1090

📠 (847) 478-1091

🌐 http://www.americantool.com/

American Tool Companies makes the famous Vice-Grip locking hand tools, along with the several other well-known brands:

- Chesco-hex tools
- Hanson-drill bits, taps, and dies
- Irwin-wood-boring and cutting tools
- Marathon-saw blades
- Prosnip-snips
- Quick-Grip-clamping tools
- Quick-Vise-portable vise
- Speedbor2000-wood-boring bits
- Unibit-step drills

Products are available through retailers; and much of the material on their Web site is marketing fluff.

However, they do offer informative "solutions and tips" for selected tools.

Craft Supplies USA

203932

1287 E. 1120 S.
Provo, UT 84606
USA

- (801) 373-0917
- (801) 377-7742
- (800) 551-8876

Construction Tools You Need

If you're fashioning your own robot frame and other mechanical parts, you need tools to cut, saw, punch, ream, file, and fasten. These are the tools you can't do without:

- *Claw hammer*, used for just about anything you can think of.
- *Rubber mallet*, for gently bashing pieces together that resist going together; also for forming sheet metal.
- *Screwdriver assortment*, including various sizes of flat-head and Philips-head screwdrivers.
- *Hacksaw*, with an assortment of blades. Coarse-tooth blades are good for wood and PVC pipe; fine-tooth blades are good for copper, aluminum, and light-gauge steel.
- *Miter box*, to cut straight lines.
- *Wrenches*, all types. Adjustable wrenches are helpful additions to the shop. Small and large Vice-Grips (that's a brand name) help you hold pieces for cutting and sanding.
- *Nut drivers*, make it easy to attach nuts to bolts.
- *Measuring tape*, a 6- or 8-foot steel measuring tape is a good choice. A cloth tape helps measure unusual shapes. For precision, purchase a small (6- to 8-inch) machinist's rule, graduated in inches, metric, or both.
- *Square (18-inch)*, for making sure that pieces you cut and assemble from wood, plastic, and metal are square.
- *File assortment*, to smooth the rough edges. Get a set for larger pieces, and another small set (called *needle files*) for miniature work.
- *Drill motor,* with variable speed control.
- *Drill bit assortment*, good, sharp ones *only*. If yours are dull, have them sharpened, or replace them. A small set with sizes from 1/16 inch to 1/4 inch is fine for starters.
- *Wire crimping tool*, for crimping lugs and connectors onto various sizes of wires. Get the kind that can handle wire from 12 to 22 gauge.
- *Vise*, for holding parts. A modest-size vice that attaches to your workbench with bolts (not a clamp; they don't hold) is sufficient.
- *Safety goggles,* for use whenever you're using tools to hammer, cut, drill, or any other time when flying debris could get in your eyes. Be sure you use the goggles.

cust@woodturnerscatalog.com

 http://www.woodturnerscatalog.com/

See listing under **Materials-Other**.

Elenco Electronics 202139

150 W. Carpenter Ave.
Wheeling, IL 60090
USA

((847) 541-3800
(847) 520-0085
elenco@elenco.com
 http://www.elenco.com/

See listing under **Kits-Electronic**.

S*K Hand Tool Corporation 202304

3535 W. 47th St.
Chicago, IL 60632
USA

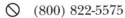 (800) 822-5575

http://www.skhandtool.com/

Wrenches, ratchets, and other great hand tools. Buy a set of these when you're young, and they'll grow old with you.

Uniplast, Inc. 203818

616 111th St.
Arlington, TX 76011
USA

((817) 640-3204
(817) 649-7095
(800) 444-9051
 customerservice@uniplastinc.com
 http://www.uniplastinc.com/

Uniplast makes hot glue guns and water-soluble adhesive sticks. They offer industrial and consumer models of glue guns. Glue sticks are available in traditional clear or in colors, as well as glow-in-the-dark.

Construction Tools You Should Have

If you want a well-stocked robot lab, consider these optional tools:

- A *drill press* helps you drill better holes, because you have more control over the angle and depth of each hole. Be sure to use a drill press vice to hold the pieces. *Never* use your hands!
- A *table saw* or *circular saw* makes cutting through large pieces of wood and plastic easier. Use it with a guide fence for straight cuts.
- A *scroll saw* or *saber saw* lets you cut out shapes. A decent scroll saw is not expensive and provides good accuracy.
- A motorized *hobby tool*, like those made by Dremel and Weller, makes short work out of a number of chores. The latest models have adjustable speed controls, which you need. Be sure to select the right bit for the job. A more powerful alternative to the hobby tool is the *Roto-Zip* (that's its trade name).
- *Hot-melt glue guns* let you apply fast-setting glue to most any surface. They are available at most hardware and hobby stores and are available in a variety of sizes.
- With a *nibbling tool* you can "nibble" small chunks from metal and plastic pieces. Use the tool to cut channels and enlarge holes.
- A *tap and die set* lets you thread holes and shafts to accept standard-size nuts and bolts.
- A *brazing tool* or *small welder* allows you spot-weld or braze two metal pieces together. Use these for small work only, and only if you're old enough to pay for fire insurance.

Brookstone Company, Inc.

http://www.brookstone.com/

Hard to find hand tools

CraftWoods

http://www.craftwoods.com/

Tools and supplies for manual woodworking crafts; hand tools

Jointech, Inc.

http://www.jointech.com/

Precision woodworking tools (for cabinetry, etc.)

Klein Tools

http://www.kleintools.com/

Hand tools (pliers, cutters, screwdrivers, etc.)

WASSCO   202568

12778 Brookprinter Pl.
Poway, CA 92064
USA

☏   (858) 679-8787

📠   (858) 679-8909

🚫   (800) 492-7726

✉   sales@wassco.com

🌐   http://www.wassco.com/

WASSCO distributes production soldering materials and supplies. Offered are solder, cleaning chemicals, abrasives, soldering tools, static control products, test and measurement gear, and hand tools.

Woodcraft Supply Corp.,   203218

P.O. Box 1686
Parkersburg, WV 26102-1686
USA

☏   (304) 428-4866

📠   (304) 428-8271

🚫   (800) 225-1153

✉   custserv@woodcraft.com

🌐   http://www.woodcraft.com/

Woodworkers tools and supplies. Be sure to check out their extensive line of plywoods (if you're building a robot base using wood). Of course, they offer the regular hand tools, like drills, saws, and planes, for working

with wood. Order online or visit one of their 61 retail locations within the U.S.

Woodsmith Store   203219

2625 Beaver Ave.
Des Moines, IA 50310
USA

☏   (515) 255-8979

🚫   (800) 444-7002

✉   woodstor@augusthome.com

🌐   http://www.woodsmithstore.com/

Woodsmith is dedicated to the home woodcrafter. They sell a lot of craft plans, but sadly, none on building robots. They do, however, offer a limited line of woodworking hand tools, bits, saws, router tables and fences, and shop jigs. Their emphasis is on precision woodworking (cabinetry, inlays, etc.), so their tools are designed for that extra measure of accuracy. Expect prices to match.

🔧 TOOLS-MACHINERY

Machinery tools include mills and lathes. A mill is used to carve material to create complex 3D shapes. It can also be used as a very accurate drill press (in fact, a mill looks a lot like a drill press). A lathe is used to cut against spinning rod or square material and is used for such jobs as cutting grooves in wheel axles.

The resources in this section are primarily aimed at desktop machinery, most of which is affordable for home workshops. See also **Tools-CNC** for computerized versions of mills and lathes.

Asteg Sales Pty Ltd.   203677

134 Pine Creek Circuit,
St Clair New South Wales 2759
Sydney
Australia

☏   +61 2 9834 6034

📠   +61 2 9834 6105

✉   sales@asteg.com

🌐   http://www.asteg.com/

Manufacturers of EMCO lathes and milling machines.

For educational products, see also:

http://www.emco.co.uk/

Clisby Miniature Machines   203075

12 Norton Summit Road
Magill, South Australia 5072
Australia

☎ +61 8 8332 5944

📠 +61 8 8332 5944

🌐 http://www.clisby.com.au/

See listing under **Tools-Precision & Miniature**.

CoolTool GmbH   203676

Fabriksgasse 15
2340 Moedling\
Austria

☎ +43 (0) 2236 892 666

📠 +43 (0) 2236 892 666 18

✉ info@unimat.at

🌐 http://www.unimat.at/

Manufacturers of midpriced motorized tools for miniature and model work, including:

- Unimat 1 (not the same as the old Unimat, but a new plastic version)
- Playmat
- Syro-Cut
- UniTurn and UniMill (higher precision lathe and mill than Unimat)

All are available through a variety of distributors and retailers. Web site is in German, Spanish, and English.

Desktop Machine Shop.com   203684

RD 2 Box 1982
Mansfield, PA 16933
USA

A Quick Look at Desktop Mills, Lathes, and CNC Routers

Model makers have used so-called *desktop* mills, lathes, and CNC routers for years. These tools are similar in features and function as the behemoths you see at large machine shops and on manufacturing floors, but are scaled down in both size and price—an ideal mix for amateur robot builders.

First, some definitions:

- A *mill* is like a vertical drill press. Instead of a cutting bit that just goes up and down, on a mill the work piece itself can be moved horizontally and laterally. This allows the mill to produce complex shapes, instead of just holes.

- A *lathe* is used to rotate a part against a cutting tool. It is typically used to contour round or cylindrical material—threads on a rod, for example. The material turns on a horizontal bed; the cutting tool is brought up against the material.

- A *CNC router* is a high-speed cutting tool, like a wood router, that is attached to a mechanism that moves the router along the X, Y, and Z axes. This movement is managed by a computer.

A CNC router is inherently a computer-controlled device. Mills and lathes can be completely manual affairs, or they, too, can be hooked up to a computer. With most models, you can purchase a manual desktop mill and lathe today, and sometime down the road retrofit it for computer control.

A desktop mill can be used to create miniature parts for your robot. Photo: Sherline Products Inc.

Desktop lathes can be used to create wheel hubs, shafts, specialty fasteners, and more. Photo: Sherline Products Inc.

((570) 549-3044

✉ marty@desktopmachineshop.com

🌐 http://www.desktopmachineshop.com/

Reseller of TAIG mills and lathes.

🌐 💻

Harbor Freight Tools ⚲   202607

3491 Mission Oaks Blvd.
P.O. Box 6010
Camarillo, CA 93011-6010
USA

Harbor Freight online.

Makers of Desktop Machinery

There are a number of makers of desktop mills, lathes, and CNC routers. Among the more common manufacturers are:

- Cool Tools
- FlashCut CNC
- ISMG
- Techno-Isel
- Liberty Enterprises
- Many/EasyCut
- MAXNC
- MicroKinetics
- Minicraft
- Minitech
- Sherline
- Super-Tech
- TAIG

Prices and features from these makers vary. If you're interested in acquiring a desktop mill, lathe, or CNC router, you're well advised to get information on as many of them as possible. The typical starting price for the better-made tool is $500, so consider research part of your investment.

Not all desktop tools are created equally. Some, like those from Cool Tools, are designed for garage shop tinkerers on a budget. They're fine for working with lightweight materials like soft plastics and thin woods, but don't think you can use them to produce highly accurate complex shapes from stainless steel! Price goes up based on accuracy, power, and size, so plan your purchase accordingly. If you need to work with pieces up to 20 inches, don't settle for a machine with a maximum cutting size of just 18 inches.

One way to save money on a desktop mill or lathe is to purchase it used. The better machines fetch good prices on eBay and other online auctions, but you may have good luck snagging a steal simply by going to garage sales and checking the local newspaper classified ads.

(805) 445-4912

 (800) 423-2567

 http://www.harborfreight.com/

Harbor Freight built a business on selling value-priced tools, much of it "off-brand," but still perfectly workable. (I still regularly use the Chinese-made drill press I bought from Harbor Freight over 20 years ago.) They offer hand and power tools, pneumatic tools, and even metal mill and lathes.

Retail stores in selected areas of North America; check the Web site for a store locator.

ISMG   203687

5151 Oceanus Dr., Bldg. 109
Huntington Beach, CA 92649
USA

 (714) 379-1380

 (714) 379-1385

(800) 575-2843

sales@ismg4tools.com

http://www.ismg4tools.com/

ISMG makes a line of desktop and full-size lathes and mills. Available from dealers worldwide.

Liberty Enterprises   203686

645 Liberty Cemetery Rd.
Morrison, TN 37357
USA

(931) 728-8984

(931) 728-3605

(800) 354-4604

sales@libertycnc.com

http://www.libertycnc.com/

Liberty makes and sells desktop mills and lathes (their mills are the gantry router type). They also sell separately a three-axis CNC motion control kit so you can build your own mill.

MicroKinetics Corporation   203685

2117-A Barrett Park Dr.
Kennesaw, GA 30144
USA

 (770) 422-7845

 (770) 422-7854

 http://www.microkinetics.com/

Desktop mills and lathes, as well as full-size production machines. Stepper motors, servomotors, and motor controllers for CNC.

Nature Coast Hobbies   203679

6773 S. Hancock Rd.
Homosassa, FL 34448
USA

 (352) 628-3990

 (352) 628-6778

(800) 714-9478

arf@naturecoast.com

http://www.naturecoast.com/

Primarily shipbuilding hobby materials and kits, but also miniatures tools, including Unimat I, Minicraft, and TAIG.

Royal Products   204012

200 Oser Ave.
Hauppauge, NY 11788
USA

(631) 273-1010

(631) 273-1066

(800) 645-4174

info@royalprod.com

http://www.royalprod.com/

Machine tools accessories (things like lathe chucks, dead centers, countersinks, etc.).

TOOLS-POWER

The job goes faster if you use power tools. This section lists sources for electrical and pneumatic power tools. The intermediate and advanced robot workshop includes at least one power saw (table saw, radial arm saw, and band saw), plus a motorized hand drill or drill press, sander, grinder, and perhaps a cut-off saw, scroll saw, jigsaw, and nut driver.

Delta Machinery

http://www.deltawoodworking.com/

Power tools, mainly for woodworking

RIDGID/Emerson Tool Company

http://www.ridgidwoodworking.com/

Maker of power tools for contractors and well-equipped home shop woodworkers

---

Campbell Hausfeld   203207

100 Production Dr.
Harrison, OH 45030
USA

⊘   (800) 543-6400

🌐   http://www.chpower.com/

There's nothing like working with air tools out in the garage. Everyone in the neighborhood can hear you work, and the "bripp, bripp" of the drill or saw or sander tells all you're hard at work building your latest invention. These tools are also fun to use, and pneumatic tools cost less than comparable electric tools.

Campbell Hausfeld is one of the premier names in home and shop air tools. They sell compressors that develop the necessary air pressure, the hoses, regulators, and dryers to condition the air, the tools themselves, and other accessories. Products are sold through retailers such as Home Depot and other hardware/home improvement stores. Some online sales of selected (often factory-serviced) products.

Campbell Hausfeld's air tools include air hammers, drills (including reversing type), ratchets, screwdrivers, sanders, nailers, and staplers. I use two reversing drills, keeping either drill bits or screwdriver bits in them, depending on what I'm doing. Quick disconnect couplers allow for fast tool changes.

Coastal Tool & Supply   202758

510 New Park Ave.
West Hartford, CT 06110
USA

📞   (860) 233-8213

🖰   (860) 233-6295

⊘   (877) 551-8665

✉   sales@coastaltool.com

🌐   http://www.coastaltool.com/

Specializing in name brand power tools, both electric and pneumatic.

Coastal Tool & Supply online

Care and Feeding of Pneumatic Tools

Water destroys air tools. Oiling your tools every day that you use them will greatly prolong their life and helps dispel trapped water. Even if you have an air dryer attached to your compressor, be sure to add oil to the tool (just squirt a dab or two into the air intake) the first time you use it that day.

You may also wish to invest in an inline oiler, which adds oiled air to the tool as you use it. The oiler connects to the hose line from the compressor to the tool.

One final word: Wear earplugs (available at any drug store) when using air tools. Your eardrums will thank you for it.

Cyber Woodworking Depot   202760

P.O. Box 80376
Springfield, MA 01138-0376
USA

✆ (413) 782-6625

📠 (413) 782-3075

🌐 http://www.toolcenter.com/

Power woodworking tools: bandsaws, drill presses, lathes, miter saws, routers, and more.

RB Industries, Inc   202761

1801 Vine St.
P.O. Box 369
Harrisonville, MO 64701
USA

✆ (816) 884-3534

⊘ (800) 487-2623

🌐 http://www.rbiwoodtools.com/

RBI makes scroll saws, wood planers, and drum sanders. Their Hawk brand scroll saws are the notable product here, as they can be used to cut out precision parts in wood, plastic, and even metal. I have had one of their 16-inch scroll saws for over a decade, and it's one of my most cherished tools. These things are expensive, but they last a lifetime.

📄 🌐 🏭

Shopsmith, Inc.   203209

6530 Poe Ave.
Dayton, OH 45414-2591
USA

✆ (937) 898-6070

📠 (937) 890-5197

⊘ (800) 543-7586

✉ customerservice@shopsmith.com

🌐 http://www.shopsmith.com/

Shopsmith makes the famous all-in-one woodworking tool that combines table saw, sander, lathe, horizontal boring machine, and drill press.

TOOLS-PRECISION & MINIATURE

Many mobile robots are small and require tools to match. In this section, you'll find specialty tools for precision metal, plastic, and wood work. Many of the tools are miniature and are ideally suited for making small parts. Included are miniature machinery tools (mills, lathes, etc.), rotary saws, and other hand-operated motorized tools.

SEE ALSO:

TOOLS-CNC AND TOOLS-MACHINERY: For additional machining tools

Best Little Machine Tool Company   204055

P.O. Box 536
Hurricane, WV 25526
USA

✆ (304) 562-3538

⊘ (800) 872-6500

✉ blueridgemachine@worldnet.att.net

🌐 http://www.blueridgemachinery.com/

Mills and lathes, both desktop and brutes, at decent prices. Printed catalog available. Among their desktop products are:

- Emco Compact 5 Lathe
- Prazi lathes and mills
- Sherline lathes and mills

Gyros Precision Tools, Inc

http://www.gyrostools.com/

GYROS offers a complete line of quality miniature tools and accessories

ModelTool.com

http://www.modeltool.com/

Small precision tools for modelers; also metal supplies (K&S; sold in packs), X-Acto, Zap (glues), Dremel

Papa John's Toolbox

http://www.hobbytool.com/

Small precision tools

Cardstone Pty Ltd.   204092

30 Industrial Ave.
Molendinar, Queensland 4214
Australia

 +61 7 5539 6388

 +61 7 5539 6188

✉ inquire@cardstone.com

🌐 http://www.cardstone.com/

Cardstone sells primarily to the dollhouse maker, but the products have equal use in robotics. This include miniature tools, saws, and drills, and Cir-Kit doll house electrical wiring and lamps (supersmall lightbulbs).

Clisby Miniature Machines   203075

12 Norton Summit Road
Magill, South Australia 5072
Australia

 +61 8 8332 5944

 +61 8 8332 5944

🌐 http://www.clisby.com.au/

Small precision lathes and milling machines (for wood and metal) at low prices. Useful for working with light-weight materials such as brass, aluminum, and milling plastics. These are small and are well suited for machining little parts, like couplers and linkages, for your robot. They aren't made for rebuilding the crankshaft for a 1955 Chevy.

Clisby metal lathe. Photo Clisby Industries Pty Ltd

FDJ On Time   203246

1180 Solana Ave.
Winter Park, FL
USA

🚫 (800) 323-6091

✉ info@fdjtool.com

🌐 http://www.fdjtool.com/

For the jewelry maker, FDJ On Time is a one-stop shop for miniature jewelry tools, casting equipment, soldering supplies, electroplating gear, cleaners, mold-making equipment, and wax-working tools. Offers an extensive list of investment casting supplies (furnaces, investment, etc.).

Separate printed catalogs are available and are free for the asking:

- Tool Catalog, lists all the tools, casting, and mold-making products
- Display Catalog, contains display boxes for jewelry

L.R. Miniatures Ltd.   202754

1088 Fairfax Circle West
Boynton Beach, FL 33436
USA

📞 (561) 965-7280

📠 (561) 965-3759

✉ lrminiatures@aol.com

🌐 http://www.lrminiatures.com/

Tools for working with small parts and miniatures. Has a very large selection.

Micro-Mark   202609

340 Snyder Ave.
Berkeley Heights, NJ 07922-1595
USA

📞 (908) 464-2984

📠 (908) 665-9383

✉ info@micromark.com

🌐 http://www.micromark.com/

Micro-Mark is about precision tools. They sell precision and miniature tools of all descriptions, including desktop mills and lathes; as well as hand tools, bits, and other accessories; small hand-operated motorized tools; casting supplies; and raw metal, plastic, and wood (well, it's not really "raw"; it's in sheet, tube, or other manufactured form).

MicroProto Systems   203691

12419 E. Nightingale Ln.
Chandler, AZ 85249
USA

((602) 791-0219

((480) 895-9648

 info@microproto.com

(http://www.microproto.com/

See listing under **Tools-CNC**.

Minicraft 203680

N1246 Thrush Dr.
Greenville, WI 54942
USA

((920) 757-1718

((920) 757-1718

 info@minicrafttools.com

(http://www.minicrafttools.com/

Sellers of Minicraft precision power tools (sanders, saws), as well as hand tools.

Minicraft Tools USA, Inc. 203681

1861 Ludden Dr.
Cross Plains, WI 53528
USA

⊘ (888) 387-9724 Ext. 305

 customer_service@rotozip.com

(http://www.minicraftusa.com/

Miniature precision tools, including RotoZip rotary saw.

See also:

http://www.minicraft.co.uk/

Moody Tools, Inc. 203250

60 Crompton Ave.
East Greenwich, RI 02818
USA

((401) 885-0911

((401) 885-4565

⊘ (800) 866-5462

 infodesk@moodytools.com

(http://www.moodytools.com/

Manufacturer of precision miniature tools. Says the Web site, "Moody Tools, Inc. has been recognized as the premier manufacturer and one of the leading suppliers of precision miniature tools for over 50 years. Moody is proud to manufacture miniature hand tools and sets available through distributors nationwide."

Nature Coast Hobbies 203679

6773 S. Hancock Rd.
Homosassa, FL 34448
USA

((352) 628-3990

((352) 628-6778

⊘ (800) 714-9478

 arf@naturecoast.com

 http://www.naturecoast.com/

Primarily shipbuilding hobby materials and kits, but also miniatures tools, including Unimat I, Minicraft, and TAIG.

Tools and Supplies for Wire Wrapping

Wire wrapping is one of the fastest methods of creating permanent and near-permanent circuits, without fussing with creating circuit boards. Wire wrapping is a point-to-point wiring system that uses a special tool and extra-fine 28- or 30-gauge wrapping wire. Wire-wrapped circuits are as sturdy as soldered circuits, when properly constructed, and you can change your designs without resoldering.

You can purchase either a manual or a motorized wrapping tool. Unless you plan on wire wrapping all day, the manual tool is sufficient. To use, insert one end of the stripped wire into a slot in the tool, and place the tool over a square-shaped wrapping post. Give the tool 5 to 10 twirls. The connection is complete. To remove the wire, you use the other end of the tool and undo the wrapping.

Sherline Products   202290

3235 Executive Ridge
Vista, CA 92083-8527
USA

☎ (760) 727-5857
☏ (760) 727-7857
⊘ (800) 541-0735
✉ sherline@sherline.com
🌐 http://www.sherline.com/

Sherline is a premier maker of miniature "desktop" lathes and vertical mills. They're a staple in home machinery shops, and there is an active trade in parts and accessories on eBay.com and other online auction Web sites. Sherline offers some CNC versions or retrofits of their products (and they sell them "CNC ready"), but many other companies offer retrofit kits. So, you can purchase a manually operated lathe or mill now and upgrade it to CNC should you wish to automate your production.

Check out Sherline's "Robot Warriors Links" page:

http://www.sherline.com/robot.htm

Sherline vertical mill. Photo: Sherline Products Inc.

TAIG Tools   203678

12419 E. Nightingale Ln.
Chandler, AZ 85249
USA

☎ (408) 895-6978
☏ (480) 895-9648
✉ sales@TaigTools.com
🌐 http://www.taigtools.com/

TAIG Tools makes of small desktop ("micro") mills and lathes. Versions of the machines can be manually operated or connected to your computer for CNC. Sold through dealers.

Be sure to check out their regular Internet specials. You can get a nice mill or lathe for less than you think. Note that several companies offer CNC retrofits for the TAIG line; TAIG also provides CNC versions of some of their tools.

For CNC versions of TAIG mills and lathes, see also MicroProto Systems:

http://www.microproto.com/

Truebite, Inc.   202753

2590 Glenwood Rd.
Vestal, NY 13850-2936
USA

Shopping Surplus for Used Surgical Tools

A useful tool for robotics construction is the hemostat, locking, fine-toothed pliers that are used in various surgical procedures. They're expensive to buy new, but are available on the surplus market. The hemostats have been autoclaved (disinfected using high heat), so while "used" they are perfectly safe.

Most of the surplus hemostats you'll find are made of stainless steel, but you might also run into plastic versions. They're lighter weight but not quite as powerful as the metal variety.

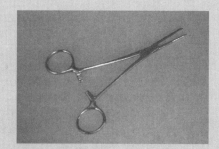

Stainless steel hemostats make good miniature pliers.

(607) 785-7664

(607) 785-2405

(800) 676-8907

truebite@truebite.com

http://www.truebite.com/

Specialty bits and accessories for precision tools (Dremel). Web site provides useful how-to articles on the proper use of the tools for common applications.

Woodsmith Store   203219

2625 Beaver Ave.
Des Moines, IA 50310
USA

(515) 255-8979

(800) 444-7002

woodstor@augusthome.com

http://www.woodsmithstore.com/

Woodsmith is dedicated to the home woodcrafter. They sell a lot of craft plans, but sadly, none on building robots. They do, however, offer a limited line of woodworking hand tools, bits, saws, router tables and fences, and shop jigs. Their emphasis is on precision woodworking (cabinetry, inlays, etc.), so their tools are designed for that extra measure of accuracy. Expect prices to match.

X-Actoblades.com   203496

P.O. Box 73
Merrick, NY 11566
USA

(516) 485-5544

(516) 489-3899

customerservice@x-actoblades.com

http://www.x-actoblades.com/

Distributors of X-Acto brand blades and tools.

Zona Tool Company   203172

16 Stony Hill Rd.
Bethel, CT 06801
USA

(203) 792-8622

(203) 790-9832

info@zonatool.com

http://www.zonatool.com/

Small precision tools for hobbies and crafts. Product line includes clamps, pin vices, small hacksaws, jeweler's saws, metal and oxide shaping tools (replacement bits for Dremel and similar tools), and sanding blocks.

 **TOYS**

Some of the earliest working robots were toys . . . though not many were designed to be played with. Rather, these robots, which followed patterns set in cams, were for timekeeping—clock making with mechanicals had been a real art for several centuries in Europe—and for entertainment. One of the most famous "show-and-tell" robots was a doll that wrote its signature while seated at a desk.

In this section, and the several to follow, are listings of toys with a robotic flavor. Some are robotic in shape, while others can be used to hack robots of your own design. Except for the very high end robot toys, such as the Sony AIBO, most are priced low enough that you won't worry about gutting them to get to their parts.

This main section lists general toy repositories, such as the venerable Toys"R"Us and the famous FAO Schwarz in New York City. The next sections tackle specific types of toys suitable for robotics:

- **Construction**-Designed for making things, these toys teach mechanical interactions, and many (such as LEGO and K'NEX) and can be used as elements.
- **Electronics**-Some electronics toys can be hacked for their sound chips, voice chips, and other components. There are many on the market; this section lists a few manufacturers that make popular brands.
- **Robots**-Robot toys include old tin windups and sophisticated microprocessor-controlled electronic pets.

Hasbro, Inc.

http://www.hasbro.com/

Toy manufacturer: Hasbro and Tiger lines; also sells replacement parts of many of their toys (some interesting robotic parts ideas)

Imagine the Challenge

http://www.imaginetoys.com/

Construction and electronic learning toys; mostly for younger children

Noveltytoys.com

http://www.noveltytoys.com

Hit toy du jour, which almost always includes a robot or two

Tiger Electronics

http://www.tigertoys.com/

Makes iCybie, Furby, and other high-tech toys

SEE ALSO:

KITS-ELECTRONIC and **KITS-ROBOTIC:** Electronics and robotic kits you can combine to make your own toys

LEGO (VARIOUS): LEGO Technic and LEGO Mindstorms construction sets

RETAIL-HOBBY & KIT: Additional robots to try

Amazon.com   202586

http://www.amazon.com/

Also sells toys, in partnership with Toys"R"Us. See listing under **Books**.

FAO Schwarz   202422

767 Fifth Ave.
New York, NY 10153
USA

📞 (212) 644-9400
📠 (212) 688-6053
🚫 (800) 876-7867
✉ fao@your-eservice.com
🌐 http://www.fao.com/

FAO Schwarz is a famous New York City–based toy retailer, with additional retail stores in California, Texas, Florida, Washington, Illinois, and several other states. They also sell mail order from the http://www.fao.com/ site. Offers a number of robotic toys, such as Sony Aibo, Wonderborg, DJ Johnny Bot, and BIO Bugs.

A unique attribute of FAO Schwarz is their knack for selecting toys that are interesting for all ages, not just for little kids. Of course this means they carry lots of robotics toys, but they also stock science and construction sets. The store caters to an upscale crowd, so they aren't afraid of stocking the really expensive stuff.

FAO Schwarz online.

GrandParents Toy Connection 202598

31 Viaduct Rd.
Stamford, CT 06907
USA

📞 (203) 602-0442
🚫 (800) 472-6312
✉️ grandtoy@erols.com
🌐 http://www.toyclassics.com/

"Old-style" toys, including some wind-up robots. Includes a 1950s design of a robotic dog (wind up).

KBtoys.com 🎖️ 202852

1099 18th St.
Ste. 1000
Denver, CO 80202
USA

📞 (303) 228-9000
🌐 http://www.kbtoys.com/

Online mail order for KB Toys retailer. Check often for deep discounts on LEGO, K'NEX, and other brands.

KB Toys Stores corporate headquarters:
100 West St.
Pittsfield, MA 01201

📞 (413) 496-3000

Manley Toy Direct 202320

2228 Barry Ave.
Los Angeles, CA 90064
USA

📞 (310) 231-7292
📠 (310) 231-7565
🚫 (800) 767-9998
🌐 http://www.manleytoy.com/

This is the direct sales site for Manley Toys, makers of the Tekno robotic dog and other products. Minimum orders apply. The Web site also provides information about the Manley Toys products. Product includes:

- Tekno Dog
- MTV Digital Camera
- Fabric, foam, and rubber balls

See also:

http://www.manleytoyquest.com/

Manley Toy Direct.

Ohio Art Company 203945

1 Toy St.
Bryan, OH 43506-0111
USA

📞 (419) 636-3141
📠 (419) 636-7614
✉️ info@world-of-toys.com
🌐 http://www.world-of-toys.com/

Started in 1908, Ohio Art is one of the oldest toy companies around. The firm is perhaps best known for the Etch-A-Sketch, a mechanized drawing tool that uses a sealed aluminum powder as a temporary drawing surface. Computerizing an Etch-A-Sketch with stepper or servomotors is a not-uncommon challenge in college- and university-level robotics courses.

Products are sold through retailers and directly from the Web site.

Totally Fun Toys 203213

521 Main St., Ste. D
P.O. Box 83
Bloomer, WI 54724
USA

📞 (715) 568-5566
📠 (715) 568-5569
🚫 (800) 977-8697
🌐 http://www.totallyfuntoys.com/

Toys, with an emphasis on construction and activity sets for the very young. However, they also carry K'NEX, as well as Rokenbok radio-controlled sets. Rokenbok vehicles are commanded via RF from a centralized station and offer hackability potential (though they are a tad expensive).

Toys"R"Us 203887

461 From Rd.
Paramus, NJ 07652
USA

☎ (201) 262-7800

📠 (201) 262-8112

🌐 http://inc.toysrus.com/

Toys"R"Us is the largest specialty toy retailer in the world (only Wal-Mart sells more toys, but they are a general discount store, not a specialty toy store). You can find a Toys"R"Us store in most every town—sometimes two or three. The stores are generally well stocked and the prices are competitive. Stock comes and goes, which is typical in the toy retail trade, but you can count on Toys"R"Us for a steady supply of these time-honored products:

- LEGO Mindstorms
- K'NEX construction sets
- Tonka motorized tractors and bulldozers (hack 'em!)
- R/C batteries and rechargers
- Nerf pistols and balls (for lightweight fighting robots)
- Motorized cars

The main Web address listed for Toys"R"Us is their corporate site (their store holdings include Toys"R"Us, Babies"R"Us, Kids"R"Us, and Imaginarium). Check the corporate Web site for their retail store locator.

The Toys"R"Us company once attempted to get into the Internet game, but decided (at least for now) to partner with Amazon.com for online sales. You can visit the Web site using the following URLs.

http://www.toysrus.com/

http://www.amazon.com/toys/

Note that Amazon handles all of the Toys"R"Us brands, including Imaginarium.

Trendmasters, Inc. 203764

611 North 10th St.
Ste. 555
St. Louis, MO 63101
USA

🚫 (800) 771-1810

✉ customerservice@trendmasters.com

🌐 http://www.trendmasters.com/

Manufacturer of the Rumble Robots toys, as well as other robotic toys (Muy Loco, Johnny Applebot, Vendobot, and more), the C-Video wireless camera, and miscellaneous goodies. Web site has some support information and replacement parts.

Web site for Trendmasters.

VTech Holdings Limited 203943

57 Ting Kok Rd., Tai Po
New Territories
Hong Kong

☎ +85 2 2680 1000

📠 +85 2 2680 1300

🌐 http://www.vtech.com.hk

VTech manufactures electronic educational products, sold through retailers such as Toys"R"Us. Several of these toys are hackable for their speech units, music synthesizers, or sensors. Web site is in English and Chinese.

See also:

http://www.vtechkids.com/

Additional Web sites for international offices in the U.K., Germany, France, and elsewhere.

🏭

TOYS-CONSTRUCTION

See **Toys** for more information about the listing in this section.

Construction Site, The 202311

200 Moody St.
Waltham, MA 02453
USA

☎ (781) 899-7900

📠 (781) 899-6485

🚫 (866) 899-7900

BRIO Corporation

http://www.briotoy.com/

BRIO construction toys

LASY

http://www.lasy.com/

Modular construction toy

Toy Kraft

http://www.toy-kraft.com/

Sells Meccano (Entech) sets and parts; based in India

---

 foreman@constructiontoys.com

 http://www.constructiontoys.com/

Online and local retailer of construction toys.

These toys are available both online and in the retail store: Capsela; Eitech; Erector; Fischertechnik; Geofix; Geomag; K'NEX; LEGO Dacta; Roger's Connection; Rhomblocks; Rokenbok; and Zome System.

These toys are available in retail stores only: Expandagon; LASY; LEGO.

Constructive Playthings 203214

U.S. Toy Company
13201 Arrington Rd.
Grandview, MO 64030
USA

 (800) 448-7830

 http://www.constplay.com

Early childhood construction toys; use 'em for parts.

Creative Learning Systems, Inc. 202427

10966 Via Frontera
San Diego, CA 92127
USA

((858) 592-7050

(858) 592-7055

⊘ (800) 458-2880

✉ info@clsinc.com

🌐 http://www.clsinc.com/

Sells many Fischertechnik kits and à la carte parts. Also, Capsela and MOVITS kits. Other science and technical educations kits, books, and products.

David Williams: Meccano
Home Page (MeccanoNet) 203253

http://www.freenet.edmonton.ab.ca/meccano/

User Meccano support page.

Direct Advantage 202600

520 W. Oklahoma Ave.
Milwaukee, WI 53207
USA

((414) 290-1000

⊘ (800) 669-7766

🌐 http://www.directadv.com/

School equipment and supplies. I was attracted to the manipulatives, which can be used as basic small parts for building robots.

Construction Sets for Robot Building

If E.T. can make a device out of old toy junk for "phoning home" across the vastness of space, imagine the robot you can build using toy construction sets. The better toy stores (especially the mail-order ones) are full of plastic and metal put-together kits that can be used to construct robots.

Throughout this section are numerous construction sets you can use for building robots.

EduBots 202429

1501 Miners Spring Rd.
Placerville, CA 95667
USA

((530) 622-6288
🚫   (877) 843-8829
   http://www.edubots.com/

Wide selection of Fischertechnik kits, partnered with Model A Technology (http://www.techeducation.com/).

Educational Insights, Inc. 203942

18730 S. Wilmington Ave.
Rancho Dominguez, CA 90220
USA

((310) 884-2000
🚫   (800) 995-4436
✉   service@edin.com
   http://www.educationalinsights.com

Over 900 items (including Amazing Live Sea-Monkeys), for home and school. Primary emphasis is on prekindergarten to eighth grade. Products are sold internationally. Highlight: Educational Insights resells Capsela construction toys.

---

eHobbyland 202776

1810 E. 12th. St.
Ste. C
Mishawaka, IN 46544
USA

((219) 256-1364
📠   (219) 256-1213
🚫   (800) 225-6509
✉   sales@hobbylandinc.com
   http://e-hobbyland.com/

Hobbies, toys, and plastic models. Model trains.

Hobby Link Japan 204031

Tatebayashi-shi
Nishitakane-cho 43-6
Gunma, 374-0075
Japan

(+81 276 80 3068
📠   +81 276 80 3067
✉   questions@hlj.com
🌐   http://www.hlj.com/

Hobby Link is a mail-order exporter based in Japan that ships internationally. You basically pay the retail price as it is in Japan, only this tends to be much less than what the same good costs at your local store. Be very careful

---

LEGO

LEGO has become the premier construction toy, for both children and adults. The LEGO Company, parent company of the LEGO brand, has expanded the line to educational resources, making the ubiquitous LEGO "bricks" common in schools across the country and around the world. LEGO also makes the Mindstorms, a series of sophisticated computerized robots.

LEGO Technic line of LEGO sets provides additional parts for robot construction. The pieces are engineered to fit the "classic" LEGO bricks, but add useful beams, connectors, and other parts to provide greater flexibility. A LEGO Technic beam has holes drilled through its sides for attaching to connector pieces. These holes are perfect for mounting LEGO components onto the rest of your robot.

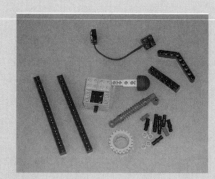

LEGO parts.

about shipping costs; you can save money buying many items at once.

They offer the full line of Tamiya Educational kits (motors, track sets, wheels, etc.), but not all kits are in stock all the time. You can arrange for your shipment to be held until most or all of the goods you want can be shipped to you.

Prices are in Japanese yen. You can convert to your currency on the Web site or at http://www.xe.com/.

Hyperbot   202425

905 South Springer Rd.
Los Altos, CA 94024-4833
USA

✆ (415) 949-2566

⊘ (800) 865-7631

✉ hyperbot@hyperbot.com

🌐 http://www.hyperbot.com/

Fischertechnik kits.

Imaginarium.com   202582

http://www.imaginarium.com/

Famous museum in San Francisco. They also sell their museum products by mail order. The Web site now under the auspices of Amazon.com.

K'NEX ⚥   203016

2990 Bergey Rd.
P.O. Box 700
Hatfield, PA 19440-0700
USA

🌐 http://www.knex.com/

K'NEX is the Rodney Dangerfield of construction toys for robotic parts: Next to LEGO, it gets no respect. Or at least, very little attention.

That's a shame, because of all the construction toy materials available, K'NEX is among the most useful in amateur robotics. K'NEX pieces snap together to make physically large yet lightweight structures. Gluing creates permanent joints, and you can use the parts, with fasteners and even nuts and machine screws, to create frames and other parts for your robots. K'NEX sets tend to be cheaper and have more parts than the average LEGO set.

K'NEX online.

Fischertechnik

The Fischertechnik kits, made in Germany, are favored in higher-ed classes and even some colleges and universities. More than "toys," Fischertechnik kits offer a snap-together approach to making working electromagnetic, hydraulic, pneumatic, static, and robotic mechanisms.

All the Fischertechnik parts are interchangeable and attach to a common plastic base plate. You can extend the lengths of the base plate to just about any size you want, and the base plate can serve as the foundation for your robot. You can use the motors supplied with the kits or use your own motors with the parts provided.

Fischertechnik parts.

Specialty robotic products in the K'NEX line include the CyberK'NEX machines (Ultra, Cybots) remote-control robot construction sets, MechWarrior construction sets, and Cool Machines, where you can build various kinds of vehicles with plates, beams, and girders.

You can buy extra K'NEX parts and sets online; toy stores carry the complete sets. Available parts items sold online include a motor pack, miniature spring motor, tires, and gears. The gears are so inexpensively priced you'll want to get them even if you don't use any other K'NEX parts (but I urge you to consider doing so).

See also the following Web sites:
http://www.cyberknex.com/-CyberK'NEX
http://www.knexeducation.com/-K'NEX education

LEGO Shop-at-Home 202583

 (800) 453-4652

 http://shop.lego.com/

See listing under **LEGO-General**.

Meccano Sources 202309

http://www.freenet.edmonton.ab.ca/
    meccano/mecsou.html

Where to get Meccano (Erector Set) kits and parts.

K'NEX

K'NEX uses unusual half-round plastic spokes and connector rods to build everything from bridges to Ferris wheels to robots. You can build a robot with just K'NEX parts or use the parts in a larger, mixed-component robot. For example, the base of a walking robot may be made from a thin sheet of aluminum, but the legs might be constructed from various K'NEX pieces.

A number of K'NEX kits are available, from simple starter sets to rather massive special-purpose collections (many of which are designed to build robots, dinosaurs, or robot-dinosaurs). Several of the kits come with small gear motors so you can motorize your creation. The motors are also available separately.

KNEX parts.

Erector Set

Erector Sets, now sold by Meccano, were once made of all metal. Now, they commonly contain both metal and plastic pieces, in various sizes, and are generally designed to build specific vehicles or other projects. Useful components of the kits include pre-punched girders, plastic and metal plates, tires, wheels, shafts, and plastic mounting panels. You can use any as you see fit, assembling your robots with the hardware supplied with the kit or with 6/32 or 8/32 nuts and bolts.

Over the years the Erector Set brand has gone through many owners. Parts from old Erector Sets may not fit well with new parts, including but not limited to differences in the threads used for the nuts and bolts. If you have a very old Erector Set (such as the ones made and sold by Gilbert), you're probably better off keeping them as collector's items, rather than raiding them for robotic parts. Similarly, today's Meccano sets are only passably compatible with the English-made Meccano sets sold decades ago. Hole spacing and sizes have varied over the years, and "mixing and matching" is not practical or desirable.

Metallus

202313

Postfach 1153
48478 Spelle
Germany

☏ +49 (0) 5977 93990
📠 +49 (0) 5977 939923
✉ rdt@redig.de
🌐 http://www.metallus.de/

Manufacturer of Marklin- and Meccano-compatible parts, including perforated strips, angled girders, flexible plates, and motors. Web site is in English and German. Ships worldwide.

Model A Technology

202450

2420 Van Layden Way
Modesto, CA 95356
USA

☏ (209) 575-3445
📠 (209) 575-2750
✉ jbmodels@inreach.com
🌐 http://www.techeducation.com/

Master distributor for Fischertechnik in the U.S. See also:

http://www.fischertechnik.com/
http://www.edubots.com/

Robotix

The Robotix kits, originally manufactured by Milton-Bradley and now sold by Learning Curve, are specially designed to make snap-together walking and rolling robots. Various kits are available, and many of them include at least one motor; you can buy additional motors if you'd like. You control the motors using a central switch pad.

The structural components in the Robotix kits are molded high-impact plastic. You can connect pieces together to form just about anything. You can cement the pieces together to provide a permanent construction.

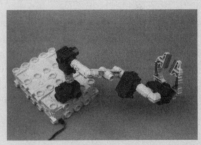

Robotix parts.

Capsela

Capsela is a popular snap-together motorized parts kit that uses unusual tube and sphere shapes. Capsela kits come in different sizes and have one or more gear motors that can be attached to various components. The kits contain unique parts that other put-together toys don't, such as plastic chain and chain sprockets/gears. Advanced kits come with remote control and computer circuits. All the parts from the various kits are interchangeable.

The links of the chain snap apart, so you can make any length chain you want. Combine the links from many kits and you can make an impressive drive system for an experimental lightweight robot.

Capsela parts.

Model A Technology.

OnlyTOYS.com 202312

ONLY TOYS, Inc.
6074 Apple Tree Dr, Ste. 4
Memphis, TN 38115
USA

((901) 866-0123

⊘ (800) 737-0123

✉ info@onlytoys.com

🌍 http://www.briotoys.com/

OnlyTOYS carries metal Erector sets; most are for building vehicles, and some (like the Steam Engine) are quite elaborate. The company also sells Rokenbok radio-controlled toys, which—though a bit expensive—offer hacking potential.

ozBricks 203116

3 Waltham Place
Avondale Heights
Melbourne, Victoria, 3034
Australia

(+61 3 8309-4034

🖅 +61 3 9337-4655

✉ info@ozbricks.com

🌍 http://www.ozbricks.com/

LEGO and LEGO Mindstorms, Down Under. Selection includes numerous Technic and Bionicle sets.

Pitsco/Dacta 203131

P.O. Box 1707
Pittsburg, KS 66762
USA

⊘ (800) 362-4308

🌍 http://www.pitsco-legodacta.com/

The Pitsco/LEGO Dacta catalog is an educator's resource for the world of LEGO and science. They offer hands-on instructional products for Robotics, Flight, Mechanisms, Early Learning, Energy, Tool, Resources, Parts, and Structures; the Robotics section offers LEGO Mindstorms and RoboLab software.

The Tools, Resources, Parts sections allow you to buy select LEGO pieces in quantity and at a discount. The parts categories are:

- Axles & Extenders
- Bands
- Beams
- Bricks
- Building Cards
- Connectors & Bushings
- Gears
- Miscellaneous
- Motors
- Plates
- Pulleys
- Special Element Sets
- Wheels & Hubs

The online store can be found directly at:
http://www.pldstore.com/

See also:
http://www.pitsco.com/

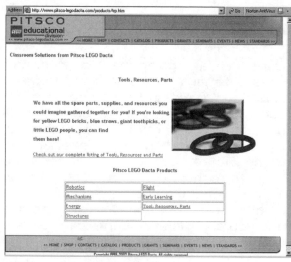

Web page for Pitsco/Dacta.

Polymorf, Inc. 204096

11500 NE 76th St.
#A-3, #309
Vancouver, WA 98682
USA

((360) 449-3024

✉ morfun@polymorf.net

🌐 http://www.polymorf.net/

Polymorf is a construction toy using hinged connectors. Seven different part designs provide triangles, squares, and rectangles, allowing for most any kind of 3D movable design. The parts are snapped together with a "pinge," which allows the pieces to swivel and move.

Though the hinged pieces are not powered, the Polymorf sets are useful as raw materials for building robots. You can glue or fasten other pieces of your robot to the Polymorf plastic. You can also use Polymorf to test different articulated designs. Different sets are available, with 230 to 1,000 pieces. In this cases, "pieces" means panels and pinges. The 1,000-piece set has 364 panels, 52 each of the seven shapes.

SmartTechToys.com 204095

301 Newbury St.
Dept. 131
Danvers, MA 01923
USA

🚫 (800) 658-5959

✉ CustomerService@smarttechtoys.com

🌐 http://smarttechtoys.com/

Robots and more robots (LEGO Mindstorms and K'NEX); toys to hack (R/C cars, Rokenbok vehicles); lasers and laser pointers.

Terrapin Software 202424

10 Holworthy St.
Cambridge, MA 02138
USA

((617) 547-5646

🖎 (617) 492-4610

🚫 (800) 774-5646

✉ info@terrapinlogo.com

🌐 http://www.terrapinlogo.com/

Developers of the Terrapin Logo programming language for home computers and LEGO Mindstorms RCX. The company also sells some LEGO and Fischertechnik kits.

Inventa

U.K.-based Valiant Technologies offers the Inventa system, a reasonably priced construction system aimed at the educational market. Inventa is a good source for gears, tracks, wheels, axles, and many other mechanical parts. Beams used for construction are semiflexible and can be cut to size. Angles and brackets allow the beams to be connected in a variety of ways. It is not uncommon—and in fact it's encouraged—to find Inventa creations intermixed with other building materials, including balsa wood, LEGO pieces, you name it. Inventa products are available from distributors, which are listed on the Valiant Web site at:

http://www.valiant-technology.com/

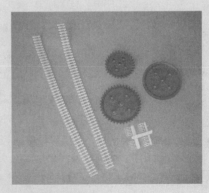

Inventa parts.

Timberdoodle Company   202423

E 1510 Spencer Lk Rd.
Shelton, WA 98584
USA

   (360) 426-0672

✉   mailbag@timberdoodle.com.

🌐   http://www.timberdoodle.com/

Timberdoodle specializes in home education products. They offer a good selection of Fischertechnik kits at good prices. Also sells Capsela, K'NEX, and electronics learning labs.

Be sure to check their "swan song" closeout deals.

Timberdoodle Web site.

Valiant Technology Ltd.   203813

Valiant House
3 Grange Mills, Weir Road
London
SW12 0NE
UK

   +44 (0) 2086 732233

   +44 (0) 2086 736333

✉   info@valiant-technology.com

🌐   http://www.valiant-technology.com/

Valiant manufactures the Inventa construction set, flexible plastic mechanical learning set that has many advantages over its more famous competitors such as LEGO and K'NEX. Inventa sets contain beams, gears, pulleys, and other mechanical parts. They also sell science class–oriented products, such as datalogger, Logo turtle robot, classroom mechanical engineering packs, electronic building block kits. Product is available through distributors and retailers.

Web Ring: Meccano   202310

http://www.meccanoweb.com/meccring/

Caters to the fans of the Meccano (or Erector Set) construction toys.

Other Construction Toys

There are numerous other construction toys that you may find handy. Try the nearest well-stocked toy store or a toy retailer on the Internet for the following:

- Expandagon Construction System (Hoberman).
- Fiddlestix Gearworks (Toys-N-Things)
- Gears! Gears! Gears! (Learning Resources)
- PowerRings (Fun Source)
- Zome System (Zome System)
- Construx (no longer made, but sets may still be available for sale)
- Fastech construction sets (no longer made, but were among the best parts assortment you could buy)

TOYS-ELECTRONICS

See **Toys** for more information about the listing in this section.

DSI Toys, Inc. 203944

1100 W. Sam Houston Pkwy. North
Houston, TX 77043
USA

- (713) 365-9900
- (713) 365-9911
- http://www.dsitoys.com/

Manufacturer of several hackable toys for robotics, including the e-BRAIN, a digital voice/music recorder with lightwave communications with a host PC. Using the Timex datalink technology, e-BRAIN receives its data from the monitor as a portion of it flashes, like a semaphore signal. Products are available at toy retailers.

Toymax Inc. 203659

125 East Bethpage Rd.
Plainview, NY 11803
USA

- (516) 391-9898
- (516) 391-9151
- (800) 222-9060
- http://www.toymax.com/

Toymax is a manufacturer of electronic toys, including small robotic toys: Rad4 and Battledrones. They also sell toy R/C cars, water pistols, and laser challenge guns (the lasers aren't real, of course, but they do contain focused optics useful for hacking).

TOYS-ROBOTS

See **Toys** for more information about the listings in this section.

ABoyd Company, LLC, The 202772

P.O. Box 4568
Jackson, MS 39296
USA

- (601) 948-3479
- (888) 458-2693
- info@aboyd.com
- http://www.aboyd.com/

Here you'll find all kinds of science fiction and monster movie memorabilia, including a number of robot toys. There's a heavy emphasis on classic Universal horror movies (of the Frankenstein and Dracula ilk), but there's also some rad robotic toys, including tin wind-ups (the 8-inch Gort is one of my favorites) and robots from movies and television.

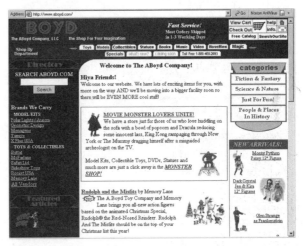

Web site for ABoyd Co.

Acme Vintage Toys & Animation Gallery 202773

9976 Westwanda Dr.
Beverly Hills, CA 90210
USA

- (310) 276-5509
- (310) 276-1183
- acmetoys@yahoo.com
- http://www.acmetoys.com/

Toys and animation art. Tin toy robots, Robby the Robot, Star Wars and Star Trek character sets, Mr. Atomic, Gigantor, and other robot figurines. Most are collectables and are very expensive.

CyberToyz 202774

Attn: Larry Waldman
2705 Wadsworth Rd.
Shaker Heig, OH 44122
USA

✉ Obiwall@aol.com

🌐 http://www.cybertoyz.net/

Collectable (and therefore fairly expensive) tin and wind-up robotic toys.

🌐 🖥

Manley Toy Quest   202659

2228 Barry Ave.
Los Angeles, CA 90064
USA

☎ (310) 231-7292

📠 (310) 231-7565

🌐 http://www.manleytoyquest.com/

Tekno robotic cats, dogs, birds, and dinosaurs; wooden Tonka; Britney Spears. Manley's products are sold in retail stores like Toys"R"Us, KMart, Wal-Mart, and others.

For a Web site dedicated to robotic toys, see also:

http://www.tekno-robot.com/

🏭

Manley Toy Quest online.

Big Life Toys

http://www.gna.com/blt/

Reproductions of old toys from the 50s and 60s, including tin robots

BioBugs

ttp://www.wowwee.com/biobugs/biointerface.html

Where BioBugs come from

Bot Collector.com

http://www.botcollector.com/

Toys and robot collectables

Cyclonians E.G.G. BOTS

http://www.bandai.com/

Makers o Cyclonians, wind up/break up spinning battle robots

Gadget Universe

http://www.gadgetuniverse.com/

Consumer electronics, including some high-tech robot toys

John's Collectible Toys & Gifts

http://www.johns-toys.com/

Robby the Robot Windup; Dr. Who Dalek; Lost in Space Robot B-9; Day The Earth Stood Still Gort Wind-Up; Terminator Endoskull Lifesize Replica

Lilliput Motor Company

http://www.lilliputmotorcompany.com/

Wind-up robotic toys

Mark Bergin Toys

http://www.bergintoys.com/

Tin robots

MetalToys.Com

http://www.metaltoys.com/

"Classic" toys, including metal reproductions of tin robots of the 50s and 60s

Robozone

http://www.robozone.com/

Anime and robot toys, collectables; ships worldwide

Sci-Fi Toys.com

http://www.sci-fi-toys.com/

Science fiction toys including tin and windup robots

Stardust Toys

http://www.tintoys.uk.com/

Specializes in "olf fashioned" tin toys, though not all may be old

Tiger Electronics UK

http://www.tigertoys.co.uk/

TigerToys in the UK

ToyTent

http://www.toytent.com/

Antique and collectable toys; some robotic, such as Robby the Robot from the movie Forbidden Planet

MGA Entertainment 203030

16730 Schoenborn St.
North Hills, 91343-6122
USA

 (800) 222-4685

 customerservice@mgae.com

 http://www.mgae.com/

MGA makes several robotic-type toys, including Insecto-Bots and the CommandoBot, a voice-active robot.

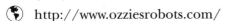

Museum of Unnatural Mystery 202775

http://www.unmuseum.org/

See listing under **Entertainment-Art**.

Ozzie's Robots Toys & Collectibles 202777

11 Cedar Ave.
Miller Place, NY 11764
USA

((631) 642-2105

 http://www.ozziesrobots.com/

Collectable robots. Some affordable reproductions; some of the true collectable stuff costs thousands of dollars. So, don't leave your robots at your friend's house when you're done playing with them.

Ray Rohr's Cosmic Artifacts 202073

P.O. Box 1001
Snoqualmie, WA 98065
USA

((425) 396-5741

 (425) 396-5742

 http://www.rayrohrtoys.com/

Wind-up robots, including Robby (from Forbidden Planet) and R2-D2 (from some movie named Star Wars). Also reference books. Mostly vintage and limited-edition toys from Japan—which means you better have a good job to pay for this stuff!

Retrofire-Robot & Space Toy Collectibles 202067

http://www.retrofire.com/

Interesting "high-tech" toys, including tin robots, space ray guns, and more.

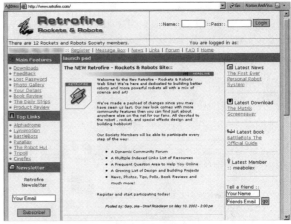

Online collectibles at Retrofire.

Robotix 203400

Learning Curve International
314 West Superior St.
6th Floor
Chicago, IL 60610-3537
USA

((312) 981-7000

 (312) 981-7500

 (800) 704-8697

cs@learningcurve.com

http://www.robotix.net/

See listing under **Robots-Hobby & Kit**.

Robots & Wind Me Up Toys 204103

713 Hampton
Tipp City, OH 45371
USA

((937) 341-1199

(208) 692-0274

http://www.windmeuptoys.com/

Robots and space toys: battery power, friction, or wind-up.

Rocket USA

202075

7775 Van Buren
Forest Park, IL 60130
USA

 (708) 358-8888

 toys@rocketusa.com

http://www.rocketusa.com/

Rocket USA is a wholesaler of collectable action toys. Products include friction-powered miniature robots and space vehicles, based on the original 1950s and 1960s toys.

Futurama toys at Rocket USA

SmartTechToys.com

204095

301 Newbury St.
Dept. 131
Danvers, MA 01923
USA

 (800) 658-5959

 CustomerService@smarttechtoys.com

http://smarttechtoys.com/

Robots and more robots (LEGO Mindstorms and K'NEX); toys to hack (R/C cars, Rokenbok vehicles); lasers and laser pointers.

Toy Rayguns

202778

http://www.toyraygun.com/

Ray guns and robots go together. This Web site is a compendium of ray guns and sci-fi art from days long gone. Not a robotics page per se, but still quite fun. (And if you look closely, there's a robot lurking here and there.) Web site also provides a user-to-user forum (although it was overrun by children playing games when I last looked), links, and an exhaustive list of books for the ray gun connoisseur.

Toy Robot Parts.com

202069

Philip De Gruchy
14 Lightfoot Street Mont Albert
Victoria 3127
Australia

lighteng@burwood.hotkey.net.au

http://www.toyrobotparts.com/

Reproduced metal and plastic parts for toy (mostly collectable Japanese) toy robots. Not inexpensive, due to the specialty nature of the products.

 **USER GROUPS**

You can share thoughts, ideas, plans, and frustrations with other robot builders at these user groups. Many provide monthly meetings (or at other regularly scheduled times); most meetings are informal, and show-and-tell and member presentations are encouraged. A few user groups are virtual, and meetings are on the Internet only.

If you like the idea of a robot user group that meets regularly, but one isn't near you, consider starting one yourself! All it takes is a desire and at least one other member. Meetings can be held at local schools, in the rec rooms of understanding companies, or even members' homes or workshops.

Art & Robotics Group (ARG) 202535

http://www.interaccess.org/arg/
Toronto, Ontario, Canada.

User group, discussion board, and latest news on the artistic side of robotics.

Atlanta Hobby Robot Club 202517

http://www.botlanta.org/
Atlanta, Ga.

Austin Robot Group 202536

http://www.robotgroup.org/
Austin, Tex.

B9 Club 203403

http://www.b9.org/index.html

User group with a difference: This one specializes in reconstruction of Robot B9 from the old Lost in Space television series.

B9 Robot Builders Club 203407

http://www.b9robotbuildersclub.com/

Dedicated to building the Robot from the Lost in Space television series. Impressions of Dr. Smith ("Oh, the pain!") are optional.

Central Illinois Robotics Club 202449

http://circ.mtco.com/
Peoria, Ill.

Says the Web site, "The Central Illinois Robotics Club was founded . . . in an effort to promote, educate, explore, and compete in the field of hobby robotics. The club is located in the greater Peoria area and meets monthly."

Central Jersey Robotics Group 203365

http://dpein.home.netcom.com/
New Jersey.

Chicago Area Robotics Group 202537

http://www.chibots.org/
Chicago, Ill.

Connecticut Robotics Society 202238

http://www.ctrobots.org/
Hartford, Conn.

As per the Web site: "We are a unique group of friends, experimenters, and mad scientists who meet monthly in Hartford, Connecticut. . . . Our interests are in electronics, mechanics, fun, and the sciences involved in automation and homebuilt robots."

Cybot Builder.com 202081

http://www.cybotbuilder.com/

CybotBuilder.com is an independent online resource for readers of Real Robots Magazine, a U.K. publication that features instructions and robot parts in each issue. See also:

http://www.realrobots.co.uk/

Web page for Cybot Builder.com.

Dallas Personal Robotics Group (DPRG)   202538

http://www.dprg.org/
Dallas , Tex.

Projects, tutorials, articles. The DPRG also sponsors the well-received RoboRama competitions. Events include line following, sumo, firefighting, and others. See also:

http://www.dprg.org/dprg_contests.html

Denver Area Robotics Club   204213

http://www.ranchbots.com/club.html
Denver, Colo.

East Bay Builders Group   203359

http://www.buildcoolstuff.com/ebg/
Berkeley, Calif.

Edmonton Area Robotics Society (EARS)   202090

http://www.ualberta.ca/~nadine/ears.html
Edmonton, Alberta, Canada.

User Groups Provide Help, Guidance

When I began to build my first robot in the late 1970s, I didn't know anyone else who was interested in my hobby. There were only a few books on the subject, and no magazines except the general-interest electronics periodicals. The world has changed—for the better—thanks primarily to the Internet. With the Internet, it's possible to locate others who have the same interests as you. With just a modicum of effort, you can find people all over the world with whom to exchange ideas. You might even be able to find folks in your hometown who share your interests.

Such are user groups, where like-minded people can get together—in person or over a modem—to discuss their interests. There are user groups for robotics, electronics, artificial intelligence, electronics, mechanics, metalworking and woodworking, programming, microcontroller and embedded systems, CNC machine tools, and much more. Some meet locally; others meet virtually on the Internet.

The Internet meetings are usually held through an online bulletin board. An example of an all-online user group is TRCY, which stands for "The Robotics Club of Yahoo." Here, some 2,000 members from around the globe use the bulletin board and chat features of Yahoo! to discuss the robot-building art. Libraries with photos, schematics, and building plans are provided.

Not all robotics user groups are virtual, of course. Many are held in business parks, school auditoriums, and people's homes. Some premier "real-life" user groups include the Seattle Robotics Society, Dallas Personal Robotics Group, HomeBrew Robotics Club, and Twin Cities Robotics Group. Those groups that provide local meeting places are indicated in the **User Groups** section by city and state.

EFREI Robotique 203368

http://assos.efrei.fr/robot/
France.

Finnish Robotics Association 203367

http://www.psavolainen.net/robotics/index.html
Tampere, Finland.

Front Range Robotics 202108

http://www.frontrangerobotics.com/
Northern Colorado.

HCC Robotica gg 203369

http://members.tripod.com/~hccrobotica/
The Netherlands.

HomeBrew Robotics Club 202211

http://www.hbrobotics.org/
San Jose, Calif.

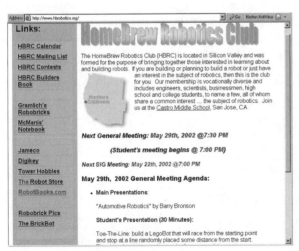

HomeBrew Robotics Club.

ISU Robotics Club 203361

http://www.ee.iastate.edu/~cybot/
Ames (Iowa State University).

KISS Institute for Practical Robotics (KIPR) 202540

http://www.kipr.org/

In the words of the Web site, "KISS Institute for Practical Robotics (KIPR) is a private non-profit community-based organization that works with all ages to provide improved learning and skills development through the application of technology, particularly robotics. We do this primarily by providing supplementary, extra-curricular and professional development classes and activities. KISS Institute's activities began in 1993."

KIPR also sponsors the annual Bot Ball tournament for middle and high school students.

Laboratory Robotics Interest Group 202541

http://lab-robotics.org/index.html
New Jersey.

Long Island NewYork Ametaur Robotics 203364

http://members.aol.com/rich924/html/club.html
New York.

MERG: Model Electronic Railway Group 202393

http://www.merg.org.uk/

Model railroad group specializing in convergence of model railroads, computers, and electronics.

Mobil Robotics Research Group 203370

http://www.dai.ed.ac.uk/groups/mrg/MRG.html
University of Edinburgh, Scotland.

Model Railroad Club 202394

http://www.ritmrc.org/

A collective of model railroad enthusiasts. The site provides some good background information on the latest in electronics for model railroading. See in particular the articles on DCC and DCC conversion.

Nashua Robot Club
203363

http://www.tiac.net/users/bigqueue/others/robot/homepage.htm
Nashua, N.H.

Northern New Mexico Robotics
202542

http://www.cbc.umn.edu/~mwd/
robot/NNMR.html
Los Alamos, N.M.

Northeastern PA Robotics Society
202063

http://www.nepars.org/

Says the Web site, "NePARS" is an informal group of hobbyists, experimenters, and robot builders who meet monthly in Wilkes-Barre, Pennsylvania. Our membership is a diverse group composed of people with many different backgrounds and experience levels. Our interests include, but are not limited to, electronics, mechanics, fun, and the science related to automation and homebuilt robotics."

Ottawa Robotics Enthusiasts (O.R.E.)
202092

http://www.ottawarobotics.org/
Ottawa, Ontario, Canada.

Phoenix Area Robot Experimenters
203357

http://www.parex.org/
Phoenix, Ariz.

Portland Area Robotics Society
202091

http://www.portlandrobotics.org/
Portland, Ore.

From the Web site, "The Portland Area Robotics Society is a club formed to help those interested in learning about and building robots. The club involves professionals, amateurs, students, college professors, engineers, artists, hobbyists, and tinkerers. PARTS will help explore all aspects of robotics for its members, and work toward expanding communication between robot enthusiasts. PARTS members share ideas, experience and enthusiasm for building robots."

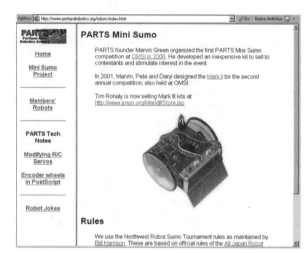

One of many informational pages at PARTS.

RoboFesta-Europe
203951

http://www.robofesta-europe.org/

RoboFesta-Europe is part of an international movement to promote interest throughout Europe in science and technology, including robotics. Sponsors Olympics-style competition events.

See also the international RoboFesta Web site at:

http://www.robofesta.net/

RoboFesta-International
203952

http://www.robofesta.net/

RoboFesta is an international movement to promote interest throughout the world in science and technology, including robotics. Sponsors Olympics-style competition events. Web site in Japanese, English, and French.

See also:

http://www.robofesta-europe.org/

RoboRama

See Dallas Personal Robotics Group (this section).

Robot Club of Traverse City 203362

http://www.wdweb.com/robotclub/
Traverse City, Mich.

Robot Group, The 202990

http://www.robotgroup.org/
Austin, Tex.

From the Web site, "The Robot Group was founded in the Spring of 1989 by a small group of Austin, TX, artists and engineers who shared a common vision: utilizing technology to provide and explore new mediums for art. Through the synergy of fusing art and technology, The Robot Group has stimulated the public into a playful interest in high technology, and art now has new vehicles for effecting culture."

See also a menagerie of robots from members of The Robot Group:

http://www.robotgroup.org/projects/

RobotBuilders.Net 203780

http://www.robotbuilders.net

Umbrella Web site for various specialty Internet-based robot-building clubs.

- B9 Club—http://www.b9.org/
- Robot Club—http://www.robotclub.org/

Robot B-9 builder's club.

- R2-D2 Builders Club—
  http://www.robotbuilders.net/r2/
- The Drone Room (Silent Running) —
  http://www.robotbuilders.net/droneroom/

Robotics Society of Southern California 203358

http://www.dreamdroid.com/
Fullerton, Calif.

Rockies Robotics Group 202543

http://www.rockies-robotics.com/
Aurora, Colo.

San Diego Robotics Society 202080

http://sdrobotics.tripod.com/
San Diego, Calif.

See also:

http://groups.yahoo.com/group/sdrs-list/

San Francisco Robotics Society of America 202544

http://www.robots.org/
San Francisco, Calif.

The San Francisco Robotics Society of America also sponsors the annual Robot Sumo conference.

Seattle Robotics Encoder 202988

http://www.seattlerobotics.org/encoder/

Semiregular online newsletter with articles, tutorials, and news about amateur robotics. Some of the articles get fairly technical; all of it is good.

Seattle Robotics Society 202014

http://www.seattlerobotics.org/
Seattle, Wash.

SRS has a major presence on the Internet and publishes the Encoder, an online technical journal on amateur robot building.

Seattle Robotics Society online.

Titan Robotics Club   203752

http://www.titanrobotics.net/

International School Robotics Club. For high-schoolers and middle-schoolers who are interested in robotics.

TRCY   202545

http://members.tripod.com/RoBoJRR/

Home page for The Robotics Club of Yahoo!.

See also:

http://members.tripod.com/RoBoJRR/

Triangle Amateur Robotics   202020

http://triangleamateurrobotics.org/
Raleigh, N.C.

Twin Cities Robotics Group   202460

http://www.tcrobots.org/
St. Paul, Minn.

Self-described as "a loose affiliation of people interested in robots, located in the Twin Cities metro area." The site hosts a number of useful resource pages, includes articles (identified by skill level), useful links, and colorful photos of the monthly meeting-see people and robots in action.

U.K. Cybernetics Club   203479

http://www.cybernetic.demon.co.uk/

For U.K.-based fans of robotics, but membership is open to everyone everywhere.

UCF Robotics   202550

http://clubs.cecs.ucf.edu/auvs/

University of Central Florida ongoing project on constructing unmanned (or unwomanned, for that matter) vehicles. That means autonomous robots.

Union College Robotics Club   203366

http://www.vu.union.edu/~robot/
Schenectady, N.Y.

Vancouver Island Robotics   202089

http://www.vancouverislandrobotics.org/

Vancouver Island, British Columbia, Canada. Sponsors workshops and day camps.

Wichita Robotic Club   202539

http://www.robot-club.org/

Meets in Wichita, Kans.. Sponsors several robotics contests.

Yahoo Groups: Homebrew_PCBs   203964

http://groups.yahoo.com/group/Homebrew_PCBs/

All about making your own PCBs (printed circuit boards) using laser printers, ink-jet printers, silk screening, smelly chemicals, you name it.

Yahoo Groups: Kansas City Robotics Society   204223

http://groups.yahoo.com/group/KCRS/

Online gathering point for all hobby or professional roboticists in the Kansas City area.

Yahoo Groups: San Diego Robotics Society   202055

http://groups.yahoo.com/group/sdrs-list

Online discussion group for the San Diego Robotics Society.

 VIDEO

Low-cost video cameras have paved the way to an increased use of vision systems in robotics. The sources in the following sections manufacture, sell, or support a variety of video equipment for allowing your robot to see. You may use video to process imagery for true video vision analysis, or you may merely broadcast a video signal from your robot to your desktop. Either way, you'll find a number of affordable products to suit your needs.

The video sections are:
- **Cameras**—Self-contained digital or analog cameras that are small enough to mount on the typical mobile robot. Digital cameras can be directly connected to a microprocessor or computer for vision analysis.
- **Imagers**—For those who wish to make their own vision systems from scratch. The imagers require interface electronics. This is not something most will want to do; rather, the listings in the section are really for their online datasheets and application notes documentation.
- **Programming & APIs**—API stands for "Application Program Interface." Examples and products for processing images. Also includes sources for capturing video popular personal computer platforms.
- **Transmitters**—Resources of battery-powered transmitters (and their associated receivers) for relaying color and black-and-white video images from one location to another. Note that these transmitters are low power only, and range is limited.

 VIDEO-CAMERAS

See the description in **Video** for more information about this section.

123securityproducts.com 203553
387 Canal St.
New York, NY 10013
USA
- ⊘ (866) 440-2288
- ✉ info@123securityproducts.com
- 🌐 http://www.123securityproducts.com/i

Black-and-white board cameras for security; video transmitters and receivers. Boasts an extremely small (but also fairly expensive) 900-MHz wireless camera/transmitter smaller than a 9-volt battery.

A3J Engineering, Inc. 203197
15344 E. Valley Blvd.
Ste. C
City of Industry, CA 91746
USA
- ☎ (626) 934-7600
- 📠 (626) 934-7609
- ✉ info@3jtech.com
- 🌐 http://www.3jtech.com/

See listing under **Communications-RF**.

🏭

Amazon Electronics/Elecronics123 202506
14172 Eureka Rd.
P.O. Box 21
Columbiana, OH 44408-0021
USA
- ☎ (330) 549 3726
- 📠 (603) 994 4964
- ⊘ (888) 549-3749
- ✉ amazon@electronics123.com
- 🌐 http://www.electronics123.com/

See listing under **Retail-General Electronics**.

CCTV Outlet 202195
1376 N.W. 22 Ave.
Miami, FL 33125
USA
- ☎ (305) 635-7060
- 📠 (305) 635-3175
- ⊘ (800) 323-8746
- ✉ sales@cctvco.com
- 🌐 http://www.cctvoutlet.com/

Cameras, lenses, RF transmitters and receivers.

CMUcam Vision Sensor ☿ 202514

http://www-2.cs.cmu.edu/~cmucam/

The CMUcam is a low-cost yet fully functional miniature digital camera; it is to robotic vision as the Basic Stamp was to microcontrollers. The Web site shows how to make one from available parts and kits you can find on the Internet (the camera itself is sold by Seattle Robotics, at http://www.seattlerobotics.com/, and Acroname, at http://www.acroname.com/). I'll let the Web site do the rest of the talking:

"CMUcam is a new low-cost, low-power sensor for mobile robots. You can use CMUcam to do many different kinds of on-board, real-time vision processing. Because CMUcam uses a serial port, it can be directly interfaced to other low-power processors such as PIC chips."

At 17 frames per second, CMUcam can do the following:

- Track the position and size of a colorful or bright object
- Measure the RGB or YUV statistics of an image region
- Automatically acquire and track the first object it sees
- Physically track using a directly connected servo
- Dump a complete image over the serial port
- Dump a bitmap showing the shape of the tracked object

Web home of the CMUcam.

Cricklewood Electronics 203856

40-42 Cricklewood Broadway
London
NW2 3ET
UK

- (+44 (0) 2084 520161
- (+44 (0) 2082 081441
- (http://www.skyelectronics.co.uk

General electronics; CCTV. Sells popular components (active andpassive) but also many hard to get and obsolete parts.

Matco Inc. 202207

2246 North Palmer Drive
Unit 103
Schaumburg, IL 60173
USA

- ((847) 303-9700
- ((847) 303-0660
- ((800) 719-9605
- (info@matco.com
- (http://www.matco.com/

MATCO sells CCD security cameras and CCTV surveillance products. They also offer security domes (plastic hemispheres of 3 to 6 inches in diameter) for real and fake security cameras.

Also in Canada:

4028 Cote Vertu.
St.Laurent, QC, H4R-1V4, Canada

Sales: (877) 720-9222 Canada/USA

Tech Support: (514) 340-9222

Fax: (775) 659-6544

matco-canada@matco.com

Micro Video Products 203998

One Mill Line Rd.
Bobcaygeon, Ontario
K0M 1A0
Canada

- ((705) 738-1755
- ((705) 738-5484
- ((800) 213-8111
- (info@microvideo.ca
- (http://www.microvideo.ca/

Micro Video sells miniature black-and-white and color video cameras (with integrated lens, though many are removable and interchangeable), bullet cameras, and wireless video systems. All cameras are NTSC (for color) or EIA (for black and white) compatible.

Newton Research Labs, Inc. 203762

441 SW 41st St.
Renton, WA 98055
USA

- ((425) 251-9600
- ((425) 251-8900

CMUCam, Omnivision, and Some Definitions

A popular example of an all-digital video camera is the CMUcam Vision Sensor, developed by Carnegie Mellon University (see http://www-2.cs.cmu.edu/~cmucam/ for more information).

The CMUCam uses an Omnivision OV6620 single-chip CMOS color imager. More on Omnivision and its products can be found at http://www.ovt.com/. Be sure to check out the spec sheets and application notes available at the Web site. At the time of this writing, the company's available digital cameras were listed in a handy product matrix, comparing such things as light sensitivity (expressed in lux), signal-to-noise ratios, and data format.

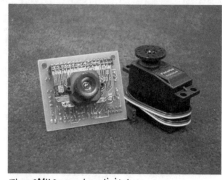

The CMUCam color digital camera.
Photo Illah Nourbakhsh.

Specifications like the ones provided for the Omnivision products are common, so for the uninitiated, here's a rundown of some of the terminology you may encounter:

• Lux—is the illumination of 1 lumen (an international unit of measurement) distributed uniformly over an area of 1 square meter. By comparison, an imager that produces an acceptable picture at 0.1 lux is considered very low light sensitive; an imager that needs 3 to 5 lux is not considered low-light sensitive.

• CCIR-601—is a data format adopted for digital television in many countries. One positive aspect of following this standard is that more and more video electronics gear are coming available that support it.

• 4:2:2—refers to sampling of the digital signal, and means that a pair of color-difference signals is sampled at half the frequency of the brightness (luminance) signal.

• Signal-to-noise ratio—is the amount of unwanted noise in relation to the desired signal. Noise in a video signal is seen as "snow." It is expressed in decibels (dB), where the higher the value, the better. With lower values (say under 35 to 40 dB), the noise starts to overcome the signal, making the image difficult to see and very hard to process electronically.

• YUV and RGB—are referred to as colorspace definitions. They define how colors are represented as a signal. Both involve some math we won't get into here, but YUV (of which there are many variants, and which is sometimes also referred to as YCbCr) is engineered to require lower bandwidth of the data transmission. This makes it ideal for slower processors.

• Resolution—can be either the number of pixels in the solid-state imaging array, or the number of pixels in the output signal of the camera. Be watchful of this specification: A solid-state camera may have an image resolution of 640 by 480, yet its imager contains fewer pixels (say, 352 by 288). The difference is made up by duplicating some pixels from the imager. Obviously, the higher the number of pixels in the imager itself, the better.

sales@newtonlabs.com

http://www.newtonlabs.com/

High-end machine vision.

In the words of the Web site, "Newton Labs develops and manufactures full turnkey machine vision systems, specializing in high speed and high resolution. Newton also manufactures robotics systems, diode laser products, and academic/research products. Newton Labs powerful, easy to use, and industrially rugged systems provide solutions for wide ranging vision and robotics applications for virtually every industry."

Pelikan Industry, Inc.   202994

555-A West Lambert Rd.
Brea, CA 92821
USA

Board camera from Pelikan. Photo Pelikan Industry, Inc.

((714) 672-0333

(714) 672-0360

 sales@pelikancam.com

http://www.pelikancam.com/

Pelikan provides video surveillance systems, including cameras and video transmitters.

Photon Vision Systems, Inc.   203724

Finger Lakes Business and Technology Park
One Technology Place
Homer, NY 13077
USA

((607) 756-5200

(607) 756-5319

 sales@photon-vision.com

http://www.photon-vision.com/

Photon Vision Systems designs and supplies CMOS imaging sensors and single-chip camera systems. Products are intended for OEM applications, but demonstration kits are available.

Plantraco Ltd.   204229

1105 8th St. East
Saskatoon, Saskatchewan
S7H 0S3
Canada

((306) 955-1836

(306) 931-0055

Better Pictures with Better Lenses

If you're interested in exploring vision systems for your robot creations, be sure to consider the quality of the lens (some solid-state imagers don't even come with a lens; you must add your own). Many of the smallest solid-state cameras were designed as electronic peepholes for doors and so have a very wide field of view—almost a "fisheye" appearance—to see as much outside as possible.

With wide views comes distortion of the image. That can make it hard to process the image to detect shapes and objects. For robotics work, you'll want the best lens you can afford, preferably one that provides a normal field of view. This will reduce the so-called barrel distortion common in wide-angle lenses.

The least expensive solid-state cameras come with a nonremovable lens, so you get what you get. The lens is interchangeable on the better units. A few are designed to work with standard C-mount lenses, which any CCTV reseller can provide.

Better Eyes to See You With

Most 'bot eyes are far simpler than the ocular sensors we humans have, yet they function quite admirably despite their lack of complexity. Here are the most common devices used for robot eyes.

- *Photoresistors*, typically a cadmium sulfide (Cds) cell (often referred to simply as a *photocell*). A Cds cell acts like a light-dependent resistor: The resistance of the cell varies depending on the intensity of the light striking it. When no light strikes the cell, the device exhibits very high resistance, typically in the high hundreds of K ohms, or even megohms. Light reduces the resistance, usually significantly (a few hundreds or thousands of ohms). Cds cells are very easy to interface to other electronics, but they are somewhat slow reacting and are unable to discern when light flashes more than 20 or 30 times per second. This trait actually comes in handy, as it means Cds cells basically ignore the on/off flashes of AC-operated lights.

- *Phototransistors* are very much like regular transistors, with their metal or plastic top removed. A glass or plastic cover protects the delicate transistor substrate inside. Unlike Cds cells, phototransistors are very quick acting, able to sense tens of thousands of flashes of light per second. The output of a phototransistor is not "linear"; that is, there is a disproportionate change in the output of a phototransistor as more and more light strikes it. A phototransistor can become easily "swamped" with too much light. Even as more light shines on the device, the phototransistor is not able to detect any more change.

- *Photodiodes* are the simpler diode versions of phototransistors. Like phototransistors, they are made with a glass or plastic cover to protect the semiconductor material inside them. And like phototransistors, photodiodes are very fast acting and can become "swamped" when exposed to a certain threshold of light. One common characteristic of most photodiodes is that their output is rather low, even when fully exposed to bright light. This means that to be effective, the output of the photodiode must usually be connected to a small amplifier of some type.

- *Pyroelectric* sensors—commonly referred to as *PIR*, for "passive infrared"—see by detecting changes in heat. They can be purchased new or salvaged from an existing motion detector. The most common use of pyroelectric infrared (or PIR) sensors is in burglar alarms and motion detectors. Because PIR sensors see only objects that move, they can be used to distinguish between animate and inanimate objects.

- *Ultrasonic* sensors send out a beam of sound, and then wait for the return echo. The difference in time between the burst of sound and its return echo indicates distance. By taking many measurements while slowly moving the ultrasonic sensor, it's possible for the robot to "map" the topography of its immediate environment.

- *Radar* is like ultrasonics, but it uses radio waves instead of sound. The advantage of radar is that the resolution—the ability to determine objects of small size—is far better than ultrasonics. The disadvantage is that workable radar detectors for robotics are quite expensive.

- *Video cameras* can be directly connected to computers, and the image they receive can be analyzed by software. They are the most like "real" eyes, and while video cameras are now relatively inexpensive and produce excellent detail, the limitation is in the software. It's not an easy task to break down and analyze a moving scene in an unpredictable environment, particularly in real time. No doubt this area of robotics will undergo massive improvements in the years to come and will someday give the ability for robots to see just as humans do.

ufoman@plantraco.com

 http://www.plantraco.com/

Sellers of upscale radio-controlled toys, including blimps and little tracked vehicles. Their Desktop Rover tracked vehicle can be controlled via a handheld remote or by software running on your computer. The company also sells a miniature wireless camera for use on its R/C products.

Polaris Industries   202033

470 Armour Dr. NE
Atlanta, GA 30324-3943
USA

((404) 872-0722

Polaris Industries micro "lipstick" camera. Photo Polaris Industries.

(404) 872-1038
(800) 752-3571
sales@polarisusa.com
http://www.polarisusa.com/

Polaris supplies security cameras and wireless transmitters for video.

QuickCam   204049

http://www.quickcam.com/

All about the Logitech QuickCam, including drivers.

Super Circuits   202732

One Supercircuits Plaza
Liberty Hill, TX 78642
USA

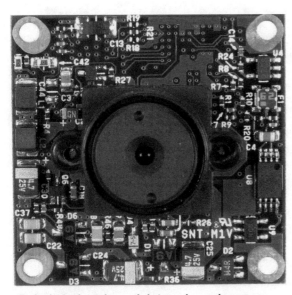

Polaris Industries miniature board camera. Photo Polaris Industries.

Google.com Search Phrases for Video and Robotics

Try these for locating more information about using cameras in robotics:

intitle:"basic stamp" camera

intitle:AVR camera

intitle:microcontroller camera

intitle:PIC camera

video robot camera

vision robot camera

webcam robot OR roboticsrobot vision mapping OR map

Super Circuits Web site.

 (800) 335-9777

 http://www.supercircuits.com/

Super Circuits sells wired and wireless video cameras, miniature (including the size of a shirt button) cameras, and video transmitters and receivers. Good stuff at reasonable prices.

URS Electronics Inc.   203615

123 N.E Seventh Ave.
Portland, OR 97293-0040
USA

Alternative Sources for Video Cameras

You don't need to buy a specialty solid-state video camera to give eyes to your robot. There are a few alternative sources for low-cost imagers:

- *Hacked wireless security camera.* You've seen ads for wireless security cameras you can "put anywhere." The camera, which operates at 9 to 12 volts DC, transmits its signal in the 2.4-gigahertz range. You need a video receiver to pick up the signal. Quality varies greatly, especially under low-light condition. Test a sample camera first before taking it apart. And be sure to turn off any microwave ovens that are operating nearby, as they emit a signal close to the operating frequency of these cameras.

- *Game Boy camera.* These were available for a time—but are now discontinued—for the venerable Nintendo Game Boy handheld game machine. You can still find them at swap meets, used computer stores, and at online auctions, such as eBay (the Game Boy device is not needed to hack the camera). The resolution is fairly low—128 by 123 pixels—but the camera is cheap and worth the hacking effort. Here are two sources for hacking the camera:

  http://pages.zoom.co.uk/andyc/camera.htm.

  http://homepages.paradise.net.nz/~vkemp/gbcam.htm

  Google.com search: *+gameboy +camera hacking*

- *Older model digital still cameras.* Digital cameras get better all the time. Ones with USB or serial interfaces to transfer the pictures to a PC are good candidates for hacking.

- *USB serial Webcams.* Many of the latest Webcams are cheap and are inherently digital, as they connect to the PC (or Macintosh or whatever) via a USB serial link. With some work, you may be able to hack a Webcam for robotics. But take note: While the data may be transferred via USB, the format of that data is likely propriety and varies from one manufacturer to the other. Some Webcam makers provide a software developer's kit or publish the data format specifications for their products, but others do not. Use Google.com to look for resources for Webcam writing drivers for Linux and other non-Windows operating systems. These pages will explain the data format of the camera. A most helpful resource is Sourceforge:

  http://www.sorceforge.net/.

  Google.com search phrase: *+CCD*

((503) 233-5341
((503) 232-3373
⊘ (800) 955-4877
✉ sales@ursele.com
🌐 http://www.ursele.com/

General industrial electronics; wire and cable; video cameras; test equipment. Local store in Portland, Ore..

Wm. B. Allen Supply Company, Inc.   202887

301 N. Rampart St.
New Orleans, LA 70112-3105
USA
((504) 525-8222
((504) 525-6361
⊘ (800) 535-9593
✉ info@wmballen.com
🌐 http://www.wmballen.com/

Wm. B. Allen supplies 35,000 items from over 150 manufacturers. For the robot builder, their electronics, video, and tools sections will be of most interest. The company is primarily in the alarm, security, and controlled access business.

🌐 💻

X10 Wireless Technology, Inc.   203076

15200 52nd Ave. South
Seattle, WA 98188
USA

((206) 241-3285
((206) 242-4644
⊘ (800) 675-3044
✉ sales@mail.x10.com
🌐 http://www.x10.com/

Home automation products for wireless remote control. Also, line of moderate-quality color and black-and-white security cameras. The main corporate Web site of X10 is at http://www.x10wti.com/.

🌐 💻

Zoomkat: Web Based Control Panel   202229

http://www.geocities.com/zoomkat/

Web-based control of servos (via a Mini SSC II servo controller). Details and programming examples for using servo control with a Web cam.

VIDEO-IMAGERS

See the description in **Video** for more information about this section.

Foveon, Inc.   204150

2820 San Tomas Expressway
Santa Clara, CA 95051
USA

CCD and CMOS Imagers

When I started robotics, consumer video cameras used a high-voltage imaging tube called a Vidicon to render a picture. Vidicons tended to be large and required a high-voltage power supply. The whole thing was heavy; with a lens, the typical video camera was the size of a lunchbox and weighed several pounds.

These days, cameras are the size of a cockroach—a regular-sized cockroach, not those brutes you encounter in New York City!—and require only minimal voltage to run. The cameras are cheap, too. You can buy a decent miniature black-and-white video camera for under $50; color versions are more, but still reasonable. These cameras, which use only solid-state components, are intended for security applications. The better ones operate under fairly low light conditions.

Solid-state video cameras use two kinds of imaging electronics: CCD or CMOS. *CCD* imagers (the *CCD* stands for "charged coupled device") is generally considered to offer a sharper picture and offers better low-light sensitivity. *CMOS* imagers cost less, but on the whole, don't deliver the sharpest pictures. While there are some CMOS imagers intended for low-light operation, most are not as sensitive as their CCD brethren.

Video Images: Analog or Digital

The vast majority of solid-state video cameras available today provide an analog output signal (specifically, it is a *composite* signal that combines synchronization and luminance signals, and with a color system, chrominance signals as well). Depending on where in the world you buy your camera, this signal is compatible with your TV and VCR.

For example, color cameras sold in the United States are compatible with the NTSC color standard; black-and-white cameras, with the RS-170 standard. Cameras for sale in many parts of Europe follow the PAL standard. Some cameras are selectable between NTSC and PAL or are available in either version.

For the most part, cameras with composite signal outputs are only modestly useful in robotics. The signal must be processed before the electronics (computer, microcontroller) on the robot can use it. The processing might be in the form of a frame capture card, or it might be some homemade sync separator and analog-to-digital comparator circuit. However, a minority of solid-state cameras provides a digital output—usually 8-bit parallel, but also USB serial. A parallel digital camera can be directly connected to your robot's computer or controller; software running on the robot reads the video signal by processing the digital data.

((408) 350-5100

◯ (877) 436-8366

🌐 http://www.foveon.com/

Makers of high-resolution CCD color imagers. Some technical documents available on the Web site.

OmniVision Technologies, Inc.   202281

930 Thompson Place
Sunnyvale, CA 94085
USA

((408) 733-3030

📠 (408) 733-3061

🌐 http://www.ovt.com/

Single-chip CMOS black-and-white and color imagers—"Single-Chip CMOS Image Sensors (Camera-on-a-Chip)."

The company designs single-chip image sensors for capturing and converting images for cameras. Their imagers are used in a number of products and designs, including the CMUcam (see http://www-2.cs.cmu.edu/~cmucam/).

VIDEO-PROGRAMMING & APIS

See the description in **Video** for more information about this section.

AVR + GameBoy™ Camera = Fun   202277

http://pages.zoom.co.uk/andyc/camera.htm

Detailed information, circuits, and sample programming (for the Atmel AVR microcontroller) for using the Gameboy camera for crude machine vision.

🗣

CMUcam Vision Sensor   202514

http://www-2.cs.cmu.edu/~cmucam/

See listing under **Video-Camera**.

🗣

CORAL Group's Color Machine Vision Project   202513

http://www-2.cs.cmu.edu/~jbruce/cmvision/

Says the Web site, their mission is to "create a simple, robust vision system suitable for real time robotics applications. The system aims to perform global low level color vision at video rates without the use of special purpose hardware."

Game Boy Camera Parallel Port Interface   203656

http://homepages.paradise.net.nz/
~vkemp/gbcam.htm

How to connect a Nintendo Gameboy camera (no longer made, but still available from some quarters) to a PC parallel port. Then how to program the PC to know how to read the data the camera is sending it. Includes circuit diagrams, how-to, and program code.

Java Media Framework API   203001

http://java.sun.com/products/
    java-media/jmf/index.html

Using Java for media and vision. Says the Web site, "The Java Media Framework API (JMF) enables audio, video and other time-based media to be added to Java applications and applets. This optional package, which can capture, playback, stream and transcode multiple media formats, extends the multimedia capabilities on the J2SE platform, and gives multimedia developers a powerful toolkit to develop scalable, cross-platform technology."

The JMF is capable of capturing, processing, and displaying data from most any analog camera. Can be used with any computer that supports Java, including Linux and the PC.

DVSec

http://MotionDetectionSoftware.com/
Video motion detection software for Windows

VideoScript Inc.

http://www.videoscript.com/
Video motion detection software for Windows and Macintosh

Linux support for Philips USB webcams   202660

http://www.smcc.demon.nl/webcam/

Site provides information and downloads for the Linux driver for Philips USB Web cams. According to the Web site, the driver also supports some cameras from Askey, Logitech, Samsung, and Creative Labs.

Newton Research Labs, Inc.   203762

441 SW 41st St.
Renton, WA 98055
USA

((425) 251-9600

℧ (425) 251-8900

✉ sales@newtonlabs.com

🌐 http://www.newtonlabs.com/

High-end machine vision. See listing under **Video-Camera**.

Open Source Computer Vision Library   202512

http://www.intel.com/research/mrl/
    research/opencv/

The Web site's aim is to "aid commercial uses of computer vision in human-computer interface, robotics, monitoring, biometrics and security by providing a free and open infrastructure where the distributed efforts of the vision community can be consolidated and performance optimized."

Programming Video for Windows   202032

http://ej.bantz.com/video/detail/

Informational Web site on using a video camera with Windows and writing programming code to capture frames with the Video for Windows application interface. For machines with Windows, obviously. (Note: Later versions of Windows incorporate video playback and capture services in the DirectX interfaces.)

QCUIAG (QuickCam and Unconventional Imaging Astronomy Group) 203657

http://www.astrabio.demon.co.uk/

Tips and techniques for using the QuickCam and other CCD/CMOS cameras for astrophotography. Many of the same techniques can be used for robot vision.

QuickCam Third-Party Drivers 204180

http://www.crynwr.com/qcpc/

Information about drivers for various computer platforms, such as Linux, for the QuickCam brand of video cameras.

RobotVision2 202983

http://hammer.prohosting.com/~vision4/RobotVision/rv2/rv2.htm

According to the Web site, "RobotVision2 (Rv2) is a real-time image processing software, using Video for Window (VFW) compatible camera, such as QuickCam, as the image source. Rv2 supports 1, 4, 8, 16, 24 and 32 bit-per-pixel, all uncompressed."

WebRemote 202251

http://www.WebRemote.co.uk/

The WebRemote Device Controller allows you to control devices across a Microsoft NetMeeting connection. Using just the NetMeeting software, you can control a Web camera (connected to the appropriate servo pan-tilt head, of course) from anywhere in the world. Use it to command the video on your robot—or even your entire robot—from a Web browser. Free download.

VIDEO-TRANSMITTERS

See the description in **Video** for more information about this section.

CCTV Outlet 202195

1376 N.W. 22 Ave.
Miami, FL 33125
USA

✆ (305) 635-7060

✆ (305) 635-3175

⊘ (800) 323-8746

✉ sales@cctvco.com

🌐 http://www.cctvoutlet.com/

Cameras, lenses, RF transmitters and receivers.

Matco Inc. 202207

2246 North Palmer Dr.
Unit 103
Schaumburg, IL 60173
USA

✆ (847) 303-9700

✆ (847) 303-0660

⊘ (800) 719-9605

✉ info@matco.com

🌐 http://www.matco.com/

See listing under **Video-Camera**.

Micro Video Products 203998

One Mill Line Rd.
Bobcaygeon, Ontario
K0M 1A0
Canada

✆ (705) 738-1755

✆ (705) 738-5484

⊘ (800) 213-8111

✉ info@microvideo.ca

🌐 http://www.microvideo.ca/

See listing under **Video-Camera**.

Polaris Industries 202033

470 Armour Dr. NE
Atlanta, GA 30324-3943
USA

✆ (404) 872-0722

✆ (404) 872-1038

(800) 752-3571

sales@polarisusa.com

http://www.polarisusa.com/

Polaris supplies security cameras and wireless transmitters for video.

Micro-miniature video transmitter. Photo Polaris Industries.

Transmitter (with integrated camera) and receiver pair. Photo Polaris Industries.

Super Circuits   202732

One Supercircuits Plaza
Liberty Hill, TX 78642
USA

(800) 335-9777

http://www.supercircuits.com/

Super Circuits sells wired and wireless video cameras, miniature (including the size of a shirt button) cameras, and video transmitters and receivers. Good stuff at reasonable prices.

X10 Wireless Technology, Inc.   203076

15200 52nd Ave. South
Seattle, WA 98188
USA

(206) 241-3285

(206) 242-4644

(800) 675-3044

sales@mail.x10.com

http://www.x10.com/

Home automation products for wireless remote control. Also line of moderate-quality color and black-and-white security cameras. The main corporate Web site of X10 is at http://www.x10wti.com/.

WHEELS AND CASTERS

If it rolls, it's in this section where you'll find many resources for wheels and casters (even tank-like treads) suitable for mobile robotics. Products include tubeless wheels designed for wheelchairs, but infinitely usable in robotics; omnidirectional wheels that rotate like a normal wheel but also allow perpendicular motion; spherical casters that don't swivel; ball transfers that act as unidirectional casters; and much more.

Several of the sources in this section manufacture wheels and casters and do not sell them directly to the public. They can be purchased from distributors, as indicated on the company Web site. Other sources do provide for direct purchase, usually through an online Web store, but also through mail order.

SEE ALSO:

MATERIALS: A number of general materials sources also offer "materials handling" products, which include wheels and casters

POWER TRANSMISSION: Shafts, bearings, pulleys, chains, and other components for driving wheels

RETAIL-HARDWARE & HOME IMPROVEMENT: Source for low-cost wheels (e.g., replacement lawn mower) and small casters

RETAIL-SURPLUS MECHANICAL: Common source for new and used wheels and casters, usually at a good discount

AIRTRAX Inc. 202101

P.O. Box 1237
Hammonton, NJ 08037
USA

☎ (609) 567-7800
📠 (609) 567-7895
🚫 (877) 2478728
✉ airtrax1@aol.com
🌐 http://www.airtrax.com/

AIRTRAX manufactures and distributes the DireXtional (omnidirectional) Wheel, intended for heavy-duty materials handling, and wheelchair wheels.

Falcon Wheel

http://www.falconwheel.com/

Natural rubber industrial press-on and pneumatic tires

Albion 204091

800 N. Clark St.
Albion, MI 49224
USA

☎ (517) 629-9501
🚫 (800) 835-8911
✉ email@albioninc.com
🌐 http://www.albioninc.com/

Albion is a manufacturer of industrial casters. Wheel choices range from cast iron to supersoft rubber.

American Airless 202342

7302 E. Alondra Blvd.
Paramount, CA 90723
USA

☎ (562) 633-7743
📠 (562) 633-4701
🚫 (800) 248-4737
✉ sales@americanairless.com
🌐 http://www.americanairless.com/

Manufacturer of tubeless (no air loss) tires, for wheelchairs and similar applications. Sizes range from 6 to 26 inches. The tires use inflated foam for pneumatic-like performance.

American Caster & Material Handling, Inc. 204209

2603 NE Industrial Dr. North
Kansas City, MO 64117
USA

☎ (816) 283-3815
📠 (816) 283-3819
🚫 (800) 688-0677
✉ custservice@americancaster.com
🌐 http://www.americancaster.com/

Sells casters and wheels for material handling; most are larger sizes for heavier loads. Lots of choices in wheels with different size hubs.

American Surplus Inc. 203593

1 Noyes Ave.
East Providence, RI 02916
USA

(401) 434-4355

(401) 434-7414

(800) 989-7176

info@American-Surplus.com

http://www.american-surplus.com/

New and used industrial surplus. See listing under **Retail-Surplus Mechanical**.

Applied Industrial Technologies   203445

One Applied Plaza
Cleveland, OH 44115-5053
USA

(216) 426-4189

(216) 426-4820

(877) 279-2799

products@apz-applied.com

http://www.appliedindustrial.com/

Industrial bearings, linear slides, gears, pulleys, pneumatics, hydraulics, and other mechanical things. Also hosts

Maintenance America, online reseller of industrial maintenance supplies and general industrial supplies (wheels, casters, fasteners, and more), tools, paints, and adhesives.

Caster Connection   203834

745 South St.
Chardon, OH 44024
USA

(800) 544-8978

http://casterconnection.com/

Casters and wheels. When I last visited I didn't stay long because of the irritating Web site design with all sorts of popup message boxes.

Cheapcasters.com   202530

Sellstar
1411 E. Edinger Ave.
Santa Ana, CA 92705
USA

Selecting the Right Caster

Robots with coaxial drives—two drive wheels on either side—need something on the front and/or the back to prevent them from tipping over. A common approach is to use casters. If the robot's drive wheels are mounted along the center line of the robot, then two casters are used. One is placed toward the front of the robot, and the other toward the back. If the drive wheels are placed at one end of the robot, then only one balancing caster need be used.

Casters add balance to two-wheeled robots.

For small roving robots, the caster is fairly small. The smallest commonly available caster has a wheel of about 1 5/8 inches; smaller models are available, but are not as easy to find. The 1 5/8-inch caster will work with robots that have 2- to 3-inch-diameter drive wheels. The larger the drive wheel, the larger the caster. (This doesn't always hold true, depending on whether the motors and wheels are mounted on the top of the robot's base, or below it. When mounted on top, there is less ground clearance, and the caster must be smaller.)

A problem with the very small casters is that, well, most aren't made very well. They are designed for extremely light duty, so their ball bearing swivel mechanism is cheaply made. This can cause the swivel to bind up or stick when used with a light load, such as your robot. When selecting a caster, spin its swivel. If it catches or binds up at all, select another model or brand (or another caster in case you simply got a bad one).

If your robot is small and light (under a few pounds), you should also avoid casters where the swivel is stiff because of heavy grease. Otherwise, the caster may not turn readily under the light weight of the robot.

(800) 835-6561

 sales@cheapcasters.com

 http://www.cheapcasters.com/

Discount casters and wheels. All sizes from 2 inches on up.

Home page of Cheapcasters.com

Cisco-Eagle Inc. 203263

2120 Valley View Lane
Dallas, TX 75234
USA

 (972) 406-9330

 (972) 406-9577

(800) 441-1162

24hours@cisco-eagle.com

http://www.cisco-eagle.com/

Manufacturer of materials-handling equipment and parts. Listed here mainly for their casters, wheels, and ball transfers.

Colson Caster Corporation 203446

3700 Airport Rd.
Jonesboro, AR 72401
USA

(800) 643-5515 Ext 1195

info@colsoncaster.com

http://www.colsoncaster.com/

Wide assortment of casters and wheels.

Swivel caster. Photo courtesy Colson Caster Corp.

Pneumatic wheel. Photo courtesy Colson Caster Corp.

Cruel Robots 202533

32547 Shawn Dr.
Warren, MI 48088
USA

Dan@cruelrobots.com

http://www.cruelrobots.com/

Performance materials and products for combat robots. Colson wheels (some are "combat ready" with heavy-duty hub already attached), axles, reducers, hubs, sprocket and chain, weapons, and casters.

Dave Brown Products 202133

4560 Layhigh Rd.
Hamilton, OH 45013
USA

(513) 738-1576

(513) 738-0152

sitemstr@dbproducts.com

http://www.dbproducts.com/

Dave Brown makes a long list of items for radio-control models. Products are available in hobby stores or online. The lines ideally suited to robotics include:

- Adhesives/additives
- Carbon fiber building materials
- Covering materials
- Lite Flite wheels

The Lite Flite wheels are a common staple of robot builders (I must be on my twentieth pair by now). The wheels are light and fairly inexpensive, and they can be attached to a variety of axles and motor shafts. By gluing a servo horn to the rim of the wheel, you can attach it to a servo that has been modified for continuous rotation.

Doug Mockett & Company, Inc. 202671

P.O. Box 3333
Manhattan Beach, CA 90266
USA

☎ (310) 318-2491
📠 (310) 376-7650

 (800) 523-1269
✉ info@mockett.com
 http://www.mockett.com/

Hardware for fine furniture; table legs, casters. High-tech and sleek looking.

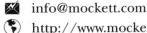

Douglas Wheel, Inc. 202171

4040 Avenida De La Plata
Oceanside, CA 92056
USA

☎ (760) 758-5560
✉ info@douglaswheel.com
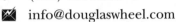 http://www.douglaswheel.com/main.htm

Douglas Wheel makes aluminum wheels for go-carts, scooters, dune buggies, golf cars, and other vehicles. Cart wheels are available in 5-inch and 6-inch sizes.

You can see the full line of products in their printed catalog-I've tried to get one several times, but nothing has ever come. You might have better luck.

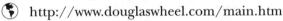

Wheels, Tires, Hubs—Oh, My!

Rolling robots use wheels. Or maybe they use tires (or *tyres,* if you're in the United Kingdom). Or maybe they use hubs and tires. Which one is right? They all are, and the terms are often used interchangeably. Yet it helps to define exactly which is which, so here's a short glossary on what these terms *should* mean.

Wheels are made up of tires, mounted on hubs. A tire is rubber, plastic, metal, or some other material, and the hub is the portion that attaches to the shaft of the axle or motor. (Similarly, the wheels on a car are made of the tire on the outside, and the hub, or rim, on the inside.)

Some wheels for robots are molded into one piece. Others, such as the Dave Brown Lite-Flight wheels, are composed of two separate pieces assembled at the plant. The Lite-Flight wheels use a plastic hub that attaches to the motor shaft or axle, and onto the hub is mounted a foam tire.

Some more terms:

- *Load wheels* are wheels with a metal or heavy-duty-plastic hub. They are for heavy loads.
- *Press-on tires* are wheels without hubs; you press the tires (sometimes they're just an O-ring) onto hubs.

Wheels are tires mounted on hubs.

Du-Bro Products, Inc.   202134

P.O. Box 815, 480 Bonner Rd.
Wauconda, IL 60084
USA

    http://www.dubro.com/rcproducts.html

Du-Bro is a leading manufacturer of hardware, wheels, and accessories for radio-controlled models. See listing under **Radio Control-Hardware**.

Model R/C tires, from Du-Bro. Photo Du-Bro Products, Inc.

Edmond Wheelchair Repair & Supply   203777

1604 Apian Way
Edmond, OK 73003
USA

(@listing-phonefaxweb: (405) 359-5006
🚫 (888) 343-2969
✉ Sales@edmond-wheelchair.com
 http://www.edmond-wheelchair.com/

Wheelchair and scooter parts, including motors, wheels, and batteries.

EL-COM   203827

12691 Monarch St.
Garden Grove, CA 92841
USA

((714) 230-6200
(📠 (714) 230-6222
🚫 (800) 228-9122
 http://www.elcomhardware.com/

Fasteners and hardware, mainly for cabinetry. Also casters, aluminum extrusions (squares, channels, bars), plastic laminates, and foam products.

Engine Trix   202168

1350 N. Acacia Dr. #2
Gilbert, AZ 85233
USA

((480) 633.8577
(📠 (480) 633.8578
✉ fasttrix@enginetrix.com
 http://www.enginetrix.com/

Sells billet wheels with tires-possible uses with robotics. Also sells Go-peds, engines, and decks.

ErgoTech Inc.   203821

Commerce Park
11 Old Newtown Rd.
Danbury, CT 06810
USA

Using Ball Transfers as Casters

Ball transfers are primarily designed to be used in materials processing— conveyor chutes, and the like. Ball transfers are made of a single ball, either metal, plastic, or rubber, held captive in a housing. They function as an omnidirectional caster for your robot.

The size of the ball varies from about 11/16 to over 3 inches in diameter. Look for ball casters at mechanical surplus stores and also at industrial supply outlets, such as Grainger, McMaster-Carr, Outwater Plastics, MSC Industrial Direct, and others.

Ball transfers can be used as omnidirectional casters.

(203) 790-4100

(203) 790-4445

sales@ergotechinc.com

http://www.ergotechinc.com/

Manufacturer of Roll-Flex multidirectional wheel. Use 'em for three-wheeled robots that can travel and turn in any direction. The Roll-Flex rollers are available as simplex or duplex (one or two sets of "tires") in diameters up to 4.72 inches.

Flexel (UK) Ltd

202343

Brackley Sawmills
Northampton Road
Brackley
Northants.
NN13 7DL
UK

+44 (0) 1280 704141

+44 (0) 1280 706373

enquiries@flexelmobility.co.uk

http://www.flexelmobility.co.uk/

Suppliers of tubeless wheelchair tires. Also sells casters, batteries, and hub brakes for wheelchairs.

J. W. Winco

202717

P.O. Box 510035
New Berlin, WI 53151-0035
USA

(262) 786-8227

(262) 786-8524

(800) 877-8351

J. W. Winco.

http://www.jwwinco.com/

Industrial components (casters, handles, cranks, hand-wheels, knobs, and O-rings), available in both metric and inch sizes. Check out the technical section for helpful engineering tidbits.

KC Marketing Group, Inc.

202531

CasterSupply.com
10 Dubelbeiss Lane
Rochester, NY 14622
USA

(716) 266-0130

(716) 266-0311

(800) 456-9340

sales@castersupply.com

http://www.castersupply.com/

Casters, wheels, drawer slides, and ball transfers. All sizes.

KIK Tire, Inc.

203774

590 Airport Rd.
Oceanside, CA 92054
USA

(760) 967-2777

(760) 967-4071

(888) 545 8473

kiktire@aol.com

http://www.kiktire.com/

KIK makes airless tires of all sizes. They're intended for things like scooters and wheelchairs, but your robot won't mind.

Kornylak Wheel Division

202102

400 Heaton St.
Hamilton, OH 45011
USA

(513) 863-1277

(513) 863-7644

(800) 837-5676

kornylak@kornylak.com

http://www.omniwheel.com/

Kornylak produces wheels for conveyors and materials handling. Among their products is the Transdisc omnidirectional caster, heavy-duty omniwheel, and Transwheels. The Transwheels are notable: They are lightweight plastic, available in 2- and 4-inch-diameter models, can be used as omnidirectional wheels, and are considerably less expensive compared to other omniwheels.

Kornylak Transdisc. Photo Kornylak Corp.

Kornylak Omniwheel. Photo Kornylak Corp.

Linco Inc.   203394

10749 East Rush St.
South El Monte, CA 91733
USA

Omniwheels Go Your Way

Imagine a wheel that spins like any other wheel, yet also allows for sideways motion. That's an *omnidirectional wheel*, *multidirectional wheel*, or *omniwheel*—an idea that goes back to about 1910. The wheel is a series of small wheels or rollers, mounted around the circumference of a larger main wheel. They're popular in materials-handling applications; the wheels are mounted in rows on top of tables or conveyors. Boxes or other goods glide effortlessly along the wheels and are allowed movement in any direction.

For robotics, omniwheels have two principle applications:

Omniwheels can be used as casters or as drive wheels. Photo Acroname, Inc.

- As drive wheels in three-wheeled robots. An example robot that uses this design is the PalmPilot Robot Kit (PPRK), sold by Acroname and several others. Rather than use two drive wheels positioned opposite one another, the PPRK uses three wheels in a triangular configuration. Only two motors propel the robot at a time, but the machine is able to move in any direction by applying power to specific motors.

- As free-wheeling casters, allowing for low-friction turning. A problem with swivel casters is that the swivel may not spin freely in turns. This causes the robot to lose accurate tracking and steering (if the robot is light enough, a caster that isn't pointed in the right direction will cause the little critter to veer off to one side!).

Omniwheels are available in sizes ranging from about 40mm (about 1.5 inch) to over 150mm (about 6 inches). The wheel material is rubber or polyurethane. Alas, omniwheels are rather expensive, but for what they do, they do it well.

(626) 448-6155

linco@linco-inc.com

http://www.linco-inc.com/

Casters. Local retail stores in southern California.

Kyosho Nitro Blizzard Tracks

Building a small robot with tracks instead of wheels? So, you know the hardest part is finding suitable tracks (or treads) that won't slip off or get fouled up traveling over carpet. You can always construct your own tracks using an automotive timing belt, a synchronous drive belt, or even the treads from a LEGO Technic set. But for mechanical reliability at an affordable price, it's hard to beat replacement tracks for the Kyosho Nitro Blizzard R/C racing car.

The replacement pack provides 28 inches of track in 7-inch segments; you screw the segments together to make the whole track loop. The track is 3 1/2 inches wide, which is fairly substantial; you can run your robot indoors or out, over a variety of surfaces, including grass, dirt, even sand. You can also purchase drive chain and sprocket wheels that mate with the track. The sprocket wheels are a little over 2 inches in diameter.

You can purchase these and other replacement parts for the Kyosho Nitro Blizzard online from Tower Hobbies (http://www.towerhobbies.com/), using their Parts Express service. The URL (subject to change) for the Nitro Blizzard is:

http://www2.towerhobbies.com/listxprs/kyoc0424.html

If this doesn't work for you, find the Parts Express link on the main Tower Hobbies home page, then scroll down to the Kyosho Nitro Blizzard listing.

National Power Chair 202169

4851 Shoreline Dr.
P.O. Box 118
Mound, MN 55364
USA

(952) 472-1511

(952) 472-1512

(800) 444-3528

info@npcinc.com

http://www.npcinc.com/

See listing under **Actuators-Motors**.

Airless tire from NPC. Photo National Power Chair

Nighthawk Manufacturing Inc. 203775

2 51331 Rande Rd. 224
Sherwood Park, Alberta
T8C 1H3
Canada

(780) 464-2856

(780) 464-2607

(800) 661-6247

nthawk@connect.ab.ca

http://www.nthawk.com/

Supplier of wheelchair parts, including airless and standard tires.

North American Roller Products, Inc. 203994

P.O. Box 2142
Glen Ellyn, IL 60138
USA

((630) 858-9161

📠 (630) 858-9103

🌐 (none specified)

Makers of popular omnidirectional wheels, intended for conveyor systems, but used a lot to make robots.

📦 $ 💻

NPC Robotics

See listing for National Power Chair under **Actuators-Motors**.

Reid Tool Supply Co. 🎖   203820

2265 Black Creek Rd.
Muskegon, MI 49444
USA

🚫 (800) 253.0421

📧 mail@reidtool.com

🌐 http://www.reidtool.com/

Reid is an all-purpose industrial supply resource. See listing under **Power Transmission**.

📄 🌐 💻

Skyway Machine, Inc.   203776

4451 Caterpillar Rd.
Redding, CA 96003
USA

((530) 243-5151

📠 (530) 243-5104

📧 sales@skywaytuffwheels.com

🌐 http://www.skywaytuffwheels.com/

Maker of specialty wheels: BEAD-LOK, bicycle, caster, utility, wheelchair, and standard hubs.

🌐 🏭

Omniwheel Sources

Here are some key sources for omniwheels, should you want to try some out:

Acroname Inc. (specialty retailer)
http://www.acroname.com/

AIRTRAX Inc. (manufacturer; heavy-duty and wheelchair applications)
http://www.airtrax.com/

Atlantic Conveying Equipment Ltd.
http://www.atlanticgb.co.uk/

Budget Robotics (specialty retailer)
http://www.budgetrobotics.com/

Kornylak Wheel Division (heavy-duty models and the low-cost Transwheel)
http://www.omniwheel.com/

Mr. Robot (specialty retailer)
http://www.mrrobot.com/

North American Roller Products (manufacturer of commonly used wheels)
*(see listing in **Wheels and Casters** for address)*

Phantasm (combat robot builder; makes custom omniwheels)
http://phantasm1.com/

Reid Tool Supply Co. (materials supply)
http://www.reidtool.com/

Superior Tire & Rubber Corporation

203264

P.O. Box 308
Warren, PA 16365
USA

☏ (814) 726-0740
🚫 (800) 289-1456
✉ sales@SuperiorTire.com
🌐 http://www.superiortire.com/

Manufacturer of industrial-size (forklifts and larger) wheels, tires, and casters. Here's some of what they offer:

- Polyurethane and rubber press-on tires
- Polyurethane and rubber load wheels
- Superior STR casters
- Custom-engineered tires

- Polyurethane and rubber track pads
- Custom molded elastomer products
- Cushomatic solid pneumatic replacement tires

Tracks USA

204233

10341-67th Ave. SE
Lake Lillian, MN 56253
USA

☏ (320) 382-6128
🌐 http://www.tracksusa.com/

Tracks USA sells replacement snowmobile tracks. Snowmobile tracks can be used with large outdoor robots and for combat robots. Sells the popular Camoplast tracks.

APPENDIX A

Yellow Pages—First Line of Defense

Someone at the phone company forgot to trademark the name "Yellow Pages," and so now there are hundreds of Yellow Pages—in print form, and on the Internet. Your own local merchants are among the best choices for parts and supplies for your robot: because they're nearby, you can visit them yourself and pick up what you need. There are no shipping costs to contend with, and you know what you're getting.

The Yellow Pages are broken up into categories, and over time, the categories have become somewhat standardized. This categorization has also extended to the numerous online Yellow Pages and business directories now available. These are free, of course, because the Yellow Pages are really just a massive classified ad section.

As useful as the Yellow Pages are, their sheer size can make finding the right thing at the right time hard. Following are the major categories for parts, supplies, and services for amateur robot building. The ones shown *in italics* are primary sources; study these well to find out what's near you.

Note: Some phone companies and Yellow Pages publishers offer their printed books on CD-ROM, and the disc includes a search engine. Check this out; it can help you find what you need.

Phone Book Yellow Pages

A

Abrasives	Grits, grinding stones, and wheels polishers
Alternators & Generators-Automotive	Car electrical system parts
Aluminum	Shapes and sheets, selection of alloys
Antiques-Dealers	Old tin and plastic robot toys, memorabilia
Antiques-Repairing & Restoring	Obsolete components, gears, pulleys
Appliances-Dealers & Service	Service dept: gears, belts, pulleys, motors
Appliances-Repair	Gears, belts, pulleys, motors
Aquariums & Aquarium Supplies-Dealers	Tubing, air fittings, small pumps
Artists' Materials & Supplies	Arts and crafts supplies, paints, precision tools
Assembly & Fabricating Service	PCB production (see also Printed & Etched Circuits)
Auctioneers	Local auction representatives (see also Liquidators)
Audio-visual Equipment	Video cameras and transmitters (see also Security)
Automated Teller Machines	Money to buy robot stuff
Automotive Dismantling & Recycling	"Junk yards": windshield and other car motors (see also Automobile Wrecking)
Automotive Electric Service	Car batteries, generators, alternators, motors
Automobile Parts, Supplies & Accessories	Car parts: switches, belts, batteries, wiring, tools (see also Alternators & Generators-Automotive)

Automobile Wrecking	More "junk yards"; (see also Dismantling & Recycling)

B

Bar Code Scanning Equipment Sales & Service	Bar code wands, scanners, printers, labels, software
Batteries-Dry Cell *Batteries-Storage*	Batteries, both chargeable and non-rechargeable
Bearings	Roller and linear bearings, gears, chain, sprocket, and usually other power transmission stuff
Bicycles-Dealers Bicycles-Repairing	Bike parts: chain, sprocket, cable, tubing, bearings
Blueprinting	Specialized printing (including large scale), laminating
Book Dealers-Retail	Books about robots
Brass	Shapes and sheets
Brewers' Equipment & Supplies	Brass tubing, fittings
Building Materials-Retail	Lumber, insulating foam, hardware, fasteners, etc.

C

Cabinet Makers	Scraps of wood (ask nicely, and most will sell or even give it to you)
Casters & Guides	Casters and wheels
Chains	Roller chain, and usually also sprocket, gears, and other power transmission
Clocks-Service & Repair	Small gears and other parts
Computers & Computer Equipment	Dealers and service shops for computer and printer parts
Copying & Duplication Service	Look for those who print on special media, like decals, laminating sheets, etc.
Copper	Shapes and sheets
Copying Machines & Supplies-Sales & Service	Used copier parts (small gears, motors, belts)
Costumes-Masquerade & Theatrical	Latex, molds, and casting materials
Craft Supplies-Retail	Arts and crafts

D

Decals	Pre-printed, custom, and printable decals
Dental Equipment & Supplies	Small tools, dental plaster, centrifugal casting
Department Stores	General merchandise, sometimes at discount
Discount Stores	Cheap general merchandise
Display Designers & Producers	Ask for odd parts: tubing, plastic scraps
Display Fixtures & Materials	Tubing, slatwall
Doughnuts	Yummm..donuts...
Drafting Room Equipment & Supplies	Printing substrates (plastic, sticky-back labels)

E

Electric Equipment & Supplies-Wholesale	Local distributors, primarily for industrial electronics
Electric Motors	Sales and repair

Electronic Equipment & Supplies-Dealers	Retailers, like Radio Shack; may include surplus, or may be in separate section
Electronic Mfrs' Representatives	Local distributors and reps, primarily for industrial electronics
Engravers-Metal Engravers-Plastic, Wood, Etc.	Engraving on metal and other materials; ask if they will engrave from an AutoCAD or DXF file you provide

F

Fabric Shops	Cloth materials, (common) craft supplies
Factory Outlets	Manufacturers' dumping ground for seconds and discontinued items; not always a good deal, but worth a look
Fasteners-Industrial & Construction	Fasteners (machine screws, bolts) and fastener systems (see also Screw Machine Products)
Fishing Tackle Dealers	Replacement graphite rods, high-strength nylon line
Flag Poles Flags & Banners	Sellers of flag pole parts: poles (metal or fiberglass), and flag pole balls, which are screwed to the top of the pole; use the balls to make wheels or bodies for your robot
Floor Materials	Rubber and foam underlayment
Foam & Sponge Rubber	Mostly for industry
Fountains-Garden, Display, Etc.	Tubing, fluid fittings, pumps
Furniture Repairing & Finishing	Fasteners, casters

G

Games & Game Supplies	Look for pinball repair; solenoids, springs, other interesting parts

H

Hardware-Retail	Hardware, fasteners, adhesives, tools, etc.
Hardwoods	Specialty wood and lumber
Hearing Aids	Small batteries, micro-miniature audio amplifiers
Hobby & Model Construction Supplies	R/C, train, and other hobbies
Hose Couplings & Fittings Hose & Tubing	Industrial grade tubing, values, couplers, and fittings; includes pneumatics and hydraulics (see also Power Transmission Equipment)
Housewares-Retail	Plastic and rubber bins (for parts), kitchen tools that might be useful for some odd robot design
Hydraulic Equipment & Supplies	Medium- and heavy machinery hydraulics

I

Industrial Equipment & Supplies	Parts for industrial applications; may include motors, gears, chain, and other power transmission

J

Jewelers' Supplies	Small tools, metal casting

K

Karts-Motorized	Go-Karts and parts

L

Labels	Labels and decals (see also Decals)
Laminating Equipment & Supplies	Tools and supplies for plastic lamination
Laminations-Plastic, Paper, Etc.	Composite materials with Wood, metal, and other material
Lasers	Industrial lasers, laser cutting services
Latex & Latex Products	Latex rubber for casting and mold making
Libraries-Public	Yes, they still have these...
Light Bulbs & Tubes Lighting Fixtures	Look for unusual lighting ideas
Liquidators	Estate and industrial sales (see also Auctioneers)
Locks & Locksmiths	Security cameras and transmitters (see also Security)
Lumber-Retail	A place to buy wood; look for shops with high-grade hardwood plywood

M

Machine Shops	Jobbers who will do metal work for you (see also Metal Fabricators)
Machinery	Industrial-grade machinery tools like lathes and mills; some may also carry desktop tools
Metal Fabricators Metal Stamping	Jobbers who will do metal work for you (see also Machine Shops, Sheet Metal Work)
Metals	Shapes and sheets (see also Aluminum, Brass, Copper, Steel Distributors & Warehouses)
Model Makers	Ask for scraps and throw-aways
Molds	Not the growing kind, but the making kind
Motorcycle Machine Shop Service Motorcycles & Motor Scooters	Heavy-duty chains, sprockets, cables
Musical Instruments-Dealers Musical Instruments-Repairing	Small gear parts (guitar tuners), metal and nylon strings, electronic musical devices

N

Notions-Retail	Odds and ends that can be useful in robot construction (e.g., heat fusing fabric tape)

P

Packaging Materials Packing Materials-Shipping Paper Products	Foam, bubble pack, cardboard
Party Supplies	Cheap items (like small plastic sticks of 3-foot long drinking straws) useful for basic construction materials
Pawnbrokers	Used stuff; watch out for unrealistic prices
Pharmacies	Bandage and gauze materials,
Pipe	PVC for irrigation and furniture
Pizza	A robot builder's gotta eat!
Plastics-Extruders	Plastic extruded into special shapes, such as tubes, squares, and rods

Plastics-Fabricating, Finishing & Decorating	Ask for scraps and surplus ends
Plastics-Foam	Foam in various thicknesses and densities; one use is to make strong composites
Plastics-Molders	Mold makers (some will sell you supplies)
Plastics-Products-Finished-Retail	Plastic sheets and other shapes
Plastics-Reinforced	Plastic composites and laminates
Plastics-Rods, Tubes, Sheets, Etc.-Supply Centers	Plastic sheets and other shapes
Plastics-Scrap	Look for small pieces
Plastics-Pressure & Vacuum Forming	Thermoplastics formed to shapes (some will sell you supplies)
Plumbing	Pipe and pipe fittings; PVC for irrigation (listing is also sometimes for plumbers)
Plumbing Fixtures, Parts & Supplies	Pipe and pipe fittings, PVC for irrigation
Power Transmission Equipment	Gears, pulleys, shafts, bearings, chain, sprocket rollers, belts, and other power transmission products
Printed & Etched Circuits	PCB production (see also Assembly & Fabricating Service)
Printing Supplies	Metal plates, plastic plates
R	
Restaurant Equipment & Supplies	Look for stainless steel bowls and other unusual products for robot materials
Rubber Products	Raw materials and construction of rubber materials (mats, gaskets, etc.)
S	
School Supplies	Educational materials; possible source for science kits, gears, and similar mechanicals
Scrap Metals	Surplus metal; tends to be big material
Screw Machine Products	Fasteners
Security Control Equipment & Systems	Video cameras, video transmitters
Sewing Machines-Service & Repair	Small parts, gears, cams
Sheet Metal Work	Sheet metal jobbers, who do projects for you (see also Metal Fabricators)
Signs	Makers of neon, plastic, foam, and other sign types; ask for scrap pieces of plastic
Skateboards & Equipment Skating Equipment & Supplies	Wheels, bearings
Sporting Goods	Fishing line, small balls (e.g. table tennis), rubber grips
Steel Distributors & Warehouses Steel-Used	Shapes and sheets
Surplus & Salvage Merchandise	May have some materials (metal, plastic) or mechanical (gears, motors), but most have tents and old Army boots
T	
Theatrical Equipment & Supplies Theatrical Make-Up	Colored gels, latex
Thrift Shops	Old stuff; good prices
Tools Tools-Electric Tools-Hand Tools-Pneumatic	Hand, power, and other tools and supplies

Tools-Electric-Repairing	Cordless drill motors and batteries
Toys-Retail	There's more than Toys"R"Us out there...
Toys-Wholesale-Mfrs	Wholesale distributors and manufacturers of toys
V	
Variety Stores	By any other name, the old five-and-dime
Video Equipment-Dealers	Video cameras
W	
Washing Machines & Dryers-Service & Repair	Gears, belts (including timing belts), solenoid valves
Water Jet Cutting	Cut materials in plastic and metal with high-powered water jet
Welding Supplies & Service	Welding rigs, welding rod and supplies
Wheel Chairs & Scooters Wheel Chairs & Scooters-Repairing	Wheelchair motors, batteries, airless tires
Wheels	Wheels and casters

Internet Yellow Pages

There are literally thousands of so-called Yellow Pages on the Internet, many of them catering to specific locales, product lines, and markets. Here are several general-purpose Yellow Pages available on the Internet; some include regional listing, but others also provide mail order/catalog links.

North America/Global

MSN Yellow Pages-*http://yellowpages.msn.com/*

QuestDex-*http://www.qwestdex.com/*

Smart Pages-*http://www.smartpages.com/*

Superior Business Network-*http://www.sbn.com/*

SuperPages-*http://www.bigbook.com/*

Switchboard-*http://www.switchboard.com/*

Telephone Directories on the Web-*http://www.teldir.com/*

Yahoo Yellow Pages-*http://yp.yahoo.com/*

Yellow Book-*http://www.yellowbook.com/*

Yellow Pages Superhighway-*http://www.bestyellow.com/*

Yellow Pages, Inc.-*http://www.yellowpagesinc.com/*

Yellow Pages.com-*http://www.yellowpages.com/*

Europe

EuroPages-*http://www.europages.com/*

Yell.com-*http://www.yell.co.uk/*

Asia and Australia/New Zealand

Chinese Yellow Pages-*http://www.chinabig.com/*

Japan Yellow Pages-*http://www.yellowpage-jp.com/*

New Zealand Yellow Pages-*http://www.yellowpages.co.nz/*

Yellow Pages Online-*http://www.yellowpages.com.au/*

COMPANY REFERENCE

The following sources provide sales or other commerce related to amateur robotics. Not listed here are Internet-only informational Web sites, newsgroups, mailing lists, etc.

12 Volt Fasteners
    Fasteners
123securityproducts.com
    Video-Cameras
3M Worldwide
    Manufacturer-Materials
    Manufacturer-Glues & Adhesives
4QD
    Motor Control
80/20 Inc.
    Machine Framing
99 Cents Only Stores
    Retail-Discount & Department
A K Peters, Ltd.
    Books-Technical
A. G. Tannenbaum
    Books-Technical
A. O. Smith
    Actuators-Motors
A-1 Electronics
    Kits-Electronic
    Retail-General Electronics
A-2-Z Solutions, Inc.
    Retail-Surplus Electronics
A3J Engineering, Inc.
    Communications-RF
    Video-Cameras
Aaeon Electronics, Inc.
    Computers-Single Board Computers
Aaron's General Store
    Fasteners
Aaron's Machine Screws
    Fasteners
ABACOM Technologies
    Communications-RF
Abatron, Inc.
    Supplies-Casting & Mold Making
AbleTronics
    Retail-General Electronics
ABoyd Company, LLC, The
    Toys-Robots
    Entertainment-Art
    Entertainment-Books & Movies
ABRA Electronics Corp.:
    Retail-General Electronics
Ace Hardware
    Retail-Hardware & Home
Improvement
Ace Hardware Hobbies
    Retail-Train & Hobby
Ace Mart Restaurant Supply

Retail-Other Materials
ACK Electronics
    Retail-General Electronics
Acme Vintage Toys & Animation Gallery
    Entertainment-Art
    Toys-Robots
Acroname Inc.
    Retail-Robotics Specialty
Action Electronic Wholesale Company
    Retail-General Electronics
Activa Products , Inc.
    Retail-Arts & Crafts
Active Electronics
    Retail-General Electronics
    Distributor/Wholesaler-Industrial
    Electronics
Active Electronics Components Depot
    Retail-General Electronics
Active Surplus
    Retail-Surplus Electronics
    Retail-General Electronics
ActiveWire, Inc.
    Computers-I/O
ActivMedia Robotics, LLC
    Robots-Industrial/Research
ActivMedia: AmigoBots
    Robots-Educational
ActivMedia: Mobile Robots
    Programming-Telerobotics
Addison lctronique Lte.
    Retail-General Electronics
Admiral Metals
    Materials-Metal
Advance Auto Parts, Inc.
    Retail-Automotive Supplies
Advanced Battery Systems
    Batteries and Power
Advanced Component Electronics (ACE)
    Retail-General Electronics
Advanced Design, Inc. / Robix
    Robots-Educational
Advanced Digital Logic
    Computers-Single Board Computers
Advanced Micro Devices (AMD)
    Manufacturer-Semiconductors
Advanced Plastics, Inc.
    Materials-Plastic
    Materials-Foam
Advantage Distribution
    Materials-Plastic
    Materials-Paper & Plastic Laminates

AE Associates, Inc.
    Retail-General Electronics
    Retail-Surplus Electronics
AeroComm
    Communications-RF
Aerospace Composite Products
    Materials-Fiberglass & Carbon
Composites
    Materials-Foam
Agilent Technologies, Inc.
    Manufacturer-Semiconductors
    Sensors-Encoders
    Sensors-Other
Air America, Inc.
    Actuators-Pneumatic
Air Dynamics
    Materials-Fiberglass & Carbon
Composites
Airline Hobby Supplies
    Radio Control-Accessories
    Supplies-Paints
Airparts, Inc.
    Materials-Metal
    Fasteners
    Tools-Hand
Airpot Corporation
    Actuators-Pneumatic
AIRTRAX Inc.
    Wheels and Casters
Airtronics Inc.
    Radio Control
Al Lasher's Electronics
    Retail-General Electronics
Albion
    Wheels and Casters
Alcoa
    Materials-Metal
Alfa Aesar
    Supplies-Chemicals
Alfa Electronics
    Test and Measurement
All American Semiconductor, Inc.
    Distributor/Wholesaler-Industrial
    Electronics
All Effects Company, Inc.
    Entertainment-Art
    Batteries and Power
All Electronics Corp.
    Retail-General Electronics
    Retail-Surplus Electronics
All Industrial Systems, Inc.

Computers-Single Board Computers
All Metals Supply, Iinc.
    Materials-Metal
    Fasteners
Allegro Micro Systems, Inc.
    Manufacturer-Semiconductors
Allied Devices
    Power Transmission
    Actuators-Motion Products
Allied Electronics
    Distributor/Wholesaler-Industrial
    Electronics
Allmetric Fasteners, Inc.
    Fasteners
Alloy Frame Systems
    Machine Framing
Alltronics
    Retail-General Electronics
    Retail-Surplus Electronics
Alpha Store Fixtures, Inc.
    Materials-Store Fixtures
ALSTOM Schilling Robotics
    Robots-Industrial/Research
Altera Corp.
    Manufacturer-Semiconductors.
Altium Limited / CircuitMaker
    Electronics-PCB-Design
Altium Limited / Protel
    Electronics-PCB-Design
Altronics Distributors Pty Ltd
    Retail-General Electronics
Alumilite Corporation
    Supplies-Casting & Mold Making
Amazon Electronics/Elecronics123
    Retail-General Electronics
    Video-Cameras
    Kits-Electronic
Amazon.com
    Books
    Entertainment-Books & Movies
    Toys
Ambroid / Graphic Vision
    Manufacturer-Glues & Adhesives
America II Corp.
    Distributor/Wholesaler-Industrial
    Electronics
    Electronics-Obsolete
American Airless
    Wheels and Casters
American Art and Clay Co., Inc. (Amaco)
    Supplies-Casting & Mold Making
American Bolt and Screw Manufacturing
Corporation
    Fasteners
American Caster & Material Handling,
Inc.
    Wheels and Casters
American Flag Store.com
    Materials-Metal
American Microsemiconductor
    Electronics-Obsolete
American Musical Supply Inc.
    Electronics-Sound & Music

American Plastics
    Materials-Plastic
American Science & Surplus
    Retail-Surplus Mechanical
    Retail-Opticals & Lasers
American Society of Mechanical
Engineers
    Professional Societies
American Surplus Inc.
    Retail-Surplus Mechanical
    Wheels and Casters
American Tool Companies, Inc. /
Vice-Grip
    Tools-Hand
America's Hobby Center
    Retail-Train & Hobby
Ametron Electronic Supply, Inc.
    Retail-General Electronics
Anaheim Automation
    Actuators-Motion Products
    Actuators-Motors
Analog Devices, Inc.
    Sensors-Tilt & Accelerometer
    Manufacturer-Semiconductors
Analytical Scientific, Ltd.
    Retail-Science
    Kits-Robotic
Anderson Power Products
    Electronics-Construction-Connectors
Android Hand and Arm Prototype
    LEGO-Mindstorms
Angela Instruments
    Electronics-Sound & Music
Angelus Research
    Robots-Educational
    Robots-Industrial/Research
Animatics Corporation
    Actuators-Motors
Antique Electronic Supply
    Retail-Other Electronics
Antratek Electronics
    Microcontrollers-Hardware
    Microcontrollers-Software
AOL Yellow Pages
    Internet-Search
AP Circuits
    Electronics-PCB-Production
APEX Electronics
    Retail-Surplus Mechanical
Apex Jr.
    Retail-Surplus Electronics
Apogee Components, Inc
    Retail-Other
Apple Speech Recognition
    Programming-Platforms & Software
Appleton Electronic Distributors, Inc.
    Distributor/Wholesaler-Industrial
    Electronics
Applied Industrial Technologies
    Actuators-Pneumatic
    Supplies-Paints
    Tools
Applied Micro Circuits Corp.

Manufacturer-Semiconductors
Applied Motion Products
    Actuators-Motion Products
Arbor Scientific
    Retail-Science
Arcade Electronics, Inc.
    Retail-General Electronics
Archie McFee & Co.
    Outside-of-the-Box
Arcom Control Systems, Inc.
    Computers-Single Board Computers
ARM Ltd.
    Manufacturer-Semiconductors
Armaverse Armatures
    Retail-Armatures & Doll Parts
Armorcast
    Supplies-Casting & Mold Making
Arrick Robotics
    Robots-Educational
    Actuators-Motion Products
ARRL: Home Page
    Professional Societies
Arrow Electronics, Inc.
    Distributor/Wholesaler-Industrial
    Electronics
Arrow Fastener Company, Inc.
    Manufacturer-Tools
ARSAPE
    Actuators-Motors
Art of Electronics, The
    Books-Electronics
Art's Hobby
    Materials-Fiberglass & Carbon
Composites
    Radio Control-Hardware
ArtSuppliesOnline.com
    Retail-Arts & Crafts
AS&C CooLight
    Materials-Lighting
ASA Micros
    Microcontrollers-Hardware
ASAP Source
    Materials-Metal
    Materials-Plastic
Aspects, Inc.
    Materials-Plastic
Asteg Sales Pty Ltd
    Tools-Machinery
Astro Flight Inc.
    Actuators-Motors
ASW-Art Supply Warehouse
    Retail-Arts & Crafts
    Materials-Paper & Plastic Laminates
AT&T Labs Natural Voices Text-to-Speech
Engine
    Electronics-Sound & Music
Athena Microsystem Solutions
    Microcontrollers-Hardware
    Computers-I/O
Atlantic Fasteners
    Fasteners
Atlas Metal Sales
    Materials-Metal

Atmel
  Manufacturer-Semiconductors
Atmel: Dream Sound Synthesis-Datasheet
  Electronics-Sound & Music
Aubuchon Hardware
  Retail-Hardware & Home
Improvement
AutoGraphics of California
  Materials-Transfer Film
Automat
  Machine Framing
Automating Mining Systems, Inc.
  Robots-Industrial/Research
Automation Sensors
  Sensors-Ultrasonic
Automationdirect.com
  Actuators-Motion Products
  Sensors-Encoders
Aveox, Inc.
  Actuators-Motors
Avnet Inc. (Avnet Electronics)
  Distributor/Wholesaler-Industrial
  Electronics
AWC / Al Williams
  Microcontrollers-Programming
Axiom Manufacturing, Inc.
  Computers-Single Board Computers
  Microcontrollers-Hardware
Ax-Man Surplus
  Retail-Surplus Electronics
  Retail-Surplus Mechanical
B&B Motor & Control Corp.
  Actuators-Motion Products
B&Q
  Retail-Hardware & Home
Improvement
B.G. Micro
  Retail-Surplus Electronics
  Retail-General Electronics
B.T.W. Electronic Parts
  Distributor/Wholesaler-Industrial
  Electronics
Badger Air-Brush Co.
  Supplies-Paints
Baldor Electric Company.
  Actuators-Motors
  Motor Control
Ball Screws & Actuators Company, Inc.
  Actuators-Motion Products
Balsa Products
  Retail-Train & Hobby
Banner Engineering Corp.
  Sensors-Other
Bar Code Discount Warehouse, Inc.
  Barcoding
BARA / British Automation and Robot
Association
  Professional Societies
Barber Colman
  Actuators-Motors
Barcode Direct
  Barcoding
Barcode Mall

Barcoding
Barcode Store
  Barcoding
Barcode Warehouse
  Barcoding
barcode-barcode / pmi
  Barcoding
BarcodeChip.com
  Barcoding
BarcodeHQ / Data Worth
  Barcoding
Bare-Metal Foil Co.
  Supplies-Casting & Mold Making
Barkingside Co.
  Retail-Science
Barnes & Noble.com
  Books
  Entertainment-Books & Movies
Barnes Products Pty Ltd.
  Supplies-Casting & Mold Making
Barnhill Bolt Co.,Inc.
  Fasteners
Barrington Automation
  Actuators-Motion Products
Baseplate
  LEGO-General
Basic Micro, Inc.
  Microcontrollers-Hardware
  Microcontrollers-Software
BasicX
  Microcontrollers-Hardware
Batteries America
  Batteries and Power
Batteries Plus
  Batteries and Power
Battery Mart
  Batteries and Power
Battery Specialties
  Batteries and Power
Battlebots
  Competitions-Combat
Baumer Electric Ltd
  Sensors
Bay Plastics Ltd,
  Materials-Plastic
Baynesville Electronics
  Retail-General Electronics
Bayside Automation Systems and
Components
  Actuators-Motion Products
  Power Transmission
BBC Robots
  Robots-Educational
  Robots-Hobby & Kit
BCD Electro Inc.
  Retail-Surplus Electronics
Bear Ingredients
  Retail-Armatures & Doll Parts
Bearing Belt Chain
  Power Transmission
Bearing Headquarters Co.
  Power Transmission
BEI Technologies

  Actuators-Motion Products
Bel Inc.
  Materials-Transfer Film
Belden Inc.
  Materials-Other
Belt Corporation of America
  Power Transmission
Best Little Machine Tool Company
  Tools-Precision & Miniature
BestRC / Hobbico, Inc.
  Radio Control
BICRON Electronics Company
  Actuators-Other
Big 5 Sporting Goods
  Retail-Other
Big Boys Toys
  Actuators-Pneumatic
Big Briar, Inc.
  Electronics-Sound & Music
Big Lots
  Retail-Discount & Department
BigTray, Inc.
  Retail-Other Materials
Bike World
  Sensors-GPS
Bimba Manufacturing Company
  Actuators-Pneumatic
BioPlastics
  Materials-Plastic
Birdfeeding.com
  Materials-Plastic
Bishop Wisecarver
  Machine Framing
  Actuators-Motion Products
Bison Gear & Engineering
  Actuators-Motors
Bitworks Inc.
  Electronics-Display
BJ's Wholesale Club, Inc.
  Retail-Discount & Department
BK Precision
  Test and Measurement
Black & Decker
  Manufacturer-Tools
Black Feather Electronics
  Retail-General Electronics
  Materials-Lighting
  Retail-Opticals & Lasers
Blue Bell Designs Inc.
  Robots-Hobby & Kit
Blue Point Engineering
  Retail-Robotics Specialty
BMI Surplus
  Retail-Surplus Electronics
  Retail-Opticals & Lasers
BobCAD CAM Inc.
  Tools-CNC
Boca Bearing
  Power Transmission
Bodine Electric Company
  Actuators-Motion Products
  Motor Control
Boeing Surplus Store

Retail-Surplus Electronics
Retail-Surplus Mechanical
Bolt Depot
Fasteners
BoltsMART
Fasteners
Boondog Automation
Electronics-Circuit Examples
Internet-Plans & Guides
Borland Software Corporation
Programming-Languages
Bosch Automation Products
Machine Framing
Boston Gear
Power Transmission
Botanical Science
Supplies-Casting & Mold Making
Bot-Kit
LEGO-Mindstorms
Bourns, Inc.
Manufacturer-Components
Sensors-Encoders
BRECOflex Co., L.L.C.
Power Transmission
BrickLink
LEGO-General
Brigar Electronics
Retail-General Electronics
Retail-Surplus Mechanical
Brikksen Company
Fasteners
Brookshire Software
Motor Control
Brown's Home Kitchen Center &
Restaurant Supply
Retail-Other Materials
Brunning Software
Microcontrollers-Programming
Budget Robotics
Retail-Robotics Specialty
Buehler Motor GmbH
Actuators-Motors
Bug'N'Bots
Robots-Hobby & Kit
Bull Electrical
Retail-General Electronics
Burden's Surplus Center
Retail-Surplus Mechanical
Burman Industries, Inc.
Supplies-Casting & Mold Making
Materials-Metal
Materials-Foam
Bytecraft Ltd.
Programming-Languages
C & H Sales
Retail-Surplus Mechanical
Retail-Surplus Electronics
C & S Sales
Test and Measurement
Kits-Robotic
Kits-Electronic
CadSoft Computer GmbH
Electronics-PCB-Design

CAIG Laboratories, Inc.
Supplies-Chemicals
Cal Plastics and Metals
Materials-Metal
Materials-Plastic
Caligari Corporation / trueSpace
Tools-CNC
Caltronics
Distributor/Wholesaler-Other
Components
Campbell Hausfeld
Tools-Power
Canadian Bearings Ltd.
Power Transmission
Canadian Tire
Retail-Other
Cardstone Pty Ltd
Tools-Precision & Miniature
Carken Co. / Deskam
Tools-CNC
Carlo Gavazzi Holding AG
Sensors
Motor Control
Carl's Electronics Inc.
Kits-Electronic
Carol Wright Girfts
Outside-of-the-Box
Carolina Science & Math
Retail-Science
CARQUEST Corporation
Retail-Automotive Supplies
Carvin Guitars
Electronics-Sound & Music
Castcraft
Supplies-Casting & Mold Making
Caster Connection
Wheels and Casters
C-AVR
Microcontrollers-Programming
CCTV Outlet
Video-Cameras
Video-Transmitters
Celebrity Standups
Entertainment-Art
Central Utah Electronics Supply
Retail-General Electronics
Cermark
Radio Control
Channellock Inc.
Manufacturer-Tools
Chatsco Distributions
Retail-Armatures & Doll Parts
Cheapcasters.com
Wheels and Casters
Chef's Depot
Retail-Other Materials
Childcraft Education Corp.
Retail-Educational Supply
ChipCenter-QuestLink
Internet-Research
ChipDocs
Internet-Research
Chip-Sources.com

Retail-Other Electronics
Chuck Hellebuyck Electronics
Microcontrollers-Hardware
Microcontrollers-Software
CID Inc.
Retail-General Electronics
Cimarron Technology, Inc.
Electronics-PCB-Production
Cinefex
Entertainment-Art
Circuit Cellar
Journals and Magazines
Circuit Specialists, Inc.
Retail-General Electronics
Cisco-Eagle Inc.
Wheels and Casters
Clarostat Manufacturing Co.
Sensors-Encoders
ClassroomDirect.com
Retail-Educational Supply
Clean Sweep Supply, Inc.
Supplies-Glues & Adhesives
Clever Gear
Retail-Other Electronics
Clippard Minimatic, Inc.
Actuators-Pneumatic
Clisby Miniature Machines
Tools-Precision & Miniature
Tools-Machinery
Clotilde, Inc.
Retail-Arts & Crafts
CNC Retro-Fit Links
Tools-CNC
CNCez PRO
Tools-CNC
cncKITS.com
Tools-CNC
Coastal Tool & Supply
Tools-Power
CodeVisionAVR
Microcontrollers-Software
CogniToy LLC / MindRover
Programming-Robotic Simulations
Cole Instrument Corp.
Sensors-Encoders
Cole-Parmer Instrument Company
Retail-Science
Colobot
Programming-Robotic Simulations
Colson Caster Corporation
Wheels and Casters
Communications Specialists, Inc.
Communications-RF
compact technik gmbh
Machine Framing
Competition-Robotics
Robots-Hobby & Kit
Compleat Sculptor, Inc., The
Supplies-Casting & Mold Making
Component Kits LLC
Kits-Electronic
Composite Store, The
Materials-Fiberglass & Carbon

Composites
    Supplies-Casting & Mold Making
    Materials-Foam
Compumotor / Parker Hannifin Corp.
    Actuators-Motors
    Actuators-Motion Products
Compumotor Engineering Reference
Guides
    Motor Control
Computronics Corporation Ltd
    Retail-General Electronics
    Electronics-Soldering
    Communications-RF
Constantines Wood Center
    Materials-Other
Construction Site, The
    Toys-Construction
Constructive Playthings
    Toys-Construction
Contact East, Inc.
    Distributor/Wholesaler-Other
Components
Control Technology Corp.
    Motor Control
Cool Neon / Funhouse Productions
    Materials-Lighting
CoolTool GmbH
    Tools-Machinery
Cooper Industries, Inc.
    Manufacturer-Tools
    Electronics-Soldering
CopyTag Limited
    Sensors-RFID
CornerHardware.com
    Retail-Hardware & Home
Improvement
Costco
    Retail-Discount & Department
Covington Innovations
    Kits-Electronic
CPE, Inc.
    Materials-Foam
Craft Supplies USA
    Materials-Other
    Tools-Hand
    Tools-Accessories
Crafter's Market
    Retail-Arts & Crafts
Creative Learning Systems, Inc.
    Toys-Construction
    Robots-Hobby & Kit
Cricklewood Electronics
    Video-Cameras
Cross Automation
    Actuators-Motion Products
Crossbow Technology, Inc.
    Sensors
Crownhill Associates Limited
    Microcontrollers-Hardware
    Microcontrollers-Software
Cruel Robots
    Power Transmission
    Wheels and Casters

Crystalfontz America Inc.
    Electronics-Display
CSK Auto Inc.
    Retail-Automotive Supplies
CSMicro Systems
    Microcontrollers-Hardware
    Microcontrollers-Software
CTC Control: Literature and Resources
    Motor Control
CTR Surplus
    Retail-Surplus Electronics
CTS Corporation
    Manufacturer-Components
CUI Inc.
    Manufacturer-Components
Custom Computer Services, Inc
    Microcontrollers-Programming
Custom Sensors Inc.
    Barcoding
Cutting Tool Mall, The
    Tools-Accessories
CVP Products
    Retail-Train & Hobby
CWIKship.com
    Retail-Hardware & Home
Improvement
Cyber Woodworking Depot
    Tools-Power
Cyberbond
    Supplies-Glues & Adhesives
Cyberbotics
    Programming-Robotic Simulation
CyberToyz
    Toys-Robots
Cynthia's Bar & Restaurant
    Retail-Other
Dal-Craft, Inc
    Retail-Arts & Crafts
Dallas Semiconductor
    Manufacturer-Semiconductors
Danaher Controls
    Manufacturer-Components
Danaher Motion MC
    Actuators
    Power Transmission
Dan's Small Parts And Kits
    Retail-General Electronics
Darice, Inc.
    Materials-Other
Data Hunter
    Communications-RF
DATAQ Instruments, Inc
    Computers-Data Acquisition
DATEL, Inc.
    Computers-Data Acquisition
Dave Brown Products
    Wheels and Casters
    Materials-Fiberglass & Carbon
Composites
Davidson Measurement Pty. Ltd.
    Sensors-Strain & Load Cells
Davis INOTEK
    Sensors

DC Electronics
    Communications-RF
Debco Electronics, Inc.
    Retail-General Electronics
    Kits-Electronic
Decade Engineering
    Electronics-Display
Delcam plc / MillWizard
    Tools-CNC
Delft Spline Systems / DeskProto
    Tools-CNC
Del-Tron Precision
    Actuators-Motion Products
Densitron Technologies plc
    Actuators-Other
    Actuators-Motors
DesignNotes.com
    Electronics-Circuit Examples
Designtech Engineering Co.
    Electronics-Display
Desktop Machine Shop.com
    Tools-Machinery
    Tools-CNC
DevX
    Programming-Tutorial & How-to
    Programming-Examples
DEWALT Industrial Tool Co.
    Manufacturer-Tools
Dexis Corporation
    Retail-Surplus Electronics
Diamond Systems Corporation
    Computers-Single Board Computers
Dick Blick Art Materials
    Retail-Arts & Crafts
Dick Smith Electronics
    Retail-General Electronics
Didel
    Robots-Hobby & Kit
Digi-Key
    Retail-General Electronics
    Distributor/Wholesaler-Industrial
Electronics
Digitrax
    Retail-Train & Hobby
Dimensional Designs
    Entertainment-Art
Dinsmore Instrument Co.
    Sensors-Other
Direct Advantage
    Toys-Construction
Directed Perception, Inc.
    Actuators-Motion Products
Disco Joints and Teddies
    Retail-Armatures & Doll Making
Discount Art Supplies
    Retail-Arts & Crafts
Discount Lasers.com
    Retail-Opticals & Lasers
Discount Package Supply, Inc.
    Outside-of-the-Box
Discount Train and Hobby
    Retail-Train & Hobby
Discount-Tools.com

Tools-Accessories
Discovery Mart .com
   Retail-Educational Supply
Display Electronics
   Retail-General Electronics
Display Warehouse
   Materials-Store Fixtures
Diverse Electronics Services
   Motor Control
   Communications-RF
Diversified Enterprises
   Robots-Hobby & Kit
Diversi-Tech Inc.
   Tools-CNC
Dixie Art & Airbrush Supplies
   Retail-Arts & Crafts
DJ Hobby
   Retail-Train & Hobby
DK Models, Inc
   Retail-Train & Hobby
DLR Kits
   Kits-Electronic
Do it Best / FixitCity.com
   Retail-Hardware & Home
Improvement
Donovan Micro-Tek
   Actuators-Motors
   Sensors-Encoder
Dontronics, Inc.
   Microcontrollers-Hardware
   Microcontrollers-Software
Dontronics: PIC List
   Microcontrollers-Hardware
Doug Mockett & Company, Inc.
   Wheels and Casters
Douglas and Sturgess, Inc.
   Supplies-Casting & Mold Making
   Materials-Paper & Plastic Laminates
   Materials-Foam
Douglas Electronics Inc.
   Electronics-PCB-Design
Douglas Wheel, Inc.
   Wheels and Casters
DoveBid Inc.
   Retail-Auctions
Dr. Dobb's
   Journals and Magazines
Draganfly Innovations
   Radio Control
Dream Catcher Hobby, Inc.
   Materials-Fiberglass & Carbon
Composites
   Retail-Train & Hobby
Dremel
   Manufacturer-Tools
Drives, Incorporated
   Power Transmission
DRMS
   Retail-Auctions
DRU Industries, Inc.
   Supplies-Casting & Mold Making
DSI Toys, Inc.
   Toys-Electronics

Du-Bro Products, Inc.
   Radio Control-Hardware
   Wheels and Casters
Du-Mor Service & Supply Co.
   Fasteners
   Materials-Metal
Dura-Belt, Inc.
   Power Transmission
Dustbots
   Internet-Informational
Dyna Art
   Materials-Transfer Film
   Electronics-PCB-Production
Dynalloy, Inc.
   Actuators-Shape Memory Alloy
Dynasys
   Sensors-IRFD
Dyna-Veyor
   Power Transmission
EA Electronics
   Actuators-Motors
   Motor Control
Eager Plastics, Inc.
   Supplies-Casting & Mold Making
Earth Computer Technologies, Inc
   Electronics-Display
EASY START: PIC microcontroller proto-
type boards
   Microcontrollers-Programming
Easy Step'n, An Introduction to Stepper
Motors for the Experimenter
   Motor Control
EasyBot
   Robots-Hobby & Kit
eBatts.com
   Batteries and Power
eBay
   Retail-Auctions
Eberhard Faber GmbH
   Materials-Other
ECD, Inc. / PCBexpress
   Electronics-PCB-Design
   Electronics-PCB-Production
ECD, Inc. / PCBPro
   Electronics-PCB-Design
   Electronics-PCB-Production
ECG Electronics / NTE Electronics, Inc.
   Internet-Research
ECN Magazine
   Journals and Magazines
Edlie Electronics
   Retail-General Electronics
Edmond Wheelchair Repair & Supply
   Wheels and Casters
   Actuators-Motors
   Batteries and Power
Edmund Industrial Optics
   Retail-Opticals & Lasers
Edmund Scientific
   Retail-Science
   Retail-Educational Supply
   Retail-Opticals & Lasers
EDN Access

Journals and Magazines
EduBots
   Toys-Construction
Educational Experience
   Retail-Educational Supply
Educational Innovations, Inc.
   Retail-Science
Educational Insights, Inc.
   Retail-Educational Supply
   Toys-Construction
EduRobot
   Robots-Educational
eduRobotics.com
   Motor Control
EEM.com
   Internet-Research
EFFECTive ENGINEERING
   Motor Control
Efston Science
   Retail-Science
eFunda
   Internet-Research
eHobbies
   Retail-Train & Hobby
eHobbyland
   Retail-Train & Hobby
   Toys-Construction
EIO.com
   Retail-Surplus Electronics
EK JAPAN CO.,LTD
   Kits-Robotic
E-Lab Digital Engineering, Inc.
   Motor Control
   Electronics-Display
ELAM Electroliminescent Industries Ltd.
   Materials-Lighting
Elan Microelectronics Corp.
   Microcontrollers-Hardware
Elantec Semiconductor, Inc.
   Manufacturer-Semiconductors
EL-COM
   Fasteners
   Wheels and Casters
   Materials-Plastic
   Materials-Metal
Electric Motor Warehouse
   Actuators-Motors
Electro Mavin
   Retail-Surplus Electronics
   Retail-Surplus Mechanical
Electro Sonic Inc.
   Retail-General Electronics
Electrocomponents plc
   Distributor/Wholesaler-Industrial
Electronics.
ElectroDynamics
   Radio Control
Electronic Depot Inc.
   Distributor/Wholesaler-Industrial
Electronics
Electronic Design
   Journals and Magazines
Electronic Dimensions

Retail-Surplus Electronics
Electronic Goldmine
    Retail-General Electronics
    Kits-Electronic
    Retail-Surplus Electronics
Electronic Model Systems
    Radio Control
Electronic Parts Outlet
    Retail-General Electronics
Electronic Products
    Journals and Magazines
Electronic Rainbow, Inc.
    Kits-Electronic
Electronic School Supply, Inc. (ESS)
    Kits-Electronic
Electronic Supply Center
    Distributor/Wholesaler-Industrial
    Electronics
Electronic Surplus Co.
    Retail-Surplus Electronics
Electronics Parts Center
    Distributor/Wholesaler-Other
Components
    Power Transmission
Electronics Plus
    Retail-General Electronics
    Retail-Surplus Electronics
Electronique Pratique
    Journals and Magazines
Electronix Express
    Retail-General Electronics
Elektor Electronics
    Journals and Magazines
Elektronikladen Mikrocomputer GmbH
    Microcontrollers-Hardware
Elenco Electronics
    Kits-Electronic
    Test and Measurement
    Electronics-Soldering
    Tools-Hand
ElextronixOnline
    Distributor/Wholesaler-Other
Components
Elite Speed Products
    Actuators-Motors
    Batteries and Power
Ellis Components
    Actuators-Motors
Elsema Pty Ltd
    Communications-RF
EMAC, Inc.
    Computers-Single Board Computers
Embedded Acquisition Systems
    Computers-Data Acquisition
    Microcontrollers-Hardware
Embedded Systems Design Website
    Microcontrollers-Hardware
Embedded Systems Programming
    Journals and Magazines
Embedded Systems, Inc.
    Microcontrollers-Hardware
    Microcontrollers-Software
Emerson Power Transmission

Manufacturing
    Power Transmission
    Actuators-Motion Products
EMJ Embedded Systems
    Computers-Single Board Computers
    Electronics-Display
Enco Manufacturing Co
    Tools
    Materials
    Supplies
Encoder Products Group
    Sensors-Encoders
Energizer Holdings, Inc.
    Batteries and Power
Energy Sales
    Batteries and Power
Engine Trix
    Wheels and Casters
Enigma Industries
    Actuators-Motors
    Programming-Robotic Simulations
Entran Devices, Inc.
    Sensors
Environmental Technology, Inc.
    Supplies-Casting & Mold Making
EOL Surplus
    Retail-Surplus Mechanical
    Retail-Surplus Electronics
ErgoTech Inc.
    Wheels and Casters
ESR Electronic Components
    Retail-General Electronics
Estimation of Motor Torque
    Actuators-Motors
EVdeals
    Actuators-Motors
    Batteries and Power
Everyday Practical Electronics
    Journals and Magazines
Evolution Robotics, Inc.
    Robots-Industrial/Research
Ewave, Inc. / Electrowave
    Communications-RF
eWellness
    Supplies-Casting & Mold Making
Exar Corp.
    Manufacturer-Semiconductors
Excess Solutions
    Retail-Surplus Electronics
Exploratorium
    Retail-Science
ExpressPCB / Engineering Express
    Electronics-PCB-Design
    Electronics-PCB-Production
EZ Flow Nail Systems
    Outside-of-the-Box
EZ Pose Flexible Doll Bodies
    Retail-Armatures & Doll Parts
Fair Radio Sales
    Retail-Surplus Electronics
Fairchild Semiconductor
    Manufacturer-Semiconductors
FAO Schwarz

Toys
Far West Materials
    Supplies-Casting & Mold Making
FARMTEK
    Retail-Other
Farnell
    Retail-General Electronics
Fascinating Electronics, Inc.
    Sensors-Ultrasonic
Fastech of Jacksonville, Inc.
    Retail-Arts & Crafts
Fastenal Company
    Fasteners
Fastener Barn, LLC
    Fasteners
Fastener-Express
    Fasteners
Fastenerkit.com
    Fasteners
Faulhaber Group
    Actuators-Motors
FDJ On Time
    Tools-Precision & Miniature
    Supplies-Casting & Mold Making
FedSales.gov
    Retail-Auctions
FEIN Power Tools Inc.
    Manufacturer-Tools
Felio Parking Sensor
    Sensors-Ultrasonic
FerretTronics
    Motor Control
Fibraplex
    Materials-Fiberglass & Carbon
Composites
Fibre Glass Developments Corporation
    Materials-Fiberglass & Carbon
Composites
    Supplies-Casting & Mold Making
Fiero Fluid Power Inc.
    Actuators-Pneumatic
    Machine Framing
Figaro USA Inc.
    Sensors-Other
Fire Mountain Gems
    Materials-Other
Fisher Scientific
    Supplies-Chemicals
FlagandBanner.com
    Materials-Metal
Flagpole Components, Inc.
    Materials-Metal
Flashcut CNC
    Tools-CNC
FLAX Art & Design
    Retail-Arts & Crafts
Flexel (UK) Ltd
    Wheels and Casters
Flexible Industrial Systems, Inc.
    Machine Framing
Flinn Scientific, Inc.
    Retail-Science
FloorBotics

Robots-Personal
FMA Direct
  Retail-Train & Hobby
Focus Group Limited
  Retail-Hardware & Home
Improvement
  Retail-Arts & Crafts
Fonix Corporation
  Electronics-Sound & Music
Forest City Surplus
  Retail-Surplus Mechanical
  Retail-Surplus Electronics
FORTH, Inc.
  Programming-Languages
Foveon, Inc.
  Video-Imagers
Frame World
  Machine Framing
Fred Barton Productions, Inc.
  Robots-Hobby & Kit
Free Software Foundation, Inc
  Programming-Languages
Freeman Manufacturing & Supply Co.
  Supplies-Casting & Mold Making
Friendly Robotics
  Robots-Personal
Frigid North Company
  Retail-General Electronics
Frontline Hobbies
  Retail-Train & Hobby
Fry's Electronics
  Retail-General Electronics
FSMLabs, Inc.
  Microcontrollers-Programming
FuelCellStore.com
  Batteries and Power
Fuller Metric Parts
  Fasteners
Fun For All! Toys
  Outside-of-the-Box
Futaba
  Radio Control
Future Electronics
  Distributor/Wholesaler-Industrial
  Electronics
  Retail-General Electronics
Future-Bot Components
  Retail-Robotics Specialty
G & G Restaurant Supply
  Retail-Other Materials
Gardner Bender
  Manufacturer-Components
Garmin Ltd.
  Sensors-GPS
Gates Rubber Co.
  Power Transmission
Gateway Electronics, Inc.
  Retail-Surplus Electronics
  Retail-Surplus Mechanical
GC/Waldom, Inc.
  Supplies-Chemicals
GE Industrial Systems
  Actuators-Motors

Motor Control
Manufacturer-Materials
GeckoSystems, Inc.
  Robots-Industrial/Research
General Robotics Corporation
  Robots-Educational
Generic Slides
  Actuators-Motion Products
Gibbons Fiberglass
  Supplies-Casting & Mold Making
Gibson Tech Ed
  Kits-Electronic
Gillette Company, The
  Batteries and Power
Gilway Technical Lamp
  Materials-Lighting
Gleason Research
  Microcontrollers-Hardware
Global Hobby Distributors
  Radio Control-Servos
Global Sources
  Internet-Links
Globe Electronic Hardware, Inc.
  Materials-Other
Globe Motors
  Actuators-Motors
Glolab Corp.
  Communications-RF
  Sensors-Optical
Glowire
  Materials-Lighting
Go Kart Supply
  Power Transmission
GoldenWest Manufacturing
  Supplies-Casting & Mold Making
  Materials-Plastic
  Materials-Paper & Plastic Laminates
Golf Car Catalog, The
  Actuators-Motors
  Materials-Other
Goodfellow
  Materials-Metal
Google Catalog Search
  Internet-Search
GPS City
  Sensors-GPS
GPS Warehouse
  Sensors-GPS
GPS World Supply
  Sensors-GPS
Grainger (W.W. Grainger)
  Materials
  Actuators-Motors
  Supplies
Grandma T's
  Retail-Arts & Crafts
  Retail-Armatures & Doll Parts
GrandParents Toy Connection
  Toys
Graybar Electric Company, Inc.
  Materials
  Supplies
Grayhill, Inc.

Sensors-Encoders
Graymark International, Inc.
  Kits-Electronic
  Kits-Robotic
Great Neck Saw Manufacturers, Inc.
  Manufacturer-Tools
Greenweld Limited
  Retail-General Electronics
Grizzly Industrial, Inc.
  Tools
GSA Auctions
  Retail-Auctions
Gurley Precision Instruments
  Sensors-Encoders
Gyration, Inc.
  Sensors-Tilt & Accelerometer
H&R Company, Inc. (Herbach and
Rademan)
  Retail-Surplus Mechanical
Hadwareshop
  Retail-Hardware & Home
Improvement
Hamamatsu Corp.
  Sensors-Optical
Hammacher Schlemmer & Company, Inc.
  Retail-Other Electronics
Hammad Ghuman
  Radio Control-Hardware
Hamtronics, Inc.
  Communications-RF
Hansen Corporation
  Actuators-Motors
Harbor Freight Tools
  Tools-Machinery
Hawker Energy Products Inc.
  Batteries and Power
Hawkes Electronics
  Retail-Other Electronics
Haydon Switch and Instrument, Inc
  Actuators-Motors
Haynes Enterprise
  Robots-Industrial/Research
Hdb Electronics
  Distributor/Wholesaler-Industrial
  Electronics
HDS Systems, Inc.
  Electronics-Display
Heartland America
  Outside-of-the-Box
Helical Products Co., Inc.
  Power Transmission
Henrys Electronics Ltd.,
  Kits-Robotic
  Kits-Electronic
Heyco Products, Inc.
  Manufacturer-Components
HGR Industrial Surplus
  Retail-Surplus Mechanical
High Performance Alloys
  Materials-Metal
High Tech Chips, Inc
  Electronics-Specialty
High-TechGarage.com

Microcontrollers-Hardware
Hirose & Yoneda Lab
    Robots-Research
Hitec/RCD
    Radio Control-Servos
Hiwin Technologies Corporation
    Actuators-Motion Products
    Actuators-Motors
HMC Electronics
    Retail-Other Electronics
Hobbees.com
    Radio Control
Hobbico, Inc
    Radio Control
Hobby Barn, The
    Retail-Train & Hobby
Hobby Club
    Retail-Train & Hobby
Hobby Horse Wisconsin Inc.
    Radio Control-Servos
Hobby Link Japan
    Toys-Construction
Hobby Lobby
    Retail-Arts & Crafts
Hobby Lobby International, Inc.
    Retail-Train & Hobby
Hobby Maker
    Retail-Train & Hobby
Hobby People
    Retail-Train & Hobby
Hobby Shack
    Retail-Train & Hobby
Hobby Stuff Inc.
    Retail-Train & Hobby
    Radio Control-Accessories
Hobbybox
    Retail-Train & Hobby
HobbyCNC
    Tools-CNC
Hobbyco Pty Ltd
    Retail-Train & Hobby
Hobbylinc.com
    Retail-Train & Hobby
    Kits-Electronic
    Kits-Robotic
Hobby's
    Retail-Train & Hobby
Hobbytron
    Kits-Robotic
    Kits-Electronic
Hohner Corp.
    Sensors-Encoders
Home Automator Magazine
    Journals and Magazines
Home Depot-Maintenance Warehouse
    Retail-Hardware & Home
Improvement
Home Depot, The
    Retail-Hardware & Home
Improvement
Home Hardware Stores Limited
    Retail-Hardware & Home
Improvement

Home Lumber Company
    Tools
Homebase
    Retail-Hardware & Home
Improvement
Honeywell International Inc.
    Sensors
Hope Education
    Retail-Educational Supply
Horizon Hobby
    Radio Control-Servos
Hosfelt Electronics
    Retail-General Electronics
    Retail-Opticals & Lasers
Hot Melt City
    Supplies-Glues & Adhesives
House of Batteries
    Batteries and Power
House of Tools
    Tools
HPS Papilio
    Materials-Transfer Film
HR Meininger Company
    Retail-Arts & Crafts
HSC Electronic Supply / Halted
    Retail-General Electronics
    Retail-General Surplus
HST Materials, Inc.
    Materials
    Supplies-Glues & Adhesives
HTH / High Tech Horizon
    Microcontrollers-Hardware
    Microcontrollers-Software
Hub Material Company
    Retail-Other Electronics
Huco Engineering Industries Ltd
    Power Transmission
Hurst Manufacturing
    Actuators-Motors
HUT Products
    Materials-Other
HVW Technologies Inc.
    Retail-Robotics Specialty
Hygloss Products, Inc.
    Retail-Arts & Crafts
Hyperbot
    Robots-Hobby & Kit
    Toys-Construction
iButton
    Microcontrollers-Hardware
IC Master
    Internet-Research
ICO RALLY
    Manufacturer-Components
IDAutomation.com, Inc.
    Barcoding
IDEA ELETTRONICA
    Retail-Science
IEE / Institution of Electrical Engineers
    Professional Societies
IEEE Robotics and Automation Society
    Professional Societies
IEEE Spectrum

Journals and Magazines
IFI Robotics
    Motor Control
Iguana Robotics
    Robots-Research
igus GMBH
    Power Transmission
    Actuators-Motion Products
Image Solutions
    Materials-Transfer Film
Imagecraft Software
    Microcontrollers-Software
Images SI Inc.
    Retail-Robotics Specialty
    Retail-Science
Imaginarium.com
    Retail-Science
    Toys-Construction
Indigo Instruments
    Retail-Science
Industrial Encoders Direct
    Sensors-Encoders
Industrial Links Ltd
    Power Transmission
Industrial Metal Supply Co.
    Materials-Metal
Industrial Profile Systems
    Machine Framing
Industrologic, Inc.
    Computers-Single Board Computers
Infineon Technologies AG
    Manufacturer-Semiconductors
Infinity Composites
    Materials-Fiberglass & Carbon
Composites
Informal Education Products
    Retail-Science
Information Unlimited
    Kits-Electronic
Inland Empire Components
    Distributor/Wholesaler-Industrial
Electronics
Innotech Systems Inc.
    Communications-Infrared
Innotek, Inc.
    Electronics-Miscellaneous
Innovation First
    Motor Control
InstrumentWarehouse.com
    Test and Measurement
Intec Automation Inc.
    Microcontrollers-Hardware
Intel Corporation
    Manufacturer-Semiconductors
Interlink Electronics, Inc.
    Sensors-Strain & Load Cells
International Components Corporation
    Distributor/Wholesaler-Other
Components
International Paper
    Materials-Paper & Plastic Laminates
International Rectifier
    Manufacturer-Semiconductors

International Robotics
    Robots-Industrial/Research
International Tool Corporation
    Tools
Intersil
    Manufacturer-Semiconductors
Invensys Plc
    Actuators
    Power Transmission
IPS Corp.
    Supplies-Glues & Adhesives
iRobot
    Robots-Industrial/Research
ISMG
    Tools-Machinery
ITW Chemtronics
    Supplies-Chemicals
Ivex Design International, Inc.
    Electronics-PCB-Design
J R Kerr
    Motor Control
J. A. LeClaire
    Materials-Lighting
J. W. Winco
    Wheels and Casters
    Materials-Other
J.C. Whitney, Inc.
    Retail-Automotive Supplies
J.L. Hammett Co.
    Retail-Educational Supply
Jameco Electronics
    Retail-General Electronics
Java Media Framework API
    Video-Programming & APIs
Jaycar Electronics
    Retail-General Electronics
    Kits-Electronic
Jbro Batteries / Lexstar Technologies
    Batteries and Power
JCM Inventures / JCM Electronic Services
    Robots-BEAM
JDR Microdevices
    Retail-General Electronics
Jensen Tools, Inc.
    Tools
    Test and Measurement
Jepson Tool Power
    Manufacturer-Tools
Jewelry Supply
    Materials-Other
Jim Allred Taxidermy Supply
    Materials-Foam
    Retail-Armatures & Doll Parts
JJC & Associates
    Power Transmission
JK Electronics
    Retail-General Electronics
JKmicrosystems, Inc.
    Computers-Single Board Computers
    Computers-I/O
Johnson Electric Holdings Limited
    Actuators-Motors
Johnson-Smith Co.

Outside-of-the-Box
Johuco Ltd.
    Robots-Hobby & Kit
Joker Robotics
    Robots-Hobby & Kit
JStamp
    Microcontrollers-Hardware
J-Tron Inc.
    Kits-Electronic
    Retail-General Electronics
J-Works, Inc.
    Computers-I/O
K&S Engineering
    Materials-Metal
Kadtronics
    Robots-Hobby & Kit
Kanda Systems Ltd.
    Microcontrollers-Hardware
    Microcontrollers-Software
Kanya AG/SA/Ltd.
    Machine Framing
Karting Distributors Inc.
    Power Transmission
Kaufman Company
    Tools
KBtoys.com
    Toys
KC Marketing Group, Inc.
    Wheels and Casters
Kee Industrial Products, Inc.
    Machine Framing
Kelvin
    Retail-Robotic
    Retail-Science
    Kits-Electronic
Ken Bromley Art Supplies
    Retail-Arts & Crafts
Kerk Motion Products, Inc.
    Actuators-Motion Products
Kevin Ross
    Microcontrollers-Hardware
Keystone Electronics Corp.
    Distributor/Wholesaler-Other
Components
Keystronics
    Retail-Surplus Electronics
KidsWheels
    Actuators-Motors
    Batteries and Power
KIK Tire, Inc.
    Wheels and Casters
Kindt-Collins Company, The
    Supplies-Casting & Mold Making
Kinetic Composites
    Materials-Fiberglass & Carbon
Composites
Kingston Vacuum Works, The
    Supplies-Casting & Mold Making
Kit Guy
    Kits-Electronic
Kits R Us
    Kits-Electronic
Kmart

Retail-Discount & Department
K'NEX
    Toys-Construction
Kodenshi Korea Corp.
    Sensors-Optical
Kornylak Wheel Division
    Wheels and Casters
Kronos Robotics
    Retail-Robotics Specialty
KRP Electronic Supermarket
    Retail-Surplus Electronics
K-Surplus Sales Inc.
    Fasteners
    Materials-Other
L.R. Miniatures Ltd.
    Tools-Precision & Miniature
Lab Electronics
    Motor Control
LabJack Corporation
    Computers-Data Acquisition
LabX
    Retail-Surplus Electronics
Laird Plastics
    Materials-Plastic
Lakeshore Learning Materials
    Retail-Educational Supply
Lakeview Research
    Books-Electronics
Laser Products
    Manufacturer-Tools
Lasermotion
    Retail-Opticals & Lasers
LAWICEL
    Microcontrollers-Hardware
Lazertran Limited
    Materials-Transfer Film
Learning Resources, Inc.
    Retail-Educational Supply
Ledex
    Actuators-Other
LedVision Holding, Inc.
    Electronics-Display
Lee Valley Tools Ltd.
    Tools
LEESON Electric Corporation
    Actuators-Motors
LEGO Dacta
    LEGO-General
LEGO Mindstorms-Home Page
    LEGO-Mindstorms
LEGO Shop-at-Home
    LEGO-General
    Toys-Construction
Leica Disto
    Sensors-Distance & Proximity
Lemos International
    Communications-RF
Leviton Mfg. Company Inc.
    Manufacturer-Components
Liberty BASIC
    Programming-Languages
Liberty Enterprises
    Tools-Machinery

Life-casting.com / Artmolds
Supplies-Casting & Mold Making
Lightweight Backpacker, The
Materials-Fiberglass & Carbon
Composites
Outside-of-the-Box
Linco Inc.
Wheels and Casters
Lindsay's Technical Books
Books-Technical
Linear Industries, Ltd.
Actuators-Motion Products
Linear Integrated Systems
Manufacturer-Semiconductors
Linx Technologies, Inc.
Communications-RF
Lite-On, Inc.
Manufacturer-Components
Little Shop of Hobbies
Retail-Train & Hobby
LNL Distributing Corp.
Retail-General Electronics
Loctite Corp
Supplies-Glues & Adhesives
Logiblocs Ltd
Electronics-Specialty
Retail-Educational Supply
Lovejoy Inc.
Power Transmission
Lowe's Companies, Inc.
Retail-Hardware & Home
Improvement
Lowrance Electronics, Inc.
Sensors-GPS
LPS Laboratories
Supplies-Chemicals
Lumitex, Inc.
Electronics-Displays
Lynxmotion, Inc.
Retail-Robotics Specialty
M.G. Chemicals
Supplies-Chemicals
M2L Electronics
Microcontrollers-Hardware
Mabuchi Motor America Corp.
Actuators-Motors
Machine Systems Ltd
Actuators-Motion Products
Magellan / Thales Navigation
Sensors-GPS
Magenta Electronics Ltd.
Kits-Electronic
Magmotor Corporation
Actuators-Motors
Main Hobby Center Inc., The
Retail-Train & Hobby
Maintenance America
Materials
Major Hobby
Retail-Train & Hobby
Malcom Company, Inc.
Tools
Manley Toy Direct

Toys
Manley Toy Quest
Toys-Robots
Manufacturer's Supply Inc.
Power Transmission
Many CNC System / EasyCut
Tools-CNC
Maplin Electronics
Retail-General Electronics
Mark Hannah Surplus Electronics
Retail-Surplus Electronics
Marlin P. Jones & Assoc. Inc
Retail-General Electronics
Retail-Surplus Mechanical
Marsh Electronics Inc.
Distributor/Wholesaler-Industrial
Electronics
Marshall Electronics, Inc.
Distributor/Wholesaler-Industrial
Electronics
MarVac Electronics
Retail-General Electronics
Maryland Metrics
Power Transmission
Fasteners
Material-Hardware
Massa Products Corp.
Sensors-Ultrasonic
Matco Inc.
Video-Cameras
Video-Transmitters
MATHWorks, Inc.
Programming-Robotic Simulations
Maxim Integrated Products
Manufacturer-Semiconductors
MAXNC Inc.
Tools-CNC
Maxon Motor AG
Actuators-Motors
MaxStream, Inc.
Communications-RF
Maxx Products International, Inc.
Radio Control
MazeBots
Programming-Robotic Simulations
McFeely's Square Drive Screws
Fasteners
McGonigal Paper & Graphics
Materials-Transfer Film
Retail-Arts & Crafts
McKenzie Taxidermy Supply
Materials-Foam
Retail-Armatures & Doll Parts
MCM Electronics
Retail-General Electronics
McMaster-Carr Supply Company
Materials
Actuators-Motors
Supplies
McNichols Company
Materials-Metal
MCS Electronics
Microcontrollers-Software

Measurement Specialties, Inc.
Sensors
Measurement Systems, Inc.
Sensors-Strain & Load Cells
Mecarobo: Educational Robots
Robots-Hobby & Kit
MechWars
Competitions-Combat
MECI-Mendelson Electronics Company,
Inc.
Retail-General Electronics
Medonis Engineering
Radio Control-Servo Control
Mekatronix, Inc.
Robots-Hobby & Kit
Sensors-Ultrasonic
Memory-Metalle GmbH
Actuators-Shape Memory Alloy
Memry Corp.
Actuators-Shape Memory Alloy
Menard, Inc.
Retail-Hardware & Home
Improvement
Mending Shed, The
Retail-Other
Meredith Instruments
Retail-Opticals & Lasers
Merkle-Korff Industries
Actuators-Motors
Merlin Systems Corp, Ltd.
Actuators-Motion Products
Sensors
Metal Supermarkets International
Materials-Metal
Metal Suppliers Online
Materials-Metal
Metallus
Toys-Construction
MetalMart.Com
Materials-Metal
MetalsDepot
Materials-Metal
Methode Electronics, Inc.
Electronics-Construction-Connectors
Metric Specialties, Inc.
Fasteners
Metric Test Equipment
Test and Measurement
Metrowerks, Inc. / Codewarrior
Programming-Languages
MGA Entertainment
Toys-Robots
Michael Burnett Productions, Inc.
Supplies-Casting & Mold Making
Michaels Stores, Inc.
Retail-Arts & Crafts
Michigan Fiberglass
Materials-Fiberglass & Carbon
Composites
Micro Computer Specialists, Inc.
Computers-Single Board Computers
Micro Engineering Labs
Microcontrollers-Hardware

Microcontrollers-Software
Micro Fasteners
    Fasteners
Micro Magazine.com
    Journals and Magazines
Micro Parts & Supplies, Inc.
    Retail-Other Materials
Micro Plastics, Inc.
    Fasteners
Micro Video Products
    Video-Cameras
    Video-Transmitters
Microbtica, S.L.
    Robots-Hobby & Kit
Microchip Technology
    Manufacturer-Semiconductors
    Microcontrollers-Hardware
    Sensors-RFID
MicroKinetics Corporation
    Tools-Machinery
    Tools-CNC
Micro-Mark
    Tools-Precision & Miniature
    Supplies-Casting & Mold Making
Micromech
    Actuators-Motors
    Motor Control
Micromint, Inc.
    Computers-Single Board Computers
MicroMo Electronics, Inc.
    Actuators-Motors
MicroProto Systems
    Tools-CNC
    Tools-Precision & Miniature
Microrobot NA Inc.
    Robots-Hobby & Kit
    Competitions-Other
Midcom (UK) Ltd
    Electronics-Obsolete
Midori America Corporation
    Manufacturer-Components
Midwest Laser Products
    Retail-Opticals & Lasers
Midwest Products Co., Inc.
    Materials-Other
Migatron Corp.
    Sensors-Ultrasonic
Milford Instruments Ltd
    Retail-Robotics Specialty
Milo Associates, Inc.
    Retail-General Electronics
Minarik Corporation
    Power Transmission
    Actuators-Motion Products
    Actuators-Motors
Miniature Bearings Australia Pty.
    Power Transmission
Miniature Molds.com
    Supplies-Casting & Mold Making
Minicraft
    Tools-Precision & Miniature
Minicraft Tools USA, Inc.
    Tools-Precision & Miniature

Minitech Machinery Corp.
    Tools-CNC
Mister Computer
    Radio Control-Servo Control
MisterArt
    Retail-Arts & Crafts
MJ Designs
    Retail-Arts & Crafts
MMS Online
    Journals and Magazines
Mode Electronics Ltd.
    Distributor/Wholesaler-Other
Components
Model A Technology
    Toys-Construction
Model Airplane News
    Journals and Magazines
Model Aviation
    Journals and Magazines
Model Builders Supply
    Manufacturer-Train & Hobby
Model Rectifier Corporation
    Manufacturer-Train & Hobby
Molon Motor and Coil Company
    Motors-Actuators
Mondo-tronics Inc.
    Actuators-Shape Memory Alloy
Mondo-tronics, Inc.
    Retail-Robotics Specialty
Moody Tools, Inc.
    Tools-Precision & Miniature
Moore Medical
    Supplies-Casting & Mold Making
Motion Control Buyer's Guide
    Motor Control
Motion Group, Inc., The
    Actuators-Motors
Motion Industries
    Power Transmission
Motion Systems Corporation
    Actuators-Motion Products
MotionShop.com
    Motor Control
    Portal-Other
Motorola Semiconductor Products Sector
    Manufacturer-Semiconductors
Mouser Electronics
    Retail-General Electronics
Movie Goods
    Entertainment-Art
Movie Poster Shop, The
    Entertainment-Art
Mr. Robot
    Retail-Robotics Specialty
MSC Fasteners
    Fasteners
MSC Industrial Direct Co., Inc.
    Power Transmission
    Tools
    Supplies-Glues & Adhesives
    Materials
    Actuators
M-Tronics Inc.

    Retail-General Electronics
Multi-Craft Plastics, inc.
    Materials-Plastic
Murata Manufacturing Co.
    Sensors
Murphy's Electronic & Industrial Surplus
Warehouse
    Retail-Surplus Mechanical
MWK Laser Products Inc.
    Retail-Opticals & Lasers
MyToolStore.com
    Tools
N R Bardwell, Ltd. / Bardwells
    Retail-General Electronics
Nancy's Notions
    Retail-Arts & Crafts
Nanomuscle, Inc.
    Actuators-Shape Memory Alloy
NAPSCO (North American Parts Search
Company)
    Power Transmission
Nasco
    Retail-Educational Supply
    Retail-Arts & Crafts
National Control Devices
    Microcontrollers-Hardware
    Radio Control-Servo Control
National Gypsum Company
    Supplies-Casting & Mold Making
National Hobby Supply
    Retail-Train & Hobby
National Power Chair (NPC)
    Actuators-Motors
    Wheels and Casters
National Semiconductor Corp.
    Manufacturer-Semiconductors
Nature Coast Hobbies
    Tools-Precision & Miniature
    Tools-Machinery
Navtech Seminars and GPS Supply
    Sensors-GPS
NEC-Tokin America
    Sensors-Tilt & Accelerometer
NEET-O-RAMA
    Entertainment-Art
Nelson Appliance Repair, Inc.
    Retail-Other
Nerd Books
    Books-Technical
NetMedia Inc. / BasicX
    Microcontrollers-Hardware
NetMedia Inc. / Siteplayer
    Computers-I/O
NetMedia Inc. / Web-Hobbies
    Radio Control-Servo Control
New Method Lasers
    Retail-Opticals & Lasers
New Micros, Inc.
    Microcontrollers-Hardware
New Scientist
    Journals and Magazines
Newark Electronics
    Distributor/Wholesaler-Industrial

Electronics
Newton Research Labs, Inc.
  Video-Cameras
  Video-Programming & APIs
Nexergy
  Batteries and Power
NiCd Lady Company, The
  Batteries and Power
Nick Carter's Taig Lathe Pages
  Tools-CNC
Nighthawk Manufacturing Inc.
  Wheels and Casters
Nitinol Devices & Components
  Actuators-Shape Memory Alloy
Nolan Supply Corporation
  Materials-Metal
  Tools
Nook Industries
  Actuators-Motion Products
Nordex, Inc.
  Power Transmission
Norland Research
  Robots-Hobby & Kit
North American Roller Products, Inc.
  Wheels and Casters
North Country Radio
  Communications-RF
Northern Tool & Equipment Co,
  Tools
  Power Transmission
Notions Marketing
  Materials-Other
NPC Robotics
  Actuators-Motors
  Wheels and Casters
NSK
  Power Transmission
NTE Electronics, Inc.
  Distributor/Wholesaler-Industrial
  Electronics
Nu Horizons Electronics Corp.
  Distributor/Wholesaler-Industrial
  Electronics
Nuts and Volts Magazine
  Journals and Magazines
Oatley Electronics
  Retail-General Electronics
  Kits-Electronic
Ocean State Electronics
  Test and Measurement
Office Depot
  Retail-Office Supplies
Office Max
  Retail-Office Supplies
Ohio Art Company
  Toys
Ohmark Electronics
  Radio Control
Ohmite Mfg. Co.
  Manufacturer-Components
Olimex Ltd.
  Electronics-PCB- Production
  Microcontrollers-Hardware

Olmec Advanced Materials Ltd.
  Materials-Lighting
  Materials-Fiberglass & Carbon
Composites
OMEGA Engineering, Inc
  Sensors-Strain & Load Cells
OmniVision Technologies, Inc.
  Video-Imagers
OMRON Corp
  Sensors-RFID
ON Semiconductor Corporation
  Manufacturer-Semiconductors
Online Metals
  Materials-Metal
OnlyTOYS.com
  Toys-Construction
OnRobo.com
  Portal-Robotics
OPAMP Technical Books
  Books-Technical
Optek Technology, Inc.
  Sensors-Infrared
Orchard Supply Hardware
  Retail-Hardware & Home
Improvement
OrderMetals.com
  Materials-Metal
O'Reilly Perl.com
  Programming-Languages
Oriental Motors
  Actuators-Motors
Oriental Trading Company
  Outside-of-the-Box
Ortho Cast, Inc.
  Supplies-Casting & Mold Making
Out of this World
  Retail-Science
Outwater Plastics Industries Inc.
  Materials
ozBricks
  Toys-Construction
Ozitronics
  Kits-Electronic
Ozzie's Robots Toys & Collectibles
  Toys-Robots
Pacific Fasteners
  Fasteners
Pacific Scientific
  Actuators-Motors
PAiA Electronics, Inc.
  Electronics-Sound & Music
Paintball-Online.com
  Actuators-Pneumatic
Parallax, Inc.
  Microcontrollers-Hardware
  Robots-Hobby & Kit
Parker Hannifin, Inc.
  Actuators-Motion Products
PartMiner Inc.
  Distributor/Wholesaler-Industrial
  Electronics
  Internet-Research
Parts Express

Retail-General Electronics
Parts for Industry
  Retail-Surplus Mechanical
  Retail-Surplus Electronics
Parts on Sale / Solatron Ttechnologies
Inc.
  Retail-General Electronics
PC/104 Embedded Solutions Magazine
  Journals and Magazines
PCB Milling
  Tools-CNC
PCBexpress
  Electronics-PCB-Production
PCB-Pool / Beta LAYOUT GmbH
  Electronics-PCB-Production
Pearl Paint
  Retail-Arts & Crafts
Pelikan Industry, Inc.
  Video-Cameras
Penn State Industries
  Tools
Pep Boys
  Retail-Automotive Supplies
PERIM-ALERT III Fence Sensor
  Sensors-Other
Perl Mongers / Perl.org
  Programming-Languages
Perma-Flex Mold Co., Inc.
  Supplies-Casting & Mold Making
Personal Robot Technolgies Inc.
  Books-Robotics
Peter H. Anderson
  Microcontrollers-Hardware
PetSafe
  Electronics-Miscellaneous
Philcap Electronic Suppliers
  Retail-General Electronics
Philips Semiconductor
  Manufacturer-Semiconductors
Phoenix Fastener Company, Inc.
  Fasteners
Photon Micro-Light
  Electronics-Specialty
Photon Vision Systems, Inc.
  Video-Cameras
PIC Design
  Power Transmission
  Actuators-Motion Products
Picard Indistries
  Actuators-Motors
  Sensors
Pico Electronics, Inc.
  Electronics-Specialty
  Batteries and Power
PID Without the Math
  Motor Control
  Books-Technical
Piezo Systems Inc.
  Electronics-Specialty
Pioneer-Standard Electronics, Inc.
  Distributor/Wholesaler-Industrial
  Electronics
Pitsco

Retail-Science
  Supplies-Casting & Mold Making
Pitsco/Dacta
  Toys-Construction
Pittman
  Actuators-Motors
Plaid Enterprises, Inc.
  Supplies-Paints
Planet Battery
  Batteries and Power
Plantraco Ltd.
  Radio Control
  Video-Cameras
Plaster Master Industries
  Supplies-Casting & Mold Making
  Materials-Foam
Plastic Products, Inc.
  Materials-Plastic
  Power Transmission
Plastic Specialties Inc.
  Materials-Plastic
Plastic World
  Materials-Plastic
Plastics Technology Online
  Journals and Magazines
Plasti-kote
  Materials-Plastic
Plastruct, Inc.
  Materials-Plastic
PMB Electronics
  Microcontrollers-Hardware
PNI Corp. / Precision Navigation
  Sensors-Tilt & Accelerometer
  Sensors-Other
Polaris Industries
  Video-Cameras
  Video-Transmitters
Polaroid/SensComp
  Sensors-Ultrasonic
Polymer Clay Express
  Retail-Arts and Crafts
Polymorf, Inc.
  Toys-Construction
  Materials-Plastic
Polytek Development Corp.
  Supplies-Casting & Mold Making
Pontech
  Radio Control-Servo Control
  Actuators-Motors
Popular Mechanics
  Journals and Magazines
Popular Science
  Journals and Magazines
Port Plastics
  Materials-Plastic
Porter Cable Corporation
  Manufacturer-Tools
Portescap
  Actuators-Motors
Powell's Books
  Books-Technical
Power Sonic Corp.
  Batteries and Power

PowerBASIC, Inc.
  Programming-Languages
Prairie Digital, Inc.
  Computers-Data Acquisition
Precision MicroDynamics, Inc.,
  Motor Control
Premier Farnell plc
  Distributor/Wholesaler-Industrial
  Electronics
Premium Supply Company
  Retail-Other Materials
Probotics, Inc. / Cye
  Robots-Personal
ProfBooks Robotics
  Books-Technical
Protean Logic
  Microcontrollers-Hardware
  Computers-Single Board Computers
PTG / Patios To Go
  Materials-Plastic
Public Missiles, Ltd.,
  Materials-Paper & Plastic Laminates
  Materials-Foam
Purity Casting Alloys Ltd.
  Materials-Metal
Putnam Precision Molding, Inc.
  Power Transmission
PVC Store, The
  Materials-Plastic
Quadravox, Inc.
  Electronics-Sound & Music
Quality Kits / QKits
  Kits-Electronic
  Retail-General Electronics
Quality Transmission Components
  Power Transmission
Quantum CNC
  Tools-CNC
Quasar Electronics Limited
  Kits-Electronic
  Kits-Robotic
QuickCam
  Video-Cameras
QUINCY
  Retail-Arts & Crafts
R & D Electronic Parts, Inc.
  Distributor/Wholesaler-Industrial
  Electronics
R & J Sign Supply
  Materials-Foam
  Materials-Paper & Plastic Laminates
R.C. Scrapyard
  Radio Control-Accessories
R.L.C. Enterprises, Inc.
  Computers-Single Board Computers
R/C Car Action
  Journals and Magazines
R/C Car Action
  Journals and Magazines
Rabbit Semiconductor
  Microcontrollers-Hardware
RACO Industries / ID Warehouse
  Barcoding

Sensors-RFID
Radar, Inc.,
  Retail-General Electronics
Radio Fence Distributors, Inc.
  Electronics-Miscellaneous
Radio Model Supplies
  Retail-Train & Hobby
Radio Shack
  Retail-General Electronics
Radiometrix Ltd
  Communications-RF
Radiotronix
  Communications-RF
RAE Corp.
  Actuators-Motors
RAF Electronic Hardware
  Distributor/Wholesaler-Other
Components
Rage
  Competitions-Entrant
Ramsey Electronics, Inc.
  Kits-Electronic
  Communications-RF
Rapid Electronics
  Distributor/Wholesaler-Industrial
  Electronics
Ray Rohr's Cosmic Artifacts
  Toys-Robots
RB Industries, Inc
  Tools-Power
RC Systems, Inc.
  Electronics-Sound & Music
RC Yellow Pages
  Radio Control
Real Robots (Cybot)
  Journals and Magazines
  Robots-Hobby & Kit
Reconn's World
  Portal-Robotics
Regal Plastics
  Materials-Plastic
Reid Tool Supply Co.
  Power Transmission
  Wheels and Casters
  Machine Framing
  Fasteners
Resources Un-Ltd.
  Retail-Other Electronics
  Retail-Opticals & Lasers
Retrofire-Robot & Space Toy Collectibles
  Entertainment-Art
  Entertainment-Books & Movies
  Toys-Robots
Reuel's
  Retail-Arts & Crafts
Reynolds Electronics
  Microcontrollers-Hardware
  Communications-RF
RF Digital Corporation
  Communications-RF
RF Monolithics, Inc.
  Communications-RF
RF Parts Company

Retail-General Electronics
RFID Components Ltd.
Sensors-RFID
RFID, Inc.
Sensors-RFID
RG Speed Control Devices Ltd.
Power Transmission
Actuators-Motion Products
Rho Enterprises
Microcontrollers-Hardware
Ridout Plastics
Materials-Plastic
Right-Tool.com
Tools
RMB Roulements Miniatures SA
Power Transmission
Actuators-Motors
Robo Systems
Tools-CNC
Roboblock System Co., Ltd
Robots-Hobby & Kit
Microcontrollers-Hardware
Microcontrollers-Software
Robodyssey Systems LLC
Robots-Hobby & Kit
Roboforge
Programming-Robotic Simulations
Robologic
Retail-Robotics Specialty
RoboProbe Technologies, Inc.
Robots-Industrial/Research
RoboRama
User Groups
Robot Cafe
Portal-Robotics
Robot Channel, The
Portal-Robotics
Robot Electronics
Sensors
Motor Control
Robot Power
Motor Control
Robot Science & Technology
Journals and Magazines
Robot Store
Retail-Robotics Specialty
Robot Store (HK)
Retail-Robotics Specialty
Robot Zone
Retail-Robotics Specialty
Robotbooks.com
Books-Robotics
Robotic Power Solutions
Batteries and Power
Actuators-Motors
Robotics Building Blocks
Microcontrollers-Hardware
Robotikits Direct
Robots-Hobby & Kit
Robotix
Robots-Hobby & Kit
Toys-Robots
RobotLogic

Motor Control
Radio Control-Servo Control
RoboToys
Retail-Robotics Specialty
RobotOz
Retail-Robotics Specialty
RobotPartz.com
Retail-Surplus Mechanical
Robots & Wind Me Up Toys
Toys-Robots
Robots Wanted: Dead or Alive, Whole or Parts
Robots-Hobby & Kit
Robots@War
Competitions-Combat
Robotwars
Competitions-Combat
Rockby Electronic Components
Retail-General Electronics
Rocket USA
Toys-Robots
Rockford Ball Screw
Actuators-Motion Products
Rockler Woodworking and Hardware
Tools
Retail-Hardware & Home Improvement
Rockwell Automation
Motor Control
Actuators-Motion Products
Roto Zip Tool Corporation
Manufacturer-Tools
Royal Products
Tools-Machinery
RP Electronics
Distributor/Wholesaler-Industrial Electronics
RS Components Ltd
Retail-General Electronics
Power Transmission
S*K Hand Tool Corporation
Tools-Hand
Saelig Co. Inc
Test and Measurement
Sage Telecommunications Pty Ltd
Microcontrollers-Hardware
Sager Electronics
Distributor/Wholesaler-Industrial Electronics
Sam Schwartz Inc.
Materials-Metal
Materials-Plastic
SAMS Technical Publishing
Books-Technical
San Diego Plastics, Inc.
Materials-Plastic
Sanyo Energy
Batteries and Power.
Sarcos Inc.
Robots-Experimental
Satco Supply
Tools
Savage Innovations / OOPic

Microcontrollers-Hardware
Savko Plastic Pipe & Fittings, Inc.
Materials-Plastic
Sax Art's & Crafts
Retail-Arts & Crafts
Retail-Educational Supply
SAYAL Electronics
Distributor/Wholesaler-Industrial Electronics
Schaevitz
Sensors
Schenz Theatrical Supply, Inc.
Supplies-Casting & Mold Making
School Specialty, Inc.
Retail-Educational Supply
School-Tech Inc.
Retail-Science
SCI FI
Entertainment-Art
Science & Hobby
Retail-Science
Science City
Retail-Science
Science Daily
Journals and Magazines
Science Experience
Retail-Science
Science Kit & Boreal Laboratories
Retail-Science
Retail-Opticals & Lasers
Science Magazine
Journals and Magazines
Science News Online
Journals and Magazines
Science Source, The
Retail-Science
ScienceKits.com, Inc.
Retail-Robotics Specialty
Scott Edwards Electronics, Inc.
Electronics-Display
Radio Control-Servo Control
Screwfix Direct Ltd.
Fasteners
Tools
Sculpture House Casting
Supplies-Casting & Mold Making
Retail-Armatures & Doll Parts
Sealevel Systems
Computers-I/O
Sears, Roebuck, & Co.
Tools
Seattle Robotics
Robots-Hobby & Kit
Vision-Cameras
Secs, Inc.
Power Transmission
Actuators-Motion Products
Seitz Corp.
Power Transmission
Actuators-Motion Products
Selectronic
Retail-Robotics Specialty
Semtech Corporation

Sensors-Strain & Load Cells
Senix Corp
   Sensors-Ultrasonic
Sensoray
   Motor Control
Sensorland.com
   Sensors
Sensors Online
   Journals and Magazines
Sensors, Inc.
   Sensors
Sensory Inc.
   Electronics-Sound & Music
Servo City
   Radio Control-Servos
   Radio Control-Accessories
Servo Systems Co.
   Actuators-Motion Products
   Retail-Surplus Mechanical
Serv-o-Link
   Power Transmission
Servo-Tek Products Co., Inc.
   Sensors-Encoders
Shadow Robot Company Ltd.
   Robots-Industrial/Research
Shape Memory Applications, Inc.
   Actuators-Shape Memory Alloy
Sharper Image, The
   Retail-Other Electronics
Sheldons Hobbies
   Retail-Train & Hobby
Sherline Products
   Tools-Precision & Miniature
Shopsmith, Inc.
   Tools-Power
SICK, Inc.
   Sensors
Sign Foam
   Materials-Foam
Silicon Chip Magazine
   Journals and Magazines
Silicones, Inc.
   Manufacturer-Materials
Siliconix Incorporated
   Manufacturer-Semiconductors
Silly Universe
   Outside-of-the-Box
Simple Step L.L.C
   Motor Control
Sine Robotics
   Robots-Industrial/Research
Skycraft Parts & Surplus Inc.
   Retail-Surplus Electronics
Skyway Machine, Inc.
   Wheels and Casters
Slice of Stainless, Inc.
   Materials-Metal
Small Parts Inc.
   Power Transmission
   Materials
   Fasteners
Smart Pages
   Internet-Search

Smarthome, Inc.
   Communications-RF
SmartTechToys.com
   Toys-Construction
   Toys-Robots
   Retail-Opticals & Lasers
Smith Fastener Company
   Fasteners
Smooth-On
   Supplies-Casting & Mold Making
Smoovy / RMB Group
   Actuators-Motors
SoftVoice Text-to-Speech System
   Electronics-Sound & Music
Solar Mower
   Robots-Personal
Solarbotics Ltd.
   Robots-BEAM
Solutions Cubed
   Motor Control
Spare Bear Parts
   Retail-Armatures & Doll Parts
Special Effect Supply Co.
   Supplies-Casting & Mold Making
   Supplies-Paints
   Materials
Special Metals Corporation
   Actuators-Shape Memory Alloy
Specialty Motions, Inc.
   Actuators-Motion Products
Specialty Resources Company
   Materials-Plastic
Specialty Tool & Bolt
   Fasteners
Spectron
   Sensors-Tilt & Accelerometer
Spectrum Educational Supplies
   Retail-Educational Supply
Spencer Gifts, Inc.
   Materials-Lighting
Sports Authority, Inc.
   Retail-Other
SpringWalker
   Robots-Industrial/Research
Square One Electronics
   Books-Electronics
ST Microelectronics
   Manufacturer-Semiconductors
Standard Supply Electronics
   Distributor/Wholesaler-Industrial
   Electronics
Standees.com
   Entertainment-Art
Stanley Works, The
   Manufacturer-Tools
Staples, Inc.
   Retail-Office Supplies
Stark Electronic
   Retail-General Electronics
Starlink Incorporated.
   Sensors-GPS
State Electronics, Inc.
   Distributor/Wholesaler-Other

Components.
Stegmann, Inc.
   Sensors-Encoders
Stepper Motor Archive, The
   Motor Control
Stiquito
   Actuators-Shape Memory Alloy
   Books-Robotics
Stock Drive Products
   Power Transmission
Stockade Wood & Craft Supply
   Materials-Other
Subtech
   Radio Control
Sullivan Products
   Actuators-Motors
   Radio Control-Accessories
Sun Equipment Corp.
   Test and Measurement
Sun Java
   Programming-Languages
Sunx Sensors USA
   Sensors
Super Battery Packs
   Batteries and Power
Super Circuits
   Video-Cameras
   Video-Transmitters
Super Tech & Associates
   Tools-CNC
SuperCal Decals
   Materials-Transfer Film
Superior Electric
   Actuators-Motors
Superior Tire & Rubber Corporation
   Wheels and Casters
Supertronix Inc.
   Retail-General Electronics
Supply Chain Systems Magazine
   Journals and Magazines
Supply Depot Inc.
   Materials
Surelight
   Materials-Lighting
Surplus Sales of Nebraska
   Retail-Surplus Electronics
Surplus Shed
   Retail-Opticals & Lasers
Surplustronics Trading LTD.
   Retail-Surplus Electronics
   Retail-General Electronics
SWS Electronics
   Retail-General Electronics
Symbol Technologies, Inc.
   Barcoding
Syn-Air Corporation
   Supplies-Casting & Mold Making
Synaptics, Inc.
   Sensors-Strain & Load Cells
Synergy Systems, LLP
   Sensors-GPS
Systronix
   Microcontrollers-Hardware

T N Lawrence & Son Ltd.
  Retail-Arts & Crafts
  Materials-Metal
  Materials-Transfer Film
Tab Electronics Build Your Own Robot Kit
  Robots-Hobby & Kit
Tadiran U.S. Battery Division
  Batteries and Power
TAIG Tools
  Tools-Precision & Miniature
  Tools-CNC
Tamiya America, Inc. — Educational
  Actuators-Motors
  Robots-Hobby & Kit
Tanner Electronics, Inc.
  Retail-General Electronics
TAP Plastics
  Materials-Plastic
  Supplies-Casting & Mold Making
Target Stores
  Retail-Discount & Department
Team Delta Engineering
  Actuators-Motors
  Motor Control
Team Whyachi LLC
  Actuators-Motors
  Competitions-Entrant
Tebo Store Fixtures
  Materials-Store Fixtures
TECEL
  Microcontrollers-Hardware
Tech America
  Retail-General Electronics
Techmaster Inc.
  Machine Framing
  Actuators-Motion Products
Techni-Tool, Inc.
  Electronics-Soldering
  Tools
Techno Profi-Team
  Machine Framing
Techno-Isel
  Tools-CNC
Technological Arts
  Microcontrollers-Hardware
Tech-supplies.co.uk
  Retail-Robotics Specialty
TechTools
  Microcontrollers-Programming
Teddy Bear Stuff
  Retail-Armatures & Doll Parts
Telelink Communications
  Communications-RF
Tern, Inc.
  Computers-Single Board Computers
Terrapin Software
  Programming-Languages
  Toys-Construction
tesa AG
  Manufacturer-Glues & Adhesives
Test Equipment Connection Corp.
  Test and Measurement
Testors Corporation

Supplies-Paints
Texas Instruments
  Manufacturer-Semiconductors
That's Cool Wire / Solution Industries
  Materials-Lighting
Thermo Centrovision
  Sensors-Optical
Think & Tinker, Ltd.
  Electronics-PCB-Production
Thinker Toys
  Robots-Hobby & Kit
  Retail-Science
Thomas Distributing
  Batteries and Power.
Thomas Register Online
  Internet-Research
Thomson Industries, Inc.
  Actuators-Motion Products
Timberdoodle Company
  Toys-Construction
Timeline Inc.
  Retail-Surplus Electronics
Tin10.Com
  Robots-Hobby & Kit
TiNi Alloy Company
  Actuators-Shape Memory Alloy
TM Technologies
  Materials-Metal
TNR Technical
  Batteries and Power
Toki Corp.
  Actuators-Shape Memory Alloy
Tool Peddler
  Tools
Toolsforless.com
  Tools
ToolSource.com
  Tools
Tools-Plus
  Tools
Toro Company, The
  Manufacturer-Tools
Torque Conversion Calculator
  Internet-Calculators & Converters
Torque Speed Applet
  Internet-Calculators & Converters
  Motor Control
Total Robots Ltd
  Retail-Robotics Specialty
Totally Fun Toys
  Toys
Tower Fasteners Co. Inc.
  Fasteners
  Materials-Other
  Power Transmission
Tower Hobbies
  Retail-Train & Hobby
Toy Robot Parts.com
  Toys-Robots
Toymax Inc.
  Toys-Electronics
Toys "R" Us
  Toys

TP Tools & Equipment
  Tools
Tracks USA
  Wheels and Casters
TrainTown Hobbies and Crafts
  Retail-Train & Hobby
Transtronics, Inc.
  Kits-Electronic
  Retail-Surplus Electronics
Travers Tool Co., Inc.
  Materials
Trendmasters, Inc.
  Toys
Tri-State Electronics
  Distributor/Wholesaler-Industrial
  Electronics
Truebite, Inc.
  Tools-Precision & Miniature
TruServ Corporation
  Retail-Hardware & Home
Improvement
TS Racing, Inc.
  Power Transmission
TTI Inc.
  Distributor/Wholesaler-Industrial
  Electronics
Tucker Electronics
  Test and Measurement
U.S. Composites, Inc.
  Materials-Fiberglass & Carbon
Composites
  Supplies-Casting & Mold Making
Ubicom / Scenix
  Manufacturer-Semiconductors
u-blox ag
  Sensors-GPS
ULINE
  Materials-Paper & Plastic Laminates
  Materials-Foam
UltraTechnology
  Programming-Languages
Unicorn Electronics
  Retail-General Electronics
Uniplast Inc.
  Supplies-Glues & Adhesives
  Tools-Hand
United States Plastic Corp.
  Materials-Plastic
Unitrode Corp.
  Manufacturer-Semiconductors
Uptown Sales Inc.
  Retail-Train & Hobby
Urethane Supply Company
  Supplies-Casting & Mold Making
  Materials-Plastic
URS Electronics Inc.
  Retail-General Electronics
  Video-Cameras
US Digital
  Sensors-Encoders
US Government Printing Office
  Books-Technical
Valiant Technology Ltd.

Toys-Construction
Electronics-Specialty
Van Aken International, Inc.
Supplies-Casting & Mold Making
Supplies-Paints
Van Dykes Taxidermy
Materials-Foam
Retail-Armatures & Doll Parts
Vantec
Radio Control
Vaughn Belting
Power Transmission
Vector Electronics and Technology, Inc.
Electronics-PCB-Production
Velleman Components NV
Kits-Electronic
Vermont American Corporation
Tools-Accessories
Vesta Technology
Computers-Single Board Computers
Victor Machinery Exchange, Inc.
Tools
Viking Office Products
Retail-Office Supplies
Vikon Technologies
Computers-Single Board Computers
Microcontrollers-Hardware
Vishay Intertechnology, Inc.
Manufacturer-Semiconductors
Sensors
Visual Communications
Materials-Transfer Film
Vitel Electronics
Distributor/Wholesaler-Industrial
Electronics
Viva Robotics, Inc.
Robots-Industrial/Research
VTech Holdings Limited
Toys
W. L. Fuller Co. Inc.
Tools-Accessories
W. M. Berg / Invensys
Power Transmission
W.J. Ford Surplus Enterprises
Retail-Surplus Electronics
W.P. Notcutt Ltd.
Supplies-Casting & Mold Making
Wacky Willy's
Retail-General Surplus
Test and Measurement
Wahl Clipper Corp.
Electronics-Soldering
Walker Electronic Supply Co.
Retail-General Electronics
Wal-Mart
Retail-Discount & Department
Walther's Model Railroad Mall
Manufacturer-Train & Hobby
Wangrow
Manufacturer-Train & Hobby
Wany Robotics
Robots-Educational
Warner Electric
Actuators-Motion Products
Wasp Bar Code Technologies

Barcoding
WASSCO
Electronics-Soldering
Supplies-Chemicals
Tools-Hand
Watson Industries, Inc.
Sensors-Other
Weeder Technologies
Computers-I/O
Weird Stuff Warehouse
Retail-Surplus Electronics
Wenzel Associates, Inc.: Technical Library
Electronics-Circuit Examples
WESCO International, Inc.
Materials
Supplies
Western Test Systems
Test and Measurement
Westrim Crafts
Materials-Other
Westward Plastics
Materials-Plastic
Wholesale Bearing & Drive Supply
Power Transmission
Wicks Aircraft Supply
Fasteners
Materials-Metal
Wiha Quality Tools
Tools
Wilde EVolutions, Inc.
Batteries and Power
Actuators-Motors
Wilke Technology GmbH
Microcontrollers-Hardware
Microcontrollers-Software
Computers-Single Board Computers
Wiltronics
Retail-General Electronics
Kits-Electronic
Win Systems
Computers-Single Board Computers
Winbond Electronics Corporation
Manufacturer-Semiconductors
Winford Engineering
Computers-I/O
Wirz Electronics
Retail-General Electronics
Wizard Devices, Inc.
Retail-Robotics Specialty
Wm. B. Allen Supply Company, Inc.
Retail-Other Electronics
Video-Cameras
Woodcraft Supply Corp.,
Tools-Hand
Materials-Other
Woodsmith Store
Tools-Hand
Tools-Precision & Miniature
Woodworker Online
Internet-Links
Woodworker's Warehouse
Tools
Woodworking Pro
Portal-Other
Journals and Magazines

Woolworths PLC
Retail-Discount & Department
Workshop Publishing
Books-Technical
World of Plastics, Inc.
Materials-Plastic
World of Robotics Online
Retail-Robotics Specialty
Worth Marine Inc.
Retail-Train & Hobby
WR Display & Packaging
Retail-Store Fixtures
Wrights
Materials-Other
X10 Wireless Technology, Inc.
Video-Cameras
Video-Transmitters
X-Actoblades.com
Tools-Precision & Miniature
XBeams, Inc.
Machine Framing
Xenoline
Materials-Lighting
Xilor inc.
Communications-RF
Communications-Infrared
Sensors-Other
XPress Metals
Materials-Metal
Yobotics
Robots-Industrial/Research
You-Do-It Electronics Center
Retail-General Electronics
Young Explorers
Retail-Educational Supply
Zagros Robotics
Robots-Hobby & Kit
Zany Brainy
Retail-Science
ZapWorld.com
Actuators-Motors
Batteries and Power
Zarlink Semiconductor Inc.
Manufacturer-Semiconductors
Zaytran Automation
Robots-Industrial/Research
ZEBEX America, Inc.
Barcoding
Zebra Technologies Corporation
Barcoding
Sensors-RFID
Zilog
Manufacturer-Semiconductors
Zircon Corporation
Manufacturer-Tools
Zona Tool Company
Tools-Precision & Miniature
Zorin Microcontroller Products
Microcontrollers-Hardware
Z-World, Inc.
Computers-Single Board Computers
Zygo Systems Limited / Chips2Ship
Electronics-Obsolete
Zykronix, Inc.
Computers-Single Board Computer

WHERE TO FIND IT

Locate the most popular categories for several hundred common (and a few not-so-common) parts and components used in robot building.

Accelerometers
  Sensors-Tilt & Accelerometer
Acrylic plastics
  Materials-Plastics
  Retail-Hardware & Home
Improvement
Actuators, materials and hardware
  Actuators-Motion Products
  Machine Framing
Adhesive tapes
  Materials
  Supplies-Glues & Adhesives
Adhesives, glues
  Manufacturer-Glues & Adhesives
  Materials
  Materials-Plastics
  Retail-Arts & Crafts
  Retail-Hardware & Home
Improvement
  Retail-Train & Hobby
  Supplies-Glues & Adhesives
Alumalite (laminate material)
  Materials-Foam
  Materials-Paper & Plastic Laminates
Alumilite (casting material)
  Supplies-Casting & Mold Making
Aluminum
  Materials-Metal
Aluminum profile extrusions
  See Machine framing.
Armatures
  Retail-Arts & Crafts
  Retail-Armatures & Doll Parts
  Supplies-Casting & Mold Making
Arts and crafts
  Retail-Arts & Crafts
  Retail-Other
  Supplies-Casting & Mold Making
Auctions
  Retail-Auctions
Automotive supplies (wires, switches, etc.)
  Retail-Automotive
  Retail-Discount & Department
AVR microcontroller
  Microcontrollers-Hardware
Ball transfers
  Wheels and Casters
Ballscrews, leadscrews
  Actuators-Motion Products
  Retail-Surplus Mechanical
Barcodes
  Barcoding
Basic electronics

Internet-Informational
Internet-Plans & Guides
Internet-Web Ring
Basic Stamp microcontroller
  Microcontrollers-Hardware
  Retail-Robotics Specialty
Batteries and battery holders
  Batteries and Power
  Radio Control
  Retail-Automotive Supplies
  Retail-General Electronics
  Retail-Hardware & Home
Improvement
  Retail-Surplus Electronics
BEAM robots
  Robots-BEAM
  Robots-Hobby & Kit
Bearings
  See Power transmission.
Belts and pulleys
  See Power transmission.
Blue foam
  See Foam.
Bolts, nuts
  See Fasteners.
Books and other literature
  Books (various)
  Internet-Informational
  Magazines and Journals
  Manufacturers (various)
Bulletin boards
  Internet-Bulletin Board/Mailing List
  Internet-Usenet Newsgroups
  Internet-Web Ring
  User Groups
Cables
  See Components, electronics.
Calculators, conversion
  Internet-Calculators & Conversion
Cameras, video
  Video (various)
Capacitors
  See Components, electronics.
Carbon composites
  See Composites.
Casters
  Materials (various)
  Retail-Hardware & Home
Improvement
  Wheels and Casters
Casting (plastic, etc.)
  Books-Technical
  Materials-Plastics

Supplies-Casting & Mold Making
Materials-Plastic
Retail-Arts & Crafts
Celastic
  Materials-Plastics
Chemicals, electronics and cleaning
  Electronics-Soldering
  Distributor/Wholesaler-Industrial
Electronics
  Retail-General Electronics
  Supplies-Chemicals
Circuit building components
  Distributor/Wholesaler-Industrial
Electronics
  Retail-General Electronics
  Retail-Surplus Electronics
Circuit making chemicals
  Distributor/Wholesaler-Industrial
Electronics
  Retail-General Electronics
  Supplies-Chemicals
Circuits, examples
  Electronics-Circuit Examples
  Internet-Informational
  Kits-Electronics
  Microcontrollers-Hardware
CNC tools
  Tools-CNC
  Tools-Machinery
  Tools-Precision & Miniature
Collectables (includes robots)
  Entertainment-Art
  Toys
Combat robot
  Competitions-Combat
  Competitions-Entrant
Communications links
  Communications-Infrared
  Communications-RF
  Electronics-Circuit Examples
  Manufacturer-Components
  Retail-General Electronics
  Video-Transmitters
Competitions, robot
  Competitions (various)
  Internet-Bulletin Board/Mailing List
  Internet-Usenet Newsgroups
Components, electronics (resistors, capacitors, ICs, switches, relays, etc.)
  Distributor/Wholesaler-Industrial
Electronics
  Manufacturer-Components
  Manufacturer-Semiconductors

Retail-General Electronics
Retail-Surplus Electronics
Composites
  Materials-Fiberglass & Carbon
Composites
  Materials-Paper & Plastic Laminates
  Materials-Plastic
  Radio Control
  Supplies-Casting & Mold Making
Computers, single board
  Computers-Single Board Computers
  Microcontrollers-Hardware
Connectors
  Distributor/Wholesaler-Industrial
Electronics
  Retail-General Electronics
  Retail-Surplus Electronics
Construction materials (frame)
  Machine Framing
  Materials-Metal
  Materials-Plastics
  Materials-Store Fixtures
  Retail-Hardware & Home
Improvement
Construction tools
  See Tools, Tools, hand and power
Controllers, motor
  Motor Control
Copper
  Materials-Metal
Corrugated cardboard
  Materials-Paper & Plastic Laminates
Couplers
  See Power transmission.
  Also: Fasteners
Craft supplies
  Retail-Arts & Crafts
  Retail-Hardware & Home
Improvement
  Retail-Train & Hobby
  Supplies (various)
Data acquisition
  Computers-Data Acquisition
Datasheets, electronics
  Manufacturer (various)
  Internet-Search
  Internet-Research
DC motors
  See Motors.
Decals
  Materials-Transfer Film
  Retail-Arts & Crafts
Diodes
  See Components, electronics.
Doll-making parts (eyes, notes)
  See Armatures.
Domes, plastic
  Materials-Plastic
  Retail-Hardware & Home
Improvement
Drill bits and accessories
  Tools-Accessories
eBay

Retail-Auctions
Educational materials
  Retail-Educational Supply
  Retail-Science
Electroluminescent wire
  Materials-Lighting
Electronics kits
  Kits-Electronic
Electronics, learning
  Internet-Informational
  Internet-Plans & Guides
  Internet-Web Ring
Embedded systems
  Computers-Single Board Computers
  Microcontrollers-Hardware
Expanded (foamed) PVC
  See PVC.
Expanded foam
  See Foam.
Fasteners (nuts, bolts, etc.)
  Distributor/Wholesaler-Other
Components
  Fasteners
  Retail-Automotive Supplies
  Retail-Hardware & Home
Improvement
Fiberglass
  Materials-Fiberglass & Carbon
Composites
  Retail-Arts & Crafts
  Retail-Hardware & Home
Improvement
Foam
  Materials-Foam
  Materials-Paper & Plastic Laminates
  Materials-Plastic
  Retail-Arts & Crafts
  Supplies-Casting & Mold Making
Foam core
  See Laminates.
Formula calculators
  Actuators-Motors
  Internet-Calculators & Conversion
Free plans
  Electronics-Circuit Examples
  Internet-Informational
  Internet-Plans & Guides
Gatorboard
  Materials (various)
Gear motors
  Actuators-Motors
  Retail-Surplus Mechanical
Gears
  See Power transmission.
Global positioning satellite (GPS)
  Sensors-GPS
Glues, adhesives
  Materials
  Materials-Plastics
  Retail-Arts & Crafts
  Retail-Hardware & Home
Improvement
  Retail-Train & Hobby

Supplies-Glues & Adhesives
Golf cart parts
  Actuators-Motors
  Batteries and Power
  Materials-Other
GPS (global positioning satellite)
  Sensors-GPS
Ham fests
  Fests and Shows
Handy Board
  Microcontrollers-Hardware
Hardware (construction)
  Materials (various)
  Retail-Hardware & Home
Improvement
  Retail-Surplus Mechanical
Hardware, electronics construction (e.g.
standoffs, handles, knobs)
  See Components, electronics.
H-bridges
  See Components, electronics.
  Also: Motor Control
Incremental encoders
  Sensors-Encoders
Infrared communications
  Communications-Infrared
  Electronics-Circuit Examples
  Retail-General Electronics
Infrared components (LEDs, sensors,
etc.)
  Distributor/Wholesaler-Industrial
Electronics
  Kits-Electronic
  Manufacturer-Components
  Retail-General Electronics
  Retail-Surplus Electronics
  Sensors-Optical
Input/output ports
  Computers-I/O
  Internet-Circuit Examples
  LEGO-Mindstorms
Integrated circuits
  See Components, electronics.
Jewelry findings
  Materials-Other
  Retail-Arts & Crafts
  Retail-Other
Kits
  Kits-Electronic
  Robots-Hobby & Kit
L293D H-bridge controller IC
  See Components, electronics.
Ballscrews, leadscrews
  Actuators-Motion Products
  Retail-Surplus Mechanical
Laminates
  Materials-Foam
  Materials-Metal
  Materials-Paper & Plastic Laminates
  Materials-Plastic
  Retail-Arts & Crafts
Lasers
  Retail-Opticals & Lasers

Retail-Other
Retail-Surplus Mechanical
Retail-Surplus Electronics
LCD panels, serial and parallel
Electronics-Circuit Example
Electronics-Display
Retail-General Electronics
Retail-Robotics Specialty
Retail-Surplus Electronics
LEDs, standard and super-bright
Distributor/Wholesaler-Industrial
Electronics
Kits-Electronic
Retail-General Electronics
Retail-Surplus Electronics
LEGO
Kits-Robotics
LEGO (various)
Robots-Hobby & Kit
Toys-Construction
Toys-Robotics
Lighting effects
Materials-Lighting
Linear actuators
Actuators-Motion Products
Linear stages
Actuators-Motion Products
Machine framing
Machine Framing
Materials-Metal
Retail-Store Fixtures
Machine screws
See Fasteners.
Magazines and other literature
Books (various)
Internet-Informational
Magazines and Journals
Manufacturers (various)
Mail lists
See Bulletin boards.
Materials handling equipment
Materials
Wheels and Casters
Maxim power management ICs
See Components, electronics.
Maze robot competitions
Competitions-Maze
Metals
Machine Framing
Materials-Metal
Materials-Store Fixtures
Retail-Hardware & Home
Improvement
Retail-Train & Hobby
Meters, volt-ohm
See Test equipment
Microcontrollers, hardware and software
Computers-Single Board Computers
Microcontrollers (various)
Retail-General Electronics
Mindstorms
See LEGO.
Mold making

Books-Technical
Materials-Plastics
Supplies-Casting & Mold Making
Materials-Plastic
Retail-Arts & Crafts
MOSFET transistors
See Components, electronics.
Motor speed control
Motor Control
Motors (includes standard, gear, and stepper, and servo)
Actuators-Motors
Distributor/Wholesaler-Industrial
Electronics
Motor Control
Retail-Surplus Mechanical
Movie robots
Books (various)
Entertainment-Art
Toys
Movies and books about robots
Books-Robotics
Entertainment- Books & Movies
Muscle Wire
See Shape memory alloy.
Ni-cad and NiMH batteries
See Batteries and battery holders.
Nitinol
See Shape memory alloy.
Notions, cloth
Retail-Arts & Crafts
Novelty items
Outside-of-the-Box
Obsolete electronic parts
Electronics-Obsolete
Internet-Search
Optical sensors
Retail-Robotics Specialty
Sensors-Optical
Optics (lenses, etc.)
See Lasers
Oscilloscopes
See Test equipment
OWI kits
Kits (various)
Paints and painting supplies
Retail-Arts & Crafts
Retail-Hardware & Home
Improvement
Supplies-Paints
Parts, locating
Electronics-Obsolete
Internet-Search
Internet-Research
PC/104 single board computers
Computers-Single Board Computers
Phototransistors, photodiodes
See Components, electronics.
Also: Sensors-optical
PICmicro microcontroller
Microcontrollers-Hardware
Piezo elements
See Components, electronics.

Pipe (plastic and metal)
Machine Framing
Materials-Metal
Materials-Plastic
Materials-Store Fixtures
Retail-Hardware & Home
Improvement
PIR (passive infrared)
Sensors-Optical
Plastic hemispheres (plastic domes)
See Domes, plastic.
Plastics
Materials-Plastics
Materials-Store Fixtures
Retail-Hardware & Home
Improvement
Retail-Train & Hobby
Supplies-Casting & Mold Making
Pneumatic cylinders and other air parts
Actuators-Pneumatic
Pneumatic tools
Tools
Tools-Power
Polymer clay
Materials-Other
Supplies-Arts & Crafts
Supplies-Casting & Mold Making
Potentiometers, trimmer pots
See Components, electronics.
Power sources (such as batteries)
Batteries and Power
Power transmission components (gears, bearings, shafts, couplers, etc).
Power Transmission
Retail-Surplus Mechanical
Printed circuit boards
Electronics-PCB-Design
Electronics-PCB-Production
Programming
Microcontrollers-Programming
Programming-Examples
Programming-Languages
Programming-Platforms & Software
PVC (plastic)
Materials-Plastics
Materials-Store Fixtures
Supplies-Casting & Mold Making
Quadrature encoders
Sensors-Encoders
Radio communications
Communications-RF
Radio Control
Retail-General Electronics
Video-Transmitters
Radio control (R/C) components (includes hardware, electrical, etc.)
Manufacturer-Train & Hobby
Radio Control (various)
Radio control (R/C) servo motors
Radio Control-Servos
Relays
See Components, electronics.
Remote control

Communications (various)
  Programming-Telerobotics
  Radio Control (various)
  Retail-Automotive Supplies
Research robots
  Internet-Edu/Government Labs
  Robots-Experimental
  Robots-Industrial/Research
Resins (casting)
  See Casting.
Resistors
  See Components, electronics.
RF transmitters/receivers
  See Communications links.
RFID sensors and equipment
  Sensors-RFID
Robot competitions
  Competitions (various)
  Internet-Bulletin Board/Mailing List
  Internet-Usenet Newsgroups
Robot kits
  Internet-Plans & Guides
  Kits-Robotics
  Retail-Robotics Specialty
  Robots-Hobby & Kit
  Toys-Robotics
Robots to buy
  Robots (various)
Rollers
  Power Transmission
  Wheels and Casters
Rotary encoders
  Retail-Surplus Mechanical
  Sensors-Encoders
Saw blades and accessories
  Tools-Accessories
Science and science fair parts and kits
  Retail-Educational Supply
  Retail-Science
Search engines
  Internet-Search
  Internet-Research
  Portal-Robotics
Semiconductors
  Distributor/Wholesaler-Industrial
Electronics
  Manufacturer-Semiconductors
  Retail-General Electronics
  Retail-Surplus Electronics
Sensors (all types)
  LEGO-Mindstorms
  Retail-Robotics Specialty
  Sensors (various)
Servo motors
  See Motors.
Shaft encoders
  Sensors-Encoders
Shafts
  See Power transmission.
Shape memory alloy
  Actuators-Shape Memory Alloy
  Books-Robotics
  Retail-Robotics Specialty

Robots-Hobby & Kit
Sign Foam
  Materials-Foam
Single board computers
  Computers-Single Board Computers
  Microcontrollers-Hardware
Sintra (trade name)
  Materials-Plastics
Software, programming and operating systems
  Microcontrollers-Software
  Programming-Platforms & Software
Soldering tools and supplies
  See Tools, electronics.
Solenoids
  Actuators-Other
  Retail-General Electronics
  Retail-Surplus Mechanical
Speech, synthesis and recognition
  Electronics-Sound & Music
  Manufacturer-Semiconductors
  Retail-Robotics Specialty
Sprockets and chain
  See Power transmission.
Stainless steel
  Materials-Metal
  Materials-Other
Standees (life-size cardboard figures)
  Entertainment-Art
Stepper motors
  See Motors.
Stiquito
  See Shape memory alloy.
Strain gauges
  Retail-Robotics Specialty
  Sensors-Strain Gauges & Load Cells
Sumo robot competitions
  Competitions-Sumo
  LEGO
Superglue
  Retail-Hardware & Home
Improvement
  Retail-Train & Hobby
  Supplies-Glues & Adhesives
Surplus, electronics
  Retail-Surplus Electronics
Surplus, mechanical and other parts
  Retail-Surplus Mechanical
Swap meets
  Fests and Shows
Switches
  See Components, electronics.
Tamiya motors and components
  Actuators-Motors
  Kits-Electronics
Taxidermy supplies (foam, eyes, etc.)
  Materials-Foam
  Retail-Armatures & Doll Parts
Test equipment (meters, oscilloscopes, etc.)
  Test and Measurement
  Kits-Electronic
  Retail-General Electronics

Retail-Surplus Electronics
Tilt sensors
  Sensors-Tilt & Accelerometer
Tools, electronics (soldering iron, test equipment, etc.)
  Distributor/Wholesaler-Industrial
Electronics
  Retail-General Electronics
  Tools
  Test and Measurement
Tools, hand and power
  Manufacturer-Tools
  Retail-Hardware & Home
Improvement
  Tools (various)
Tools, machining and CNC
  Tools-CNC
  Tools-Machinery
  Tools-Precision & Miniature
Tools, pneumatic
  Tools
  Tools-Power
Toys, construction
  LEGO-General
  Toys-Construction
  Toys-Robotics
Toys, general
  LEGO (various):
  Retail-Hobby & Kit
  Toys (various)
Transfer films for copiers and printers
  Materials-Transfer Film
  Retail-Arts & Crafts
Transistors
  See Components, electronics.
Tubing, plastic and rubber
  Materials-Plastic
  Radio Control-Hardware
  Retail-Hardware & Home
Improvement
Ultrasonic sensors
  Retail-Robotics Specialty
  Sensors-Ultrasonic
University/college robot labs
  Internet-Edu/Government Labs
  Internet-Informational
User groups
  Internet-Bulletin Board/Mail List
  Internet-Usenet Newsgroups
  Portal-Robotics
  Professional Societies
  User Groups
VCR replacement parts
  Distributor/Wholesaler-Other
Components
Velleman kits
  Kits-Electronics
Video
  Video (various)
Vision
  Video (various)
Voltage regulators
  See Components, electronics.

Volt-ohm meters
  See Test equipment
Walking robots
  Internet-Edu/Government Labs
  Robots-Walking
Wheelchair motors
  Actuators-Motors

Retail-Surplus Mechanical
Wheels
  Materials (various)
  Retail-Surplus Mechanical
  Retail-Hardware & Home
Improvement
  Wheels and Casters

Wire and cable
  See Components, electronics.
  Also: Materials-Other
Woodworking tools and supplies
  Materials-Other
  Tools (various)

ABOUT THE AUTHOR

Gordon McComb is an avid electronics hobbyist and robot enthusiast. His books include the best-selling *Robot Builder's Bonanza, Gordon McComb's Gadgeteer's Goldmine,* and *Lasers, Ray Guns, and Light Cannons.* For 13 years he wrote a weekly nationally syndicated newspaper column on computers, and is the founder of the Robotics Workshop column in *Poptronix Magazine.*